Volkswagen

Station Wagon / Bus

Service Manual Type 2

1968, 1969, 1970, 1971, 1972, 1973, 1974, 1975, 1976, 1977, 1978, 1979

Selected Books and Repair Information From Bentley Publishers

Motorsports

Alex Zanardi: My Sweetest Victory *Alex Zanardi and Gianluca Gasparini* ISBN 978-0-8376-1249-2

The Unfair Advantage *Mark Donohue and Paul van Valkenburgh* ISBN 978-0-8376-0069-7

Equations of Motion - Adventure, Risk and Innovation *William F. Milliken* ISBN 978-0-8376-1570-7

Engineering

The Hack Mechanic Guide to European Automotive Electrical Systems *Rob Siegel* ISBN 978-0-8376-1751-0

Physics for Gearheads *Randy Beikmann* ISBN 978-0-8376-1615-5

Bosch Automotive Handbook Updated 9th Edition *Robert Bosch GmbH* ISBN 978-0-8376-1732-9

Bosch Fuel Injection and Engine Management *Charles O. Probst, SAE* ISBN 978-0-8376-0300-1

Maximum Boost: Designing, Testing, and Installing Turbocharger Systems *Corky Bell* ISBN 978-0-8376-0160-1

Supercharged! Design, Testing and Installation of Supercharger Systems *Corky Bell* ISBN 978-0-8376-0168-7

Audi

Audi A4 Service Manual: 2002-2008, 1.8L Turbo, 2.0L Turbo, 3.0L, 3.2L *Bentley Publishers* ISBN 978-0-8376-1574-5

Audi TT Service Manual: 2000-2006, 1.8L turbo, 3.2 L *Bentley Publishers* ISBN 978-0-8376-1625-4

Audi A6 (C5 platform) Service Manual: 1998-2004 *Bentley Publishers* ISBN 978-0-8376-1670-4

BMW

Memoirs of a Hack Mechanic *Rob Siegel* ISBN 978-0-8376-1720-6

BMW 3 Series (F30, F31, F34) Service Manual: 2012-2015 *Bentley Publishers* ISBN 978-0-8376-1752-7

BMW X3 (E83) Service Manual: 2004-2010 *Bentley Publishers* ISBN 978-0-8376-1731-2

BMW X5 (53) Service Manual: 2000-2006 *Bentley Publishers* ISBN 978-0-8376-1643-8

BMW 3 Series (E90, E91, E92, E93) Service Manual: 2006-2011 *Bentley Publishers* ISBN 978-0-8376-1723-7

BMW 3 Series (E46) Service Manual: 1999-2005 *Bentley Publishers* ISBN 978-0-8376-1657-5

BMW 3 Series (E36) Service Manual: 1992-1998 *Bentley Publishers* ISBN 978-0-8376-1709-1

BMW 3 Series (E30) Service Manual: 1984-1990 *Bentley Publishers* ISBN 978-0-8376-1647-6

BMW Z3 (E36/7) Service Manual: 1996-2002 *Bentley Publishers* ISBN 978-0-8376-1617-9

BMW 5 Series (E60, E61) Service Manual: 2004-2010 *Bentley Publishers* ISBN 978-0-8376-1689-6

BMW 5 Series (E39) Service Manual: 1997-2003 *Bentley Publishers* ISBN 978-0-8376-1672-8

Corvette

Corvette: America's Star-Spangled Sports Car *Karl Ludvigsen* ISBN 978-0-8376-1659-9

Zora Arkus-Duntov: The Legend Behind Corvette *Jerry Burton* ISBN 978-08376-0858-7

Porsche

Porsche 911 (996) Service Manual: 1999-2005 *Bentley Publishers* ISBN 978-0-8376-1710-7

Porsche 911 (993) Service Manual: 1995-1998 *Bentley Publishers* ISBN 978-0-8376-1719-0

Porsche Boxster Service Manual: 1997-2004 *Bentley Publishers* ISBN 978-0-8376-1645-2

Porsche 911 Carrera Service Manual: 1984-1989 *Bentley Publishers* ISBN 978-0-8376-1696-4

Porsche 911 SC Service Manual: 1978-1983 *Bentley Publishers* ISBN 978-0-8376-1705-3

Porsche: Excellence Was Expected *Karl Ludvigsen* ISBN 978-0-8376-0235-6

Porsche — Origin of the Species *Karl Ludvigsen* ISBN 978-0-8376-1331-4

Volkswagen

Volkswagen Rabbit, GTI Service Manual: 2006-2009 *Bentley Publishers* ISBN 978-0-8376-1664-3

Volkswagen Jetta Service Manual: 2005-2010 *Bentley Publishers* ISBN 978-0-8376-1616-2

Volkswagen Jetta, Golf, GTI Service Manual: 1999-2005 *Bentley Publishers* ISBN 978-0-8376-1678-0

Volkswagen Jetta, Golf, GTI: 1993-1999, Cabrio: 1995-2002 Service Manual *Bentley Publishers* ISBN 978-0-8376-1660-5

Volkswagen GTI, Golf, Jetta Service Manual: 1985-1992 *Bentley Publishers* ISBN 978-0-8376-1637-7

Volkswagen Corrado Repair Manual: 1990-1994 *Bentley Publishers* ISBN 978-0-8376-1699-5

Volkswagen Passat, Passat Wagon Service Manual: 1998-2005 *Bentley Publishers* ISBN 978-0-8376-1669-8

MINI Repair Manuals

MINI Cooper Service Manual: 2007-2013 *Bentley Publishers* ISBN 978-0-8376-1730-5

MINI Cooper Service Manual: 2002-2006 *Bentley Publishers* ISBN 978-0-8376-1639-1

Mercedes-Benz

Mercedes-Benz C-Class (W202) Service Manual 1994-2000 *Bentley Publishers* ISBN 978-0-8376-1692-6

Mercedes-Benz Technical Companion *Staff of The Star and members of Mercedes-Benz Club of America* ISBN 978-0-8376-1033-7

BentleyPublishers®
Automotive Reference

Bentley Publishers has published service manuals and automobile books since 1950. For more information, please contact Bentley Publishers at 1734 Massachusetts Avenue, Cambridge, MA 02138 USA, or visit our web site at `BentleyPublishers.com`

Volkswagen

Station Wagon/Bus

Service Manual Type 2

1968, 1969, 1970, 1971, 1972, 1973, 1974, 1975, 1976, 1977, 1978, 1979

Volkswagen of America, Inc.

BENTLEY PUBLISHERS® | Automotive Reference™

Bentley Publishers, a division of Robert Bentley, Inc.
1734 Massachusetts Avenue
Cambridge, MA 02138 USA
800-423-4595 / 617-547-4170

Information that makes the difference®

BentleyPublishers® .com

Technical contact information

We welcome your feedback. Please submit corrections and additions to our technical discussion forum at:

`http://www.BentleyPublishers.com`

Updates and corrections

We will evaluate submissions and post appropriate editorial changes online as updates or tech discussion. Appropriate updates and corrections will be added to the book in future printings. Check for updates and corrections for this book before beginning work on your vehicle. See the following web address for additional information:

`http://www.BentleyPublishers.com/updates/`

WARNING—Important Safety Notice

Do not use this manual for repairs unless you are familiar with basic automotive repair procedures and safe workshop practices. This manual illustrates the workshop procedures for some maintenance and service work. It is not a substitute for full and up-to-date information from the vehicle manufacturer or for proper training as an automotive technician. Note that it is not possible for us to anticipate all of the ways or conditions under which vehicles may be serviced or to provide cautions as to all of the possible hazards that may result.

We have endeavored to ensure the accuracy of the information in this manual. Please note, however, that considering the vast quantity and the complexity of the service information involved, we cannot warrant the accuracy or completeness of the information contained in this manual.

FOR THESE REASONS, NEITHER THE PUBLISHER NOR THE AUTHOR MAKES ANY WARRANTIES, EXPRESS OR IMPLIED, THAT THE INFORMATION IN THIS MANUAL IS FREE OF ERRORS OR OMISSIONS, AND WE EXPRESSLY DISCLAIM THE IMPLIED WARRANTIES OF MERCHANTABILITY AND OF FITNESS FOR A PARTICULAR PURPOSE, EVEN IF THE PUBLISHER OR AUTHOR HAVE BEEN ADVISED OF A PARTICULAR PURPOSE, AND EVEN IF A PARTICULAR PURPOSE IS INDICATED IN THE MANUAL. THE PUBLISHER AND AUTHOR ALSO DISCLAIM ALL LIABILITY FOR DIRECT, INDIRECT, INCIDENTAL OR CONSEQUENTIAL DAMAGES THAT RESULT FROM ANY USE OF THE EXAMPLES, INSTRUCTIONS OR OTHER INFORMATION IN THIS MANUAL. IN NO EVENT SHALL OUR LIABILITY WHETHER IN TORT, CONTRACT OR OTHERWISE EXCEED THE COST OF THIS MANUAL.

Before attempting any work on your Volkswagen, read **ix Warnings and Cautions** and any **WARNING** or **CAUTION** that accompanies a procedure in the manual. Review the **WARNINGS** and **CAUTIONS** each time you prepare to work on your Volkswagen.

Your common sense and good judgment are crucial to safe and successful service work. Read procedures through before starting them. Think about whether the condition of your car, your level of mechanical skill, or your level of reading comprehension might result in or contribute in some way to an occurrence which might cause you injury, damage your car, or result in an unsafe repair. If you have doubts for these or other reasons about your ability to perform safe repair work on your car, have the work done at an authorized Volkswagen dealer or other qualified shop.

Part numbers listed in this manual are for identification purposes only, not for ordering. Always check with your authorized Volkswagen dealer to verify part numbers and availability before beginning service work that may require new parts.

Special tools required to perform certain service operations are identified in the manual and are recommended for use. Use of improper tools may be detrimental to the car's safe operation as well as the safety of the person servicing the car.

The vehicle manufacturer will continue to issue service information updates and parts retrofits after the editorial closing of this manual. Some of these updates and retrofits will apply to procedures and specifications in this manual. We regret that we cannot supply updates to purchasers of this manual.

This manual is prepared, published and distributed by Bentley Publishers, 1734 Massachusetts Avenue, Cambridge, Massachusetts 02138 USA. All information contained in this manual is based on the information available to the publisher from Volkswagen at the time of editorial closing. Volkswagen has not reviewed and does not vouch for the accuracy or completeness of the technical specifications and work procedures described and given in this manual.

This manual is simultaneously printed in Canada.

ISBN 978-0-8376-1635-3 **Job code: V279-22**

Library of Congress Catalog Card No. for the paperback version 78-75038
VWoA Inc. Part No. LPV 997 288

Manufactured in the United States of America.

Lot #202507R003

FOREWORD

Service to VW owners is of top priority to the Volkswagen organization and has always included the continuing development and introduction of new and expanded services. In line with this purpose, Volkswagen of America, Inc. has introduced this Volkswagen Official Service Manual.

This Type 2 Manual covers the Station Wagon, the Panel Truck, the Pickup Truck, the Kombi, and the Campmobile of Model Years 1968, 1969, 1970, 1971, 1972, 1973, 1974, 1975, 1976, 1977, 1978, and 1979. (Cars of each Model Year are usually introduced in August of the preceding year.)

NOTE ——
This Manual covers neither the automatic transmission used on 1976 and later models, nor the manual transmission used on 1977 and later models and all but the first 1976 models.

The chassis numbers assigned to the Station Wagon, the Panel Truck, the Pickup Truck, the Kombi, and the Campmobile for the Model Years covered in this Manual are:

Station Wagon

1968: 228 000 001 to 228 220 000
1969: 229 000 001 to 229 300 000
1970: 220 2000 001 to 220 2300 000
1971: 221 2000 001 to 221 2300 000
1972: 222 2000 001 to 222 2300 000
1973: 223 2000 001 to 223 2300 000
1974: 224 2000 001 to 224 2300 000
1975: 225 2000 001 to 225 2300 000
1976: 226 2000 001 to 226 2300 000
1977: 227 2000 001 to 227 2300 000
1978: 228 2000 001 to 228 2300 000
1979: 229 2000 001 to 229 2300 000

Kombi and Campmobile

1968: 238 000 001 to 238 220 000
1969: 239 000 001 to 239 300 000
1970: 230 2000 001 to 230 2300 000
1971: 231 2000 001 to 231 2300 000
1972: 232 2000 001 to 232 2300 000
1973: 233 2000 001 to 233 2300 000
1974: 234 2000 001 to 234 2300 000
1975: 235 2000 001 to 235 2300 000
1976: 236 2000 001 to 236 2300 000
1977: 237 2000 001 to 237 2300 000
1978: 238 2000 001 to 238 2300 000
1979: 239 2000 001 to 239 2300 000

Panel Truck

1968: 218 000 001 to 218 220 000
1969: 219 000 001 to 219 300 000
1970: 210 2000 001 to 210 2300 000
1971: 211 2000 001 to 211 2300 000
1972: 212 2000 001 to 212 2300 000
1973: 213 2000 001 to 213 2300 000
1974: 214 2000 001 to 214 2300 000
1975: 215 2000 001 to 215 2300 000
1976: 216 2000 001 to 216 2300 000
1977: 217 2000 001 to 217 2300 000
1978: 218 2000 001 to 218 2300 000

Pickup Truck

1968: 268 000 001 to 268 220 000
1969: 269 000 001 to 269 300 000
1970: 260 2000 001 to 260 2300 000
1971: 261 2000 001 to 261 2300 000
1972: 262 2000 001 to 262 2300 000
1973: 263 2000 001 to 263 2300 000
1974: 264 2000 001 to 264 2300 000
1975: 265 2000 001 to 265 2300 000
1976: 266 2000 001 to 266 2300 000
1977: 267 2000 001 to 267 2300 000
1978: 268 2000 001 to 268 2300 000

The chassis number is found in three places: behind the front passenger's seat, on the right-hand engine cover plate, and on the dashboard. This Manual is organized so that changes from model year to model year are noted, and if a change within one model year is made, the chassis number of the first Volkswagen with the change is given.

For the Volkswagen owner with mechanical skills and for independent garages, this Manual gives Volkswagen-approved specifications and procedures. In addition, a Volkswagen owner who has no intention of working on his car will find that reading and owning this Manual will enable him or her to discuss repairs intelligently with a professional mechanic.

The aim throughout has been simplicity, clarity, and completeness with step-by-step procedures and accurate specifications. We have endeavored to ensure the highest degree of accuracy possible. When the vast array of data presented in this Manual is taken into account, however, no claim to infallibility can be made.

The Volkswagen owner intending to do maintenance and repairs should have a set of metric wrenches, a torque wrench, screwdrivers, and feeler gauges, since these basic hand tools will be used in accomplishing a majority of the repairs described in this Manual. Usually, there will be a caution in the text when a repair requires special tools or special skills.

If you are a professional mechanic already working on imported cars, you may have some Volkswagen special tools that are shown in some of the illustrations in this Manual. If you have previously worked only on American-manufactured cars, you will not have to replace your expensive micrometers, vernier calipers, and other precision tools because specifications are given both in millimeters and in inches, except when special Volkswagen metric tools are indispensable (such measurements are given only in millimeters).

Volkswagens are constantly being improved and sometimes such changes—both in parts and specifications—are made applicable to older Volkswagens. Thus, a replacement part to be used on an older Volkswagen may not be the same as the part used in the original installation. Such changes are noted in this Manual. If a specification given in this Manual differs from one in an earlier source, disregard the earlier specification. The specifications in this Volkswagen Official Service Manual are accurate as of the publication date of this Manual.

Volkswagen offers an extensive warranty. Therefore, before deciding to repair a Volkswagen that is covered by the new-car warranty, consult your Authorized Volkswagen Dealer. You may find that he can make the repair either free or at minimum cost. Regardless of its age, or whether it is under warranty, your Volkswagen is both an easy car to service and an easy car to get serviced. So if at any time a repair is needed that you feel is too difficult to do yourself, a trained Volkswagen mechanic is ready to do the job for you.

Volkswagen of America, Inc.

Please read these warnings and cautions before proceeding with maintenance and repair work.

WARNING—

- Some repairs may be beyond your capability. If you lack the skills, tools and equipment, or a suitable workplace for any procedure described in this manual, we suggest you leave such repairs to an authorized Volkswagen dealer service department, or other qualified shop.

- Volkswagen is constantly improving its cars. Sometimes these changes, both in parts and specifications, are made applicable to earlier models. Therefore, before starting any major jobs or repairs to components on which passenger safety may depend, consult your authorized Volkswagen dealer about Technical Bulletins that may have been issued since the editorial closing date of this manual.

- Do not re-use any fasteners that are worn or deformed in normal use. Some fasteners are designed to be used only once, and are unreliable and may fail if used a second time. This includes, but is not limited to, nuts, bolts, washers, circlips and cotter pins. Always follow the recommendations in this manual to replace these fasteners with new parts where indicated, and any other time it is deemed necessary by inspection.

- Never work under a lifted car unless it is solidly supported on stands designed for the purpose. Do not support a car on cinder blocks, hollow tiles or other props that may crumble under continuous load. Never work under a car that is supported solely by a jack. Never work under the car while the engine is running.

- If you are going to work under a car on the ground, make sure that the ground is level. Block the wheels to keep the car from rolling. Disconnect the battery negative (–) terminal (Ground strap) to prevent others from starting the car while you are under it.

- Never run the engine unless the work area is well ventilated. *Carbon monoxide kills.*

- Tie long hair behind your head. Do not wear a necktie, a scarf, loose clothing, or a necklace when you work near machine tools or running engines. If your hair, clothing, or jewelry were to get caught in the machinery, severe injury could result.

- Do not attempt to work on your car if you do not feel well. You increase the danger of injury to yourself and others if you are tired, upset or have taken medicine or any other substance that may impair you from being fully alert.

- Illuminate your work area adequately but safely. Use a portable safety light for working inside or under the car. Make sure the bulb is enclosed by a wire cage. The hot filament of an accidentally broken bulb can ignite spilled fuel or oil.

- Catch draining fuel, oil, or brake fluid in suitable containers. Do not use food or beverage containers that might mislead someone into drinking from them. Store flammable fluids away from fire hazards. Wipe up spills at once, but do not store the oily rags, which can ignite and burn spontaneously.

- Always observe good workshop practices. Wear goggles when you *operate machine* tools or work with battery acid. Gloves or other protective clothing should be worn whenever the job requires working with harmful substances.

- Friction materials such as brake or clutch discs may contain asbestos fibers. Do not create dust by grinding, sanding, or by cleaning with compressed air. Avoid breathing asbestos fibers and asbestos dust. Breathing asbestos can cause serious diseases such as asbestosis or cancer, and may result in death.

- Finger rings, bracelets and other jewelry should be removed so that they cannot cause electrical shorts, get caught in running machinery, or be crushed by heavy parts.

- Disconnect the battery negative (–) terminal (Ground strap) whenever you work on the fuel system or the electrical system. Do not smoke or work near heaters or other fire hazards. Keep an approved fire extinguisher handy.

- Batteries give off explosive hydrogen gas during charging. Keep sparks, lighted matches and open flame away from the top of the battery. If hydrogen gas escaping from the cap vents is ignited, it will ignite gas trapped in the cells and cause the battery to explode.

- Connect and disconnect battery cables, jumper cables or a battery charger only with the ignition switched off, to prevent sparks. Do not disconnect the battery while the engine is running.

- Do not quick-charge the battery (for boost starting) for longer than one minute. Wait at least one minute before boosting the battery a second time.

- Do not allow battery charging voltage to exceed 16.5 volts. If the battery begins producing gas or boiling violently, reduce the charging rate. Boosting a sulfated battery at a high charging rate can cause an explosion.

- Some aerosol tire inflators are highly flammable. Be extremely cautious when repairing a tire that may have been inflated using an aerosol tire inflator. Keep sparks, open flame or other sources of ignition away from the tire repair area. Inflate and deflate the tire at least four times before breaking the bead from the rim. Completely remove the tire from the rim before attempting any repair.

- The air-conditioning system is filled with chemical refrigerant, which is hazardous. The A/C system should be serviced only by trained technicians using approved refrigerant recovery/recycling equipment, trained in related safety precautions, and familiar with regulations governing the discharging and disposal of automotive chemical refrigerants.

- Do not expose any part of the A/C system to high temperatures such as open flame. Excessive heat will increase system pressure and may cause the system to burst.

- Greases, lubricants and other automotive chemicals contain toxic substances, many of which are absorbed directly through the skin. Read manufacturer's instructions and warnings carefully. Use hand and eye protection. Avoid direct skin contact.

CAUTION—

- Volkswagen part numbers listed in this manual are for identification purposes only, not for ordering. Always check with your authorized Volkswagen dealer to verify part numbers and availability before beginning service work that may require new parts.

- Before starting a job, make certain that you have all the necessary tools and parts on hand. Read all the instructions thoroughly, do not attempt shortcuts. Use tools appropriate to the work and use only replacement parts meeting Volkswagen specifications. Makeshift tools, parts and procedures will not make good repairs.

- Use pneumatic and electric tools only to loosen threaded parts and fasteners. Never use these tools to tighten fasteners, especially on light alloy parts. Always use a torque wrench to tighten fasteners to the tightening torque specification listed.

- Be mindful of the environment and ecology. Before you drain the crankcase, find out the proper way to dispose of the oil. Do not pour oil onto the ground, down a drain, or into a stream, pond, or lake. Consult local ordinances that govern the disposal of wastes.

Letters from Manual users suggest that the following rules need to be emphasized.

• **The air cleaner** should always be installed before you adjust the ignition timing, the idle speed, or the idle mixture.

• Always check and adjust the ignition timing while the engine is running, using a **stroboscopic timing light**. Do not attempt to adjust the timing with the engine turned off, which was the procedure used on early VWs.

• Contrary to the mistaken instruction given in the **1972 Owner's Maintenance and Repair Guide**, it is never necessary to make an additional timing mark on the crankshaft pulley before you adjust the ignition timing. Set the ignition timing using the existing V-shaped notch.

• During routine maintenance, adjust the **idle speed only**. The idle mixture does not require adjustment unless (1) exhaust emissions are excessive; (2) the engine has had major repairs; (3) the carburetor has been replaced or rebuilt on a 1970 through 1974 model; or (4) the intake air sensor has been replaced or repaired on a 1975 or later model.

• Never try to adjust **the idle speed** until you have regapped or replaced the spark plugs, adjusted the valves, and set the ignition timing to specifications. The engine must be thoroughly warmed up when you adjust the idle speed.

• Never try to adjust the idle speed by turning the **throttle valve adjustment screw** (the procedure used on most other cars and on earlier model VWs). Doing this will upset the spark advance settings. Use only the idle speed adjustment procedure described in the **FUEL SYSTEM** or **FUEL INJECTION** sections of this Manual.

• After you have adjusted the idle speed, **recheck the ignition timing.** If the idle speed is correct and the ignition timing has not changed, you have adjusted the idle speed correctly.

• Before you adjust the idle mixture with an **exhaust gas analyzer**, always disconnect the evaporative emission control hose that connects the charcoal canister to the air cleaner. Otherwise, fuel tank vapors may upset the accuracy of the analyzer readings.

• Do not adjust the **valves** to less than the 0.15-mm (.006-in.) clearance specified in this Manual. The older, outdated clearance specification, which may be given on an engine decal or in the Owner's Manual, has been superseded.

• Always adjust the valves while the engine is **cold** (oil temperature no more than 50°C (122°F)). Never attempt to adjust the valves with the engine hot or running.

• Never raise a VW by placing a **jack** under the engine or under the transmission. Doing this may ruin, or seriously damage, the light alloy castings from which these components are made.

Directions for using torque wrenches calibrated in newton meters

In adopting the SI *(Systeme International)* units of measure, which constitute the Modernized Metric System, tool manufacturers are beginning to introduce torque wrenches that are calibrated in newton meters. As metrication proceeds, torque specifications given in foot *pounds (ft. lb.)* and meter kilograms (mkg) will eventually be replaced by torque specifications given in newton meters (N·m or Nm).

At present, there are in use too few torque wrenches calibrated in newton meters to justify the inclusion of newton meter torque specifications in this Manual. Nevertheless, if you purchase a new torque wrench, we recommend that you try to obtain one that is calibrated in newton meters. Such a tool can easily be used with this Manual by converting the meter kilogram specifications to newton meters.

To convert meter kilograms (mkg) to newton meters, simply disregard the decimal point. For example, 3.5 mkg would become 35 Nm. To convert centimeter kilograms (cmkg) to newton meters, point-off the one place with a decimal. For example, 50 cmkg would become 5.0 Nm. These conversions are not mathematically precise (3.5 mkg actually equals 34.3 Nm) but they are adequate for normal workshop purposes.

BODY AND FRAME

Contents

Body and Frame

VW Type 2 vehicles have a full-length, ladder-type frame that provides mounting points for the axles, engine, and controls. The frame is welded to the all-steel unit construction body so that the body and frame together form one rigid assembly. Despite detail variations, this basic design has remained unchanged during the model years covered by this Manual.

The Station Wagon, Kombi, Campmobile, Delivery Van, and Pick-Up Truck share the same frame. The frame consists of two channel-section sidemembers, a number of crossmembers, supports, outriggers, and stiffeners—all joined by electric welding. The sidemembers are approximately 50 × 100 mm (2 × 4 in.) in cross section with a wall thickness of about 2 mm (.079 in.). The arched portions of the sidemembers, located above the axles, are considerably deeper and have welded plates to form a box section. On vehicles with the sliding steel roof, Campmobile roof, or sliding doors on both sides, the body structure is further strengthened by reinforcement plates welded beneath the floor panel.

The seat box, integral with the frame, is built up of sheet steel pressings welded to one another and to the frame sidemembers. The body floor panels, seat mounting platform, wheel housings, cab rear panel, and jack brackets are welded to the frame—in addition to being welded to the front, rear, or side panels of the body. This form of construction provides maximum strength with a minimum possibility for leaks and rattles.

The bodies of the Station Wagon, Kombi, Campmobile, and Delivery Van differ only in their window arrangements and trim. The Pick-Up and Double Cab Pick-Up bodies employ construction similar to that of the fully-enclosed models but with additional spot-welded box-section strengthening pillars for the cab. Also, the Pick-Up's cargo platform is made of ribbed 1-mm (.040 in.) steel panels, whereas the floor panels on fully-enclosed models are 0.76 and 0.89 mm (.030 and .035 in.) thick.

While many repairs described in this section can be carried out by car owners, a number of the procedures may be of practical value only to professional mechanics. If you lack the skills, tools, or the workshop necessary for making body adjustments and repairs, we suggest that you leave such work to an Authorized VW Dealer or other qualified shop. We especially urge you to consult your Authorized VW Dealer before attempting repairs on a car still covered by the new-car warranty.

1. GENERAL DESCRIPTION

The Type 2 body (Fig. 1-1) is a welded steel assembly that includes the front and rear ends, the inner and outer side panels, and the roof and floor panels. The sill panels are welded to the side panels, forming two box-section sidemembers.

© 1974 VWoA—3024

Fig. 1-1. Type 2 body, welded to frame. Doors, windows, and working parts have not been installed.

Doors and Windows

An exceptionally large variety of door and window arrangements are available on Type 2 vehicles. On all models, the front passenger doors are attached to the front body pillar by two hinges. The Double Cab Pick-Up has two rear passenger doors that are mounted on hinges on the door pillar. Sliding side doors are available on all fully-enclosed models and can be obtained on both sides of the Delivery Van. In addition, the fully-enclosed models have a large rear lid for access to the rear luggage compartment.

Seats

The driver's seat is individually mounted and adjustable front to rear. The angle of its seat back is also adjustable. Head restraints are available as an optional addition to all seats. The seats in the rear passenger compartment can be removed so that large objects can be carried inside the vehicle.

Heating and Ventilation

Fresh air is drawn in by the engine cooling fan and is heated as it passes through the fins of the heat exchangers. The flow of heated air into the passenger compartment is regulated by flaps at the front of each heat exchanger and at the heat outlets. The flaps are controlled by cables that link them to three levers located on the instrument panel.

A fresh-air box in the front of the vehicle collects outside air for cool-air ventilation. The air enters the interior through two vents below the windshield and two outlets in the dashboard. Two additional fresh-air discharge vents are located on the partition between the front seats and the rear passenger compartment. They are individually adjustable and provide fresh air ventilation toward the rear.

The air that enters the interior of the vehicle via the fresh air circulation system is drawn out through openings in the front door frames. The air flow can be regulated by levers in the inside panel of the front doors. Both an air conditioning system and an auxiliary heater are available as optional equipment.

Interior Trim and Sliding Roof

Those portions of the floor panel and wheel housings inside the passenger compartment are soundproofed with thermoplastic damping material, which also insulates against road heat. The floor is covered with synthetic rubber floormats. The upholstery is easily cleaned vinyl that is perforated to improve air circulation.

A sliding steel roof (sun roof) is optional equipment on several Type 2 models. Opening the sliding roof creates a clear space above the driver and front passenger seats. A hand crank controls the roof, which can be adjusted to any position from fully open to fully closed. For safety, the crank should be folded into its recess when not in use.

The interior trim panel for the sliding roof is made of the same perforated vinyl used for the headliner. The trim panel cannot be removed until the sliding roof and one of its side runners have been removed. The sliding roof can be adjusted, however, without completely removing the trim panel from the vehicle.

Campmobile Equipment

A wide variety of special equipment is available for Campmobile models. The center side window is a louvered, knob-controlled, jalousie-type window with removable mosquito net screening. The roof trim is plywood. Wood-grained material is used for the closets, bench seat frame, storage shelf, and other interior appointments. Campmobile information given in this Manual is directly applicable only to 1968 through 1973 models.

2. MAINTENANCE

Only one maintenance operation, lubrication of the door and lid hinges and locks, is required at regular mileage intervals. This procedure is covered in **LUBRICATION AND MAINTENANCE**. Care of the body, trim, upholstery, and windows is also described briefly in **LUBRICATION AND MAINTENANCE.**

3. BUMPERS

The bumpers on 1968 through 1972 models are similar and mounted in the same way. The bumpers and their mountings are different on 1973 and later vehicles and will be covered separately from those of the earlier models.

3.1 Removing and Installing Front Bumper
(1968 through 1972)

Thoroughly clean the exposed threads on all bumper mounting bolts before you attempt to remove them. If corrosion is evident, or if the nuts or bolts are difficult to turn, apply penetrating oil to the threads.

To remove:

1. Remove the two side support bolts located just ahead of each front wheel as shown in Fig. 3-1.

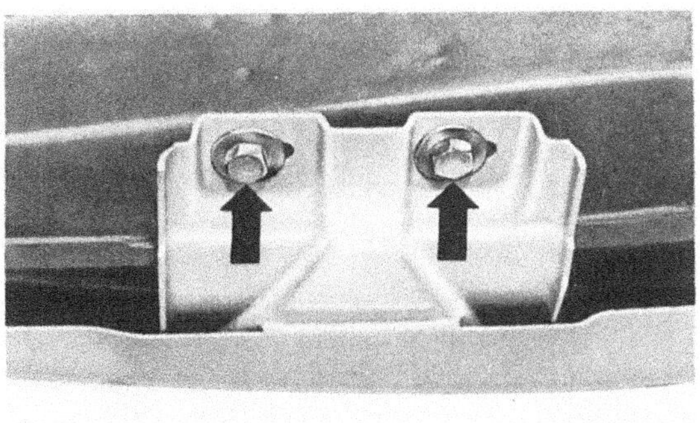

Fig. 3-1. Bolts and washers (arrows) that hold the front bumper side supports on the body.

2. Remove the bolts indicated in Fig. 3-2 from one bumper bracket.

Fig. 3-2. Bolts that hold bracket to frame sidemember.

NOTE —
On later models the towing eye is held by the bracket bolts and will come off when the bracket is unbolted from the body.

3. Place a support under the unbolted end of the bumper or have someone hold it. Then remove the two bolts from the other bracket and take the bumper off the car. 1
4. To remove the bracket from early models, remove the nuts and bolts indicated in Fig. 3-3; on later models, remove the nuts and bolts indicated in Fig. 3-4.

Fig. 3-3. Nuts and bolts (arrows) that hold early bracket on bumper.

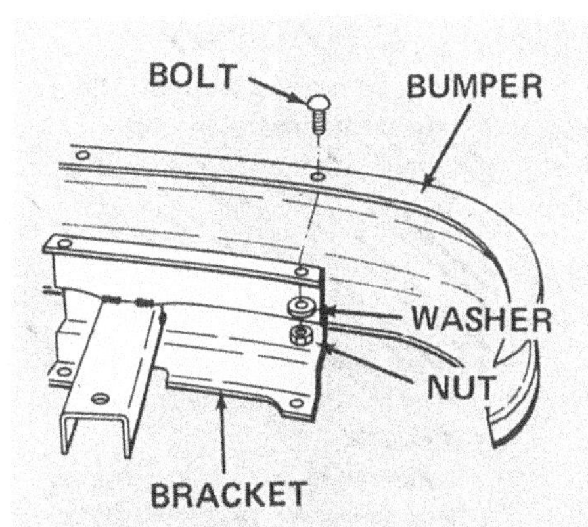

Fig. 3-4. Nuts and bolts that hold late bracket on bumper.

5. To remove the end pieces, take out the three nuts and bolts that hold them on the bumper. Be careful not to damage the rubber seal.

Installation is the reverse of removal. Loosely install the assembled bumper and brackets on the vehicle. Then, by sliding the brackets and side supports on their elongated bolt holes, obtain a uniform gap between the bumper and the body. When the bumper is properly aligned, torque the bolts to 3.5 mkg (25 ft. lb.).

3.2 Removing and Installing Front Bumper
(from 1973)

The one-piece front bumper installed on 1973 and later models is bolted on an energy-absorbing support. The bumper is removed by taking out the four bolts, one of which is indicated in Fig. 3-5. Installation is the reverse of removal. Torque the bolts to 3.5 mkg (25 ft. lb.).

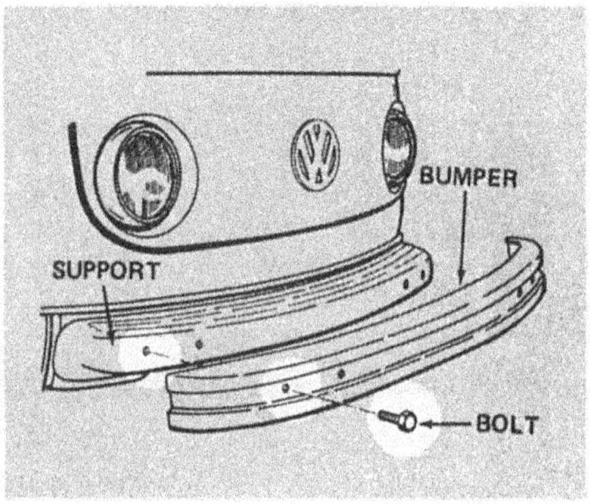

© 1974 VWoA—3690

Fig. 3-5. Front bumper removal and installation on 1973 and later models.

3.3 Removing and Installing Rear Bumper
(1968 through 1972)

The gravel guard on late rear bumpers can be removed and installed without removing the bumper. See Fig. 3-6.

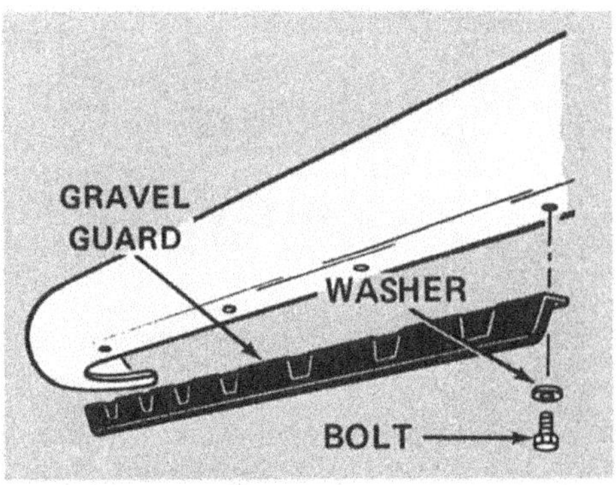

© 1974 VWoA—3691

Fig. 3-6. Removal and installation of gravel guard.

To remove bumper:

1. Remove the bolts and nuts at both sides of the vehicle that hold the cover plate to the body (Fig. 3-7).
2. Remove the two bolts that hold one of the bumper brackets to the frame sidemember.

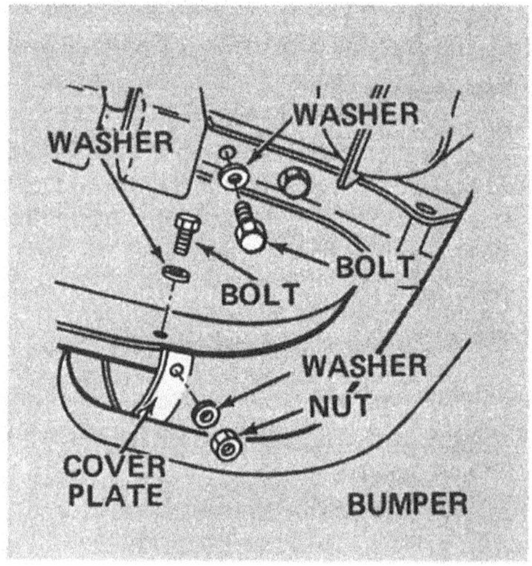

© 1974 VWoA—3692

Fig. 3-7. Nuts and bolts that hold rear bumper.

3. Place a support under the unbolted end of the bumper or have someone hold it. Then remove the two bolts from the other bracket and take the bumper off the car.
4. To remove the brackets, remove the nuts shown in Fig. 3-8.

NOTE —
Thoroughly clean the exposed threads. If corrosion is evident, or if the nuts are hard to turn, apply penetrating oil.

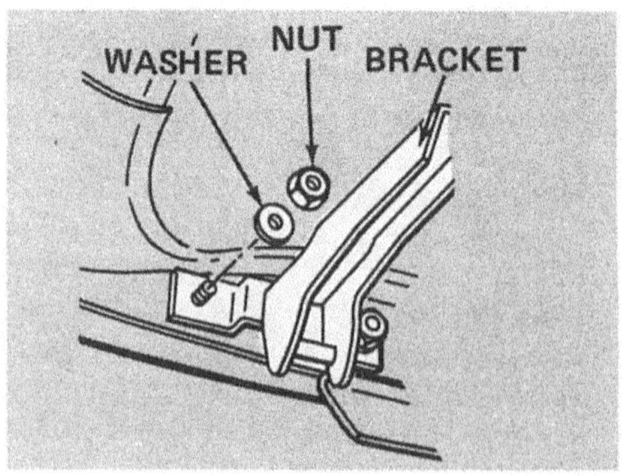

© 1974 VWoA—3693

Fig. 3-8. Bracket being removed from rear bumper.

5. To remove the cover plate, remove the nuts and bolts shown in Fig. 3-9.

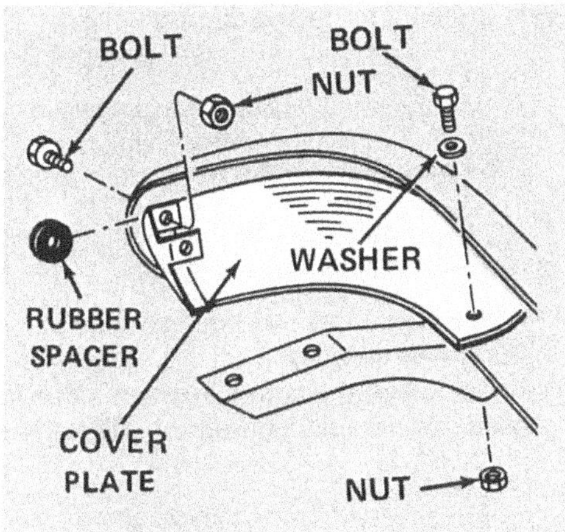

© 1974 VWoA—3694

Fig. 3-9. Cover plate being removed or installed on rear bumper.

Installation is the reverse of removal. Be careful not to damage the rubber spacers by overtightening the bolt on the cover plate. Loosely install the assembled bumper and brackets on the vehicle. Then, by sliding the brackets and cover plate on their elongated bolt holes, obtain a uniform gap between the bumper and the body. When the bumper is properly aligned, torque the bracket bolts to 3.5 mkg (25 ft. lb.). Do not overtighten the bolts that hold the cover plates on the body.

3.4 Removing and Installing Rear Bumper (from 1973)

To remove the rear bumper from 1973 and 1974 vehicles, remove the two bracket-securing bolts (see Fig. 3-10).

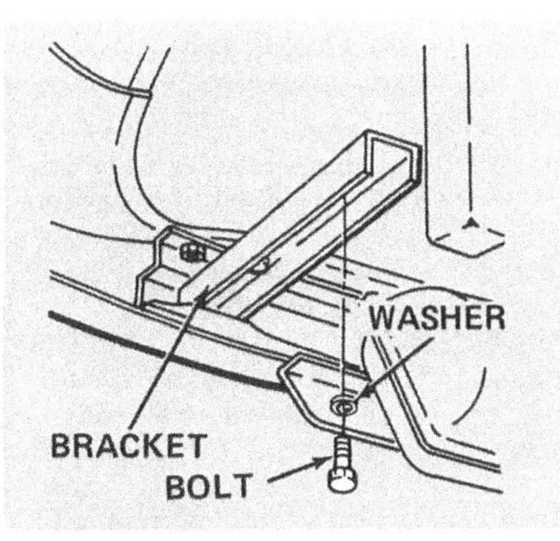

© 1974 VWoA—3695

Fig. 3-10. One of the two bolts that hold the rear bumper bracket on the frame sidemember.

Place a support under the unbolted end of the bumper or have someone hold the bumper. Then remove the two bolts from the other bracket and take off the bumper. If necessary, remove the brackets as shown in Fig. 3-11.

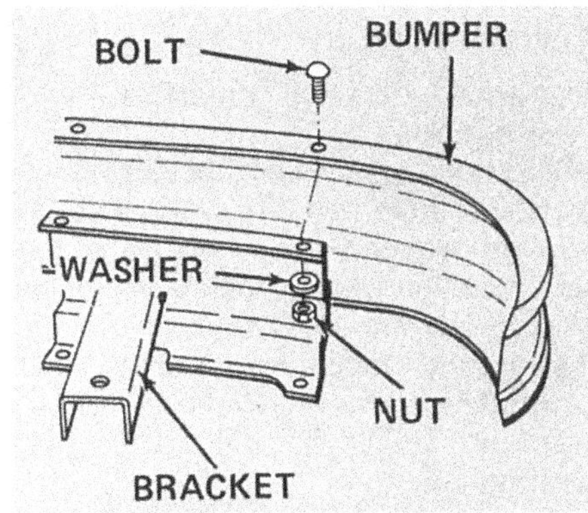

© 1974 VWoA—3696

Fig. 3-11. Bracket being removed or installed on rear bumper of 1973 or 1974 vehicle.

Installation is the reverse of removal. Loosely install the assembled bumper and brackets on the vehicle. Then, by sliding the brackets on their elongated bolt holes, obtain a uniform gap between the bumper and the body. When the bumper is properly aligned, torque the bracket bolts to 3.5 mkg (25 ft. lb.).

4. FUEL FILLER FLAP

The fuel filler flap does not lock and is discontinued on 1974 and later models. It can be removed by taking out the two Phillips head screws indicated in Fig. 4-1. The rubber buffers can also be pried out and replaced. During installation, make sure that the flap contacts the body evenly all around before tightening the screws.

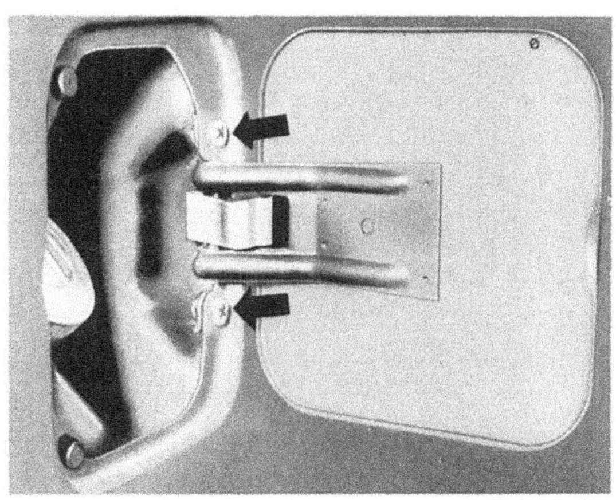

© 1974 VWoA—3031

Fig. 4-1. Screws (arrows) that hold fuel filler flap.

5. DOORS

In addition to the front doors and the sliding side door, the door locks and windows are also covered here.

5.1 Removing and Installing Front Door

If the same door is to be reinstalled, mark the location of the hinges on the body. This will prevent your having to align the door later. From Chassis No. 236 2126 632, new hinges (Part No. 281 831 405) are used on the front doors. When installing a new door on an earlier vehicle, you must also install the new-type hinges, which are maintenance-free and do not require oiling. However, if you press out the hinge pins, you must be careful not to damage the teflon bushing when the pin is reinstalled. Otherwise the hinge will jam and may then require periodic lubrication.

To remove:

1. Take the circlip off the pin for the door check strap. Remove the pin and disconnect the strap.
2. With the door solidly supported so that the upper hinge will not be bent or broken by the door's weight, remove the two Phillips head (or socket head) screws that hold the lower hinge to the hinge pillar (Fig. 5-1). If the screws are rusted tight, loosen them with an impact driver.
3. With the door supported, remove the two Phillips head (or socket head) screws from the upper hinge and remove the door together with its hinges.

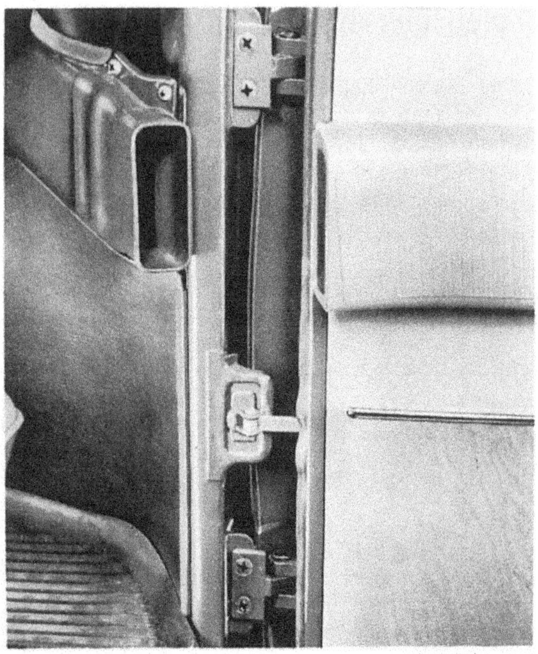

© 1974 VWoA—3032

Fig. 5-1. Door check strap and front door hinges. Late models have socket head hinge screws.

To install:

1. Inspect the rubber weatherstrip around the door. If it is cracked or deformed, replace it.

   NOTE —
   Before replacing a weatherstrip, clean away all the old adhesive with solvent. Install the new weatherstrip with trim cement.

2. If a new door is being installed, remove the lock striker plate from the lock pillar.
3. If the original door is being installed, mount the hinges with reference to the marks you made before removing the door. If a new door is being installed, do not fully tighten the hinge mounting.

   NOTE —
   The hinges are screwed to movable threaded plates. This makes it possible to shift the position of the door in its opening for alignment purposes.

4. To align a new door, position the door in the door opening so that it contacts the weatherstrip evenly all around and the door's trim molding is in line with the trim molding on the side of the body.
5. After the door is aligned, tighten the hinge mounting screws. Set Phillips head screws firmly with an impact driver.
6. Attach the door check strap to its bracket with the pin and circlip.
7. Install the lock striker plate and adjust it as described in **5.2 Adjusting Door Striker Plate.**

5.2 Adjusting Door Striker Plate
(all doors and lids)

After installing a replacement door, adjust the striker plate so that the rear edge of the door aligns with the body. The striker plate should also be adjusted if the door rattles or requires excessive force to close and lock. A door rattle that persists even after you have made all possible adjustments indicates a worn rubber wedge. You can correct the condition either by installing a 0.50-mm to 1.50-mm (.020-in. to .060-in.) shim between the wedge and the striker plate, or by replacing the rubber wedge. The first three steps in the adjustment procedure will tell you whether such adjustment or replacement is necessary.

To adjust:

1. Take out the two large screws and remove the striker plate. Check the door alignment and, if necessary, correct it by moving the hinges as described in **5.1 Removing and Installing Front Door.**

2. Insert the striker plate, bottom first, in the latch. Press the latch down into its fully locked position. Then turn the striker plate as indicated in Fig. 5-2.

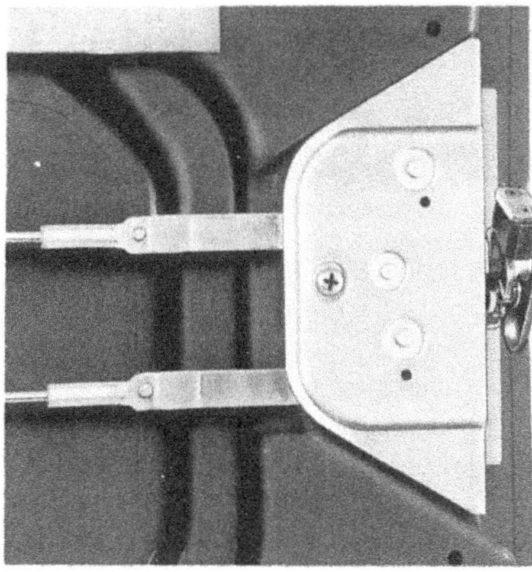

© 1974 VWoA—3034

Fig. 5-2. Striker plate being positioned on door latch. Push in at bottom, then rotate in at top.

3. If you can move the striker plate vertically as indicated in Fig. 5-3, the rubber wedge is worn. Either add shims between the rubber wedge and the striker plate or replace the rubber wedge. If lack of lubrication has caused the pin on the striker plate to wear, replace the entire striker plate.

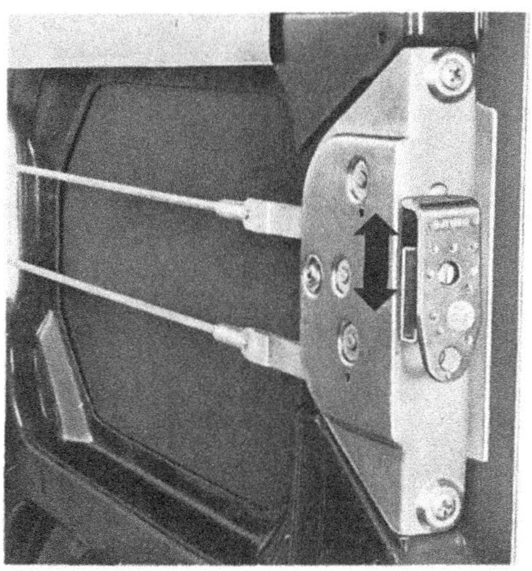

© 1974 VWoA—3035

Fig. 5-3. Undesirable vertical movement (arrow).

NOTE —
To install shims, remove the two Phillips head screws in the angled part of the striker plate. Insert the shim as shown in Fig. 5-4, then reassemble.

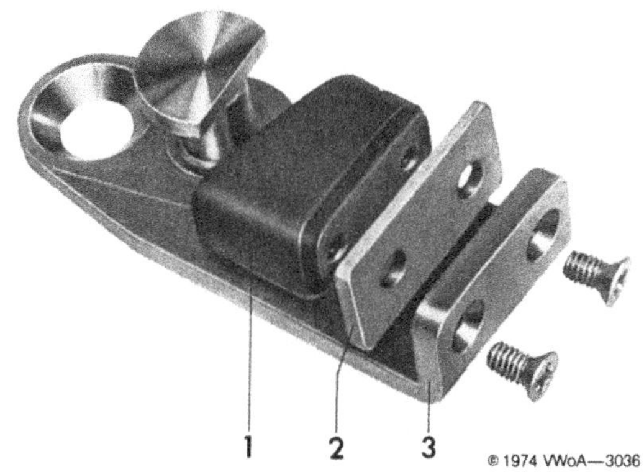

© 1974 VWoA—3036

1. Rubber wedge   2. Shim   3. Striker plate

Fig. 5-4. Position of shim in striker plate.

NOTE —
To correct either a misalignment between the door and front body or a lack of uniformity in the gap between the door and body, adjust the hinges and not the striker plate. Only misalignment (vertical or horizontal) between the door and rear body should be corrected by adjusting the striker plate.

4. After correcting any excessive wedge play, correct any misalignment between the door and rear body. To do this, install the striker plate in its centered position on the lock pillar.

5. Close the door. Then see if the door aligns with the rear body. If necessary, adjust the striker plate as indicated in Fig. 5-5.

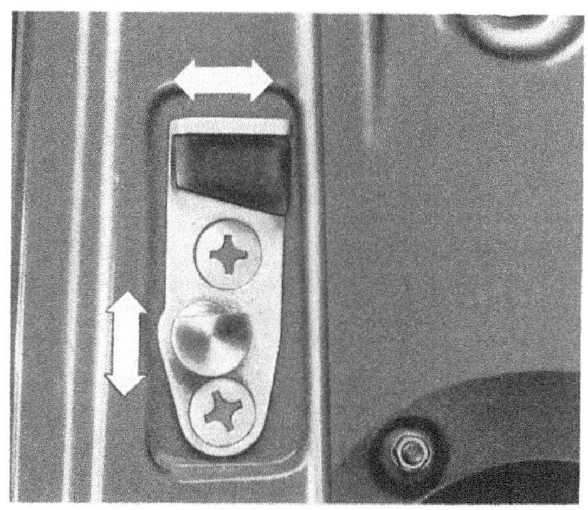

© 1974 VWoA—3037

Fig. 5-5. Adjustments for correcting door misalignment. Move striker plate as indicated by top double arrow until door closes flush with side of body. Move the plate as indicated by the side double arrow to align the door trim with the rear body trim.

6. After aligning the door, grasp the door handle and feel for play between the lock and the striker plate. If there is play or if the door will not latch, rotate the striker plate as indicated in Fig. 5-6. If the door is hard to close or if the handle works stiffly, rotate the striker plate in the opposite direction.

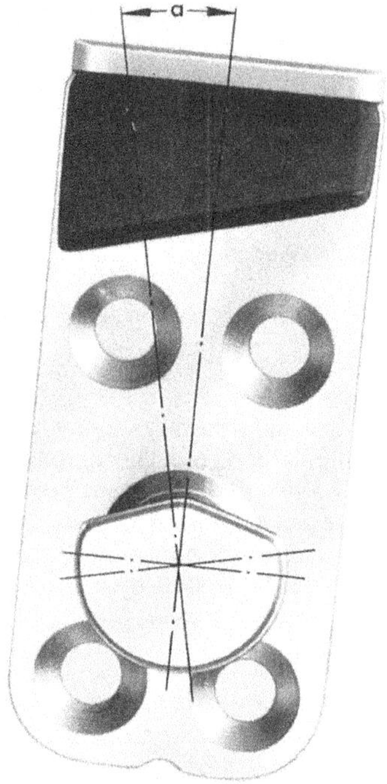

© 1974 VWoA—1756

Fig. 5-6. Striker plate turned from vertical as indicated by arc **a**. This moves the wedge closer to the door latch. Notice the point at the center of the striker plate lug about which the striker plate is rotated.

NOTE —
If the striker plate has been set too high, the door will be hard to open with the press button and will drop slightly as it swings open. If the plate has been set too low, the door will spring out of its locked position when slammed or merely engage the secondary latch position.

5.3 Removing and Installing Front Door Trim Panel

Only the air duct installed on new vehicles since April 1971 (beginning with chassis No. 221 2197 035) is available as a replacement part. If the late duct must be installed on an earlier model, the trim panel must be reworked as indicated in Fig. 5-7. Replacement trim panels already have the correct holes for the new air duct.

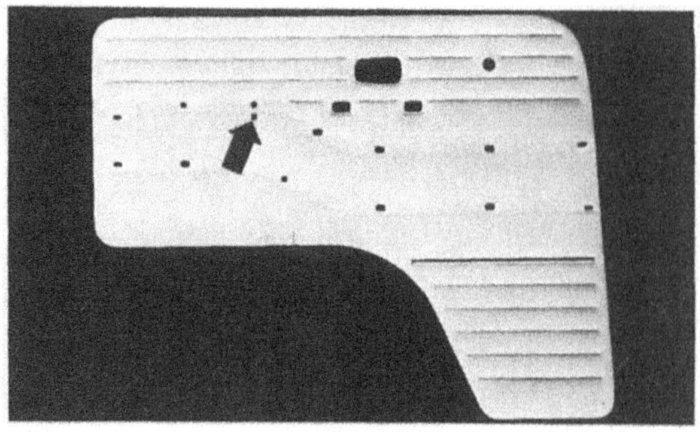

© 1974 VWoA—3227

Fig. 5-7. Trim panel modification. A round hole must be drilled as indicated by the arrow. The other holes have been elongated, where necessary, to make possible the installation of the late-type air duct.

To remove trim panel:

1. Remove the screws indicated in Fig. 5-8. Then remove the two similar screws at the opposite end of the air duct.

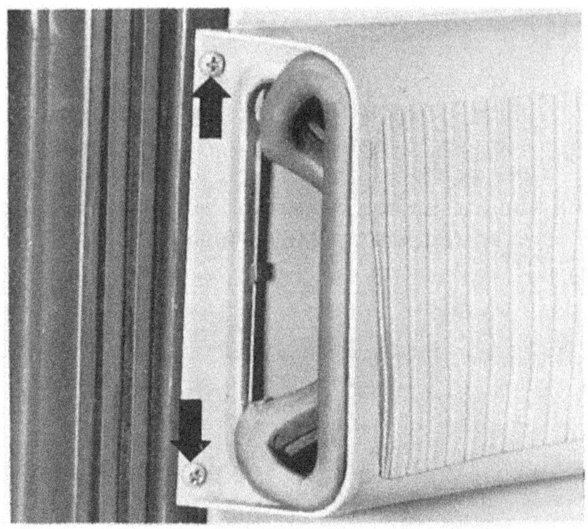

© 1974 VWoA—3038

Fig. 5-8. Screws in edge of air duct (arrows). The rubber seal has been pushed aside for access to the screws.

2. Remove the two screws that hold the door pull handle, then remove the pull handle.
3. Pry the plastic trim off the window crank. Remove the Phillips head screw that holds the crank on the window lifter shaft and remove the crank and the escutcheon.

NOTE —
The screw is held by Loctite® and requires a properly fitting Phillips screwdriver to snap it free.

4. Pry the finger plate out of the lock release lever escutcheon. Then remove the screw indicated in Fig. 5-9.

Fig. 5-9. Screw that holds lock release lever bracket.

5. Using a wooden wedge, pry off the trim panel all around the door.
6. Remove the trim panel. On some vehicles, there is a hook that engages a hole in the door (Fig. 5-10). When removing the trim panel, pull it away slightly, then lift the panel to unhook it.

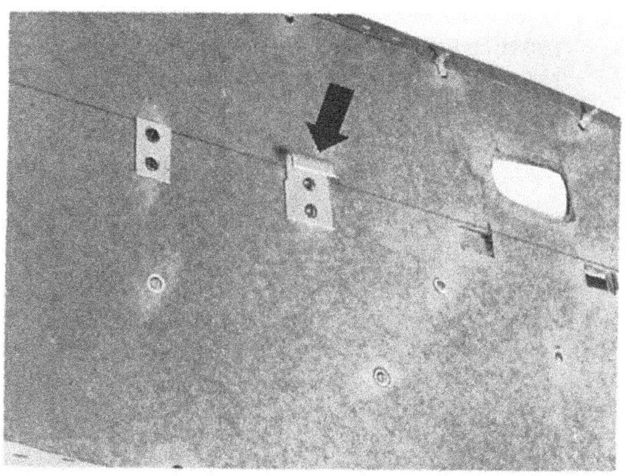

Fig. 5-10. Hook (arrow) on the back of trim panel.

7. If necessary, pull the plastic sheet off the inner door panel and pry out the seals for the trim panel clips.
8. If necessary, the air duct and the armrest can be dismounted from the trim panel by removing the screws.

Installation is the reverse of removal. If either the plastic sheet over the door inner panel or the rubber seal for the release lever have been removed, reglue them. Install the window crank screw with Loctite®. The crank should point forward and up with the window closed.

If necessary, replace the seal at the front of the air duct. Bend up the eight metal tabs on the frame (Fig. 5-11) to free the seal from the duct. Replacement seals are available in two thicknesses: 12 mm and 17 mm ($^{15}/_{32}$ and $^{43}/_{64}$ in.).

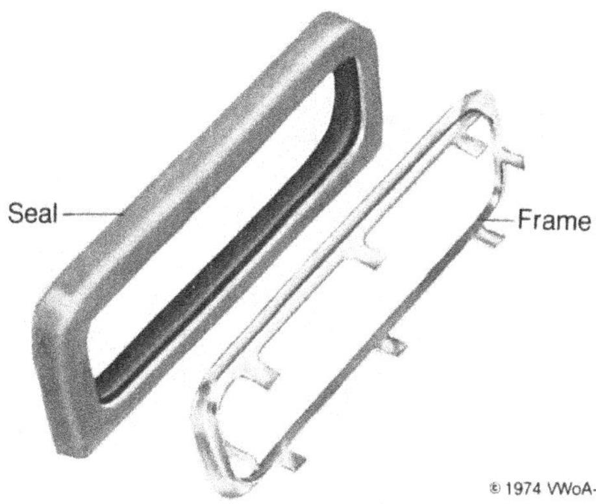

Fig. 5-11. Air duct seal and mounting frame.

5.4 Removing and Installing Door Check

If any part of the door check is faulty, the whole unit must be replaced. Individual components are not supplied. However, replacement pins, circlips, and socket head screws are available.

To replace the door check, remove the circlip and take out the check strap pin. Remove the door trim panel as described in **5.3 Removing and Installing Front Door Trim Panel**. Then remove the socket head screws indicated in Fig. 5-12. Installation is the reverse of removal. Lightly lubricate the check strap with multipurpose grease and the various joints with engine oil.

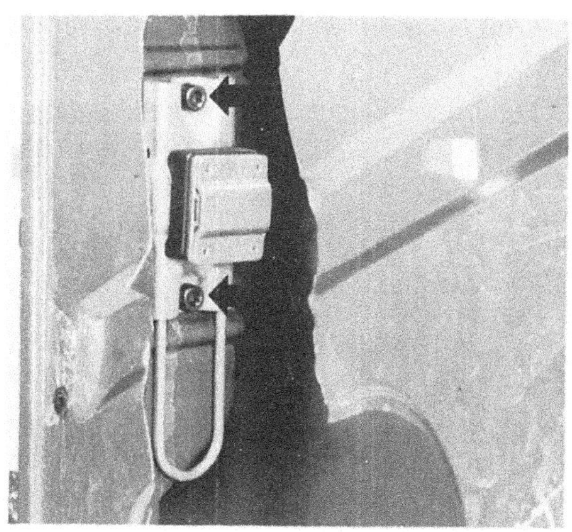

Fig. 5-12. Socket head screws that hold the door check. Remove them with a 5-mm Allen key.

5.5 Removing and Installing Door Handle

To remove the door handle, first remove the trim panel as described in **5.3 Removing and Installing Front Door Trim Panel**. Partially remove the plastic sheet from the door inner panel. Then, using a 5-mm Allen key, remove the two socket head screws indicated in Fig. 5-13.

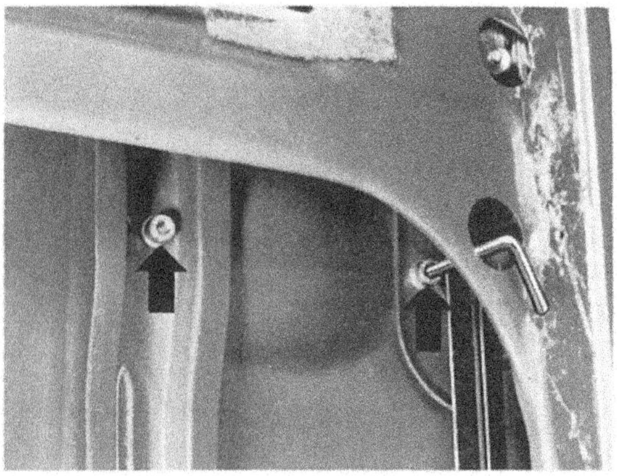

Fig. 5-13. Socket head screws (arrows) being removed.

When installing the door handle, clean the lock and the door release trigger, then apply door and lock lubricant. Make sure the rubber gaskets between the handle and outer door panel are in good condition and that they seat properly when you mount the door handle on the door.

5.6 Removing and Installing Window Rear Guide Channel and Run Channel

The window rear guide channel is held at the top by a retaining clip and at the bottom by the bolt indicated in Fig. 5-14.

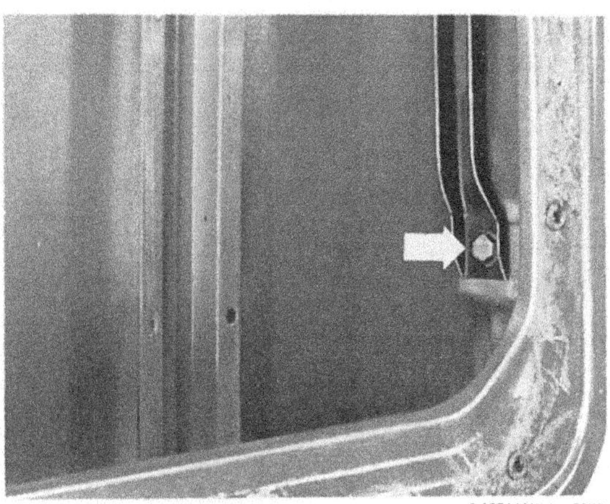

Fig. 5-14. Bolt (arrow) that holds guide channel.

Though the rear guide channel seldom requires replacement, it is necessary to remove it for access to the door lock. To remove the guide channel, take out the bolt and pull the guide channel downward.

Before installing, make sure that the retaining clip is tight. Detach the run channel slightly near the window slot in order to see that the guide channel engages the clip properly. The run channel must be knocked into the clip with a blunt wedge of wood or plastic and not with a sharp tool such as a screwdriver.

The run channel should be replaced if it is worn. The rear run channel is held in the window slot with hardened steel clips. The front run channel is simply pressed into the guide channel. The window will operate stiffly after a new run channel has been installed. To correct this, dust the run channel with talcum powder.

5.7 Removing and Installing Front Door Lock, Release Lever, and Pull Rod

The lock release lever and the pull rod must be removed before you can remove the door lock mechanism.

To remove release lever and pull rod:

1. Remove the door trim panel as described in **5.3 Removing and Installing Front Door Trim Panel**. Then remove the two bolts indicated in Fig. 5-15.

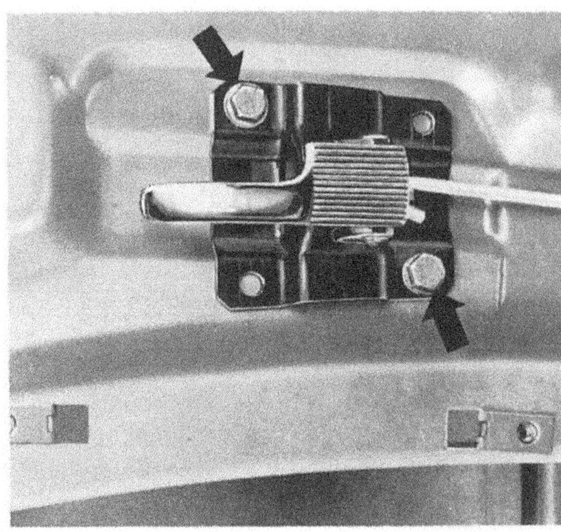

Fig. 5-15. Bolts (arrows) that hold lock release lever.

2. Swing the release lever downward at the front and unhook it from the pull rod. On 1969 through 1973 models, also unhook the rod for the locking lever.

3. Working through the opening in the door inner panel, remove the spring clip indicated in Fig. 5-16, then unhook the rod from the door lock.

© 1974 VWoA—3057

Fig. 5-16. Spring clip for rod on door lock.

Installation is the reverse of removal. Place the rubber packing on the pull rod before you install the lock release lever and do not forget the washers for the screws that hold the release lever on the door. After the rod(s) and release lever have been installed, check the security of the anti-rattle felt on the pull rod. If necessary, move the felt to the correct location and reglue it.

Check to see that the lock operates properly before installing the trim panel. If you install a new pull rod, it is often necessary to correct the length of the rod by bending it. This can be done with a pair of pliers or the bending tool shown in Fig. 5-17.

© 1974 VWoA—3058

Fig. 5-17. Bending tool being used to alter the effective length of the pull rod.

The door lock (Fig. 5-18) can only be removed after taking off the door handle and the rear guide channel as described under the two preceding headings and then removing the lock release lever and pull rod(s) as just described. On 1968 and on 1974 and later models, you must also unscrew the locking knob at the rear of the window. Remove the three Phillips screws, place the latch in its vertical position, then take the lock out downward complete with its plastic cover.

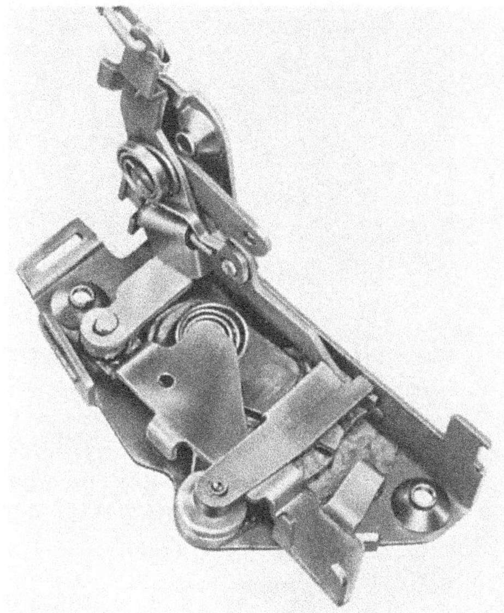

© 1974 VWoA—3054

Fig. 5-18. Door lock removed from the door.

Installation is the reverse of removal. Check the lock and the plastic cover. If the lock is worn or the cover does not fit properly, replace the faulty part. Prior to installation, thoroughly lubricate all moving parts of the lock with door and lock lubricant. Check the operation of the lock after you have installed the pull rod(s) and lock release lever.

5.8 Removing and Installing Window Lifter and Door Window Glass

The door window glass can only be removed after you have removed the window lifter mechanism. The window lifter is a crank-operated flexible cable that is carried inside curved tubes. The window lifter is attached to the door panel at three points.

To remove lifter and glass:

1. Remove the door trim panel as described in **5.3 Removing and Installing Front Door Trim Panel.**
2. Remove the plastic sheet that is beneath the door trim panel.

3. Remove the bolt indicated in Fig. 5-19.

© 1974 VWoA—3060

Fig. 5-19. Bolt for the window front guide channel.

4. Lower the window. Then remove the two bolts indicated in Fig. 5-20 and disconnect the window lifter cable bracket from the window lifter channel.

© 1974 VWoA—3061

Fig. 5-20. Bolts that hold the window lifter cable bracket to the window lifter channel.

5. Working near the bottom of the door inner panel, disengage the plastic cable guide tube from the tab on the door inner panel.

NOTE —
To disengage the plastic guide tube, use a screwdriver at the point indicated in Fig. 5-21 to slightly bend the tab away from the plastic guide tube. Then move the plastic guide tube sideways to disengage it.

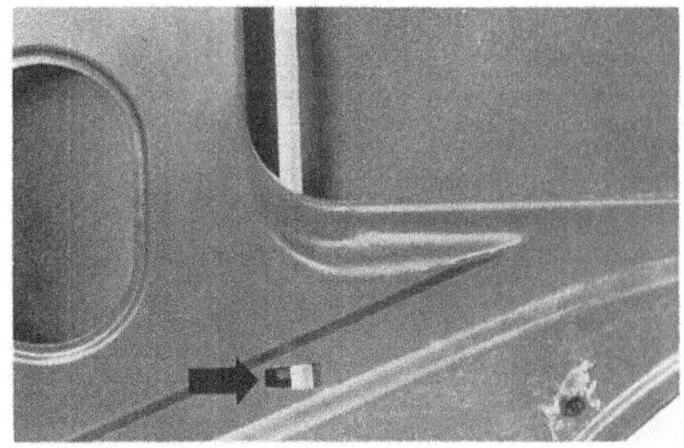

© 1974 VWoA—3062

Fig. 5-21. Metal tab (arrow) that must be pried away to free plastic guide tube from door inner panel.

6. Remove the two bolts from the cable drive gear. (The bolts are located on either side of the shaft to which the window lifter crank attaches.)
7. Remove the four bolts indicated in Fig. 5-22.

© 1974 VWoA—3063

Fig. 5-22. Bolts (arrows) that hold the window lifter assembly to the door panel.

8. Press the window front guide channel forward off its bracket. Then, through the gap thus formed, pull the window lifter assembly downward and out of the door.
9. If necessary, slide the door window together with its lifter channel downward and out through the same opening.
10. If the glass must be replaced, knock the lifter channel off using a rubber hammer and a wooden or plastic block. Inspect the rubber in the lifter channel. If it is in poor condition, it too should be replaced.

11. When installing new glass in the lifter channel, the dimension indicated in Fig. 5-23 must be the same on both sides of the glass.

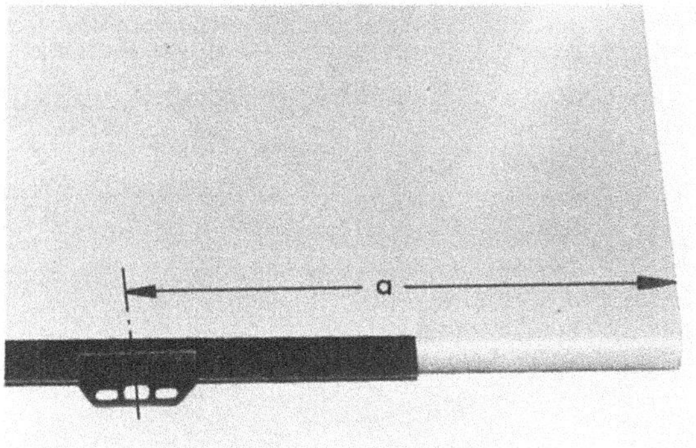

© 1974 VWoA—3059

Fig. 5-23. Glass fitting dimension. Dimension **a**, measured from the center of the lifter channel, must be the same to both edges of the glass.

To install:

1. Check to see that the lifter mechanism works easily. If necessary, lubricate the spiral cable with molybdenum grease. Replacement cables are available and can be installed as described under the next heading. However, if other parts of the lifter are worn or damaged, replace the lifter as a unit.
2. To prevent rattles, make sure that the plastic foam strips at the top of the window lifter tubes are fixed firmly in position.

NOTE —
If the cable rattles in the metal guide tube, carefully squeeze the guide tube together at the appropriate places.

3. If removed earlier, install the window glass together with the lifter channel.
4. Press the window front guide channel off its bracket for clearance, then insert the window lifter mechanism through the gap thus formed.
5. Using the bolts removed earlier, attach the window winder and the front guide channel. Make sure that you install the dished washers with their convex surface upward.
6. Engage the plastic guide tube in the metal tab on the door inner panel. Then bend the tab around the guide tube and out into the slot in the panel.
7. Pull the window down and loosely attach the cable bracket to the lifter channel. To center the bracket on the lifter channel, temporarily install the crank and raise and lower the window several times. Then tighten the bolts that hold the bracket on the lifter channel.
8. Assemble the remaining parts of the door.

Replacing Window Lifter Cable

Replacement cables for the door window lifter have recently become available. In replacing the cable, the window winder mechanism need not be removed completely from the door. However, it must be unbolted as described in the procedure just given so that the window lifter mechanism can be tilted inside the door. To hold the window fully up while you are changing the cable, insert a punch or screwdriver through one of the openings in the door inner panel.

Push the cable drive (the shaft to which the crank is attached) out of the door inner panel. Then tilt and lower the window lifter assembly until the cable bracket hangs below the bottom edge of the door and the cable drive is accessible through the large opening in the door inner panel. To remove the old cable, temporarily install the crank and wind the cable out as far as possible. Bend up the sheet metal tabs on the guide tube at the point where the cable emerges from the guide tube's lower end. Then pull the old cable out of the guide tube.

Lubricate the new cable with molybdenum grease and insert it into the guide tube until it is engaged by the cable drive. Use the window lifter crank to fully pull in the cable. Then bend down the sheet metal tabs on the end of the guide tube. The remainder of installation is the reverse of removal.

5.9 Removing and Installing Vent Wing and Vent Wing Frame

The vent wing and frame should be removed from the door before attempting to remove the vent wing from the frame.

To remove:

1. Remove the door trim panel as described in **5.3 Removing and Installing Front Door Trim Panel.**
2. Remove the plastic sheet that is under the door trim panel.
3. Working through the openings in the upper part of the door inner panel (Fig. 5-24), remove the one bolt that holds the front guide channel and the one bolt that holds the rear guide channel. Then push the door window fully down.

NOTE —
The window must be up for access to the bolts. If you cannot push the door window down after you have removed the bolts, temporarily install the window winder crank and lower the window in the normal way.

1

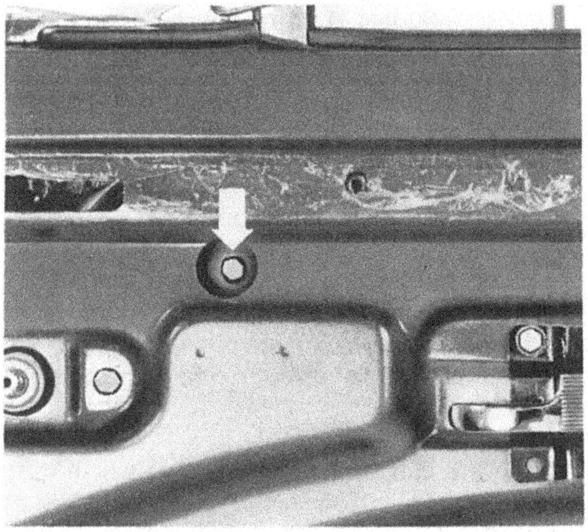

© 1974 VWoA—3064

Fig. 5-24. Opening (arrow) for access to guide channel bolt.

4. Pull the rear run channel out of its slot at the point adjacent to the top of the front guide channel (Fig. 5-25). Then remove the Phillips head screw that holds the top of the front guide channel.

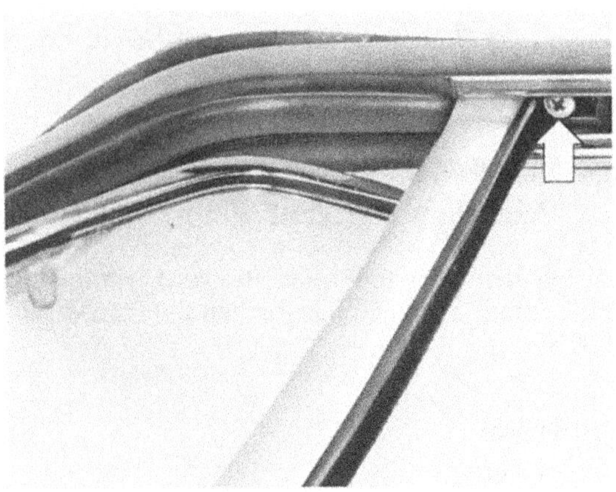

© 1974 VWoA—3065

Fig. 5-25. Phillips head screw that holds top of door front guide channel. Rear channel must be pulled out for access to this screw.

5. Pull out the window slot weatherstrip at the point adjacent to the front guide channel (Fig. 5-26).
6. Pull the vent wing together with its frame toward the door lock (toward the rear edge of the door).
7. Tilt the top part of the vent wing frame toward the inside of the vehicle. Then, being careful not to damage the paint in the window slot, lift the vent wing together with its frame out of the door.

© 1974 VWoA—3066

Fig. 5-26. Wooden wedge being used to pry the window slot weatherstrip out of the door panel. The spring wire clip that holds the weatherstrip is shown by the dotted line.

8. To remove the vent wing from its frame, grind off the bottom of the rivet indicated in Fig. 5-27. Then, using a punch, drive the rivet out upward.

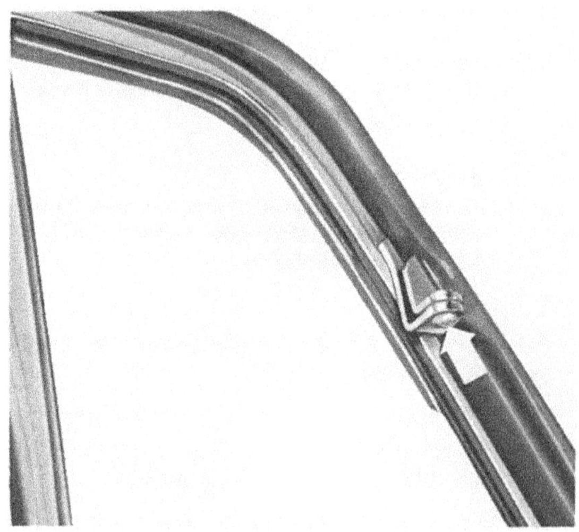

© 1974 VWoA—3067

Fig. 5-27. Rivet that holds the vent wing to the vent wing frame. Grind the end off the rivet at the point indicated by the arrow.

9. Loosen the clamp screw. Tilt the vent wing outward, then pull the spindle on its lower edge up and out of the clamp in the frame.
10. To install the vent wing in the frame, insert the vent wing in the rubber slot and locate the spindle in the clamp on the frame.

11. Align the upper bracket on the vent wing with the bracket on the frame. Then, making sure that the washers are installed, install a new rivet as shown in Fig. 5-28.

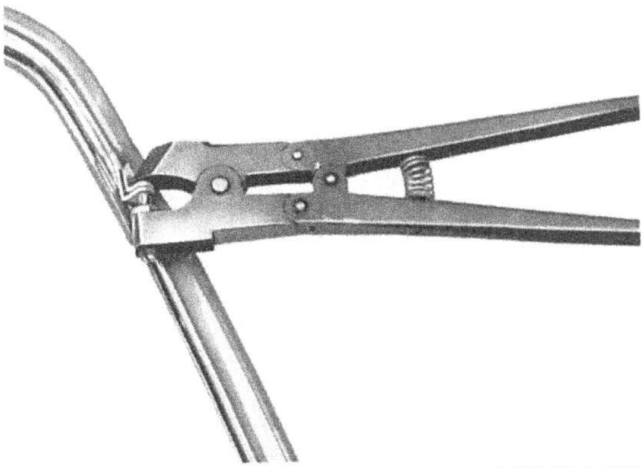

Fig. 5-28. New rivet being installed.

12. Adjust the opening/closing friction of the vent wing by turning the clamp screw indicated in Fig. 5-29.

NOTE —
This screw is also accessible with the vent wing and its frame installed in the door. If the vent wing will not stay open, or requires excessive force to open and close, remove the door trim panel. Insert a screwdriver through the opening in the door inner panel and adjust the screw.

Fig. 5-29. Location of adjusting clamp screw (arrow).

13. Using the reverse of the removal procedures, insert the vent wing and frame in the door. If necessary, adjust its position as described in the following procedure before installing the remainder of the door parts.

To adjust vent wing:

1. Lightly coat the vent wing weatherstrip with glycerine, talcum powder, or silicone spray. Then open and close it to check for jamming.
2. If the vent wing jams at side **A**, as given in Fig. 5-30, pull the lower weatherstrip out of the vent wing frame as far as the pivot point. Then place a plastic or wooden wedge on the exposed frame. Using a hammer, lightly tap the frame deeper into the window slot.
3. If the vent wing jams at side **B**, as given in Fig. 5-30, fully open the door window. Loosen the bolt for the front guide channel and pull the vent wing frame and the front guide channel slightly to the rear. Then tighten the bolt.

NOTE —
It may be necessary after making the preceding adjustment to adjust the rear guide channel slightly rearward. To do this, loosen the bolt at the bottom of the channel.

4. If the vent wing glass is not parallel with the front window frame at side **C**, as illustrated in Fig. 5-30, open the vent wing to 90°. Then carefully align the front guide channel with the door window glass. If the vent wing still is not parallel along side **C**, slightly bend the upper hinge brackets, as necessary.

Fig. 5-30. Points for checking vent wing alignment.

5.10 Removing and Installing Sliding Door Outside Runner Cover

The outside runner cover (Fig. 5-31) must be removed before you can remove the sliding door.

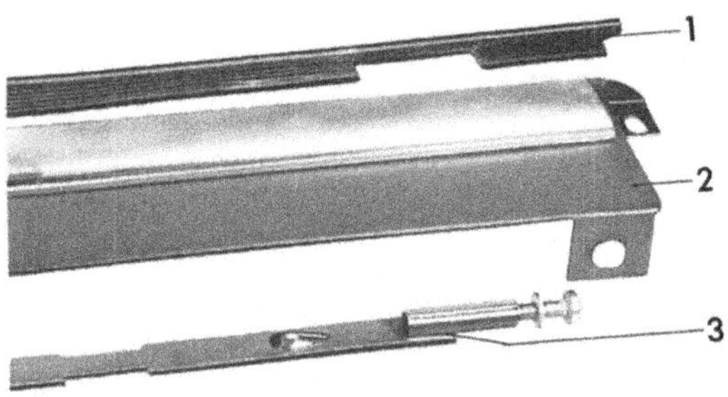

© 1974 VWoA—3071

1. Beading 2. Cover 3. Retaining strip

Fig. 5-31. Components of sliding door outside runner cover.

To remove:

1. Remove the three Phillips head screws that hold the cover on the body. Their locations are shown in Fig. 5-32 and Fig. 5-33.

NOTE —
On vehicles manufactured after September 1967, you must also remove a cover-securing nut and bolt from inside the passenger compartment.

© 1974 VWoA—3072

Fig. 5-32. Phillips head screws (arrows) at the front of the outside runner cover.

© 1974 VWoA—3073

Fig. 5-33. Phillips head screw (arrow) at the rear of the outside runner cover.

2. Fully open the sliding door.
3. Loosen the Phillips head screw on the retaining strip (Fig. 5-34) by about 15 turns.

© 1974 VWoA—3074

Fig. 5-34. Phillips head screw (arrow) that is threaded into the end of the retaining strip.

4. Place a punch against the head of the Phillips head screw. Then, using a hammer, tap the punch sharply to drive the retaining strip toward the rear of the body.
5. Starting at the rear of the sliding door, lift the outside runner cover up and out of the retaining strip.
6. After taking off the cover, fully remove the Phillips head screws from the ends of the retaining strip and take the retaining strip off the body.

To install:

1. Loosely install the retaining strip. Press the cover into the gap between the body and the retaining strip from above, then insert the beading.

   NOTE —
   Place a small amount of plastic sealing compound ("dum-dum") between the retaining strip and the body to maintain a gap while the cover and beading are being installed.

2. Install the two Phillips head screws that screw in from below.
3. By turning the Phillips head screw in the front end of the retaining strip, tension the retaining strip while making sure that the beading is correctly positioned.
4. Secure the cover at the lock pillar end with the remaining Phillips head screw. On vehicles manufactured after September 1967, install the nut and bolt from inside the passenger compartment.

5.11 Removing and Installing Sliding Door

Before you can remove the sliding door, you must remove the outside runner cover as described in **5.10 Removing and Installing Sliding Door Outside Runner Cover.**

To remove door:

1. Remove the outside runner cover. Then push the door far enough to the rear so that the guide piece and the roller on the hinge link can be lifted sideways out of the recess in the center runner. See Fig. 5-35.

© 1974 VWoA—3075

Fig. 5-35. Roller (arrow) aligned with recess in the top of the center runner.

2. Push the door fully to the rear and lift it until the upper roller can be lifted out of the top runner (Fig. 5-36).

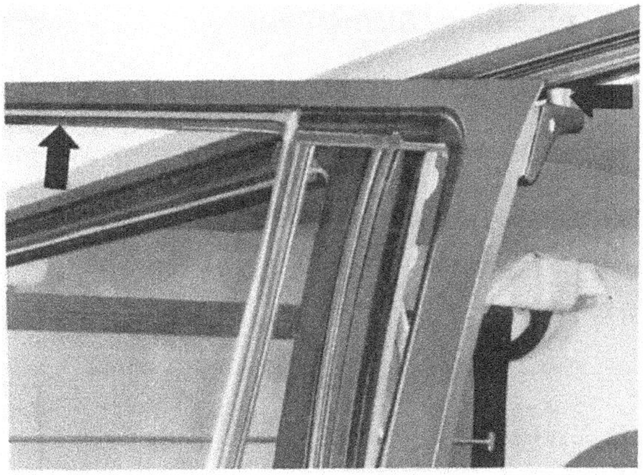

© 1974 VWoA—3076

Fig. 5-36. Upper part of sliding door being unhooked. Lift the door as indicated by the left arrow. Then lift the roller (right arrow) out of the top runner.

3. Swing the door slightly outward and pull the lower rollers out of the break in the bottom runner (Fig. 5-37) and remove the door.

© 1974 VWoA—3077

Fig. 5-37. Door being swung outward (arrow) so that the lower rollers can be unhooked from the bottom runner.

To install:

1. Inspect the runners and, if necessary, straighten them. Check the guide and support rollers on the door. Replace damaged rollers. Lubricate the rollers with multipurpose grease if they turn stiffly.
2. Inspect the rubber weatherstrip around the door opening in the body. If necessary, remove the old weatherstrip, clean away all old adhesive with solvent, then use trim cement to glue a new weatherstrip into place.
3. Insert the door first into the bottom runner and then into the top runner.
4. Push the door forward until the roller and guide can be inserted in the break in the center runner.

5.12 Adjusting Sliding Door

The sliding door is properly adjusted if the gap between the door and the door opening is even all around. The trim or waistline on the door must align with the trim or waistline on the body and the door surface must be flush with the surface of the body.

To adjust:

1. If the door is not in line at the bottom, adjust the lower roller. To do this, loosen the Phillips head screw and the two socket head screws that hold the lower roller bracket. Then insert or remove shims as indicated in Fig. 5-38.

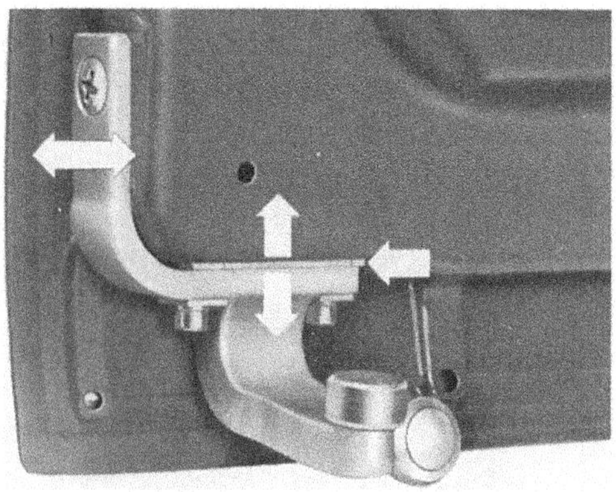

© 1974 VWoA—3078

Fig. 5-38. Lower roller adjustment. The roller bracket can be repositioned horizontally or vertically as indicated by the double arrows. Shims are inserted at the point indicated by the right-hand arrow.

2. If the door is not flush with the body at the top, loosen the nut on the top roller shaft. Then adjust the roller as indicated in Fig. 5-39.

© 1974 VWoA—3079

Fig. 5-39. Top roller adjustment. Position the roller on the bracket as needed (arrows) to align the door with the body.

3. To prevent excessive vertical door movement, loosen the three Phillips head screws that hold the top roller bracket. Then raise the bracket until the clearance between the roller and runner is as small as possible.
4. To adjust the door gap to a uniform width, close the door. Then loosen the four bolts on the hinge housing and adjust the angle of the hinge link.
5. To check for excessive latch play, press the door firmly near the hinge link. If there is any detectable play, adjust the striker plate (Fig. 5-40).

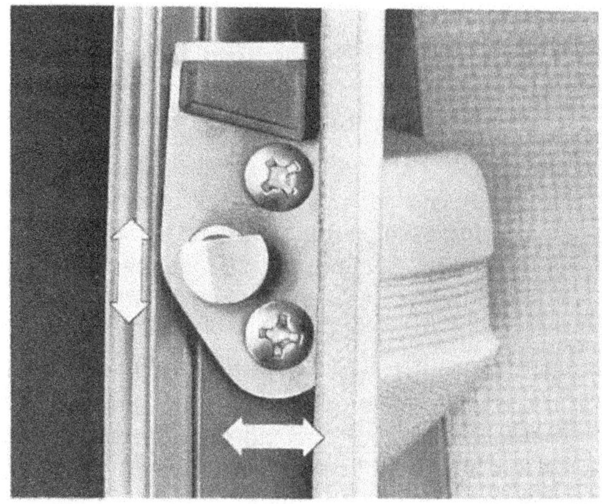

© 1974 VWoA—3081

Fig. 5-40. Striker plate adjustment. Loosen the Phillips head screws. Then relocate the striker plate as indicated by the double arrows until all play is removed from the door.

6. To align the door retainer with the rear bracket, loosen the Phillips head screws and shift the position of the retainer as indicated in Fig. 5-41.

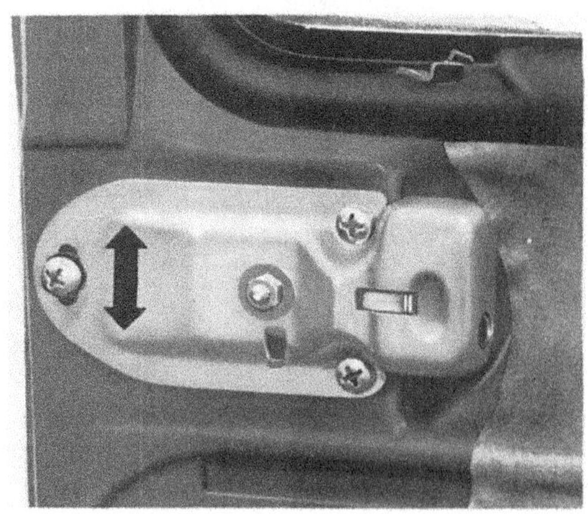

© 1974 VWoA—3082

Fig. 5-41. Door retainer adjustment. Shift position of retainer as indicated by double arrow.

7. To adjust the locking plate (Fig. 5-42) for the remote control lock, slightly loosen the two locking plate bolts and then close the door to center the locking plate. Open the door and tighten the bolts. If necessary, up to two spacer shims can be placed behind the locking plate.

© 1974 VWoA—3083

Fig. 5-42. Locking plate adjustment (double arrow).

5.13 Removing and Installing Sliding Door Retainer

To remove the retainer, remove the three Phillips head screws that hold it on the door panel. Then push the connecting rod off the pin on the door retaining catch and withdraw the retainer as shown in Fig. 5-43. Following installation, adjust the retainer as described in **5.12 Adjusting Sliding Door.**

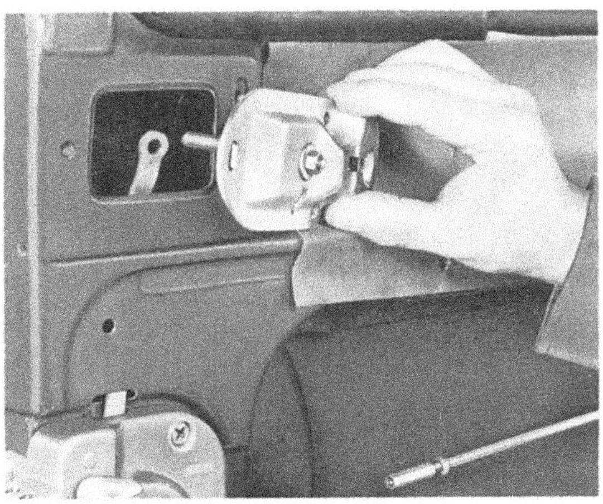

© 1974 VWoA—3085

Fig. 5-43. Retainer being removed. Note relationship of the pin and the eye in the door retaining catch.

5.14 Removing and Installing Remote Control Lock and Center Lock

The door must be removed as described in **5.11 Removing and Installing Sliding Door.** Remove the retainer as described in **5.13 Removing and Installing Sliding Door Retainer.** Then, using a wooden wedge, carefully pry off the door trim panel.

To remove remote control lock:

1. On 1968 through 1970 models, loosen the locknuts on the connecting rods. Then unscrew the connecting rod sleeves from the threaded rods on the lock. Remove the screws indicated in Fig. 5-44.

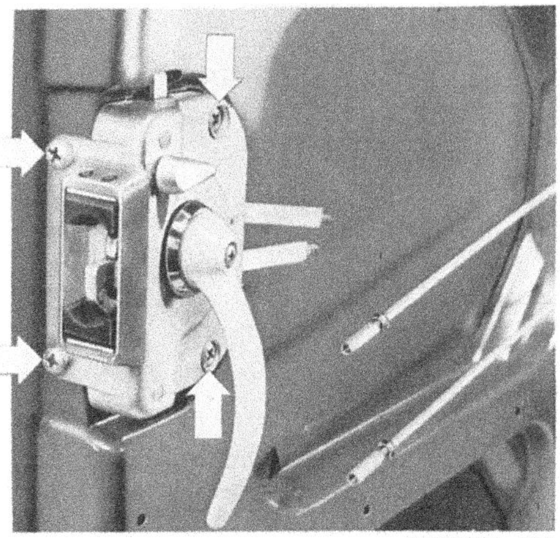

© 1974 VWoA—3086

Fig. 5-44. Lock removal (1968 through 1970 models). Unscrew connecting rods, then take out screws (arrows).

NOTE —
1971 and later models have one cable and one connecting rod. See Fig. 5-45.

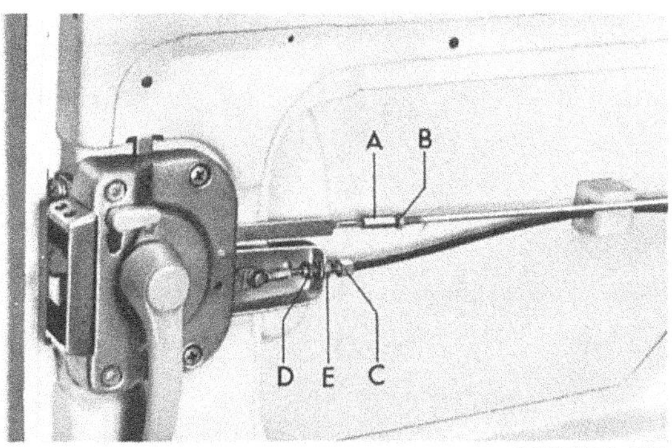

© 1974 VWoA—3098

A. Rod sleeve
B. Rod locknut
C. Cable adjusting nut
D. Cable locknut
E. Cable adjuster

Fig. 5-45. Remote control lock with cable.

2. On 1971 and later models, remove the circlip that holds the cable eye on the lock. Loosen the cable locknut, then screw the cable adjuster in toward the remote control lock. Push the cable adjuster out of the opening in the remote control lock. Disconnect the connecting rod as on earlier models, then remove the screws indicated in Fig. 5-46.

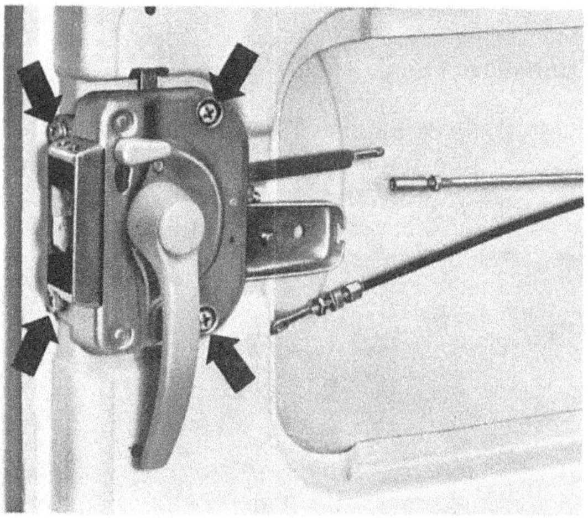

© 1974 VWoA—3099

Fig. 5-46. Lock removal (from 1971). Disconnect cable and rod, then take out the screws (arrows).

3. Pull the remote control lock down until its upper control rod is out of the door inner panel, then remove the lock. If necessary, the remote control lock can be disassembled as shown in Fig. 5-47 or Fig. 5-48. The components shown are those available as replacement parts.

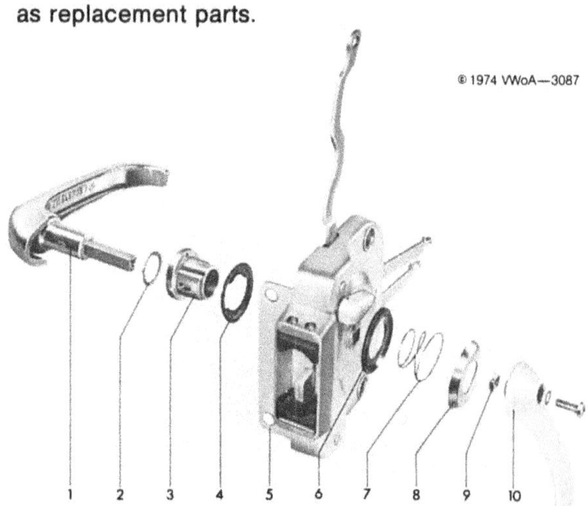

1. Outer handle
2. Sealing ring
3. Outer escutcheon
4. Gasket
5. Remote control lock
6. Gasket
7. Spring
8. Inner escutcheon
9. Guide piece
10. Inner handle

Fig. 5-47. Components of early lock (1968-1970).

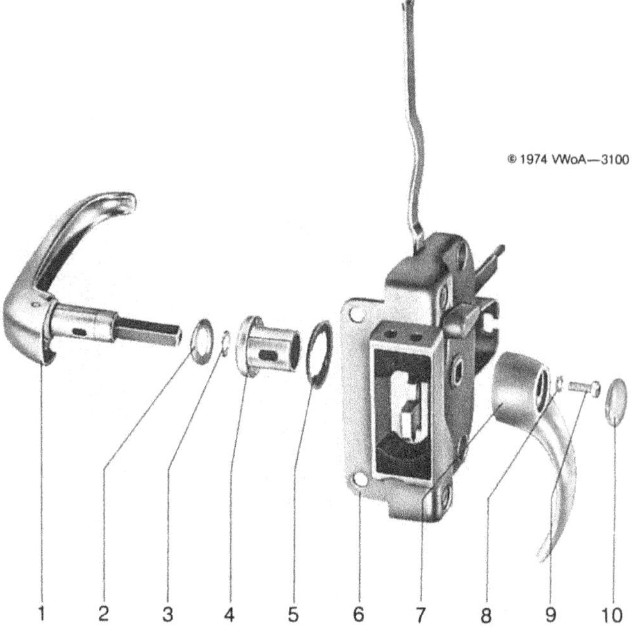

1. Outer handle
2. Sealing ring
3. Guide washer
4. Escutcheon
5. Grommet
6. Remote control lock
7. Inner handle
8. Dished washer
9. Phillips head screw
10. Cap

Fig. 5-48. Components of late lock (from 1971).

Installation is the reverse of removal. Before installing the trim panel, adjust the remote control lock as described in **5.15 Adjusting Remote Control Lock and Center Lock**. Also adjust the retainer as described in **5.12 Adjusting Sliding Door**.

To remove center lock:

1. Remove the remote control lock. Then remove the Phillips head screws indicated in Fig. 5-49.

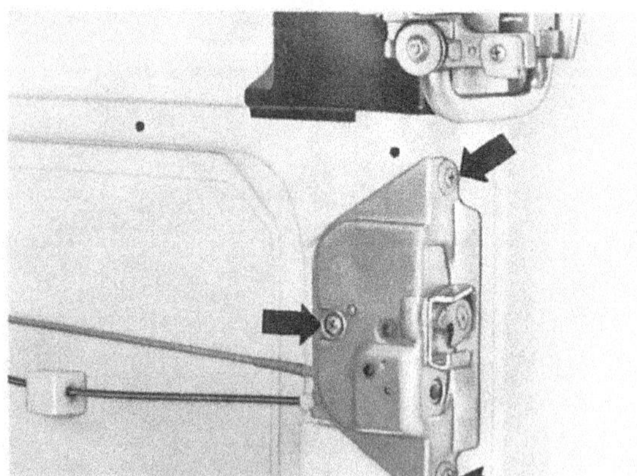

© 1974 VWoA—3102

Fig. 5-49. Screws that hold center lock on door.

2. Remove the center lock from the door. On 1971 and later models, pull the lock out to the side so that the cable is withdrawn from the rubber guide on the door.
3. If necessary, the connecting rod(s) and cable can be removed from the center lock by prying off the E-clips as indicated in Fig. 5-50.

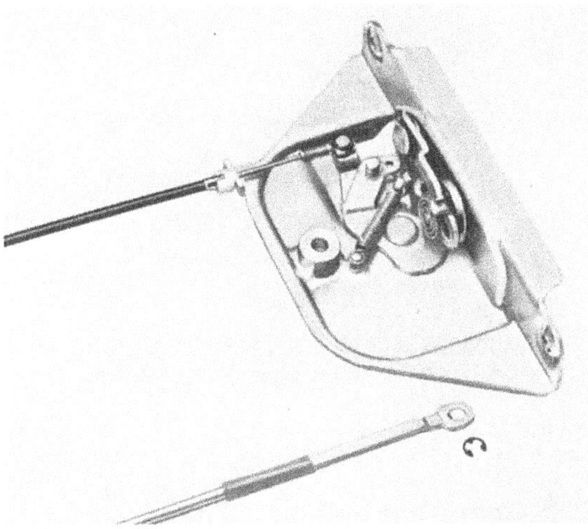

© 1974 VWoA—3104

Fig. 5-50. E-clip and connecting rod removed.

Installation is the reverse of removal. Inspect all parts, particularly the rubber buffers indicated in Fig. 5-51, and replace any parts that are worn or damaged. Lubricate all moving parts with door and lock lubricant. Before installing the trim panels, adjust the center lock together with the remote control lock as described in **5.15 Adjusting Remote Control Lock and Center Lock**. Also adjust the retainer as described in **5.12 Adjusting Sliding Door.**

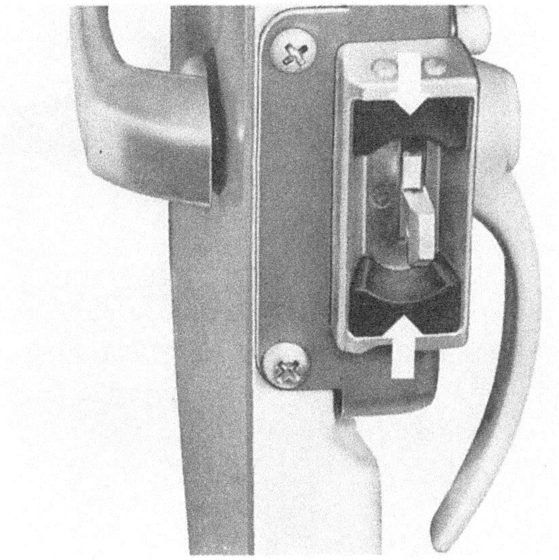

© 1974 VWoA—3103

Fig. 5-51. Rubber buffers in center lock.

5.15 Adjusting Remote Control Lock and Center Lock

If the door cannot be opened and closed or locked properly after carrying out the adjustments described in **5.12 Adjusting Sliding Door**, the adjustment of the connecting rods should be checked.

1

The procedure for adjusting 1968 through 1970 locks is slightly different from the procedure for adjusting 1971 and later locks. The two procedures are covered separately.

To adjust locks (through 1970):

1. Secure the center lock mechanism with two 4-mm screws as shown in Fig. 5-52, or use suitable punches or nails for the same purpose. Screws, however, can be threaded into the operating levers.

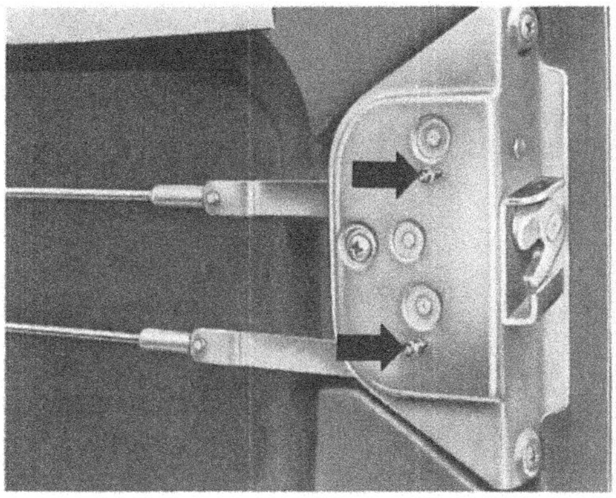

© 1974 VWoA—3092

Fig. 5-52. Screws (arrows) being used to secure the operating levers inside the center lock.

2. Similarly secure the remote control lock by installing a 4-mm screw or other suitable object through the lock housing and into the lower pull rod.
3. Install the connecting rods on the threaded ends of the remote control lock pull rods. Turn the adjusting sleeves until the connecting rods are just barely tensioned, then tighten the locknuts. Remove the 4-mm screws from the locks.

Check the lock operation. Owing to overlapping tolerances, it may be found that the door locking mechanism does not work satisfactorily despite careful adjustment. In such cases, shortening or lengthening the upper and lower connecting rods will correct the problem. To open the door, it should be necessary to move the outer door handle at least 40 mm (1⁹⁄₁₆ in.).

To prevent rattles, make sure that the foam rubber pad at the center of the connecting rods is securely glued to the door inner panel.

To adjust late locks:

1. Secure the center lock mechanisms with a 4-mm screw as shown in Fig. 5-53, or use suitable punches or nails for the same purpose. A screw, however, can be threaded into the operating lever.

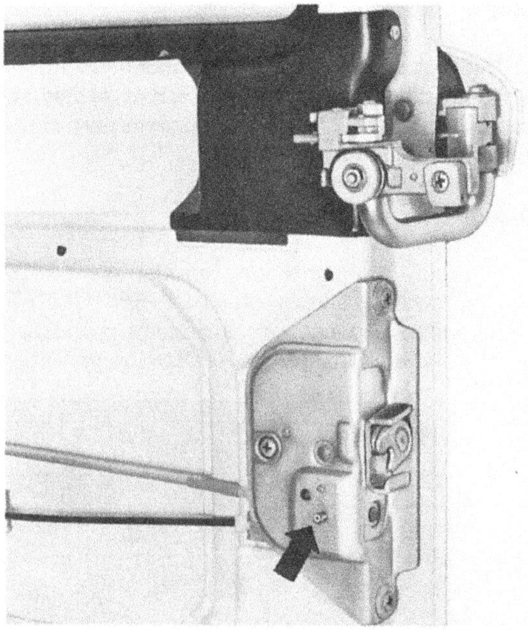

© 1974 VWoA—3105

Fig. 5-53. Screw (arrow) being used to secure the operating lever inside the center lock.

2. Install the cable eye on the operating lever and secure it with the circlip. Then install the cable adjuster in the notch in the remote control lock.
3. To adjust the cable, hand-turn the adjusting nut until it is finger-tight. Then tighten the locknut.
4. To adjust the connecting rod, pull the rod tight. Screw the threaded sleeve as far as the stop on the lock pull rod. Then tighten the locknut.
5. Remove the 4-mm screw from the center lock.

Check the lock operation. Owing to overlapping tolerances, it may be found that the door locking mechanism does not work satisfactorily despite careful adjustment. In such cases, shortening or lengthening the connecting rod and cable will correct the problem. To open the door, it should be necessary to move the outer door handle at least 40 mm (1$\frac{9}{16}$ in.); however, it should not be necessary to move the handle more than 50 mm (2 in.).

To prevent rattles, make sure that the foam rubber blocks and the plastic clips are positioned as shown in Fig. 5-54. They should also be properly glued to the door panels.

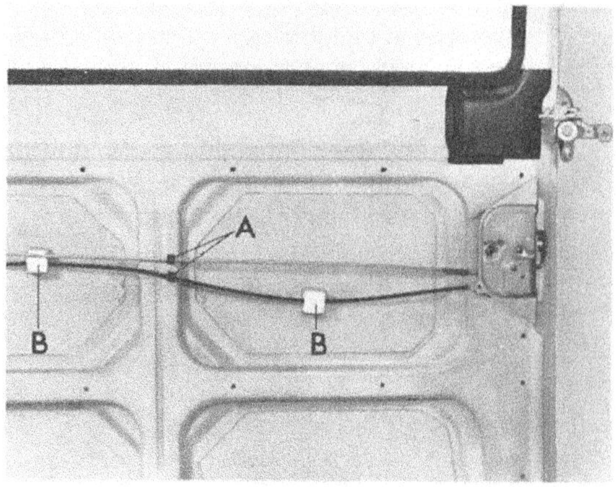

© 1974 VWoA—3109

A. Plastic clips B. Foam rubber blocks

Fig. 5-54. Proper position for foam rubber blocks and plastic clips on the door panels.

5.16 Removing and Installing Sliding Door Retainer Rear Bracket

The sliding door retainer rear bracket is located at the extreme rear of the center runner. The rear bracket is the part that the retainer latches onto when the door is fully opened.

For access, you must remove the sliding door as described in **5.11 Removing and Installing Sliding Door**. To remove the retainer rear bracket, partially detach the rear trim panel. Then remove the Phillips head screw as indicated in Fig. 5-55.

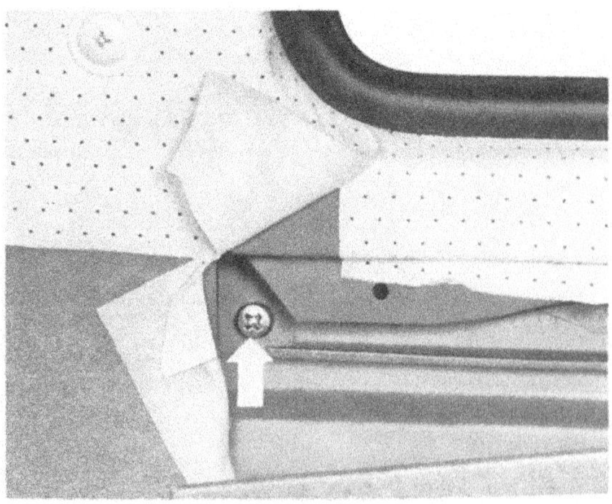

© 1974 VWoA—3110

Fig. 5-55. Screw (arrow) that holds the retainer rear bracket from inside the passenger compartment.

Working outside the vehicle, remove the two bolts and take off the bracket. The bracket, bolts, and replaceable components of the bracket are shown in Fig. 5-56.

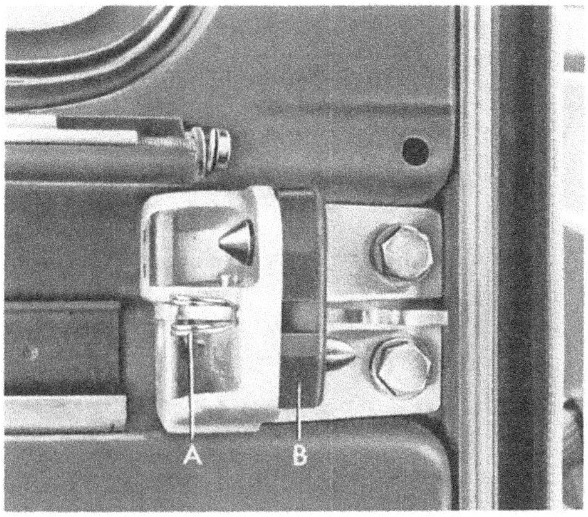

© 1974 VWoA—3111

A. Spring B. Rubber buffer

Fig. 5-56. Sliding door retainer rear bracket.

Installation is the reverse of removal. Inspect the spring and rubber buffer. Replace worn or damaged parts. The retainer rear bracket should be carefully aligned on the panel. After installing the sliding door, it may be necessary to adjust the door retainer as described in **5.12 Adjusting Sliding Door.**

5.17 Removing and Installing Hinge

The components of the hinge are shown in Fig. 5-57. Most of them are available as individual replacement parts.

1. Return spring
2. Hinge link
3. M 6 nut
4. Lock washer
5. Roller bracket
6. Guide block
7. Roller and bearing
8. Phillips head screw
9. Spring washer
10. Thin M 6 nut
11. Housing
12. Circlip
13. Pivot
14. Pin
15. Operating cam
16. Spring washer
17. M 8 nut
18. Washer
19. Lower locking lever
20. Spacer
21. Spring
22. Upper locking lever
23. Pin
24. E-clip

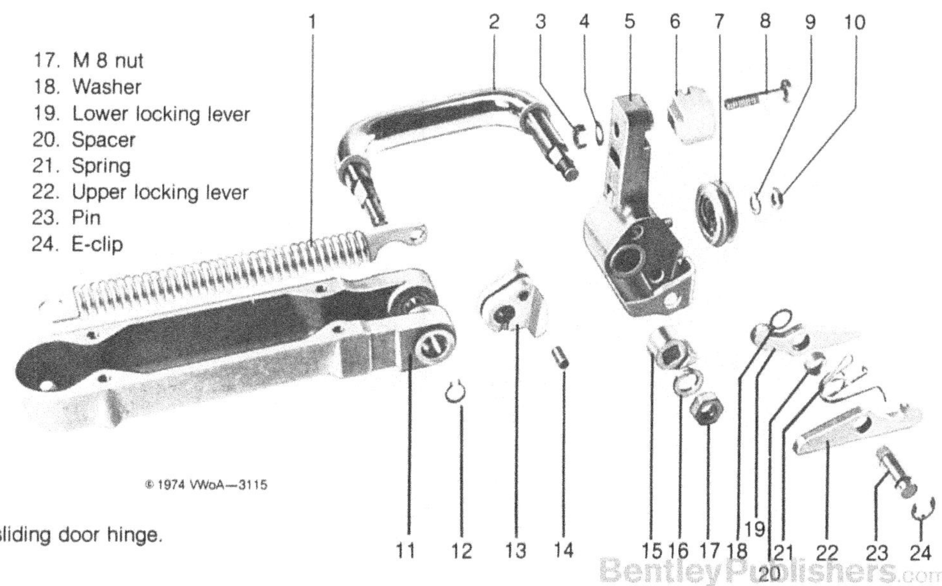

© 1974 VWoA—3115

Fig. 5-57. Components of sliding door hinge.

The sliding door must be removed as described in **5.11 Removing and Installing Sliding Door** before you can remove the hinge. To remove the hinge from the door, take out the four bolts shown in Fig. 5-58.

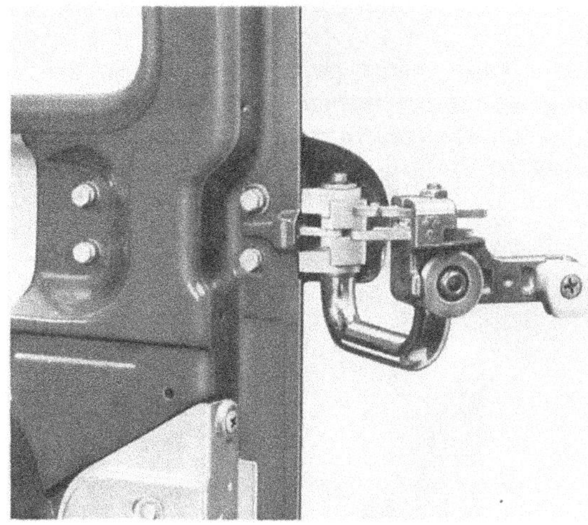

© 1974 VWoA—3112

Fig. 5-58. Bolts that hold hinge on door panel.

Before you install the hinge, apply engine oil to the pivot points for the hinge link. There are grooves in the pivot so that oil can easily be introduced between the pivot and the housing. Also oil the washers and other parts of the pivot points, but do not oil the roller and bearing or the return spring. Lightly coat the return spring with multipurpose grease. Apply the same lubricant to the roller and bearing, working as much grease as possible into the bearing itself.

Installation is the reverse of removal. Following installation, adjust the door as described in **5.12 Adjusting Sliding Door.**

5.18 Removing and Installing Sliding Door Vent Wing

The vent wing pivot spindles are mounted in a stamped steel frame. If the vent wing opens and closes too easily, or if it is difficult to move, the clamp for the lower pivot should be adjusted. To do this, raise the weatherstrip as shown in Fig. 5-59. Then turn the Phillips head adjusting screw until the correct degree of friction is obtained.

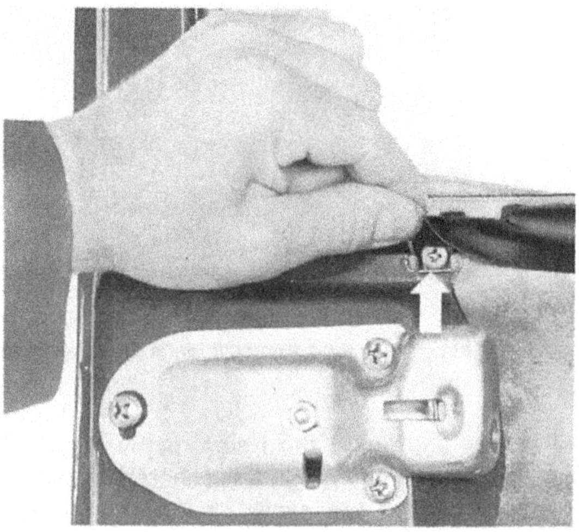

© 1974 VWoA—3116

Fig. 5-59. Vent wing adjusting screw (arrow).

To remove vent wing:

1. Bend up the two metal tabs (Fig. 5-60) on the vent wing frame. With the vent wing installed, these will be found in the end face of the door.

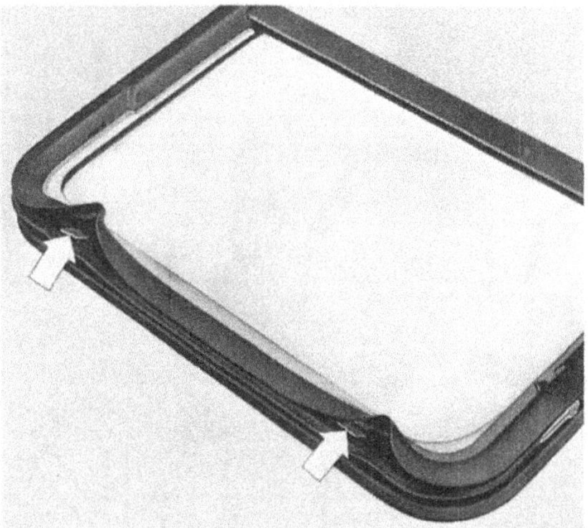

© 1974 VWoA—3117

Fig. 5-60. Metal tabs that hold the vent wing frame in the sliding door.

2. Hand-press the vent wing, vent wing frame, and weatherstrip out of the window opening.
3. If necessary, remove the Phillips head adjusting screw (Fig. 5-61) from the pivot spindle clamp. Then take the vent wing out of the frame.

> ***CAUTION —***
> *Proceed carefully so that you do not accidentally damage the paint on the frame.*

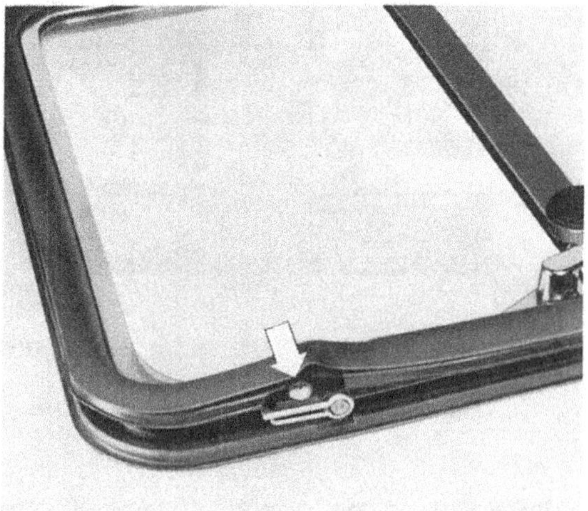

© 1974 VWoA—3118

Fig. 5-61. Screw (arrow) that must be removed before you can take the vent wing out of its frame.

To install:

1. If the vent wing was removed from the vent wing frame, place a glass installer's cord in the outer lip of the weatherstrip. Then install the vent wing in its frame while pulling out the cord.
2. Check to see that the weatherstrip is properly positioned on the frame and against the vent wing. Then install the adjusting screw and tighten it until the correct degree of pivot friction is obtained.
3. Install the vent wing and frame from the outside of the sliding door. Starting at the top, insert the frame behind the sealing flange in the window opening.

> **NOTE —**
> Installation is easier if you first remove the large window from the sliding door.

4. Pull the weatherstrip lip over the flange with a glass installer's cord, making sure that the weatherstrip fits properly all around.
5. To secure the frame and vent wing in the door, bend down the two metal tabs that are on the frame.

6. SEATS AND INTERIOR TRIM

The seats in the rear passenger compartment can be removed by unbolting their retainers from the floor. To remove the driver's seat, fully lift the adjusting lever, then slide the seat forward and out of the runners. If the seat is hard to move, check the runners on the seat frame. Straighten the runners and remove burrs as necessary. To remove the front passenger seat, lift the front edge of the seat until the hook on the backrest disengages from the retainer on the partition panel. The seat is riveted to the backrest and the two parts must be removed as a unit. Lifting the front of the seat causes the backrest to slide down, thus unhooking it.

When you install the front passenger seat, first engage the seat on the two rear brackets that are welded to the floor. Then lower the front edge of the seat until the backrest hooks on its retainer. If the backrest does not engage the retainer, or if the front of the seat cannot be pushed fully down, loosen the two Phillips head screws and move the retainer loop up or down on the partition panel.

Removing and Installing Partition Panel Trim

Fig. 6-1 shows the trim panel removed. To remove the trim panel, first squeeze the handle together and pull it out (right-hand panel only). Then, using a wooden wedge, carefully pry the trim panel off the partition.

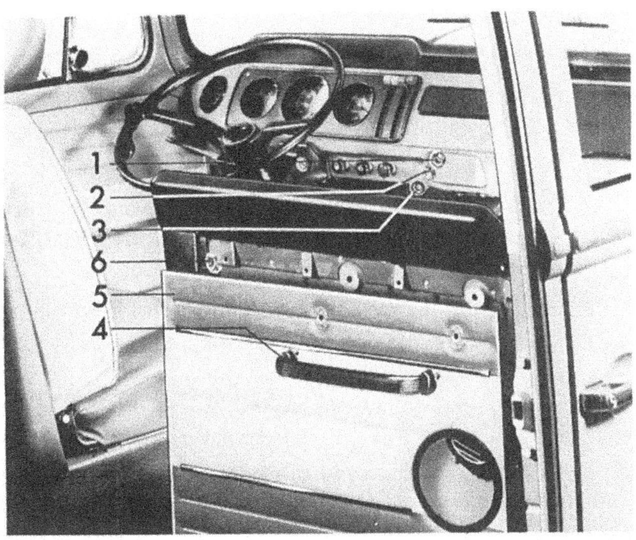

© 1974 VWoA—3129

1. Nut (2)
2. Spring washer (2)
3. Flat washer (2)
4. Handle
5. Trim panel
6. Steel clips (6)

Fig. 6-1. Trim panel removed.

Installation is the reverse of removal. When you install the handle, squeeze it together, push it through the trim panel, then release it. Check to see that it has hooked solidly to the partition.

Once the right-hand trim panel is off, you can replace the padded strips and the seat mounting trim. However, you must also remove the front passenger seat before you can remove the Phillips head screws indicated in Fig. 6-2 and Fig. 6-3.

1

© 1974 VWoA—3120

Fig. 6-2. Screws that hold top padded strip.

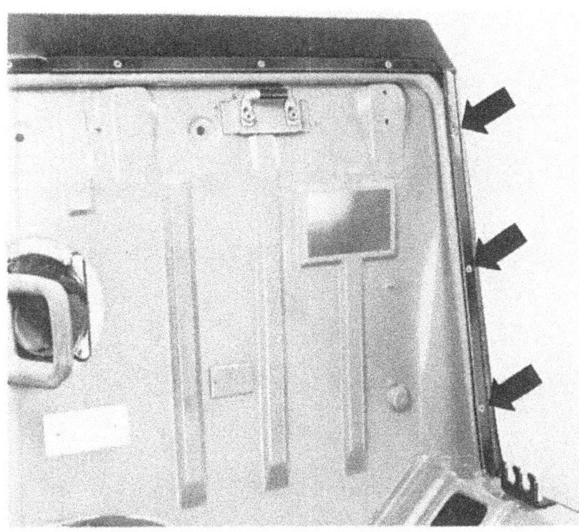

© 1974 VWoA—3124

Fig. 6-3. Screws that hold side padded strip.

NOTE —
After you have removed the Phillips head screws, take off the plastic trim strip that is installed between the screws and the padded strips.

To remove padded strips, drill out the rivets as shown in Fig. 6-4.

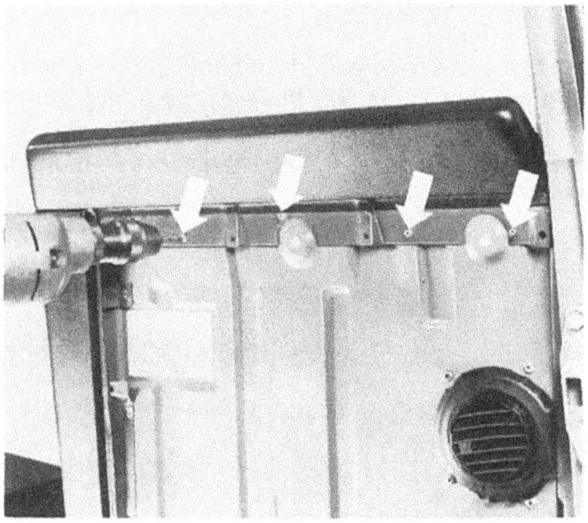

© 1974 VWoA—3121

Fig. 6-4. Rivets for padded strips being drilled out. Four similar rivets hold the side padded strip.

Install the new padded strips with pop rivets as shown in Fig. 6-5.

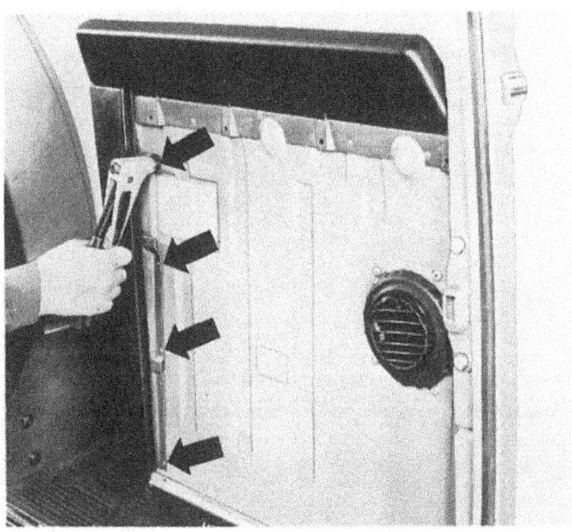

© 1974 VWoA—3126

Fig. 6-5. Pop rivets being installed.

NOTE —
The seat mounting trim strip, which is mounted at the bottom of the partition, is held at the top by three rivets. To replace the trim strip, drill out the rivets as described for the padded strips. Then install the new seat mounting trim strip with pop rivets.

Removing and Installing Side Trim Panels

The side trim panels for the rear passenger compartment and for the luggage compartment are shown removed in Fig. 6-6 and Fig. 6-7. The passenger compartment side trim panel is attached with steel clips and is also held by the armrest and its screws. The luggage compartment side trim panel is held by Phillips head screws. The heads of the screws are covered by plastic caps.

To remove the passenger compartment side trim panel, carefully pry the plastic caps out of the holes for the armrest screws. Remove the screws and take off the armrest.

NOTE —
The screws, which are not easily seen after the caps have been pried off, are installed at an angle.

Using a wooden wedge, pry off the trim panel. Insert the wedge near each of the 19 steel clips, being careful not to damage the vinyl covering on the trim panel or the paint on the body.

Installation is the reverse of removal. During installation, make sure that the side trim panel is flush with the rear side trim panel.

NOTE —
The rear side trim panels, located at each end of the rear seat, are also installed with steel clips. These panels can be removed and installed using a procedure similar to that just described for the side trim panel.

To remove luggage compartment side trim:

1. When removing the left-side trim panel, remove the spare wheel cover and the spare wheel. Then remove the rubber buffers for the spare wheel.
2. Detach the rear seat backrest brackets on both sides of the rear seat, then push the backrest forward.
3. Pry out the plastic caps and remove the five Phillips head screws.
4. Pull the side trim panel forward and out of the retaining channel on the rear body pillar.

Installation is the reverse of removal. Before installing the screws, make sure that the front edge of the luggage compartment side trim panel is flush with the edge of the rear passenger compartment rear side trim panel.

CAUTION —
Be careful not to burr or otherwise damage the luggage compartment side trim panel as you insert it in the retaining channel on the rear body pillar.

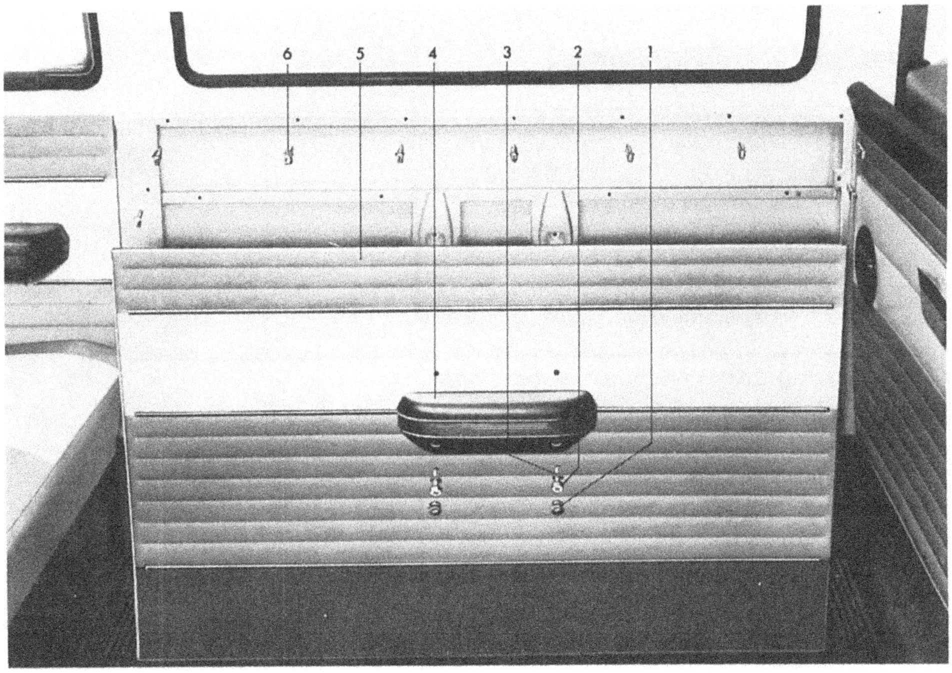

© 1974 VWoA—3130

1. Caps for armrest screws (2)
2. Armrest screw (2)
3. Spring washer (2)
4. Armrest
5. Side trim panel
6. Steel clips (19)

Fig. 6-6. Rear passenger side trim panel removal.

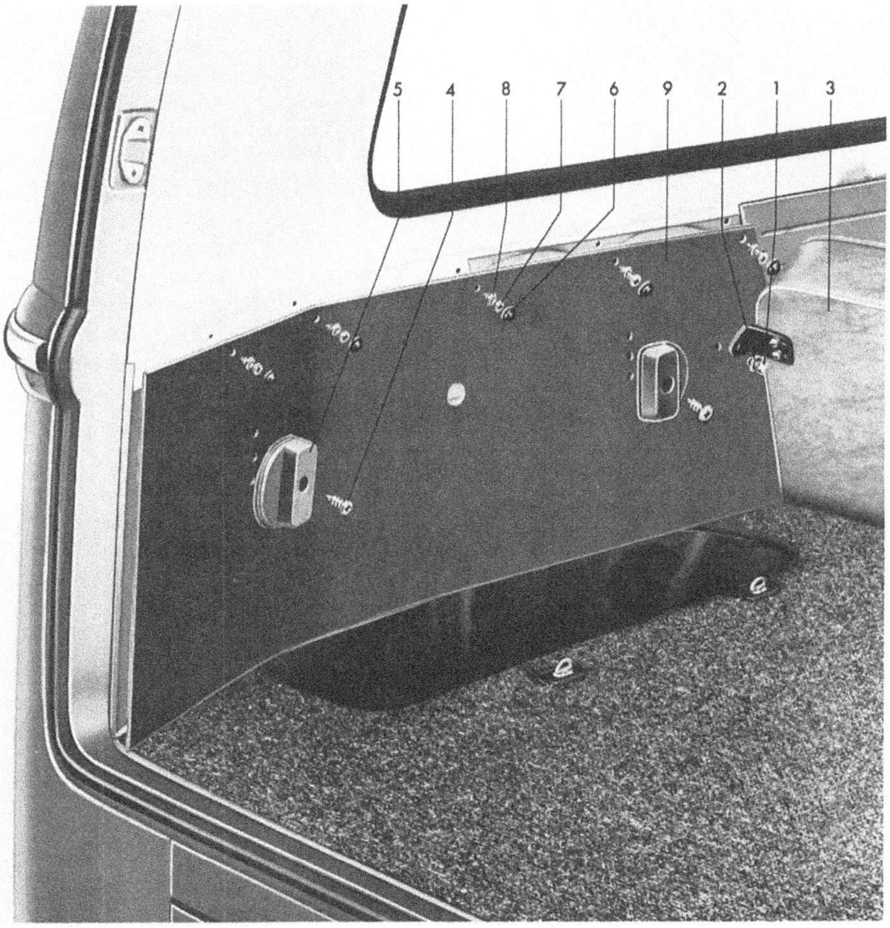

© 1974 VWoA—3131

1. Bolt (2)
2. Spring washer (2)
3. Rear seat backrest
4. Screw (2)
5. Rubber buffer (2)
6. Cap (5)
7. Screw (5)
8. Washer (5)
9. Trim panel

Fig. 6-7. Luggage compartment side trim panel removal.

7. CAMPMOBILE EQUIPMENT

The procedures given here apply mainly to 1968 through 1973 Campmobiles. The publisher regrets that service information for 1974 and later Campmobiles is unavailable from the vehicle's manufacturer. To determine whether a particular procedure applies to a late model Campmobile, please compare the equipment found on the vehicle with the pictures of Campmobile equipment that appear on this page and on following pages.

Removing and Installing Icebox Cabinet

The icebox cabinet is mounted against the partition behind the front passenger seat. To remove the icebox cabinet, remove the front passenger seat. Then remove the two hand nuts indicated in Fig. 7-1.

© 1974 VWoA—3158

Fig. 7-1. Hand nuts (arrows) behind front seat.

Pull the cabinet away from the partition, tilt it slightly, and then lift it up to withdraw the drain hoses from the hole in the floor (Fig. 7-2).

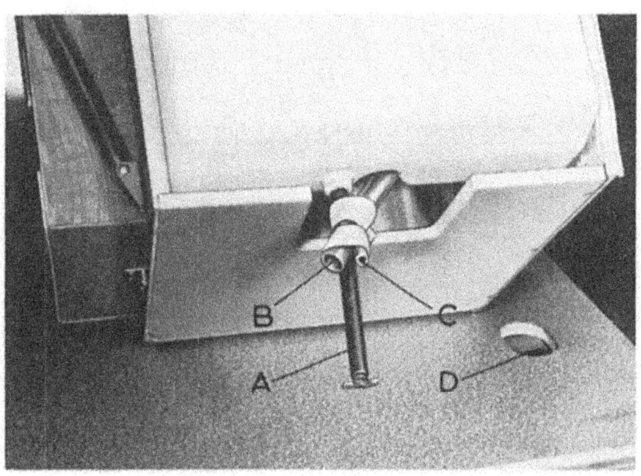

© 1974 VWoA—3159

A. Hose with drain tap for water container
B. Drain hose for sink
C. Drain hose for drip tray
D. Hole in floor

Fig. 7-2. Hoses under icebox cabinet.

Installation is the reverse of removal. Replace the rubber seal (Fig. 7-3) if it is worn or damaged.

© 1974 VWoA—3160

Fig. 7-3. Rubber seal **A** located under floor panel.

Checking, Removing, and Installing Water Pump

If the pump fails to deliver water though the water container is full, remove the drawer from the icebox cabinet. Check to see that the hose from the water container is attached to the pump. (The pump is in the plastic housing that is integral with the base of the spigot.) If the hose is correctly connected, remove the hand nuts (shown earlier in Fig. 7-1) that hold the icebox cabinet on the partition. Check to see that the hose is connected to the water container. If the hose connections are in order, but no water is delivered, remove the two screws from the sides of the spigot housing and remove the pump as shown in Fig. 7-4 for inspection, repair, or replacement. Installation is the reverse of removal.

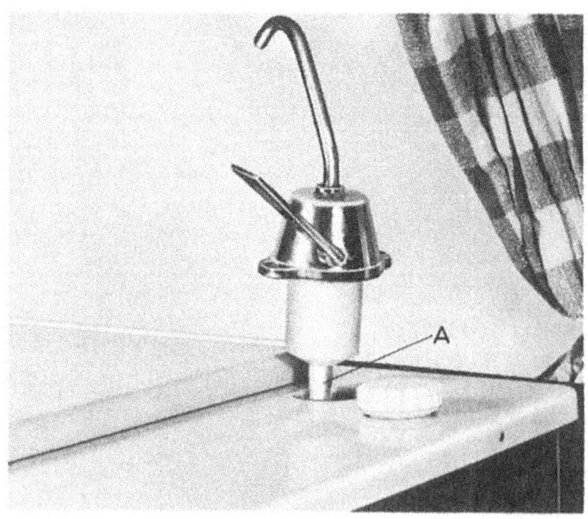

© 1974 VWoA—3706

Fig. 7-4. Pump removal. Detach hose at **A.**

1

Removing and Installing Front Bench Seat

To remove the front bench seat (which is mounted back-to-back with the driver's seat), first lift out the seat cushion. Then take out the four bolts that hold the seat retaining plates to the floor. See Fig. 7-5.

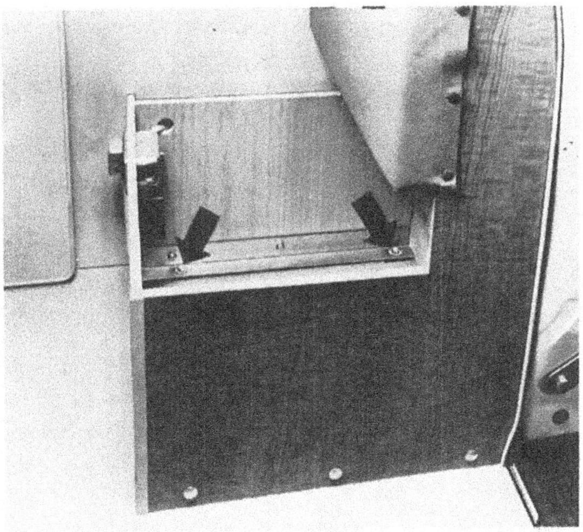

© 1974 VWoA—3174

Fig. 7-5. Two of four bolts (arrows) that hold seat.

Remove the two screws in the face of the electrical receptacle and pull out the receptacle cover. Remove the screw and the wires indicated in Fig. 7-6. Attach tags to the wires as you remove them so that they can be reinstalled on the correct terminals. The seat may then be removed from the vehicle.

Installation is the reverse of removal. When installing the receptacle cover, hold the housing to keep the screws from pushing it off the inside of the seat box.

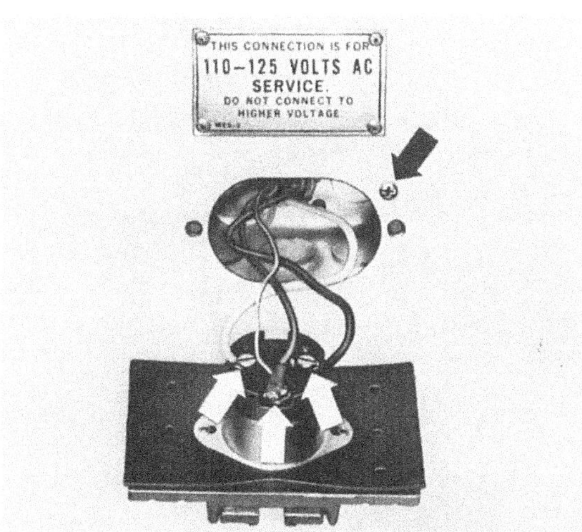

© 1974 VWoA—3173

Fig. 7-6. Screw and wires that hold receptacle.

Removing and Installing Rear Bench Seat

The rear bench seat must be removed together with the storage cabinet that is under it. But before it can be removed, you must remove the front bench seat and the folding table.

To remove rear bench seat:

1. Release the rear bench seat cushion, then lift it up and secure it with the support. (The catches that hold the cushion are inside the storage cabinet and are accessible after you open the cabinet door.)
2. Remove the nuts indicated in Fig. 7-7. Then, working under the vehicle, pull the bolts out of the floor panel. If necessary for access to the left-side bolt, temporarily remove the heater pipe extension.

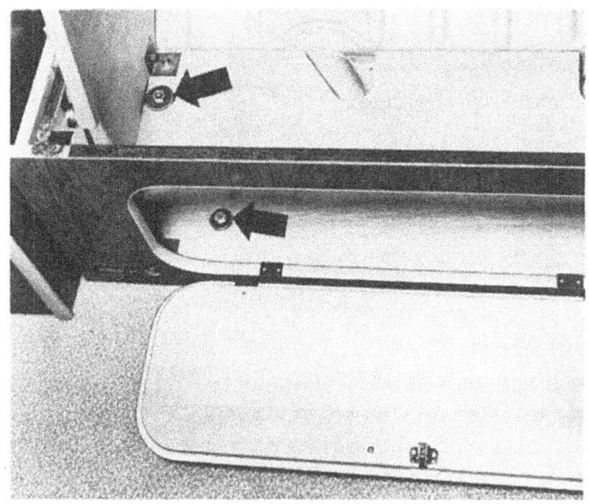

© 1974 VWoA—3175

Fig. 7-7. Nuts and bolts that hold cabinet bottom.

3. Remove the nuts indicated in Fig. 7-8. Then, working under the vehicle, remove the studs and plate.

© 1974 VWoA—3176

Fig. 7-8. Bracket on wheel housing.

4. To detach the seat brackets from the engine cover panel, remove the two Phillips head screws and the bolt indicated in Fig. 7-9 from the right-hand bracket. Remove the two similar screws (but no bolt) from the left-hand bracket.
5. Take the seat out forward.

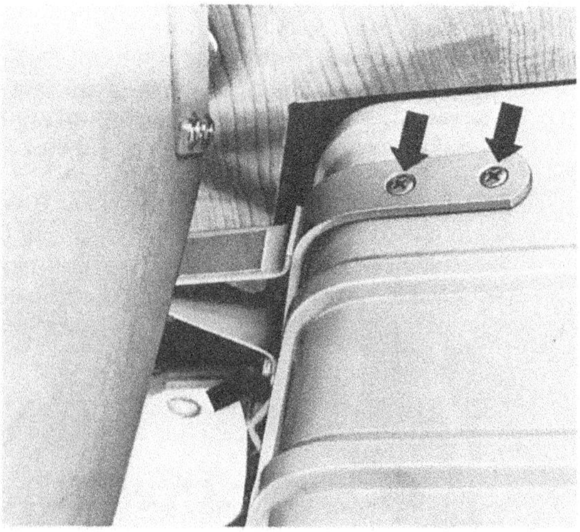

© 1974 VWoA—3177

Fig. 7-9. Screws and bolt (arrows) that hold the right-hand bracket.

Installation is the reverse of removal. The stud plate for the bracket on the wheel housing is shown in Fig. 7-10. Use new self-locking nuts during installation. Also, do not forget the rubber seal since leaks will occur if you do. If the rubber seal is worn or damaged, replace it. Following installation, check to see that seat cushion hinges work properly.

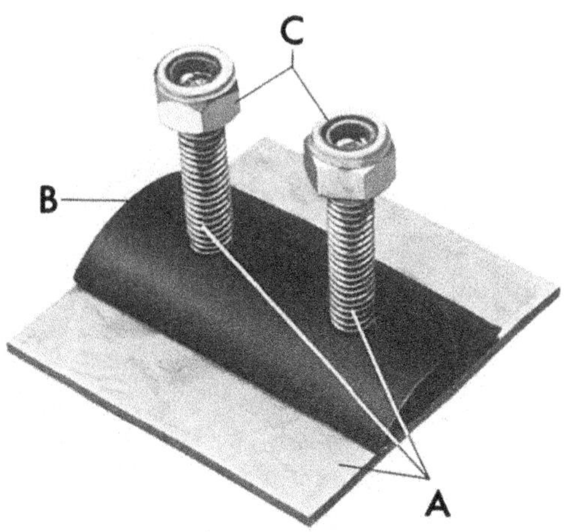

© 1974 VWoA—3178

Fig. 7-10. Stud plate assembly **A**. The rubber seal is at **B**; the self-locking nuts are at **C**. The studs are welded to the plate, but the rubber seal can be removed.

Removing and Installing Linen/Clothes Closet

You must remove the icebox cabinet, the storage shelf, the front bench seat, and the rear bench seat before you can remove the closet. To remove the closet, take out the two Phillips head screws indicated in Fig. 7-11.

© 1974 VWoA—3163

Fig. 7-11. Screws (arrows) that hold the closet front brackets to the body. The sliding door must be opened for access to the screws.

After removing the hand nut just inside the linen closet door (Fig. 7-12), you can swing the closet sideways, pull it forward, and lift it out of the car.

© 1974 VWoA—3164

Fig. 7-12. Hand nut (arrow) at rear of closet.

Installation is the reverse of removal. Install the two Phillips head screws at the front before you secure the rear of the closet. Insert the rear bracket through the hole in the cabinet and engage it over the edge of the rear side panel as shown in Fig. 7-13. Install the hand nut. If necessary, tighten the locknut against the outside of the closet after the hand nut is installed.

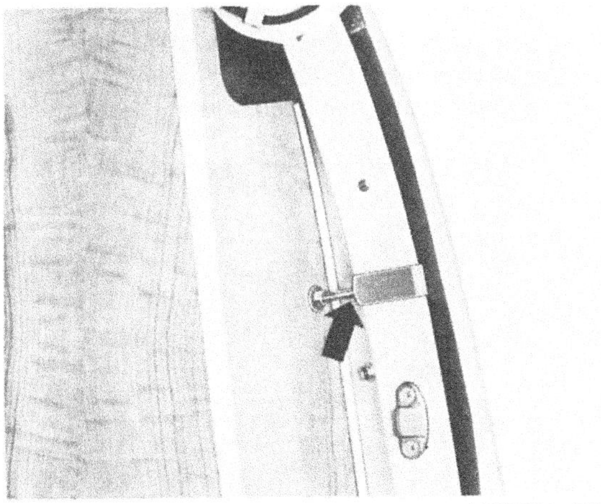

© 1974 VWoA—3165

Fig. 7-13. Position of rear bracket (arrow) on closet and on body rear side panel. The luggage compartment lid must be opened for access.

Eliminating Closet Door Rattle

If the closet doors rattle, use trim cement to stick small pieces of foam rubber on the door edges as shown in Fig. 7-14. If the rattle persists, move the latch plate slightly inward (Fig. 7-15).

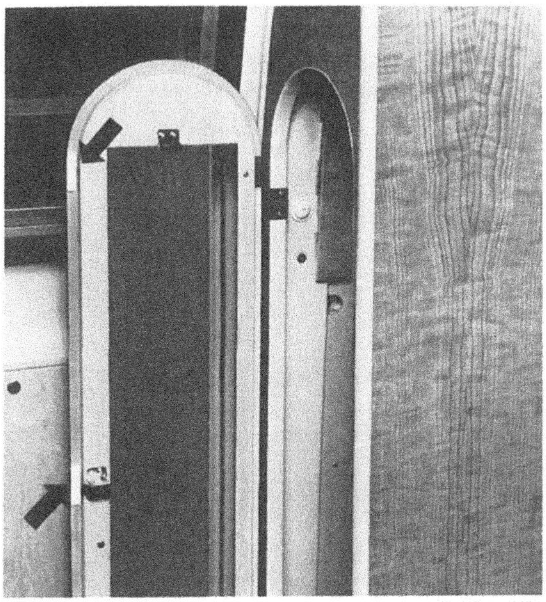

© 1974 VWoA—3170

Fig. 7-14. Foam rubber strips (arrows) applied to door.

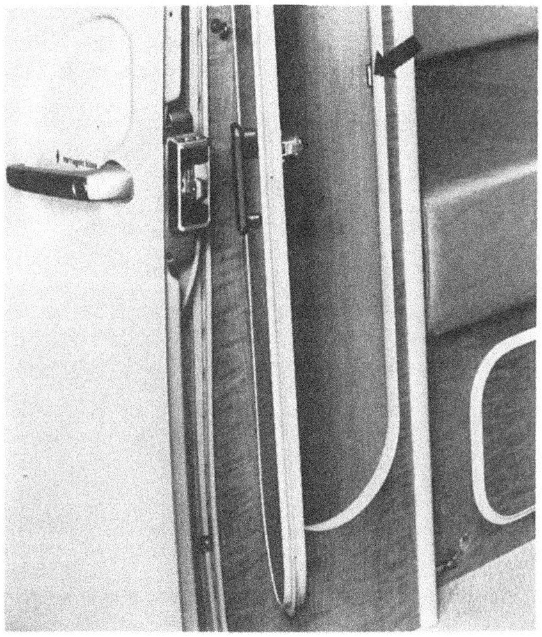

© 1974 VWoA—3171

Fig. 7-15. Latch plate. Move deeper into closet.

Removing and Installing Storage Shelf

The storage shelf, located at the rear of the Campmobile, must be removed before you can remove the clothes/linen closet. It is helpful to have someone support the storage shelf after it is partially unbolted, freeing you to complete the removal procedures. If you attempt the job by yourself, be careful that the shelf does not fall or shift its position in such a way that it damages the interior trim.

To remove shelf:

1. Remove the two screws indicated in Fig. 7-16.

© 1974 VWoA—3166

Fig. 7-16. Screws (arrows) that hold left side of shelf.

2. Remove the nut, washer, and carriage bolt at the right-hand side of the storage shelf. Push the bolt back into the linen/clothes closet (Fig. 7-17).

© 1974 VWoA—3167

Fig. 7-17. Bolt at left-hand side of storage shelf.

3. Remove the five screws with press-button retainers (Fig. 7-18).

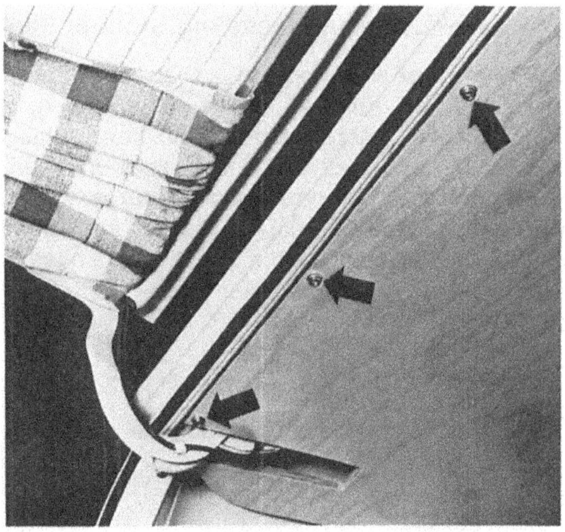

© 1974 VWoA—3168

Fig. 7-18. Three of the five screws (arrows) that hold shelf to the roof frame. Luggage compartment lid must be raised for access to the screws.

4. Using a screwdriver at the left-hand side of the storage shelf, pry the bracket down toward the side window. (This is the bracket you removed the screws from in Step 1.) Then carefully take the storage shelf out downward.

Installation is the reverse of removal. Again pry down the left-side bracket while you lift the storage shelf into position.

Removing and Installing Plywood Roof Trim

Though it makes the job easier, it is not absolutely necessary to remove the side roof trim panel over the sliding door before you remove the front roof trim. However, both side trim panels must be removed before you can remove the rear roof trim.

To remove the side roof trim panel above the sliding door, pull the panel out of its retaining and locating channels as shown in Fig. 7-19. Then take the locating channels off the roof trim. Installation is the reverse of removal.

© 1974 VWoA—3183

Fig. 7-19. Panel above sliding door being removed. Locating channel is at **A**, retaining channel at **B**, and roof trim panel at **C**.

The side roof trim panel above the side window is shown in Fig. 7-20. To remove the panel, first remove the interior lamp and disconnect it from its wire. To remove the panel, pull it out of the retaining and locating channels. Then take the locating channels off the roof trim. Installation is the reverse of removal.

© 1974 VWoA—3184

Fig. 7-20. Panel above side window and related parts. Roof trim panel is at **A**, locating channel at **B**, and retaining channel at **C**.

To remove front roof trim:

1. Remove the roof side trim panel over the side window.
2. Slightly loosen the four Phillips head screws that hold the retaining plate for the folding cot. Then fully remove the two self-locking nuts and screws.
3. Pull the beading (Fig. 7-21) off the roof frame until the beading no longer contacts the front roof panel.

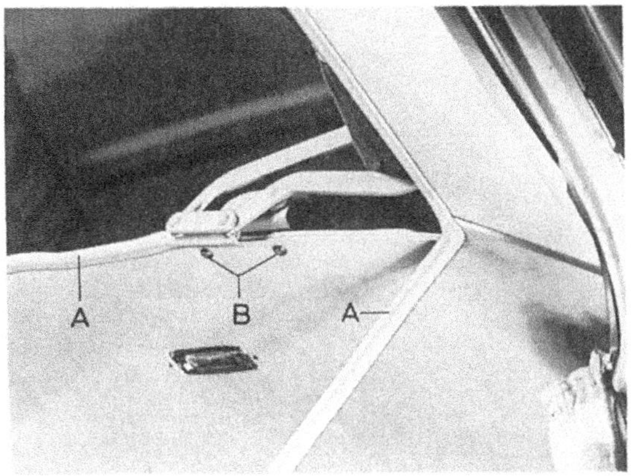

Fig. 7-21. Screws with self-locking nuts **B** that must be removed before you can remove beading **A.**

4. Remove the sun visors together with their retainers, the rear view mirror, and the interior lamp. Then remove the Phillips head screws indicated in Fig. 7-22.

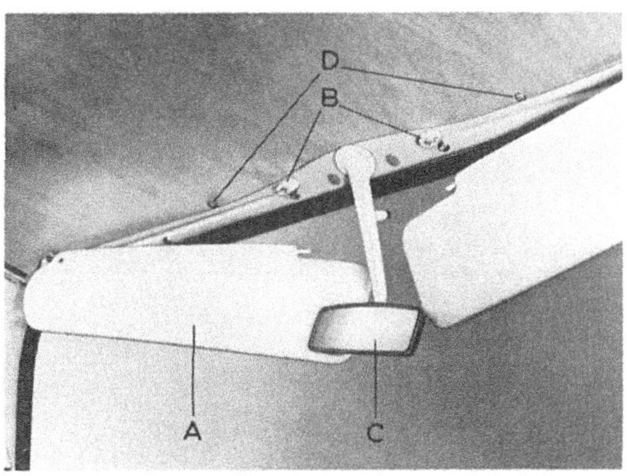

Fig. 7-22. Sun visors **A**, retainers **B**, rear view mirror **C**, and Phillips head screws **D.**

5. Pull down the roof front trim panel in the center as shown in Fig. 7-23. Press it out of the retaining channel on one side or the other, then pull the panel out of the retaining channel on the opposite side and remove the panel.

Fig. 7-23. Roof front panel being removed.

Installation is the reverse of removal. Install the trim panel in one retaining channel, bend the panel down in the center, then locate it in the other retaining channel. Lift the plate for the folding cot while you install the beading on the edge of the roof frame. Use new self-locking nuts for the cot screws.

To remove roof rear trim panels:

1. Remove the roof side trim panels over the sliding door and the side window. Then remove the storage shelf as previously described or remove the luggage net on vehicles that are so equipped.
2. Pull the beading off the front of the roof frame until it no longer contacts the trim panel. Then remove the screws and self-locking nuts indicated in Fig. 7-24.

Fig. 7-24. Folding cot screws with self-locking nuts (arrows).

3. Remove the three Phillips head screws that are near the rear edge of the trim panel. Also remove the securing bracket below the screws (Fig. 7-25).

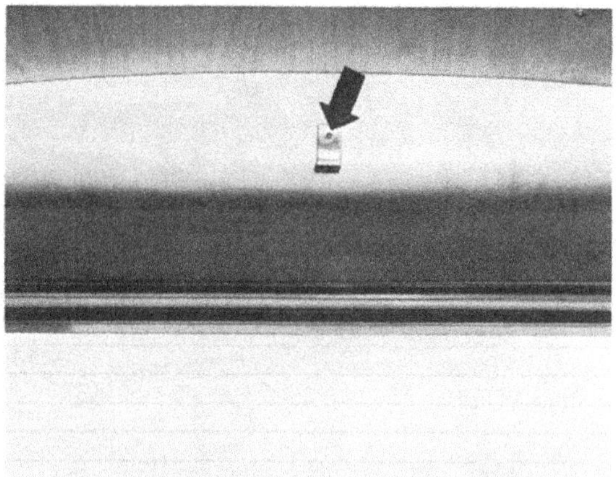

© 1974 VWoA—3187

Fig. 7-25. Securing bracket (arrow). Phillips head screws in roof panel are just above the bracket.

4. Pull the panel down in the center, then carefully lift its edges out of the retaining channels. Remove the trim panel from the vehicle.
5. If necessary, raise the luggage compartment lid, then remove the Phillips head screws with press-button fasteners indicated in Fig. 7-26. Then remove the end part of the roof trim.

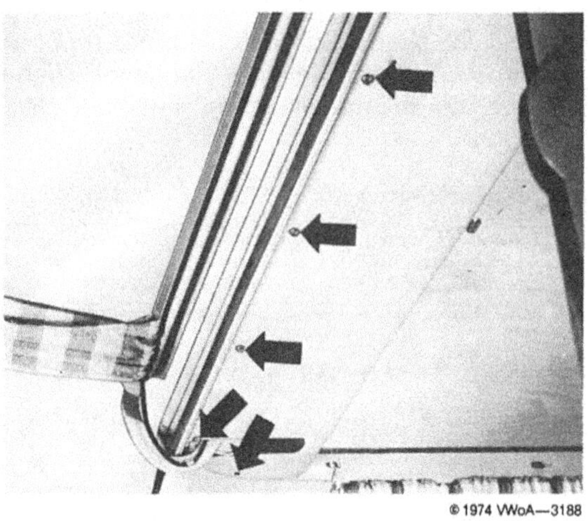

© 1974 VWoA—3188

Fig. 7-26. Five screws with press-button fasteners.

Installation is the reverse of removal. Install the trim panel in one retaining channel, bend the panel down in the center, then locate it in the other retaining channel. Use new self-locking nuts on the screws for the folding bed retaining plate.

Removing and Installing Louver (jalousie) Window

To remove the jalousie side window, first remove the curtain and curtain rail by taking out the screws. Push the mosquito net screen up about 5 mm ($\frac{7}{32}$ in.), then pull the screen down and out of the frame. Pull off the rubber weatherstrip as shown in Fig. 7-27.

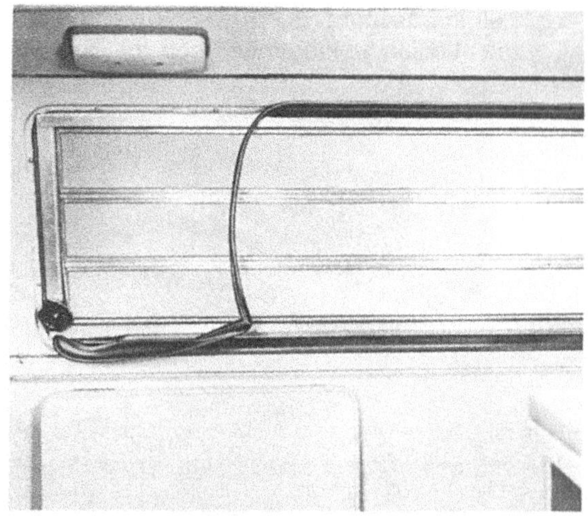

© 1974 VWoA—3189

Fig. 7-27. Rubber weatherstrip partially removed.

Remove the 15 Phillips head screws around the window frame (Fig. 7-28), then carefully hand-press the jalousie window out of its opening in the body.

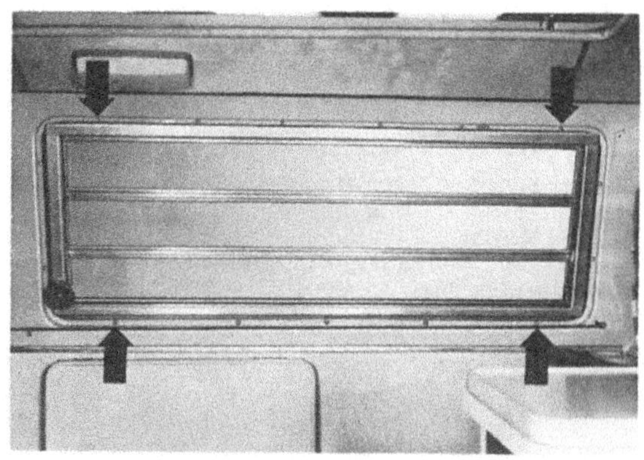

© 1974 VWoA—3190

Fig. 7-28. Four of 15 screws (arrows) that hold the jalousie window in body.

Installation is the reverse of removal. Make sure that the rubber weatherstrip on the window's outer seat is properly positioned, then install the window and frame upper edge first. Install the four screws indicated in Fig. 7-28, and then the remaining 11 screws. When installing the inner rubber seal, do so with the wide part of the seal along the bottom of the window.

1

7.1 Front Hinged Roof

The front hinged roof, available on Campmobiles, provides space for a folding cot. A Campmobile with the front hinged roof raised is shown in Fig. 7-29.

© 1974 VWoA—3199

Fig. 7-29. Front hinged roof in its raised position.

Removing and Installing Front Hinged Roof

The front hinged roof must be in its raised position before you can remove it.

To remove:

1. Remove the screws and detach the strips that hold the bellows on the roof panel (Fig. 7-30).

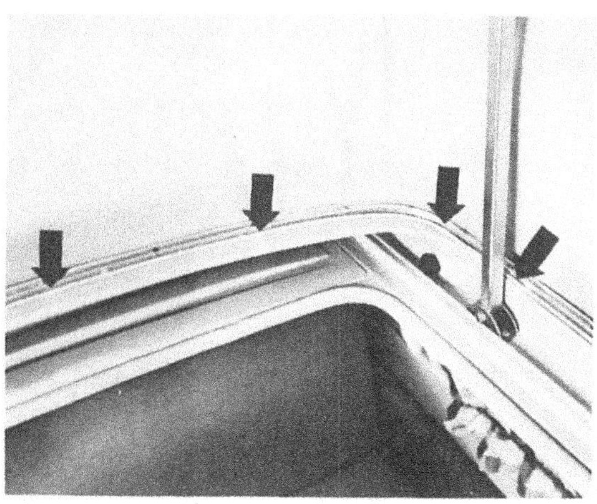

© 1974 VWoA—3200

Fig. 7-30. Strips that hold the bellows on the body. Remove screws at arrow.

2. Remove the hinge mounting bolts indicated in Fig. 7-31. Have someone hold the roof firmly while you remove the bolts on the opposite side.

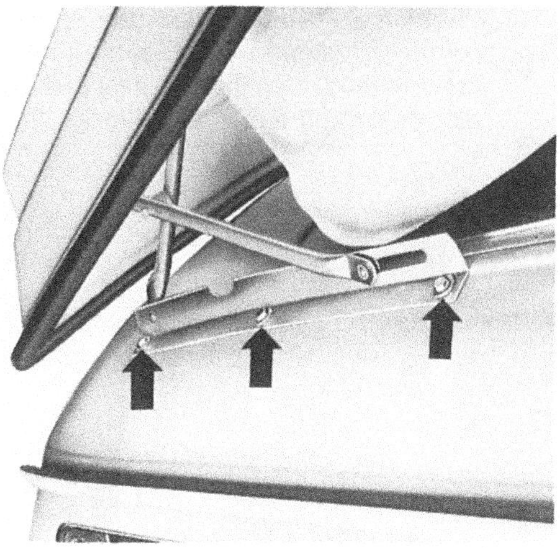

© 1974 VWoA—3201

Fig. 7-31. Bolts (arrows) that hold hinge on roof panel. Be careful not to let the unbolted hinge scratch the paint on the roof panel.

3. Remove the two Phillips head screws indicated in Fig. 7-32 from both roof support retaining plates. Then, while holding the hinges and supports to avoid damaging the paint, carefully lift off the roof.

© 1974 VWoA—3202

Fig. 7-32. Screws (arrows) that hold the roof supports.

To install:

1. Place the roof in position and loosely install the hinge mounting bolts.
2. Carefully align the roof with the ribs in the luggage pan at the rear part of the vehicle roof. Then tighten the hinge mounting bolts.
3. Using the Phillips head screws, install both roof supports on the roof frame.
4. Carefully lower the roof.

5. Check the position of the rear rubber seal at the points indicated in Fig. 7-33. If necessary, reposition the hinges so that the seal makes proper contact all along the rear roof edge.

© 1974 VWoA—3204

Fig. 7-33. Points where seal contact should be checked (arrows). Check on opposite side also.

6. Using trim cement, glue the plastic molding to the vehicle roof. Then position the bellows on the strip.
7. Using the screws, install the bellows-securing strips in the following order: front, rear, and sides.

Removing and Installing Roof Seal

To remove the rubber seal, raise the hinged roof, then pull off the seal as shown in Fig. 7-34.

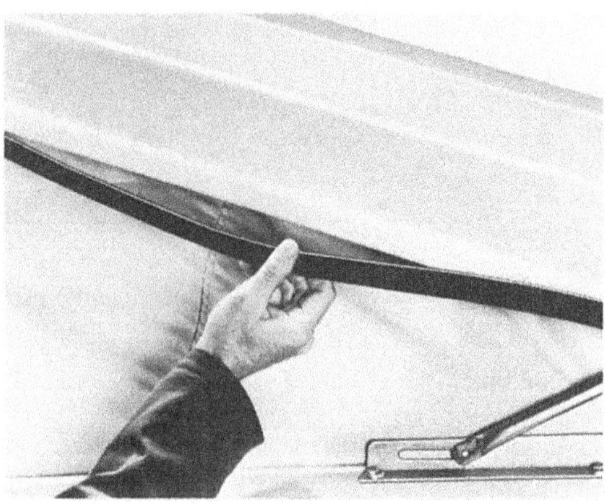

© 1974 VWoA—3704

Fig. 7-34. Seal being removed from front hinged roof.

The cross section of the rear seal is different from the cross section of the seals used at the front and sides. This difference is shown in Fig. 7-35.

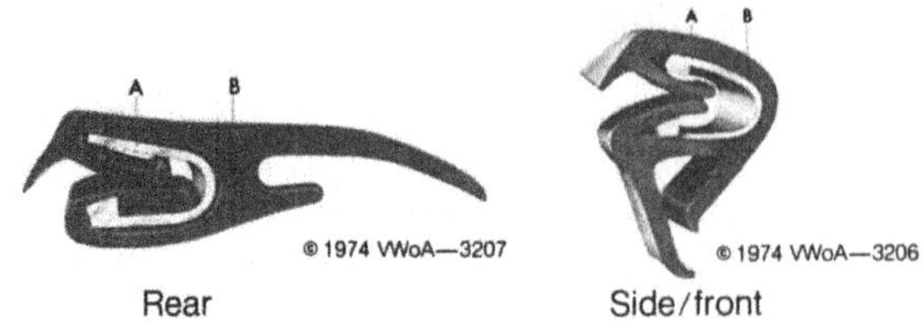

Fig. 7-35. Cross section of rear seal (left); cross section of front and side seals (right). The rubber seals are at **B**; the steel clips at **A.**

Coat the new seal with glycerine or silicone spray. Then hand-press the seal over the lip on the hinged roof. Fully drive the seal over the lip as shown in Fig. 7-36.

NOTE —
If the steel clips fit loosely, squeeze them together with pliers to restore their tension. Pull the seal off at appropriate points while doing this.

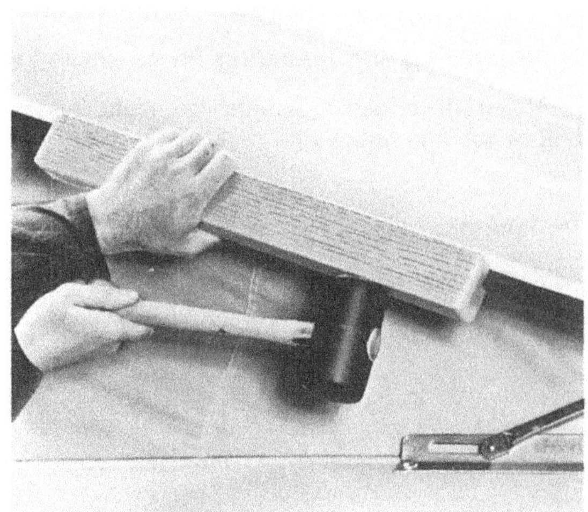

© 1974 VWoA—3705

Fig. 7-36. A wooden block and rubber mallet being used to drive the seal fully into place.

Removing and Installing Bellows

Though the front hinged roof is shown removed in the illustrations, you do not need to remove it in replacing the bellows. If the existing bellows is not to be reinstalled, cut the material off just above the vehicle roof so that the screws in the securing strips are more readily accessible.

To remove the bellows, remove the screws and securing strips around the lower edge of the bellows. Pull out the staples that hold the plastic strips around the upper edge of the bellows and remove the plastic strips as shown in Fig. 7-37. Then remove the additional staples that hold the bellows to the wooden frame and remove the bellows from the front hinged roof.

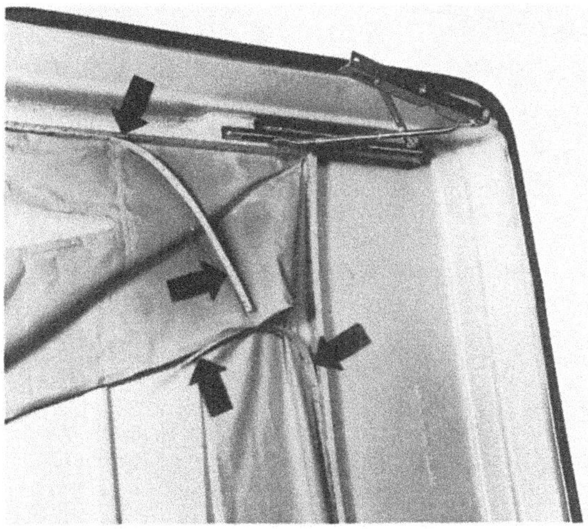

© 1974 VWoA—3208

Fig. 7-37. Plastic strips being removed. The top and right-hand arrows indicate the staples. The two central arrows indicate the plastic strips.

When you install the bellows, first tack or staple it to the wooden frame at the four corners. Then tack or staple the material to the wooden frame all around and staple on the plastic strips as shown in Fig. 7-38.

© 1974 VWoA—3210

Fig. 7-38. Staple gun being used to attach the bellows and plastic strips to the wooden frame.

To install the lower edge of the bellows on the vehicle roof, first use trim cement to glue the plastic molding to the vehicle roof. Then position the bellows on the strip. Using the screws, install the bellows-securing strips in the following order: front, rear, and sides.

Removing and Installing Hinges and Roof Supports

The hinged roof supports are held on the vehicle roof by two Phillips head screws, as shown earlier in the procedures for removing and installing the front hinged roof. A single bolt with a self-locking nut holds each support on its top bracket. Use new self-locking nuts during installation. Install the supports first at the top and then at the bottom.

When you remove a hinge, support the unbolted side of the roof so that it does not shift and bend the hinge on the opposite side. If both hinges are being replaced, replace them one at a time.

To remove the hinge, raise the front hinged roof and then remove the three bolts and the two self-locking nuts indicated in Fig. 7-39. Be careful not to scratch the paint as you remove the hinge from the vehicle.

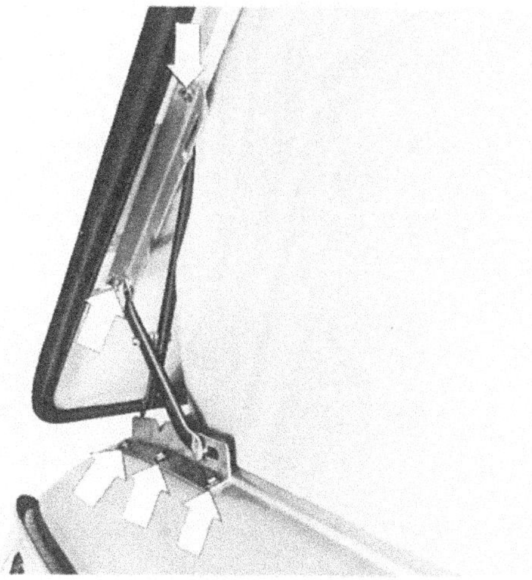

© 1974 VWoA—3211

Fig. 7-39. Bolts (lower three arrows) and self-locking nuts (upper two arrows).

Installation is the reverse of removal. Use new self-locking nuts on the bolts in the hinged roof. Loosely install the three bolts that hold the hinge on the vehicle roof, then carefully close the roof. Align the hinged roof along its rear edge, then tighten the three bolts that hold the hinge on the vehicle roof.

Removing and Installing Folding Cot

You do not need to fully remove the folding cot in order to replace the canvas cover and side tubes. The removal of the canvas cover and side tubes is described following the regular removal and installation procedure.

1

The front hinged roof must be raised before the cot can be unfolded into its sleeping position as shown in Fig. 7-40.

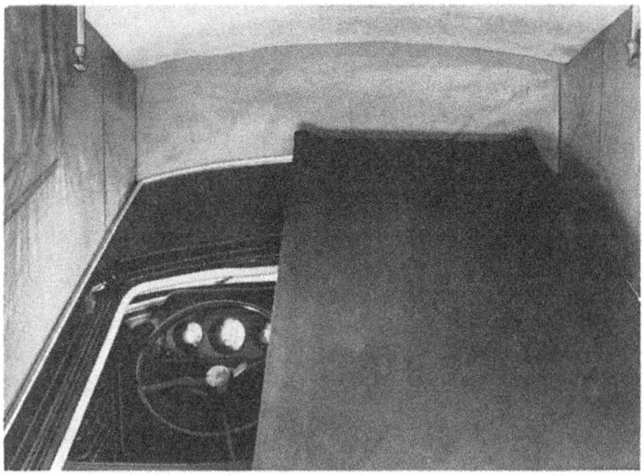

© 1974 VWoA—3212

Fig. 7-40. Folding cot ready for use.

To remove folding cot:

1. Raise the front hinged roof, but do not unfold the cot.
2. Remove the four Phillips head screws that hold the front folding cot hinge to the vehicle roof (Fig. 7-41).

© 1974 VWoA—3213

Fig. 7-41. Phillips head screws that hold the folding cot front hinge to the vehicle roof.

3. Remove the two screws indicated in Fig. 7-42 from the front hinge.

NOTE —
Use a wrench from above to hold the self-locking nuts.

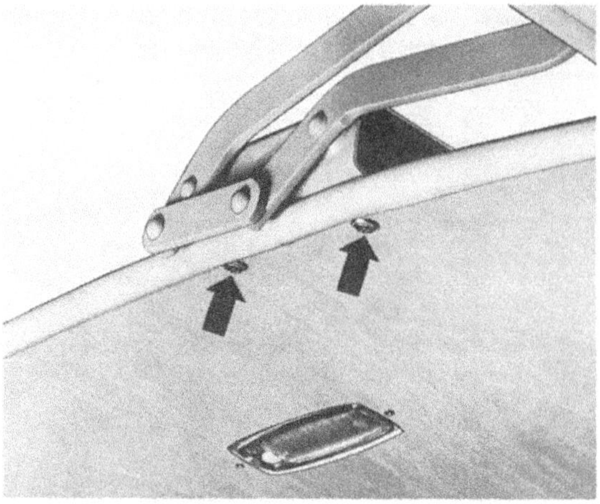

© 1974 VWoA—3214

Fig. 7-42. Phillips head screws with self-locking nuts. These screws hold the front hinge on the trim panel and roof pan.

4. Remove the four Phillips head screws and two self-locking nuts indicated in Fig. 7-43. Then carefully remove the folding cot from the vehicle.

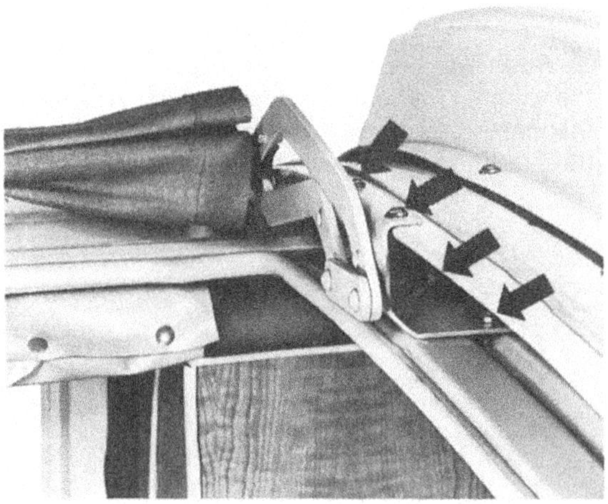

© 1974 VWoA—3215

Fig. 7-43. Screws that hold rear hinge. The screws indicated by the left two arrows are threaded into the roof frame. The screws indicated by the right two arrows have self-locking nuts (visible).

Installation is the reverse of removal. Use new self-locking nuts when you install the screws that pass through the trim panels and roof pan.

To remove the folding cot tube and canvas:

1. Remove the self-locking nuts from the bolts in the front and rear hinges. This may be done with the cot installed inside the vehicle.

2. Push the bolts out of the holes in the hinge arms and connecting tube brackets (Fig. 7-44).

Fig. 7-44. Connecting tube mounting on hinge. The connecting tube with bracket is at **A**, the bolt at **B**, and the self-locking nut at **C.**

3. Carefully lower the tubes and canvas into the vehicle interior. Then withdraw the tubes from the canvas.

Installation is the reverse of removal. Use new self-locking nuts on the bolts that hold the tubes on the hinges.

Removing and Installing Roof Luggage Rack

To remove the luggage rack, first remove the four Phillips head screws from the front edge of the rack (Fig. 7-45).

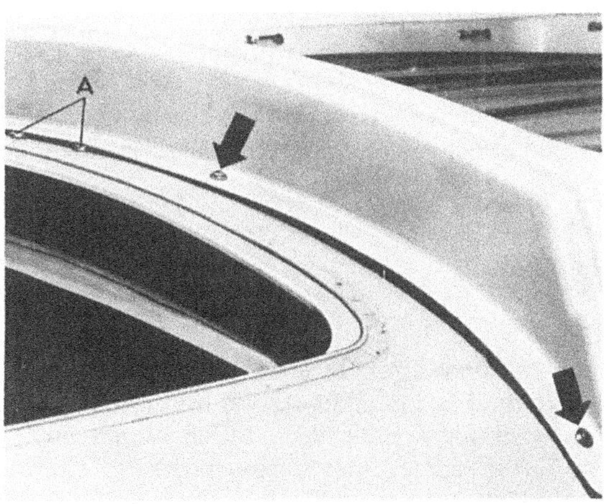

Fig. 7-45. Screws at front edge of luggage rack. Arrows indicate two of the four Phillips head screws. There are two rubber spacer washers at **A**. Two plastic caps must be pried off for access to the two center screws.

Remove the three Phillips head screws at the rear edge of the luggage rack (Fig. 7-46). Lift the luggage rack off the vehicle and remove the three rubber spacers so that they will not be lost.

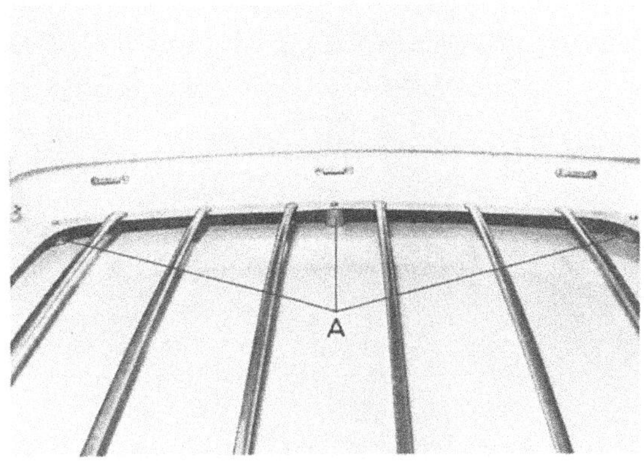

Fig. 7-46. Phillips head screws with rubber spacers **A** that hold the rear of the luggage rack.

Installation is the reverse of removal. Do not forget the two rubber spacer washers that go between the front of the luggage rack and the center of the roof. Carefully press the plastic caps over the two front center screws after they have been installed. Tighten the screws at the rear of the luggage rack until they hold the rack firmly but not so tightly that the rubber spacers are compressed.

8. SLIDING ROOF

(sun roof)

The sliding steel roof (sun roof) is optional equipment on the vehicles covered by this Manual. The trim panel cannot be removed until you have removed the sliding roof itself and have removed the left side runner.

8.1 Removing and Installing Sliding Roof

The sliding roof can be adjusted without removing it from the vehicle. If the sliding roof does not operate properly, read **8.4 Adjusting Cables** and **8.5 Adjusting Sliding Roof Height** before deciding whether it is necessary to remove the sliding roof.

To remove sliding roof:

1. Using the hand crank, slightly open the sliding roof.

NOTE —
A gap of from 125 to 150 mm (5 or 6 in.) should be adequate.

1

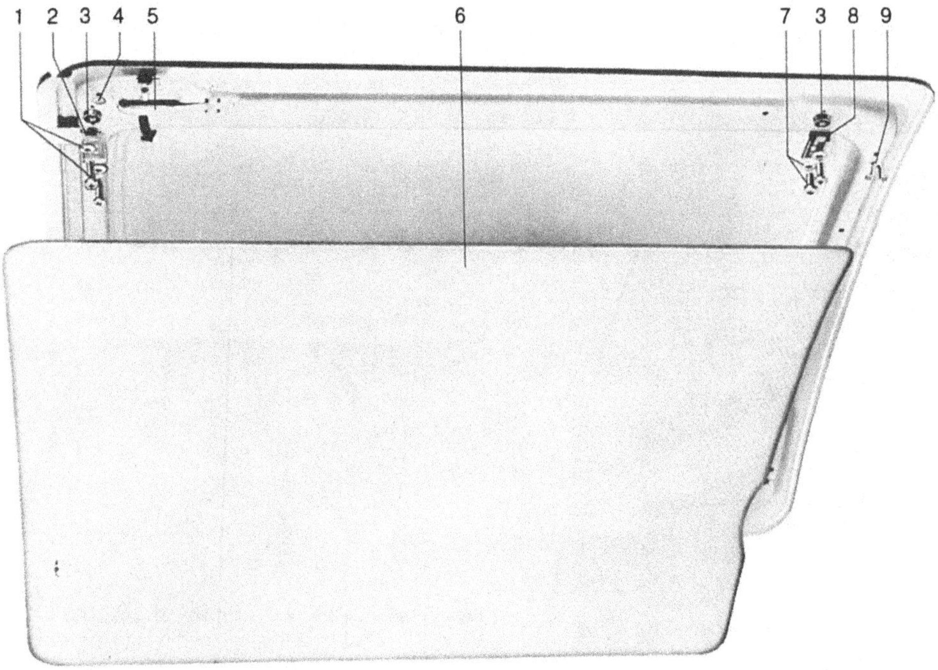

1. Phillips head screw with spring washer (4)
2. Rear guide (2)
3. Height adjusting screw (4)
4. Panel plug (2)
5. Leaf spring (swings aside as indicated by arrow) (2)
6. Trim panel
7. Phillips head screw with spring washer (4)
8. Front guide (2)
9. Clip for trim panel (5)

Fig. 8-1. Components of sliding roof panel (viewed from right-hand side).

2. Using a wooden wedge, carefully pry the front edge of the trim panel off the sliding roof panel. Work as closely as possible to each of the five clips (see Fig. 8-1).
3. Push the trim panel to the rear as far as it will go.
4. Using the hand crank, close the sliding roof until it is only open about 50 mm (2 in.). Then take out the four Phillips head screws with spring washers and remove the front guides.
5. Unhook the leaf springs from the lifters. Then swing the springs inward as indicated by the arrow in Fig. 8-1.
6. Pull the left and right lifters—located at the rear on each cable—out of the brackets on the roof, then turn the lifters down.
7. Taking care not to damage the seal or scratch the paint, lift the sliding roof panel out through the top of the roof opening. (The rear guides must come out through the recesses in the side runners.)

To install:

1. Insert the sliding roof panel into the roof opening rear edge first and engage the rear guides in the runners.
2. Slowly push the sliding roof panel to the rear while gradually lowering the front edge into position.
3. Pull the sliding panel fully forward. Insert the left and right lifters—located on the ends of the cables—into the brackets on the sliding roof. Then hook the leaf springs over the lifters.
4. Using the Phillips head screws and spring washers, install the front guides.
5. Adjust the sliding roof as described in **8.4 Adjusting Cables** and **8.5 Adjusting Sliding Roof Height.**
6. Using the hand crank, open the sliding roof halfway.
7. Pull the trim panel forward and attach it to the sliding roof by hand-pressing in the five clips.

8.2 Removing and Installing Sliding Roof Trim Panel

The sliding roof trim panel should be removed only if the trim panel itself must be replaced or repaired.

To remove:

1. Remove the sliding roof panel as described in **8.1 Removing and Installing Sliding Roof.**
2. Working through the roof opening, remove the eight Phillips head screws and the single countersunk screw that hold the left runner on the body. (The runners and related parts are shown in Fig. 8-2.)
3. Pull the left runner and its rear retainer forward out of the bracket and lift it out through the top of the roof opening.
4. Carefully take out the trim panel—left side first—through the top of the roof opening.

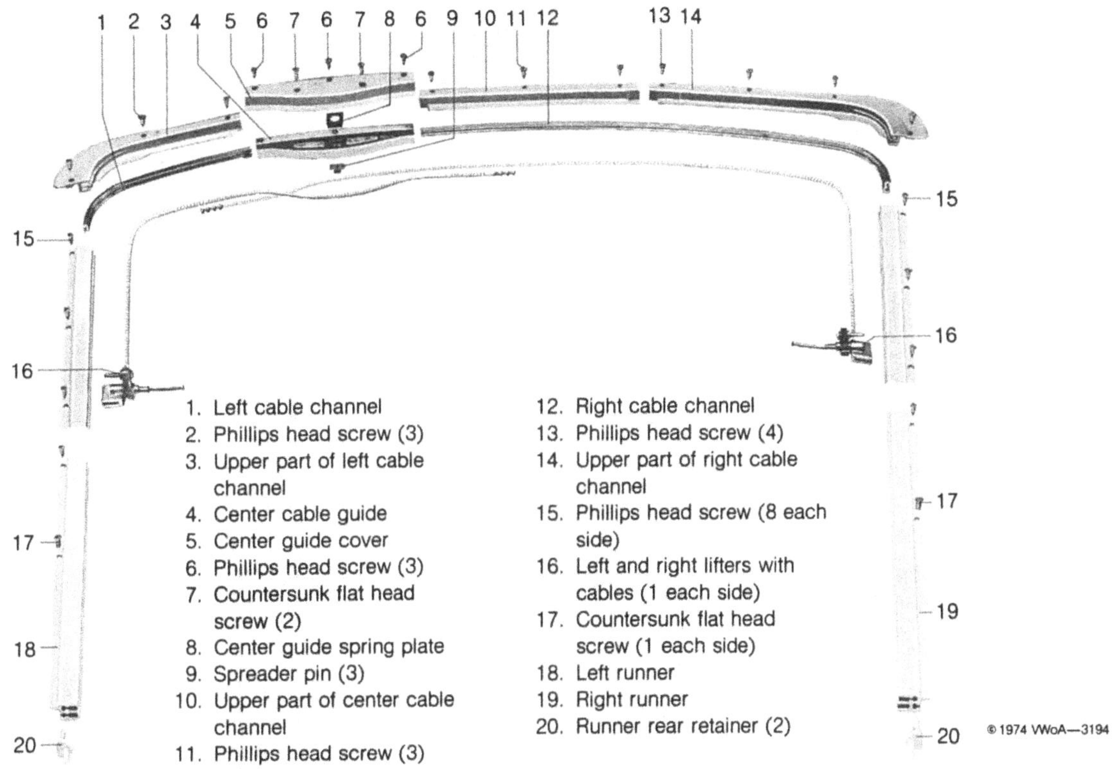

Fig. 8-2. Runners, cables, and related parts.

To install trim panel:

1. Inspect the trim panel frame, trim panel, and the steel clips. Replace worn or damaged parts. Dirty trim panels can be cleaned with plastic cleaner.
2. Place the trim panel in the runners. Then install the left runner.
3. Install and adjust the sliding roof panel. See **8.1 Removing and Installing Sliding Roof.**
4. Pull the trim panel forward and attach it to the sliding roof by hand-pressing in the five clips.

8.3 Removing and Installing Runners and Cables

In removing and installing the runners, cables, and other parts shown in Fig. 8-2, be careful not to scratch the paint on the roof. To prevent rust, touch up any paint scratches that you find.

To remove:

1. Remove the sliding roof as described in **8.1 Removing and Installing Sliding Roof.** It is not necessary to remove trim panel.
2. Remove the center guide cover. To do this, remove the three Phillips head screws and the two countersunk flat head screws.
3. Remove the upper part of the center cable channel by taking out three screws.
4. Remove the upper part of left cable channel by taking out three screws.
5. Remove the upper part of the right cable channel by taking out three screws.
6. Remove the center guide spring plate.
7. To remove the left and right side runners, remove the 16 Phillips head screws and the two countersunk flat head screws. Pull each runner together with its rear retainer forward out of the brackets, and then out through the top of the roof opening.
8. Lift out the center cable guide and the left and right cable channels.
9. Pull the cables and lifters off the runners.

To install:

1. Reversing the removal procedures, loosely install the side runners and lifters with cables in the sliding roof opening.
2. If previously removed, install the trim panel before installing the left runner.

3. Carefully align the side runners. Then tighten the screws.

CAUTION —
The runner rear retainers must fit tightly in their brackets. If improperly aligned, the side runners will shift sideways as the roof is opened and closed, thus preventing the roof from working smoothly.

4. Install the center cable guide and the left and right cable channels.
5. Position the cables as shown in Fig. 8-2.
6. Install the left, right, and center cable channel upper parts. Do not forget the center guide spring plate and the spreader pin.
7. Install the sliding roof as described in **8.1 Removing and Installing Sliding Roof**; adjust the roof as described under the following two headings.

8.4 Adjusting Cables

The cables should be adjusted after removing and installing the sliding roof panel and whenever the sliding roof fails to open and close evenly at both sides.

To adjust:

1. Using the hand crank, slightly open the sliding roof.

NOTE —
A gap of from 125 to 150 mm (5 or 6 in.) should be adequate.

2. Using a wooden wedge, carefully pry the front edge of the trim panel off the sliding roof panel. Work as close as possible to each of the five clips. (See **8.1 Removing and Installing Sliding Roof.**)
3. Push the trim panel to the rear as far as it will go.
4. Using the hand crank, fully close the sliding roof.
5. Carefully pry the plastic cap off the hub of the hand crank. Then remove the Phillips head screw and take off the hand crank and escutcheon. (See Fig. 8-3.)
6. Loosen by approximately six turns each the two Phillips head screws that hold the cable drive gear assembly on the roof.
7. Pull the cable drive gear assembly down until the gear no longer engages the cable. (When disengaged, the gearshaft can easily be finger-turned.)

1. Cable drive gear assembly
2. Phillips head screw (2)
3. Escutcheon
4. Hand crank
5. Short Phillips head screw
6. Plastic cap

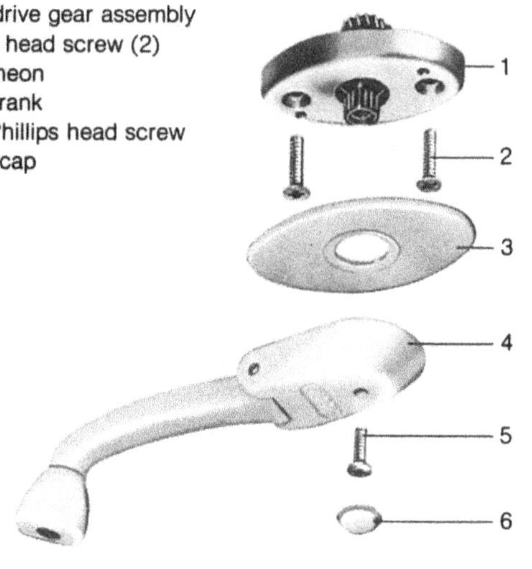

Fig. 8-3. Components of hand crank and cable drive gear.

8. Check the position of the sliding roof in the roof opening. If necessary, hand-shift the roof until it is square with the opening.
9. Place the lifters—located at the rear of each cable—in a vertical position and, if necessary, adjust the roof height as described under the following heading.
10. Finger-turn the drive gearshaft clockwise as far as it will go, then turn it counterclockwise one-half turn.
11. Hand-press the cable drive gear assembly upward until the gear engages the cables, then tighten the two Phillips head screws.
12. Install the escutcheon, hand crank, short Phillips head screw, and plastic cap.
13. To check the cable adjustment and crank position, open and close the sliding roof several times. If necessary, reposition the hand crank on the drive gearshaft so that the crank can be folded into the recess when the sliding roof is fully closed.
14. If the sliding roof does not open and close evenly on both sides, repeat the adjustment. When the adjustment is correct, pull the trim panel forward and attach it to the sliding roof by hand-pressing in the five clips.

8.5 Adjusting Sliding Roof Height

The height of the sliding roof should be adjusted after removing and installing the sliding roof panel and whenever the top of the closed sliding roof does not lie flush with the roof of the vehicle.

To adjust height:

1. Using the hand crank, slightly open the sliding roof.

   NOTE —
   A gap of from 125 to 150 mm (5 or 6 in.) should be adequate.

2. Using a wooden wedge, carefully pry the front edge of the trim panel off the sliding roof panel. Work as close as possible to each of the five clips. (See **8.1 Removing and Installing Sliding Roof.)**
3. Push the trim panel to the rear as far as it will go.
4. Using the hand crank, fully close the sliding roof.
5. Loosen the two Phillips head screws with spring washers in each of the two front guides (Fig. 8-4).
6. To adjust the height of the front edge of the sliding roof panel, turn the adjusting screw one way or the other on each of the two front guides.

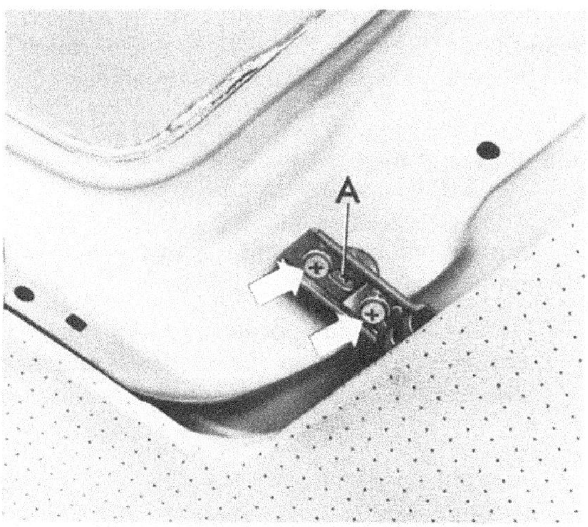

© 1974 VWoA—3196

Fig. 8-4. Front guide. The arrows indicate the Phillips head screws with spring washers. The adjusting screw is at **A.**

7. To adjust the height of the rear edge of the sliding roof panel, open the roof slightly. Then unhook the leaf springs from both rear lifters and swing the springs to the position shown in Fig. 8-5.
8. Loosen the nut and bolt indicated in Fig. 8-5 on both rear lifters.
9. To correct the height, move the lifter pins in their elongated holes. When the height is correct, tighten the nuts and bolts.
10. Swing the leaf springs back to their original position and hook them over the lifters.

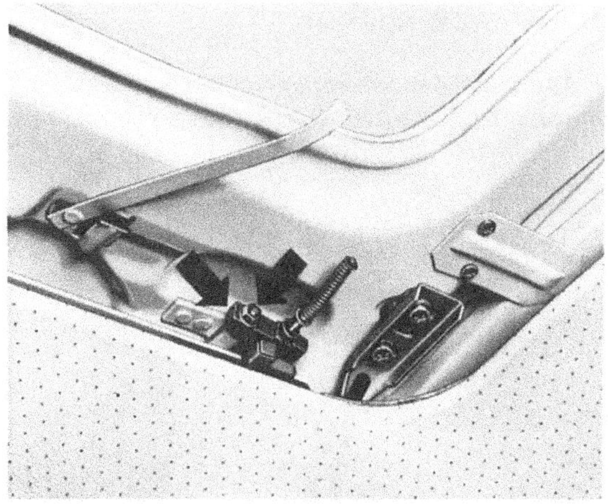

© 1974 VWoA—3197

Fig. 8-5. Rear lifter with leaf spring moved aside. The arrows indicate the nut and bolt that lock the position of the lifter pin.

11. Loosen the two Phillips head screws with spring washers on each rear guide. Then, by turning screw **A** indicated in Fig. 8-6, adjust the height of each rear guide so that it fits in the recess in the runner when the roof is being opened or closed.

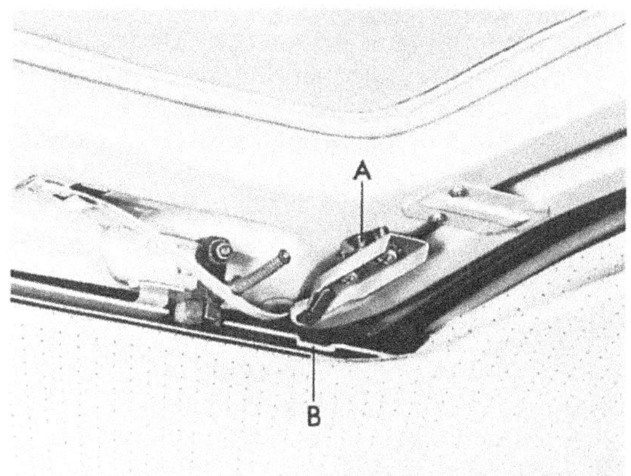

© 1974 VWoA—3198

Fig. 8-6. Guide adusting screw **A** and recess in side runner **B.**

NOTE —
If the rear guide is not properly adjusted, the roof will rattle.

12. Tighten the Phillips head screws with spring washers in each rear guide. Then check to see that the roof opens and closes evenly on both sides. If not, adjust the cables as described in **8.4 Adjusting Cables.**
13. Pull the trim panel forward and attach it to the sliding roof by hand-pressing in the five clips.

9. Outside Mirror

The outside mirror mounting has two pivots so that the mirror can be adjusted to any angle. You can remove the outside mirror as shown in Fig. 9-1. Use a 19-mm (¾ in.) wrench to unscrew the retainer from the door.

© 1974 VWoA—3191

Fig. 9-1. Outside mirror being removed.

Before you install the mirror, make certain that the two pivots move freely. Their freedom of movement can be altered by tightening or loosening the mirror retainer or the cap nut. If necessary, disassemble the mirror as shown in Fig. 9-2 so that you can clean and lubricate the pivots.

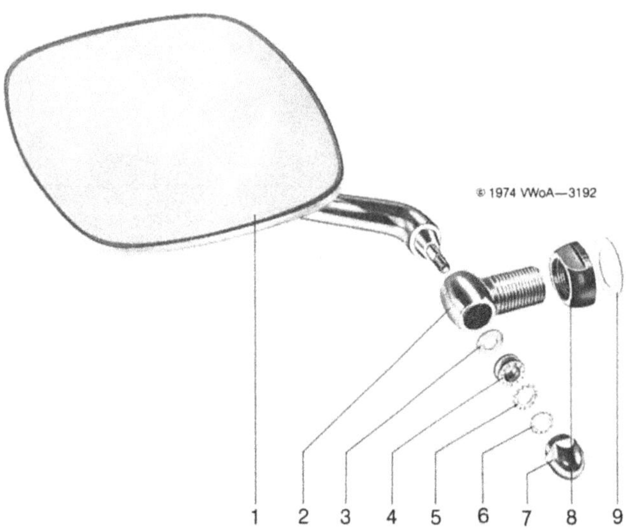

1. Mirror
2. Mirror bracket
3. Washer with hexagonal hole
4. Coil spring
5. Serrated washer
6. Washer
7. Self-locking cap nut
8. Mirror retainer
9. Plastic washer

Fig. 9-2. Components of the outside mirror.

During assembly, use multipurpose grease to lubricate the taper and thread on the mirror, the taper inside the mirror bracket, and the coil spring. If the plastic washer or the serrated washer is damaged, replace it. Do not forget to install the plastic washer under the mirror retainer when you install the mirror on the vehicle.

10. Heating and Ventilation

No special instructions are necessary for the removal of many heating and ventilation components. An examination of the components will quickly reveal the locations of the screws that hold them to the vehicle body. Though there have been detail modifications to some parts of the heating and ventilation system during the model years covered by this Manual, the basic design of its components has remained unchanged.

10.1 Removing and Installing Fresh Air and Heating Controls

The fresh air and heating controls are located behind the dashboard. The control knobs project through slots in the right-hand side of the instrument panel.

To remove controls:

1. Press out the four plastic plugs that hold the lever end pieces to the control levers. The locations of the plugs are shown in Fig. 10-1.
2. Pull the lever ends out through the slots in the instrument panel. Then remove the leaf springs that are also held to the levers by the plastic plugs.

© 1976 VWoA—3700

Fig. 10-1. Locations of plastic plugs (arrows). Although the instrument panel is removed in this illustration, you cannot remove it until after you have pulled the lever ends out of the slots in the panel.

3. Working behind the dashboard, pry off the spring clips that hold the adjustable connecting links to the control levers and to the relay levers. The parts are identified in Fig. 10-2.

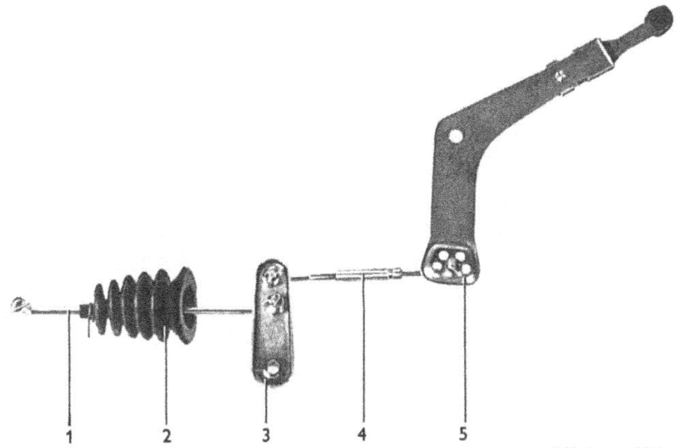

© 1976 VWoA—3702

1. Connecting rod
2. Rubber boot
3. Relay lever
4. Adjustable connecting link
5. Control lever

Fig. 10-2. Components of ventilation control linkage.

4. Remove the three Phillips head screws indicated in Fig. 10-3. Then pull the control downward, disconnect the heater cables, and remove the controls from the vehicle.

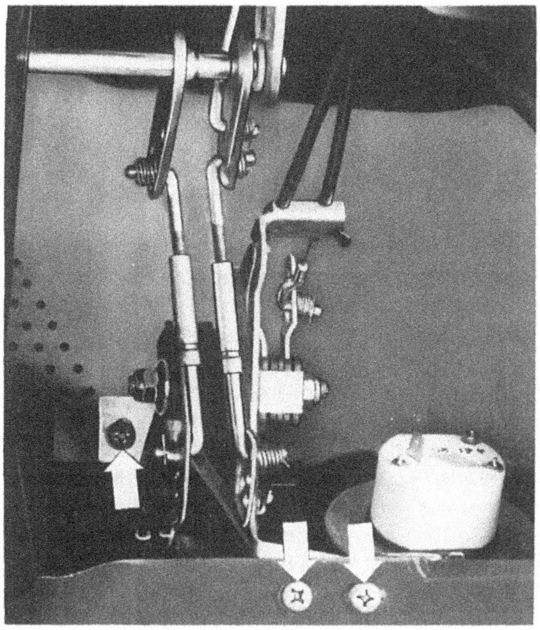

© 1976 VWoA—3701

Fig. 10-3. Screws (arrows) that hold controls to dashboard.

To disassemble the control lever assembly, remove the self-locking nuts. Then remove the levers and the friction washers from the pivot pins. Be sure to take notice of the order in which the levers and the friction washers are installed on the pins. If any friction washer is worn or damaged, replace it. During assembly, you can adjust the friction by loosening or tightening the self-locking nuts.

Installation is the reverse of removal. Following installation, make sure that the fresh air flaps are closed with the control lever in its highest position. If the flaps fail to close, or close too soon, you can adjust the linkage by installing the adjustable connecting link in a different hole in the control lever or by altering the length of the adjustable connecting link.

10.2 Removing and Installing Fresh Air Flaps and Linkage

It is not necessary to remove the control linkage in order to remove the fresh air flaps. However, if you must remove the linkage, it is necessary to remove the air intake grille and mesh panel from the front of the vehicle so that you can disconnect the connecting rods from the fresh air flaps.

To remove:

1. Remove the six Phillips head screws that hold the air intake grille to the body. Three of these screws are indicated in Fig. 10-4.
2. Remove the air intake grille and the mesh panel that is behind the grille.

© 1976 VWoA—3697

Fig. 10-4. Phillips head screws (arrows) that hold air intake grille and mesh panel.

3. Open the fresh air flap. Carefully pry off the spring clip that holds each connecting rod to its flap. Then disconnect the connecting rod from the flap.

4. Remove the two screws indicated in Fig. 10-5. Then remove the fresh air flap from the vehicle.

© 1976 VWoA—3698

Fig. 10-5. Screws (arrows) that hold fresh air flap to vehicle body.

5. Working beneath the dashboard, carefully pry off the spring clips that hold the connecting rods and the adjustable connecting links to the relay levers. Then disengage the rods and links from the relay levers. (These parts were previously identified in Fig. 10-2.)
6. Working outside the vehicle, push the rubber boots and the connecting rods out of the front panel toward the interior of the car.
7. Remove the Phillips head screws that hold the linkage brackets to the body (Fig. 10-6). Then remove the linkage from the vehicle.

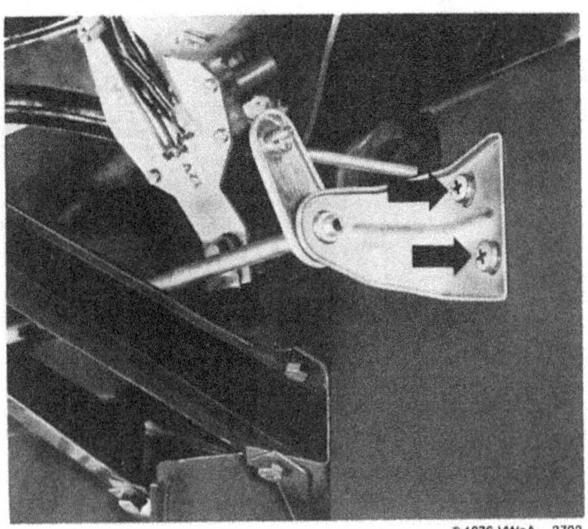

© 1976 VWoA—3703

Fig. 10-6. Screws (arrows) that hold linkage to body.

Before you reinstall the flap(s) or the linkage, inspect the plastic bearings and replace any that are worn or damaged. Also replace worn or damaged rubber boots and replace the flap seals if they are in any way faulty.

Installation is the reverse of removal. Make sure that the rubber boots for the connecting rods are installed correctly (Fig. 10-7). Following installation of all the components, check that the flaps and control levers operate smoothly. The flaps should be closed when the control levers are in their highest positions. If the flaps close too soon, or fail to close, adjust the linkage by installing the adjustable connecting links in different holes in the control levers or by altering the lengths of the adjustable connecting links.

© 1976 VWoA—3699

Fig. 10-7. Rubber boot and connecting rod installed in body. Arrows indicate holes for flap bearing mounting screws.

10.3 Heat Ducts and Outlets

Except for the front heater outlets (defroster ducts), the removal of the various heat outlet components will be apparent by looking at them. Replacement cables are available for the rear heater outlets.

To remove the front heater outlets (defroster ducts), remove the two screws that hold each defroster trim plate to the top of the dashboard. Then, working beneath the dashboard, squeeze the slots of the plastic outlet pipes together so that you can push the outlet pipes up slightly—through the holes in the top of the dashboard. This will make it possible to pull the Y-shaped branch pipe off the heat pipe. Separate the Y-shaped branch pipe from the two outlet pipes. Then, while squeezing their slots together, pull the plastic outlet pipes down and out of the dashboard. Installation is the reverse of removal.

Section 2

FRONT AXLE

2

Contents

Front Axle

Prior to the models covered by this Manual, the Type 2 front axle was similar to the front axle of the Volkswagen Beetle. However, beginning with the 1968 models, the front axle for Type 2 vehicles was completely redesigned. Although it employs the same operating principles as the earlier axle, every part of the new unit differs in size and shape. The worm and peg steering used on earlier models has likewise been revised and is replaced by worm and roller steering on the 1973 and later models.

The front axle has three functions: springing (or suspension), steering, and wheel alignment. Tire wear and vehicle handling are good indicators of how well these functions are being fulfilled. However, it is always best to maintain the front axle in serviceable condition so that abnormal tire wear and poor handling never affect the car.

Independent front wheel suspension is by trailing links and transverse torsion bars. This system not only prevents the transferal of road shocks from one front wheel to the other, but also provides excellent resistance to their transferal to the suspension parts, chassis, and passengers. Vertical travel of the front suspension is limited by progressively acting rubber stops. The greater the impact of the torsion arms (the trailing links) against the rubber stops, the greater their springing and energy absorbing reaction. But because the trailing link system minimizes the suspension deflection caused by severe bumps, the stops are seldom called upon during normal highway driving.

Ball joints at the ends of the torsion arms provide a flexible mounting for the steering knuckle. These joints not only permit the free vertical movement of the front wheels during bump and rebound, but also allow the wheels to be turned around a vertical axis for steering. Front wheel camber adjustments are accomplished by turning an eccentric bushing incorporated in the upper ball joint mountings. And, since the joints are lubricated at the factory, they rarely require further lubrication.

Many repair procedures require special equipment that car owners and small repair shops may not have—a hydraulic press and precision measuring jigs, for example. If you lack the skills, tools, or a clean workshop for servicing the front axle and steering, we suggest you leave such repairs to an Authorized VW Dealer or other qualified shop. We especially urge you to consult your Authorized VW Dealer before attempting repairs on a car still covered by the new-car warranty.

1. GENERAL DESCRIPTION

The front axle is a rigid beam with pivoting members that provide suspension movement, steering movement, and rotational movement of the wheels. The steering gearbox is coupled to the front wheels by a linkage that is an inherent part of the front axle. Therefore, all steering adjustments and repairs are covered in this section of the Manual. The names and positions of the front axle's parts are shown in Fig. 1-1.

Axle Beam

The axle beam itself is an electrically welded assembly consisting of two large parallel steel tubes and a number of heavy-gauge steel stampings. The axle is mounted on the vehicle by bolting its end plates, which incorporate the upper shock absorber mountings, solidly to the frame sidemembers. The axle beam tubes contain the torsion bars, torsion bar mounts, and torsion arm bearings. The bearings for the steering relay lever shaft are housed in a mounting welded to the front of the lower axle tube.

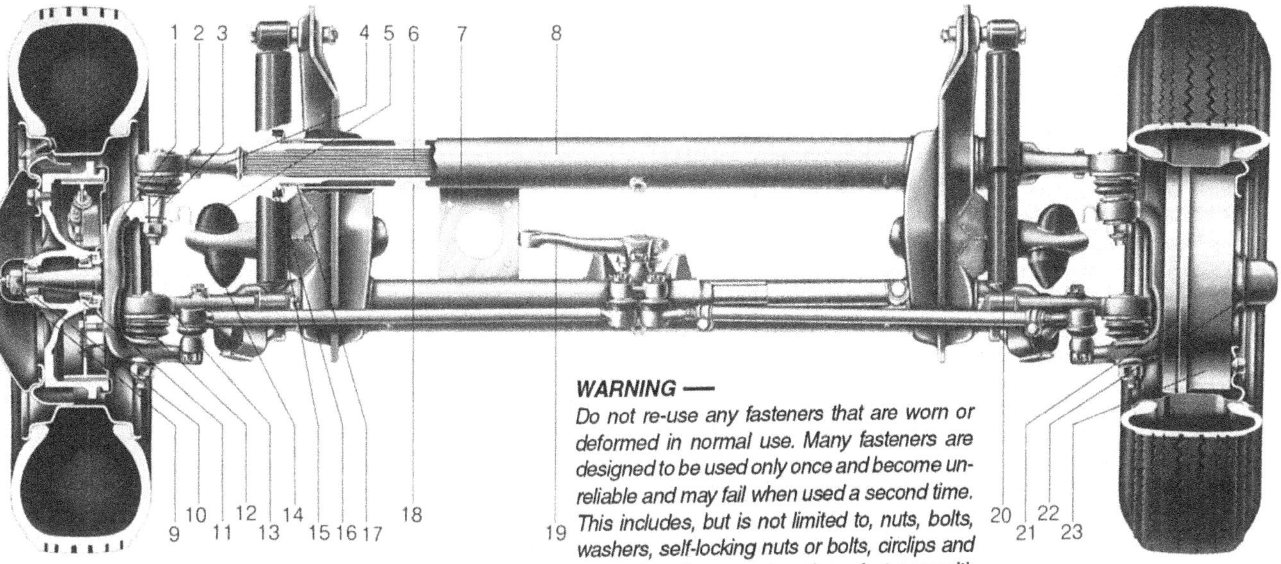

WARNING —
Do not re-use any fasteners that are worn or deformed in normal use. Many fasteners are designed to be used only once and become unreliable and may fail when used a second time. This includes, but is not limited to, nuts, bolts, washers, self-locking nuts or bolts, circlips and cotter pins. Always replace these fasteners with new parts.

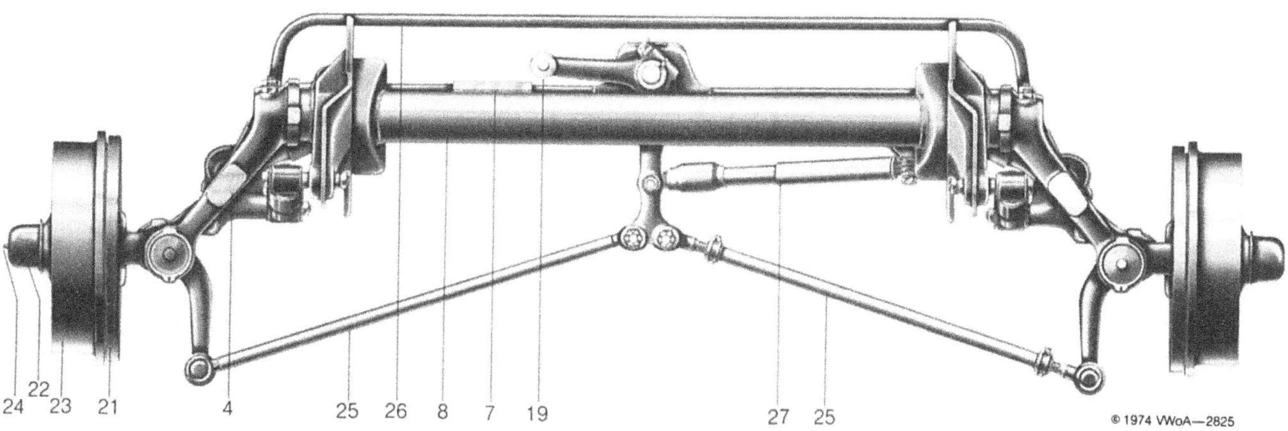

1. Steering knuckle
2. Upper ball joint
3. Eccentric camber adjusting bushing
4. Upper torsion arm
5. Upper torsion arm stop
6. Torsion bar
7. Brake servo mounting plate
8. Front axle beam
9. Wheel bearing clamp nut
10. Outer wheel bearing
11. Inner wheel bearing
12. Lower ball joint
13. Lower torsion arm
14. Lower torsion arm stop
15. Torsion arm seal retainer
16. Torsion arm seal
17. Torsion arm needle bearing
18. Plastic seat and metal bushing
19. Relay lever
20. Shock absorber
21. Brake backing plate
22. Dust cap
23. Brake drum
24. Speedometer cable
25. Tie rods
26. Stabilizer bar
27. Steering damper

Fig. 1-1. Cutaway and top views of the front axle.

Suspension

Multi-leaf torsion bars provide the springing. There are two bars, one in the upper axle tube and one in the lower. Setscrews hold the bars stationary in their center bushings. Each half of each torsion bar has a torsion arm mounted at its outer end. The lower torsion arms are joined by a stabilizer bar that increases the front suspension's resistance to roll during cornering.

Steering

Worm and peg steering is used on vehicles built from August 1967 through July 1972. The 1973 and later models have worm and roller steering. Both types can be adjusted, if necessary, to compensate for minor wear.

The ball-joint tie rod ends do not require lubrication during their service life. Since only the right-hand tie rod has threaded ends, you adjust the front wheel toe simply by loosening the locknuts and turning the right-hand tie rod. A hydraulic steering damper is linked to the right-hand tie rod to minimize the road shock transmitted to the steering wheel.

Wheel Bearings

Tapered-roller wheel bearings are used on all vehicles covered by this Manual. The inner race for each of the four bearing assemblies consists of a cone surrounded by a number of caged tapered rollers. The outer races are a press fit in the wheel hub bore. The hub is an integral part of the front wheel brake drum, but is separate from the front wheel brake disc used on 1971 and later models. Tapered-roller bearing adjustment is possible by turning the clamp nut on the steering knuckle's stub axle.

2. MAINTENANCE

The diagnosis and maintenance steps that must be performed at regular mileage intervals are listed here. Lubrication and checking procedures are described fully in **LUBRICATION AND MAINTENANCE** or under the listed headings in this section of the Manual.

1. Lubricating the front axle
2. Lubricating and adjusting the front wheel bearings
3. Checking the dust seals and plugs on the suspension ball joints and the tie rod ends (see **7. Torsion Arms, Suspension Ball Joints, and Torsion Bars** and **9. Steering**)
4. Checking the ball joint play (see **7. Torsion Arms, Suspension Ball Joints, and Torsion Bars**)
5. Checking the steering play (see **9. Steering**)
6. Checking the front wheel camber and toe (see **3. Front Wheel Alignment**).

3. FRONT WHEEL ALIGNMENT

Only camber and toe are adjustable. Caster angle and kingpin inclination are determined by the manufactured dimensions of the suspension parts, so damaged parts must be replaced to correct these alignment factors.

The following preparatory steps are essential to accurate alignment measurements:

1. Have the car on a level surface.
2. Inflate the tires to specifications and unload the car except for the spare wheel and a full fuel tank. Then jounce the car several times and let it settle into its normal position.
3. Check the adjustment of the steering gear and the front wheel bearings. Adjust them if necessary. See **5.2 Adjusting Front Wheel Bearings** and **9.1 Adjusting Worm and Peg Steering** or **9.2 Adjusting Worm and Roller Steering**.
4. Make sure there is no play in the tie rod ends, relay lever, or other parts of the steering linkage.

Measuring wheel camber and toe requires suitable gauges. Professional-grade instruments may cost several hundred dollars, but the modestly priced gauges available from mail order houses are adequate for home use. Instructions are supplied by the manufacturer.

3.1 Checking and Adjusting Camber

Camber is the angle at which wheels depart from the true vertical when viewed from directly in front of the car. If the tops of the wheels lean outward slightly, they are said to have positive camber; if they lean inward, they have negative camber.

To check:

1. After placing the car on a level surface with its front wheels pointing straight ahead, apply a bubble protractor as shown in Fig. 3-1.

© 1974 VWoA—2196

Fig. 3-1. Bubble protractor applied against front wheel.

2. Using chalk, mark the wheel at the points where it is *contacted* by the protractor.
3. Turn the spirit level carrier on the protractor until the bubble is centered, then read the camber angle on the scale.

NOTE —
If you are using some other type of gauge, follow the manufacturer's instructions.

4. Roll the car forward a half-turn on the wheels, then repeat the measurement at the chalk-marked points.
5. Take the new reading and average it with the one you obtained earlier. The result is the camber angle, corrected for wheel runout.
6. Repeat the entire procedure on the opposite wheel.

The front wheels should have 40′ ± 20′ of positive camber. Also, the difference in camber between the wheels should not vary more than 30′. If not within specifications, the camber of each wheel should be adjusted to as near 40′ as possible.

To adjust camber:

1. Loosen the self-locking nut on the upper ball joint stud. Make certain the notch in the eccentric camber adjusting bushing is roughly toward the front of the vehicle, as indicated in Fig. 3-2.

NOTE —
The vehicle must be standing on its wheels while adjustments are being made.

© 1974 VWoA—2856

Fig. 3-2. Notch in eccentric bushing (wheel removed for clarity). Adjust camber by turning the bushing a maximum of 90° to either side.

2. Set the spirit level carrier on the protractor to the specified angle of 40′ positive camber.
3. Turn the eccentric camber adjusting bushing (Fig. 3-3) until the bubble in the protractor is centered when the protractor is applied to the chalk-marked points.

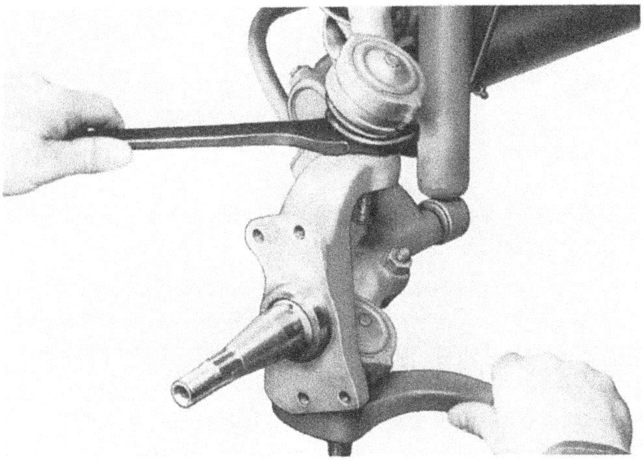

© 1974 VWoA—2848

Fig. 3-3. Eccentric camber adjusting bushing being turned. (Wheel and brake removed for clarity.)

4. Torque the self-locking nut on the ball joint stud to 10 mkg (72 ft. lb.). Recheck the camber and, if necessary, repeat the adjustments to bring camber within specifications.
5. Check the toe and adjust it if necessary.

3.2 Checking and Adjusting Toe

All Type 2 vehicles are designed to operate with a small amount of toe-in. This means that the front edges of the tires are slightly closer together than the rear edges. Shops with optical aligning devices should follow the equipment manufacturer's instructions to obtain a toe-in angle of 15′ ± 15′—or 5′ ± 15′ after adding 12 to 18 kg (27 to 39 lb.) of extra weight above the wheel. The maximum toe change should not exceed 25′.

Most small shops and individual car owners check the toe with a track gauge. This device is used to measure first the distance between two marked points at the front edges of the rims, then the distance between the same two points after the car has been rolled forward until the marks are at the rear. The measurement made at the rear should be 0.0 to 3.3 mm (0.000 to 0.130 in. or approximately 0 to ⅛ in.) greater.

NOTE —
These specifications apply only when the wheels are in their straight-ahead position. Toe specifications with the wheels turned appear in **10. Front Axle Technical Data**.

2

To check wheel toe:

1. Turn the steering to its center position.
2. Check the toe. If a track gauge is used, mark the measuring points with chalk. This allows you to make measurements between the same two points after you roll the car ahead a half-turn of the wheels.
3. Adjust the toe if it is not within specifications.

To adjust:

1. Loosen the clamp bolts on both of the clamps on the right-hand tie rod.
2. Correct the toe-in by turning the right-hand tie rod. Rotate the top of the tie rod toward the front of the car to increase the toe-in; to the rear to decrease it. Check toe measurements frequently.
3. When the toe-in is correct, position the tie rod so that the ball joints are not angled. Torque the clamp bolts to 1.5 mkg (11 ft. lb.).

NOTE —
Following adjustments, the steering wheel will normally be centered when the front wheels are in their straight-ahead position. If not, check for bent steering knuckles or bent steering linkage parts, then adjust the steering wheel position as described in **9. Steering**.

3.3 Checking and Adjusting Steering Lock

A provision for adjusting steering lock has been incorporated on all Type 2 vehicles beginning with chassis No. 211 2066 397, manufactured in November 1970. It consists of two adjusting bolts with locknuts as indicated in Fig. 3-4. The bolt heads limit the travel of the relay lever on both sides.

Fig. 3-4. Bolts used to adjust steering lock. On vehicles manufactured before November 1970, steering lock is limited by the position in which the brackets are welded to the lower axle tube.

To check and adjust steering lock:

1. Raise the vehicle—preferably on a hydraulic lift.
2. Turn the front wheels fully to the right and to the left. In either position, the clearance between the tires and the stabilizer bar should be 20 mm (3/4 in.) as indicated in Fig. 3-5.

Fig. 3-5. Clearance **a**, between tire and stabilizer bar, should be 20 mm (3/4 in.).

3. If clearance is outside the prescribed range, loosen the locknut on the steering lock adjusting bolt(s).
4. Turn the bolt(s) in or out until the prescribed clearance between the tire and stabilizer bar is obtained, then tighten the locknut.

4. SHOCK ABSORBERS

Make sure you install only shock absorbers intended for the front of the vehicle. Mismatched shock absorbers will impair handling and ride. It is not necessary, however, to replace both shock absorbers if only one is defective. Also, front axle shock absorbers of different manufacture can be combined as long as their damping characteristics are identical. If the vehicle is to be subjected to heavy loads, rough roads, or extremely high temperatures, it may be advisable to install heavy-duty shock absorbers in the interest of longer service—despite their effect on the riding qualities of the car.

CAUTION —
Install heavy-duty shock absorbers on all four wheels at the same time. Otherwise, handling will be adversely affected.

4.1 Checking Shock Absorbers

You can quick-check the shock absorbers by grasping the front bumper and rocking the car vigorously. When you let go, the vehicle should rebound only once and then settle into its normal attitude. If the car continues to rock or bob, the shock absorbers are worn. Excessive bobbing on the highway also signals defective shock absorbers. Badly worn shock absorbers often make knocking noises when the vehicle is driven.

You can hand-check a shock absorber by extending and compressing it while holding it in its installed position. It should operate smoothly and with uniform resistance throughout its entire stroke. If possible, compare the used shock absorber with a new one. New shock absorbers that have been in storage may have to be pumped several times before they reach full efficiency.

An adequate supply of fluid is placed in the shock absorbers during *manufacture* to compensate for small leaks. Minor traces of fluid are acceptable if the shock absorber still functions efficiently.

4.2 Replacing Shock Absorbers

Shock absorbers cannot be repaired or serviced and should be replaced if faulty.

To replace:

1. Raise the vehicle and remove the front wheel.
2. Remove the M 12 nut and bolt that hold the upper end of the shock absorber to the front axle end plate.
3. Pull the top of the shock absorber to the rear, then remove the M 10 nut that holds the lower end of the shock absorber on the threaded stud in the lower torsion arm.
4. Remove the old shock absorber.
5. Install the lower end of the new shock absorber on the stud in the lower torsion arm.
6. Incline the top of the shock absorber about 30° to the rear, and then torque the nut to 2.5 to 3.5 mkg (18 to 25 ft. lb.) while the shock absorber is in this position.

NOTE —
This procedure will prevent the lower rubber bushing from twisting as the suspension moves upward. If the vehicle is standing on its wheels while the new shock absorber is being installed, do not incline the shock absorber 30° to the rear—keep it upright.

7. Move the top end of the shock absorber into line with the upper mounting hole. Install the nut and bolt with a torque of 5.0 mkg (36 ft. lb.).

If only the rubber shock absorber bushings are faulty, it is possible to replace them without replacing the entire shock absorber.

To replace shock absorber bushing:

1. Press the metal sleeve and rubber bushing out of the shock absorber as shown in Fig. 4-1.

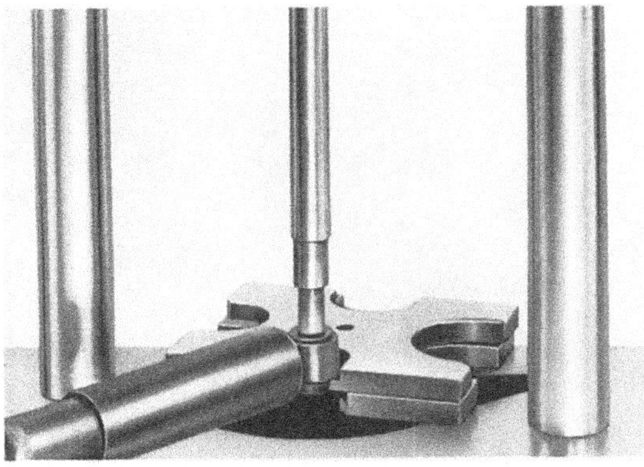

© 1974 VWoA—2887

Fig. 4-1. Metal sleeve and rubber bushing being pressed out of the shock absorber.

2. Coat the new rubber bushing with talcum powder or silicone spray. Using a vise, press the rubber bushing into the shock absorber until the bushing's shoulder contacts the shock absorber eye.
3. Using the setup shown in Fig. 4-2, press the metal sleeve in from the same side.
4. If necessary, correct the position of the rubber bushing using the same setup used to remove the old sleeve and bushing.

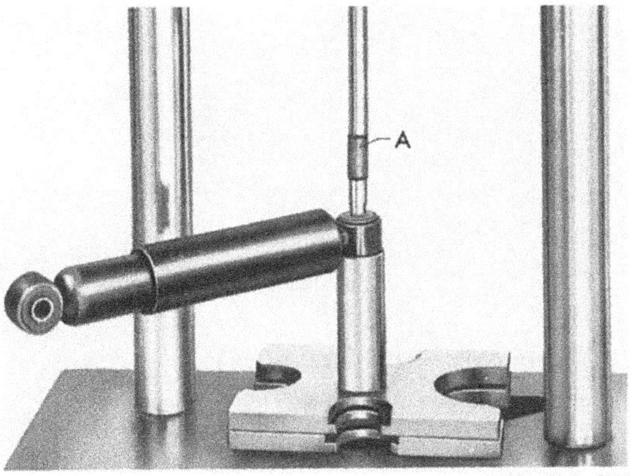

© 1974 VWoA—2888

Fig. 4-2. Metal sleeve being pressed in.

2

5. FRONT WHEEL BEARINGS

There are two tapered-roller bearing assemblies at each front wheel. Each bearing has a solid steel outer race and an inner race that consists of a steel cone and a number of caged tapered rollers. The bearings and related parts are shown in Fig. 5-1.

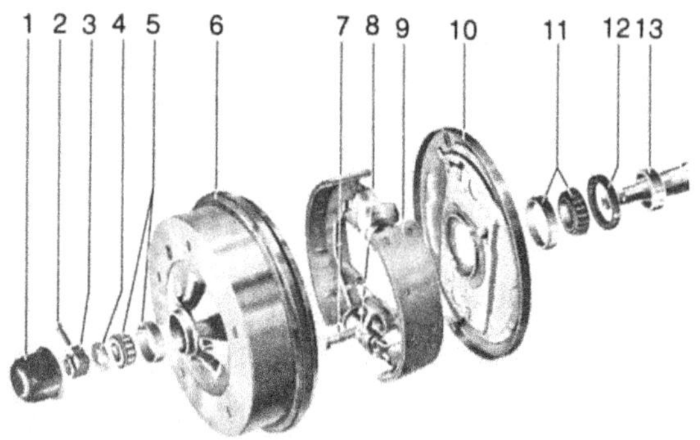

WARNING —
Do not re-use fasteners such as nuts, bolts, washers, self-locking nuts, circlips, cotter pins that are worn or deformed in normal use. Always replace.

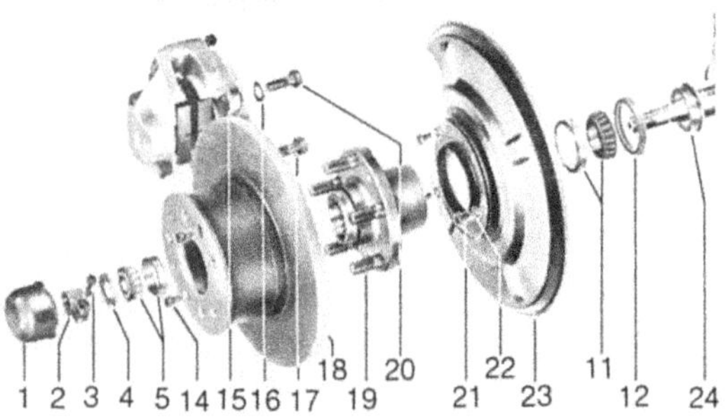

1. Dust cap
2. Socket head clamp screw
3. Clamp nut
4. Thrust washer
5. Outer tapered-roller bearing
6. Brake drum
7. Wheel cylinder mounting bolt
8. Lock washer
9. Front brake shoe
10. Brake backing plate
11. Inner tapered-roller bearing
12. Grease seal
13. Spacer ring (drum brakes)
14. Socket head screw
15. Brake caliper
16. Spring washer
17. Fitted bolt with shoulder
18. Brake disc
19. Wheel hub
20. M 12 bolt
21. Bolt
22. Lock washer
23. Splash shield
24. Spacer ring (disc brakes)

Fig. 5-1. Wheel bearings and related parts.

5.1 Removing and Installing Front Wheel Bearings

Before you can remove the brake disc and wheel bearings from vehicles with disc brakes, the brake caliper must be removed as described in **6.1 Removing and Installing Steering Knuckle**. You can remove the inner races with common hand tools. The solid steel outer races, however, are pressed into the brake drum or the wheel hub and should be removed with a hydraulic press and appropriate mandrels.

To remove inner races:

1. Remove the road wheel or, on 1971 and later models, the brake caliper. Then pry off the dust cover.
2. Using a 7-mm (9/32-in.) hex key, loosen the socket head screw in the clamp nut. Then unscrew the clamp nut from the stub axle on the steering knuckle.

   NOTE —
   The clamp nut for the left front wheel has a left-hand thread.

3. Pull off the brake drum or disc, being careful not to let the thrust washer and the outer tapered-roller bearing inner race fall out and onto the floor.
4. With the brake drum or disc on the workbench, carefully remove the thrust washer and the outer bearing's inner race. Store them in a clean, dust-free place.

   NOTE —
   On vehicles with disc brakes, you can separate the disc from the wheel hub by taking out the socket head screws, then tapping the wheel studs with a rubber mallet.

5. Pry the grease seal out of its recess in the rear of the drum or wheel hub. Then lift out the inner tapered-roller bearing's inner race. Store it with the outer bearing.

To install inner races:

1. Carefully clean the inner bearing races with solvent, then dry them with compressed air.

   CAUTION —
   Do not use solvents such as gasoline because they remove all lubrication. Also, do not let blasts of compressed air spin the races. Unlubricated bearings can be damaged by rapid movement.

2. Inspect the inner bearing races. Replace them if they are worn, burred, rough, or heat-blued.
3. Clean the brake drum or wheel hub and inspect the outer bearing races. Replace them if they are worn, burred, rough, or heat-blued.
4. Pack the inner bearing's inner race with multipurpose grease, as described in **LUBRICATION AND MAINTENANCE**. Carefully place it inside the hub.
5. Press a new grease seal into place.
6. Inspect the stub axle for burrs or blued areas. Check the bearing seat dimensions as described in **6.2 Checking Steering Knuckle**. If satisfactory, lightly coat the stub axle with multipurpose grease.

7. Carefully slide the brake drum or wheel hub, with or without the brake disc, onto the stub axle so that the grease seal or bearing races are not accidentally damaged by the sharp threads.
8. Pack the outer bearing's inner race with multipurpose grease. Then carefully slide it onto the stub axle and into the hub.
9. Install the thrust washer and the clamp nut. Tighten the clamp nut on the stub axle until the bearings just contact their outer races.
10. Adjust the bearings as described in **5.2 Checking and Adjusting Front Wheel Bearings**. Install the dust cap, the brake caliper (where fitted), and the road wheel.

To remove outer races:

1. Support the brake drum or wheel hub, outside down, on the press bed. Using a suitable driving mandrel, press out the outer bearing's outer race.
2. Turn over the brake drum or wheel hub.
3. Using a suitable driving mandrel, press out the inner bearing's outer race.

Installation is the reverse of removal. The inner bearing's outer race, being the larger, is pressed out last but the sequence is unimportant to installation. Make certain, however, that the hub recesses are clean and free of burrs or pressure marks that could prevent the races from seating completely.

5.2 Checking and Adjusting Front Wheel Bearings

Wheel bearings should turn smoothly and not have excessive axial play. If the bearings feel gritty, have tight spots, or make noises when the wheel turns, they probably need to be replaced. Excess axial play, though, can be corrected by adjusting.

To adjust bearings:

1. Raise the wheel, then pry off the dust cap.
2. If the bearings have just been installed, torque the clamp nut to about 1.0 mkg (7 ft. lb.) while you hand-turn the brake disc or drum.

> ***CAUTION —***
> *Never torque the clamp nut to more than 1.3 mkg (9.5 ft. lb.). Doing so will damage the bearing races.*

3. To measure the bearing axial play, install a dial indicator on one of the wheel lugs or in place of a wheel bolt (or use a dial indicator with a magnetic base).
4. Position the dial indicator pin against the end of the stub axle as shown in Fig. 5-2. Then move the wheel in and out by hand, turning the clamp nut one way or the other until the axial play is between 0.03 and 0.12 mm (.001 and .005 in.).

© 1974 VWoA—2837

Fig. 5-2. Dial indicator being used to measure wheel bearing axial play (indicated by arrow).

> **NOTE —**
> Turn the wheel and repeat the measurement at several different points. The readings should not vary greatly and their average should fall within the prescribed range. Replace bearings that will not adjust properly.

5. Torque the socket head clamp screw to 1.5 to 2.0 mkg (11 to 14 ft. lb.).
6. Install the dust cap and lower the wheel to the ground.

6. STEERING KNUCKLES

The steering knuckles, with their integral steering arms, are held onto the torsion arms by ball joints. For safety as well as drivability, it is important that the steering knuckles are not bent. Check them carefully after an accident or other severe impact.

6.1 Removing and Installing Steering Knuckle

When removing the steering knuckle, take off the brake assembly only if you are planning to replace the steering knuckle itself. The brake assembly can be left in place if only the ball joints require attention.

2

To remove steering knuckle:

1. On vehicles with drum brakes, remove the brake drum and the wheel bearings as described in **5.1 Removing and Installing Front Wheel Bearings**.
2. On cars with disc brakes, remove the brake caliper from the steering knuckle by removing the two bolts and the brake hose clamp indicated in Fig. 6-1.

> ***CAUTION —***
> *The caliper must have cooled to room temperature before you remove it. Otherwise, it may be damaged by heat distortion.*

© 1974 VWoA—2843

Fig. 6-1. Bolts and brake hose clamp that hold disc brake caliper on steering knuckle.

3. Using a stiff wire hook, suspend the brake caliper from the car body so that it is not being supported by the brake hose.

> ***WARNING —***
> *Never let a brake caliper or brake backing plate assembly hang by the brake hose. Doing so could weaken the hose and cause subsequent brake failure.*

4. On cars with drum brakes, remove the four M 10 bolts that hold the brake backing plate assembly on the steering knuckle. Remove the backing plate assembly and suspend it from the car body so that its weight is not supported by the brake hose.
5. Remove the cotter pin and the castellated nut from the tie rod end stud. Then press out the tie rod end as shown in Fig. 6-2.

> ***CAUTION —***
> *Do not hammer out the tie rod end. Doing so will ruin the threads and make reinstallation impossible.*

© 1974 VWoA—2841

Fig. 6-2. Tie rod end being pressed out. The tool shown prevents damage to the threads on the tie rod end stud.

6. On cars with disc brakes, remove the three bolts that hold the splash shield on the steering knuckle. Then remove the splash shield.
7. Remove the M 18 self-locking nut from the lower suspension ball joint stud. The install an M 18 × 1.5 cap nut and press the ball joint stud loose as shown in Fig. 6-3.

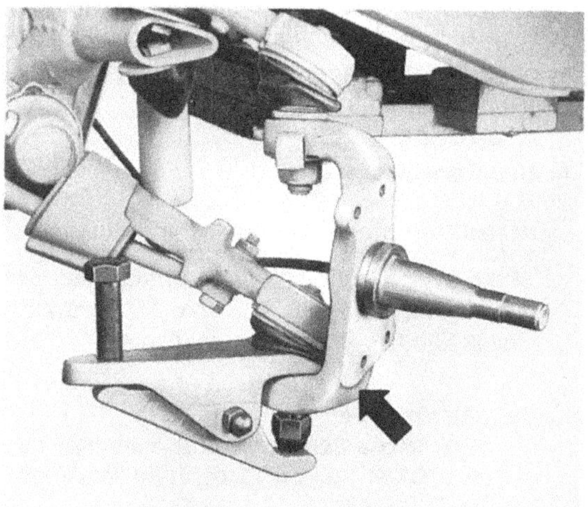

© 1974 VWoA—2847

Fig. 6-3. Ball joint being pressed loose in steering knuckle (arrow).

8. Remove the M 18 self-locking nut from the upper ball joint stud. Then turn the eccentric camber adjusting bushing to free the steering knuckle from the upper ball joint. See Fig. 6-4.

© 1974 VWoA—2848

Fig. 6-4. Eccentric camber adjusting bushing being turned. Support the steering knuckle so that it does not drop off suddenly.

NOTE —
If the steering knuckle is being removed so that the upper ball joint can be repaired or replaced, press the ball joint stud out of the eccentric camber adjusting bushing. See **7.4 Repairing and Replacing Ball Joints**.

To install:

1. Loosely attach the steering knuckle to the lower ball joint.
2. Using a tool similar to the one shown in Fig. 6-5, lift the lower torsion arm until the upper ball joint stud enters the steering knuckle.

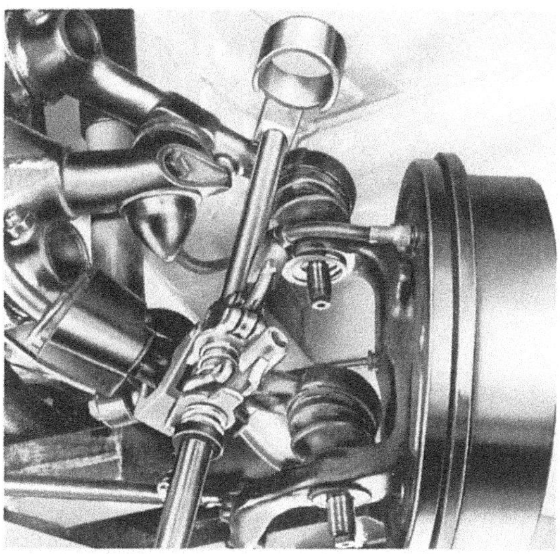

© 1974 VWoA—2855

Fig. 6-5. Special tool used to lift torsion arm.

3. Position the eccentric camber adjusting bushing so that its notch points forward as shown in Fig. 6-6.

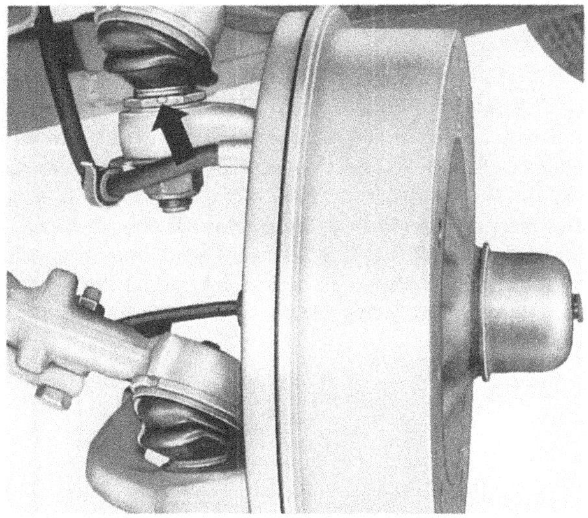

© 1974 VWoA—2856

Fig. 6-6. The notch on the eccentric camber adjusting bushing.

4. Install new self-locking nuts on the ball joint studs. Torque only the nut on the lower ball joint stud to 10 mkg (72 ft. lb.).
5. Install the tie rod end in the steering knuckle. Torque the castellated nut to 3.0 mkg (22 ft. lb.). Advance the nut, if necessary, to uncover the cotter pin hole, then install a new cotter pin.
6. On vehicles with drum brakes, install the brake backing plate assembly. Torque the M 10 bolts to 5.5 to 6.0 mkg (40 to 43 ft. lb.).
7. On vehicles with disc brakes, install the splash shield. Torque the bolts to 1.0 mkg (7.0 ft. lb.).
8. Install the brake drum or disc. Then adjust the front wheel bearings as described in **5.2 Checking and Adjusting Front Wheel Bearings**.
9. On vehicles with disc brakes, install the brake caliper. Torque the M 12 bolts used on 1968 through 1972 models to 10 mkg (72 ft. lb.). Torque the M 14 bolts used on 1973 and later models to 16 mkg (116 ft. lb.).
10. Install the road wheel. Torque the wheel bolts used on 1968 through 1970 models to 13 mkg (94 ft. lb.). Torque the wheel nuts used on 1971 and later models to 12 to 14 mkg (87 to 101 ft. lb.).

WARNING —
- *Tighten the wheel bolts or nuts with the vehicle on the ground. The leverage needed for this job is enough to topple a car off the lift.*
- *Do not re-use fasteners such as nuts, bolts, washers, self-locking nuts, circlips, cotter pins that are worn or deformed in normal use. Always replace.*

2

11. Adjust the camber and toe as described in **3. Front Wheel Alignment**. Torque the self-locking nut on the upper ball joint stud to 10 mkg (72 ft. lb.).

6.2 Checking Steering Knuckle

The steering knuckle can be checked either on or off the car. Measure the stub axle at the three points indicated in Fig. 6-7. The diameter of the outer bearing seat (**A**) should be 19.03 to 19.05 mm (.7491 to .7499 in.). The diameter of the inner bearing seat (**B**) should be 31.73 to 31.75 mm (1.2492 to 1.2499 in.). The spacer ring seat (**C**) should have a diameter of 38.04 to 38.08 mm (1.4976 to 1.4992 in.). Check the dimensions with a micrometer or a vernier caliper.

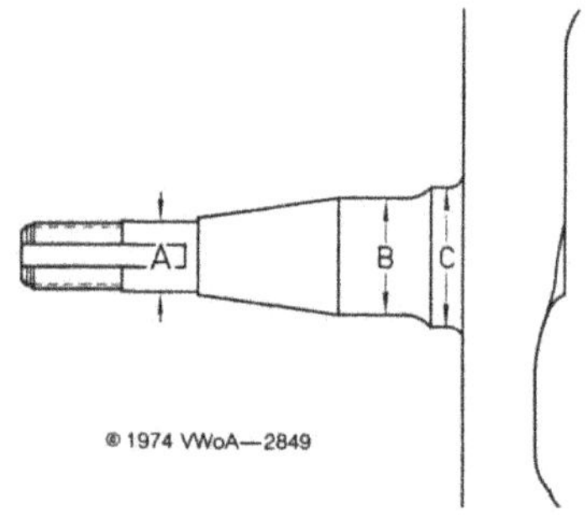

Fig. 6-7. Stub axle measuring points. Dimensions are given in preceding text.

You can check the steering knuckle's stub axle for bending using a vernier caliper and a machinist's square as shown in Fig. 6-8. Make your measurements at not less than three points around the stub axle. The difference between any two measurements should not exceed 0.25 mm (.010 in.).

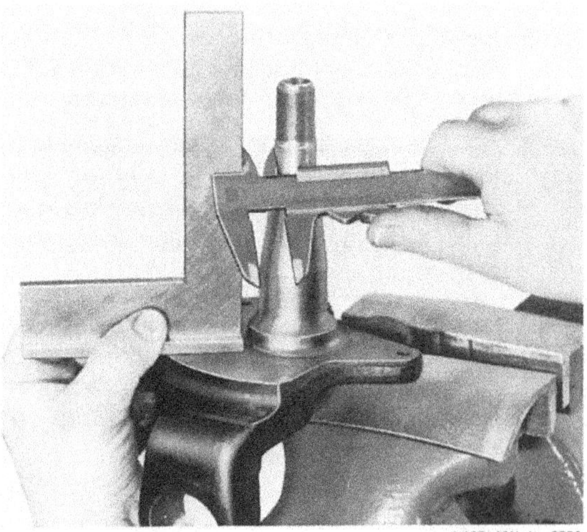

© 1974 VWoA—2850

Fig. 6-8. A vernier caliper and machinist's square being used to check the stub axle for bending.

On vehicles with drum front brakes, use tool VW 258f/2 to check the steering arm for bending. The steering knuckle is within tolerances when the hole in the steering arm is in line with another hole in the gauge, and when the top face of the steering arm eye is parallel to the surface of the gauge as indicated by **a** in Fig. 6-9.

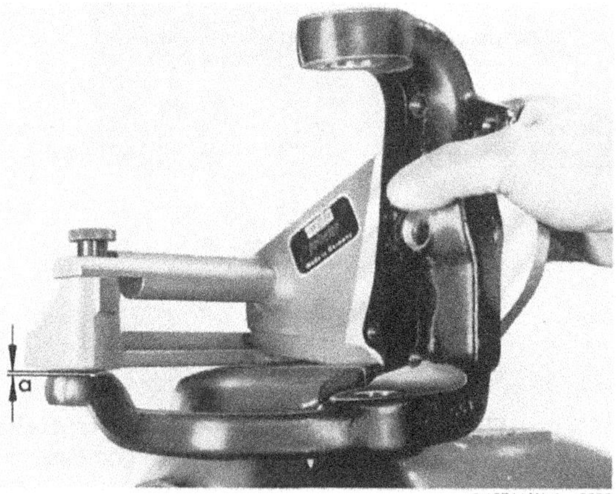

© 1974 VWoA—2852

Fig. 6-9. Special gauge in position with its center plate slipped over the stub axle. Check the steering arm at **a**, as described in the text.

On vehicles with disc front brakes, use a straightedge and vernier caliper to check the steering arm for bending (Fig. 6-10). The distance from the straightedge to the outer edge of the tie rod hole should be 110.50 to 111.50 mm (4.350 to 4.389 in.).

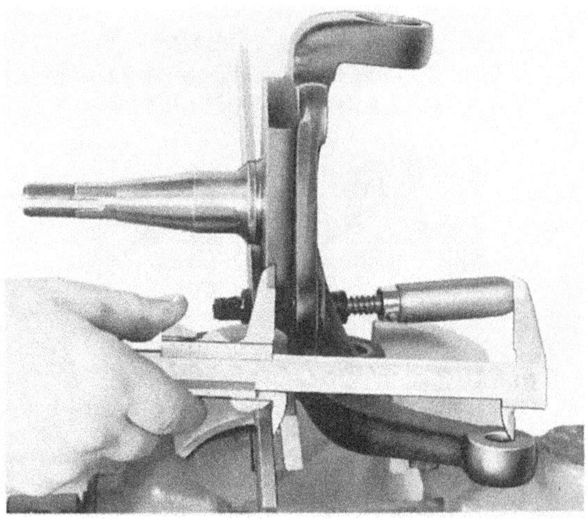

© 1974 VWoA—2853

Fig. 6-10. Steering arm being checked for bending.

CAUTION —
Bent steering knuckles must be replaced and not straightened. Bending them back to their original shape will seriously weaken them structurally.

7. TORSION ARMS, SUSPENSION BALL JOINTS, AND TORSION BARS

The torsion bars that provide springing for the front wheels are housed inside the axle beam. Socket head setscrews hold the torsion arms on the outer ends of the bars. The suspension ball joints are a press fit in the torsion arms.

7.1 Checking Ball Joints

A special lever and vernier caliper should be used to check the ball joint play. If the steering knuckle has been removed, check the ball joints as described in **7.4 Repairing and Replacing Ball Joints**.

To check:

1. Lift the car. Then turn the steering to one side.
2. Install the special lever as shown in Fig. 7-1.
3. Place a vernier caliper over the ball joint with one jaw on the steering knuckle and the other on the torsion arm. Note the reading.
4. Pull down on the lever in order to raise the torsion bar and expand the ball joint. Note the new reading on the vernier caliper.

© 1974 VWoA—2861

Fig. 7-1. Lever and vernier caliper in position. Lever should draw the two torsion arms toward one another when the handle is raised.

5. Compare the two readings. The difference between them is the ball joint play.

NOTE —
Neither the upper or lower ball joint should exceed 2.00 mm (.080 in.) play. Joints that exceed this must be replaced.

7.2 Removing and Installing Torsion Arm

Each torsion arm is held on its torsion bar by a socket head setscrew and locknut.

To remove:

1. Remove the steering knuckle complete with brake assembly.
2. If the lower torsion arm is to be removed, disconnect the stabilizer bar by driving the retainers off the two stabilizer bar rubber mounting clamps.

NOTE —
It is necessary to bend down the locking tabs on the retainers before driving them off. Obtain new retainers for use during assembly.

3. Remove the nut from the stabilizer bar mounting bolt on the lower torsion arm, then take off the stabilizer bar.
4. Loosen the locknut on the socket head setscrew that holds the torsion arm on the torsion bar. Then remove the setscrew.
5. Remove the torsion arm from the end of the torsion bar.

Installation is the reverse of removal. Inspect the rubber seal for the torsion arm. Replace the seal if it is worn, cracked, or loose-fitting.

7.3 Checking Torsion Arms

The torsion arms can be checked for bending only after they have been removed from the car. A special measuring jig, VW 282d, is required for this job.

CAUTION —
If you lack the skills or special tools needed for checking the torsion arms or replacing the suspension ball joints, we suggest you leave such repairs to an Authorized VW Dealer or other qualified shop. We especially urge you to consult your Authorized VW Dealer before attempting repairs on a car still covered by the new-car warranty.

Suspension ball joints used prior to the 1973 model year have screw-in plastic plugs in their end plates. The latest type ball joints do not have these plugs. Instead, there is a small depression in the end plate with a 6-mm (.236-in.) diameter test surface. This modification has necessitated a minor change in the test jig. When checking torsion arms with the late-type ball joints, the test plates on the test jig must be changed.

The ball joint without a plastic plug is shown in Fig. 7-2.

© 1974 VWoA—2653

Fig. 7-2. Late-type ball joint with test surface (arrow) instead of plastic plug.

To check torsion arms:

1. Carefully clean the torsion arm and ball joint.
2. Inspect the bearing surfaces on the torsion arm. If they are worn, replace the torsion arm complete with the ball joint and also the needle bearing and metal bushing in the axle tube.
3. On early-type ball joints, remove the plastic plug from the end plate and install the test point in its place as shown in Fig. 7-3.

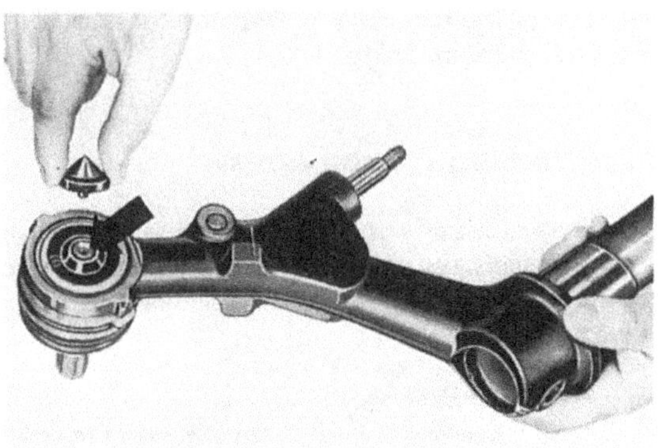

© 1974 VWoA—2862

Fig. 7-3. Test point being installed. Screw the point into the hole for the plastic plug.

4. With the late-type ball joints, install the proper measuring plates on the test jig and screw the measuring pin into the appropriate hole in the measuring plate.
5. Using the correct bushings from the set supplied with the VW 282d, install the torsion arm in the test jig. Use bushing VW 282d/12 for the torsion arm's inner bearing surface and spacer ring VW 282d/21 for the outer bearing surface.
6. See whether the test point contacts the test jig measuring plate (or the point on the measuring plate contacts the test surface on the ball joint) as shown in Fig. 7-4 or Fig. 7-5.

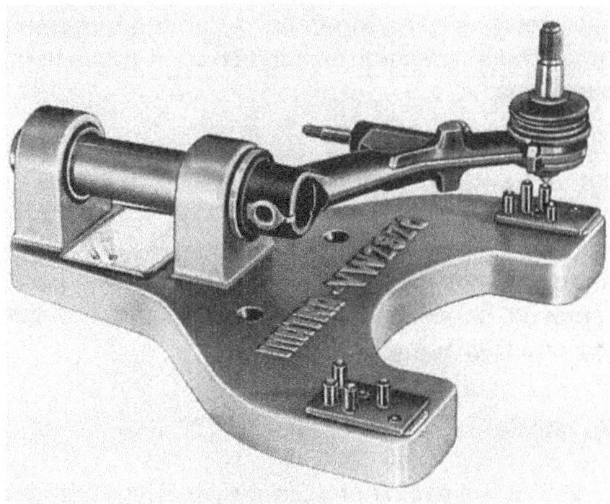

© 1974 VWoA—2863

Fig. 7-4. Early-type ball joint. Test point should contact the boss on test plate.

NOTE —
The lower torsion arms should contact the small, outer boss on the test jig, as shown in Fig. 7-4. Upper torsion arms should contact the taller, outer boss that is just to the left of the small boss being contacted in the illustration. The threaded holes for the test point used with late-type ball joints are in the same relative locations as the bosses used with the early ball joints.

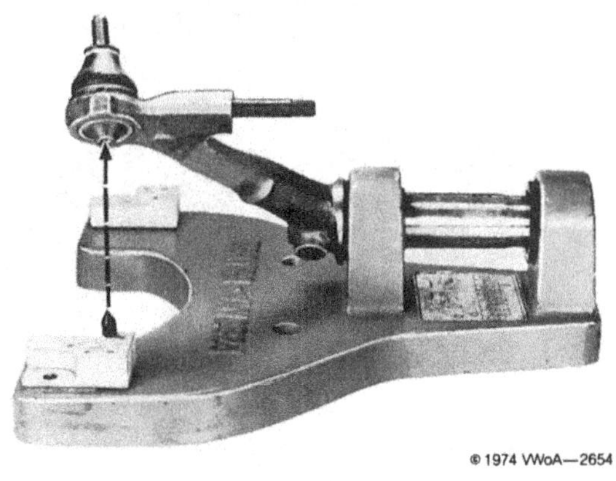

© 1974 VWoA—2654

Fig. 7-5. Late-type ball joint. Point on jig should contact test surface on ball joint. Torsion arm shown is not for a Type 2 vehicle.

CAUTION —
Bent torsion arms must be replaced complete with ball joints and not straightened. Bending them back to their original shape will seriously weaken them structurally.

7. On early-type ball joints, remove the test point and screw in a new plastic plug.

7.4 Repairing and Replacing Ball Joints

To check the ball joint play after the torsion arm has been removed, place a vernier caliper over the ball joint as shown in Fig. 7-6. Press the stud in all the way and note the measurement. Then pull the stud out and take another measurement. Compare the two readings. The difference between them is the ball joint play.

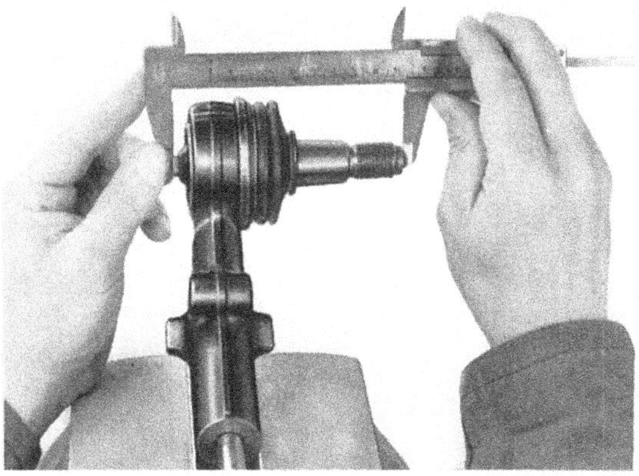

© 1974 VWoA—2864

Fig. 7-6. Ball joint play being checked. Make measurements with the stud all the way in and all the way out.

NOTE —
New ball joints must not exceed 0.30 mm (.012 in.) play. Used ball joints must not exceed 2.00 mm (.080 in.) play. Used ball joints that are at or near this wear limit must be replaced.

Damaged ball joint seals can be replaced. However, the late-type ball joints without plastic plugs must be replaced if dirt has entered them. The early-type ball joints should be thoroughly cleaned in solvent after the damaged seal has been removed. Then remove the plastic plug from the tapped hole and install a grease fitting. Force multipurpose grease through the joint until all traces of dirt have been expelled.

Install the small steel retaining ring on the new dust seal. Then fill the new dust seal with approximately 15 g (½ oz.) of multipurpose grease.

Install the dust seal on the ball joint. Then place a conical sleeve over the seal as shown in Fig. 7-7. Slide the new retaining ring over the conical sleeve and off its large end onto the ball joint seal. The conical sleeve will keep the retaining ring from accidentally puncturing the new seal.

© 1974 VWoA—2865

Fig. 7-7. Retaining ring being slipped off large end of conical sleeve and onto dust seal. The retaining ring groove must be grease-free.

After the new dust seal has been installed, remove the grease fitting, then seal the tapped hole with a new plug.

CAUTION —
Do not hammer in the plug, screw it in. Otherwise it will soon fall out.

Depending on the tolerance range of the torsion arm, either a standard size ball joint or an oversize ball joint is installed. The torsion arms with oversize holes—0.30 mm (.012 in.) larger—are identified by a letter **B** stamped in the locations shown in Fig. 7-8.

© 1974 VWoA—2867

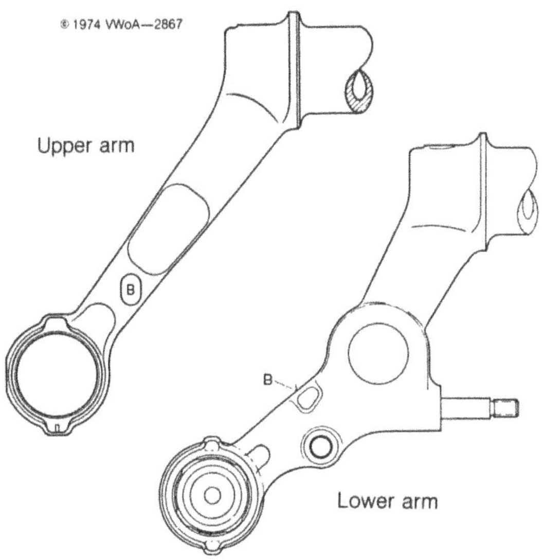

© 1974 VWoA—2868

Fig. 7-8. Marking locations for torsion arms with oversize ball joint holes.

The oversize ball joints are 0.30 mm (.012 in.) larger in diameter and are identified by two additional V-shaped notches as indicated in Fig. 7-9. These notches are located 45° from the two square installation-position grooves found on both standard and oversize ball joints.

© 1974 VWoA—2866

Fig. 7-9. V-notches (arrows) that identify an oversize ball joint.

The ball joints are a press fit in the torsion arms. They must fit properly and, once pressed out, must never be reinstalled in a torsion arm.

To replace ball joint:

1. Press the old ball joint out of the torsion arm as shown in Fig. 7-10.

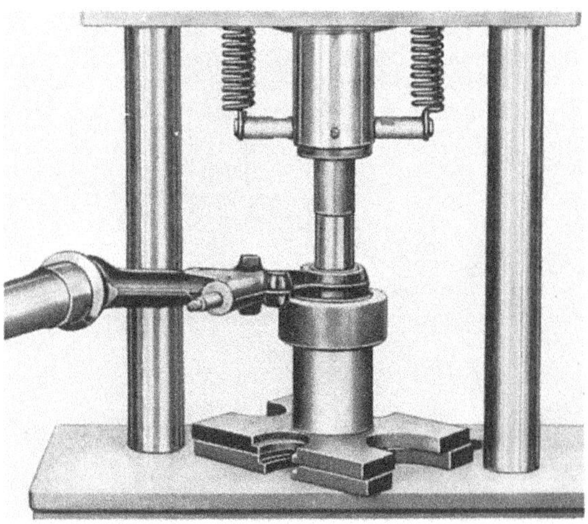

© 1974 VWoA—2869

Fig. 7-10. Hydraulic repair press being used to press the ball joint out of the torsion arm.

2. If necessary, press the eccentric camber adjusting bushing off the ball joint for an upper torsion arm as shown in Fig. 7-11.

© 1974 VWoA—2870

Fig. 7-11. Ball joint being pressed loose from eccentric camber adjusting bushing. Note the nut installed on the ball joint stud. This nut will keep the ball joint from flying out suddenly.

3. Align the installation-position groove with the boss on the torsion arm (Fig. 7-12).

WARNING —
Never reinstall used ball joints. They will not fit tightly and could come out of the torsion arm while the car is being driven.

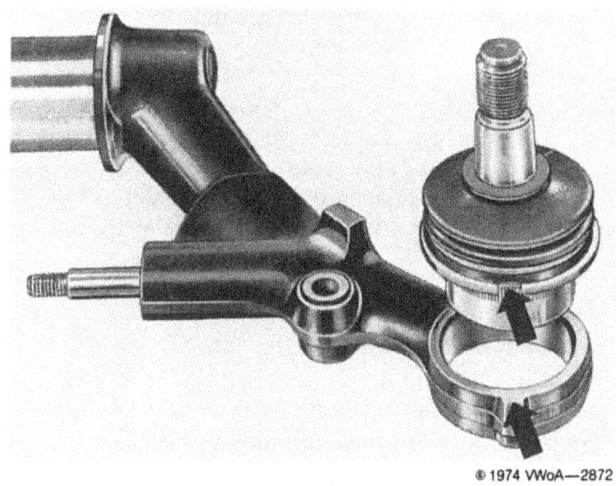

© 1974 VWoA—2872

Fig. 7-12. Groove in ball joint (top arrow). When you install the ball joint, align the groove with the boss on the torsion arm (bottom arrow).

4. Being careful to keep the groove and the forged boss aligned, press the new ball joint into the torsion arm. Use a sleeve-type driving mandrel that will apply pressure to the outer part of the ball joint only.

NOTE —
Press the ball joint in from the bottom of the torsion arm.

5. Using suitable mandrels and adaptors, install peening tool VW 471 on the press ram. Peen the ball joint cover to obtain the pattern indicated by **a** in Fig. 7-13.

NOTE —
The peening tool must be applied three times with a pressure of 6 tons to ensure that the joint fits tightly in the arm.

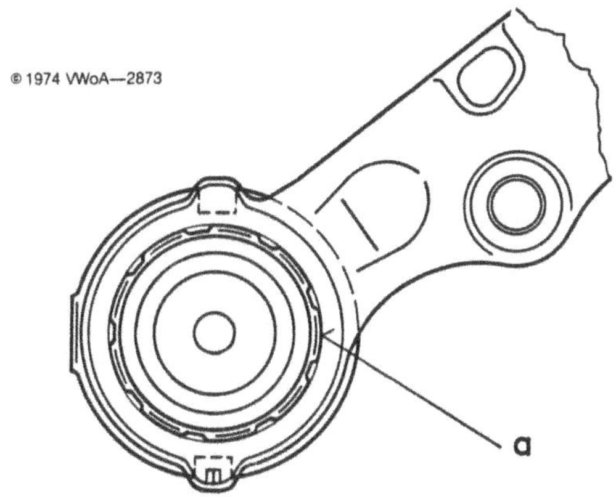

Fig. 7-13. Peened ball joint cover. Cover material is expanded as indicated by **a**.

Replacing Shock Absorber Stud

An oversize shock absorber stud is available to replace studs that are bent, broken, or corroded. First drive out the dowel pin, then extract the shock absorber stud. If the stud is broken, centerpunch the remaining part and drill a 3-mm (1/8-in.) pilot hole. Then drill out the stud with a 10.75-mm (7/16-in.) drill. The broken part will usually come out with the last few turns of the drill. Enlarge the hole to 12.3 mm (31/64 in.); then ream the hole with a 12.455 to 12.482-mm (.4904 to .4914-in.) reamer. (Alternately, you can grind the oversize stud down to obtain a 0.01 to 0.05-mm [.0004 to .002-in.] press fit.) Press in the oversize stud until it projects 45.00 to 45.50 mm (1.771 to 1.791 in.) as shown by dimension **b** in Fig. 7-14. Drill a 4.00 to 4.08-mm (.157 to .161 or about 5/32-in.) hole at dimension **a**. Then drive in the dowel pin.

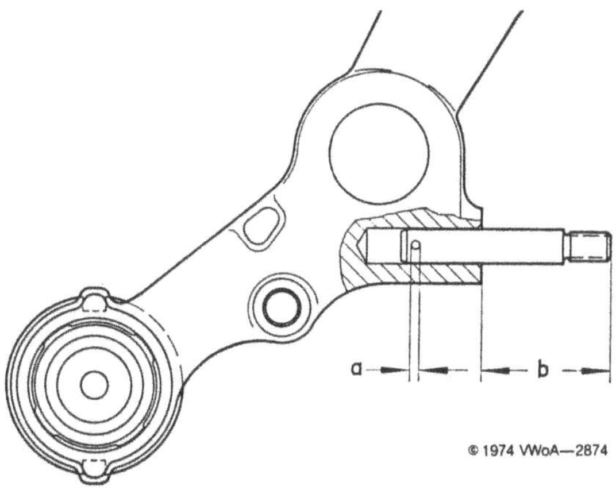

Fig. 7-14. Oversize stud fitting dimensions (see text).

7.5 Removing and Installing Torsion Bars

From August 1969, the torsion bars have a white paint spot on one end. This end must be on the left side of the car. Use only unmarked bars on earlier models.

To remove:

1. Remove both steering knuckles, complete with brake assemblies, as described in **6.1 Removing and Installing Steering Knuckle**. Suspend the steering knuckles from the axle tubes with heavy wire hooks.

   WARNING —
   Never let the knuckle hang by the brake hose. Doing so could weaken the hose and cause subsequent brake failure.

2. On lower torsion bars only, remove the shock absorber. On upper torsion bars only, detach the front gearshift rod at the shift rod coupling, then push the front gearshift rod to one side.
3. Remove the torsion arm from one end of the torsion bar, as described in **7.2 Removing and Installing Torsion Arm**.
4. Loosen the locknut on the socket head setscrew. Then remove the setscrew (Fig. 7-15).

Fig. 7-15. Setscrews with locknuts (arrows) that hold the torsion bars in their center bushings.

5. Gripping the torsion arm that is still installed, pull the torsion bar out of the axle beam tube.

Installation is the reverse of removal. Clean the torsion bar, inspect it for cracks and breakage, and then coat it liberally with multipurpose grease before you install it. Align the countersunk mark in the center of the bar with the setscrew hole. Torque first the setscrew and then the locknut to 4.0 mkg (29 ft. lb.). Then install the torsion arm and both steering knuckles.

2

7.6 Replacing Needle Bearings And Metal Bushings

The needle bearings are more likely to require replacement than the metal bushings, which are subject to very little wear. However, if wear is noted on the torsion arm bearing surface, replace the metal bushing as well as the worn torsion arm.

To remove:

1. Remove the torsion arms and torsion bars as described in **7.5 Removing and Installing Torsion Bars**.
2. Using an expansion tool or a toggle washer behind the needle bearing, pull out the needle bearing with a slide hammer as shown in Fig. 7-16.

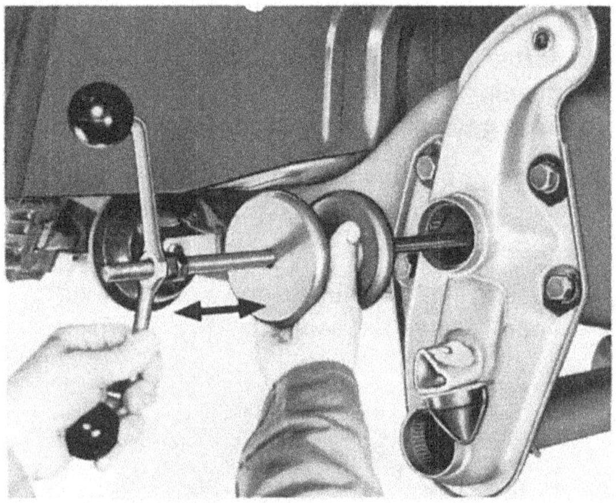

© 1974 VWoA—2879

Fig. 7-16. Needle bearing being pulled out of axle tube with slide hammer and toggle washer.

3. Measure the metal bushing for wear as shown in Fig. 7-17.

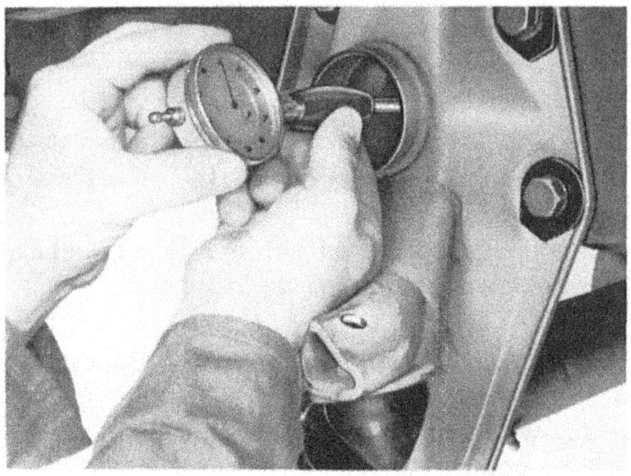

© 1974 VWoA—2878

Fig. 7-17. Internal measuring gauge being used to check inside diameter of metal bushing.

4. If the metal bushing has worn to a diameter greater than 43.40 mm (1.7086 in.), pull it out with a slide hammer and the tools shown in Fig. 7-18.

CAUTION —
Do not pull out the plastic sleeve as it is not subject to wear. Replacement sleeves are not supplied, so if you remove or damage them the entire axle beam must be replaced.

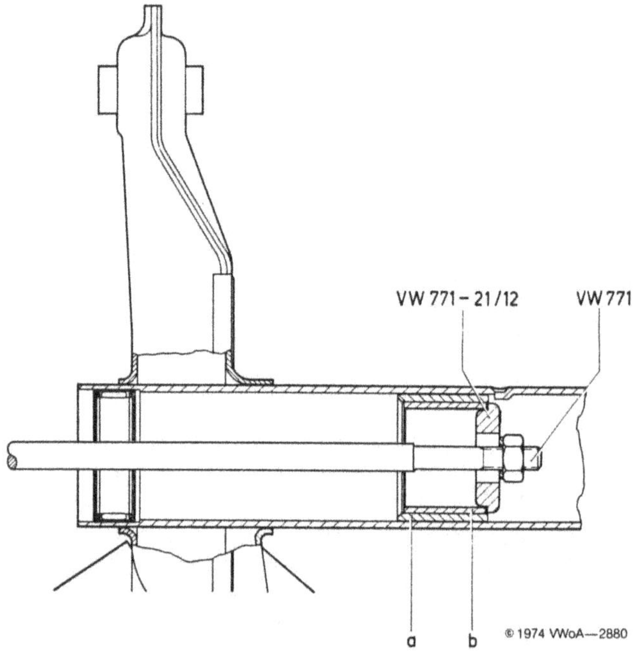

© 1974 VWoA—2880

a. Plastic sleeve b. Metal bushing

Fig. 7-18. Toggle washer in position to pull out metal bushing without removing plastic sleeve.

5. Clean the needle bearing seats. Then, using the internal measuring gauge illustrated in Fig. 7-17, check the diameter of the needle bearing seats.

NOTE —
This measurement is essential since both standard and oversize needle bearings are used.

Install standard bearings in either upper or lower tubes that measure 56.97 to 56.99 mm (2.2429 to 2.2437 in.); install oversize bearings in either upper or lower tubes that measure 57.17 to 57.19 mm (2.2507 to 2.2515 in.). Standard bearings are 56.96 to 56.99 mm (2.2425 to 2.2437 in.) in diameter; oversize bearings are 57.17 to 57.19 mm (2.2507 to 2.2515 in.) in diameter.

Oversize bearings are marked with the letter **U** on the hardened end which faces outward. If the bearing seats are no longer within tolerance, replace the axle beam.

Clean the axle tubes, particularly at the needle bearing and metal bushing seats. Then drive in the needle bearing, hardened face outward, until the shoulder on the special drift (VW 772) contacts the axle tube as shown in Fig. 7-19. Using the needle bearing as a pilot for the drift, drive the metal bushings in with VW 772.

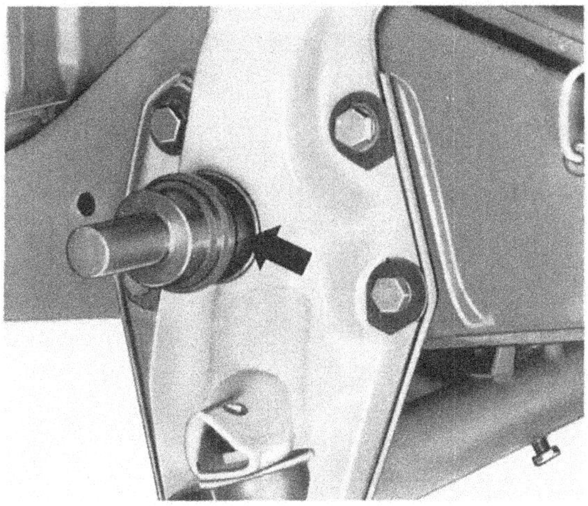

© 1974 VWoA—2883

Fig. 7-19. Drift used to drive in needle bearing and metal bushing. It should contact tube at arrow.

If the special drift is unavailable, position the needle bearing and the metal bushing as shown in Fig. 7-20. Dimension **a** is 141.00 −1.00 mm (5.551 −.040 in.). Dimension **b** is 7.00 +0.50 mm (.276 +.020 in.).

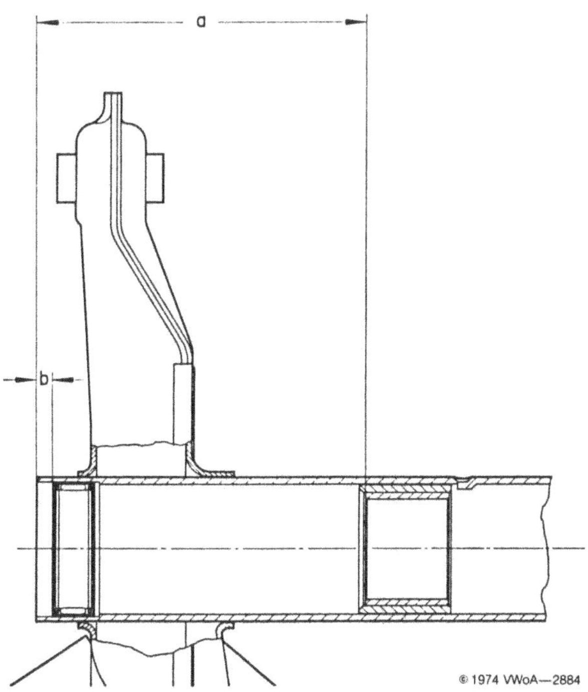

© 1974 VWoA—2884

Fig. 7-20. Needle bearing and metal bushing positions.

After the needle bearing and metal bushing are driven into place, install the seal retainers with their lugs in a vertical position as shown in Fig. 7-21.

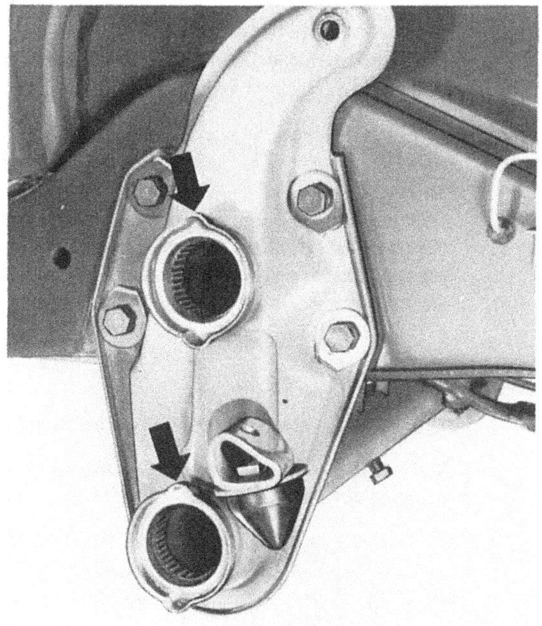

© 1974 VWoA—2885

Fig. 7-21. Seal retainers installed over needle bearings. The lugs should be vertical, as indicated by the two arrows.

NOTE —
Oversize seal retainers are available for use with oversize needle bearings. The oversize seal retainers are marked with a groove as indicated by the arrow in Fig. 7-22.

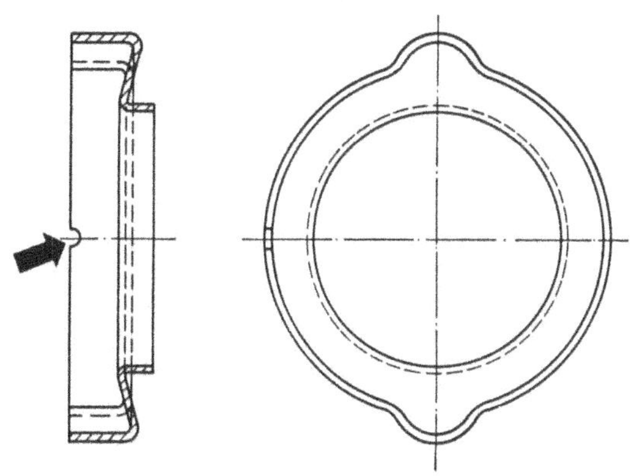

© 1974 VWoA—2886

Fig. 7-22. Oversize seal retainer marked by groove.

After the seal retainers have been correctly installed, install new rubber seals. Then reinstall the torsion bars and torsion arms, the steering knuckle and brake assembly, and other remaining parts.

8. REMOVING AND INSTALLING AXLE BEAM

Though most front axle repairs can be carried out with the axle beam mounted on the car, it may be necessary to remove the axle beam when correcting body or frame damage or to facilitate a complete front axle rebuild. If you suspect that the axle beam has been bent by accident damage, you can check the axle tubes with a straightedge.

CAUTION —
Bent axle beams must be replaced and not straightened. Bending them back to their original shape will seriously weaken them structurally.

To remove:

1. Raise the car and remove the front road wheels.
2. Disconnect the brake hoses at the brackets. Plug the brake lines with new bleeder valve dust caps.
3. Remove the cotter pin from the speedometer cable where the cable end projects from the left dust cap. Then, working behind the steering knuckle, pull out the cable complete with its housing.
4. Working under the car, remove the cover plate that is beneath the pedal cluster.
5. Engage either first or third gear. Then disconnect the gearshift rod at the coupling indicated in Fig. 8-1.

© 1974 VWoA—2828

Fig. 8-1. Coupling (arrow) where gearshift rod can be disconnected.

6. On vehicles with manual transmissions, remove the gearshift lever and front gearshift rod. On vehicles with automatic transmissions, remove the front gearshift rod by unbolting it from the lower part of the selector lever. See **TRANSMISSION AND REAR AXLE** or **AUTOMATIC TRANSMISSION**.
7. Disconnect the clutch cable from the pedal arm at the point indicated in Fig. 8-2.

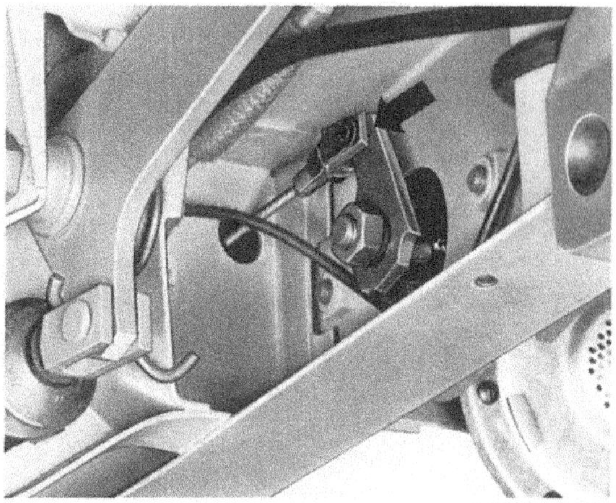

© 1974 VWoA—2829

Fig. 8-2. Clutch cable disconnecting point (arrow).

8. Disconnect the parking brake cables from the parking brake lever.
9. Remove the cotter pin and castellated nut from the drag link end stud. Then press the drag link end out of the relay lever with the tool shown in Fig. 8-3.

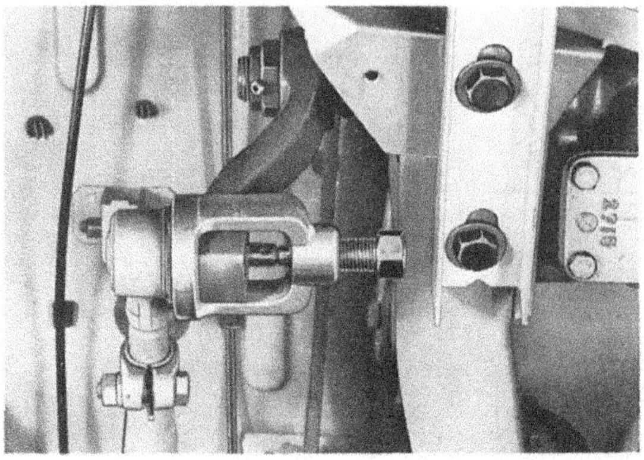

© 1974 VWoA—2830

Fig. 8-3. Drag link end being pressed out. The drag link is shown being pressed out of the steering gearbox drop arm, but the same tool should be used to press it out of the relay lever.

CAUTION —
Do not hammer out the drag link end. Doing so will ruin the threads and make reinstallation impossible.

10. So that the steering damper will not be accidentally damaged by the floor jack's supporting adapter, unbolt the damper from the bracket on the axle tube. Then swing the damper out of the way to the rear.
11. Position a floor jack with the VW 610 front axle supporting adapter under the front axle. Then raise the jack until the adapter is in firm contact with the axle beam.

> ***WARNING —***
> *If you do not have a suitable jack and adapter, have at least two helpers support the front axle while you are unbolting it. Trying to handle this job alone could lead to serious injury owing to the weight of the axle.*

12. Remove the four axle beam mounting bolts from each side plate. Then, using the floor jack, lower the axle and pull it from under the vehicle.

To Install:

1. Place the axle in the floor jack front axle adapter VW 610. Then raise the axle and position it on the vehicle.

> **NOTE —**
> If there is a gap between the axle end plates and the frame sidemembers, take up the gap with spacers, which are available in thicknesses of 0.50 and 1.00 mm (.020 and .040 in.).

2. Using new lock washers, install, but do not tighten, the four mounting bolts in each end plate.
3. To ensure that braking stresses will not loosen the mounting bolts, tilt the axle forward to take up the play between the mounting bolts and the bolt holes in the axle beam end plates. Then torque the mounting bolts to 9.0 to 12.5 mkg (65 to 90 ft. lb.).
4. Install the drag link in the relay lever. Torque the castellated nut to 3.0 mkg (22 ft. lb.). Advance the nut, if necessary, to uncover the cotter pin hole, then install a new cotter pin.
5. Using a new lockplate, install the steering damper on its bracket on the lower axle beam tube. Torque the mounting bolt to 4.0 to 4.5 mkg (29 to 32 ft. lb.).
6. Install the front gearshift rod. If the vehicle has a square head setscrew at the coupling, lock it with safety wire.

The remainder of installation is the reverse of removal. Make sure the brake hoses are not twisted and use a new cotter pin in the end of the speedometer cable. Following axle installation, bleed the brakes, adjust the parking brake, and adjust the front wheel camber and toe.

9. STEERING

The worm and peg steering gearbox used on 1968 through 1972 Type 2 vehicles is shown in Fig. 9-1.

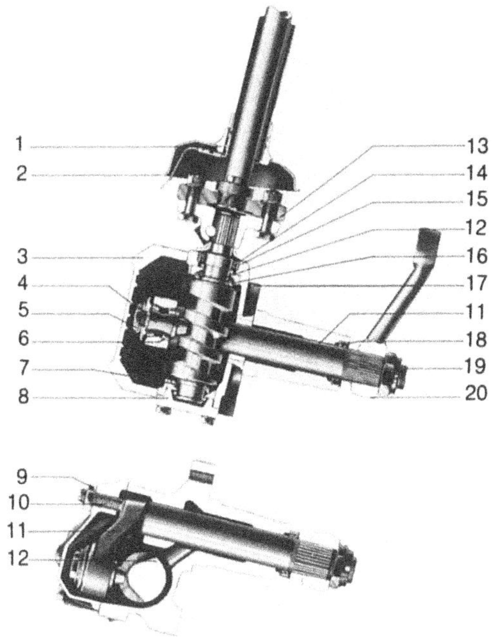

1. Column tube retaining ring
2. Column tube cover
3. Steering gearbox cover
4. Peg
5. Peg securing nut
6. Rollers (16 each race)
7. Lower bearing
8. End plate
9. Locknut for adjusting screw
10. Adjusting screw
11. Peg shaft
12. Worm spindle
13. Steering coupling
14. Worm spindle cap
15. Worm spindle oil seal
16. Upper bearing
17. Steering gearbox case
18. Peg shaft oil seal
19. Castellated nut
20. Drop arm

Fig. 9-1. Worm and peg steering gearbox.

> ***WARNING —***
> *Do not re-use fasteners such as nuts, bolts, washers, self-locking nuts, circlips, cotter pins that are worn or deformed in normal use. Always replace.*

The steering system normally requires no maintenance. However, the steering gearbox can be damaged if the vehicle is operated for a long time with improperly adjusted steering. Replacement parts are available for the worm and peg steering gearbox. Complete data for rebuilding this unit are given later in this section.

Although the removal and installation procedures given in this section are illustrated with photos of the worm and peg unit, the procedures apply equally to the worm and roller gearbox. For the most part, the worm and roller steering gearbox must be replaced as a unit if it is worn or damaged. However, the roller shaft oil seal, Part No. 211 415 273A is available as a replacement part.

2

Fig. 9-2 is a cross-section of the worm and roller steering gearbox. It shows the location of the sole replaceable component, the roller shaft oil seal.

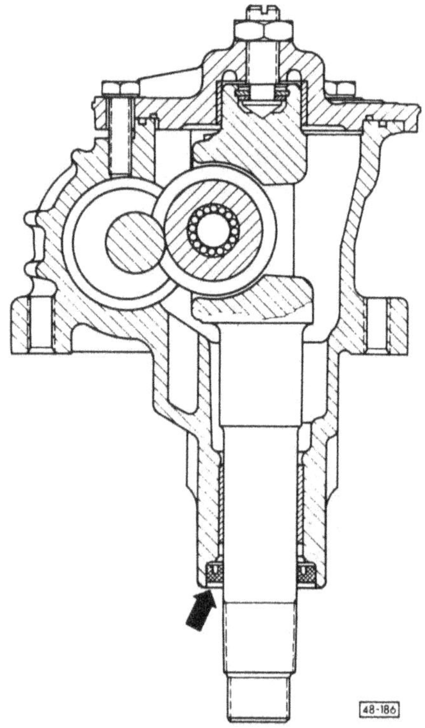

©1977 VWoA-4420

Fig. 9-2. Roller shaft oil seal (arrow) on 1973 and later steering gearbox.

To replace seal:

1. Remove the drop arm as described in **9.5 Removing and Installing Drag Link and Drop Arm.**
2. Using a sharp-edged screwdriver, pry out the faulty seal.

> ***CAUTION —***
> *Be careful not to scratch the roller shaft, as doing this will cause rapid wear of the new seal and renewed leakage.*

3. Thoroughly clean the seal recess in the steering gearbox housing and wipe clean the exposed end of the roller shaft.
4. Wrap the splines of the roller shaft with thin plastic electrical tape in order to protect the seal from being cut by the splines as the seal is installed.
5. Lightly lubricate the seal lip and pack the groove in the seal with multipurpose grease. Then slide the oil seal onto the shaft—with the seal's grooved side toward the steering gearbox.
6. Using an appropriate tool, drive the oil seal fully into position.

> **NOTE —**
> A suitable driving tool is a tube that has an inside diameter of 32 mm ($1\,^{9}/_{32}$ in.), an outside diameter of 36 mm ($1\,^{13}/_{32}$ in.), and a length of 100 mm (4 in.).

7. Reinstall the drop arm. Use a new lock plate and torque the nut to 14.0 mkg (101 ft. lb.).
8. Using a mirror, check the steering gearbox oil level. If necessary, add hypoid gear oil until the roller shaft is covered.

9.1 Adjusting Worm and Peg Steering

Check the worm and peg steering with the vehicle lifted. Turn the steering back and forth through its centered position several times. You should feel resistance as the steering passes through its centerpoint, but it should do so smoothly—without sticking. If there is no resistance or if the steering binds, adjust the *steering.*

To adjust:

1. Torque the steering gearbox mounting bolts to 3.5 to 5.0 mkg (25 to 36 ft. lb.) and the steering gearbox cover bolts to 1.5 mkg (11 ft. lb.).
2. Loosen the locknut for the adjusting screw. Have someone gradually turn the adjusting screw (Fig. 9-3) until you feel the correct resistance as the steering passes through the centerpoint.

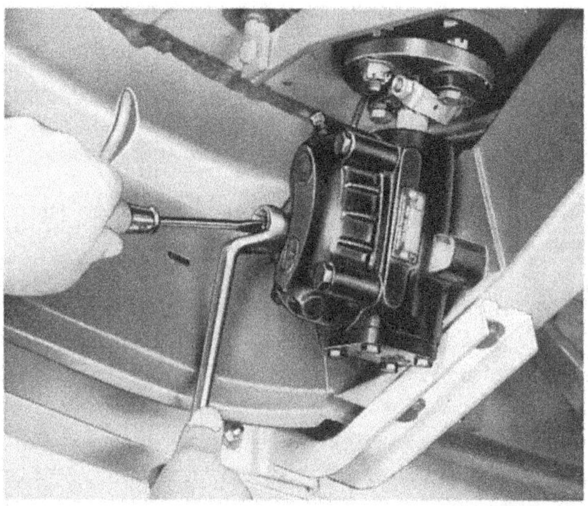

© 1974 VWoA—2921

Fig. 9-3. Worm and peg steering being adjusted.

3. If you lack an experienced helper, attach a torque gauge as shown in Fig. 9-4.

© 1974 VWoA—2922

Fig. 9-4. Torque gauge being used to check worm and peg steering adjustment.

NOTE —
Take off the horn button and disconnect the horn wire so that the torque gauge can be attached to the steering wheel nut. Also, the drag link must be disconnected from the drop arm as described in **9.5 Removing and Installing Drag Link and Drop Arm**.

4. Turn the adjusting screw one way or the other in small increments until it is possible to turn the steering through the centerpoint with a torque of 8 to 10 cmkg (7 to 9 in. lb.).

NOTE —
The torque given in Step 4 applies only to gearboxes that have had at least 3000 miles (5000 km) of service. New gearboxes—or gearboxes that have new cases, peg and worm sets, bearings, or seals—should be adjusted to a turning torque of 24 cmkg (21 in. lb.).

9.2 Adjusting Worm and Roller Steering

To check the adjustment of the worm and roller steering, lift the vehicle—or support the front axle on safety stands. With the front wheels in their straight-ahead position, grip one steering wheel spoke at its extreme outer end. Turn the steering wheel lightly in both directions. The freeplay should not exceed 15 mm ($5/8$ in.) measured at the wheel rim.

If the steering freeplay is excessive, check to see that the looseness is not caused by worn tie rod ends, a loose drop arm, or a worn relay lever shaft. Make sure that the steering gearbox is mounted firmly and that its cover bolts are torqued to 2.5 mkg (18 ft. lb.). If no faults are found, correct excessive centerpoint freeplay by making adjustments at the steering gearbox.

To adjust:

1. Disconnect the drag link from the drop arm as described in **9.5 Removing and Installing Drag Link and Drop Arm**.
2. Turn the steering wheel 180° to 200° to either the left or the right.

NOTE —
The steering is centered when the pointer on the worm spindle dust cap is in line with the 7-mm ($9/32$-in.) square boss on the worm spindle cap. For all repair operations with the steering centered, the pointer on the dust cap must be within the 7-mm ($9/32$-in.) limits of the square boss (Fig. 9-5).

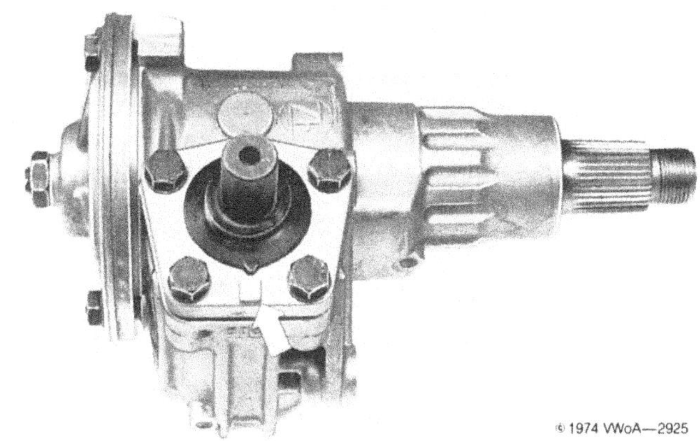

© 1974 VWoA—2925

Fig. 9-5. Square boss (arrow) that pointer on dust cap must be toward when the steering is centered.

3. Loosen the locknut for the adjusting screw. Then turn out the adjusting screw one complete turn.
4. Working under the vehicle, move the drop arm back and forth. Gradually turn in the adjusting screw until no play is felt at the drop arm.
5. Hold the adjuster in its optimum position while tightening the locknut to 5.0 to 6.0 mkg (36 to 43 ft. lb.).
6. Take off the horn button and disconnect the horn wire. Then attach a torque gauge to the steering wheel nut as shown earlier in Fig. 9-4.
7. Turn the steering wheel until the steering is nearly centered. Using the torque gauge, move the steering through its centerpoint. The steering is correctly adjusted if a turning torque of 8 to 12 cmkg (7 to 11 in. lb.) is required.
8. If necessary, correct the adjustment, then recheck it with the torque gauge.

9.3 Removing and Checking Steering Damper

The steering damper is a hydraulic cylinder mounted between an eye in the relay lever and a bracket on the axle beam. To remove it, take out the cotter pin and unscrew the M 10 nut from the bolt through the relay lever. Then remove the bolt that holds the damper on the axle beam.

Hand-check the steering damper by extending and compressing it while holding it in its installed position. It must operate with uniform resistance throughout its entire stroke. If necessary, compare the used unit with a new steering damper. Minor fluid leakage does not make replacement necessary as long as efficiency is not impaired.

To prevent steering trouble due to premature failure of the steering damper, make certain that the damper you install is the correct one for the vehicle. Check the rubber bushing and sleeves in the removed damper before reinstallation. If they are worn or damaged, replace them.

During installation, make sure that the large flat washer is between the damper bushing and the relay lever. Use a new cotter pin at the relay lever and a new lock washer under the head of the bolt that holds the damper on the axle beam.

9.4 Removing and Checking Tie Rods

To remove the tie rods, first disconnect the steering damper from the relay lever. Then take out the cotter pins in the tie rod end studs and remove the castellated nuts. Use a tool such as the one shown in Fig. 9-6 to press the tie rod ends out of the relay lever and the steering arms.

> ***CAUTION —***
> *Do not hammer out the tie rod ends. Doing so will ruin the threads and make reinstallation impossible.*

© 1974 VWoA—2841

Fig. 9-6. Tool used to press out tie rod ends.

Installation is the reverse of removal. Carefully inspect the tie rods for cracks. Bending can be detected by rolling the tie rods over a flat surface.

> ***CAUTION —***
> *Bent tie rods must be replaced and not straightened. Bending them back to their original shape will seriously weaken them structurally.*

Check the tie rod ends for play and replace any that are worn. If you cannot hand-move the tie rod end stud, replace the tie rod end. If the tie rod end boots are torn or cracked, they can be replaced. However, the entire tie rod end should be replaced if dirt has entered the ball socket. If in doubt, replace the tie rod end.

After installing the tie rod ends in the relay lever and the steering arms, torque the castellated nuts to 2.5 mkg (18 ft. lb.). Advance the nuts, if necessary, to uncover the cotter pin holes. Then install new cotter pins in all four tie rod end studs.

> **NOTE —**
> When you install the right-hand tie rod, the tie rod end with right-hand threads should be at the steering knuckle end.

Check the front wheel toe and adjust it to the specifications given in **3.2 Checking and Adjusting Toe**. Twist the right-hand tie rod as far as it will go so that the tie rod ends are parallel. Then torque the clamp bolts for the right-hand tie rod to 1.5 mkg (11 ft. lb.).

> *WARNING —*
> *Do not re-use fasteners such as nuts, bolts, washers, self-locking nuts, circlips, cotter pins that are worn or deformed in normal use. Always replace.*

9.5 Removing and Installing Drag Link and Drop Arm

The drop arm must be removed before the steering gearbox can be removed from the frame.

To remove:

1. Working under the vehicle, remove the cover plate that is beneath the pedal cluster.
2. Remove the cotter pins and then the castellated nuts from the drag link ends.
3. Using the tool illustrated earlier in Fig. 9-6, press the drag link ends out of the drop arm and the relay lever.

> ***CAUTION —***
> *Do not hammer out the drag link ends. Doing so will ruin the threads and make reinstallation impossible.*

4. Remove the cotter pin and then the castellated nut from the peg shaft or roller shaft of the steering gearbox.

5. Using a puller as shown in Fig. 9-7, remove the drop arm from the shaft.

© 1974 VWoA—2896

Fig. 9-7. Puller being used to remove drop arm from peg (or roller) shaft.

Installation is the reverse of removal. Make sure that the splines both in the drop arm and on the shaft are in good condition and that the splines in the replacement arm match those of the shaft. Align the arm with the shaft as shown in Fig. 9-8. On early models, torque the nut to 8.0 to 11.0 mkg (58 to 80 ft. lb.); on 1973 and later models, to 14.0 mkg (101 ft. lb.). Torque the castellated nuts on the drag link ends to 3.0 mkg (22 ft. lb.). Advance the nuts, if necessary, to uncover the cotter pin holes, then install the new cotter pins.

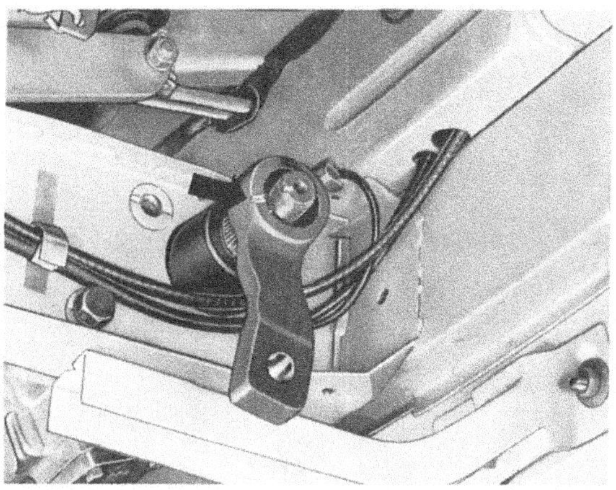

© 1974 VWoA—2897

Fig. 9-8. Mark on drop arm (arrow) that must be aligned with notch in shaft. From chassis No. 212 2038 133, there are two marks. Align the mark labeled **L** on vehicles with left-hand drive. Align the mark labeled **R** on vehicles with right-hand drive.

9.6 Removing and Installing Relay Lever, Lever Shaft, and Bushings

Play in the relay lever can usually be corrected by replacing the relay lever shaft and its bearings.

To remove relay lever and shaft:

1. Working under the vehicle, remove the cover plate that is beneath the pedal cluster.

2. Disconnect the tie rods, the steering damper, and the drag link from the relay lever. These jobs are described under the three preceding headings.

3. To prevent the spring washer from forcing the lever up and damaging the bolt, clamp the relay lever as shown in Fig. 9-9. Then remove the nut and clamp bolt from the relay lever.

4. Remove the clamp and then the relay lever.

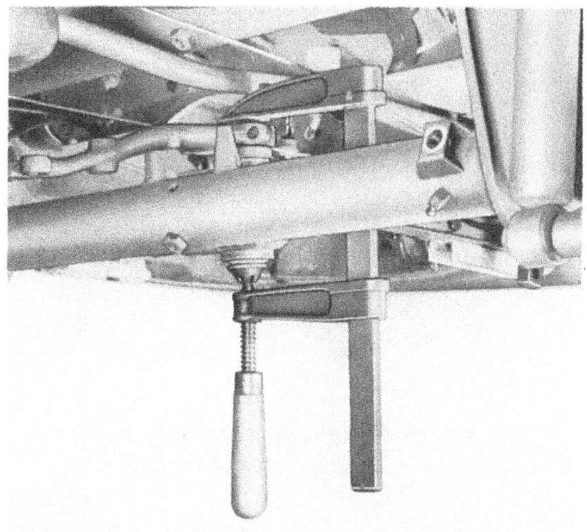

© 1974 VWoA—2892

Fig. 9-9. Clamp installed on relay lever and shaft.

5. Using a screwdriver, pry the end cap off the bottom of the relay lever shaft mounting. (Alternately, you can drive it off with a hammer and cold chisel.)

6. Pull the relay lever shaft out downward.

NOTE —
Normally, the shaft will fall out of its bushings when the end cap is removed. Because there is a head on the lower end of the shaft, it cannot be driven out upward. Check the top end of the shaft for burrs if it is stuck.

7. Inspect the end cap, the upper protection cap, and its gasket. Obtain replacements for any worn or damaged parts.

2

To replace lever shaft bushings:

1. Remove the grease nipple. Then, using a slide hammer and appropriate extraction tools, drive the old bushings out downward (Fig. 9-10).

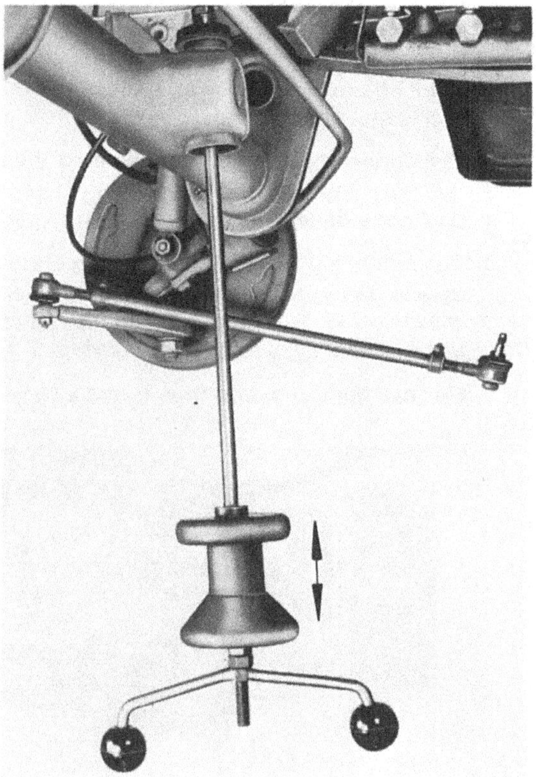

Fig. 9-10. Bushing being driven out. Move the hammer as indicated by the double arrow.

2. Using the tools shown in Fig. 9-10, drive in the new lower bushing from above until it is flush with the bottom of the mounting.
3. Drive the upper bushing in only until its upper edge projects 1.10 mm (.043 in.) above the mounting. Measure this by placing a feeler gauge between the driving tool and the top of the mount, or by measuring the projecting part of the bushing with a vernier caliper or depth micrometer.

Installation of the relay lever and shaft is the reverse of removal. Before you install the clamp bolt and nut, compress the spring washer by installing a clamp, as shown in Fig. 9-9. Torque the nut to 8.0 to 11.0 mkg (58 to 80 ft. lb.).

Install the end cap and the grease nipple. Then lubricate the relay lever shaft. Following tie rod installation, torque the castellated nuts on the tie rod ends to 2.5 mkg (18 ft. lb.) and the steering damper bolt to 4.0 to 4.5 mkg (29 to 32 ft. lb.).

9.7 Removing and Installing Steering Wheel, Steering Column, and Steering Column Tube

A new steering column was introduced on the 1975 models. This column is different in two major ways from the steering column used from 1968 through 1974. The new steering column has a different column switch assembly that is similar to the switch assembly used on late-model VW Beetles. Secondly, the new column tube mounting's support and energy absorbing bracket are welded to the steering column tube instead of being separate parts as on the earlier models. Fig. 9-11 is an exploded view of the new steering column.

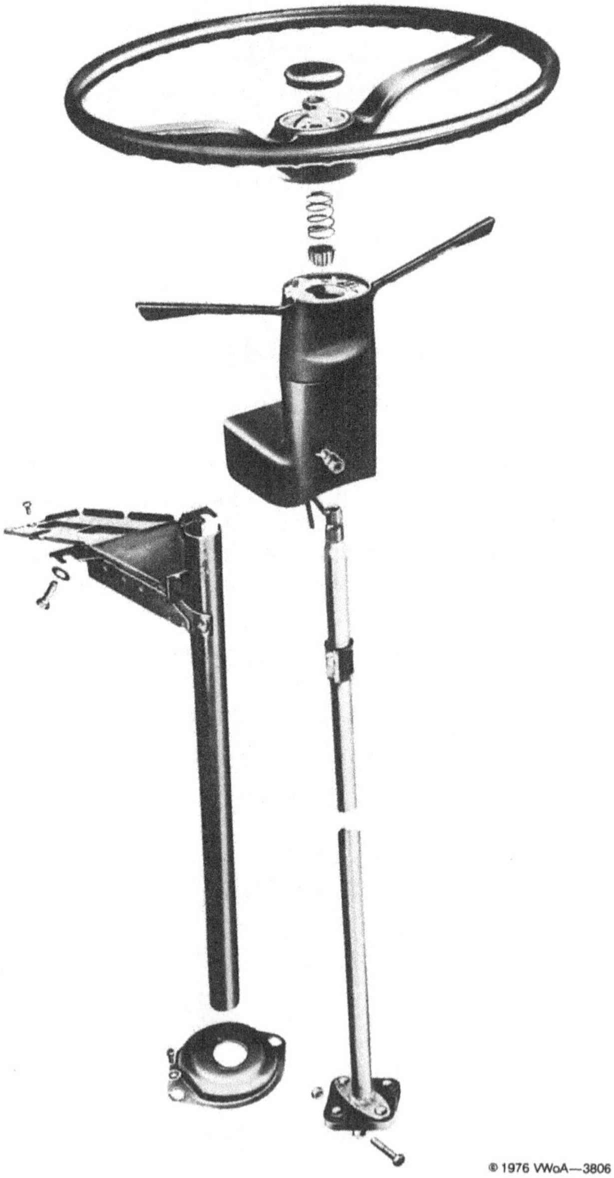

Fig. 9-11. Steering column of 1975 and later models. Identify components by comparing this illustration with Fig. 9-12.

The components of the steering column used on 1968 through 1974 vehicles are identified in Fig. 9-12. This illustration will be helpful if you must replace any of the individual parts. The illustration also gives the names and locations of the parts. The part names will appear frequently in the following procedures. The turn signal switch, which is not shown in this illustration, should be removed or installed as described in **ELECTRICAL SYSTEM**.

Fig. 9-12. Components of steering column assembly.

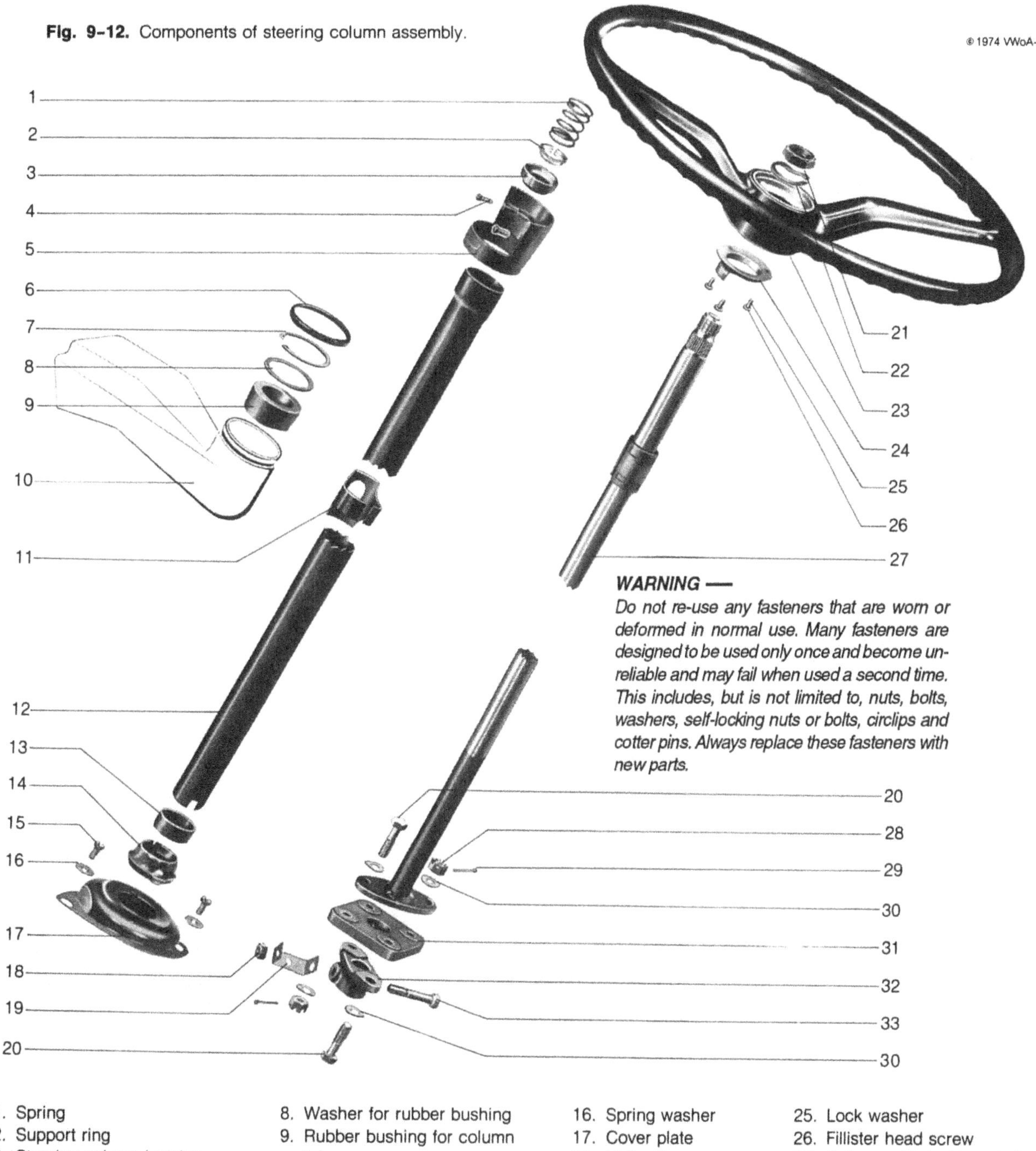

WARNING —
Do not re-use any fasteners that are worn or deformed in normal use. Many fasteners are designed to be used only once and become unreliable and may fail when used a second time. This includes, but is not limited to, nuts, bolts, washers, self-locking nuts or bolts, circlips and cotter pins. Always replace these fasteners with new parts.

1. Spring
2. Support ring
3. Steering column bearing
4. Fillister head screw
5. Retainer for turn signal switch
6. Insulating washer for ignition/steering lock
7. Circlip
8. Washer for rubber bushing
9. Rubber bushing for column tube
10. Mounting
11. Insulating ring
12. Column tube
13. Seal
14. Plastic clamping ring
15. Binding screw
16. Spring washer
17. Cover plate
18. M 8 nut
19. Lockplate
20. M 8 bolt
21. Steering wheel nut
22. Spring washer
23. Steering wheel
24. Cancelling ring
25. Lock washer
26. Fillister head screw
27. Column with locking shells
28. M 8 castellated nut
29. Cotter pin
30. Spring washer
31. Coupling disc
32. Coupling flange
33. M 8 clamp bolt

The entire steering column, together with the steering wheel and the steering column tube, can be removed from the vehicle as a unit. Individual steering column assembly components need only be removed if you are repairing the steering column.

To remove steering column assembly as a unit:

1. Disconnect the battery ground strap.
2. Disconnect the wiring of the column-mounted switches as described in **ELECTRICAL SYSTEM**.
3. Working under the vehicle, disconnect the horn ground wire from the steering column. Then unbolt the steering column coupling from the steering gearbox worm spindle.
4. Working inside the vehicle, remove the screws that hold the cover plate between the pedals.
5. Unbolt the column tube mounting's support from the dashboard. Then remove the steering column assembly from the vehicle as a unit.
6. Inspect the coupling disc. If the flexible material is worn or deteriorating, replace the disc.

> ***CAUTION —***
> *If the vehicle has been in an accident, you should also check the bracket and the support carefully. If you find cracks or bending, replace the bracket and the support. On 1975 and later models, you must replace the entire steering column tube. Do not attempt to weld or straighten the parts. Doing so will destroy the effectiveness of the energy absorbing system.*

Installation is the reverse of removal. During installation of the steering column on 1970 and later vehicles, install the plastic coated washers for the energy absorbing column tube support as shown in Fig. 9-13. The redesigned steering column tube, introduced on the 1975 models, is installed in a similar way even though the column tube mounting's support and energy absorbing bracket are integral with the steering column tube.

> **NOTE —**
> The purpose of the plastic coated washers is to let the support detach itself smoothly from the bolts in the dashboard—should sufficient stress be imposed on the mounting. The bracket then collapses at the predetermined location and the support is deflected to the side.

If the plastic coated washers are in any way damaged—or if the column tube support has been forced off the bolt by impact—you should always replace the washers.

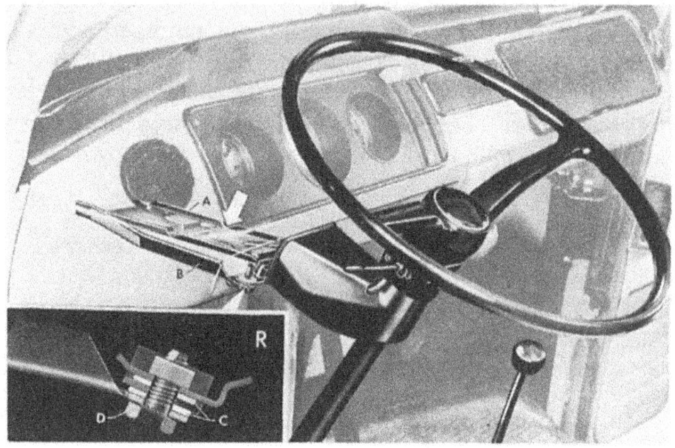

© 1974 VWoA—2827

A. Bracket
B. Support
C. Plastic coated washers
D. Self-tapping bolt

Fig. 9-13. Energy absorbing column tube support used on 1970 through 1974 vehicles. The inset (**R**) shows the proper installation position for the plastic coated washers. The arrow indicates the bracket's predetermined collapse point.

To remove steering wheel:

1. Disconnect the horn ground wire either at the horn or at the bottom of the steering column tube.
2. On 1968 through 1974 models only, remove the screws that hold the turn signal switch to the steering column tube. Move the turn signal switch aside.
3. Carefully pry out the horn button, then remove the steering wheel nut. On 1975 and later models, lift the steering wheel off the steering column.
4. On 1968 through 1974 models, use a puller, such as the one shown in Fig. 9-14, to pull the steering wheel off the steering column.

© 1974 VWoA—2900

Fig. 9-14. Puller being used to remove steering wheel from pre-1975 steering column.

5. If you intend to replace the steering wheel, take out the three screws that hold the cancelling ring to the bottom of the steering wheel hub. Then remove the cancelling ring so that it can be installed on the new steering wheel.

To install:

1. If necessary, install the cancelling ring on the steering wheel.
2. Center the steering. Then install the steering wheel so that its spokes are horizontal and the tab on the cancelling ring is to the left.
3. Install the steering wheel nut. On 1968 through 1974 models, torque the nut to 2.5 to 3.0 mkg (18 to 22 ft. lb.). On 1975 and later models, torque the nut to 5.0 mkg (36 ft. lb.).
4. Reinstall the horn button so that the crest is upright when the steering wheel's spokes are in a horizontal position.
5. On 1968 through 1974 models, loosely install the turn signal switch. Adjust the switch on the column tube to obtain a gap of from 2.00 to 3.00 mm (.080 to .120 in.) between the switch and the steering wheel. Then tighten the screws.

To remove steering column:

1. Disconnect the battery ground strap. Remove the steering wheel. On vehicles with the ignition/steering lock, turn the ignition key to its on position.
2. Working between the pedals, remove the two binding screws that hold the cover plate to the floor panel.
3. Working under the vehicle, remove the cover plate that is beneath the pedal cluster.
4. Bend the lockplate away from the nut, and then remove the M 8 clamp bolt that holds the coupling flange on the steering gearbox worm spindle.
5. Remove the steering gearbox as described in **9.8 Removing and Installing Steering Gearbox**.
6. Take the steering column out downward complete with the steering coupling.

Installation is the reverse of removal. If necessary, refer to **9.8 Removing and Installing Steering Gearbox**. Install the drop arm and drag link as described in **9.5 Removing and Installing Drop Arm and Drag Link**. Torque the steering gearbox mounting bolts to 3.5 to 5.0 mkg (25 to 36 ft. lb.). On early models, torque the drop arm nut to 8.0 to 11.0 mkg (58 to 80 ft. lb.); on 1973 and later models, torque to 14.0 mkg (101 ft. lb.). Torque the castellated nut on the drag link end to 3.0 mkg (22 ft. lb.). Advance the nut, if necessary, to uncover the cotter pin hole, then install a new cotter pin. Torque the steering wheel nut to 2.5 to 3.0 mkg (18 to 22 ft. lb.).

The column tubes of 1975 and later vehicles cannot be removed separately as described in the following procedure. Instead, remove the steering column assembly as a unit, then separate it into its individual components.

To remove 1968 through 1974 column tube:

1. Disconnect the battery ground strap. Remove the steering wheel. On vehicles with the ignition/steering lock, turn the ignition key to its on position.
2. Remove the circlip from its groove above the rubber bushing for the steering column.
3. Working between the pedals, remove the two binding screws that hold the cover plate to the floor panel.
4. Slightly lift the column tube together with the cover plate, pull the horn ground wire off the terminal on the column tube, and then straighten the terminal. (See Fig. 9-15.)

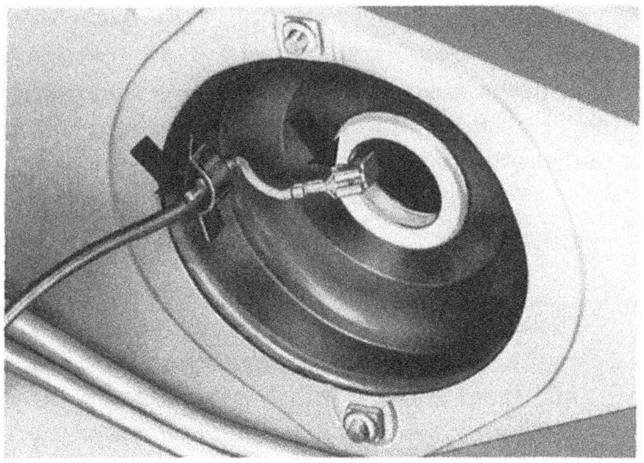

© 1974 VWoA—2904

Fig. 9-15. Horn ground wire (steering column and coupling removed for clarity). The wire passes through a retainer on the cover plate (left arrow). The terminal on the column tube (right arrow) must be bent to one side as shown when the column tube is installed.

5. Pull the cover plate off the lower end of the column tube together with the plastic clamping ring. The cover can remain in the car, attached to the horn ground wire.
6. Pull the column tube upward out of the steering/ignition lock and the mounting.

2

To install:

1. Align the insulating ring on the column tube so that both elongated holes match.
2. Coat the rubber bushing with talcum powder. Then push the column tube into the mounting and steering/ignition lock complete with the rubber bushing and the insulating ring.
3. Slide the cover plate together with the plastic clamping ring upward onto the column tube.
4. Install the horn ground wire on the terminal on the column tube. Then bend the terminal to roughly a right angle with the tube.
5. Push the column tube down until the hole in the tube is aligned with the locking pin of the steering/ignition lock. Check the lock's operation.
6. Mount the cover plate on the floor panel so that the column tube is centered with the steering column.
7. Install the washer and circlip above the rubber bushing for the column tube.
8. Inspect the steering column bearing and replace it if it is faulty. Install the bearing and the support ring with the shoulder on the support ring upward.
9. Install the steering wheel as previously described.

9.8 Removing and Installing Gearbox

Although the worm and peg steering gearbox is shown in the illustrations, the procedure for removing the late-model worm and roller gearbox is the same.

To remove:

1. On 1968 through 1974 models, remove the screws that hold the turn signal switch to the column tube, and then move the turn signal switch aside. On 1975 and later models, remove the steering column assembly as a unit.
2. Working under the vehicle, remove the cover plate that is beneath the pedal cluster.
3. Remove the drag link and drop arm as described in **9.5 Removing and Installing Drag Link and Drop Arm**.
4. Bend the lockplate away from the nut, and then remove the M 8 clamp bolt that holds the coupling flange on the steering gearbox worm spindle. (See Fig. 9-16.)

NOTE —
The clamp bolt is the horizontal bolt. It is not necessary to remove the cotter pins and then remove the castellated nuts from the bolts that hold the steering coupling together.

© 1974 VWoA—2902

Fig. 9-16. Clamp bolt (arrow) that holds steering coupling flange on worm spindle.

5. Remove the bolts indicated in Fig. 9-17. Then push the steering column upward and remove the gearbox from the frame sidemember.

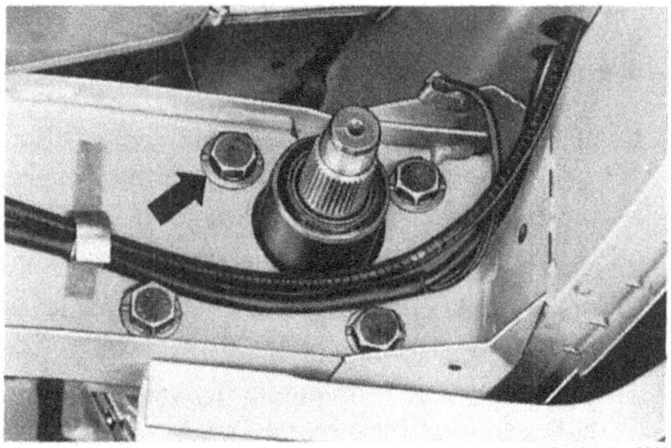

© 1974 VWoA—2903

Fig. 9-17. Bolts that hold gearbox to frame.

Installation is the reverse of removal. Make sure the gearbox is filled with oil. (See **9.9 Disassembling, Assembling, and Adjusting Steering Gearbox**.) Install the drop arm and drag link as described in **9.5 Removing and Installing Drag Link and Drop Arm**. Torque the steering gearbox mounting bolts to 3.5 to 5.0 mkg (25 to 36 ft. lb.). On early models, torque the drop arm nut to 8.0 to 11.0 mkg (58 to 80 ft. lb.); on 1973 and later models, torque to 14.0 mkg (101 ft. lb.). Torque the castellated nut on the drag link end to 3.0 mkg (22 ft. lb.). Advance the nut, if necessary, to uncover the cotter pin hole, then install a new cotter pin. During installation, adjust the pre-1975 turn signal switch on the column tube to obtain a gap of from 2.00 to 3.00 mm (.080 to .120 in.) between the switch and the steering wheel. Then tighten the screws.

9.9 Disassembling, Assembling, and Adjusting Steering Gearbox

Only the worm and peg steering gearbox (Fig. 9-18) installed on 1968 through 1972 models can be repaired. The worm and roller gearbox installed on 1973 and later models should be replaced as a unit if it is worn or damaged. Replacement gearboxes are supplied dry and must be filled with 9⅔ U.S. oz. (8 Imperial oz.; 284 ml) of hypoid oil prior to installation.

Unless you have a torque gauge such as the one shown in the illustrations, it is not possible to adjust accurately the worm and peg steering gearbox. If the bearings are too tight, they will bind and produce excessive wear; if they are too loose, there will be excessive play in the steering.

CAUTION —
If you lack the skills, tools, or a clean workshop for servicing the worm and peg steering gearbox, we suggest you leave such repairs to an Authorized VW Dealer or other qualified shop. We especially urge you to consult your Authorized VW Dealer before attempting repairs on a car still covered by the new-car warranty.

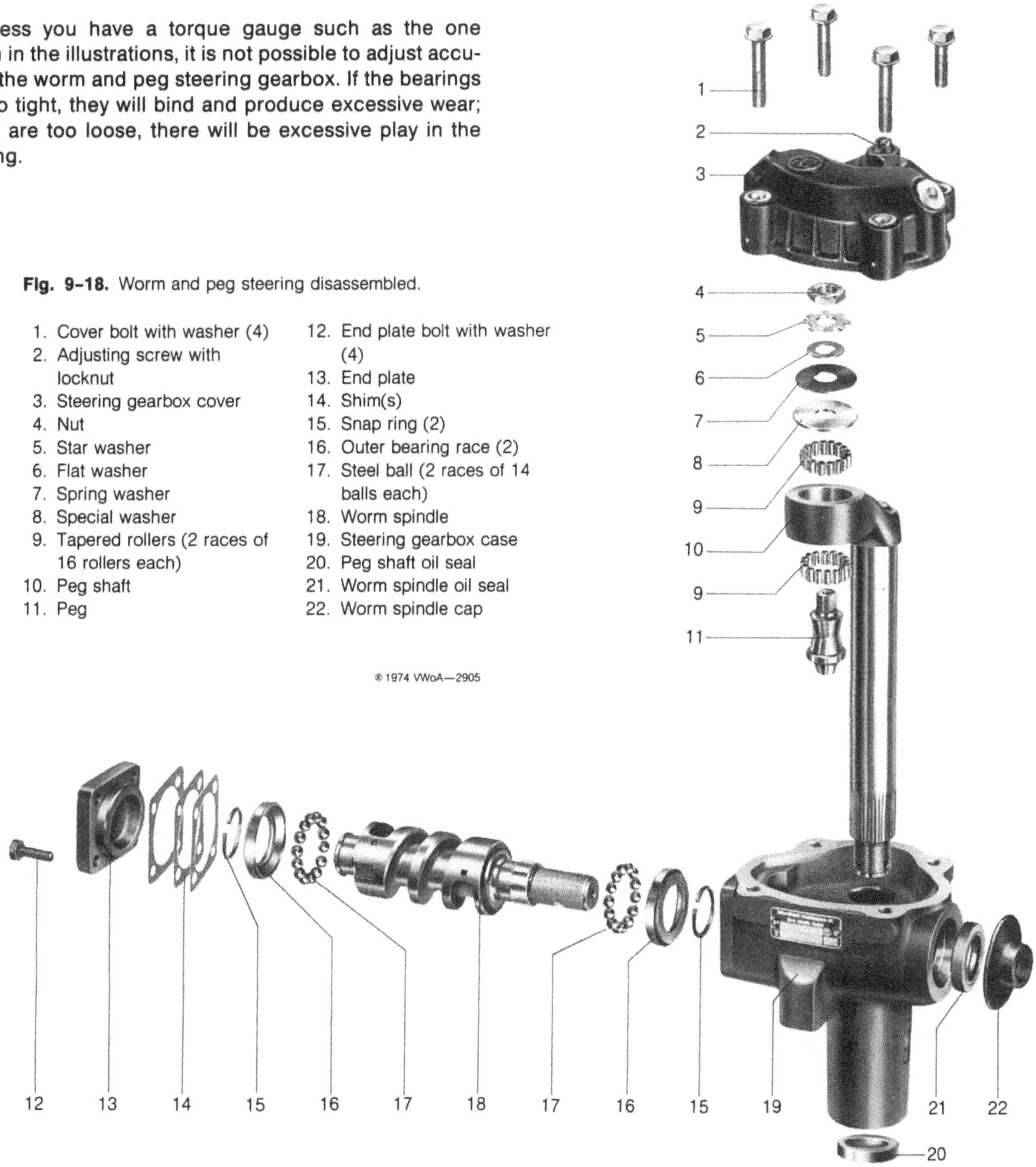

Fig. 9-18. Worm and peg steering disassembled.

1. Cover bolt with washer (4)
2. Adjusting screw with locknut
3. Steering gearbox cover
4. Nut
5. Star washer
6. Flat washer
7. Spring washer
8. Special washer
9. Tapered rollers (2 races of 16 rollers each)
10. Peg shaft
11. Peg
12. End plate bolt with washer (4)
13. End plate
14. Shim(s)
15. Snap ring (2)
16. Outer bearing race (2)
17. Steel ball (2 races of 14 balls each)
18. Worm spindle
19. Steering gearbox case
20. Peg shaft oil seal
21. Worm spindle oil seal
22. Worm spindle cap

To disassemble:

1. Thoroughly clean the outside of the gearbox—especially the parts of the worm spindle and peg shaft that project from the case.
2. Remove the four cover bolts with washers. Then remove the steering gearbox cover complete with adjusting screw and adjusting screw locknut.
3. Lift the peg shaft out upward. To remove the peg and rollers from the shaft, turn the peg clockwise as indicated in Fig. 9-19.

© 1974 VWoA—2906

Fig. 9-19. Direction peg should be turned (arrow) when removing peg and tapered rollers.

4. Remove the four end plate bolts with washers. Then remove the endplate and the shims.
5. Withdraw the worm spindle from the steering gearbox case. If necessary, remove the snap rings and disassemble the ball bearing races.

To assemble and adjust:

1. Inspect the oil seals and replace them if they are worn or cracked. Then install the worm spindle, the original shims, and the end plate. Attach a torque gauge as shown in Fig. 9-20.

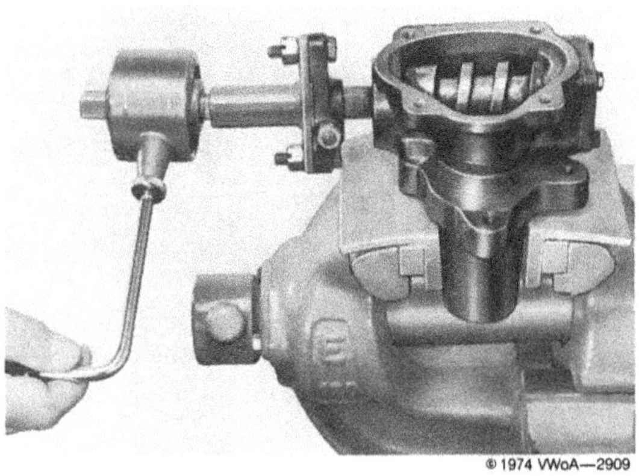

© 1974 VWoA—2909

Fig. 9-20. Torque gauge installed on worm spindle.

2. Using the torque gauge, turn the worm spindle. Turning torque should be 2.0 to 5.0 cmkg (2 to 4 in. lb.).
3. If the turning torque is too great, add shims between the end plate and the steering gearbox case (Fig. 9-21); if the turning torque is too little, remove the shims.

NOTE—
Shims are available in the following thicknesses: 0.10, 0.125, 0.15, and 0.30 mm (.004, .0049, .0059, and .0118 in.).

© 1974 VWoA—2910

Fig. 9-21. Point (arrow) at which shims are installed.

4. When worm spindle turning torque is correct, assemble the peg shaft. Use grease as an adhesive to hold the tapered rollers on the peg as shown in Fig. 9-22.

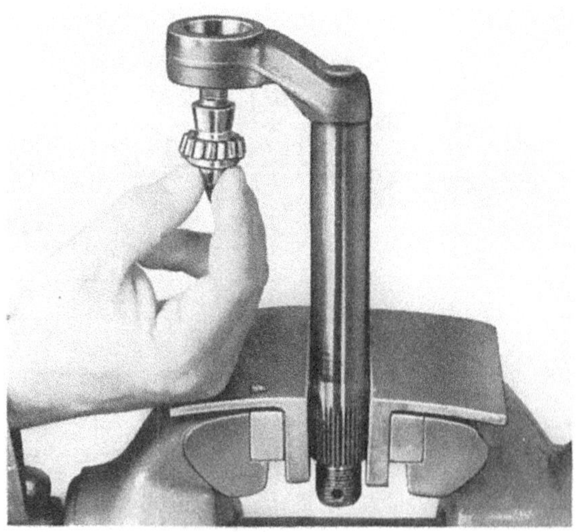

© 1974 VWoA—2911

Fig. 9-22. Peg with tapered rollers being installed in peg shaft.

5. With the cylindrical end of the peg clamped between hardwood blocks in a vise, install the special washer over the threaded end of the peg with the washer's shoulder toward the tapered rollers.
6. Install the spring washer, flat washer, star washer, and nut on the peg. Then torque the nut to 25 cmkg (22 in. lb.).
7. Remove the peg shaft from the vise. Then clamp the lower end of the shaft in the vise between soft jaws so that the peg is free to turn.
8. Install the torque gauge shown in Fig. 9-20 on the nut. Check the turning torque of the peg. It must be from 2 to 3 cmkg (or 2 to 3 in. lb.).
9. Adjust the peg's turning torque, if necessary, by tightening or loosening the nut. Then lock the nut by bending over one tab on the star washer as shown in Fig. 9-23.

Fig. 9-23. Tab on star washer (arrow) bent over against one flat of the nut.

10. Being careful not to damage the oil seal, install the peg shaft in the gearbox so that the peg engages the center of the worm.
11. Thoroughly clean the mating surfaces of the gearbox case and the gearbox cover.
12. Coat the mating surface of the case with sealing compound. Install the cover. Then torque the four cover bolts with washers to 2.5 mkg (18 ft. lb.).

> ***CAUTION —***
> *Loosen the adjusting screw before installing the cover. The cover can be damaged if the adjusting screw contacts the peg shaft before the cover is seated on the case.*

13. Fill the gearbox with 9⅔ U.S. oz. (8 Imperial oz.; 284 ml) of hypoid transmission oil.

14. Install the torque gauge as shown in Fig. 9-24. Then turn the steering away from its centerpoint. (See **9.10 Locating Steering Centerpoint**.)

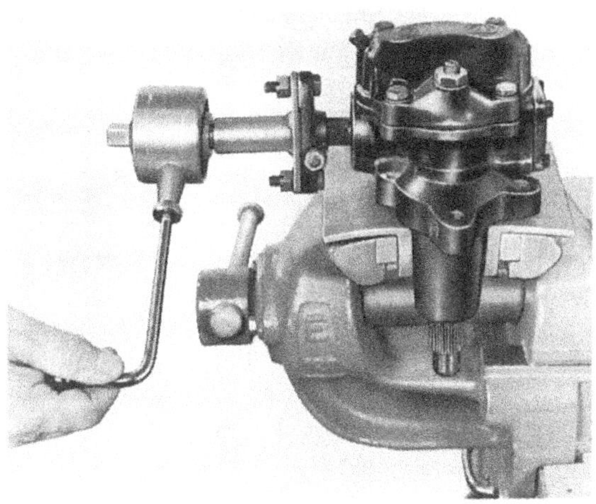

Fig. 9-24. Torque gauge installed on worm spindle.

15. Using the torque gauge, turn the worm spindle back and forth through its centerpoint. Gradually tighten the adjusting screw until it requires approximately 24 cmkg (21 in. lb.) to turn the steering through the central pressure point.
16. When the correct turning torque is obtained, hold the adjuster in its optimum position while you tighten the locknut as shown in Fig. 9-25.

> **NOTE —**
> The 24 cmkg (21 in. lb.) specification applies only if the gearbox is new or if a new case, peg and worm set, bearing, or seal has been installed. If these parts have had at least 3000 miles (5000 km) of service, adjust to 8 to 10 cmkg (7 to 9 in.lb.).

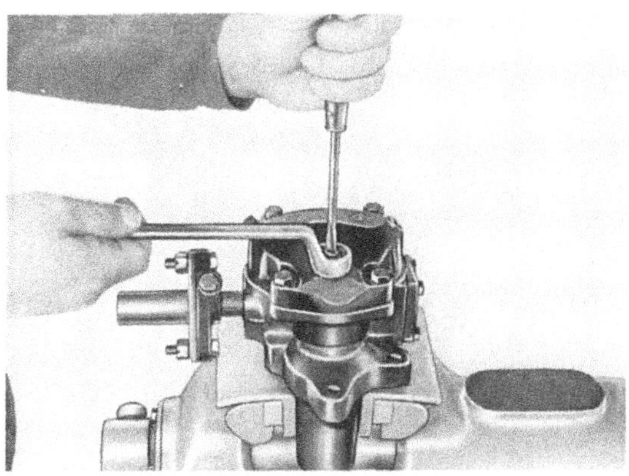

Fig. 9-25. Locknut being tightened while adjusting screw is held stationary with screwdriver.

2

9.10 Locating Steering Centerpoint

Beginning with chassis No. 211 2249 275 manufactured during June 1971, centering marks were placed on the steering gearbox. These marks, which are indicated in Fig. 9-26, are on the steering gear case and the worm spindle cap.

Fig. 9-26. Centerpoint marks (arrows) that appear on late worm and peg steering gearboxes.

Whenever the worm spindle cap has been removed during repairs, the position of the cap on the worm spindle must be determined following assembly and adjustment of the steering gearbox.

To mark centerpoint:

1. Draw a thin paint or chalk line somewhere on the worm spindle as indicated in Fig. 9-27. Then install the torque gauge on the worm spindle.

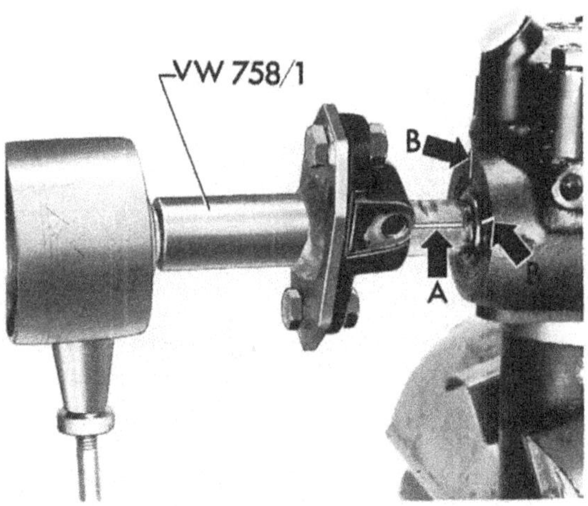

Fig. 9-27. Paint or chalk line on worm spindle, indicated by arrow **A**. The two arrows labeled **B** point to the marks that must be located with the help of the torque gauge.

2. Turning the worm spindle with the torque gauge, approach the central pressure point from both directions. Exactly at the point where the gauge starts to indicate a higher torque—more than 5 cmkg (4.4 in. lb.)—make marks on the case in line with the line on the worm spindle.
3. Set the mark on the spindle (arrow **A** in Fig. 9-28) exactly half-way between the two marks on the case (arrows **B** in Fig. 9-27).

Fig. 9-28. Line on worm spindle (**A**) centered between marks on gearbox case (arrows **B**).

4. Carefully remove torque gauge and adapter without turning the worm spindle. (Alternately, you can make a third mark on the gearbox case that is in line with the centered mark on the worm spindle.)
5. Liberally coat the underside of the worm spindle cap with multipurpose grease. Then, using a tube-type driver, push the cap onto the worm spindle so that the lug on the cap is in line with the boss on the case.

NOTE —
The alignment of the lug on the cap with the boss on the case is indicated by the dotted line in Fig. 9-28.

6. Again check the adjustment with the torque gauge to make sure that the cap has been properly installed.

10. FRONT AXLE TECHNICAL DATA

The technical data on the following pages contain all the dimensions and adjustment specifications needed to service, repair, or rebuild the front axle and steering. Further data on adjusting the alignment of the rear suspension can be found in **TRANSMISSION AND REAR AXLE**. The data given there apply equally to vehicles with manual and automatic transmission.

I. Wheel Alignment Specifications

Designation	Value
1. Wheelbase	2400 mm (94.5 in.)
2. Turning circle	12.3 m (40 ft. 4 in.)
3. Track at front	1395 mm ($54\frac{59}{64}$ in.)
4. Total toe with wheels not pressed	+15′ ± 15′ or 0.0 to 3.3 mm (0 to $\frac{1}{8}$ in.)
5. Total toe with wheels pressed	+ 5′ ± 15′
6. Pressure applied to wheels	15 ± 3 kg (33 ± 6 lb.)
7. Maximum permissible difference between toe with wheels pressed and not pressed	25′
8. Front wheel camber in straight-ahead position	+40′ ± 20′
Maximum permissible difference between sides	30′
9. Toe-out at a 20° lock to left and right (wheels not pressed)	−2° 30′ ± 30′
10. Offset between stub axles	max. 0.80 mm ($\frac{1}{32}$ in.)
11. *Caster angle of a wheel*	3° ± 40′
equals the camber difference of a wheel on a 20° lock to left and right	2° ± 25′

II. Tolerances, Wear Limits, and Settings

Designation	New Part mm (in.)	Wear Limit mm (in.)
A. Front torsion bars ... number of leaves	9	—
length	980 (38.582)	—
Setting (1968–1969) ... angle	56° ± 30′	—
Setting (1970–1974) ... angle	60° ± 1°	—
B. Axle beam and torsion arms		
1. Torsion arm bearings in axle beam		
a. Seat for upper needle bearing ... diameter	56.97–56.99 (2.2429–2.2437)	—
Needle bearing ... diameter	56.96–56.99 (2.2425–2.2437)	—
Thrust rings ... diameter	57.17–57.19 (2.2507–2.2515)	—
Oversize ... diameter	57.17–57.19 (2.2507–2.2515)	—
b. Seat for lower needle bearing ... diameter	56.97–56.99 (2.2429–2.2437)	—
Needle bearing ... diameter	56.96–56.99 (2.2425–2.2437)	—
Oversize ... diameter	57.17–57.19 (2.2507–2.2515)	—
Needle bearing ... diameter	57.17–57.19 (2.2507–2.2515)	—
2. Bushing for		
a. Torsion arm, upper	43.20–43.27 (1.7007–1.7035)	43.40 (1.7086)
b. Torsion arm, lower	43.20–43.27 (1.7007–1.7035)	43.40 (1.7086)
3. Torsion arm ... twist	max. 0.30 (.012)	—
C. Ball joints, steering knuckles		
1. Ball joints, upper ... play	max. 0.3 (.012)	2.0 (.080)
Ball joints, lower ... play	max. 0.3 (.012)	2.0 (.080)
2. Steering knuckle/stub axle measured with a vernier caliper and square ... distortion	0.40 (.016)	—
3. Wheel bearing, inner ... inside diameter	31.75–31.77 (1.2500–1.2507)	—
outside diameter	59.13–59.16 (2.3279–2.3291)	—
4. Wheel bearing, outer ... inside diameter	19.05–19.07 (.7500–.7507)	—
outside diameter	45.24–45.26 (1.7811–1.7818)	—
5. Seat for inner wheel bearing stub axle ... diameter	31.73–31.75 (1.2492–1.2500)	—
Seat in brake disc or drum ... diameter	59.09–59.12 (2.3263–2.3275)	—
6. Seat for outer wheel bearing stub axle ... diameter	19.03–19.05 (.7492–.7500)	—
Seat in brake disc or drum ... diameter	45.20–45.23 (1.7795–1.7807)	—
7. Wheel bearing ... play	0.03–0.12 (.001–.0047)	—
D. Steering (through July 1972)		
1. Steering wheel turns from lock to lock	$2\frac{3}{4}$	—
2. Steering gear ratio	15.0	—
3. Overall ratio	15.7	—
4. Worm turning torque (for axial adjustment) ... without oil seal	2.0–5.0 cmkg (1.7–4.3 in. lb.)	—
5. Tightening torque for peg securing nut	25 cmkg (21.7 in. lb.)	—
6. Peg turning torque	2.0–3.0 cmkg (1.7–2.6 in. lb.)	—
7. Total turning torque (steering gear assembled)	24 cmkg (20.8 in. lb.)	—
E. Steering (from August 1973)		
Worm and roller steering, no repair possible	—	—
Total turning torque through center point (for roller shaft adjustment)	8–12 cmkg (7–11 in. lb.)	

III. Tightening Torques for Axle and Steering

Location	Designation	mkg	ft. lb.
Front axle to frame	bolt	9.0-12.5	65-90
Upper shock absorber mounting	nut	5.0	36
Lower shock absorber mounting	nut	2.5-3.5	18-25
Ball joints to steering knuckle (use new nuts in place of nuts that have been removed)	self-locking nut	10	72
Clamp screw for wheel bearing clamp nut	socket head screw	1.5-max. 2.0	11-max. 14
Tie rods and drag link	castellated nut M 12	3.0	22
	castellated nut M 10	2.5	18
Clamps for tie rods and drag link	nut/bolt	1.5	11
Steering damper to relay lever	bolt	4.0-4.5	29-32
Steering damper to axle tube	bolt	4.0-4.5	29-32
Setscrew for torsion bar	setscrew	4.0	29
Locknut for setscrew	locknut	4.0	29
Stabilizer bar to torsion arm	M 10 nut	3.5-5.0	25-36
	M 8 nut	2.5	18
Steering (up to July 1972)			
Steering gear to side member	bolt	3.5-5.0	25-36
Drop arm to shaft	nut	8.0-11.0	58-80
Relay lever to shaft	bolt	6.0	43
Steering wheel to column	nut	2.5-3.0	18-22
Flange to steering worm	nut	2.0	14
Coupling disc to flange	castellated nut	1.5	11
Steering column to floor plate	fillister head screw	0.5	3.5
Steering gear case cover	bolt	2.5	18
Steering gear end cover	bolt	1.5	11
Steering (from August 1972)			
Drop arm to shaft	nut	14.0	101
Locknut for adjuster	nut	5.0-6.0	36-43
Steering wheel to column (1973 and 1974 models)	nut	2.5-3.0	18-22
Steering wheel to column (1975 and later models)	nut	5.0	36

See BRAKES AND WHEELS for torque specifications related to the brakes.

WARNING —
Do not re-use any fasteners that are worn or deformed in normal use. Many fasteners are designed to be used only once and become unreliable and may fail when used a second time. This includes, but is not limited to, nuts, bolts, washers, self-locking nuts or bolts, circlips and cotter pins. Always replace these fasteners with new parts.

Section 3

FUEL SYSTEM

3

Contents

2 Fuel System

TABLES

Fuel System

The fuel system, as covered in this section of the Manual, applies mainly to the 1968 through 1974 vehicles that have carburetors. The electronic fuel injection system, introduced on the 1975 models, is covered separately in **FUEL INJECTION**. The data given in this section under **3. Fuel Tank** and **7.1 Evaporative Emission Control** apply to vehicles both with fuel injection and with carburetors. You should refer to **FUEL INJECTION** for all other information related to fuel injection engines—including emission control data and air cleaner servicing. Whether they have carburetors or fuel injection, VWs covered by this Manual are designed to operate on regular (91 octane) gasoline. Fuel injection vehicles that are equipped with catalytic converters *require* lead-free gasoline.

The fuel system for dual-carburetor and single-carburetor engines handles five main tasks necessary for proper engine operation: (1) it provides storage space for the gasoline; (2) it includes the components necessary for delivering gasoline to the engine; (3) it is responsible for admitting the proper amount of filtered air to the engine; (4) it incorporates a carburetor and distribution system for mixing fuel and air in precisely controlled proportions and delivering it to the cylinders; and (5) it modifies the density of the incoming air so that the combustion process does not produce an excess of undesirable exhaust emissions. The fourth function mentioned above, that of mixing the fuel with air, is handled on 1968 through 1971 models by one single-venturi downdraft carburetor. The 1972 through 1974 models have two single-venturi downdraft carburetors.

On 1968 through 1971 models, the single carburetor is mounted atop a tubular welded-steel intake manifold that has an exhaust-warmed preheating pipe. On 1971 models, the cylinder heads have dual intake ports rather than the single siamesed intake port used in the heads of earlier engines. On the 1971 engines, the two outer ends of the intake manifold are joined to bifurcated cast aluminum intake pipes that conduct the fuel/air mixture into the cylinder head intake ports. The dual-carburetor engine introduced on the 1972 models has each carburetor mounted on a bifurcated cast aluminum intake manifold bolted to the top of each cylinder head. The dual manifolds are connected by a balance pipe and are heated solely by their contact with the cylinder heads.

Because many of the repairs and tune-up procedures described in this section of the Manual have a direct influence on exhaust emissions, they should not be undertaken unless all prescribed equipment is available. If you lack the skills, special equipment, or tools needed for servicing and adjusting the fuel system, we suggest you leave such repairs to an Authorized VW Dealer or other qualified shop. We especially urge you to consult your Authorized VW Dealer before attempting repairs on a car still covered by the new-car warranty.

4 Fuel System

1. General Description

As noted earlier, the fuel system may conveniently be divided into five subsystems, each with a separate function. For brevity, these will be called fuel storage, pump and lines, air cleaner, carburetor, and emission controls.

Fuel Storage

The fuel tank is housed beneath the rear luggage area, ahead of the engine. The fuel capacity is 15.8 U.S. gallons (13.2 Imperial gallons; 60 liters). The tank is equipped with a sending unit for the electrical fuel gauge, a pickup tube for transfer of fuel to the engine, and a vent system—routed, on 1970 and later vehicles, into the evaporative emission control system. Except on 1974 and later models, the unvented filler cap is concealed beneath a spring-loaded flap in the left rear side of the vehicle.

Pump and Lines

Three different fuel pump designs have been used on the carburetor-engined vehicles covered by this Manual. The 1968 through 1970 models have a pump with a slightly cone-shaped cover. The 1971 model has a pump with a dome-shaped stamped steel cover. Individual parts are available for the pump used on 1968 through 1970 vehicles, but the pump with the domed cover is the standard replacement part for all single-carburetor engines.

The third type of fuel pump, used on dual-carburetor engines only, is located at the front of the engine near the flywheel—just ahead of the No. 1 cylinder. This pump has a cast metal top that houses a fuel cutoff diaphragm. The built-in fuel cutoff prevents a gradual flow of fuel to the carburetors after the engine has been stopped.

On vehicles with single-carburetor engines, fuel reaches the pump through a steel line connected to the tank by a hose. On vehicles with dual-carburetor engines, the tank and pump are connected by a hose only. All models have a hose from the pump to the carburetor. However, there is a metal T-piece on dual-carburetor engines so that the single hose from the pump can serve two hoses connected to the widely-separated carburetors.

Air Cleaner

Single-carburetor engines and 1972 dual-carburetor engines have an oil bath air cleaner. Dual-carburetor engines on 1973 and later vehicles have a dry-type air cleaner with a pleated paper filter. The air cleaner used on 1975 and later vehicles is covered in **FUEL INJECTION**. There have been almost annual modifications to the air cleaner and its ducts. Most of the modifications have been to the intake air preheating system. Originally incorporated to prevent carburetor icing, intake air preheating has become an important auxiliary to the emission control system.

Carburetors

The 1968 and 1969 models have a single Solex 30 PICT-2 carburetor. The 1970 models have a single Solex 30 PICT-3 carburetor and the 1971 models a single Solex 34 PICT-3 carburetor. The 1972 through 1974 models have two carburetors. The left carburetor is a Solex 34 PDSIT-2 and the right carburetor is a Solex 34 PDSIT-3. On dual-carburetor engines, the idle mixture circuit and the vacuum ports for the distributor are in the left carburetor.

All the carburetors used on Type 2 vehicles have built-in automatic chokes. A thermostatic spring closes the choke valve when the engine is cold. When the ignition is turned on, an electric heating element warms the thermostatic spring, causing the choke to open at a predetermined rate. Whenever the throttle valve closes—as at idle or during deceleration—a vacuum diaphragm overrides the thermostatic spring, slightly opening the choke.

Emission Controls

The evaporative emission control, standard beginning with the 1970 models, prevents gasoline fumes from escaping into the atmosphere. The fuel tank is vented into a system that traps and contains fuel vapors until they can be drawn into the air cleaner and burned by the engine.

The 1973 and later models are equipped with an EGR (exhaust gas recirculation) system. The system diverts a portion of the exhaust gases into the intake manifolds below the carburetors, modifying the density and content of the incoming mixture. The recirculated exhaust gases lower the flame peaks in the combustion process, thereby reducing the formation of oxides of nitrogen, an air pollutant. The EGR system of 1975 and later models is covered in **FUEL INJECTION**. The 1973 and 1974 models also have an air injection type exhaust afterburning system. An air pump, driven by the engine, injects fresh air into the exhaust system at the point where hot gases emerge from the cylinder heads. The additional air causes unconsumed fuel to burn in the exhaust system, thereby reducing hydrocarbon emissions.

A throttle valve positioner is installed on 1970 and 1971 engines. The positioner automatically adjusts the throttle closing rate for minimum exhaust emissions and prevents the carburetor throttle valve from closing suddenly when the accelerator pedal is released—again to reduce emissions. On 1971 models, this latter function is supplemented by a dashpot installed on the carburetor. A dashpot, but no throttle valve positioner, is used on dual-carburetor engines.

2. Maintenance

There are only a few maintenance operations that must be carried out at a specified mileage or after a certain period of service. These are listed below and covered briefly in **LUBRICATION AND MAINTENANCE.**

1. Checking the throttle positioner (where fitted)
2. Servicing the air cleaner
3. Replacing the air filter (paper type)
4. Checking the exhaust gas recirculation (EGR) valve (1973 and later models only)
5. Replacing the element-type EGR filter
6. Replacing the activated charcoal canister (1970 and later models only)
7. Replacing the air filter for the exhaust afterburning air pump (1973 and 1974 models only)
8. Checking the air pump belt tension.

3. FUEL TANK

If the fuel tank must be removed for cleaning or repairs, it is important that the connections leading to the tank be reinstalled correctly and in their original locations. Installation errors can lead to fuel starvation or to the improper venting of fumes. Fig. 3-1 shows the ventilation system found on 1970 and 1971 models.

3.1 Removing and Installing Fuel Tank

To avoid the hazard of spilled fuel, the tank should be no more than half full during removal. Surplus fuel should be drained off.

WARNING —
Disconnect the battery ground strap. Do not smoke or work near heaters or other fire hazards. Have a fire extinguisher handy.

To remove:

1. Except on pickup trucks, remove the engine as described in **ENGINE AND CLUTCH**. On 1972 and later models, it is possible (barely) to remove the tank panel without removing the engine.

2. On fully-enclosed vehicles manufactured after May 1968, remove the screws that hold the panel in the front of the engine compartment. Then remove the panel (Fig. 3-2).

© 1974 VWoA—3020

Fig. 3-2. Panel that separates engine compartment from fuel tank compartment.

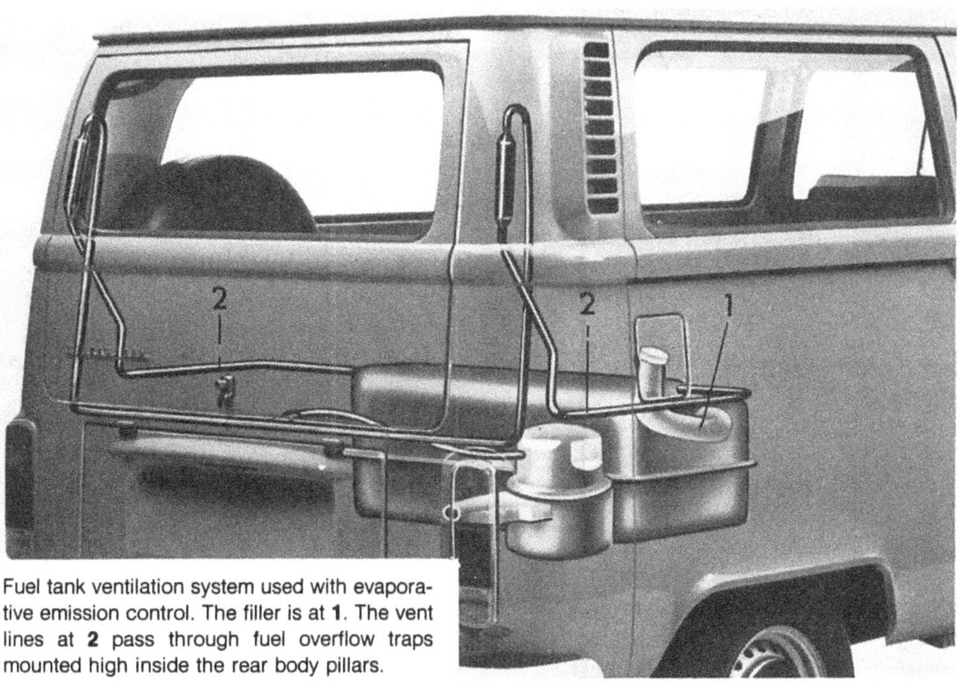

Fig. 3-1. Fuel tank ventilation system used with evaporative emission control. The filler is at **1**. The vent lines at **2** pass through fuel overflow traps mounted high inside the rear body pillars.

© 1974 VWoA—3254

3. On pickup trucks, remove both of the panels shown in Fig. 3-3.
4. Disconnect the fuel hose(s). Quickly plug the hose(s) and connection(s).

© 1974 VWoA—3021

Fig. 3-3. Two panels that cover tank compartment on pickup truck bodies.

5. Loosen the large hose clamps on the fuel filler neck hose. Then remove the hose.
6. Disconnect the wire from the fuel gauge sending unit. Detach the ventilation hose(s) from the fuel tank.
7. Remove the bolts that hold the tank retaining straps. Then take off both straps.
8. On pickup trucks, remove the tank toward the cargo area; on fully-enclosed vehicles, remove the tank toward the engine compartment.

Installation is the reverse of removal. New clamps should be used on the ventilation hose(s) and the hoses must not be installed in a twisted position or bent sharply enough to collapse them. If necessary, replace the rectangular foam rubber tank seal that goes beneath the tank and surrounds the fuel pickup.

3.2 Fuel Gauge Sending Unit And Fuel Pickup

The removal, installation, and testing of the fuel gauge sending unit is covered in **ELECTRICAL SYSTEM.**

The fuel pickup in the bottom of the tank consists of a union nut, a gasket, and a feed pipe. The inner end of the feed pipe is covered by a wire gauze filter. To remove the filter, unscrew the union nut, remove the feedpipe, and withdraw the wire gauze filter from the opening.

3.3 Treating Corroded Fuel Tank

Water is heavier than gasoline, so condensation or water from contaminated fuel will collect in the bottom of the tank where it may cause rusting. Not only can rust particles contaminate the rest of the fuel system but, given time, the fuel tank may be rusted through. Corroded tanks should be either replaced or cleaned.

> ***WARNING —***
> *Tanks containing holes caused by corrosion should not be welded or soldered using open flame. Even an empty tank contains fumes that make it a potential bomb.*

There are two standard agents for cleaning rusted fuel tanks:

A. Derusting phosphate agent (commercially available as Rodine 50 or a similar product) that is mixed either in a solution of one part agent to ten parts water or according to the manufacturer's directions.

B. Aqueous solution of hydrochloric acid (that is, industrial muriatic acid, specific gravity 1.190) in the proportion of 20 parts hydrochloric acid solution, 80 parts water, and one part inhibitor (such as Rodine 50).

Cleaning agent **A** is the preferred treatment. It is milder and leaves a protective phosphate film. Both methods require immediate rinsing with a soluble oil mixture (one part machine coolant to 20 parts water). Rust will form again if the tank is not rinsed thoroughly.

To remove rust:

1. Seal the fuel pickup and vent pipes. Then remove the fuel gauge sending unit.
2. Fill the tank with solution **A** or **B**. Make sure the solution completely fills the tank so that acid fumes will not attack the tank wall above the fluid level.

> **NOTE —**
> Leave the solution in the tank until the rust is removed. Forty minutes may be sufficient for a very small accumulation, but severe rusting may require as long as eight hours. For best results, let the solution stand overnight.

3. Pour out the solution, and pour in 4 to 5 U.S. quarts (6.5 to 8.5 Imperial pints; 4 to 5 liters) of rinsing solution. Rock the tank vigorously to slosh the rinse over the entire interior.
4. Drain the tank and dry it with compressed air.

The cleaning solution can be used 10 or 15 times. Store the solution only in a glass container with a glass or rubber stopper that will not be attacked by vapors from the acid in the solution.

4. FUEL PUMP AND LINES
(carburetor engines only)

A mechanical fuel pump is used on all single-carburetor and dual-carburetor engines. The pump on single-carburetor engines is operated by a cam on the distributor drive shaft. On dual-carburetor engines, the pump is operated by an eccentric cam at the front of the camshaft.

4.1 Fuel Pump Troubleshooting

Table a lists possible fuel pump trouble symptoms with their probable causes and remedies. When more than one probable cause or remedy is given, check them in the order in which they are listed. The numbers in bold type in the Remedy column refer to the headings in this section under which the prescribed repairs are described.

4.2 Removing and Installing Fuel Pump

On single-carburetor engines, the fuel pump is mounted near the distributor, on top of the crankcase. On dual-carburetor engines, the fuel pump is mounted at the front right-hand side of the crankcase, just ahead of the No. 1 cylinder. On single-carburetor engines, the stroke of the pump is determined by the number of gaskets installed under the black intermediate flange that goes between the pump and the engine's crankcase. It is important that the stroke be limited to its specified length.

To remove pump:

1. Pull the fuel hoses off the pump, quickly plugging them to prevent the escape of gasoline.

WARNING —
Disconnect the battery ground strap. Do not smoke or work near heaters or other fire hazards. Have a fire extinguisher handy.

2. On single-carburetor engines, remove the nuts on the pump flange; on dual-carburetor engines, remove the socket head screws. Then take off the pump.
3. Remove the pushrod from the center of the intermediate flange. Then remove the intermediate flange and gaskets from the crankcase.

3

To adjust pump stroke
(single-carburetor engine only):

1. Place two new gaskets on the pump mounting studs. Then install the intermediate flange on top of them.

CAUTION —
Always install the intermediate flange on the crankcase before you insert the pushrod. Otherwise, the pushrod may slip through the flange and into the crankcase.

Table a. Fuel Pump Troubleshooting

Symptom	Probable Cause	Remedy
1. Fuel leaking at joint faces of pump	a. Slotted screws loose b. Diaphragm cracked	a. Tighten screws. See **4.3**. b. Replace diaphragm. See **4.3**.
2. Fuel leaking at diaphragm rivets	Diaphragm damaged during assembly	Replace diaphragm. See **4.3**.
3. Fuel leaking through diaphragm itself	Diaphragm material damaged by solvent in fuel	Replace diaphragm. See **4.3**.
4. Diaphragm damaged, apparently from excessive pump stroke	Pump incorrectly installed, gasket too thin	a. Install pump correctly with additional gasket. Replace diaphragm. See **4.2**, **4.3**. b. Check pushrod stroke. See **4.2**.
5. Pump pressure low	a. Pump incorrectly installed, gasket too thick b. Spring pressure low	a. Install pump correctly, removing one gasket if necessary. See **4.2**. b. Stretch spring to lengthen or, if necessary, replace. See **4.3**.
6. Carburetor flooding	a. Pump pressure excessive, forcing needle valve down. Pump gasket too thin b. Spring pressure excessive	a. Install pump correctly. Check pushrod stroke. Add gasket if needed. See **4.2**. b. Press spring together to shorten or, if necessary, replace spring. See **4.3**.
7. Insufficient fuel delivery	Valves leaky or sticking	Free valves or replace pump. See **4.2**, **4.3**.

2. Install the pushrod in the intermediate flange with the tapered end down.
3. Turn the engine by hand until the pushrod rises to its highest point of travel.
4. Measure the distance that the pushrod projects above the intermediate flange as shown in Fig. 4-1. The pushrod should project 13 mm (½ in.). If it does not, adjust it to this dimension by removing or installing gaskets under the intermediate flange.

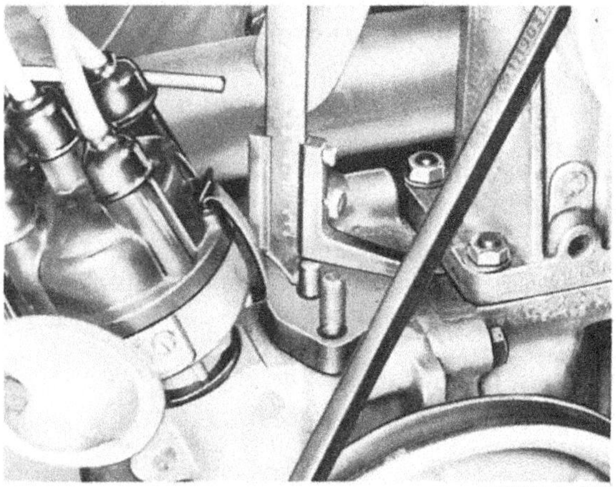

© 1974 VWoA—2394

Fig. 4-1. Depth gauge being used to check length of pushrod where it projects from intermediate flange.

To install pump:

1. Fill the lower part of the pump with universal grease. Coat the pushrod with molybdenum grease.
2. On single-carburetor engines, install a new gasket over the studs where they project from the intermediate flange. On dual-carburetor engines, install the intermediate flange with a new gasket on each side. Then insert the pushrod with the tapered end down on single-carburetor engines, or the large-diameter end innermost on dual-carburetor engines.
3. On single-carburetor engines, place the pump over the studs and install the two spring washers and nuts. Torque the nuts to 2.5 mkg (18 ft. lb.).
4. On dual-carburetor engines, install the pump with the two socket head screws.
5. Using new hose clamps, install the fuel hoses on the fuel pump.

NOTE —
If the original hoses are to be reinstalled, carefully inspect their ends to see that they have not been weakened or deformed by the previous installation of hose clamps. A new clamp may not seal hoses that have lost resiliency. In such cases, a new hose should always be installed.

6. Reconnect the battery ground strap.

Checking Fuel Pump Pressure And Delivery Capacity

You can check the fuel pump's pressure by installing a "T" fitting between the pump and the carburetor so that a pressure gauge can be installed. To check the fuel pump's delivery capacity, run the engine on an auxiliary fuel supply while you collect the fuel pump output in a container so it can be measured.

Maximum delivery pressure should be between 3.0 and 5.0 psi (0.20 and 0.35 kg/cm^2) for single-carburetor engines or 5.0 psi (0.35 kg/cm^2) for dual-carburetor engines. Minimum delivery capacity should be 400 cc per minute at 4000 rpm (3800 rpm for dual-carburetor engines).

4.3 Disassembling and Assembling Fuel Pump

Fig. 4-2 is an exploded view of the fuel pump used on dual-carburetor engines.

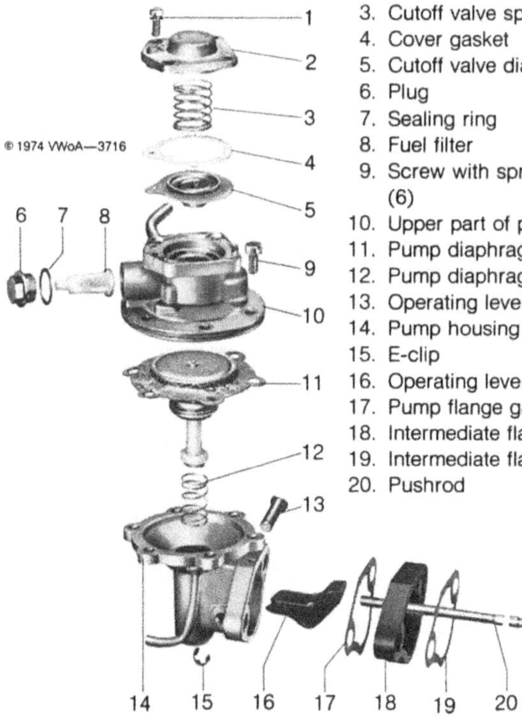

1. Screw with spring washer (4)
2. Fuel cutoff valve cover
3. Cutoff valve spring
4. Cover gasket
5. Cutoff valve diaphragm
6. Plug
7. Sealing ring
8. Fuel filter
9. Screw with spring washer (6)
10. Upper part of pump
11. Pump diaphragm
12. Pump diaphragm spring
13. Operating lever pin
14. Pump housing
15. E-clip
16. Operating lever
17. Pump flange gasket
18. Intermediate flange
19. Intermediate flange gasket
20. Pushrod

Fig. 4-2. Replaceable parts of dual-carburetor engine fuel pump.

Replacement diaphragms and a repair kit are also available for the pump used on 1968 through 1970 models (Fig. 4-3). The fuel filter can be removed for cleaning by taking off the cover. The 1971 single-carburetor engine has the same pump flange gasket, intermediate flange, and intermediate flange gasket.

Fig. 4-3. Components of fuel pump used on 1968 through 1970 models.

1. Cover bolt
2. Sealing ring
3. Cover
4. Cover seal
5. Fuel filter
6. Screw (6)
7. Spring washer (2)
8. Upper part of pump
9. Pump diaphragm
10. Pump housing
11. Pump flange gasket
12. Intermediate flange
13. Intermediate flange gasket
14. Pushrod

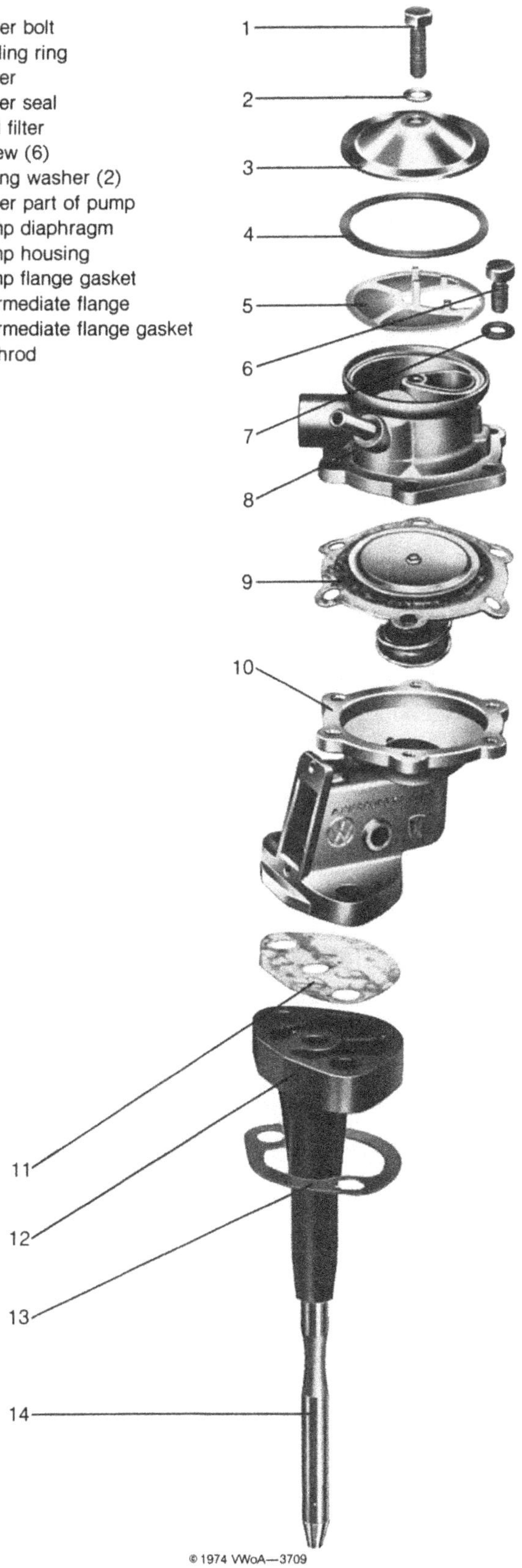

© 1974 VWoA—3709

The fuel pump used on 1971 models can be only partially disassembled, as illustrated in Fig. 4-4, to permit cleaning and replacement of the filter and its gasket. The pump should be replaced if it is faulty or fails to deliver the specified fuel pressure and capacity.

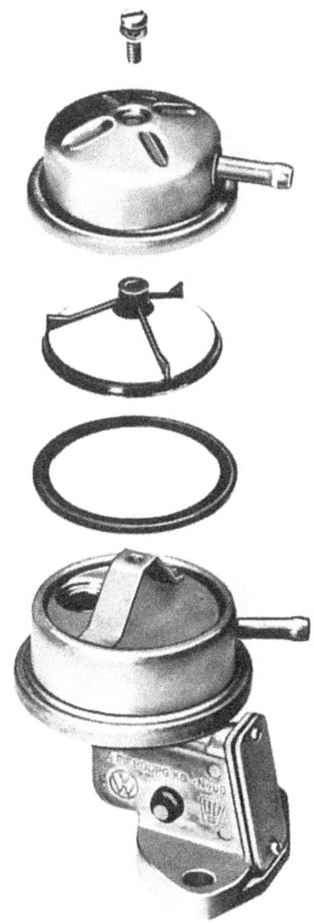

© 1974 VWoA—2408

Fig. 4-4. 1971 fuel pump disassembled.

5. CARBURETOR

The Solex 30 PICT-2 carburetor installed on 1968 and 1969 models differs from the later carburetors primarily in the design of the idle mixture circuit. The 30 PICT-2 has only a volume control screw—typical of carburetors built prior to the enactment of strict exhaust emissions standards. Later carburetors have a volume control screw with a limited range of adjustment and an idle speed adjusting screw that controls the flow of fuel and air to the idle circuit.

The Solex 30 PICT-3 carburetor installed on 1970 models and the Solex 34 PICT-3 carburetor installed on 1971 models are fundamentally the same. Some of the parts and subsystems on the later unit have been relocated. For example, the electromagnetic cutoff valve is located on the right-hand side of the 30 PICT-2 and 30 PICT-3 and on the left-hand side of the 34 PICT-3 carburetor.

The electromagnetic cutoff valve is a simple device that consists of a cutoff plunger and an electrically operated solenoid. The cutoff valve has no influence on carburetion while the engine is running, but prevents the engine from "running on" or "dieseling" after the ignition is turned off. On 34 PICT-3 carburetors, a spring inside the electromagnetic cutoff valve causes the cutoff plunger to snap out of the valve and block the idle air bypass drilling when the ignition is turned off. This prevents air and fuel from reaching the intake manifold when the driver's foot is off the accelerator.

When the ignition is turned on, a wire to the electromagnetic cutoff valve energizes the solenoid. The solenoid withdraws the cutoff plunger to the position shown in Fig. 5-1.

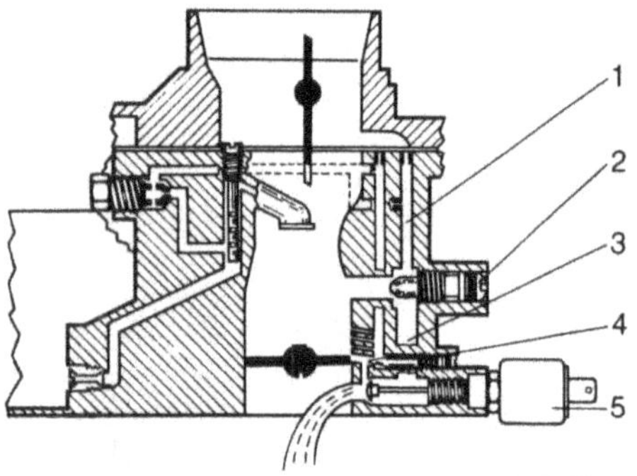

© 1974 VWoA—2412

1. Bypass air drilling
2. Idle speed adjusting screw
3. Bypass drilling
4. Idle mixture adjusting screw
5. Electromagnetic cutoff valve

Fig. 5-1. Electromagnetic cutoff valve, 34 PICT-3.

On 30 PICT-2 and 30 PICT-3 carburetors, the electromagnetic cutoff valve controls fuel delivery to the idle system as shown in Fig. 5-2.

© 1974 VWoA—2421

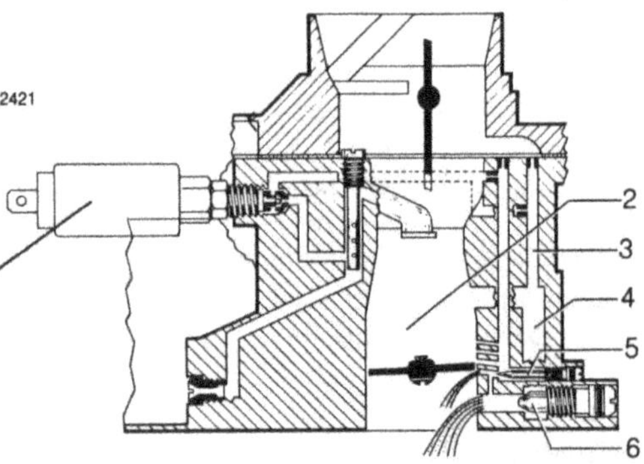

1. Electromagnetic cutoff valve
2. Carburetor throat
3. Bypass air drilling
4. Bypass drilling
5. Idle mixture adjusting screw
6. Idle speed adjusting screw

Fig. 5-2. Electromagnetic cutoff valve (30 PICT-2 and 30 PICT-3).

The 34 PDSIT-2 and 34 PDSIT-3 carburetors used on dual-carburetor engines are completely different from the carburetors used on single-carburetor engines. A central idling system is built into the 34 PDSIT (left) carburetor. A mixture distributor tube (Fig. 5-3) carries the idle mixture from the idling system on the left carburetor to the two intake manifolds.

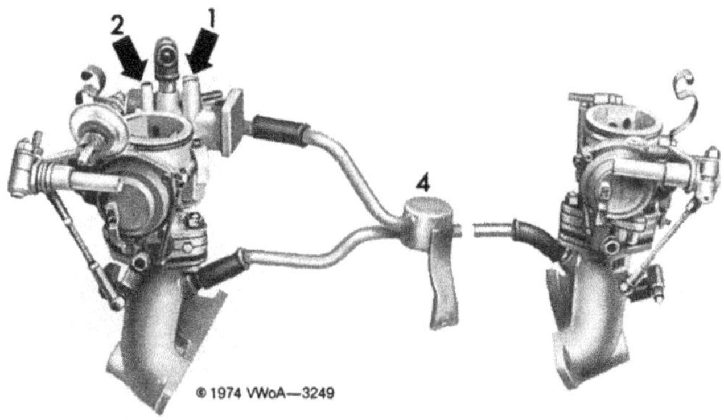

Fig. 5-3. Dual carburetor system. The idle adjustments on the central idling system are at arrows **1** and **2**. The mixture distributor is at **4**.

A schematic view of the central idling system on dual-carburetor engines is given in Fig. 5-4. Notice the position of the electromagnetic cutoff valve in the idle circuit.

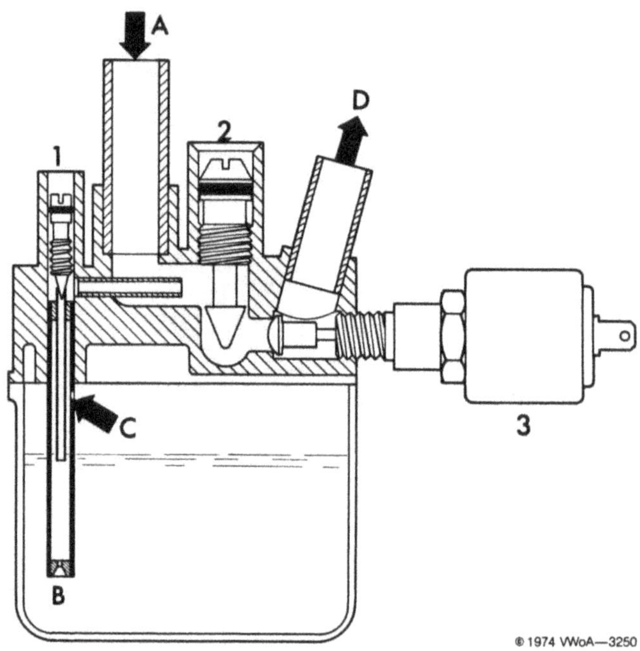

© 1974 VWoA—3250

1. Mixture control screw
2. Idle speed adjusting screw
3. Electromagnetic cutoff valve
A. Air from air cleaner
B. Emulsion tube jet
C. Emulsion tube air drilling
D. Mixture to cylinders

Fig. 5-4. Central idling system.

5.1 Carburetor Troubleshooting

Table b lists carburetor troubles with their probable causes and remedies. The numbers in bold type in the Remedy column refer to the headings in this section under which the prescribed repairs are described.

5.2 Synchronizing Dual Carburetors

If the carburetors need to be synchronized, do so before you adjust the idle.

Table b. Carburetor Troubleshooting

Symptom	Probable Cause	Remedy
1. Idle poor, resists adjustment	a. Dirt in idle system	a. Clean carburetor(s), adjust as prescribed. See **5.4**, **5.2**, **5.3**.
	b. Throttle valve gap incorrect	b. Adjust gap to 0.10 mm (.004 in.). See **5.3**.
2. Engine does not start (tank has fuel, and ignition is working)	a. Choke valve sticking	a. Apply penetrating fluid to free choke valve shaft. See **5.4**.
	b. Automatic choke not working properly	b. Check vacuum diaphragm for freedom of movement. See **5.4**.
	c. Bimetal spring unhooked or broken	c. Reconnect spring or, if broken, replace complete ceramic plate (when installing, match index marks). See **5.4**.
	d. Ceramic plate broken	d. Replace ceramic plate (when installing, match index marks). See **5.4**.
	e. Float needle valve sticking and carburetor flooding	e. Clean or replace float needle valve. See **5.4**.
3. Engine runs continuously at fast idle	a. Automatic choke not switching off	a. Check heater element and connections. See **5.4**.
	b. Choke valve or fast idle cam sticking	b. Apply penetrating fluid to free choke valve or cam. See **5.4**.
	c. Carburetors not synchronized	c. Adjust linkage, then adjust carburetors. See **5.2**.
	d. Throttle valve sticking	d. Free throttle valve lever and pull rod. A bent pull rod must be replaced. See **5.4**.
	e. Throttle valve positioner out of adjustment	e. Adjust throttle valve positioning. See **5.4**.
	f. Throttle valve positioner cannot be adjusted	f. Replace throttle valve positioner. See **5.4**.
4. Exhaust backfire when car is overrunning the engine (coasting)	a. Idle mixture slightly weak	a. Enrich mixture by turning volume control screw, or idle mixture adjusting screw, counterclockwise. See **5.3**.
	b. Throttle valve positioner out of adjustment	b. Adjust throttle valve positioner or replace it. In extreme cases, set cut-in speed to max. 1900 rpm. See **5.4**.
	c. Faulty mixture control valve on vehicles with exhaust afterburning (from 1973)	c. Replace mixture control valve. See **7.4**.
5. Engine idles unevenly or stalls (ignition dwell correct)	a. Vacuum hose to brake servo or automatic transmission disconnected or leaking	a. Replace or reconnect vacuum hose. Tighten hose clamps. See **BRAKES AND WHEELS**.
	b. Vacuum hoses between exhaust gas recirculation (EGR) valve and intake manifolds cracked, loose, or off	b. Replace hose. See **7.2**.
	c. Faulty EGR valve	c. Check valve and replace if necessary. See **7.2**.
	d. Idle adjustment incorrect	d. Adjust idle to 850 ± 50 rpm. (950 ± 50 rpm for vehicles with automatic transmission). See **5.3**.
	e. Pilot jet blocked	e. Clean jet. See **5.4**.
	f. Air leak in manifold	f. Check carburetor flange gasket, manifold sleeves, and intake pipe gaskets. See **5.4**, **ENGINE AND CLUTCH**.
6. Engine "runs on" when ignition is switched off	a. Faulty electromagnetic cutoff valve or wire	a. Test wire for current. Test solenoid for continuity. Replace faulty cutoff valves. See **5.4**.
	b. Idle speed too fast	b. Adjust idle speed. See **5.3**.
	c. Idle mixture too rich	c. Weaken idle mixture by turning volume control screw, or idle mixture adjusting screw, clockwise. See **5.3**.

Continued on next page

3

Table b. Carburetor Troubleshooting (continued)

Symptom	Probable Cause	Remedy
7. Poor transition from idle to operating speed	a. Accelerator pump passages blocked or ball sticking b. Torn accelerator pump diaphragm c. Idle adjustment incorrect d. Amount of fuel injected is incorrect	a. Clean accelerator pump and check operation. See **5.4**. b. Replace diaphragm. See **5.4**. c. Adjust idle. See **5.3**. d. Adjust accelerator pump volume. See **5.3**.
8. Engine stalls when accelerator pedal is released suddenly	Idle mixture too rich	Adjust idle. See **5.3**.
9. Engine runs unevenly (surges) with black smoke at low idle and smokes badly as idle speed increases. Spark plugs soot up quickly and misfire	a. Excessive pressure on float needle valve b. Float leaking c. Float needle valve not closing d. Automatic choke stuck closed	a. Check fuel pump pressure and reduce if necessary. See **4.2**. b. Replace float. See **5.4**. c. Clean or replace float needle valve. See **5.4**. d. Apply penetrating fluid to free choke valve shaft. See **5.4**.
10. Engine runs unevenly at full throttle, misfires, and cuts out or lacks power (ignition dwell, spark advance, and spark plugs all right)	a. Fuel starvation at driving speeds b. Low fuel level in float bowl c. Low fuel pressure d. Dirt in fuel system	a. Clean main jet. See **5.4**. b. Clean float needle valve. See **5.4**. c. Check fuel pump pressure and increase if necessary. See **4.2**. d. Clean fuel tank. Flush lines. Clean carburetor and fuel pump. See **3.3**, **4.3**, **5.4**.
11. Fuel consumption excessive	a. Jet sizes not properly matched b. Excessive pressure at float needle valve c. Float leaking d. Float needle valve not closing e. Automatic choke not working properly	a. Install correct set of jets. Check spark plugs for soot fouling. See **5.4**, **8**, **ENGINE AND CLUTCH**. b. Check fuel pump pressure and reduce if necessary. See **4.2**. c. Replace float. See **5.4**. d. Clean needle valve or replace float. See **5.4**. e. Apply penetrating fluid to free choke valve shaft. See **5.4**.

To synchronize:

1. Dismount the air cleaner ducts from both carburetors, leaving the hose attached (Fig. 5-5).

© 1974 VWoA—3253

Fig. 5-5. Air cleaner duct (**2**) detached from carburetor. Hose (**3**) must remain connected. Idle speed adjusting screw is at **4**; mixture control screw is at **5**. **1** and **6** are 1972 model synchronizing adjustments.

2. Disconnect the lower ball joint of the right-hand throttle operating rod from the lever on the right-hand carburetor's throttle valve switch.

3. On 1972 models, turn the synchronizing screw on the left carburetor (Fig. 5-6) until the right carburetor's throttle operating rod can be reconnected to the throttle arm without moving either carburetor's throttle valve from its fully closed position.

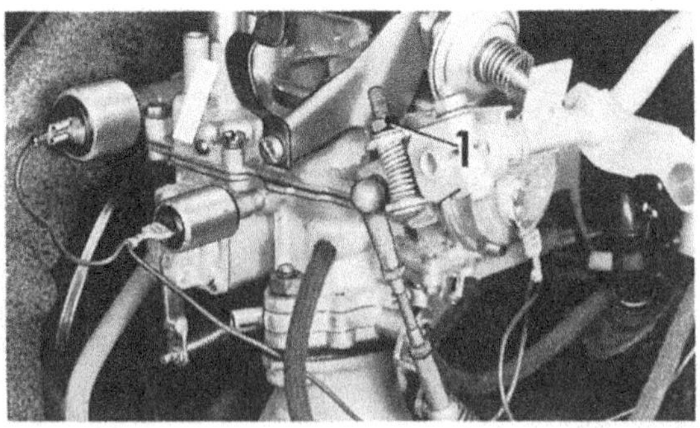

© 1974 VWoA—3252

Fig. 5-6. Synchronizing screw (at **1**) on the left-hand carburetor.

4. On 1973 and 1974 models, adjust the length of the right-hand carburetor's throttle operating rod until the rod can be reconnected to the throttle arm without moving either carburetor's throttle valve from its fully closed position. The approximate distance between the two ball joints of the operating rod should be 107 mm ($4\frac{7}{32}$ in.).
5. Insert an oil thermometer in place of the oil dipstick. Start the engine and run it until the oil temperature reaches 50° to 70°C (122° to 158°F). Check to see that both automatic chokes are fully open.
6. Run the engine at a steady speed between 2000 and 3000 rpm. Place a Unisyn®, Auto-Syn®, or other airflow-measuring synchronizing device over the throat of one carburetor and then the other. See Fig. 5-7.

CAUTION—
If you lack the skills, tools, or equipment for synchronizing the carburetor, we suggest you leave such repairs to an Authorized VW Dealer or other qualified shop. We urge you to consult your Authorized VW Dealer before attempting repairs on a car still covered by the new-car warranty.

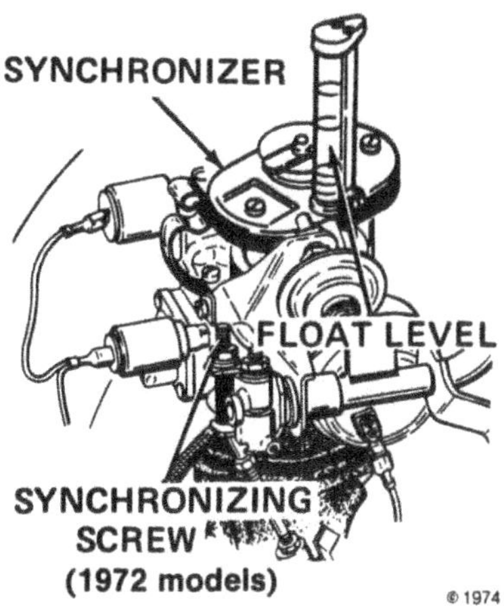

Fig. 5-7. Synchronizing device placed on carburetor. Air flow is indicated by the height to which the float rises in the graduated tube.

7. If one carburetor is passing more air than the other, adjust the synchronization. On 1972 models, turn the synchronizing screw one way or the other until the air flow is the same at both carburetors. On 1973 and 1974 models, turn the adjustment on the right-hand carburetor's operating rod (Fig. 5-8).
8. Adjust the idle.

Fig. 5-8. Adjustment on 1973 and 1974 right-hand carburetor throttle operating rod. Loosen locknuts, then use a wrench to turn the rod.

5.3 Adjusting Idle

Because even minor carburetor adjustment changes will affect exhaust emissions, please follow the instructions carefully and use the specified equipment. Professional mechanics should find out whether state authorization is required before a repair shop can make adjustments that influence exhaust emissions.

On 1970 through 1974 models, adjust only the idle speed adjusting screw during routine maintenance. Do this as described in the last step of each carburetor adjusting procedure. You should not adjust the idle mixture adjusting screw unless (1) you have installed a different carburetor; (2) you have removed, repaired, or rebuilt the carburetor; (3) you have had to alter the carburetor synchronizing adjustments; or (4) the engine is producing excessive emissions.

In troubleshooting the engine, eliminate all other possible trouble sources before you touch the carburetor adjustments. Also, adjusting the idle rpm should be the last step in a tune-up. Otherwise, valve and ignition adjustments will upset the previously made idle adjustment.

CAUTION—
If you lack the skills, tools, or test equipment for adjusting the carburetor, we suggest you leave such repairs to an Authorized VW Dealer or other qualified shop. We urge you to consult your Authorized VW Dealer before attempting repairs on a car still covered by the new-car warranty.

3

Adjusting Single Carburetor

Before you adjust the idle, clean, re-gap, or replace the spark plugs as needed. Make certain that valve clearances, ignition dwell angle, and ignition timing are correct. Connect a dwell meter/tachometer to the ignition system so that you can measure idle speed accurately. Insert a thermometer for measuring oil temperature in place of the oil dipstick.

The air cleaner must be installed while you adjust the idle mixture or idle rpm. If you use an exhaust gas analyzer during idle mixture adjustment, disconnect the evaporative emission control hose from the air cleaner (the hose that joins the air cleaner with the activated charcoal filter canister). Otherwise, the analyzer readings may be influenced by possible fuel tank vapors.

To adjust idle (30 PICT-2):

1. Start the engine and run it until the oil temperature reaches 50° to 70°C (122° to 158°F). Check to make sure that the automatic choke is fully open.
2. Adjust the idle speed by turning the idle speed adjusting screw on the throttle arm (see Fig. 5-9). The idle speed should be 800 to 900 rpm.

Fig. 5-9. Idle adjustments on 30 PICT-2 carburetor. The idle speed adjusting screw is at **1**; the volume control screw is at **2**.

3. Gradually turn the volume control screw clockwise until the engine begins to slow down. From this position, turn the volume control screw counterclockwise until the engine achieves the fastest obtainable idle.
4. If necessary, return the idle to 800 to 900 rpm by turning the idle speed adjusting screw.

To adjust idle (30 PICT-3):

1. Start the engine and run it until the oil temperature reaches 50° to 70°C (122° to 158°F). Check to make sure that the automatic choke is fully open.
2. Adjust the idle speed by turning the idle speed adjusting screw (Fig. 5-10). The idle speed should be 800 to 900 rpm.

CAUTION —
Do not adjust the idle speed on the 30 PICT-3 carburetor by turning the throttle valve stop screw. Doing so would adversely affect exhaust emissions.

1. Throttle valve stop screw
2. Idle mixture adjusting screw
3. Idle speed adjusting screw

Fig. 5-10. Adjustments influencing idle speed on 30 PICT-3 and 34 PICT-3 carburetors.

To adjust idle (34 PICT-3):

1. Start the engine and run it until the oil temperature reaches 50° to 70°C (122° to 158°F). Check to make sure that the automatic choke is fully open, then stop the engine.
2. Turn the throttle valve stop screw out until there is clearance between its tip and the fast idle cam. See Fig. 5-10 given earlier. Then turn the screw in until it just touches the fast idle cam.
3. From this position, turn the screw one-quarter turn further in.

4. Slowly turn the idle mixture adjusting screw in until it comes to a stop, and then turn it back 2½ to 3 complete turns.

5. Start the engine and note the idle rpm. If necessary, turn the idle speed adjusting screw to adjust the idle to 800 to 900 rpm.

6. By turning the idle mixture adjusting screw, adjust to the fastest obtainable idle. Then turn the idle mixture adjusting screw slowly clockwise until the engine speed drops by 20 to 30 rpm.

7. By turning the idle speed adjusting screw, reset the idle to 800 to 900 rpm.

> **NOTE —**
> If a CO tester is used to check the concentration of CO in the exhaust, follow the instrument manufacturer's instructions. With the idle speed adjusted to specifications, CO should be 3% ± 1%. If not, turn the idle mixture adjusting screw to obtain a reading in that range.

Adjusting Dual Carburetors

Before you adjust the idle, clean, re-gap, or replace the spark plugs as needed. Make certain that valve clearances, ignition dwell angle, and ignition timing are correct. Connect a dwell meter/tachometer to the ignition system so that you can measure idle speed accurately. Insert a thermometer for measuring oil temperature in place of the oil dipstick.

If you are installing a new or rebuilt left-hand carburetor, or if the regular idle adjustment procedure fails to obtain the correct results, carry out the following preparatory steps before attempting to adjust (or readjust) the idle.

> **NOTE —**
> The 0.10-mm (.004-in.) throttle valve opening, measured in the following procedure, is to be measured with the choke valve fully open. Do not confuse this dimension with the 0.60-mm or 0.70-mm (.024-in. or .028-in.) settings given in **8. Fuel System Technical Data**. The latter dimension is the fast idle setting, measured with the choke valve fully closed. See **5.4 Removing, Rebuilding, and Installing Carburetor**.

To prepare:

1. Check the throttle valve opening. With the choke valve fully open, the gap between the throttle valve and the throttle body must not exceed 0.10 mm (.004 in.). Adjust the gap (Fig. 5-11) and then secure the stop screws with plastic caps.

© 1974 VWoA—3237

Fig. 5-11. Throttle valve closing gap (dimension **a**).

2. Install the carburetor and check synchronization. Then connect the tachometer and a CO tester.

3. Pull the right-hand carburetor's throttle valve operating rod off the ball on the throttle relay shaft. Pull the vacuum hose off the retard side of the vacuum unit on the distributor.

4. On 1973 and 1974 vehicles, pull the left hose off the exhaust afterburning air injection pump and seal the hose. Disconnect the wire from the terminal on the central idling system's electromagnetic cutoff valve (Fig. 5-12).

© 1974 VWoA—3238

Fig. 5-12. Terminal (arrow) of the central idling system's electromagnetic cutoff valve.

5. Without using force, carefully turn in the volume control screws (Fig. 5-13) on both carburetors until the screws contact their seats. From this position, turn the screws out 2½ turns.

© 1974 VWoA—3239

Fig. 5-13. Volume control screw (arrow) in throttle body.

3

6. Start the engine and set the idle at 500 to 700 rpm. By turning both volume control screws uniformly in the same direction, adjust the CO level to between 3% and 5%.

7. Disconnect the wire from the terminal on the electromagnetic idling cutoff valve on one carburetor (Fig. 5-14). Read the decrease in rpm.

© 1974 VWoA—3240

Fig. 5-14. Terminal on electromagnetic idling cutoff valve.

8. Repeat Step 7 on the other carburetor. The decrease in rpm should be the same as for the first carburetor. If it is not, turn the volume control screws in opposite directions until the decrease in rpm is the same at both carburetors with CO at between 3% and 5%.

NOTE —
The objective of your adjustments is to synchronize the mixture richness of the two carburetors. The regular synchronizing procedure is for synchronizing the throttle valve openings only.

9. Reconnect all hoses and wires. Then adjust the idle as described in the following procedure.

To adjust idle:

1. Check the synchronization of the carburetors as described in **5.2 Synchronizing Dual Carburetors**.
2. Start the engine and run it until the oil temperature reaches 50° to 70°C (122 to 158°F). Check to make sure that both automatic chokes are fully open.
3. Connect an accurate tachometer to the engine, following the instructions supplied by the instrument's manufacturer.
4. Adjust the idle speed by turning the idle speed adjusting screw (Fig. 5-15). The idle speed should be 800 to 900 rpm for vehicles with manual transmissions or 900 to 1000 rpm for vehicles with automatic transmissions.

© 1974 VWoA—3251

Fig. 5-15. Idle adjustments for the dual-carburetor engine's central idling system. The idle speed adjusting screw is at **2**; the idle mixture screw is at **1**.

5. If you have a CO tester, turn the idle mixture screw to obtain 3 ± 1 volume % of CO. Then perform Step 9. If you do not have a CO tester, carry out Steps 6 through 9.
6. Slowly turn in the idle mixture screw on the central idling system until the idle speed drops noticeably.
7. Gradually turn out the idle mixture screw until the fastest obtainable idle speed is reached.
8. Slowly turn in the idle mixture screw until the idle slows by 30 to 50 rpm. Then turn out the screw ¼ turn from this point.
9. By turning the idle speed adjusting screw, reset the idle to 800 to 900 rpm on vehicles with manual transmissions or to 900 to 1000 rpm on vehicles with automatic transmissions.

NOTE —
If a satisfactory idle cannot be obtained, carry out the preparatory steps given earlier, then adjust the idle again.

Adjusting Throttle Valve Positioner

(1968 through 1971 models only)

Two throttle valve positioners have been used. The 1968 and 1969 models have a one-piece positioner; the 1970 and 1971 models have a two-piece positioner. A tachometer with a range of at least 0 to 3000 rpm must be used to adjust either kind of positioner.

To adjust one-piece positioner:

1. Start the engine and run it until the oil temperature reaches 50° to 70°C (122° to 158°F). Then adjust the idle.
2. Loosen the lock screw for the adjusting screw on the altitude corrector (Fig. 5-16).

© 1972 VWoA—41

Fig. 5-16. One-piece throttle valve positioner. The adjusting screw on the altitude corrector is at **1**.

3. With the engine running, turn the adjusting screw on the altitude corrector clockwise until the stop washer on the pull rod contacts the throttle valve positioner housing. The idle speed should then be between 1700 and 1800 rpm.
4. If the idle speed is above 1800 rpm, loosen the locknuts on the pull rod. Turn the rod to lengthen it, thereby reducing the idle speed to between 1700 and 1800 rpm. Then tighten the locknuts.
5. If the idle speed is below 1700 rpm, loosen the locknuts on the pull rod. Turn the rod to shorten it, thereby increasing the idle speed to between 1700 and 1800 rpm. Then tighten the locknuts.
6. Turn the adjusting screw on the altitude corrector counterclockwise until the engine idle speed is 800 to 900 rpm.
7. Pull the throttle valve arm to the rear until the engine is running at 3000 rpm. Then release the lever. It should take from 3 to 4 seconds for the idle speed to drop to 1000 rpm.
8. If the time taken is less than 3 seconds, correct it by turning the adjusting screw on the altitude corrector clockwise. If the time taken is more than 4 seconds, correct it by turning the adjusting screw on the altitude corrector counterclockwise.
9. Tighten the lock screw for the adjusting screw on the altitude corrector.

To adjust two-piece positioner:

1. Start the engine and run it until the oil temperature reaches 50° to 70°C (122° to 158°F). Then adjust the idle.
2. Check the fast idle speed by pulling the fast idle lever back against the adjusting screw in the fast idle lever stop as shown in Fig. 5-17. The fast idle should be 1550 rpm ± 100 rpm.

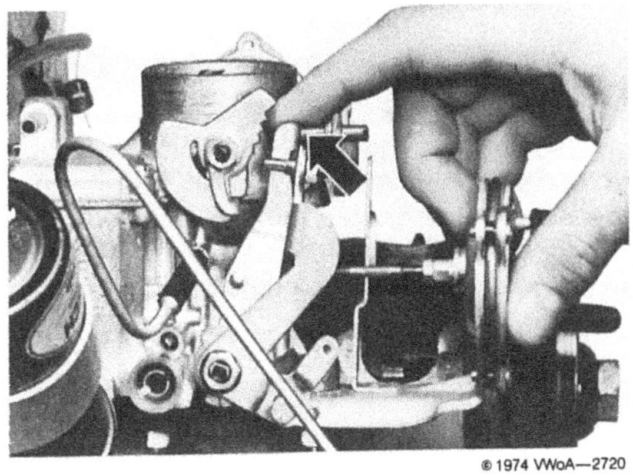

© 1974 VWoA—2720

Fig. 5-17. Fast idle lever being pulled back with fingertip until it contacts adjusting screw (arrow).

3. If the fast idle is not within specifications, turn the adjusting screw in the fast idle lever stop to bring it within the prescribed range. After a warmup drive, the fast idle should not exceed 1700 rpm.
4. Pull the throttle valve lever away from the fast idle lever until the engine is running at 3000 rpm. Then release the lever. It should take 3.5 seconds ± 1 second for the engine to return to its normal idle.
5. If the throttle valve closing time is not within specifications, loosen the lock screw for the adjusting screw on the altitude corrector. Then turn the screw on the altitude corrector (Fig. 5-18) clockwise to increase closing time or counterclockwise to decrease closing time.

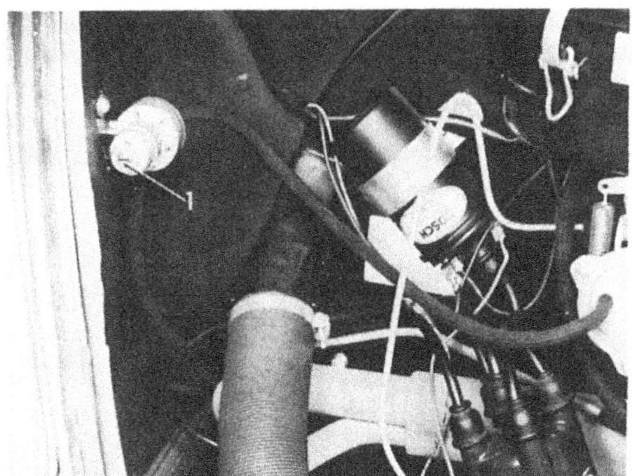

© 1974 VWoA—2722

Fig. 5-18. Adjustment screw (**1**) on the altitude corrector.

After a warmup drive, the throttle valve closing time should not exceed 6 seconds. If the adjustment produces erratic results, check the condition of the hoses between the altitude corrector and the throttle valve positioner's diaphragm unit.

Adjusting Dashpot

(1971 models only)

The 1971 models are equipped with a dashpot. It is adjusted by loosening the two locknuts and repositioning the dashpot in its mounting bracket. Dimension **a** in Fig. 5-19 should be adjusted to 1.00 mm (.040 in.) with the dashpot's plunger fully in and the throttle fully closed on the warm running position of the fast idle cam.

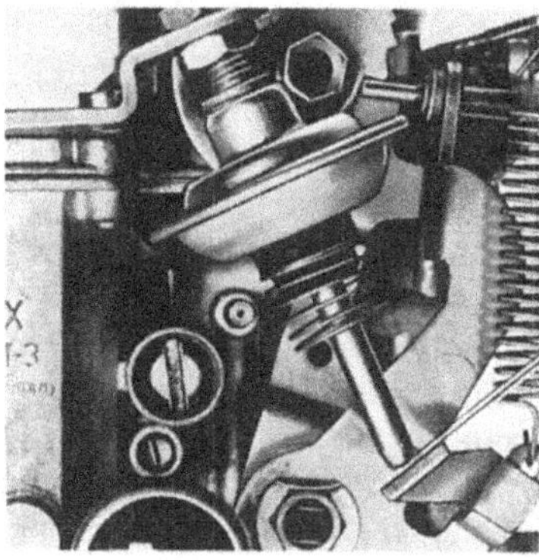

Fig. 5-19. Clearance (at dimension **a**) between the plunger and throttle valve arm with the dashpot in its closed position.

Adjusting Dashpot

(1972 through 1974 models only)

The dashpot installed on dual-carburetor engines does not normally require adjustment unless it is removed or the throttle linkage has been disassembled. The dashpot (Fig. 5-20) is adjusted by loosening the two locknuts and repositioning the dashpot in its mounting bracket. Adjust the dashpot until the clearance between its plunger and the tab on the throttle relay shaft is 0.04 mm (.0015 in.) with the dashpot's plunger held fully in.

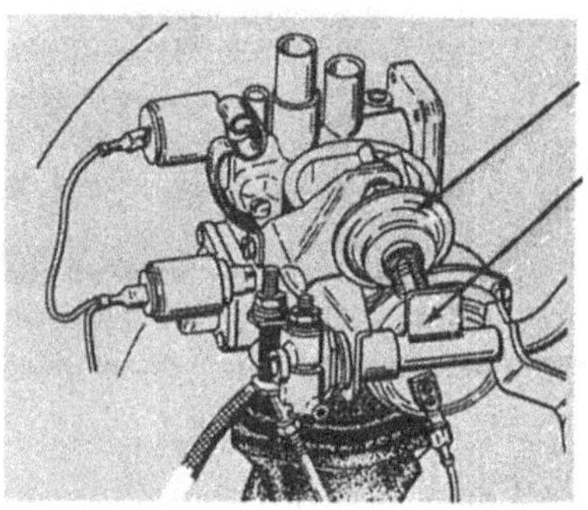

© 1974 VWoA—3711

Fig. 5-20. Dashpot mounted on left carburetor. Measure clearance between the dashpot plunger and the tab on the throttle relay shaft.

Adjusting Accelerator Pump

(single-carburetor engines)

The accelerator pump's injection quantity is adjustable. Except on 34 PICT-3 carburetors, the adjustment is made by installing the cotter pin through a different hole in the connecting link. The 34 PICT-3 carburetors have a bellcrank with an adjusting segment (Fig. 5-21).

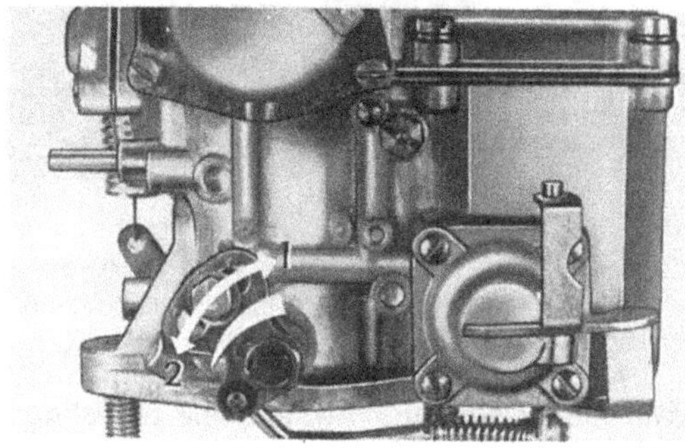

© 1974 VWoA—3010

Fig. 5-21. Injection quantity adjustment on 34 PICT-3 carburetor. Turning the segment toward **1** increases the pump output; turning the segment toward **2** decreases output.

Before you measure the injection quantity, make sure that the float bowl is filled with fuel. Then attach a length of hose or tubing to the discharge end of the accelerator pump injector so that the expelled gasoline can be caught and measured in a 25 cc glass graduate. Hold the glass graduate under the end of the tubing and operate the throttle valve rapidly exactly ten times. Divide the amount caught by ten to get the average quantity of a single injection pulse. The specified injection quantity for carburetors of a particular type and part number can be found in **8. Fuel System Technical Data**.

Except on 34 PICT-3 carburetors, move the cotter pin to the outer hole to decrease injection quantity; to the inner hole to increase injection quantity. On 34 PICT-3 carburetors, loosen the retaining screw and turn the adjusting segment clockwise to decrease injection quantity or counterclockwise to increase injection quantity. Then tighten the retaining screw to hold the adjusting segment in that position.

Check the injection quantity again after making the adjustment to see that it is within specifications. Best economy is obtained with the adjustment at the lower end of the specification range. If the injection quantity exceeds specifications, exhaust emission will be adversely affected.

Adjusting Accelerator Pump

(dual-carburetor engines)

The accelerator pump's injection quantity is adjustable. On dual carburetors, you should strive to adjust the injection quantities of the two carburetors to as near the same amount as possible. Always check the injection quantity of a new carburetor, since the factory often adjusts it to a higher amount than is economical on the engine. You should also adjust the injection quantity after rebuilding the carburetor or when there is engine hesitation during acceleration.

On carburetors used with manual transmissions, and on carburetors used with automatic transmissions during the 1972 and early 1973 model years, the accelerator pump adjustment is made by installing the cotter pin through a different hole in the connecting link. Fig. 5-22 shows the location of the cotter pin. The cotter pins of the two carburetors should always be installed in the same hole of their respective connecting links.

Fig. 5-22. Location of cotter pin in accelerator pump connecting link. There are three holes in the connecting link.

Beginning late in the 1973 model year, a new kind of accelerator pump adjustment was introduced on carburetors used with automatic transmissions. On these carburetors, there is an adjusting sleeve that can be turned in order to adjust the injection quantity. See Fig. 5-23.

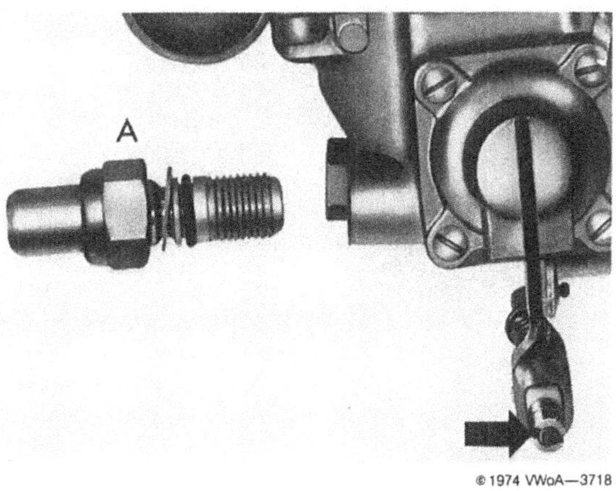

Fig. 5-23. Accelerator pump adjustment on 1974 carburetor. The thermostatic valve is at **A**. The threaded adjusting sleeve is indicated by the arrow.

On the 34 PDSIT-series carburetors installed on 1974 models with automatic transmissions, a thermostatic valve (Fig. 5-23) regulates the injection quantity of the accelerator pump. When the carburetor body temperature is below about 21°C (70°F), the accelerator pump injects approximately 1.5 cc of fuel per stroke. When the carburetor body temperature is above about 21°C (70°F), the thermostatic valve opens so that part of the fuel from the accelerator pump can flow through a relief drilling and back into the float bowl. With the thermostatic valve open, approximately 0.7 cc of fuel is injected per pump stroke.

To test the thermostatic valve, remove the carburetor upper part. Then blow into the relief drilling in the float bowl. Air should pass through the drilling at temperatures above 23°C (73°F), but not at temperatures below 19°C (66°F). You can check the carburetor temperature by inserting a thermometer into the fuel in the float bowl.

Before you measure the injection quantity, make sure that the float bowl is filled with fuel. Then attach a length of hose or tubing to the discharge end of the accelerator pump injector so that the expelled gasoline can be caught and measured in a 25 cc glass graduate. Hold the glass graduate under the end of the tubing and operate the throttle valve rapidly exactly ten times. Divide the amount caught by ten to get the average quantity of a single injection pulse.

> **NOTE —**
> When measuring the injection quantity of a carburetor that has a thermostatic valve, the carburetor body temperature must be below 19°C (66°F) so that the thermostatic valve will be closed.

The specific injection quantity for carburetors of a particular type and part number can be found in **8. Fuel System Technical Data**. Move the cotter pin to the outer hole to decrease injection quantity or to the inner hole to increase the injection quantity. If the carburetor has an adjusting sleeve, move the sleeve outward to decrease the injection quantity or inward to increase the quantity.

5.4 Removing, Rebuilding, And Installing Carburetor

If a carburetor must be replaced, it is important that the new carburetor have the same part number as the original, or that the new carburetor be the correct replacement for the car model being serviced. Always obtain replacement parts with reference to the carburetor part number and the engine number.

To remove carburetor:

1. Remove the air cleaner as described in **LUBRICATION AND MAINTENANCE**.

3

2. Disconnect all vacuum, air, and fuel hoses from the carburetor.

> ***WARNING —***
> *Disconnect the battery ground strap. Do not smoke or work near heaters or other fire hazards. Have a fire extinguisher handy.*

3. Disconnect the wires from the automatic choke heating element and the electromagnetic cutoff valve(s).
4. On single-carburetor engines, loosen the clamp screw in the accelerator cable pivot pin, then pull the cable out of the pin.
5. On dual-carburetor engines, pull the throttle operating rod ball sockets off the balls on the carburetor throttle arms.
6. Unhook the throttle return spring (where applicable).
7. Remove the two nuts from the studs that hold the carburetor on the manifold flange. Then remove the carburetor and gasket.

Removing and Installing Throttle Valve Positioner

On single-carburetor engines, the throttle valve positioner or the positioner's diaphragm unit will come off along with the carburetor. The pull rod should be disconnected from the carburetor's fast idle lever by prying off the retaining pin for the pull rod clevis and not by unscrewing the pull rod from the clevis or the diaphragm. Disconnecting the pull rod as described will avoid the need for adjusting the pull rod when the unit is reinstalled. If any part of the throttle valve positioner is replaced, installation should be followed by the adjustments described in **7.3 Throttle Valve Positioner** and in **5.3 Adjusting Idle.**

Disassembling and Assembling Carburetor

The 30 PICT-3 carburetor is shown disassembled in Fig. 5-24. Disassembly of the 30 PICT-2 carburetor is il-

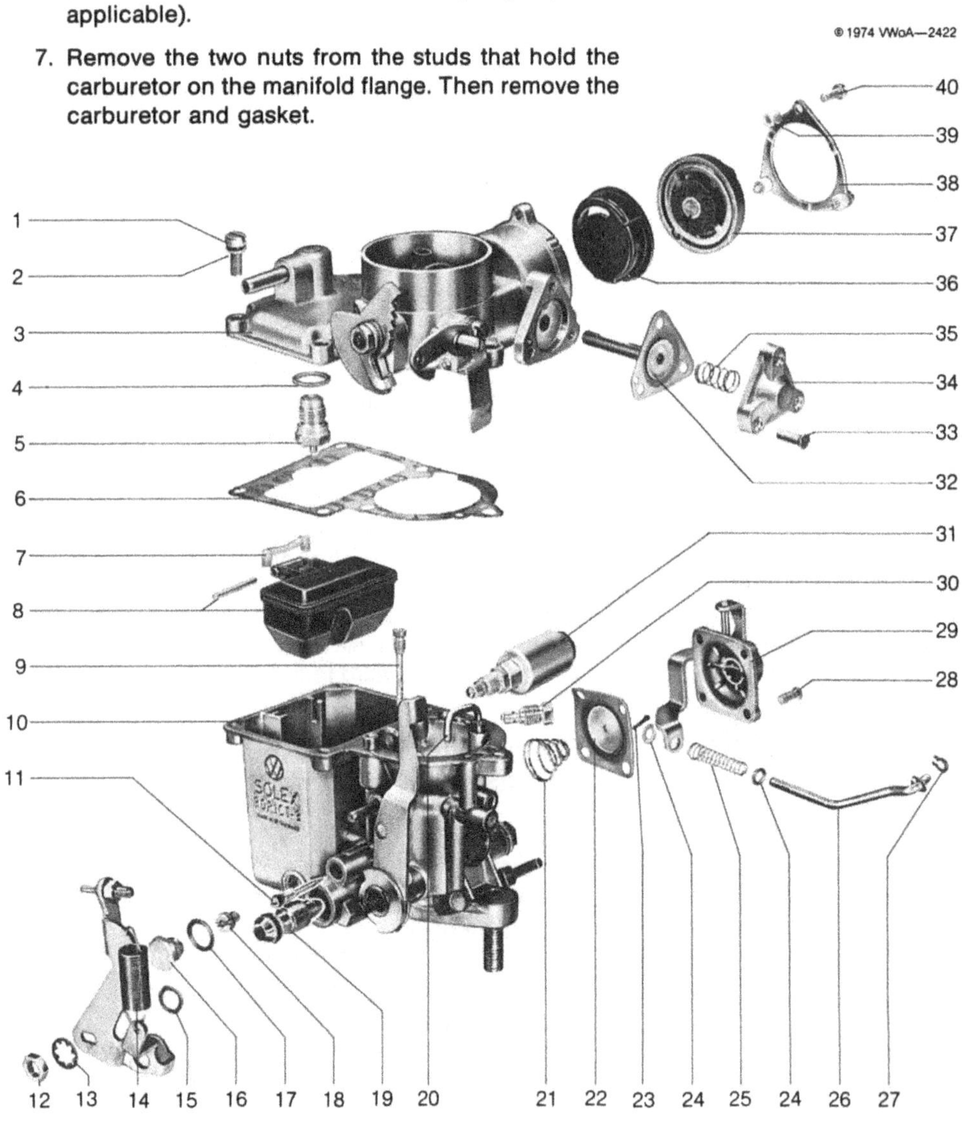

1. Fillister head screw
2. Spring washer
3. Carburetor upper part
4. Float valve washer
5. Float valve
6. Gasket
7. Float pin retainer
8. Float with pivot pin
9. Air correction jet with emulsion tube
10. Carburetor body
11. Idle mixture adjusting screw
12. Nut
13. Lock washer
14. Throttle return spring
15. Spring washer
16. Main jet cover plug
17. Main jet cover plug seal
18. Main jet
19. Idle speed adjusting screw
20. Accelerator pump injector
21. Pump diaphragm spring
22. Pump diaphragm
23. Cotter pin
24. 1-mm (.040-in.) thick washer
25. Connecting link spring
26. Connecting link
27. Clip
28. Screw
29. Pump cover
30. Pilot jet
31. Electromagnetic cutoff valve
32. Vacuum diaphragm
33. Oval head screw
34. Vacuum diaphragm cover
35. Vacuum diaphragm spring
36. Plastic cap
37. Choke heating element
38. Cover retaining ring
39. Retaining ring spacer
40. Small fillister head screw

Fig. 5-24. 30 PICT-3 carburetor disassembled.

lustrated in Fig. 5-25; disassembly of the 34 PICT-3 carburetor is illustrated in Fig. 5-27; disassembly of the 34 PDSIT-2/3 carburetor is illustrated in Fig. 5-26.

As can be seen by comparing Fig. 5-24 through Fig. 5-27, the parts of the various carburetors are quite similar. Jet sizes and other specifications for tne different units can be determined from the Table **I** in **8. Fuel System Technical Data**. The following basic disassembly sequence applies to all carburetors:

1. Remove the carburetor upper part from the carburetor body.
2. Remove the various jets and adjustment screws from the carburetor body and remove the float valve from the carburetor upper part.
3. Disassemble the accelerator pump and linkage.
4. On the 34 PDSIT-2/3 only, remove the throttle body from the carburetor body.
5. Disassemble the throttle valve shaft assembly.
6. Disassemble the automatic choke assembly.
7. On the 34 PDSIT-2 only, disassemble the central idling system.

1. Fillister head screw
2. Spring washer
3. Carburetor upper part
4. Float valve washer
5. Float valve
6. Gasket
7. Float pin retainer
8. Float with pivot pin
9. Air correction jet with emulsion tube
10. Carburetor body
11. Main jet
12. Main jet cover plug seal
13. Main jet cover plug
14. Throttle return spring
15. Volume control screw
16. Spring
17. Accelerator pump injector
18. Pump diaphragm spring
19. Pump diaphragm
20. Cotter pin
21. 1-mm (.040-in.) thick washer
22. Connecting link
23. Clip
24. Connecting link spring
25. Pump cover
26. Screw
27. Electromagnetic cutoff valve
28. Vacuum diaphragm spring
29. Oval head screw
30. Vacuum diaphragm cover
31. Vacuum diaphragm
32. Plastic cap
33. Choke heating element
34. Retaining ring spacer
35. Cover retaining ring
36. Small fillister head screw

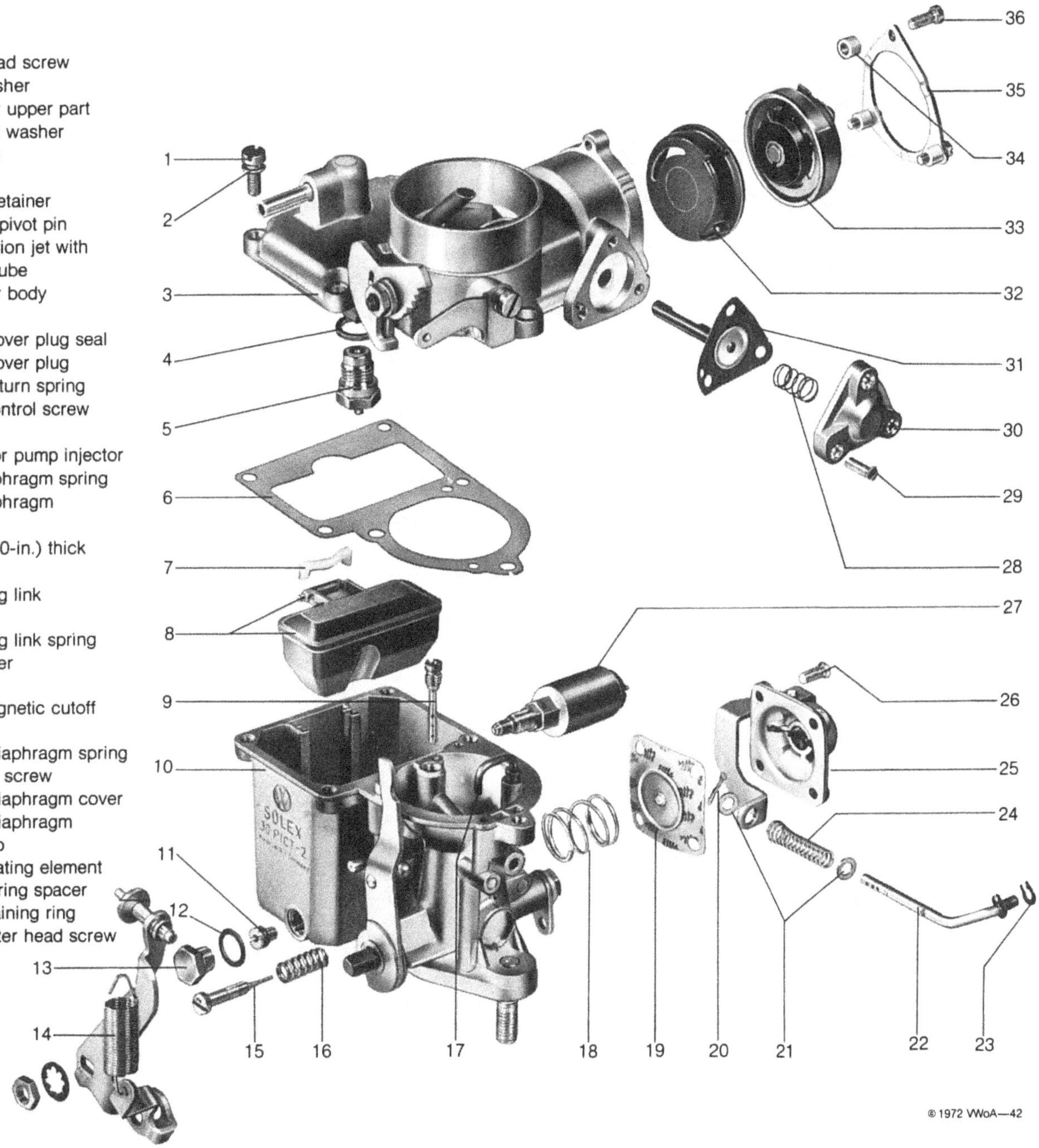

Fig. 5-25. 30 PICT-2 carburetor disassembled.

Fig. 5-26. 34 PDSIT-2/3 carburetor disassembled.

1. Screw with spring washer (6)
2. Idle mixture enrichment
3. Central idling system electromagnetic cutoff valve
4. O-ring
5. Idle mixture screw
6. O-ring
7. O-ring
8. Idle speed adjusting screw
9. Carburetor upper part
10. Screw with washer (5)
11. Dashpot with bracket
12. Gasket for choke heating element
13. Choke heating element
14. Cover retaining ring
15. Throttle shaft switch (auto. trans. 34 PDSIT-3)
16. Washer (auto. trans. 34 PDSIT-3)
17. Spring ring (auto. trans. 34 PDSIT-3)
18. Screw (auto. trans. 34 PDSIT-3)
19. Float valve washer
20. Float valve
21. Accelerator pump injector
22. Gasket
23. Float pin retainer
24. Float pin
25. Float
26. Main jet cover plug
27. Main jet cover plug seal
28. Main jet
29. Electromagnetic idling cutoff valve
30. Pump diaphragm spring
31. Pump diaphragm
32. Screw
33. Pump cover

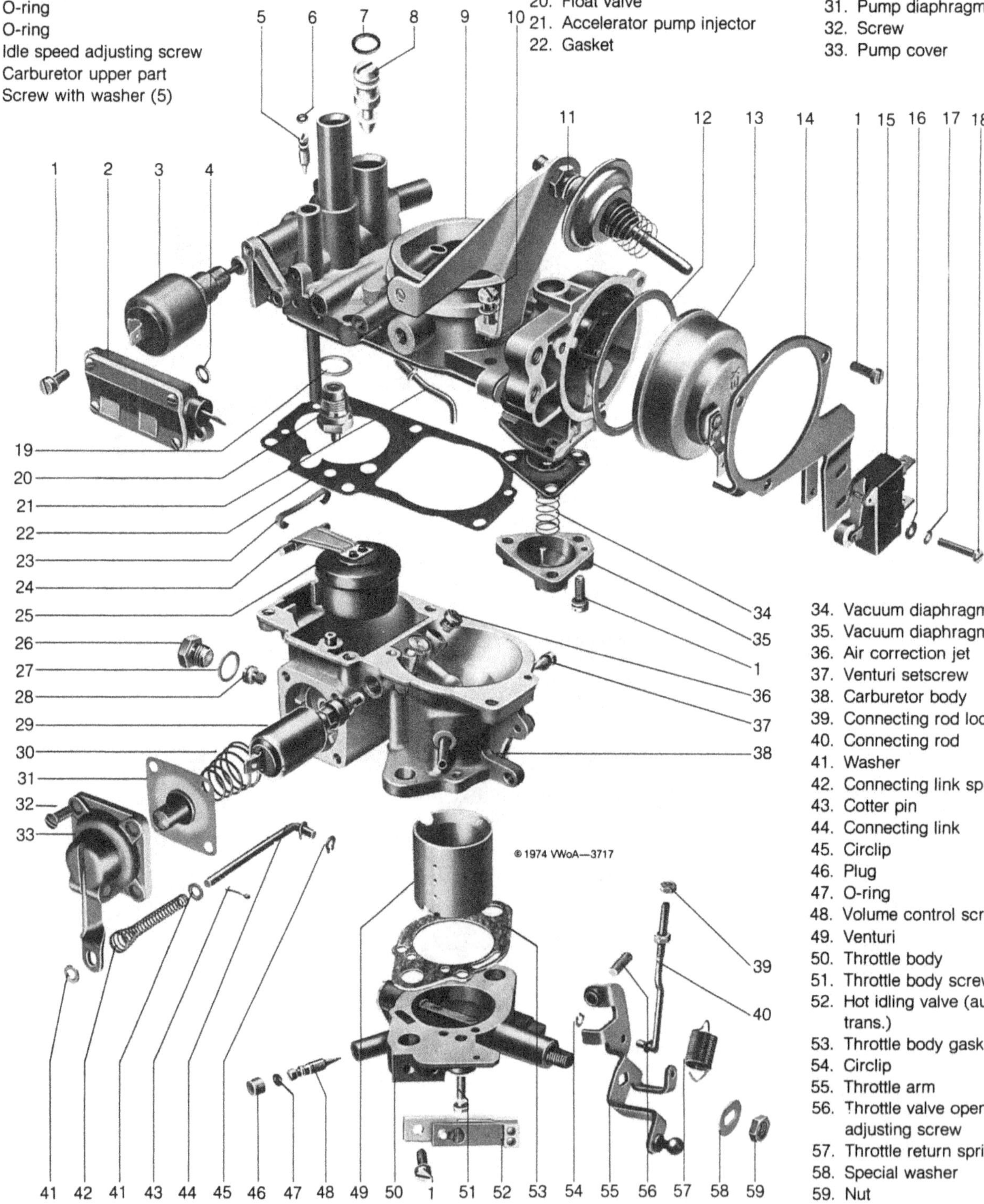

34. Vacuum diaphragm spring
35. Vacuum diaphragm cover
36. Air correction jet
37. Venturi setscrew
38. Carburetor body
39. Connecting rod locknut
40. Connecting rod
41. Washer
42. Connecting link spring
43. Cotter pin
44. Connecting link
45. Circlip
46. Plug
47. O-ring
48. Volume control screw
49. Venturi
50. Throttle body
51. Throttle body screw (2)
52. Hot idling valve (auto. trans.)
53. Throttle body gasket
54. Circlip
55. Throttle arm
56. Throttle valve opening adjusting screw
57. Throttle return spring
58. Special washer
59. Nut

Fig. 5-27. 34 PICT-3 carburetor disassembled.

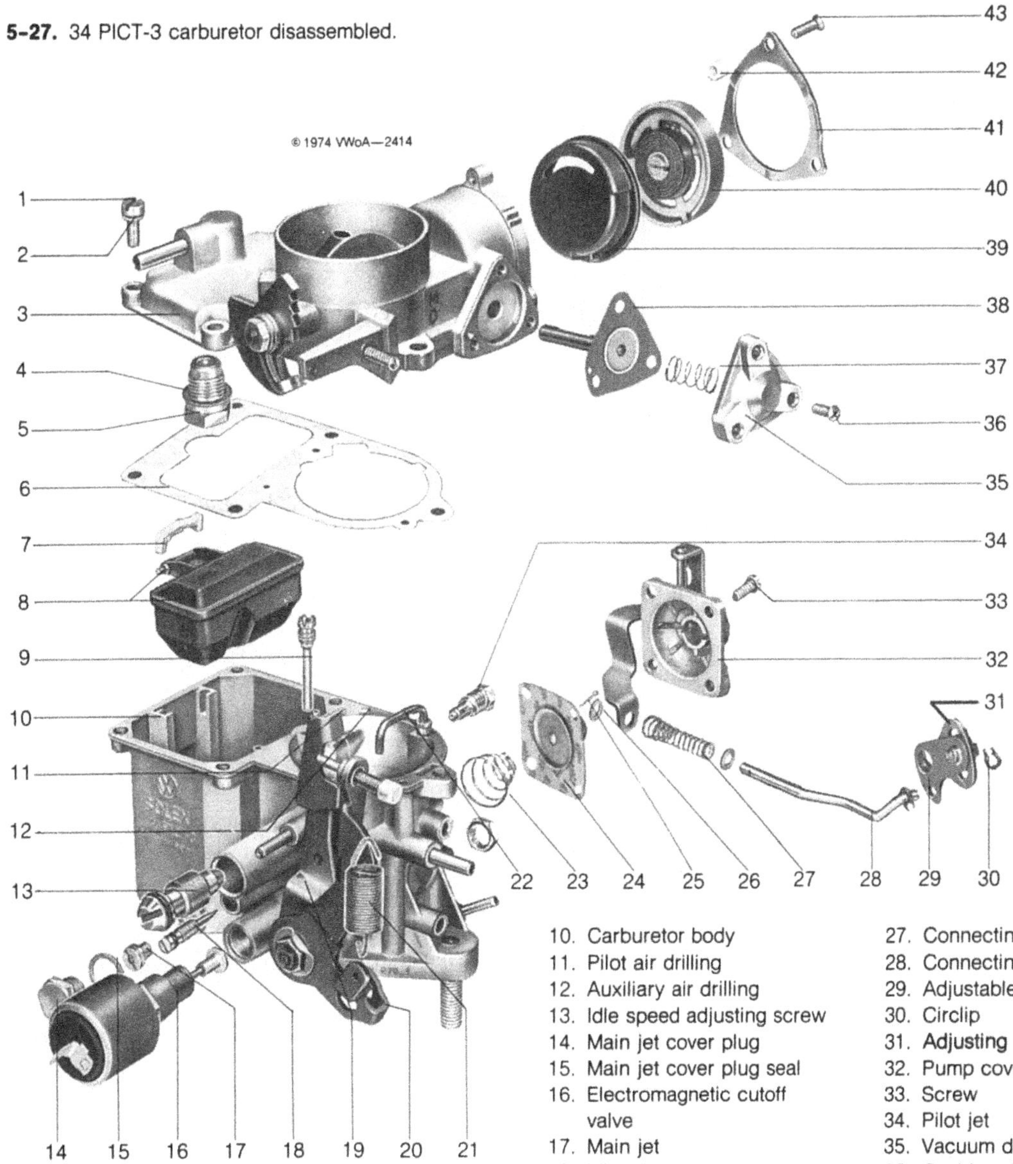

1. Fillister head screw
2. Spring washer
3. Carburetor upper part
4. Float valve washer
5. Float valve
6. Gasket
7. Float pin retainer
8. Float with pivot pin
9. Air correction jet with emulsion tube
10. Carburetor body
11. Pilot air drilling
12. Auxiliary air drilling
13. Idle speed adjusting screw
14. Main jet cover plug
15. Main jet cover plug seal
16. Electromagnetic cutoff valve
17. Main jet
18. Idle mixture screw
19. Fast idle lever
20. Throttle valve lever
21. Throttle return spring
22. Accelerator pump injector
23. Pump diaphragm spring
24. Pump diaphragm
25. Cotter pin
26. 1-mm (.040-in.) thick washer
27. Connecting rod spring
28. Connecting link
29. Adjustable bellcrank
30. Circlip
31. Adjusting segment
32. Pump cover
33. Screw
34. Pilot jet
35. Vacuum diaphragm cover
36. Oval head screw
37. Vacuum diaphragm spring
38. Vacuum diaphragm
39. Plastic cap
40. Choke heating element
41. Cover retaining ring
42. Retaining ring spacer
43. Small fillister head screw

Assembly is the reverse of disassembly. Obtain a rebuilding kit and install the components it contains. With the exception of the choke heating element, the pump diaphragm, the float, and the vacuum diaphragm, wash all old parts in lacquer thinner, acetone, or a commercial carburetor cleaner.

WARNING —
Do not smoke or work near heaters or other fire hazards. The cleaning agents are highly flammable.

CAUTION —
If you lack the skills, tools, or a suitable workshop for rebuilding the carburetor, we suggest you leave such repairs to an Authorized VW Dealer or other qualified shop. We especially urge you to consult your Authorized VW Dealer before attempting repairs on a car still covered by the new-car warranty.

Blow out all jets, valves, and drillings with compressed air. Do not clean them with pins or pieces of wire, which could upset their precise calibration. Install the choke heating element with its mark in line with the middle mark on the housing (Fig. 5-28).

Fig. 5-28. Distance **a** from upper to middle mark. Some 1970 carburetors have the element's mark at the upper housing mark, as shown, rather than at the middle mark. Check before disassembly.

The body joint gasket for PICT carburetors must be the correct one. Those for the 30 PICT-2 are marked with brown stripes; those for the 30 PICT-3 are marked with yellow stripes; those for the 34 PICT-3 are marked with black stripes. The location of these stripes is shown in Fig. 5-29.

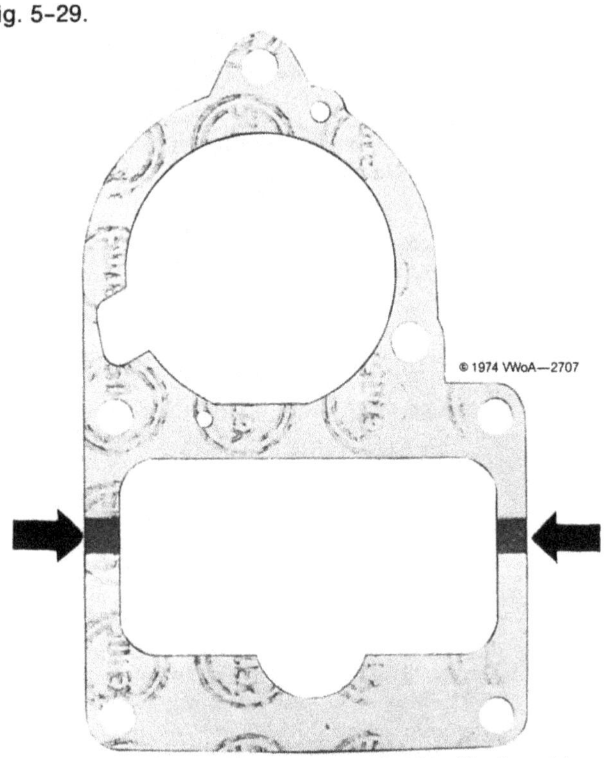

Fig. 5-29. PICT carburetor gasket identification stripes.

When assembling the accelerator pump linkage on a 30 PICT-2, 30 PICT-3, or 34 PDSIT-2/3 carburetor, install the cotter pin in its original hole in the connecting link. On carburetors of all types, the opposite end of the connecting link should have at least 0.30 to 0.50 mm (.012 to .020 in.) of axial play in the bellcrank.

Check the float valve for binding and leakage. It should not be possible to blow air through the valve while the needle is pressed lightly onto its seat. To check it for leaks, immerse the float in hot water. If bubbles appear, replace the float. Also make certain that an 8.5-gram float (PICT carburetors) or a 7.0-gram float (PDSIT carburetors) is installed and that the washer used under the float valve is the correct thickness. Select the washer as described in the procedure that follows.

On 34 PDSIT-2/3 carburetors only, you must adjust the throttle valve's fast idle opening after you have installed the carburetor upper part on the carburetor body. This opening measurement, made with the choke valve fully closed, should not be confused with the throttle valve gap which is measured with the choke valve fully open. See **5.3 Adjusting Idle**.

To adjust the throttle valve's fast idle opening, hold the choke valve fully closed. Then adjust the positions of the two nuts on the connecting rod until the throttle valve just permits a rod-type feeler gauge to be inserted as shown in Fig. 5-30. Use an 0.60-mm (.024-in.) feeler for 1972 and 1973 models; use an 0.70-mm (.028-in.) gauge for 1974 models. Tighten the nuts against the choke linkage pivot. Then recheck the throttle valve opening measurement with the choke valve fully closed.

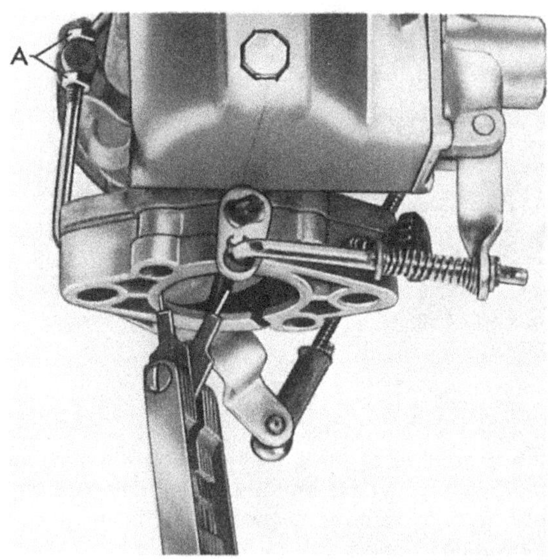

Fig. 5-30. Throttle valve fast idle opening being adjusted. Feeler gauge is inserted between throttle valve and throttle body. Connecting rod adjusting nuts are at **A**.

If you have disassembled the throttle body, or if you have disturbed the adjustment of the throttle valve stop screw, adjust the throttle valve opening with the choke valve open as described in **5.3 Adjusting Idle**.

Adjusting Fuel Level in Float Bowl

Either the carburetor must be level (carburetor removed), or the car positioned on a level surface (carburetor installed). If the carburetor is installed, idle the engine briefly to ensure that the float bowl is full; if the carburetor is not installed, fill the float bowl using a piece of hose attached to the fuel inlet pipe. Then remove the carburetor upper part and the gasket so that the fuel level can be measured as shown in Fig. 5-31.

WARNING —
Disconnect the battery ground strap. Do not smoke or work near heaters or other fire hazards. Have a fire extinguisher handy.

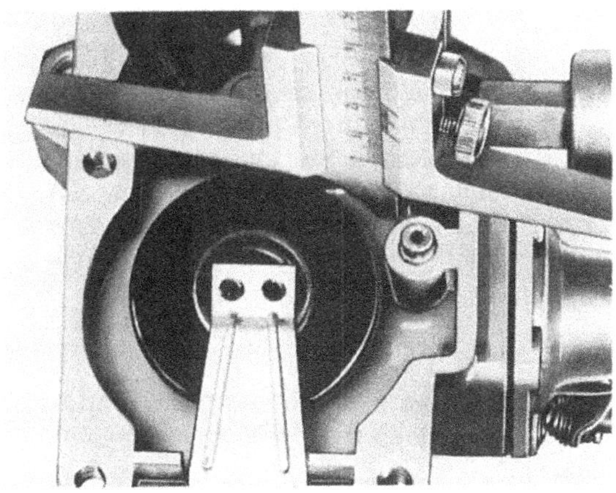

© 1974 VWoA—2724

Fig. 5-31. Fuel level being measured using a depth gauge. Gauge bridge contacts the carburetor body. Slide tip just touches fuel.

The distance from the top of the carburetor to the surface of the fuel should be 19.5 to 20.5 mm (.767 to .807 in.) for 1968 through 1970 vehicles; 17.0 to 19.0 mm (.667 to .748 in.) for 1971 models; and 12 to 14 mm (.470 to .550 in.) for dual-carburetor models. If the fuel level is too high, use a thicker washer under the float valve; if it is too low, use a thinner washer. Washers are available in thicknesses of 0.50 mm (.020 in.), 0.80 mm (.031 in.), 1.00 mm (.040 in.), and 1.50 mm (.060 in.).

Checking Electromagnetic Cutoff Valve

The electromagnetic cutoff valve can be checked while it is installed. Turn on the ignition without starting the engine, then remove the wire from the terminal on the electromagnetic cutoff valve. Touch the wire to the terminal several times. The valve should make a clicking sound each time contact is made.

The same test can be carried out with the electromagnetic cutoff valve removed from the car. Connect negative battery current to the valve's outer casing, and apply positive battery current to the terminal. On 34 PICT-3 and 34 PDSIT-2/3 carburetors, you must apply slight finger pressure to the electromagnetic cutoff valve before the plunger will retract into the solenoid.

Carburetor Installation

Lightly lubricate the choke valve shaft and throttle valve shaft with engine oil and the external linkage with molybdenum grease. Using a new gasket, install the carburetor on the intake manifold, then torque the nuts to 2.0 mkg (14 ft. lb.). Secure the fuel hose with a new hose clamp. If new parts have been installed on the throttle positioner's operating diaphragm unit (where fitted), adjust the pull rod as described in **7.3 Throttle Valve Positioner.** 3

Have someone hold the accelerator pedal to the floor while you connect the accelerator cable. On single-carburetor engines, install the cable end so that gap **a** indicated in Fig. 5-32 is 1.00 mm (.040 in.). Then adjust the carburetor as described in **5.3 Adjusting Idle**. Use the same procedure for dual-carbureator engines, but adjust the gap to 1.00 to 1.50 mm (.040 to .060 in.).

© 1974 VWoA—2387

Fig. 5-32. Full-throttle clearance between throttle valve lever and stop on carburetor body.

6. AIR CLEANER

Four different air cleaners have been installed on the carburetor-engined vehicles covered by this Manual. The intake air preheating flap on 1968 through 1970 models is regulated by a cable linked to the engine's cooling air system thermostat. On 1971 models, the flap is controlled by a separate thermostat built into the air cleaner. As on the earlier models, the 1972 models have an oil bath air cleaner. However, the air cleaner on 1972 models has a different design suited to the dual-carburetor engine. 1973 and 1974 models have a dry-type air cleaner with a pleated paper filter. On the air cleaners used on dual-carburetor engines, the intake air preheating flap is regulated by a thermostatically controlled vacuum unit, making the intake air preheating system responsive to both temperature and engine load.

The 1969 through 1971 models also have a weighted control flap on the air cleaner. It is important that you check the weighted arm for free movement after you have installed the air cleaner. If the arm is blocked by a carelessly installed hose, oil may be drawn from the crankcase into the air cleaner. Removing, installing, and servicing the air cleaner are described in **LUBRICATION AND MAINTENANCE.**

Checking and Adjusting 1970 Intake Air Preheating System

Adjust the cable after the air cleaner has been installed and before the carburetor is adjusted when performing a tune-up.

To install and adjust:

1. Push the cable housing as far as it will go into the retainer on the air cleaner intake. Then tighten the clamp screw in the retainer.
2. Place the cable eye over the cranked end on the flap shaft, then secure it with the steel clip.
3. With the engine cold, check to see that the coil spring portion of the cable is slightly compressed and that the flap is completely closed. If they are not, replace the cable. See Fig. 6-1.

© 1974 VWoA—3004

Fig. 6-1. Cable correctly installed on air cleaner.

Checking and Adjusting 1971 Intake Air Preheating System

The thermostat location on the air cleaner's intake tube is shown in Fig. 6-2.

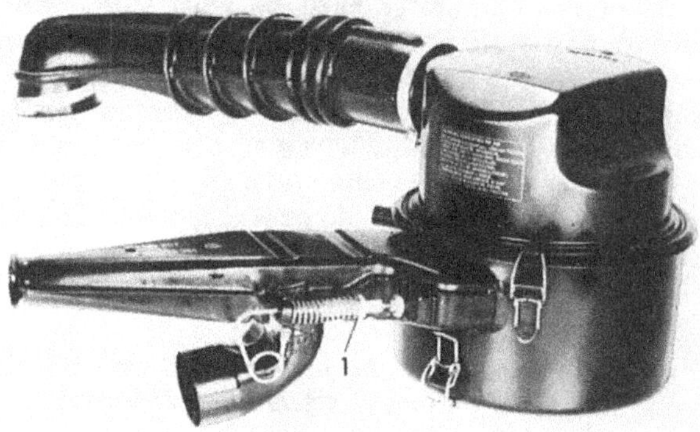

© 1974 VWoA—2407

Fig. 6-2. The 1971 air cleaner. Thermostat is located at **1.**

To check the thermostat, temporarily remove the activated charcoal filter hose from the air cleaner and install a thermometer and rubber stopper in its place. The intake air preheating flap should open between 27.5° and 32.5°C (81.5° and 90.5°F). Minor adjustments can be made by bending the looped portion of the preheating flap arm.

Checking and Adjusting 1972 and Later Intake Air Preheating System

The main parts of the temperature- and load-sensitive intake air preheating system are shown in Fig. 6-3.

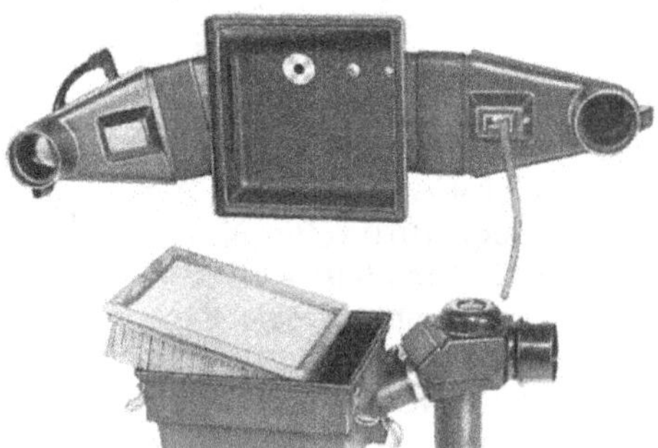

© 1974 VWoA—3562

Fig. 6-3. Air cleaner with thermostatically controlled vacuum-operated flap. The thermostatic vacuum valve is located in the upper part. The valve is the part with the hose attached to it in the illustration. The diaphragm-type vacuum unit that moves the flap is in the intake of the air cleaner lower part.

The thermostatic vacuum valve installed after December 1972 is different from that introduced on early 1972 models. It can be identified by a brass hose connection for the vacuum unit. On the early valve, both connections are black plastic. The later valve keeps the flap closed to cool air regardless of engine load during the warm-up period. The early valve allows the flap to admit cool air under heavy engine loads. If the thermostatic valve or vacuum unit is defective, it should be replaced.

7. Checking, Repairing, and Adjusting Emission Controls

The VW engine will not run properly unless the emission controls are kept in proper working order. Be especially careful not to mix up the emission control hoses when you install them. Doing so will adversely affect both performance and fuel economy.

Although all models covered are equipped with an efficient positive crankcase ventilation (PCV) system, it should be noted by those who are not familiar with the VW engine that the system does not include a PCV valve—a fixture that is common on other makes.

7.1 Evaporative Emission Control

The activated charcoal filter canister must be replaced after 30,000 mi. (48,000 km) of service. It is located in the engine compartment as shown in Fig. 7-1.

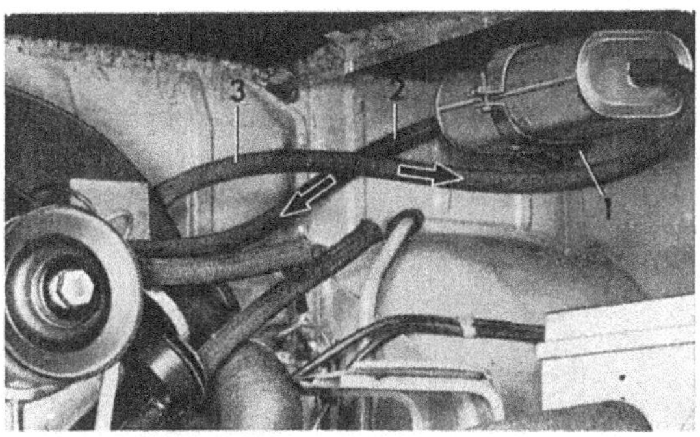

© 1974 VWoA—3712

Fig. 7-1. Activated charcoal filter canister (**1**). Air from the fan enters the canister through hose **3**. Hose **2** conducts fuel vapors to the air cleaner. (The air cleaner has been removed for clarity.)

To replace the canister, remove the hoses that are connected to it, then take out the Phillips head screw in the canister mounting bracket. Note the hose installation positions so that the hoses can be installed in their correct positions on the new canister.

7.2 Exhaust Gas Recirculation

Fig. 7-2 is a drawing of the exhaust gas recirculation (EGR) system as installed in 1973 and later models with manual transmissions. The EGR system as installed on 1973 and later models with automatic transmissions is shown in Fig. 7-3. The EGR system of 1975 and later models is covered in **FUEL INJECTION**.

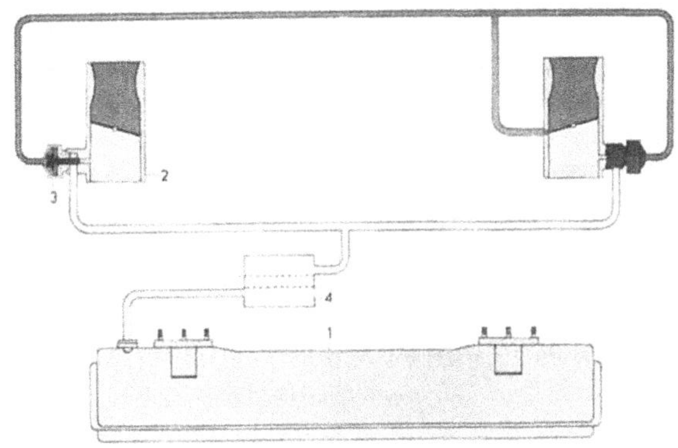

© 1974 VWoA—3244

1. Muffler
2. Carburetor and intake manifold (2)
3. Exhaust gas recirculation valve (2)
4. Element type filter

Fig. 7-2. Exhaust gas recirculation system as used on vehicles with manual transmissions. 1974 models have one central EGR valve instead of two separate valves.

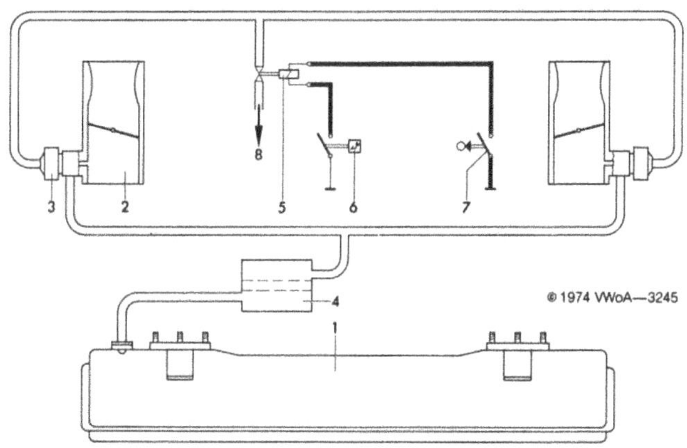

© 1974 VWoA—3245

1. Muffler
2. Carburetor and intake manifold (2)
3. Exhaust gas recirculation valve (2)
4. Element type filter
5. Solenoid-operated vacuum valve
6. Temperature-controlled switch
7. Off/on switch on throttle valve shaft
8. Vacuum hose connected to the vacuum powered brake servo system.

Fig. 7-3. Exhaust gas recirculation system as used on vehicles with automatic transmissions. 1974 models have one central EGR valve instead of two separate valves.

Exhaust gas is drawn from the muffler into a filter that cools the gas and removes particulates. The introduction of exhaust gas into the intake manifolds is controlled by vacuum operated exhaust gas recirculation valves and, on vehicles with automatic transmissions, by a valve controlled by temperature and throttle valve position.

On early 1973 vehicles with automatic transmissions, the temperature-controlled switch permits exhaust gas circulation only when the engine compartment air temperature exceeds 12°C (54°F). On late 1973 and all 1974 vehicles with automatic transmissions, the switch permits exhaust gas recirculation any time the engine cooling air temperature exceeds 85°C (185°F). The early 1973 temperature-controlled switch is located above the battery. On late 1973 and all 1974 vehicles, the switch is on the engine, between the coil and the distributor. It is operated by a tab on the cooling air control flap shaft.

Checking and Adjusting Throttle Valve Switch

(automatic transmission only)

The off/on switch is mounted on the automatic choke cover retaining ring as shown in Fig. 7-4. The switch should go on when the throttle valve has opened 10° and should remain on until 10° before the throttle valve is fully opened. If not, loosen the screws and move the switch on the elongated mounting bracket holes until the operating range is correct. Use a protractor to check the angle of the throttle valve shaft.

© 1974 VWoA—3247

Fig. 7-4. Throttle valve switch. A cam on the throttle valve shaft arm contacts a roller on the switch.

Checking and Adjusting Temperature Switch

(automatic transmission only)

The temperature switch used on early 1973 models has no adjustment. The switch on late 1973 and 1974 models has a tab that is operated by the cooling air control flap shaft. To check the switch, insert a 1.00-mm (.040-in.) feeler gauge between the switch tab and the flap shaft.

To check the switch, make sure that the cooling air flaps are closed. You should hear the switch click when you insert a 1.00-mm (.040-in.) feeler gauge between the switch tab and the flap shaft. If not, adjust the switch mounting bracket on its elongated screw hole until the switch operates correctly.

Checking Exhaust Gas Recirculation Valve

To check the 1973 exhaust gas recirculation valves, remove them from the intake manifolds and make sure that they are clean. Hand-press the valve pin to make sure that it moves freely. Then connect the valve to a vacuum hose on an engine other than that of the vehicle you are servicing. Start the engine. At 1500 to 2000 rpm the valve pin should pull in and then return to its original position when you slow down the engine. To check the 1974 valve, simply run the engine and observe whether the visible pin moves in and out proportional to rpm.

7.3 Throttle Valve Positioner

Individual parts for the throttle valve positioner can be replaced on cars equipped with this emission control device. Fig. 7-5 is an exploded view that shows the components of the unit. The gasket should be replaced whenever the altitude corrector is reassembled.

Fig. 7-5. Exploded view of the two-piece throttle valve positioner. Parts **3** through **11** are the same as on the earlier one-piece positioner.

1. Diaphragm unit
2. Diaphragm unit retaining ring
3. Control part cover
4. Gasket
5. Control part
6. Control diaphragm
7. Plastic foam filter
8. Altitude corrector bellows
9. Altitude corrector housing
10. Phillips screw (6)
11. Setscrew
12. Mounting clamp
13. Rubber mounting

When a new diaphragm unit is installed, you must adjust the pull rod length after the diaphragm unit has been installed on the carburetor. To adjust, loosen the locknuts at each end of the pull rod. One end of the pull rod has right-hand threads; the other end has left-hand threads. Its effective length can be adjusted by turning the rod in one direction or the other. Adjust its length until the fast idle lever does not touch either the carburetor body or the throttle valve lever when the throttle valve is closed (Fig. 7-6).

© 1974 VWoA—2726

Fig. 7-6. Turning pull rod to adjust its effective length. Large arrows indicate space between fast idle lever and carburetor body and between fast idle lever and throttle valve lever.

After the pull rod length has been adjusted, tighten the two locknuts to hold the pull rod in place. The throttle valve positioner should then be adjusted with the engine running as described in **5.3 Adjusting Idle**. However, it is usually best to perform a tune-up before carrying out fine throttle valve positioner adjustments.

7.4 Exhaust Afterburning

The 1973 and 1974 models are equipped with an air injection exhaust afterburning system. A schematic view of this system is given in Fig. 7-7.

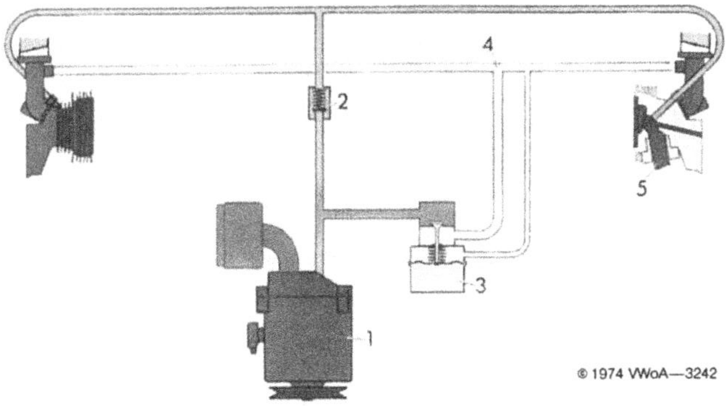

© 1974 VWoA—3242

1. Air pump
2. Check valve
3. Anti-backfire idle mixture valve
4. Vacuum connection to central idling system
5. Exhaust port in cylinder head

Fig. 7-7. Components of exhaust afterburning system.

Replacement parts are not supplied for the air pump. Defective pumps must be replaced. The pump's filter element should be replaced every 18,000 mi. (30,000 km) as described in **LUBRICATION AND MAINTENANCE**. Belt adjustment is also covered there.

If the engine backfires—especially during deceleration—check for leakage in the exhaust system, insufficient valve clearance, and incorrect ignition timing. If these faults are not present, check for leakage in the hoses attached to the anti-backfire idle mixture valve. The valve itself is seldom faulty. However, if no other cause for the backfiring can be found, check to see that the valve diaphragm is not leaking. Faulty valves must be replaced.

8. FUEL SYSTEM TECHNICAL DATA

I. Fuel Pump

Model year	Engine code letter	Part No.	Minimum delivery capacity	Maximum delivery pressure
1968 through 1970	B	211 127 025	400 cc/min. @ 3800 rpm	3.5 psi (0.25 kg/cm²)
1971	AE	113 127 025 C (D)	400 cc/min. @ 4000 rpm	3-5 psi (0.20-0.35 kg/cm²)
1972 and later	CB, CD, AW	021 127 025 A	400 cc/min. @ 3800 rpm	5 psi (0.35 kg/cm²)

II. Carburetor Settings and Jets

Model year	**1968–1969**	**1970**	**1971**	**1972**	
Carburetor type	**30 PICT-2**	**30 PICT-3**	**34 PICT-3**	**left side: 34 PDSIT-2**	**right side: 34 PDSIT-3**
Part No.	**113 129 027 H**	**211 129 029 Q**	**211 129 031 G**	**021 129 027 L**	**021 129 028 L**
From engine No.	**B 5 000 001**	**B 5 116 437**	**AE 000 001**	**CB 000 001**	
Venturi (dia.)	24 mm	24 mm	26 mm	26 mm	
Main jet	X116	X112.5	X125	X137.5	
Air correction jet	125Z	140Z	60Z	155	
Pilot jet	55	g 65	g 57.5	55	
Pilot jet air bleed	135	135	147.5	145	
Auxiliary fuel jet	—	45	42.5	45	—
Auxiliary air jet	—	130	90	0.7	—
Power fuel jet	60	100/100	95/95	—	
Float needle valve (dia.)	1.5 mm	1.5 mm	1.5 mm	1.2 mm	
Washer under float needle valve (thickness)	1.5 mm	1.5 mm	0.5 mm	0.5 mm	
Fuel level	19.5–20.5 mm	19.5–20.5 mm	17–19 mm	12–14 mm	
Float weight	8.5 g	8.5 g	8.5 g	7.0 g	
Pump injection quantity	1.3–1.6 cc/stroke	1.05–1.35 cc/stroke	1.45 ± 0.1 cc/stroke	0.8–1.0 cc/stroke	
Throttle valve gap	—	—	—	0.6 mm (choke closed)	
Gasket for carb. upper part	brown	yellow	black	—	

Model year	**1973 (Manual transmission)**		**1973 (Automatic transmission)**		**1974**	
Carburetor type	**left side: 34 PDSIT-2**	**right side: 34 PDSIT-3**	**left side: 34 PDSIT-2**	**right side: 34 PDSIT-3**	**left side: 34 PDSIT-2**	**right side: 34 PDSIT-3**
Part No.	**021 129 027 P**	**021 129 028 P**	**021 129 027 M**	**021 129 028 M**	**021 129 031 N/P/Q***	**021 129 032 N/P/Q***
From engine No.	**CB 062 001**		**CD 000 001**		**AW 000 001**	
Venturi (dia.)	26 mm		26 mm		26 mm	
Main jet	X130		X132.5		X130	
Air correction jet	140		155		175	
Pilot jet	55		50		52.5	
Pilot jet air bleed	140		130		120	
Auxiliary fuel jet	45	—	45	—	45	—
Auxiliary air jet	0.7	—	0.7	—	0.7	—
Power fuel jet	—		—		—	
Float needle valve (dia.)	1.2 mm		1.2 mm		1.2 mm	
Washer under float needle valve (thickness)	1.0 mm		1.0 mm		1.0 mm	
Fuel level	12–14 mm		12–14 mm		12–14 mm	
Float weight	7.0 g		7.0 g		7.0 g	
pump injection quantity	0.7 ± 0.1 cc/stroke		0.7 ± 0.5 cc/stroke		0.55–0.85/1.3–1.7 cc/stroke	
Throttle valve gap	0.6 mm (choke closed)		0.6 mm (choke closed)		0.7 mm (choke closed)	
Gasket for carb. upper part	—		—		—	

***N = manual trans. (exc. Calif.), Q = manual trans. (Calif.), P = auto. trans.**

Section 4

ELECTRICAL SYSTEM

Contents

4

TABLES

Electrical System

The electrical system is basically an efficient means for transmitting power from the engine to remote parts of the car. It does this with the help of a generator or alternator that converts some of the engine's mechanical energy into electrical energy. The electrical energy is carried over wires to motors that convert it back into mechanical energy or to bulbs that convert it into heat and light. The battery in the system supplies electrical power mainly when the engine is not running.

4

Every terminal in the electrical system is numbered. The terminal numbers for all major electrical connections are given in the color wiring diagrams that appear at the end of this section. The terminal number is usually stamped on the component itself as an aid to proper installation.

Though most of the electrical terminal numbers are used only once to denote a particular terminal on a particular component, there are several numbers that do not designate specific terminals and which appear in numerous locations throughout the electrical system. Such numbers identify main sources of electrical current. All terminals numbered 15 originate at the ignition switch and supply current only when the ignition switch is in its on position. Terminals numbered 30 supply positive polarity current directly from the battery with no intervening switch that can be used to turn it off. Terminals numbered 31 are ground connections between a switch and another electrical component. Terminals numbered 85 receive a ground wire that is connected directly to the chassis of the car. Terminals identified by the number 50 receive current only when the ignition switch is in its start position. A letter suffix is sometimes added to the terminal number to distinguish separate parts of the same circuit or to prevent the confusion of two circuits that have similar functions.

All electrical circuits other than those required for starting and operating the engine are protected by fuses. To prevent accidental shorts that might cause a blown fuse or damage wires and electrical components that are not protected by fuses, you should always disconnect the ground strap from the negative pole of the battery before working on the electrical system of your car. If you lack the skills or the equipment needed for testing and repairing the electrical system we suggest that you leave such work to an Authorized VW Dealer or other qualified shop. We especially urge you to consult your Authorized VW Dealer before attempting repairs on a car still covered by the new-car warranty.

1. GENERAL DESCRIPTION

A brief description of the electrical system's components is presented here for familiarization purposes. Please notice that no mention is made of the electronic fuel injection system used on 1975 and later models. Complete information on the fuel injection system, including a wiring diagram, is given in **FUEL INJECTION**.

System Voltage and Polarity

The VW has a 12-volt, negative ground electrical system. This means that the voltage regulator keeps voltage in the system at approximately the 12-volt rating of the battery and that the negative pole of the battery is connected directly to the car frame.

Battery

The six-cell, 12-volt lead-acid battery is located at the right-hand side of the engine compartment. On 1968 through 1974 models, the battery is rated at 45 ampere-hours; on 1975 and later models, the battery is rated at 54 ampere-hours.

Starting System

The starter is series-wound and has an overrunning clutch. Output is 0.7 horsepower on vehicles with manual transmissions and 0.8 horsepower on vehicles with automatic transmissions. The starter and its attached solenoid are located near the front of the engine.

Charging System

A belt-driven direct current (DC) generator is installed on 1968 through 1971 models. All later models have an alternator with a separate voltage regulator. The DC generators on 1968 models are rated at 30 amperes. The DC generator on 1969 through 1971 models is rated at 38 amperes. The alternator is rated at 55 amperes.

Ignition System

Ignition troubleshooting, maintenance, and repair are covered in **ENGINE AND CLUTCH** and in **LUBRICATION AND MAINTENANCE.** Radio suppression is by resistance built into the spark plug connectors and the distributor rotor.

Wiring

All components of the VW electrical system (except for the heavy battery cables) have push-on connectors. A system of fuses prevents short circuits or excessive current from damaging the electrical system.

Lights

The lighting system includes parking lights, side marker lights, turn signals, back-up lights, interior lighting, and sealed beam headlights. The headlight beams are dimmed or raised by pulling the turn signal lever toward the steering wheel. Actual switching is carried out by a dimmer relay mounted behind the instrument panel.

VW Computer Analysis

Many VW cars have been equipped with a separate wiring harness that serves the VW Computer Analysis system. A central socket in the engine compartment receives individual wires connected to various measuring points on the car. These connections are marked by encircled numbers in the wiring diagrams in this section.

Although they are not vital to car operation, all such connections must be kept intact if the computer analysis system is to work properly. Never connect any device other than the test plug of the VW Computer Analysis system to the test network central socket in the engine compartment. Doing so could damage the socket, the test sensors, or the vehicle components containing them.

Windshield Wipers

The blades of the two-speed windshield wiper system automatically return to their parked position when they are switched off. The wiper switch includes a windshield washer control. On late models the wipers and washers are controlled by a lever at the right side of the steering column; on early models, by a knob on the dashboard.

The washers work by compressed air. The fluid reservoir for the windshield washers is located on the right side of the dashboard. There is also an air connection under the plastic reservoir cover which permits you to pressurize the container with any standard tire pump or service station air hose.

Instruments

A flexible cable driven by the left-hand front wheel operates the speedometer. The fuel gauge is located at the left of the speedometer. The oil pressure, generator, high beam, and turn signal warning lights are mounted in the fuel gauge dial. An optional, electrically wound clock may be installed at the right side of the speedometer.

2. MAINTENANCE

No routine lubrication of the generator, starter, or other motors is required. However, the following checks are included in **LUBRICATION AND MAINTENANCE.**

1. Checking the lights and switches
2. Checking the windshield wipers and washers
3. Checking the battery
4. Testing the charging and starting systems.

3. FUSES

Most of the fuses in the electrical system are located in a fuse box under the dashboard near the steering column. Fig. 3-1 is a diagram of a typical VW fuse box, although the fuse boxes of some vehicles covered in this Manual have only ten fuses rather than the twelve shown.

Table a lists the fuse number of each electrical component on each model covered in this Manual. Always replace red (16-amp) fuses with other red fuses and white (8-amp) fuses with other white fuses. If in doubt, consult the Owner's Manual supplied with the car. Never patch a blown fuse with aluminum foil or replace it with wire or a fuse of greater capacity.

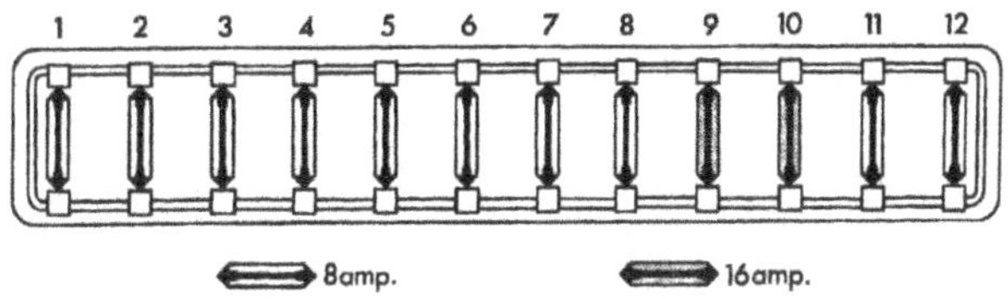

© 1974 VWoA—1363

Fig. 3-1. Typical fuse box. The location of the 16-amp fuses varies among model years.

Table a. Fuse Locations in Fuse Box

	1968 & 1969	1970	1971	1972	1973	1974	1975 & later
Auxiliary heater (switch current)			9/11	9/10	9/10	9/10	9/10
Backup lights (automatic transmission only)					11	11	11
Buzzer alarm		2	9	9	9	9	9
Emergency flasher system	9	2	8	8	8	8	8
Fuel gauge		4	12	11	11	11	11
High beam, left and high beam indicator light	3	7	6	6	5	5	5
High beam, right	4	8	5	5	6	6	6
Horn	1	3	10	12	12	12	12
Interior light, front	9	2	8	8	8	8	8
Interior light, rear	9	1	9	9	9	9	8
Kickdown (automatic transmission — except 1976)					11	11	11
License plate lights	7	12	1	2	2	2	2
Low beam, left	5	9	4	4	3	3	3
Low beam, right	6	10	3	3	4	4	4
Parking lights	7	12	1	2	2	2	2
Rear window defogger (switch current)	2	3	10	10	10	10	10
Side marker lights, front					2	2	2
Side marker light, left rear		11	2	1	1	1	1
Side marker light, right rear		12	1	2	2	2	2
Stop lights	1	3	10	12	12	12	12
Tail light, left	8	11	2	1	1	1	1
Tail light, right	7	12	1	2	2	2	2
Turn signals	1	4	12	11	11	11	11
Unassigned (accessory)	10	6	7	7	7	7	7
Warning light, brake	1	3	12	12	12	12	12
Warning light, generator or alternator		4	12	11	11	11	11
Warning light, oil pressure		4	12	11	11	11	11
Windshield wipers	2	5	11	10	10	10	10

4

Replacing In-line Fuses

In addition to the fuses located in the fuse box, there may be one or more fuses located in in-line fuse holders. Except on vehicles with automatic transmissions, the in-line fuse holder for the back-up lights is mounted on the engine, next to the ignition coil. (The fuse is in the fuse box on vehicles with automatic transmissions.)

The 16-amp fuse for the warm air blower is located in the engine compartment, near the blower. On 1971 and later models equipped with the optional auxiliary heater, the fuse for the heater is in an in-line holder next to the heater—also in the engine compartment. The locations for these in-line fuses are illustrated in the Owner's Manual.

4. BATTERY

Each of the six battery cells contains a set of brown lead oxide positive plates and gray sponge lead negative plates. The cells are connected in series by heavy lead bars and are enclosed in a plastic case having six compartments. The battery case also serves as a tank for the electrolyte—a solution of sulfuric acid diluted with water to a specific gravity of 1.285, which means that the electrolyte weighs 1.285 times as much as an equal volume of water. The battery plates that make up the cells are completely immersed in the electrolyte.

The terminal posts are labeled + and − and are further identified by the fact that the positive post is the thicker of the two. The ground strap connected to the negative post bolts to the frame of the car. The cable attached to the positive post extends to the starter solenoid where it is joined by the main line serving the rest of the electrical system. The batteries in VW cars built between June 1971 and July 1974 have an additional central terminal with a small guage wire attached to it. This is the Computer Analysis connection used to check the electrolyte level.

Discharging

The battery does not store electricity. Rather, it produces electrical current by means of a reversible electrochemical reaction. When a current is completed between the two battery posts, sulfuric acid from the electrolyte combines with the lead in the plates to produce lead sulfate, releasing a great many electrons in the process.

Charging

The electrochemical reaction by which the battery produces electrical current is reversed when direct current is sent back into the cells. The charging system of the car supplies this current. When the discharged battery plates are charged with direct current from an outside source, the lead sulfate in the plates is converted back to its original state, returning sulfuric acid to the solution in the electrolyte.

A battery can never be charged to a voltage level in excess of the voltage it is capable of producing electrochemically. As charging proceeds, the battery's voltage builds to a peak called terminal voltage. If charging is continued beyond the terminal voltage, the water in the electrolyte begins to decompose into hydrogen and oxygen. This condition is called gassing.

Temperature Effects

Temperature changes modify the efficiency of the battery as well as alter the specific gravity of its electrolyte. Low outside temperatures can create slow starting by thickening the engine and transmission oils and simultaneously reducing the battery power available for running the starter motor. The current-producing capacity of a battery chilled to −15°C (5°F) is only half what its capacity is at 20°C (68°F).

An additional winter hazard is the danger of partly-discharged batteries freezing in cold weather due to the higher proportion of water in their electrolyte. A frozen battery will produce no current but can usually be restored to service if thawed out slowly. The following list shows the safe low temperature limits for batteries in various states of change.

Specific gravity	Freezing point
1.285	−68°C (−90°F)
1.200	−27°C (−17°F)
1.120	−11°C (12°F)

4.1 Servicing and Testing Battery

The level of the electrolyte should never be allowed to fall below the tops of the plates in any cell. As water is lost through evaporation and electrolysis, fresh water must be added to maintain the electrolyte's level at the bottoms of the indicator tubes built into the battery filler openings. Use only distilled water to replenish the electrolyte. Water that is not chemically pure may have an adverse effect on battery life and efficiency.

The battery will lose more water in summer than it will in winter. In very hot weather it may be necessary to check the electrolyte level as often as once a week. Never overfill the cells. Doing so could cause the electrolyte to boil over during a long daylight drive when the load on the electrical system is light and generator output is high.

Battery terminals must be tight-fitting and free of corrosion and acid salts. If you notice even a trace of corrosion, remove the positive cable and ground strap from the battery posts and clean the posts and terminal clamps with a wire-brush battery-terminal cleaning tool. After the terminals have been cleaned and the positive cable and ground strap tightly installed on the battery posts, the terminals and posts should be coated lightly with petroleum jelly or sprayed with a commercial battery terminal corrosion inhibitor.

WARNING —
Keep sparks and open flame away from the top of the battery. Hydrogen gas from the battery could explode violently.

The top of the battery should always be kept clean. Even a thin layer of dust containing conductive acid salts can cause the battery to begin discharging. Corrosion and acid salt accumulations should be washed away with baking soda solution. Be extremely careful that none of this solution enters the cells through the vent holes. Even a drop or two will seriously impair the efficiency of the battery.

Periodic battery tests should be made to help keep track of battery condition. Such tests can also be made to help pinpoint the source of suspected battery trouble.

WARNING —
Wear goggles when you work with battery electrolyte and do not allow the liquid to contact your skin or clothing. Electrolyte is corrosive and can cause severe burns. If it should spill on your skin, flush the area of contact immediately with large quantities of water. Spilled electrolyte can be neutralized with a strong baking soda solution.

Hydrometer Testing

The simplest tool for testing the battery is a hydrometer. It consists of a glass cylinder with a freely moving float inside. When electrolyte is drawn into the cylinder by squeezing and releasing a rubber ball, the level to which the float sinks indicates the specific gravity of the electrolyte. A specific gravity scale on the float is read at the point where it intersects the surface of the electrolyte. The more dense the concentration of sulfuric acid in the electrolyte, the less the float will sink and the higher the reading. Specific gravity values for different states of charge are as follows:

State of charge	Specific gravity
Fully discharged	1.120
Half discharged	1.200
Fully charged	1.285

Voltage Testing

Total battery voltage can be tested with a special voltmeter such as the one shown in Fig. 4-1.

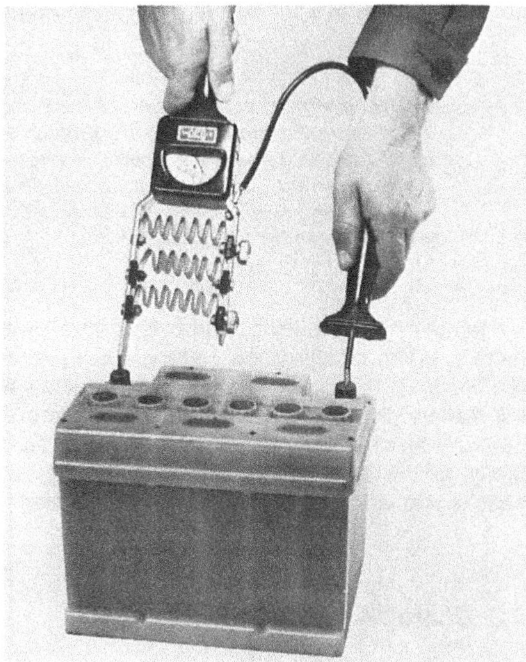

Fig. 4-1. Using voltmeter to test total voltage of battery.

The tester should consist of a voltmeter connected in parallel with a test load of approximately 110 amps. The minimum voltage indicated should not be less than 9.6 volts. If total voltage drops below this value during the 5- to 10-second test, the battery is either discharged or sulfated. A sulfated battery is one in which the plates are covered by a layer of lead sulfate that is difficult to reconvert. Sulfating is visible as a gray coating on the plates.

CAUTION —
A discharged battery should be recharged immediately. Otherwise, sulfating will lead to the loss of active plate materials and to reduced battery capacity.

The voltage of an individual cell should not vary from the others by more than 0.2 volts. This can be determined by applying one prong of the tester to the negative battery post, then dipping the other prong into the electrolyte of successive cells, and finally applying it to the positive post. The readings should be 2, 4, 6, 8, 10, and 12 volts. This test should last for no more than 10 seconds.

4

4.2 Removing and Installing Battery

The battery is fastened to the body by two clamps. One clamp is welded to the floor and the other clamp is held by a bolt. Before removing the bolt, disconnect first the ground strap and then the positive cable from the battery posts. If necessary, disconnect the wire from the central terminal that serves the VW Computer Analysis System.

> ***CAUTION —***
> *If a new ground strap must be installed on a car equipped for VW Computer Analysis, make certain the part number is correct. The wrong ground strap will cause incorrect voltage data to be fed into the computer.*

Remove the mounting clamp bolt, then remove the battery. When installing the battery, clean and install the terminals as described earlier in **4.1 Servicing and Testing Battery**. Be sure to install the wire for the VW Computer Analysis system. The battery must be mounted firmly to the body during installation to prevent road shocks and vibration from damaging the plates.

4.3 Charging with Battery Charger

Normally, a battery should be charged at no more than 10 percent of its rated capacity. For example, a charging current of 4.5 amperes would be used on a battery having 45 Ah (ampere-hours) capacity. However, a current as low as 5 percent of the rated capacity (2.25 amps for a 45 Ah battery) can be used in normal charging and should always be used the first time a new battery is charged.

In normal charging, the battery is considered fully charged when it is gassing freely and the voltage of the individual cells has risen to 2.5 to 2.7 volts (about 15 volts for the battery). An hour or so after you have switched off the charging current, use the voltmeter/tester to determine the rest voltage of the battery. This should be 2.1 to 2.2 volts per cell, or approximately 12.5 to 13.0 volts for the battery.

Quick Charging

To save time in an emergency, a higher current can be used to charge batteries in good condition. Only sound batteries that are already in service are suitable for quick charging. Neither factory-new nor sulfated batteries should ever be quick-charged.

> ***WARNING —***
> *Do not boost a sulfated battery at a high charging rate. The battery could explode.*

To quick charge:

1. Remove the battery caps, then connect a battery charger and voltmeter to the battery. Quick-charge at 40 amperes for three minutes.
2. Observe the voltmeter reading during charging. If total battery voltage exceeds 15.5 volts, the battery plates are sulfated or worn out and the battery should be replaced.
3. If the total voltage is less than 15.5 volts, test the individual cell voltages. If cell voltages vary by more than 0.1 volt, the battery plates are worn out and the battery should be replaced. If cell voltages are within 0.1 volt, measure the specific gravity and quick-charge as follows:

Specific gravity	Period of charge
1.150 or less	1 hour
1.150 to 1.175	45 minutes
1.175 to 1.225	15 minutes
above 1.225	Slow charge only

> ***WARNING —***
> *Smoking and open flames should not be permitted in a room where batteries are charged. Charging causes excess water in the electrolyte to decompose into hydrogen and oxygen, a dangerously explosive combination of gases.*

> ***CAUTION —***
> *Do not store precision tools in a room where batteries are charged. The corrosive fumes generated can severely damage the tools.*

Storing Battery

A battery that is not in use will gradually discharge itself. At room temperature it will lose about one percent of its remaining capacity a day. The rate of discharge increases with higher temperatures. If the battery is allowed to remain in a partly or fully discharged condition for long periods, it will become badly sulfated and may never be serviceable again. The following procedure is recommended to prevent this happening in a battery that is to be stored either in or out of the car.

To store battery:

1. Charge the battery. Check the electrolyte level and the specific gravity, then make corrections as necessary.
2. Store the battery in a cool, dry place.
3. Every 6 to 8 weeks, discharge the battery and recharge it.
4. Before returning the battery to service, charge it with a very low current (not over 3 amps).

5. STARTING SYSTEM

Table b lists the starters installed on Type 2 vehicles built during the years covered by this Manual. Starter 003 911 023A is installed only in cars having automatic transmissions. Bosch starters 311 911 023 B, 311 911 023 C, and 311 911 023 D are similar in construction and identical in technical specifications. The latter two (types C and D) have aluminum field coils rather than the copper field coils used in other types. They also have a modified drive mechanism and are equipped with four carbon brushes instead of two. Starter 311 911 023 D can also be distinguished from the other Bosch starters by its shorter length—31.8 mm (1¼ in.) less than the others.

Table b. Starter Types

Starter	Chassis Numbers	Output
111 911 023 A (VW)	from 218 000 001 through 210 2106 747	.7 bhp
311 911 023 B (Bosch)	from 218 000 001 through 219 2000 000	.7 bhp
311 911 023 C (Bosch)	from 210 2000 001 through 211 2300 000	.7 bhp
311 911 023 D (Bosch)	from 210 2000 001	.7 bhp
003 911 023 A (Bosch)	from 213 2000 001 (automatic transmission)	.8 bhp

5.1 Starting System Troubleshooting

Troubleshooting procedures that are applicable to cars with any type starter appear in **Table c**. The bold numbers in the Remedy column refer to the headings in this section under which the prescribed tests and repairs can be found. If more than one probable cause is listed, check them one by one in the order in which they appear.

Table c. Starting System Troubleshooting

Problem	Test and Probable Cause	Remedy
1. Starter does not operate when ignition key is turned to start position	Turn lights on for test: a. Lights are out. Loose cables or poor ground connection. Battery run down b. Lights go out when key is moved to starting position. Insufficient current due to loose connections or corroded terminals c. Lights go dim when key is moved to starting position. Battery run down d. Connect a jumper cable between starter terminals 30 and 50. If starter runs, cable 50 to ignition switch or cable 30 to lighting switch is faulty, or ignition switch is defective e. Lights stay bright, solenoid operates. Connect jumper cable from starter terminal 30 to connector strip terminal. Solenoid contacts are faulty if starter runs	a. Check battery cable terminals. Test battery. Charge if necessary. See **4.1**, **4.2**, **4.3**. b. Clean and tighten all battery cable connections. See **4.2**. c. Charge battery. See **4.3**. d. Eliminate open circuits. Replace defective parts. See **7**, **14**. e. Replace solenoid. See **5.3**.
2. Starter does not operate when battery cable is directly connected with terminal stud of connector strip	a. Brushes sticking b. Brushes worn c. Weak spring tension. Brushes do not make contact d. Commutator dirty e. Commutator rough, pitted, or burned f. Armature or field coils defective	a. Clean brushes and guides. See **5.4**. b. Replace brushes. See **5.4**. c. Replace springs. See **5.4**. d. Clean commutator. See **5.4**. e. Recondition or replace starter motor. See **5.2**, **5.4**. f. Recondition or replace starter motor. See **5.2**, **5.4**.
3. Starter turns too slowly or fails to turn engine over	a. Battery run down b. Insufficient current flow due to loose or corroded connections c. Brushes sticking d. Brushes worn e. Commutator dirty f. Commutator rough, pitted, or burned g. Armature or field coils defective	a. Charge battery. See **4.3**. b. Clean battery terminals and cable clamps; tighten connections. See **4.1**, **4.2**. c. Clean brushes and guides. See **5.4**. d. Replace brushes. See **5.4**. e. Clean commutator. See **5.4**. f. Recondition or replace starter motor. See **5.2**, **5.4**. g. Recondition or replace starter motor. See **5.2**, **5.4**.
4. Starter makes unusual sounds and cranks engine erratically or fails to crank	a. Drive pinion defective b. Flywheel ring gear defective	a. Replace drive pinion. See **5.4**. b. Replace flywheel. See **ENGINE**.
5. Drive pinion does not move out of mesh	a. Drive pinion or armature shaft dirty or damaged b. Solenoid switch defective	a. Recondition or replace starter motor. See **5.2**, **5.4**. b. Replace solenoid switch. See **5.3**.

4

5.2 Removing and Installing Starter

Starter removal will be easier if you place the vehicle on a lift and take off the right-hand rear wheel.

To remove:

1. Disconnect the ground strap from the negative post of the battery, then disconnect the wires from terminals 30 and 50 of the starter solenoid.
2. While a helper holds a nut inside the engine compartment, remove the upper starter mounting bolt.
3. Remove the lower starter mounting nut from its stud, then withdraw the starter motor.

   NOTE —
   After removing the Bosch starter motor, check the bushing in the clutch housing. If the bushing is worn or damaged, replace it.

To install:

1. Lubricate the starter shaft bushing with multipurpose grease, then apply a good sealing compound around the starter opening in the clutch housing.
2. Slide the starter onto the stud and secure it loosely with the nut. Then loosely install the bolt and nut. Working alternately, gradually torque both fasteners to 3.0 mkg (22 ft. lb.).
3. Clean the wires and terminals, then tightly install all the electrical connections.

5.3 Removing, Adjusting, and Installing Solenoid

The following procedure applies only to the Bosch starter. The VW unit is covered later.

To remove:

1. Remove the starter from the car.
2. Remove the nut and the connector strap. Then remove the two screws that hold the solenoid to the drive end plate.
3. Lift the pull rod upward and out of the engaging lever. This will be easier if you pull the pinion clockwise and outward at the same time.
4. Withdraw the solenoid.

When installing a new solenoid on a Bosch starter, be sure to adjust the pull rod length as indicated in Fig. 5-1. You should never adjust pull rod length to accommodate a defective solenoid. A defective solenoid should be replaced.

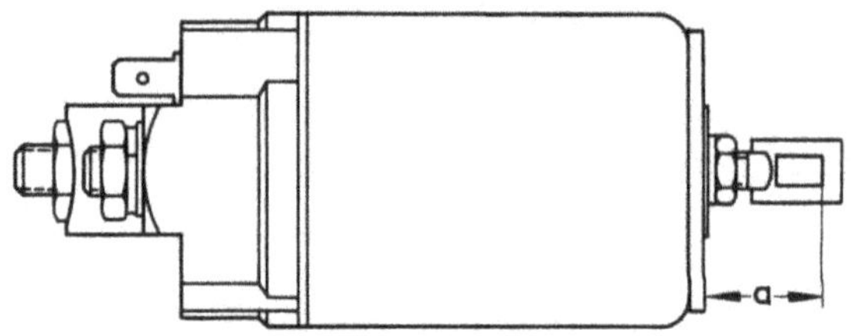

© 1974 VWoA—1261

Fig. 5-1. Pull rod adjustment. Loosen locknut, then turn rod clockwise or counterclockwise until dimension **a** is 19 mm ± 0.10 (.748 ± .004 in.). Tighten locknut and recheck dimension **a**.

To install solenoid:

1. Seat the molded rubber seal properly on the drive end plate.
2. Place a small strip of plastic sealing compound on the outer edge of the solenoid end face.
3. Withdraw the drive pinion as far as possible and install the solenoid as shown in Fig. 5-2.
4. Install the nut and connector strap.

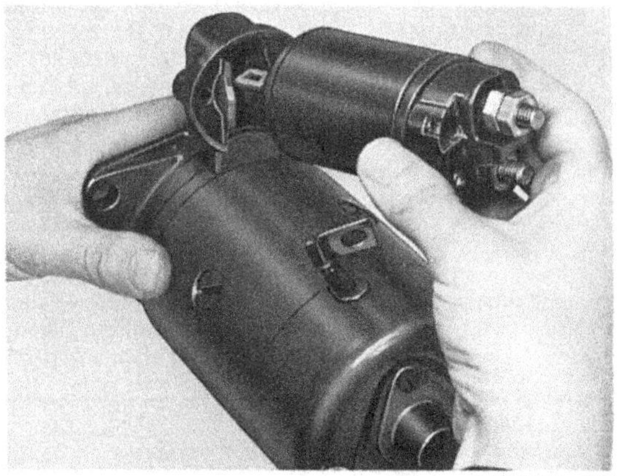

© 1974 VWoA—1263

Fig. 5-2. Solenoid being installed. Hook pull rod eye over engaging lever. Use sealing compound between solenoid end face and drive end plate.

VW Starter

The solenoid is attached to the VW starter by two nuts on studs built into the solenoid housing, a connecting strap, and an electrical connection. The solenoid core is a separate part. To remove the core from the operating lever, take off the spring clips that lock the two pins, push out the pins, remove the insulating plate, and then turn the contact plate 90°. Unlike the Bosch starter, no pull rod adjustment is called for. During assembly, use a good plastic sealing compound to seal the pin holes and the mating surfaces between the solenoid housing and the starter drive end plate.

5.4 Disassembling and Assembling Starter

Three representative starters are shown disassembled in Fig. 5-3, Fig. 5-4, and Fig. 5-5. These exploded views include the VW starter, one Bosch starter having two brushes, and another equipped with four brushes. Basic disassembly requires only that the circlips or C-washers be removed from the armature shaft and the two long through bolts be removed. Further disassembly can be done with a screwdriver.

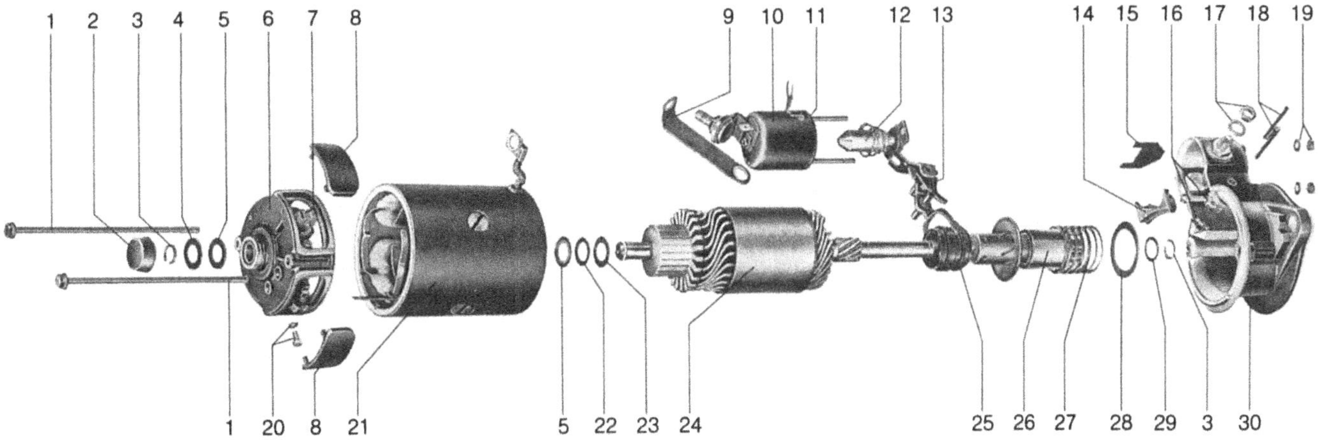

1. Through bolt
2. Cap
3. Circlip
4. Steel washer
5. Bronze washer
6. Commutator end plate
7. Brush holder and brushes
8. Brush inspection cover
9. Connecting strap
10. Solenoid housing and winding
11. Insulating disk
12. Solenoid core
13. Linkage
14. Molded rubber seal
15. Insulating plate
16. Spring clip
17. Nut and lock washer
18. Pins
19. Small nut and lock washer
20. Screw and lock washer
21. Pole housing
22. Dished washer
23. Steel washer
24. Armature
25. Connecting bushing
26. Drive pinion
27. Spring
28. Intermediate washer
29. Dished washer
30. Drive end plate

Fig. 5-3. Disassembly of VW starter 111 911 023A.

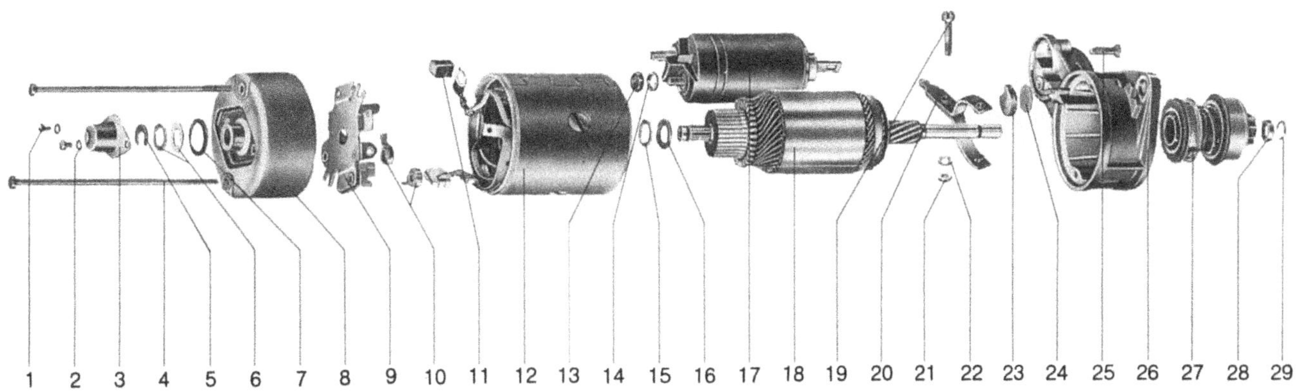

1. Screw
2. Washer
3. End cap
4. Through bolt
5. C-washer
6. Shims
7. Sealing ring
8. End plate
9. Brush holder
10. Springs
11. Rubber grommet
12. Pole housing
13. Nut
14. Lock washer
15. Insulating washer
16. Thrust washer
17. Solenoid
18. Armature
19. Pin
20. Engaging lever
21. Nut
22. Lock washer
23. Molded rubber seal
24. Disk
25. Screw
26. Drive end plate
27. Drive pinion
28. Stop ring
29. Circlip

Fig. 5-4. Disassembly of Bosch starter 311 911 023 B.

© 1974 VWoA—1315

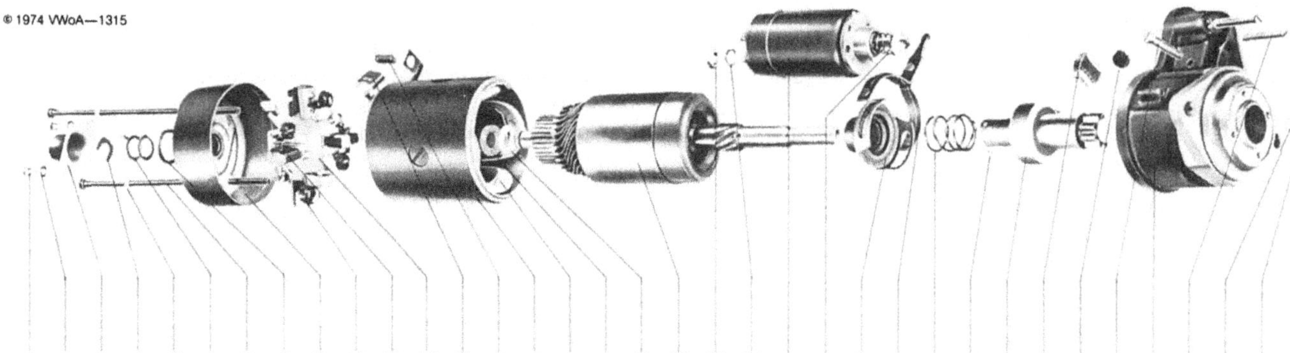

1. Screw
2. Washer
3. End cap
4. C-washer
5. Washer
6. Shims
7. Through bolt
8. Sealing ring
9. End plate
10. Retaining spring
11. Negative brush
12. Brush holder
13. Positive brush
14. Rubber grommet
15. Pole housing
16. Field winding
17. Insulating washer
18. Thrust washer
19. Armature
20. Nut
21. Lock washer
22. Solenoid
23. Solenoid return spring
24. Operating sleeve
25. Engaging lever
26. Engaging spring
27. Detent balls
28. Drive pinion
29. Molded rubber seal
30. Disk
31. Pin
32. Drive end plate
33. Screw
34. Lock washer
35. Nut

Fig. 5-5. Disassembly of Bosch starter 003 911 023A.

Remove the solenoid before disassembling the starter. The cap must also be removed from the commutator end of the starter to gain access to the C-washer or circlip on the armature shaft. After the C-washer or circlip has been removed, the through bolts can be taken out and the starter end plate and housing lifted off the armature. Note the number and position of all spacer washers to ensure proper reassembly. See Fig. 5-6 for additional information on the VW starter.

Starter Drive Servicing

The starter drive need not be disassembled if only electrical tests are to be made or the brushes replaced. If, however, the commutator must be machined or if the starter drive is defective, further disassembly is required. The drive pinion is removed from Bosch starters as in Fig. 5-7.

© 1974 VWoA—2303

Fig. 5-6. VW starter disassembly. Remove brush inspection covers and lift brushes to position shown before removing the through bolts. Note circlip groove in armature shaft.

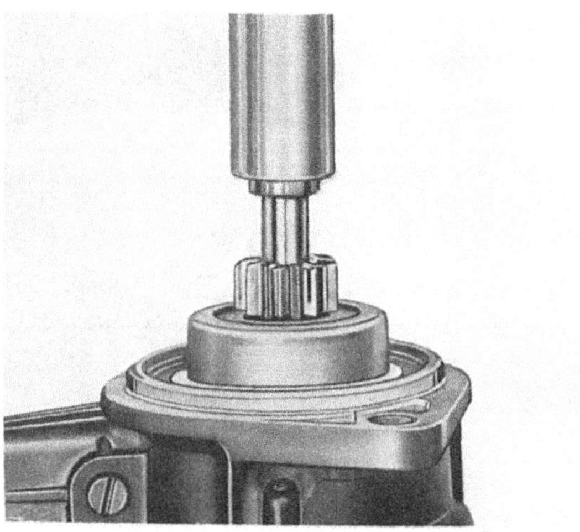

© 1974 VWoA—1278

Fig. 5-7. Removing Bosch starter pinion. Remove circlip and stop ring, file burrs from circlip groove, then press off pinion as illustrated.

To remove the armature and starter drive from the drive end plate of a VW starter, remove the pins shown in Fig. 5-8. Then remove the solenoid core as described in **5.3 Removing, Adjusting, and Installing Solenoid.**

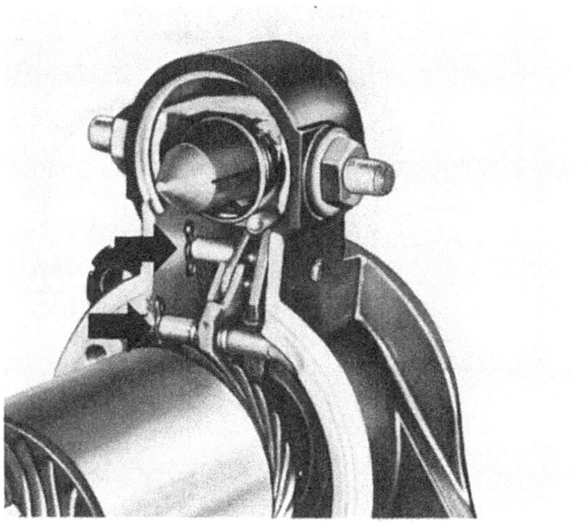

© 1974 VWoA—2304

Fig. 5-8. Removing VW starter drive. Remove the two spring clips at arrows, then push pins out of starter drive end casting.

To remove the drive pinion from the VW starter armature, take the circlip off the shaft and withdraw the connecting bushing as indicated in Fig. 5-9. Then, turn the whole drive pinion assembly çlockwise and jerk it slightly to remove it from the shaft. The connecting bushing and the five steel balls beneath it may then be taken off. During assembly, use multipurpose grease as an adhesive to hold the balls in place.

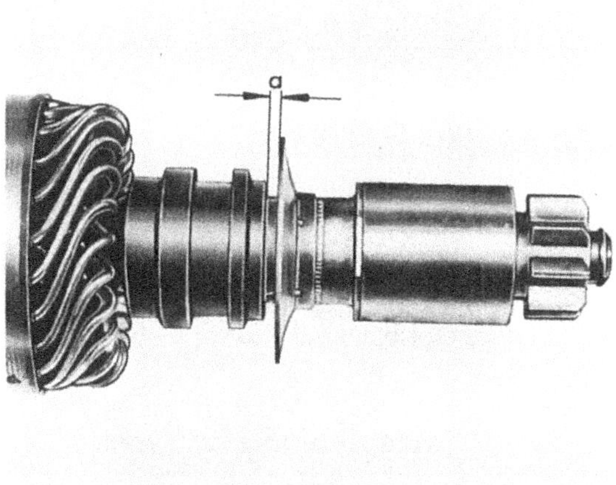

© 1974 VWoA—2305

Fig. 5-9. Removing drive pinion assembly from armature shaft of VW starter. The connecting bushing (the grooved sleeve nearest the armature windings) must be pulled back to make gap **a** approximately 3 to 6 mm (1/8 to 1/4 in.) wide.

Servicing Brushes and Commutator

You need only remove the commutator end plate from VW starters, or the end plate and brush holder from Bosch starters, to replace the brushes. However, the starter will have to be completely disassembled so the commutator can be machined if it is found to be out-of-round by more than 0.03 mm (.001 in.). The diameter of VW starter commutators must not be reduced below 33.00 mm (1.300 in.) or those of Bosch starters reduced below 34.50 mm (1.350 in.). The insulation strips should be undercut by about 0.50 mm (.020 in.) after the commutator has been turned on a lathe.

Unsolder the old brush wires and solder the new ones in their place. If the end plate bushing is worn, it can be pressed out and a replacement installed. Heat the end plate in hot oil before pressing in the new bushing.

To assemble starter:

4

Do not wash the self-oiling end plate bushing or the drive pinion assembly in solvent. This would destroy the lubricant built into these parts at the factory. Lubricate the bushings at both ends of the armature shaft with multipurpose grease. The drive mechanism must not be greased or it may jam in cold weather. During assembly, all points shown in Fig. 5-10 should be made weatherproof with a good sealer.

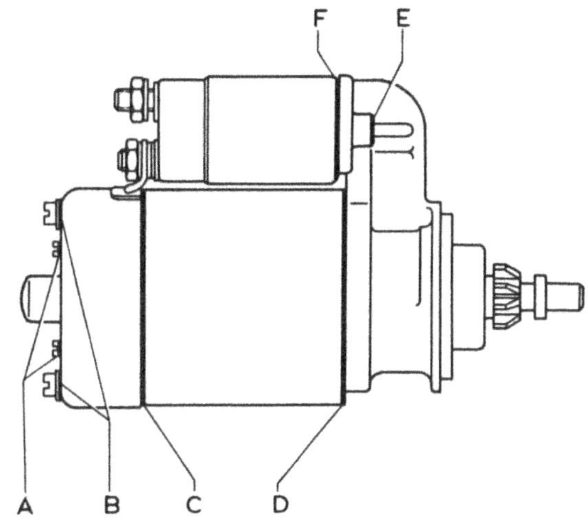

© 1974 VWoA—1314

A. Holes for end cap screws
B. Holes for through bolts
C. Joint between pole housing and end plate
D. Joint between pole housing and mounting bracket
E. Holes for solenoid mounting screws
F. Surface between solenoid switch and drive end housing

Fig. 5-10. Sealing locations on Bosch starter. On VW starters, the holes for the engagement linkage pins must also be sealed.

5.5 Starter Bench Tests

A battery, or several batteries wired in parallel, with a rated capacity of 135 ampere hours should be available for bench test. This will ensure that decreasing battery power does not influence the test readings. Automotive electrical shops that have starter motor test stands will find the data in **Table d** and **Table e** useful to determine how closely starters conform to factory specifications. Testing should be carried out in the following sequence:

1. No load test—starter motor running freely.
2. Load test—test stand flywheel braked to limit starter to the rmp given in **Table d**. This test must last no longer than 10 seconds.
3. Stall torque test—test stand flywheel braked to stop the starter under load. This test must last no longer than 5 seconds.
4. Solenoid pull-in voltage under load—starter switched on and off under light load and the pinion checked for proper engagement.

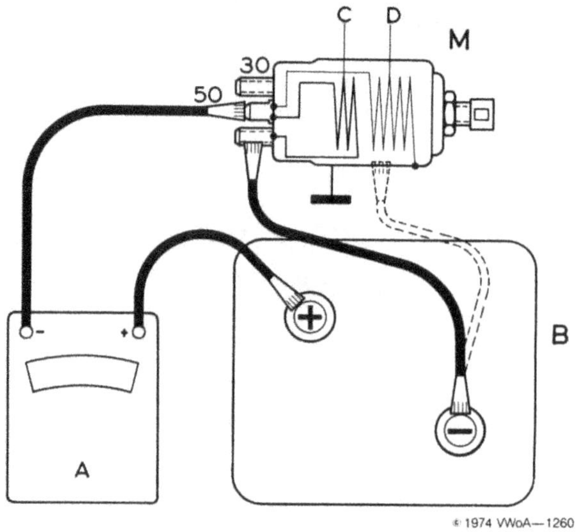

A. Ammeter
B. Battery
C. Pull-in winding
D. Holding winding
M. Solenoid

Fig. 5-11. Solenoid winding tests. Move negative connection to position shown by dotted lines to check current draw in holding winding. Bosch unit is shown, but VW solenoid uses a similar hook-up.

Table d. Starter Data

Starter Type	No-load Test			Load Test			Stall Torque Test		
	Current (amps)	Voltage	Speed (rpm)	Current (amps)	Voltage	Speed (rpm)	Current (amps)	Voltage	Solenoid pull-in voltage
311 911 023B	35–45	12	7400–9100	170–205	9	900–1300	220–260	6	7
111 911 023A	25–40	12	6200–7800	170–195	9	1050–1350	270–290	6	8
003 911 023A	35–50	12	6400–7900	160–200	9	1100–1400	250–300	6	8

Table e lists the normal current draw for the two windings of a solenoid that is in good condition. The pull-in winding is tested as shown in Fig. 5-11 and the holding winding as shown by the dotted lines in the same illustration. Faulty windings cannot be repaired. If readings vary from those listed in **Table e**, the solenoid must be replaced.

Table e. Solenoid Switch Test Data

	Current draw (amps)	
	Bosch	VW
Pull-in winding, max.	35	30
Holding winding, max.	11	12

6. CHARGING SYSTEM

Table f lists the generators installed in 1968 through 1971 Type 2 vehicles. In addition to the direct current generators listed in **Table f**, an alternator has been installed on Type 2 vehicles beginning with the 1972 models. The alternator is manufactured by Bosch.

Table f. Generator Types

Generator	From chassis No.	Through chassis No.
211 903 031 A	218 000 001	218 163 485
113 903 031 G	218 163 486	218 202 251
211 903 031 D	219 000 001	211 2276 560

The mounting for the alternator is different from the mounting for the DC generator. The electrical circuits and the regulators are also entirely different for the two types of charging systems.

> ***CAUTION —***
> *Tests and checks suitable for DC generator charging systems must not be used when working with alternator charging systems. Some DC generator tests will damage the diodes in an alternator. With the exception of data appearing in **6.8 Alternator**, all material in this section applies solely to charging systems equipped with DC generators.*

6.1 Generator Troubleshooting

Troubleshooting procedures applicable to cars with DC generators appear in **Table g**. The bold numbers in the Remedy column refer to the headings in this section under which the prescribed tests and repairs can be found. If more than one probable cause is listed, check them in the order in which they appear.

Although many of the entries in **Table g** will be seen to apply to both alternator and DC generator charging systems, the table was devised with only the DC generator in mind. There are several probable trouble causes applicable to alternators that do not occur in DC generators and a number of troubles common to DC generators that do not affect alternators.

6.2 In-car Testing of Generator And Regulator

A number of in-car tests are possible using a precision voltmeter, an ammeter, and a variable resistance. These tests will not only determine the condition of the generator and the regulator but can also determine whether the troubles that you detect are in the regulator or in the generator itself. All DC generators must meet the specifications given in **Table h**. Instructions for making the tests appear following the table.

Table h. Generator Technical Data

Nominal data		
	211 903 031 A **113 903 031 G**	**211 903 031 D**
Maximum current, amps	30	38
Mean regulating voltage	14	14
Nominal output speed, rpm	2000	2400
Test data		
	211 903 031 A **113 903 031 G**	**211 903 031 D**
Cut-in speed, generator rpm (1.9 times engine rpm)	1450	1300
Cut-in voltage	12.4–13.1	12.5–13.2
Return current, amps	2.0–7.5	5.0–11.5
No-load regulating voltage	13.5–14.5	13.5–14.2
Load regulating voltage	12.8–13.8	12.9–13.6
Load current, amps	45	35

4

Table g. Generator Troubleshooting

Problem	Probable Cause	Remedy
1. Ignition is on but warning light does not glow	a. Discharged battery b. Broken battery case or plates c. Bulb burned out d. Corroded and/or loose battery terminals e. Loose connections and/or broken wiring f. Ignition/starter switch defective g. Generator brushes do not make contact with commutator h. Regulator faulty	a. Charge battery. See **4.3**. b. Replace battery. See **4.2**. c. Replace bulb. See **13.3**. d. Clean and/or tighten terminals. See **4.1**. e. Tighten and/or repair wires. See **14**. f. Replace ignition/starter switch. See **7**. g. Free the brushes or replace them. If necessary, replace the brush springs. See **6.5**. h. Replace regulator. See **6.3**.
2. Light stays on or flickers when engine is running	a. Fan belt broken or slips badly b. Regulator faulty c. Loose connections or broken wire d. Generator faulty e. Commutator bars short-circuited	a. Replace or adjust fan belt. See **6.4**. b. Replace regulator. See **6.3**. c. Test wires. Repair as necessary. See **14**. d. Test generator. Repair as necessary. See **6.2**. e. Clean commutator with crocus cloth. See **6.5**.
3. Light goes out only at high speed	a. Generator faulty b. Regulator faulty	a. Test generator. Repair as necessary. See **6.2**. b. Replace regulator. See **6.3**.
4. Light remains on with the ignition switched off	Regulator contact points sticking (burned)	Replace regulator. See **6.3**.

To test no-load voltage:

1. Disconnect the wire from regulator terminal B+ (51), being careful not to ground the lead. Connect a voltmeter as shown in Fig. 6-1.

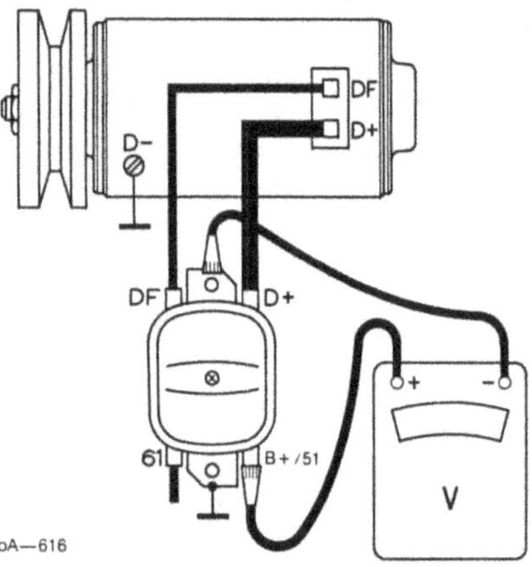

© 1972 VWoA—616

Fig. 6-1. Hook-up for no-load test. Positive voltmeter lead replaces wire normally attached to terminal B+ (51).

2. Start the engine, then slowly increase engine speed to approximately 2000 rpm. The voltmeter should indicate 13.5 to 14.5 volts.
3. When the engine is turned off, the voltmeter reading should drop from 12 to 0 volts just before the engine stops completely. This indicates that the regulator points are not sticking.
4. Check cut-in speed by temporarily reducing the engine idle to 650 to 700 rpm. The voltmeter should suddenly rise from zero as engine speed reaches approximately 685 rpm for generator 211 903 031D or 760 to 770 rpm for other generators.

Load regulating voltage can be tested as shown in Fig. 6-2. The test instruments are a voltmeter with a range of 0 to 30 volts, an ammeter with a range of −50 to +50 amps, and a rheostat that can be loaded to 50 amps.

With the engine running at approximately 1500 rpm, adjust the rheostat to a load setting of 45 amps. The voltage should be 12.9 to 13.6 for generator 211 903 031D or 12.8 to 13.8 for other generators. If it is not, the regulator should be replaced.

Use the hook-up shown in Fig. 6-3 to determine whether the trouble is in the regulator or in the generator. If there is no voltage, the regulator is faulty and must be replaced before you can make the test. Use an 8-gauge or larger jumper wire and limit the test's duration to 15 seconds. At 1500 generator rpm (790 engine rpm), the generator's voltage output should be 12 volts. At double this rpm, the voltage output should be 36 volts. If these specifications are met, the trouble is in the regulator. If not, remove and disassemble the generator for the tests described in **6.7 Testing Disassembled Generator**.

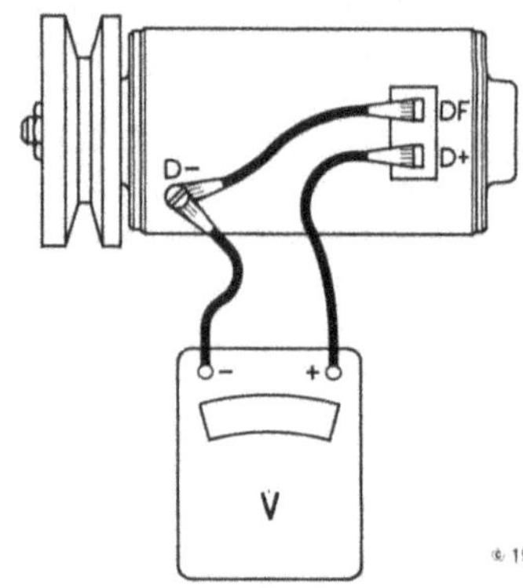

© 1972 VWoA—619

Fig. 6-3. Testing generator without regulator.

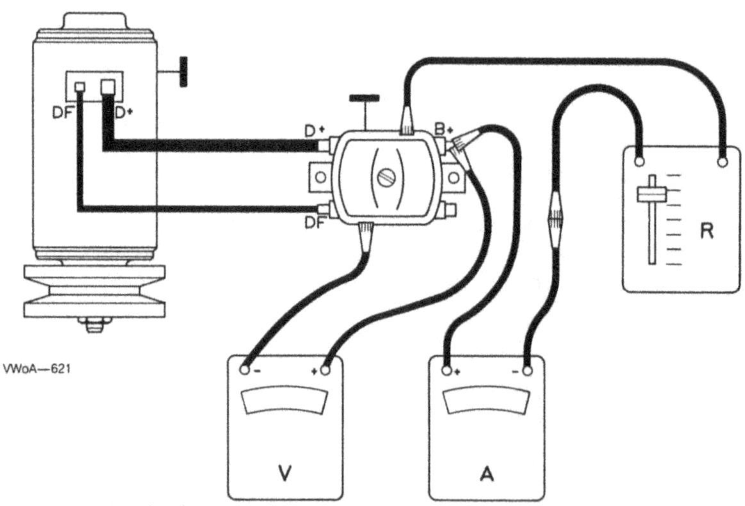

© 1972 VWoA—621

Fig. 6-2. Load regulating voltage test hook-up.

The return current test illustrated in Fig. 6-4 will detect a faulty cut-out relay in the regulator unit.

CAUTION —
The D+ terminals on the regulator and generator must not be disconnected while the engine is running. Doing so could burn out the generator's field windings.

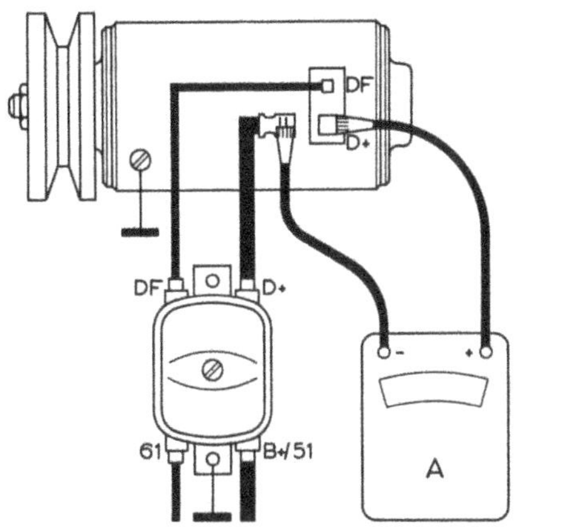

Fig. 6-4. Hook-up used for return current test.

After installing the ammeter as shown, start the engine and reduce the idle speed until the ammeter needle moves into the negative (discharge) range. Turn the engine off. The ammeter needle should jump back to zero before the engine has stopped completely. The maximum discharge reading obtained is the return current. It should be 5.0 to 11.5 amps for generator 211 903 031D or 2.0 to 7.5 amps for other generators. If the ammeter does not return to zero when the engine is completely stopped, the regulator is defective and must be replaced.

6.3 Removing and Installing Regulator

The regulator is mounted in the engine compartment, next to the battery. Always disconnect the battery ground strap before removing the regulator. Attach a tag to each wire as you remove it so that you can later identify the wires and install them on their proper terminals. Be very careful not to interchange wires DF and D+ or B+ and 61. Either error could damage the regulator, the generator, or both.

The regulator is held to the body by two Phillips head sheet metal screws. On early models, be sure that the regulator base has good ground contact. Late models have a ground wire between the regulator and the generator. If a new regulator is being installed on an early model, also add the ground wire.

6.4 Removing and Installing Generator

Before you can remove the generator, you must partially raise the fan housing as described in conjunction with cooling air system removal in **ENGINE AND CLUTCH**. With the fan housing raised, remove the four bolts indicated in Fig. 6-5, then remove the generator and the fan from the fan housing.

© 1974 VWoA—2237

Fig. 6-5. Bolts (arrows) that hold generator and fan cover to fan housing.

If necessary, you should polarize generators as shown in Fig. 6-6.

CAUTION —
Polarization of an alternator is both unnecessary and damaging to the alternator.

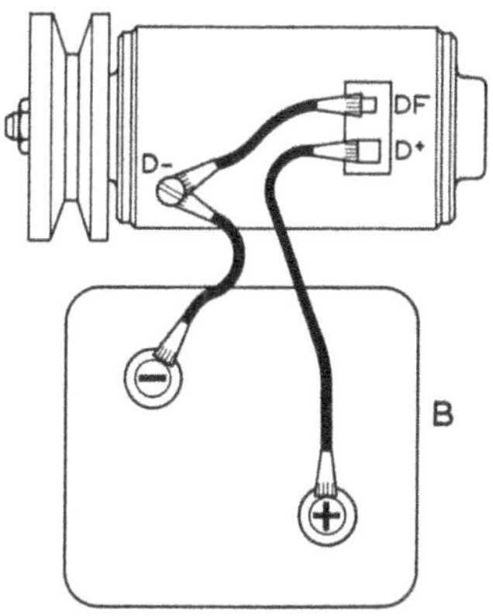

Fig. 6-6. Battery wired to polarize the generator.

You do not need to polarize the generator unless it is a new generator, a rebuilt generator, or a used generator

4

that has been stored for several weeks or more. Also polarize generators that have undergone electrical testing while disassembled. During polarization, the generator should be allowed to run briefly as a motor in order to give the pole shoes residual magnetism of the correct polarity. Without residual magnetism, the generator will not generate enough electricity to close the cut-out relay in the regulator when the engine is started.

If you must disassemble the generator before you install it, or if you must replace the fan, you can remove the fan from the shaft by taking off the special nut shown in Fig. 6-7. Two nuts hold the fan cover to the generator.

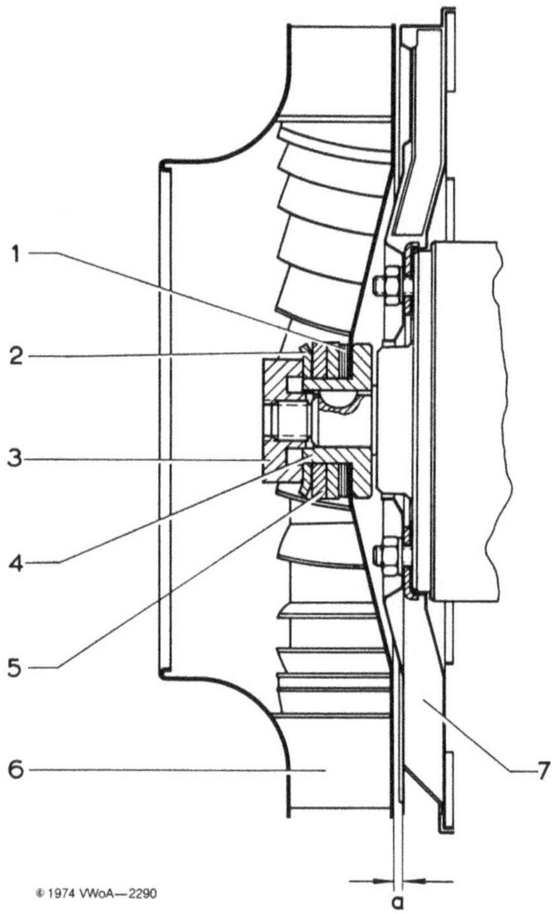

1. 0.50-mm (.019-in.) shims
2. Lock washer
3. Special nut
4. Hub
5. Carrier plate (2)
6. Fan
7. Fan cover
a. Distance between fan and fan cover

Fig. 6-7. Fan installation. During assembly, arrange shims to obtain a 2-mm (.080-in.) clearance at **a**. On generator 211 903 031 D, the clearance can be 1.20 to 2.10 mm (.047 to .082 in.). To obtain this clearance some shims must be placed between the fan and the hub flange.

Always install the fan cover in its original position on the generator. When you install the fan on the generator shaft, arrange the 0.50-mm (.020-in.) shims so that the clearance between the fan and the fan cover is 2 mm (.080 in.). Make certain that you install the lock washer so that its raised periphery is toward the special nut. Torque the special nut to 5.5 to 6.5 mkg (40 to 47 ft. lb.).

Install the generator in its original position on the fan housing. After you have installed the generator on the fan housing, hand-turn the shaft to make sure that the fan is not rubbing against the fan housing. If the shaft turns freely without noise, you have done the job correctly. If the fan rubs, either the fan housing is distorted or the fan shims were incorrectly installed. Install the fan housing and other parts as described in **ENGINE AND CLUTCH**.

6.5 Replacing Generator Brushes and Servicing Commutator

You can remove the generator brushes through the inspection ports in the generator housing. Lift the brush springs with a wire hook so that the old brushes can be slipped out of their holders. To free their connecting wires, remove the terminal screws. The generator brushes should be replaced whenever they are so worn that they no longer protrude above the tops of the brush holders.

Clean the commutator with trichloroethylene or a similar solvent before installing the new brushes. Bent or rusted brush springs should be replaced. If commutator out-of-round exceeds 0.015 mm (.0006 in.), the generator must be disassembled so that the commutator can be machined. The normal diameter of the commutator is 33.40 mm (1.315 in.). It should not be turned down below 32.80 mm (1.291 in.) when machined. After machining, undercut the insulation between the commutator bars until it is about 0.50 mm (.020 in.) below the surface of the bars themselves.

6.6 Disassembling and Assembling Generator

You may find it necessary to disassemble the generator for electrical tests, commutator servicing, or replacement of bearings or other internal parts. For basic disassembly, simply remove the two nuts that hold the fan cover to the generator and then withdraw the two long through bolts. Further disassembly can be done with only a screwdriver, although it may be necessary to use a press or bearing puller to remove the ball bearing races from the armature shaft. Complete generator disassembly is illustrated in Fig. 6-8.

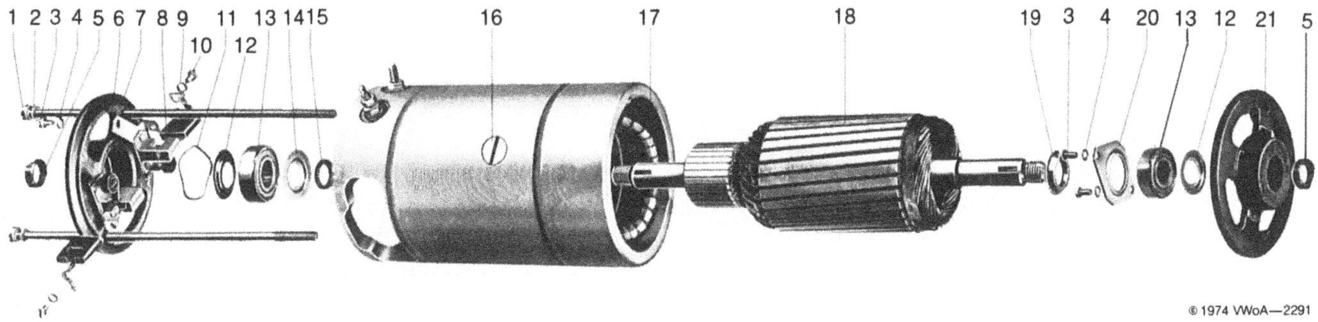

© 1974 VWoA—2291

1. Housing through bolt (2)
2. B 6 lock washer (2)
3. AM 4 × 8 fillister head screw
4. B 4 lock washer (3)
5. Spacer ring (2)
6. Brush spring (2)
7. End plate, commutator end
8. Carbon brush (2)
9. B 4 lock washer (2)
10. AM 4 × 6 fillister head screw
11. Lock washer
12. Dished washer (2)
13. Ball bearing (2)
14. Splash shield
15. Thrust washer
16. Pole shoe screw (2)
17. Field coil (2)
18. Armature
19. Splash shield
20. Retaining plate
21. End plate, fan end

Fig. 6-8. Generator disassembly. Remove through bolts to separate armature and end plate from housing.

Using a press and suitable support plates, you can press the armature out of the end plate as shown in Fig. 6-9. The cutout in the thin support plate should have a radius of 11 mm (7/16 in.) so that the bearing will not be damaged. The same support plate can be used for pressing the bearing off the commutator end of the armature.

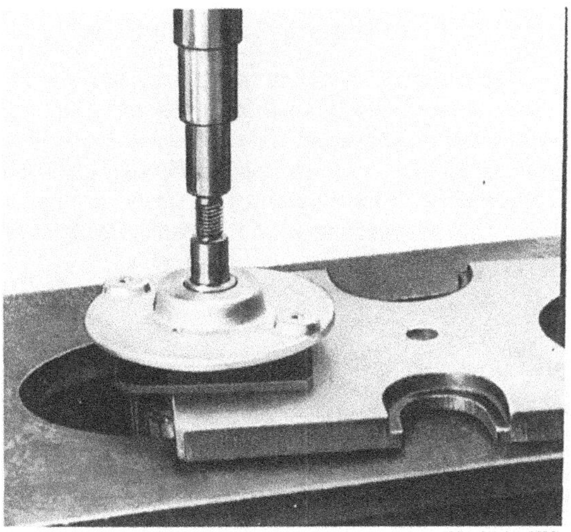

© 1974 VWoA—2293

Fig. 6-9. Pressing armature shaft out of end plate at fan end. Same support plate set-up can be used to press bearing off opposite end of armature.

To assemble:

1. Check the brushes and commutator for wear. If necessary, service them as described in **6.5 Replacing Generator Brushes and Servicing Commutator.**

NOTE —
The commutator should be smooth, grayish black, and clean. If it is greasy, clean it with trichloroethylene or a similar solvent.

2. Press the bearing onto the commutator end of the armature. Then press on the end plate at the fan end, as shown in Fig. 6-10.

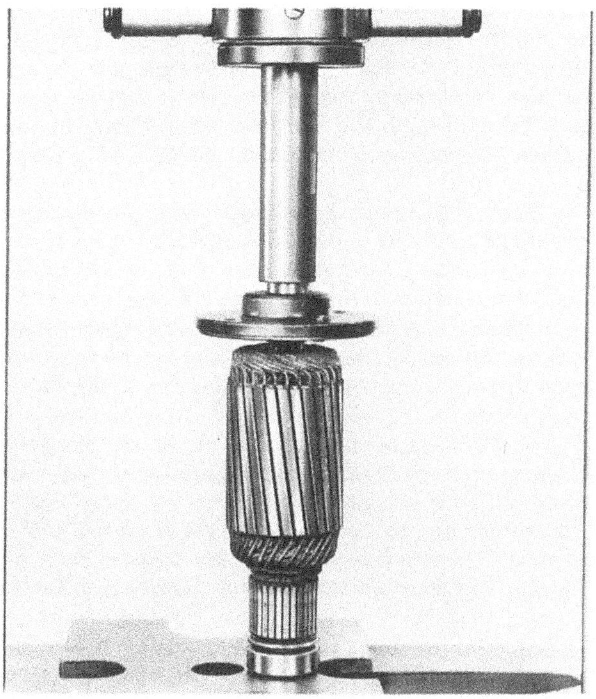

© 1974 VWoA—2295

Fig. 6-10. Pressing end plate onto armature.

3. Using a repair press, install the spacer rings.
4. To ensure alignment, fit the end plate notches over the projections on the housing as the parts are assembled. See Fig. 6-11.

© 1974 VWoA—2296

Fig. 6-11. Notch (arrow) that must fit over projection on housing. The armature must turn freely following installation of the through bolts.

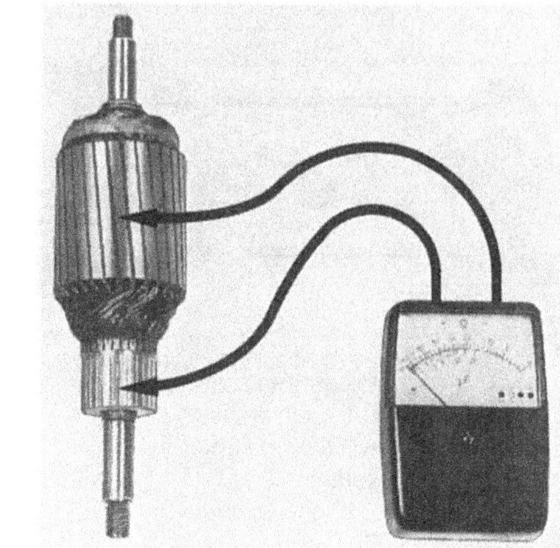

© 1972 VWoA—622

Fig. 6-12. Armature ground test. If testing between commutator and armature body or shaft produces an ohmmeter reading or causes a battery-powered test lamp to light, windings are grounded.

6.7 Testing Disassembled Generator

If you have tested the charging system and have determined that trouble lies in the generator, further tests can be made after the generator has been removed and disassembled. Several tests can be made with a simple battery-powered test light, although an ohmmeter is preferable. A resistance-measuring bridge is recommended for testing the field windings for shorts. However, an ammeter used in conjunction with a battery can also be used. To test the windings with an ammeter, compare the current draw through the two field coils. If one coil consumes 0.5 amperes or more than the other, it is shorted.

Check the armature for open circuits, shorts, and grounded windings. Burn marks between the commutator segments usually indicate open circuits, which often result from windings that have come unsoldered from their commutator bars. Such unsoldered windings are the result of overheating caused by a faulty regulator. An electrical specialist may be able to repair the armature.

Short circuits occur when the insulation between armature windings breaks down. It is generally best to replace shorted armatures. An armature tester called a growler is used to test for shorts. Growlers are available at most automotive supply shops that offer machine shop service. The shop will test an armature for you at low cost.

Armature grounds occur when insulation breaks down and allows the windings to come into electrical contact with the armature laminations or the shaft. It is easy to detect this condition with a battery-powered test light or with the ohmmeter shown in Fig. 6-12.

Occasionally, carbon dust from the brushes will short an armature. If arcing current has not permanently damaged the insulation, cleaning with trichloroethylene or a similar solvent will usually cure such shorts.

Testing Field Windings

Shorted field windings result from faulty insulation that allows one strand of wire to come into electrical contact with another. Because shorts reduce the strength of the magnetic field, they limit generator output. Shorts may be detected by the ammeter-and-battery test described earlier, or by the test shown in Fig. 6-13. For this test a resistance measuring bridge is essential.

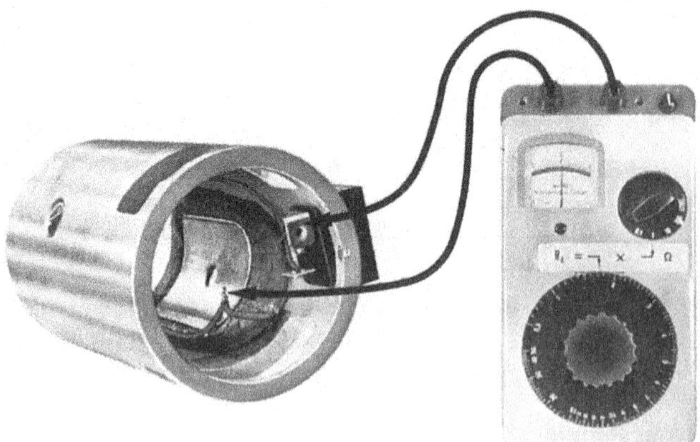

© 1972 VWoA—624

Fig. 6-13. Testing for shorted windings. A resistance measuring bridge should be used to test each field coil. If one coil has lower resistance than the other, it is shorted.

In addition to short circuits between strands of wire in the field coils, the generator windings can also become short-circuited to the generator housing. To locate such trouble, test between the field windings and the housing. These parts should not be in electrical contact. Although Fig. 6-14 shows an ohmmeter making the continuity test, a battery-powered test light will work equally well. The ohmmeter should produce no reading, and a test lamp should not light.

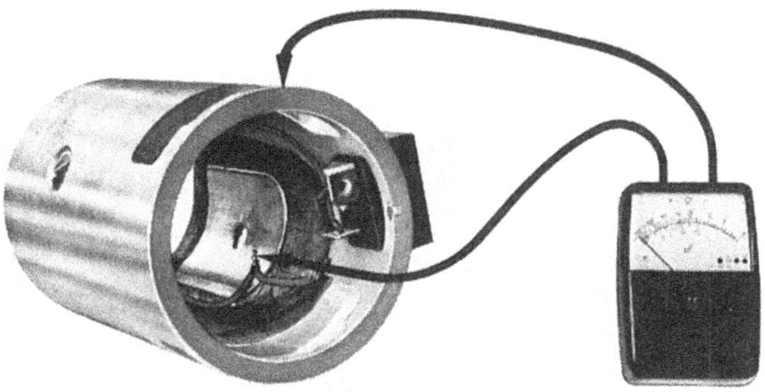

© 1974 VWoA—1197

Fig. 6-14. Testing for shorts to ground. If a test between the field coils and generator housing reveals electrical contact, the windings are shorted to ground.

If there is no charging system output, and if the tests described in **6.2 In-car Testing of Generator and Regulator** indicate that the generator is faulty, check the brushes and commutator. If they are in good condition, disassemble the generator and test for open field windings as shown in Fig. 6-15. Burned out windings usually result from excessive current loads caused by a faulty regulator or by defective charging system wiring. Repeat the in-car tests after repairing or replacing the generator to detect any remaining charging system faults.

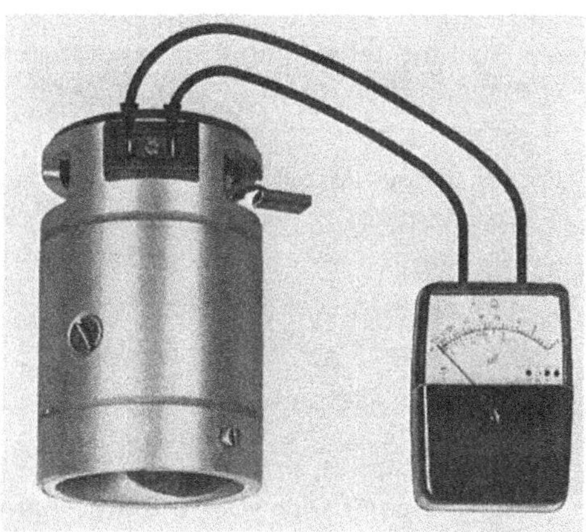

© 1972 VWoA—623

Fig. 6-15. Testing field coils for open circuits. Testing between two terminals should cause the ohmmeter to register or the test lamp to light. If it does not, the windings are open. Note that the positive brush must not touch the generator housing.

6.8 Alternator

All vehicles with the dual-carburetor engine—1972 and later models—have an alternator. Only one alternator type is used, VW part No. 021 903 023 A.

Fig. 6-16 shows the hook-up for testing alternator output. The positive battery cable must be disconnected and a battery cutout switch installed as shown with the cable reconnected to the cutout switch.

> ***CAUTION —***
> *An alternator must never be run without a battery. Doing so will severely damage the alternator, the regulator, or both.*

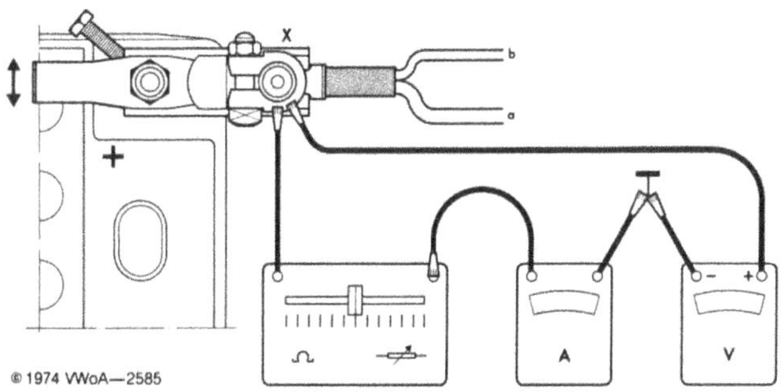

© 1974 VWoA—2585

a. To starter
b. To regulator
X. Battery cutout switch (SUN electric No. 7052-003 or similar equipment)

Fig. 6-16. Alternator charging system test. Variable resistance, ammeter, and voltmeter are connected in series as shown. Note that the ammeter lead and the negative voltmeter lead connected to it are both grounded to the chassis. The double arrow at the left indicates switch operation.

To test:

1. Make the test connections while the alternator is stationary. Then close the cutout switch and start the engine.
2. Run the engine at 975 rpm (2200 alternator rpm). Adjust the variable resistance so that the ammeter gives a reading of about 30 amps.
3. Move the cutout switch in order to cut the battery out of the test circuit. The load current is now determined by the variable resistance.
4. Readjust the variable resistance so that the ammeter reading is approximately 36 amps with the engine at about 975 rpm (2200 alternator rpm).

4

5. Read the voltage indicated on the voltmeter. It should be approximately 14 volts.

If there are deviations from the prescribed voltage under load, install a new regulator and repeat the test. If there are still deviations, replace the alternator. Though some individual replacement parts are available, it is best to replace faulty alternators as a unit.

Removing and Installing Alternator

The procedure given here covers only the removal of the alternator. The procedure for replacing the alternator belt is given in **ENGINE AND CLUTCH**.

To remove:

1. Remove the spring-ring type hose clamp from the oil filler. Then remove the oil filler neck.
2. Remove the engine oil dipstick. Slide the rubber grommet up and off the dipstick tube.
3. Remove the fillister head screws that hold the right rear part of the engine cover plate. Then slide the plate up and off the oil filler pipe together with the rubber oil filler pipe grommet.
4. Remove the oblong insert from the alternator cover plate. Then, using a 12-point driver, loosen the socket head screw indicated in Fig. 6-17.

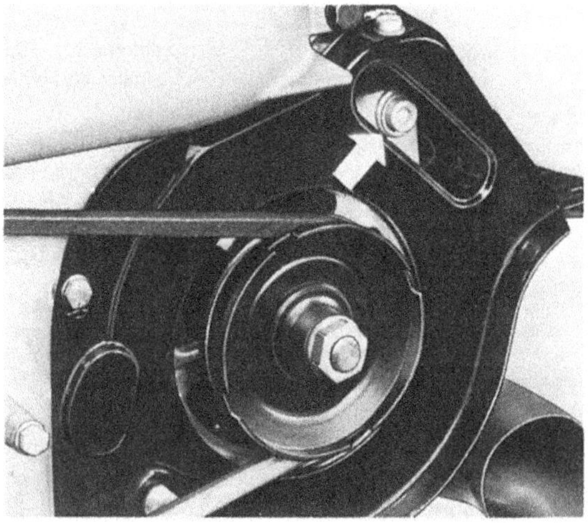

© 1974 VWoA—2937

Fig. 6-17. Socket head bolt (arrow) that holds upper side of alternator.

5. Loosen the lower alternator mounting bolt. Then swing the alternator to the left and unhook the drive belt from the alternator pulley.
6. Fully remove the lower alternator mounting bolt and then take off the alternator cover plate.
7. Remove the wiring harness and its grommet from the engine sheet metal parts only to the extent that is necessary to remove the wires from the alternator.
8. Carefully remove the alternator cooling air rubber elbow from the fan housing.
9. Remove the screws for the alternator tensioning arm bracket (Fig. 6-18). Then remove the alternator together with the tensioning arm bracket.

© 1974 VWoA—2939

Fig. 6-18. Screws that hold the alternator tensioning arm bracket.

To install:

1. Position the alternator on the engine. Make sure that the grommet for the wiring harness is correctly installed as you attach the wiring.
2. Install the tensioning arm securing bracket on the engine. Then insert the upper generator mounting bolt.
3. Temporarily install the lower mounting bolt. Insert the alternator cooling air rubber elbow into the fan housing.
4. Remove the lower mounting bolt, install the alternator cover plate, and then reinstall the bolt.
5. Hook the alternator drive belt over the alternator pulley. Adjust the alternator position until the belt yields 15 mm (5/8 in.) when you depress it firmly near the mid-point with your thumb. Then tighten both alternator mounting bolts.
6. Install the oblong insert in the alternator cover plate. Install the engine cover plate together with the rubber grommet for the oil filler pipe.
7. Install the rubber grommet for the dipstick tube. Then install the dipstick and the oil filler neck.

7. REMOVING AND INSTALLING TURN SIGNAL SWITCH AND IGNITION/STEERING LOCK

There have been a number of modifications to the ignition/steering lock during the model years covered by this Manual. Differences will be pointed out in the course of the procedures.

To remove 1968 through 1974 switch:

1. Disconnect the battery ground strap. Working under the vehicle, disconnect the horn wire from the terminal that is on the column tube, just above the steering gearbox flexible coupling.
2. Remove the two Phillips head screws that hold the turn signal switch on the steering column tube. Remove the turn signal switch and, on 1973 and later models, the windshield wiper switch. Let the switches hang by their wires. (See **10.7 Removing and Installing Wiper Switch.**)
3. Carefully pry off the horn button. Then remove the M 18 nut that holds the steering wheel on the steering column.
4. Remove the screws that hold the turn signal canceling ring on the bottom of the steering wheel. Allow the ring to slide down the steering column. Then, using a puller such as the one shown in Fig. 7-1, remove the steering wheel.

© 1974 VWoA—2954

Fig. 7-1. Steering wheel being pulled off column.

5. Remove the insulating ring, circlip, and washer that are on top of the rubber bushing. The bushing is fitted between the steering column tube and the ignition/steering lock housing.
6. On vehicles with a steering column lock, turn the ignition key to the drive position.
7. Lubricate the rubber bushing with talcum powder or silicone spray. Then, using two screwdrivers, carefully pry the bushing out of the column tube and lock.
8. On 1973 and later vehicles only, bleed the air from the windshield washer reservoir. Then disconnect the hoses from the windshield washer valve.
9. Working behind the dashboard, disconnect the wires for the turn signal switch, the ignition/steering lock and, on 1973 and later cars, the wiper switch.
10. Remove the screws and bolts indicated in Fig. 7-2. Then slide the ignition/steering lock housing up and off of the steering column tube.

© 1974 VWoA—2955

Fig. 7-2. Phillips head screws (arrows) and bolts that hold the lock housing.

11. To remove the ignition/starter switch from the lock body, remove the screw indicated in Fig. 7-3.

© 1974 VWoA—2967

Fig. 7-3. Ignition/starter switch retaining screw (arrow). On 1970 and later models, the housing cannot be turned to one side as shown, but must be removed as described in the preceding steps.

4

12. To remove the ignition/steering lock from the lock housing (1971 and later models only), remove the two screws indicated in Fig. 7-4. On earlier models, replace the lock together with the housing.

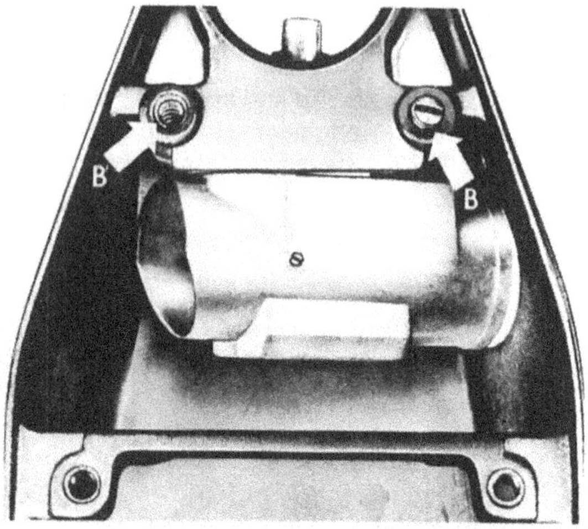

© 1974 VWoA—2966

Fig. 7-4. Screws (arrows **B**) that hold the late-type lock in the lock housing.

To install 1968 through 1974 switch:

1. If you removed the ignition/steering lock body from the housing, install the lock body in the housing and tighten the screws.
2. Place the lock housing on the column tube. Then reconnect the wires for the ignition/steering lock, the turn signal switch and, on 1973 and later models, the windshield wiper/washer switch.
3. Install the bolts and screws that hold the lock housing on the steering column mounting bracket.

> ***CAUTION —***
> *On 1970 and later models, you must torque the screws for the bracket to exactly 1.5 mkg (11 ft. lb.). Otherwise, the energy absorbing steering column will not collapse properly under impact. For additional information, see* **FRONT AXLE.**

4. Install the rubber bushing, washer, circlip, and insulating ring.
5. Attach the canceling ring to the steering wheel. Then, with the steering centered as described in **FRONT AXLE**, install the steering wheel so that the steering wheel spokes are horizontal and the tongue on the canceling ring is on the left.
6. Install the M 18 steering wheel nut and torque it to 2.5 to 3.0 mkg (18 to 22 ft. lb.).
7. Reconnect the horn wire.
8. Install the turn signal switch and, on 1973 and later vehicles, the windshield wiper/washer control. The clearance between the switch housing and the steering wheel should be 2.00 mm (.080 in.) as shown in Fig. 7-5.

© 1974 VWoA—2956

Fig. 7-5. Clearance (dimension **a**) between turn signal switch and steering wheel.

To remove 1975 and later switches:

1. Disconnect the battery ground strap. Bleed all air pressure from the windshield washer fluid reservoir.
2. Carefully pry the horn button out of the center of the steering wheel.
3. Remove the nut that holds the steering wheel to the steering column. Then lift off the steering wheel.
4. Remove the spring and the spacer from the top end of the steering column.
5. Remove the four screws that hold the turn signal switch and the wiper switch to the lugs on the steering lock housing.

> **NOTE —**
> Be careful not to lose the four spacer tubes that are between the switches. The four switch mounting screws pass through the centers of these spacer tubes.

6. Disconnect the water hoses from the water valve on the wiper switch. Disconnect the wires for the turn signal switch and the wiper switch.
7. Remove the turn signal switch, the wiper switch, and the switch housing upper trim from the steering column.
8. Locate the round hole that is in the lower left-hand side of the lower part of the switch housing trim. The hole provides access to the shear bolt (Fig. 7-6).

9. Working through the access hole, drill into the center of the shear bolt. The hole that you drill should be at least 7 mm (or ¼ in.) deep.

CAUTION —
To prevent the drill bit from wandering, which could damage the steering lock housing, carefully centerpunch the shear bolt before you attempt to drill it. Select a drill size that is appropriate to the diameter of the shear bolt and to the diameter of an available left-hand thread cutting tap.

10. Insert a left-hand thread cutting tap into the drilled shear bolt. Then turn the tap counterclockwise in order to remove the shear bolt.

11. Insert the key into the ignition lock and unlock the steering column.

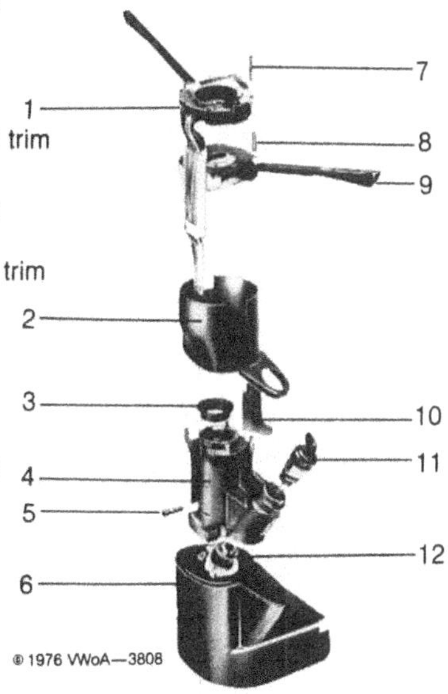

1. Turn signal switch
2. Switch housing upper trim
3. Plastic bushing
4. Steering lock housing
5. Shear bolt
6. Switch housing lower trim
7. Screw
8. Spacer tube
9. Wiper switch
10. Lock cover plate
11. Lock cylinder
12. Ignition/starter switch

Fig. 7-6. Exploded view of steering column switches.

12. Working beneath the switch housing lower trim, remove the two Phillips head screws that hold the trim to the bottom of the steering lock housing.

13. Remove the steering lock by sliding it up and off the steering column.

14. If necessary, remove the ignition/starter switch mounting screw. Then remove the ignition/starter switch from the steering lock housing.

15. To remove the lock cylinder, use pliers to pull the lock cover plate out of the steering lock housing. This will provide access to the hole for lock cylinder removal.

You may find that on 1976 and later vehicles the hole used to release the retaining spring is missing. A hole must be drilled in the switch housing, as indicated in Fig. 7-7. Use a piece of steel wire inserted in the hole to depress the spring while you remove the cylinder.

CAUTION —
Remove the turn signal and windshield wiper switches before you attempt to drill the hole. Use a drill with a 3-mm (or ⅛-in.) diameter.

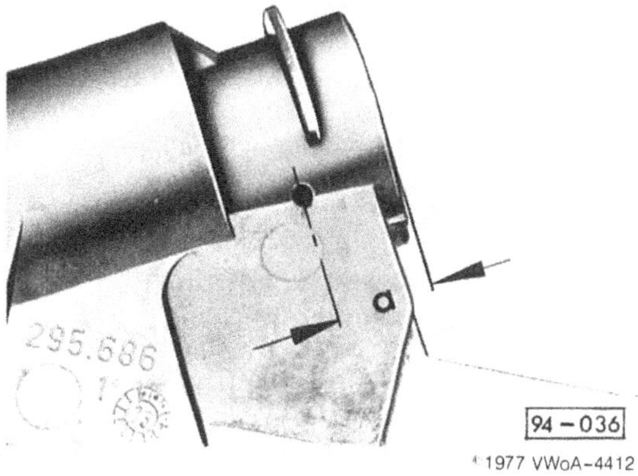

Fig. 7-7. Location for hole in ignition switch housing. Dimension **a** is 13 mm (just over ½ in.).

Installation is the reverse of removal. Loosely install a new shear bolt but do not tighten it until you have installed the steering wheel. Center the steering. Then install the steering wheel so that its spokes are horizontal. Torque the steering wheel nut to 5.0 mkg (36 ft. lb.).

4

With the steering wheel installed, adjust the gap between the bottom of the steering wheel and the face of the turn signal switch to 2 to 3 mm (¹⁄₁₆ to ⅛ in.). Then tighten the shear bolt until its head breaks off. Install the horn button so that the crest is upright when the steering wheel spokes are horizontal.

8. LIGHTS AND SWITCHES

To replace bulbs in the tail lights, license plate light, front turn signals, or side marker lights, remove the Phillips head screws in the lens. See Fig. 8-1.

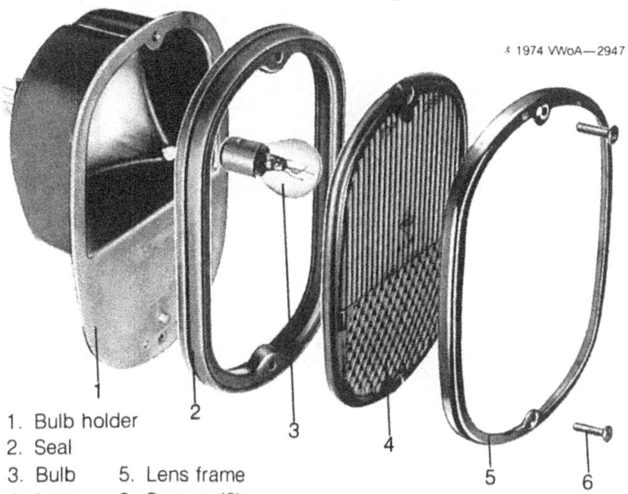

1. Bulb holder
2. Seal
3. Bulb
4. Lens
5. Lens frame
6. Screws (2)

Fig. 8-1. Typical light assembly.

When installing the lens, make certain that you install the seal properly and that it is in serviceable condition. Instrument panel light bulb replacement is covered in **13.3 Removing and Installing Instruments.**

With the exception of the interior lights, all bulbs are of the bayonet base type. To remove a bayonet base bulb, press the bulb in and turn it counterclockwise until it stops. Then pull out the bulb. Installation is the reverse of removal.

8.1 Replacing Sealed Beam Unit and Aiming Headlights

You can remove the headlight trim ring by loosening a single screw in the edge of the ring. The screw is permanently installed in the trim ring and can also be used as a handle to pull the trim ring off the car. Remove the sealed beam unit and retaining ring by taking out the three short screws. You may then pull the sealed beam unit forward and remove the cable connector from its terminals (Fig. 8-2).

CAUTION —
Do not alter the position of the long headlight adjustment screws. If you do, you will have to readjust the headlights.

Fig. 8-2. Sealed beam removal.

When you install a sealed beam unit, be sure its three glass lugs correctly engage the support ring. Install the retaining ring, then hook the trim ring over the lug at the outer side of the body recess. Make certain that the rubber gasket is properly positioned before you tighten the trim ring mounting screw. It should not be necessary to aim the headlights after changing a sealed beam unit. If you are in doubt, however, the aim should be checked.

Each headlight has two adjusting screws. Adjust lateral aim with the screw **A** shown in Fig. 8-3. Adjust vertical aim with screw **B.**

Fig. 8-3. Headlight being adjusted.

To adjust:

1. Position the car on a level surface 7.62 m (25 ft.) from a vertical wall. Have the fuel tank about half-filled and make certain that the tire pressures are correct.
2. Roll the car back and forth a few yards to settle the suspension. Then load the driver's seat with one person or a weight of 70 kg (154 lb.).
3. Remove the headlight trim rings. Turn on the low beams and cover one of the headlights. The uncovered light's upper and left edges of high intensity must be in the position shown in Fig. 8-4.

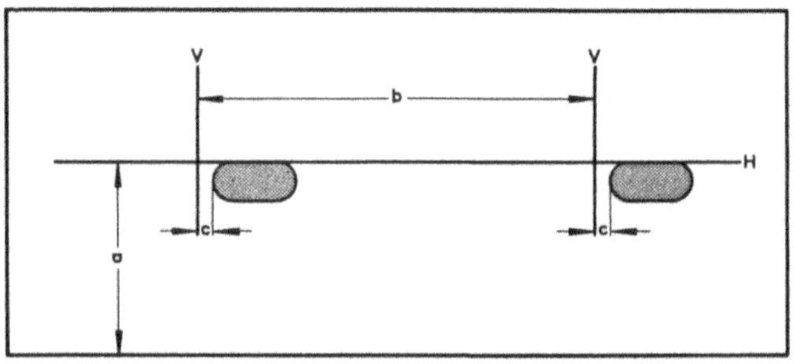

Fig. 8-4. Aiming target on vertical wall. Shaded areas are zones of high light intensity. Vertical lines (**V**) and horizontal line (**H**) intersect at the headlight centers. Dimension **a** is the headlights' height above ground and **b** is the 1080-mm (42½-in.) width between the headlights. Dimension **c** equals 50.8 mm (2 in.).

4. Repeat procedure 3 on the opposite headlight and adjust as necessary. The high beams will automatically be in adjustment once the low beams are aimed to the proper specifications.

NOTE —
Check your state laws to determine whether adjustments must be made by a licensed shop. Your state may also have laws specifying different aiming from that described here.

8.2 Removing and Installing Switches in Dashboard

The dashboard-mounted switches installed in Type 2 vehicles are either push-pull or rotary type.

To remove switch:

1. Disconnect the battery ground strap.
2. Unscrew the knob from the switch.
3. Using a special wrench, remove the escutcheon that holds the switch in the dashboard. See Fig. 8-5.
4. Reaching under the dashboard, pull the switch out of the dashboard toward the front of the car.
5. Labeling each wire so that you will be able to return it to its correct terminal during installation, disconnect all wires from the switch.

Installation is the reverse of removal. Make sure the notches in the switch body engage the projections in the hole in the dashboard.

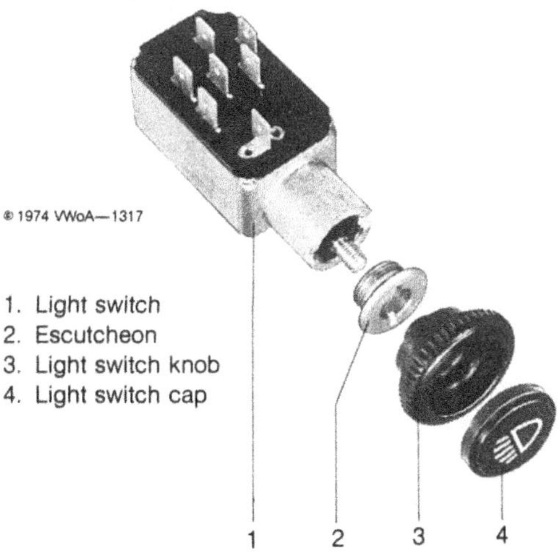

Fig. 8-5. Typical switch. This is a main lighting switch but others are removed similarly. On 1968 through 1972 models, bleed the air from the windshield washer reservoir before taking hoses off wiper switch.

Removing Door Contact Switches

You can remove the door contact switches as illustrated in Fig. 8-6.

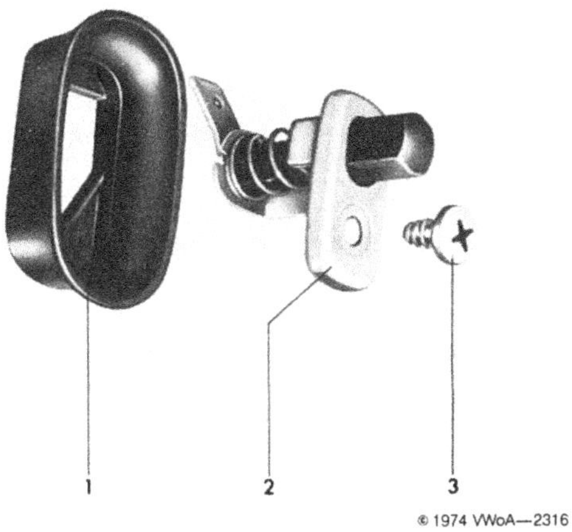

Fig. 8-6. Door contact switch removal. Remove screw (**3**), withdraw switch (**2**), and disconnect wire. Seal (**1**) comes off along with the switch.

4

8.3 Dimmer Relay

Beginning with the 1970 models, Type 2 vehicles have a fuse box with an integral relay console. To remove the dimmer relay, bend back the clip, then pull the fuse box out of its bracket complete with wires (Fig. 8-7). The dimmer relay is the larger of the two cube-shaped relays on the back of the fuse box. Obtain replacement parts with reference to the part number on the original relay.

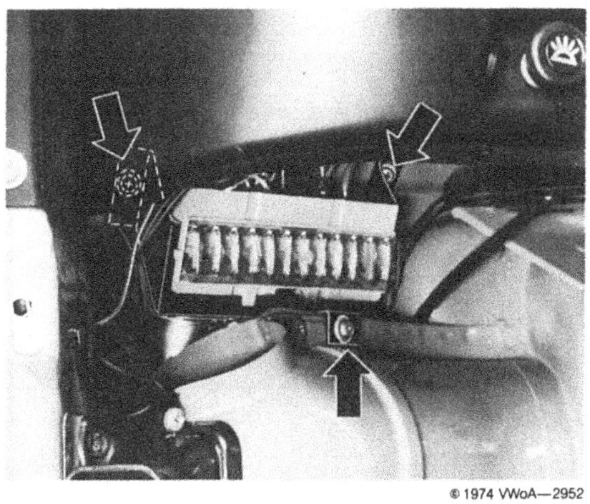

Fig. 8-7. Fuse box with relay console. Arrows indicate screws that hold the fuse box bracket on the car.

On 1968 and 1969 models, the dimmer relay is attached to the bottom of the steering column support. It is an oblong box with four wires attached to it. (See Fig. 9-1.)

9. TESTING TURN SIGNALS AND EMERGENCY FLASHERS

The turn signal switch and relay are often mistakenly blamed for troubles caused by dirty, corroded, or loose-fitting turn signal bulb contacts. Before starting to troubleshoot either the turn signal switch or the turn signal/emergency flasher relay, be sure that all bulb contacts are clean and tight.

The turn signals and emergency flashers share the same relay. If the emergency flashers work but the turn signals do not, you can be certain the relay is not faulty. The same is true if the turn signals work but the emergency flashers do not. The test described here will keep you from unnecessarily replacing a good relay.

Testing Turn Signal/Emergency Flasher Relay

With the ignition off, connect terminal + or −49 (depending on the relay manufacturer's designation) of the relay to terminal 30 of the fuse box. Operate the turn signal switch both ways. If the turn signals on both sides light up, the relay and turn signal switch are not defective. If they do not light up, test first the emergency flasher switch and then the turn signal switch.

To test emergency flasher switch:

1. Disconnect the battery ground strap.
2. Remove the emergency flasher switch and turn it on.
3. With an ohmmeter, test to make sure there is continuity between terminals 30 and + and also between terminal 49a and terminals **R** and **L**.
4. If the resistance is greater than zero ohms, replace the emergency flasher switch.

To test turn signal switch:

1. At the plug guide, disconnect the black/green/-white wire to the turn signal switch.
2. With an ohmmeter, test for continuity between terminal 54BL and terminals **R** and **L** of the turn signal switch.
3. If the resistance is greater than zero ohms, replace the turn signal switch.

Replacing Turn Signal/Emergency Flasher Relay

The relay for the turn signals/emergency flashers is located near the headlight dimmer relay. It is the largest relay on the 1970 and later fuse box/relay console (see **8.3 Dimmer Relay**). Its position on 1968 and 1969 models is shown in Fig. 9-1. The silvery-colored relay in front of it is the dimmer relay.

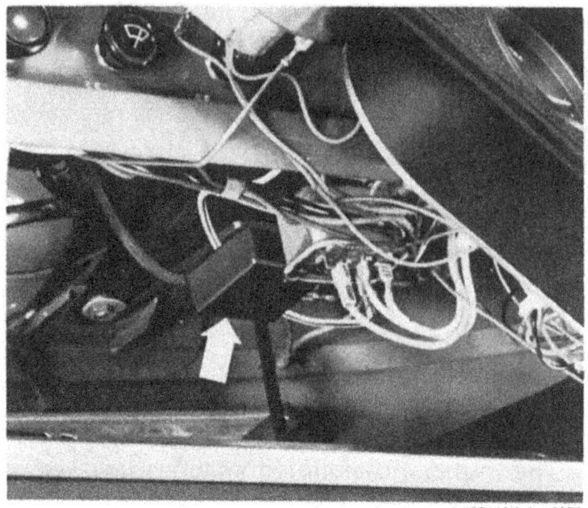

Fig. 9-1. Turn signal/emergency flasher relay (1968 and 1969 models). The arrow indicates the flasher relay. The silvery relay in front of the flasher relay is the dimmer relay.

Always disconnect the battery ground strap before you remove the relay. On early models, attach identifying tags to each wire as it is removed so that you will be able to return it to the proper terminal on the new relay. Defective relays cannot be repaired and must be replaced.

Removing Turn Signal Switch

The turn signal switch and, on 1973 and later vehicles, the windshield wiper/washer control can be removed without taking off the ignition/steering lock. However, because the turn signal switch must be removed before you can remove the ignition/steering lock, the removal of the turn signal switch is covered in **7. Removing and Installing Turn Signal Switch and Ignition/Steering Lock.**

10. WINDSHIELD WIPERS AND WASHER

Operation of the two-speed windshield wipers is controlled by turning a rotary switch on 1968 through 1971 cars, or by depressing a lever on the steering column on later models. The blades park automatically when the wipers are turned off. A push button in the center of the rotary switch actuates the windshield washers. On 1972 and later models, lifting the column-mounted lever toward the steering wheel triggers the washers. The washers operate by compressed air stored in the washer fluid reservoir.

CAUTION —
In servicing the windshield washer system, always bleed the air pressure from the fluid reservoir before disconnecting any of the hoses. Otherwise, fluid will be expelled from the reservoir.

10.1 Windshield Wiper Motor Troubleshooting

Table i is designed to help you determine the cause and remedy for wiper motor malfunctions. The numbers in the Remedy column refer to headings in this section where the suggested repairs are described.

In-car Testing

In addition to analyzing operational faults, you can evaluate the wiper motor by checking the current draw (at terminal 30) at both low and high speeds. The current draw should be about 2.5 amps at low speed and 3.5 amps at high speed. The test should be made with the wiper blades pulled away from the windshield. Otherwise, friction will cause your ammeter readings to be inaccurate.

If the wipers make squeaking noises and run slowly with a high current draw, the probable cause is inadequate bearing lubrication. If current draw is high without squeaking noises, the wiper motor's armature is probably shorted.

Armature and brush sets are available as replacement parts. If any other part of the wiper motor and drive is faulty, the entire wiper motor assembly must be replaced. When obtaining replacement armatures, make sure that the worm threads on the new armature have the same pitch and spacing as those of the original armature. Two worm thread dimensions have been used—both by the same manufacturer. If you install the wrong armature, it will damage the drive gearshaft and the entire wiper motor would have to be replaced.

10.2 Removing and Installing Wiper Motor and Frame

Though most of the components for the windshield wiper mechanism have undergone minor changes during the years covered by this Manual, the basic design remains the same.

To remove motor and frame:

1. Disconnect the battery ground strap.
2. Remove the wiper arm cap nuts. Then take off the wiper arms.
3. Remove the seals for the wiper shafts. Then remove the nuts, washers, and outer bearing seals.
4. Working under the dashboard, remove the flexible heater branch connections.
5. On 1968 through 1972 models, disconnect the wiper motor wiring harness from the wiper switch. Attach identifying tags to each wire as you remove it so that you will be able to reconnect the wire to its original terminal. On 1973 and later models, discon-

Table i. Windshield Wiper Motor Troubleshooting

Problem	Probable Cause	Remedy
1. Windshield wiper motor does not work, operates too slowly, cuts out, or comes to a standstill	a. Brushes worn b. Brush tension spring too weak c. Brushes stuck in their holders d. Commutator dirty e. Moving joints of windshield wiper linkages dry or jammed f. Battery voltage too low g. Armature burned by short circuits h. Switch faulty or wires loose	a. Replace brushes. See **10.5**. b. Replace tension spring. See **10.5**. c. Free brushes. See **10.5**. d. Clean commutator. See **10.5**. e. Thoroughly lubricate all moving joints with universal grease; eliminate jamming. See **10.3**. f. Charge battery; check cables and connections. See **4.1**, **4.3**. g. Replace motor or armature. See **10.2**, **10.4**, **10.5**. h. Replace switch; repair faulty connections. See **10.7**.
2. Windshield wiper motor continues to run or fails to return blades to parking position after manual switch is turned off	a. Contacts in cover damaged b. Contact mount (insulation plate) broken c. Contacts dirty d. Poor connection from terminal 31b via wiper switch to ground	a. Replace cover. See **10.5**. b. Replace cover. See **10.5**. c. Clean contacts. See **10.5**. d. Check connection; replace parts if necessary. See **10.7**.
3. Wiper linkage squeaks; motor operates slowly; armature is burned	a. Moving joints of windshield wiper linkages need grease b. Gear shaft lacks sufficient end clearance c. Drive housing not correctly positioned on motor	a. Thoroughly grease all moving joints with universal grease. See **10.3**. b. Adjust axial play. See **10.5**. c. Install cover properly. See **10.5**.

nect the wiper motor wiring harness from the wiper control switch wiring harness at the in-line connector.

6. Remove the mounting bolt from the wiper motor cover (early models) or the bracket on the wiper motor housing (late models). Then take the wiper motor and frame out downward and to the right (Fig. 10-1).

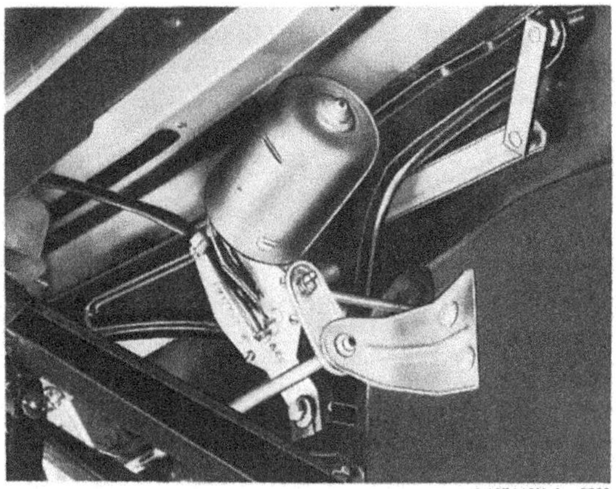

© 1974 VWoA—2329

Fig. 10-1. Wiper motor and frame in position under the dashboard. Pull the frame slightly to the rear to withdraw the wiper arm shafts from the front body panel.

Installation is the reverse of removal. Make sure that the wiper shafts are perpendicular to the surface of the windshield and that the bearing washers and seals are installed in their proper order. If in doubt, refer to Fig. 10-2. Read **10.6 Removing, Servicing, and Installing Wiper Blades and Arms** before you install the wiper arms.

When installing the wiper arms, torque the cap nuts (or the regular nut shown in Fig. 10-3) to 42 to 60 cmkg (37 to 52 in. lb.). If necessary, the correct installation of the wires can be determined by studying the wiring diagram presented elsewhere in this section.

© 1974 VWoA—2346

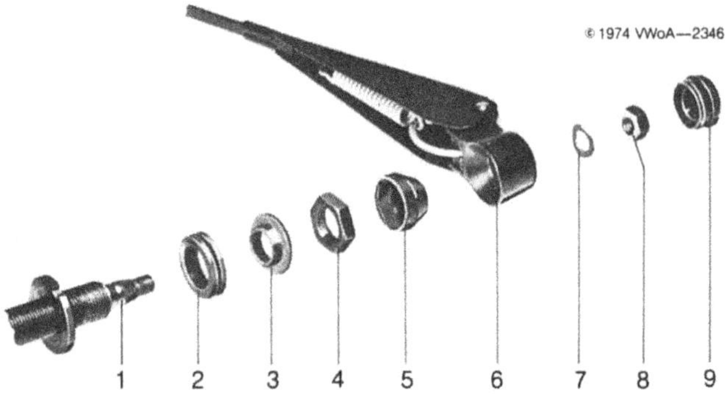

1. Wiper shaft
2. Seal
3. Guide
4. M 12 nut
5. Bearing seal
6. Wiper arm
7. Spring washer
8. M 6 nut
9. Plastic trim cap

Fig. 10-3. Wiper arm installation, 1973 and later models.

10.3 Replacing Wiper Shaft, Wiper Shaft Bearing, or Connecting Rod

You must remove the wiper frame in order to replace the bearing, shaft, or connecting rod. The wiper shaft, with its bearing and related parts, is shown in Fig. 10-2.

Fig. 10-2. Wiper shaft and bearing with related parts.

© 1974 VWoA—2976

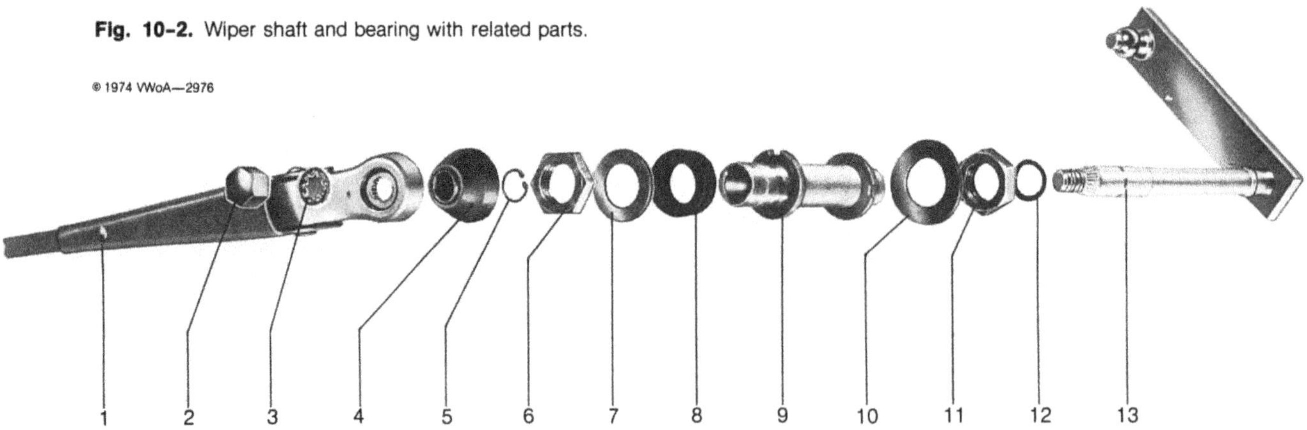

1. Wiper arm
2. Cap nut
3. Serrated washer
4. Bearing seal
5. Circlip
6. M 12 nut
7. Washer
8. Seal
9. Wiper shaft bearing
10. Large spring washer
11. Brass M 12 nut
12. Small spring washer
13. Wiper shaft

To remove:

1. Disconnect the battery ground strap, then remove the wiper frame and motor.
2. Using a screwdriver, carefully pry the connecting rod(s) off the ball pins on the wiper shaft(s) and drive crank.
3. Remove the circlip from the groove in the wiper shaft. Then withdraw the wiper shaft from the bearing complete with its small spring washer.
4. Remove the brass nut, then take the bearing out of the wiper frame.

Installation is the reverse of removal. Make sure that the groove in the bearing engages the projection on the wiper frame. Lubricate the wiper shaft ball pin and the drive crank ball pins with multipurpose grease before you install the connecting rods.

10.4 Removing and Installing Wiper Motor

You can remove the wiper motor from the frame for servicing or replacement, but only after you remove the frame—with the wiper motor still attached—from the car. Using a screwdriver, carefully pry the connecting rods off the ball pin on the drive crank. Mark the installation position of the motor on the frame, then remove the two mounting bolts and take off the motor.

During installation, align the motor with the marks. Lubricate the drive crank ball pin and spherical seats in the connecting rods with multipurpose grease. If the motor has been disassembled, or if you have removed the drive crank, adjust the parking position before you install the motor on the frame.

To adjust parking position:

1. Connect the wiper motor to the wiper switch according to the terminal designations.
2. Support the motor so that it will not fall.
3. Connect the ground terminal of the wiper motor and terminal 31 on the switch to the battery's negative pole. The parking position brake will not work unless these connections are made.
4. Connect terminal 30 on the switch to the battery's positive pole. Turn on the wiper switch and let the motor run for about 15 seconds.
5. Turn the switch off. The motor should stop in its correct parking position.
6. Install the drive crank. Position it parallel to the armature shaft when viewed from the end of the drive gearshaft (Fig. 10-4).

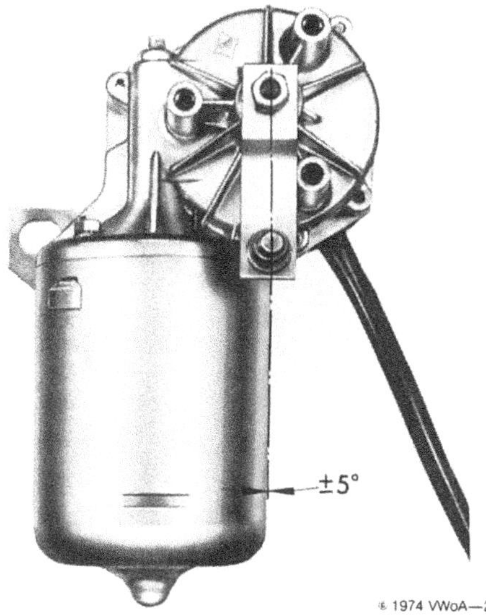

Fig. 10-4. Correct position of drive crank in the parking position. The axis of the crank can deviate from the side of the pole housing by an angle of ± 5°.

10.5 Disassembling and Assembling Wiper Motor

Armatures and brush sets are the only replacement parts available. The drive housing need not be taken apart if only the brushes and commutator require servicing.

To disassemble wiper motor:

1. Disconnect the battery ground strap, then remove the wiper frame and motor as described in **10.2 Removing and Installing Wiper Motor and Frame.**
2. Remove the motor from the frame as described in **10.4 Removing and Installing Wiper Motor.**
3. Remove the two hexagon head screws. While slowly turning the drive crank to free the worm gear, lift the drive housing off the motor.
4. Inspect the brushes. If they are oily, clean them with trichloroethylene or some similar solvent. If they are worn, replace them.

NOTE —
Although the brushes and brush springs can be removed from the brush holder for cleaning, the holder, springs, and brushes should be replaced as a unit if the brushes are worn. All necessary parts are included in the brush replacement set.

4

5. Lift the armature out of the pole housing. Inspect the commutator. See Fig. 10-5. If the commutator is oily, clean it with trichloroethylene or some similar solvent.

6. If the commutator is worn or burned, it can be turned down to a minimum diameter of 21.50 mm (.846 in.). Permissible commutator out-of-round is 0.03 mm (.001 in.). After turning, undercut the insulating strips between the commutator bars. Then clean away all metal particles to prevent shorts.

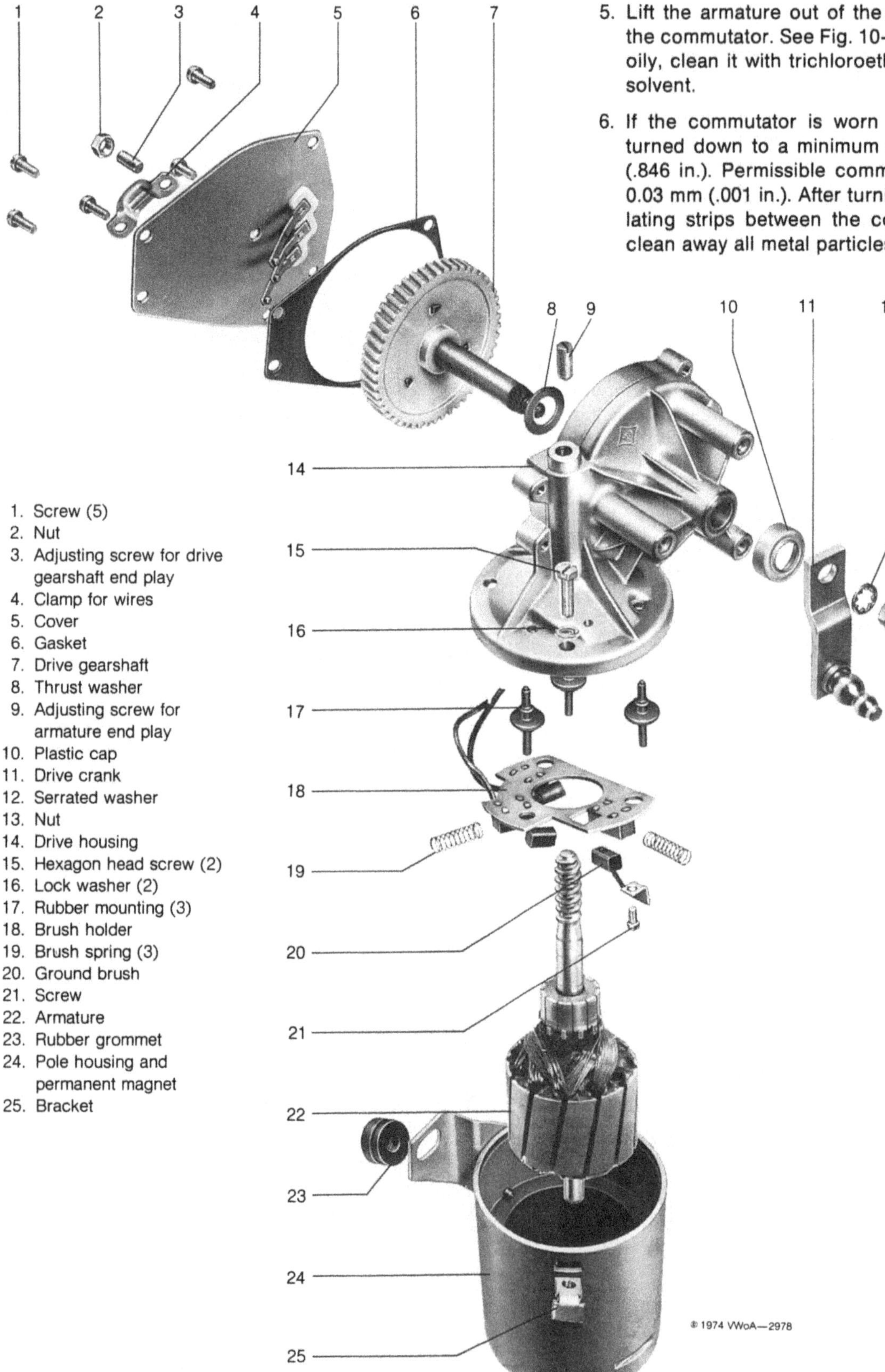

1. Screw (5)
2. Nut
3. Adjusting screw for drive gearshaft end play
4. Clamp for wires
5. Cover
6. Gasket
7. Drive gearshaft
8. Thrust washer
9. Adjusting screw for armature end play
10. Plastic cap
11. Drive crank
12. Serrated washer
13. Nut
14. Drive housing
15. Hexagon head screw (2)
16. Lock washer (2)
17. Rubber mounting (3)
18. Brush holder
19. Brush spring (3)
20. Ground brush
21. Screw
22. Armature
23. Rubber grommet
24. Pole housing and permanent magnet
25. Bracket

Fig. 10-5. Exploded view of windshield wiper motor and drive.

Because the wiper motor uses permanent magnets for its field poles, there are no field coils to test. Armatures can be checked using the test described in **6.7 Testing Disassembled Generator**. If the bearings in the pole housing are worn, replace the entire motor.

To disassemble drive:

1. Remove the five cover screws, then remove the cover and gasket.
2. Remove the drive crank nut and the serrated washer. Remove the drive crank and the plastic cap.
3. Remove the drive gearshaft together with the thrust washer.

Assembly of the wiper motor and drive is the reverse of removal. Use universal grease to lubricate the armature and drive gearshaft bearings.

The armature's end play should be 0.20 to 0.30 mm (.008 to .012 in.), and is adjusted with the adjusting screw in the drive housing near the end of the armature. The end play of the drive gearshaft should also be 0.20 to 0.30 mm (.008 to .012 in.) and is adjusted with the screw and locking nut at the end of the drive gearshaft.

Before installing the drive crank on the drive gearshaft, you must adjust the parking position as described in **10.4 Removing and Installing Wiper Motor**. If this is not done, excessive starting loads might be placed on the wiper motor.

10.6 Removing, Servicing, and Installing Wiper Blades and Arms

To remove a wiper blade, first fold the wiper arm away from the windshield. Then turn the blade as shown in Fig. 10-6 until it contacts its stop.

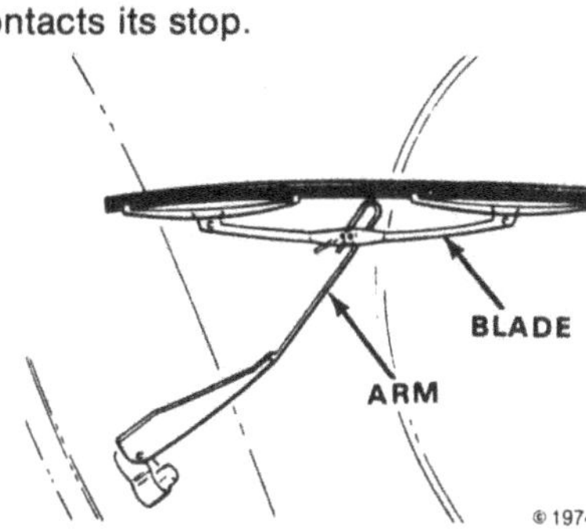

Fig. 10-6. Wiper blade turned to an angle with wiper arm.

Lift the retaining spring and slide the blade down the wiper arm until the hook of the arm is off the pivot pin. You may then lift the blade off upward. The sequence is illustrated in Fig. 10-7.

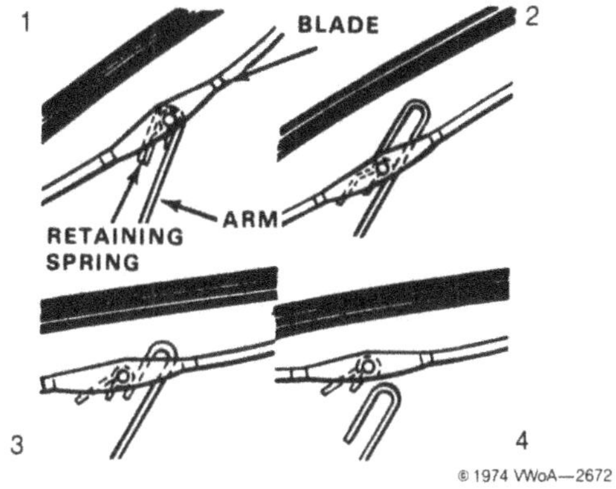

Fig. 10-7. Blade removal sequence.

To install wiper blade:

1. Position the new blade on the arm. The arm must enter through the hole in the blade frame on the side opposite the retaining spring.
2. Slide the blade down the arm until you can slip the hook on the arm over the pivot pin.
3. Pull the blade upward until the retaining spring is fully enclosed in the hook, then place the wiper against the windshield. For installation, reverse the sequence given in Fig. 10-7.

4

The rubber filler can be renewed without replacing the entire wiper blade. See Fig. 10-8.

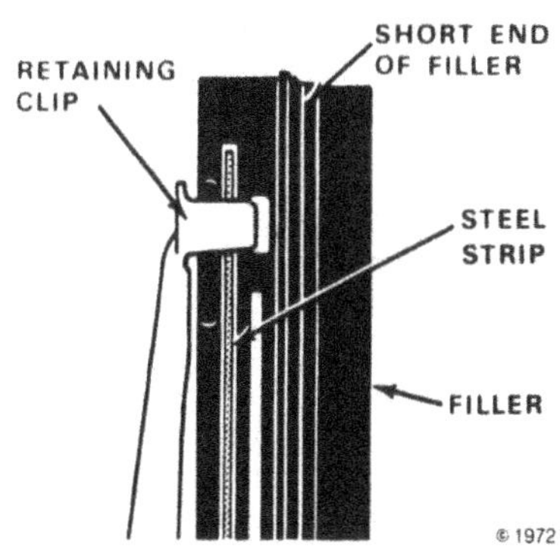

Fig. 10-8. Components of wiper blade filler assembly.

To replace filler:

1. Remove the wiper blade from the wiper arm.
2. Tightly compress the short end of the filler between your thumb and finger, then twist one side of the filler out of the retaining clip.

3. With the free side of the retaining clip resting in the groove with the steel strip, repeat the preceding step to free the other side of the filler from the retaining clip.
4. Slide the short end of the filler toward the center of the blade until the short end is completely free of the retaining clip.
5. Shift the filler sideways and unhook the steel strips from the retaining clip as indicated in Fig. 10-9. Then slide the filler out of the other retaining clips.

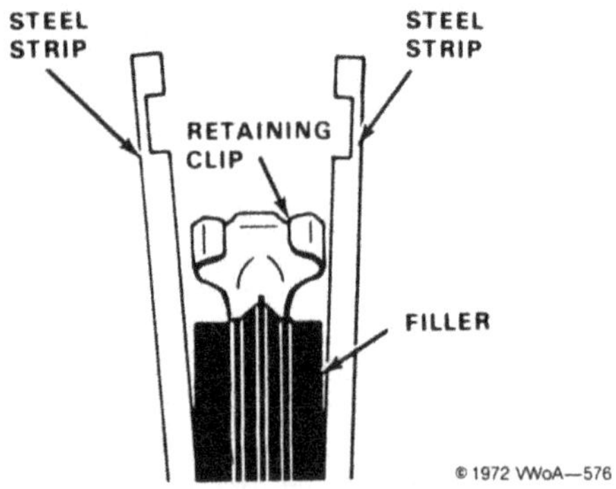

Fig. 10-9. Steel strips unhooked from retaining clip.

6. Place both steel strips in the grooves of the new filler. Make sure the notches in the steel strips face the filler and engage the projections in the filler grooves.
7. Hold the filler so that the strips are kept in the grooves. Starting at the open end of the filler, carefully slide the filler and the steel strips into the retaining clips.
8. When the closed end of the filler has reached the end retaining clip, compress the filler until the retaining clip rides over the raised edge next to the retaining clip recess in the end of the filler.
9. Make sure the retaining clip completely engages the recess in the filler. Then install the wiper blade on the wiper arm.

Removing, Installing, and Adjusting Wiper Arms

On vehicles built before model year 1973, the wiper arm is held on the wiper shaft by a cap nut. An M 6 hexagon nut is used on 1973 and later models. The nut is covered by a black plastic cap. When removing late-type wiper arms, it is necessary to pry off the cap carefully in order to gain access to the mounting nut.

On all models, proper installation requires that the spring washer or serrated washer be under the nut. It is important to adjust the angle of the wiper arm to specifications before torquing the mounting nuts. The nuts should be torqued to 42 to 60 cmkg (37 to 52 in. lb.).

The wiper arms must be correctly installed after they have been removed for replacement or repairs. Proper wiper operation under all weather conditions is possible only when the arms are adjusted accurately to specifications. The adjustment of the wiper on the wiper arms shafts is measured at the two points indicated in Fig. 10-10. Dimension **a** should be 80 mm (3$\frac{5}{32}$ in.). Dimension **b** should be 100 mm (3$\frac{15}{16}$ in.).

Fig. 10-10. Points where wiper arm adjustment is measured.

10.7 Removing and Installing Wiper Switch

Instructions applicable to rotary-type wiper switches are given in **8.2 Removing and Installing Switches in Dashboard**. The installation of the windshield washer hoses on the lever-type switch is shown in Fig. 10-11. On the rotary-type switch, the hose from the reservoir is connected to the endmost of the two connections. The lever-type switch is removed along with the turn signal switch as described in **7. Removing and Installing Turn Signal Switch and Ignition/Steering Lock**. The turn signal and

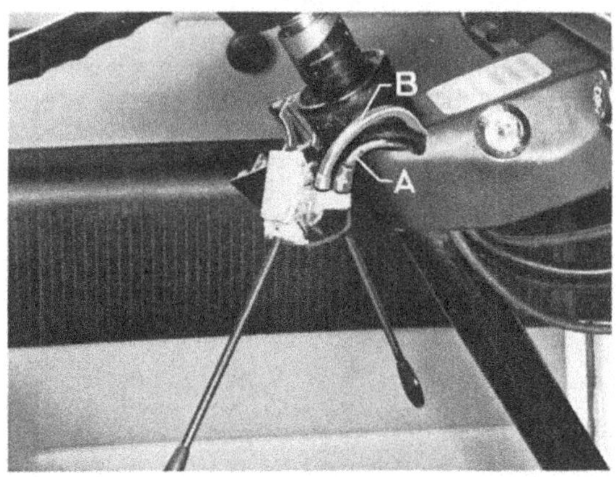

Fig. 10-11. Hose installation on lever-type switch. Hose **A** is from reservoir; hose **B** goes to jets.

wiper switches used through 1974 are shown separated in Fig. 10-12. The clamp pieces have lugs that keep the switch assembly from rotating on the column. The switch used on 1975 and later models is shown in **7. Removing and Installing Turn Signal Switch and Steering/Ignition Lock**.

When you install the column-mounted wiper switch, or when you install new washer hoses, make sure that the hoses are held by the metal clip on the bottom of the steering column mounting bracket. Do not allow the hose to hang down loosely.

Fig. 10-12. Pre-1975 column switches disassembled. Notice that the windshield washer valve can be replaced without also replacing the wiper switch.

To facilitate installation of the hoses on the washer valve, moisten the connections with water only. Other lubricants can cause windshield washer failure. Additional information on the windshield washer system is given in **10.9 Windshield Washers**.

10.8 Troubleshooting Windshield Wipers

Distorted wiper blades sometimes place an excessive load on the wiper motor, causing the fuse to blow. **Table j** lists the probable causes and suggested remedies for windshield wiper problems. The numbers in bold type in the Remedy column refer to the numbered headings in **ELECTRICAL SYSTEM** where repairs are described.

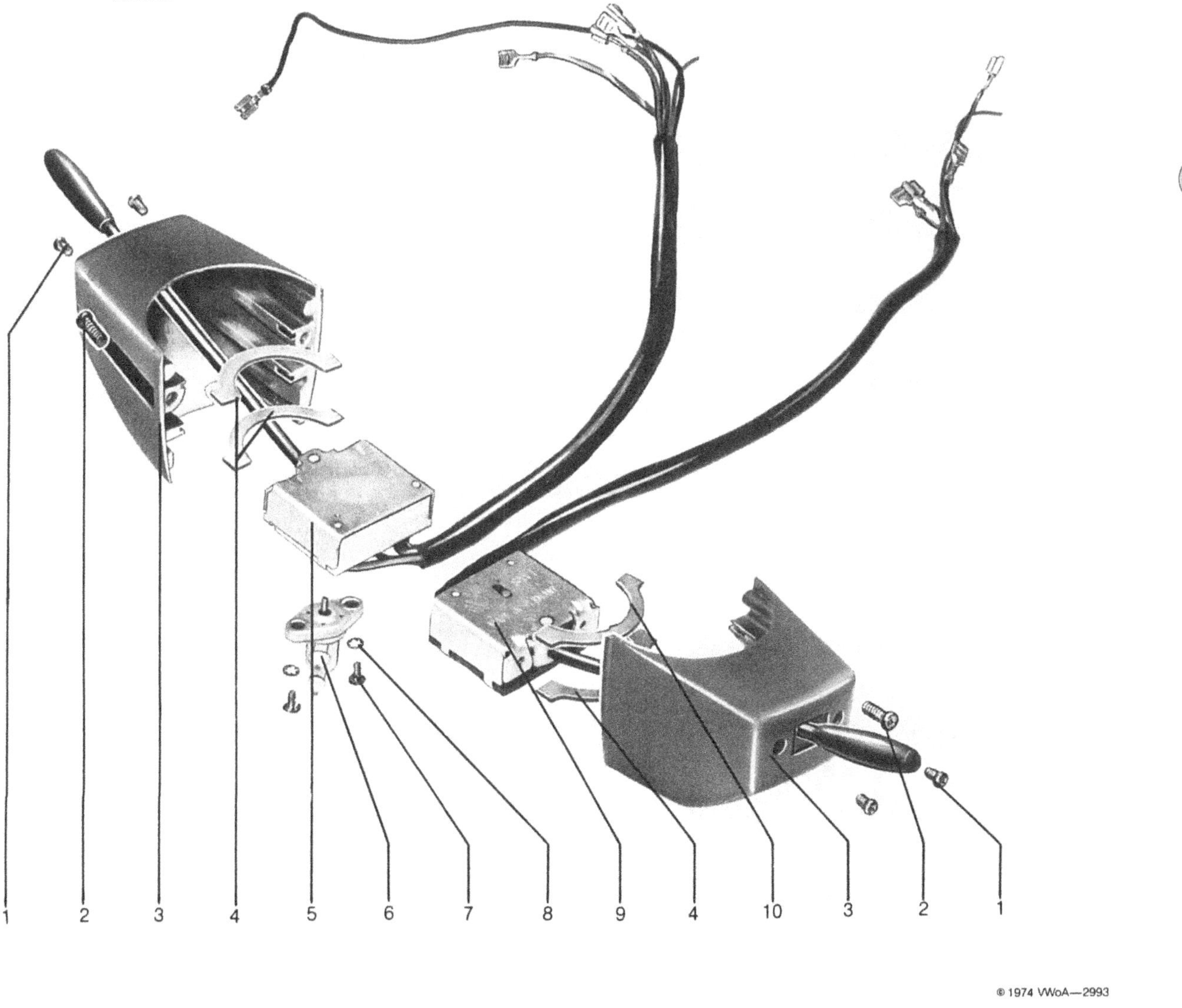

1. M 4 oval head screw
2. M 5 screw
3. Housing for switches
4. Clamp for switches
5. Wiper switch
6. Washer valve
7. M 4 binding screw
8. Serrated washer
9. Turn signal switch
10. Upper clamp with lug for turn signal switch

Table j. Windshield Wiping System Troubleshooting

Problem	Probable Cause	Remedy
 © 1972 VWoA—762 1. Smearing	a. Blade dirty b. Blade lips frayed; rubber damaged or worn out c. Blades old, blade surface cracked	a. Clean blade with hard nylon brush and soap solution or alcohol. b. Install new rubber fillers. See **10.6**. c. Install new rubber fillers. See **10.6**.
 © 1972 VWoA—763 2. Traces of water on windshield form small beads	Window soiled by paint, polish, oil, or diesel exhaust deposits	Clean windshield with clean cloth and grease/oil silicone remover.
3. Blade misses parts of windshield	a. Filler torn out of retainer b. Blade not in uniform contact with glass; spring or retainer distorted c. Pressure exerted by wiper insufficient	a. Reinstall filler carefully. See **10.6**. b. Install new blade. See **10.6**. c. Lightly lubricate arm linkage and spring or install new arm. See **10.6**.
 © 1972 VWoA—764 4. Blade wipes well on one side but badly on other, blade shudders	a. Filler distorted, no longer flips b. Wiper arm distorted; blade not vertical on the windshield	a. Clean blade with hard nylon brush and soap solution or alcohol, or install new filler. b. Twist wiper arm carefully until it is vertical.

10.9 Windshield Washers

To clean or adjust the windshield washer jets, insert a fine wire or sewing needle as shown in Fig. 10-13.

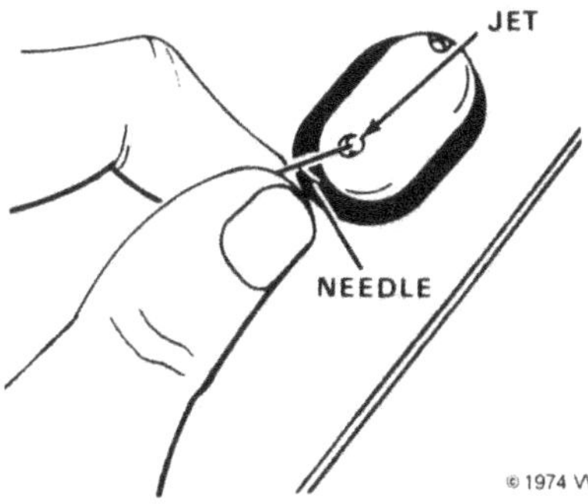

Fig. 10-13. Ball-shaped windshield washer jet being pivoted in its socket.

The reservoir for the windshield washer system is located behind the right front trim panel which is under the dashboard ahead of the front passenger seat. To remove the reservoir, first take the plastic cap off the trim panel. Then bleed the air from the system, either by depressing the air pressure valve or by loosening the filler cap on the reservoir neck.

Remove the Phillips head sheet metal screws, then take off the trim panel. If only the air pressure valve and hose are faulty, you can replace them now with no further disassembly. Just unscrew the plastic ferrule nut that holds the hose on the reservoir, then install the new hose and valve. To take out the reservoir, remove the plastic retaining nut and the three bolts indicated in Fig. 10-14. Installation is the reverse of removal.

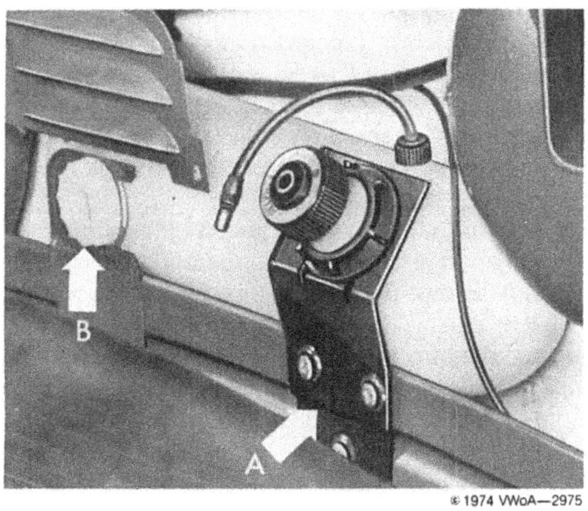

Fig. 10-14. Reservoir mounting. The plastic retaining nut is at **B**. Three bolts at **A** hold the bracket on the body. If necessary, remove the bracket from the reservoir neck by taking off the plastic ring nut.

11. HEATED REAR WINDOW

A relay in the engine compartment controls the temperature of the heated rear window. Disconnect the battery ground strap before removing either the wires or the mounting screws from the relay. Make sure there is good ground contact between the relay and the body of the car when a new relay is installed.

Before removing the rear window, disconnect the two wires from the terminals. Removal and installation of the heated rear window are the same as for any other rear window. However, such work should only be attempted by someone who is familiar with automotive glass work and has the special tools required.

If only the conductive grid is damaged, it is unnecessary to replace the window. Repair material is available at VW dealers to patch broken circuits.

To repair:

1. Apply a strip of masking tape along each edge of the broken conductor.
2. Apply the repair material evenly over the break.
3. Allow the repair to dry for one hour at room temperature. Then remove the tape and test the heating element.

4

12. HORNS AND BUZZERS

The horn is mounted under the front of the vehicle, on the frame. The brown ground wire is connected to the horn button on the steering wheel. The yellow and black hot wire is connected directly to the fuse box. Check the fuse for the horn before you begin troubleshooting.

Troubleshooting and Adjusting Horn

The best way to troubleshoot the horn circuit is with a voltmeter. Using the wiring diagrams as a guide, you can test whether or not battery voltage is reaching the horn, the fuse, and other parts of the circuit.

In particular, check for loose, dirty terminals and worn electrical contacts. If the ground (negative polarity) is weak at the brown wire to the horn, check the ground potential at the terminal on the bottom of the steering column tube. If the ground potential is weak, check the wire and the contact ring for the horn button.

If there is no electrical continuity between the two terminals on the horn, the horn is burned out. But if there is continuity, it may be possible to dissassemble and repair the horn. Look for burned or dirty contact points, a faulty condenser, or water inside the horn caused by faulty sealing.

If the horn is receiving full battery voltage, but makes only a soft clicking sound (or has a poor tone), the trouble can usually be cured by adjustment.

CAUTION —
Do not perform needless horn adjustments. Misadjustment can damage the horn.

Chip out the sealing compound over the adjusting screw. Then connect an ammeter in series with the horn. (That is, positive polarity current for the horn must pass through the ammeter.) Make repeated adjustments, striving for the clearest sound with minimum amperage draw.

CAUTION —
Do not turn the adjusting screw while the horn is sounding. Doing so could damage the horn.

If a suitable ammeter is not available, turn the adjusting screw counterclockwise until there is no sound when current is supplied to the horn. Turn the screw clockwise very gradually (about ¼ turn at a time) until the horn has a clear tone. After adjustment, seal the adjusting screw with a good sealing adhesive.

Warning Buzzer

All 1970 and later models are equipped with an ignition key warning buzzer that sounds if the ignition key is in the lock when the driver's door is opened. The buzzer, located beside the headlight dimmer relay, is connected to terminal 30 of the fuse box and, via the left-hand door contact switch, to the warning system contact in the ignition lock.

13. INSTRUMENTS

A speedometer and a fuel gauge are the only gauge-type instruments in the dashboard instrument panel. An electrically wound clock is optional. There are warning lights for oil pressure, the generator (or alternator), the high beams, and the turn signals mounted in the dial of the fuel gauge.

13.1 Replacing Speedometer Cable

The left front wheel drives the speedometer via a flexible cable. The cable is held in the front wheel bearing dust cover by a cotter pin or circlip.

To replace cable:

1. Working behind the dashboard, unscrew the knurled ferrule nut that holds the cable to the speedometer head.
2. Remove the left front hubcap. Pry off the circlip or remove the cotter pin from the square end of the speedometer cable where it projects from the wheel bearing dust cover.
3. Working from behind the wheel, pull the cable out of the steering knuckle.
4. Pull the cable out through the opening below the dashboard.

Installation is the reverse of removal. Pass the new cable through the body grommet without stretching the cable or bending it sharply. Insert the drive end in the speedometer head and tighten the ferrule nut, then insert the other end into the steering knuckle.

CAUTION —
The radius of any bend must be at least 150 mm (6 in.). Otherwise the cable will soon break at the bend.

13.2 Fuel Gauge Troubleshooting

The balancing coil type fuel gauge is controlled by an electromechanical sending unit in the fuel tank. Inaccurate fuel gauges must be replaced or returned to their manufacturer for calibration (VDO Instruments, Ltd., 116 Victor Ave., Detroit, Mich. 48203).

A fuel gauge that never moves from the 1/1 position has a grounded control circuit. Pull apart in-line wire connector T^1, which is located near the fuel tank. (See **14. Wiring Diagrams**.) If the gauge falls from the 1/1 mark, the trouble is in the sending unit. If the gauge does not fall from the 1/1 mark, the control wire is grounded somewhere between the sending unit and the gauge.

If the gauge fails to register at all, despite a full tank of fuel, pull apart in-line wire connector T^1. Then ground the control wire that goes from the connector to the gauge against a clean, unpainted metal part of the car. If the gauge moves up to 1/1, the sending unit is faulty or improperly grounded. If the gauge still fails to move, the gauge or the control wire is faulty.

Removing and Testing Fuel Gauge Sending Unit

If an ohmmeter indicates infinite ohms when it is connected to ground and to the sending unit wire (at in-line wire connector T^1), the sending unit is burned out and must be replaced. To remove the sending unit, you must first remove the fuel tank as described in **FUEL SYSTEM**.

To take out the sending unit, disconnect the wire shown in Fig. 13-1. Then remove the five bolts in the top of the sending unit.

> ***WARNING —***
> *Disconnect the battery ground strap. Do not smoke or work near heaters or other fire hazards. Have a fire extinguisher handy.*

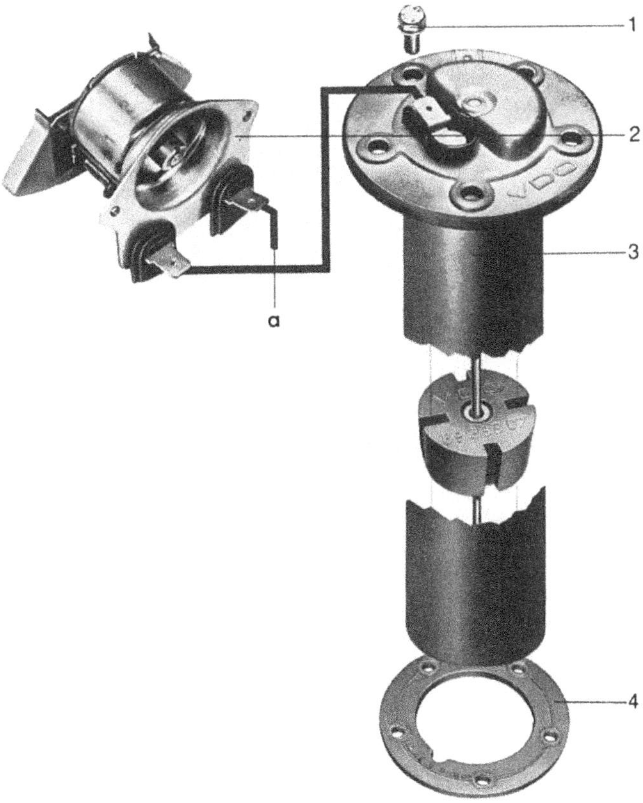

1. Bolt
2. Fuel gauge
3. Sending unit
4. Gasket

© 1974 VWoA—2982

Fig. 13-1. Fuel gauge and sending unit. The ground wire is shown connected from the sending unit to the gauge. The wire at **a** goes to terminal 15 on the fuse box.

Carefully lift out the sending unit. Then remove the gasket. You can test the sending unit by connecting a battery and voltmeter in series between the sending unit housing and the gauge wire terminal.

> ***WARNING —***
> *Do not make tests near the fuel tank. An electrical spark could cause an explosion.*

© 1976 VWoA

Observe whether or not the voltmeter reading changes continuously as the sending unit's float is moved by hand through its full range. If the voltmeter needle does not move or does not move smoothly, the sending unit is defective and should be replaced.

When installing the sending unit, be sure that the rubber gasket is positioned correctly. The ground lug on the sender must be able to slip through the cutout in the gasket and make contact with the fuel tank. The fuel tank should be clean and bright where the ground lug contacts it.

13.3 Removing and Installing Instruments

The instrument panel is held in the dashboard by four Phillips head sheet metal screws. On 1968 through 1972 models, these screws thread into brackets on the dashboard. On 1973 and later models, the screws thread into spring clips as indicated in Fig. 13-2.

© 1974 VWoA—2987

Fig. 13-2. Spring clips (arrows) that receive the sheet metal screws on 1973 and later models. Notice the openings for the bulb holders in the back of the fuel gauge.

The instrument panel must be removed from the dashboard before you can remove the instruments. The instruments are housed in a single instrument cluster. An exploded view of the instrument cluster is given in Fig. 13-3.

To remove instrument cluster:

1. Disconnect the battery ground strap.
2. Press out the plastic pins in the knobs for the fresh air control levers, then remove the knobs.
3. Take out the Phillips head screws at the four corners of the instrument panel, being careful not to lose the spring clips on 1973 and later models.

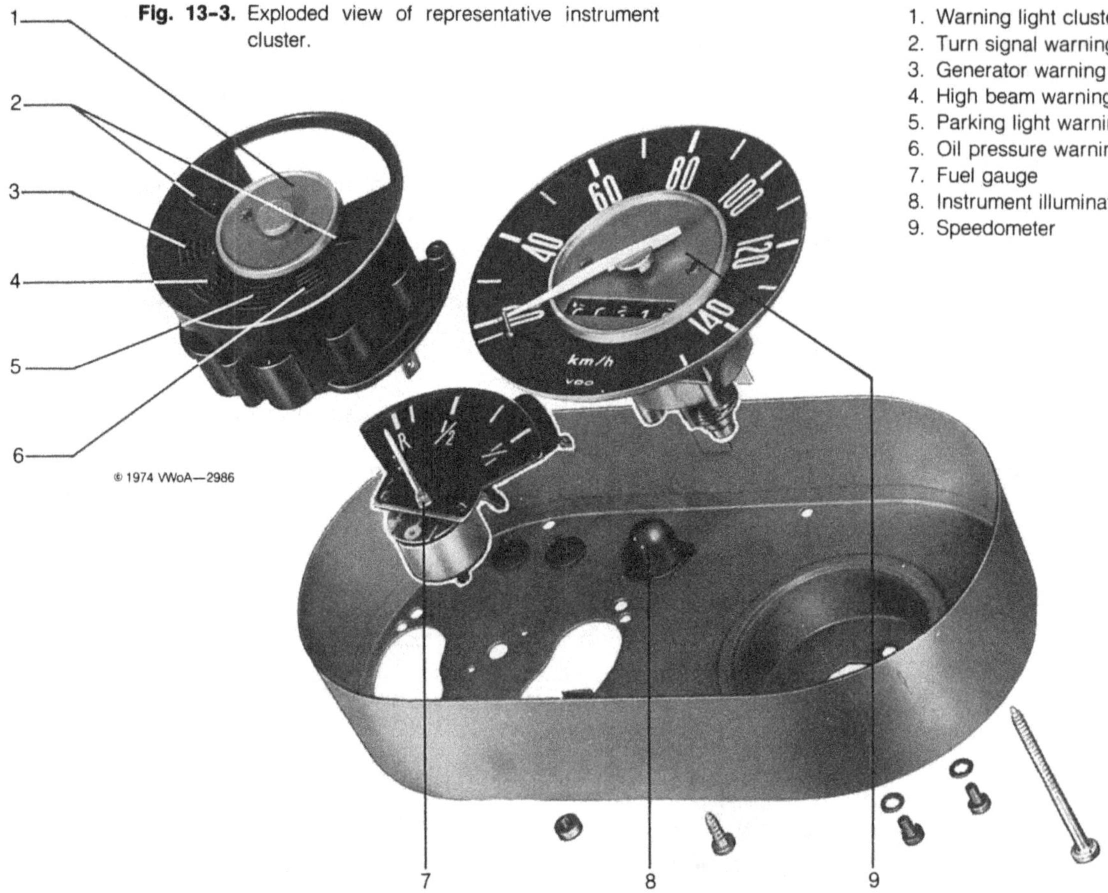

Fig. 13-3. Exploded view of representative instrument cluster.

1. Warning light cluster
2. Turn signal warning light
3. Generator warning light
4. High beam warning light
5. Parking light warning light
6. Oil pressure warning light
7. Fuel gauge
8. Instrument illuminating light
9. Speedometer

4. Unscrew the ferrule nut that holds the speedometer cable on the speedometer head. Then withdraw the cable from the speedometer.
5. Slightly lift the instrument panel out of the dashboard. Disconnect the wires from the rear of the panel. Attach tags to the wires as you remove them so that you can return them to their original terminals during installation.
6. Remove the instrument panel.
7. To remove the instrument cluster from the back of the instrument panel, take out the two long screws. One of these screws is shown at the extreme right-hand side of Fig. 13-3.
8. If necessary, remove the instruments from the instrument cluster by taking off the nut and screws shown at the bottom of Fig. 13-3.

Installation is the reverse of removal.

Replacing Instrument Panel Bulbs

It is unnecessary to remove the instrument panel to replace bulbs. The individual bulbs are housed in bulb holders that slide into the rear of the instrument cluster housing and the warning light cluster. To replace a bulb reach behind the dashboard and withdraw the bulb holder. Install the new bulb in the holder, then slip the holder back into the rear of the cluster.

14. WIRING DIAGRAMS

Two kinds of wiring diagrams are presented here. Those for 1968 through 1972 vehicles show individual components as line drawings. Diagrams for 1973 and later vehicles are of the current-flow type and represent the components schematically. Both kinds of diagrams show the wires' insulation color; the small numbers in the wire lines indicate their gauge in mm².

Note that in the wiring diagrams for cars built since January 1971, the test connections for the VW Computer Analysis system are shown. Always reconnect these wires when servicing the electrical system.

CAUTION —
Never connect any device other than the test plug of the VW Computer Analysis system to the test network central socket in the engine compartment. Other test equipment will not guarantee accurate readings and could damage the socket, the test sensors, or the vehicle components containing them.

The current-flow diagrams for 1973 and later cars have symbols for the electrical components rather than pictorial representations as in earlier diagrams. However, pictorial representations do appear separately at the bottom of the diagram to show where the wires attach to the actual component. This is an aid to troubleshooting and component replacement. The symbols used in the diagram itself are explained in Fig. 14-1. The thin black lines in the current-flow diagrams are not actual wires but ground connections via the vehicle chassis.

Along the bottom of each current-flow diagram is a yellow band containing numbers. These numbers (current tracks) will help you find electrical components easily. Appearing after each component listed in the description are numbers in a yellow legend labeled "current track." These numbers indicate which current track in the current-flow diagram contains the component.

Fig. 14-1. Schematic representation of electrical components in current-flow diagrams.

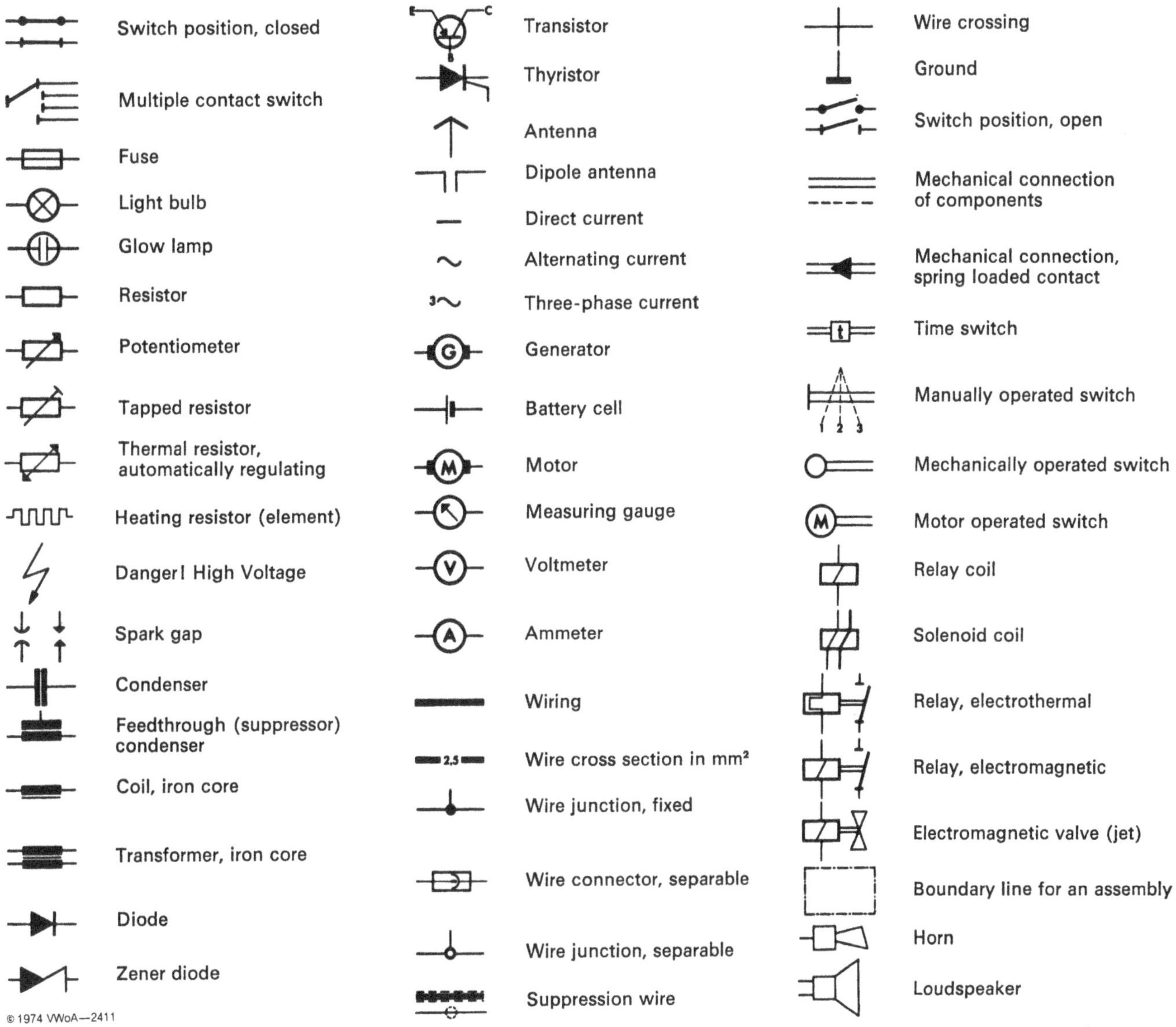

4

A. How to read current flow diagrams

Wire Color Code	
Black	-BK
Blue	-BL
Brown	-BR
Clear	-CL
Gray	-GY
Green	-G
Light Green	-LT.G
Red	-R
Violet	-V
White	-W
Yellow	-Y

Wire color (corresponds with color in vehicle)

Wire cross section in mm²

Part designation
Helps you to find, in the legend, the part which the symbol in the diagram belongs to, e.g. E^9 = Switch for fan motor.

Terminal with the designation which is on the actual part.

Symbol (here: Bulb)

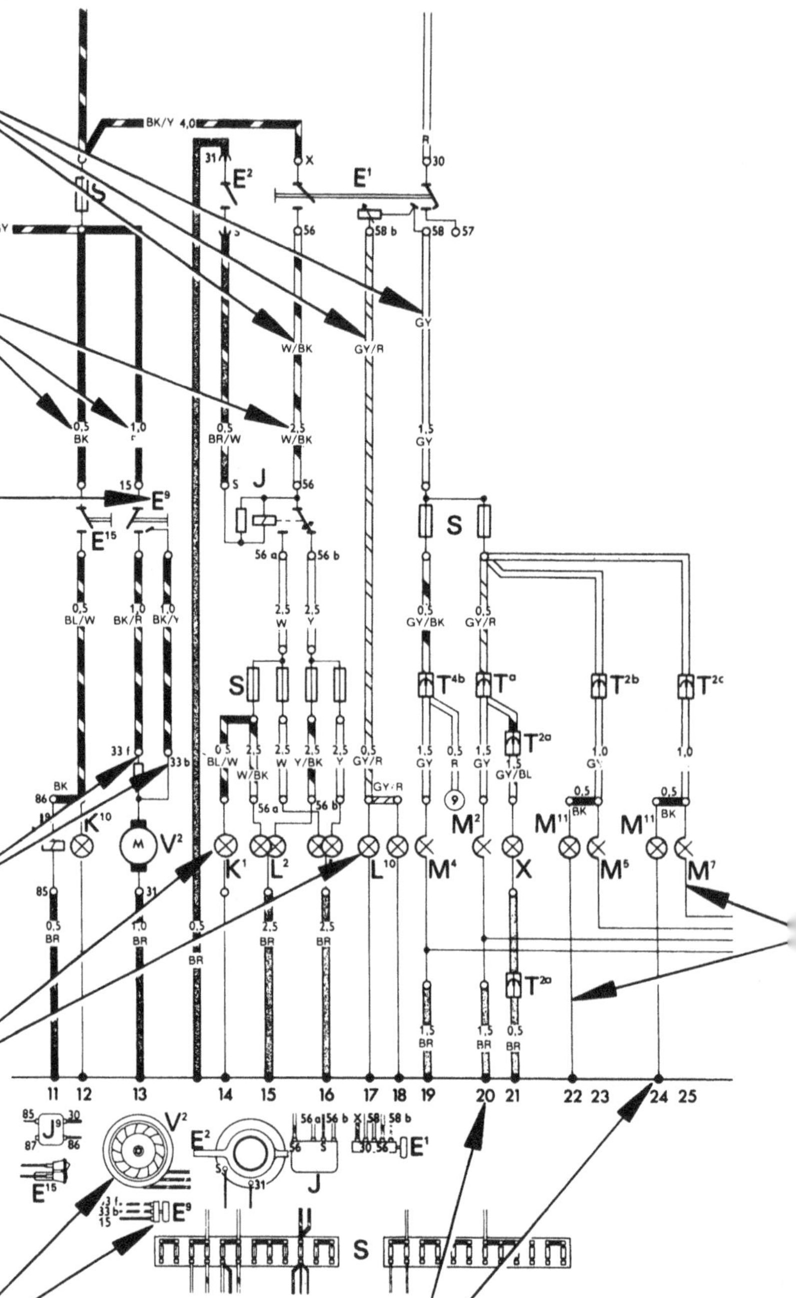

Sketch of part, the symbol for which is in the current track above it. Also shows how wires are connected.

Current tracks with track numbers to facilitate finding of part in diagram (see legend).

	Description	current track
A	– Battery	4
B	– Starter	5, 6
C	– Generator	1, 2, 3
C^1	– Regulator	1, 2, 3
D	– Ignition / starter switch	6, 7, 12
E	– Windshield wiper switch	8, 9, 10
E^1	– Light switch	16, 17, 19
E^2	– Turn signal and headlight dimmer switch	14, 38, 39
E^3	– Emergency flasher switch	38, 39, 42, 43, 44
E^9	– Fan motor switch	13
E^{15}	– Rear window defogger switch	12

Legend, consisting of:

Part designation,

Current track number,
with the aid of which you can find the part in the diagram, e.g. turn signal switch E^2, current track 14.

T^2 – Wire connector, double
- a – in engine compartment lid
- b – in luggage compartment, front, left
- c – in luggage compartment, front, right

T^3 – Wire connector, triple
- a – in luggage compartment, front, left
- b – behind the engine compartment insulation, right

Explanation (where you can find a wire connector, for example, on the vehicle).

Internal connections (thin lines).
These are **not** actual wires but ground connections which go through parts such as bulb holders for example.

Please note:
All switches and contacts are in a mechanically neutral position.
The various contacts in switches are shown in the current track in which they operate.

B. VW Type 2—from August 1967 (1968 and 1969 models)

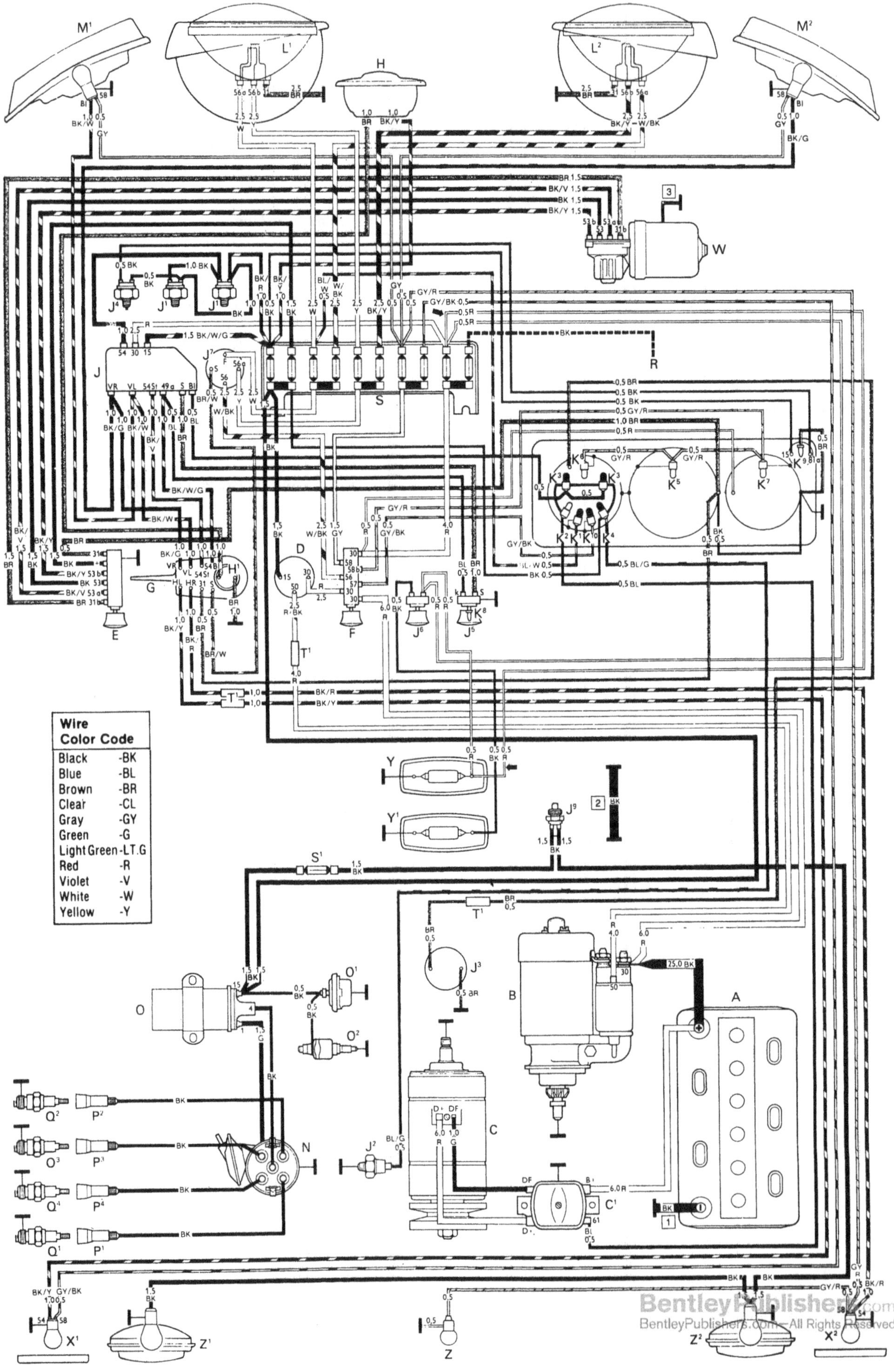

Wire Color Code	
Black	-BK
Blue	-BL
Brown	-BR
Clear	-CL
Gray	-GY
Green	-G
Light Green	-LT.G
Red	-R
Violet	-V
White	-W
Yellow	-Y

B. VW Type 2—from August 1967 (1968 and 1969 models)

A – Battery
B – Starter
C – Generator
C^1 – Regulator
D – Ignition/starter switch
E – Windshield wiper switch
F – Light switch
G – Turn signal switch and hand dimmer
H – Horn
H^1 – Horn button
J – Emergency flasher relay
J^1 – Brake light switch (2×)
J^2 – Oil pressure switch
J^3 – Fuel gauge sender unit
J^4 – Warning switch for brakes
J^5 – Emergency flasher switch
J^6 – Interior light switch
J^7 – Dimmer relay
J^9 – Back-up light switch
K^1 – High beam warning light
K^2 – Generator and fan warning light
K^3 – Turn signal warning lights
K^4 – Oil pressure warning lights
K^5 – Speedometer light
K^6 – Fuel gauge light
K^7 – Clock light
K^8 – Emergency flasher warning light
K^9 – Brake system warning light
L^1 – Sealed beam unit, left
L^2 – Sealed beam unit, right
M^1 – Parking light, left
M^2 – Parking light, right
N – Distributor
O – Ignition coil
O^1 – Automatic choke
O^2 – Electro-magnetic pilot jet
P^1 – Spark plug connector, No. 1 cylinder
P^2 – Spark plug connector, No. 2 cylinder
P^3 – Spark plug connector, No. 3 cylinder
P^4 – Spark plug connector, No. 4 cylinder
Q^1 – Spark plug for No. 1 cylinder
Q^2 – Spark plug for No. 2 cylinder
Q^3 – Spark plug for No. 3 cylinder
Q^4 – Spark plug for No. 4 cylinder
R – Radio
S – Fuse box
S^1 – Back-up light fuse
T^1 – Cable connector
U^1 – Turn signal, front left
U^2 – Turn signal, front right
W – Windshield wiper motor
X^1 – Brake, turn signal and tail light, left
X^2 – Brake, turn signal and tail light, right
Y – Interior light, front
Y^1 – Interior light, rear
Z – License plate light
Z^1 – Back-up light, left
Z^2 – Back-up light, right

[1] – Battery to body ground strap
[2] – Transmission to body ground strap
[3] – Windshield wiper motor ground cable

Black dotted lines = optional extras
All fuses: 8 amps.

➤ – for Pick up only

C. Fresh Air Fan Circuit and Rear Window Defogger Circuit for VW Type 2—from August 1969 (1970 Models)

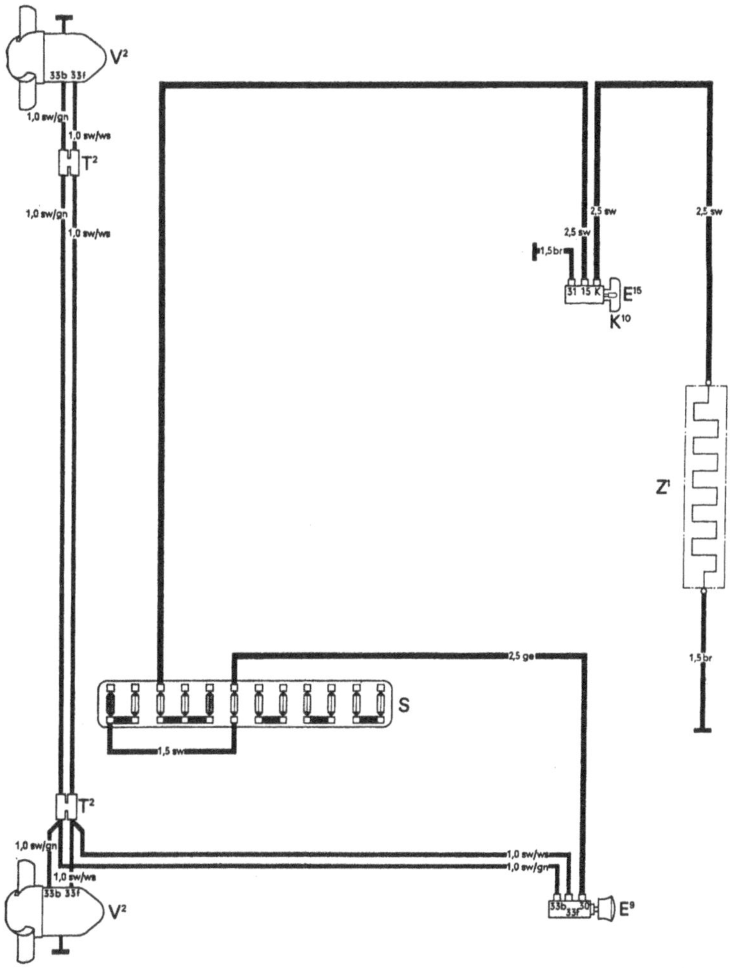

E^9 – Switch for fan motors
E^{15} – Rear window defogger switch
K^{10} – Rear window defogger warning light
S – Fuse box
T^2 – Cable connector, double
Y^2 – Fan motor, front
Z^1 – Rear window defogger heating element

Color of cables:

sw – black
ge – yellow
br – brown
gn – green
ws – white

D. VW Type 2—from August 1969 (1970 Models)

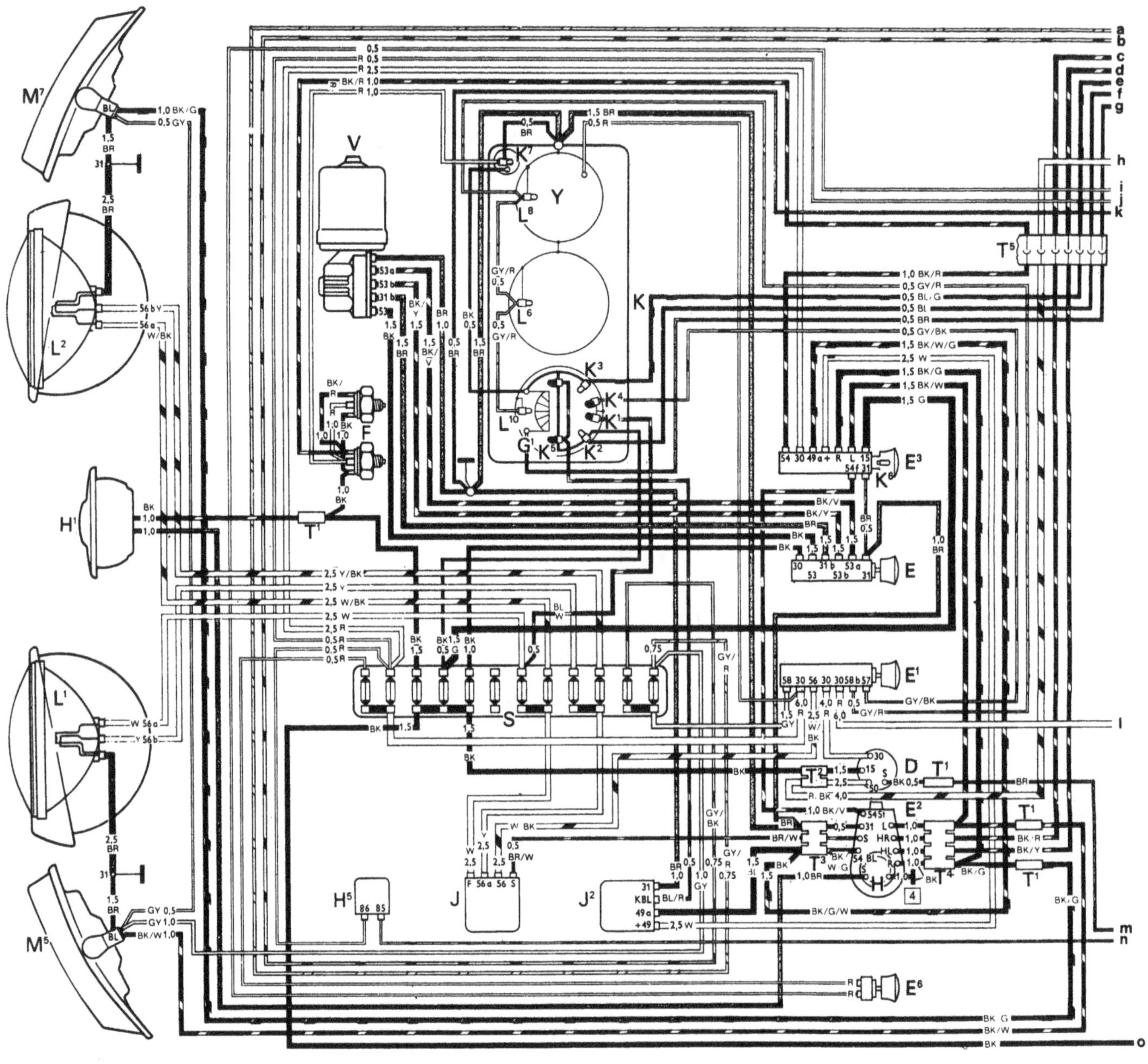

A – Battery
B – Starter
C – Generator
C' – Regulator
D – Ignition / starter switch
E – Windshield wiper switch
E' – Light switch
E² – Turn signal and headlight dimmer switch
E³ – Emergency flasher switch
E⁶ – Switch for interior light, rear
F – Brake light switch
F' – Oil pressure switch
F² – Door contact switch, left with contact for buzzer H⁵
F³ – Door contact, switch right
F⁴ – Switch for back-up lights
G – Fuel gauge sending unit
G' – Fuel gauge
H – Horn button
H' – Horn
H⁵ – Ignition key warning buzzer
J – Dimmer relay
J² – Emergency flasher relay
K – Instrument panel insert
K' – High beam warning light
K² – Generator charging warning light
K³ – Oil pressure warning light
K⁴ – Parking light warning light
K⁵ – Turn signal warning light
K⁶ – Emergency flasher warning light
K⁷ – Dual circuit brake system warning light
L' – Sealed beam light, left headlight
L² – Sealed beam light, right headlight
L⁶ – Speedometer light
L⁸ – Clock light
L¹⁰ – Instrument panel light
M⁵ – Turn signal and parking light, front, left
M⁷ – Turn signal and parking light, front, righ
M⁹ – Tail / brake / turn signal light, left
M¹⁰ – Tail / brake / turn signal light, right
M¹² – Side marker lights, rear
N – Ignition coil
N' – Automatic choke
N³ – Electro-magnetic pilot jet
O – Distributor
P' – Spark plug connector, No. 1 cylinder
P² – Spark plug connector, No. 2 cylinder
P³ – Spark plug connector, No. 3 cylinder
P⁴ – Spark plug connector, No. 4 cylinder
Q' – Spark plug, No. 1 cylinder
Q² – Spark plug, No. 2 cylinder

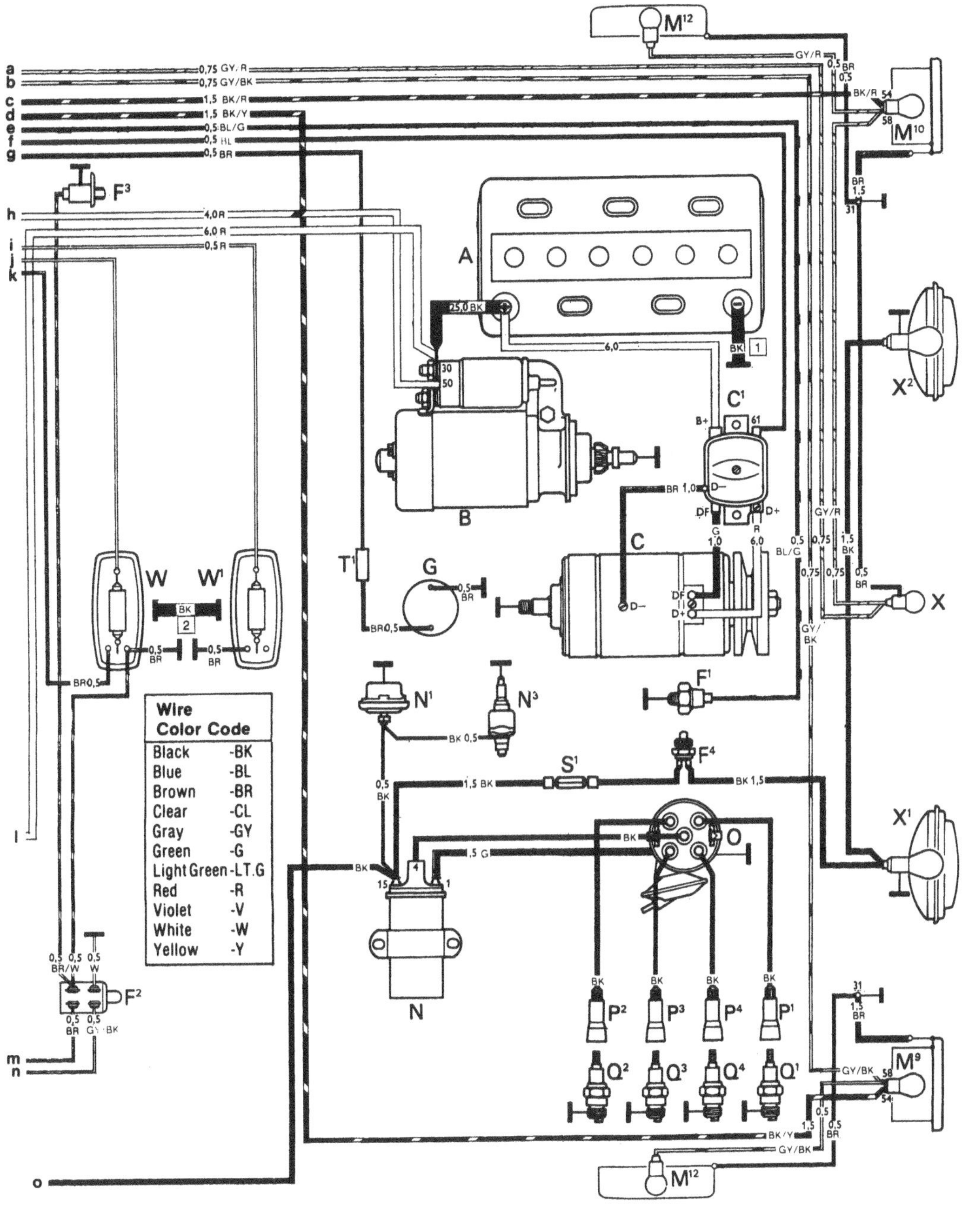

Wire Color Code	
Black	-BK
Blue	-BL
Brown	-BR
Clear	-CL
Gray	-GY
Green	-G
Light Green	-LT.G
Red	-R
Violet	-V
White	-W
Yellow	-Y

4

- Q³ – Spark plug, No. 3 cylinder
- Q⁴ – Spark plug, No. 4 cylinder
- S – Fuse box
- S¹ – In-line fuse for back-up lights
- T¹ – Cable connector, single
- T² – Cable connector, double
- T³ – Cable connector, triple
- T⁴ – Cable connector (four connections)
- T⁵ – Push-on connector
- V – Windshield wiper motor
- W – Interior light, front
- W¹ – Interior light, rear
- X – License plate light
- X¹ – Back-up light, left
- X² – Back-up light, right
- Y – Clock

- [1] – Ground strap from battery to frame
- [2] – Ground strap from transmission to the frame
- [4] – Ground cable from horn button to steering coupling

E1. VW Type 2—from August 1970 (1971 Models)

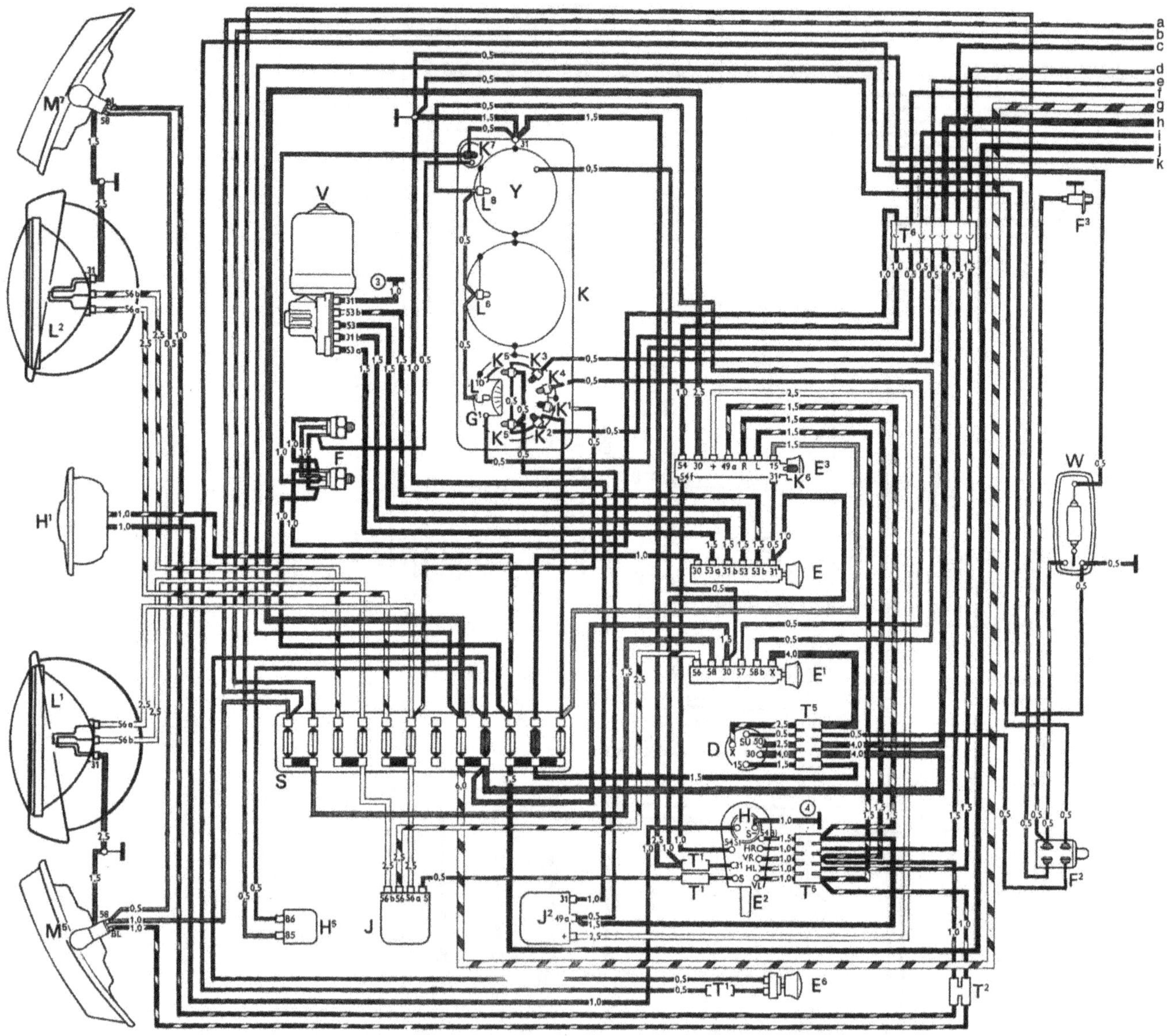

A – Battery
B – Starter
C – Generator
C' – Regulator
D – Ignition / starter switch
E – Windshield wiper switch
E¹ – Light switch
E² – Turn signal and headlight dimmer switch
E³ – Emergency flasher switch
E⁶ – Interior light switch, rear
F – Brake light switch with warning switch
F¹ – Oil pressure switch
F² – Door contact switch, left, with contact for buzzer
F³ – Door contact switch, right
F⁴ – Back-up light switch
G – Fuel gauge sending unit
G¹ – Fuel gauge
H – Horn button
H¹ – Horn
H⁵ – Ignition key warning buzzer
J – Dimmer relay
J² – Emergency flasher relay
K – Instrument panel insert
K¹ – High beam warning light
K² – Generator charging warning light
K³ – Oil pressure warning light
K⁴ – Parking light warning light
K⁵ – Turn signal warning light
K⁶ – Emergency flasher warning light
K⁷ – Dual circuit brake system warning light
L¹ – Sealed beam unit, left headlight
L² – Sealed beam unit, right headlight
L⁶ – Speedometer light
L⁸ – Clock light
L¹⁰ – Instrument panel light
M⁵ – Turn signal and parking light, front, left
M⁷ – Turn signal and parking light, front, right
M⁹ – Tail-, brake-, turn signal light, left
M¹⁰ – Tail-, brake-, turn signal light, right
M¹² – Sidemarker light, rear
N – Ignition coil
N¹ – Automatic choke
O – Ignition distributor
P¹ – Spark plug connector, No. 1 cylinder
P² – Spark plug connector, No. 2 cylinder
P³ – Spark plug connector, No. 3 cylinder
P⁴ – Spark plug connector, No. 4 cylinder

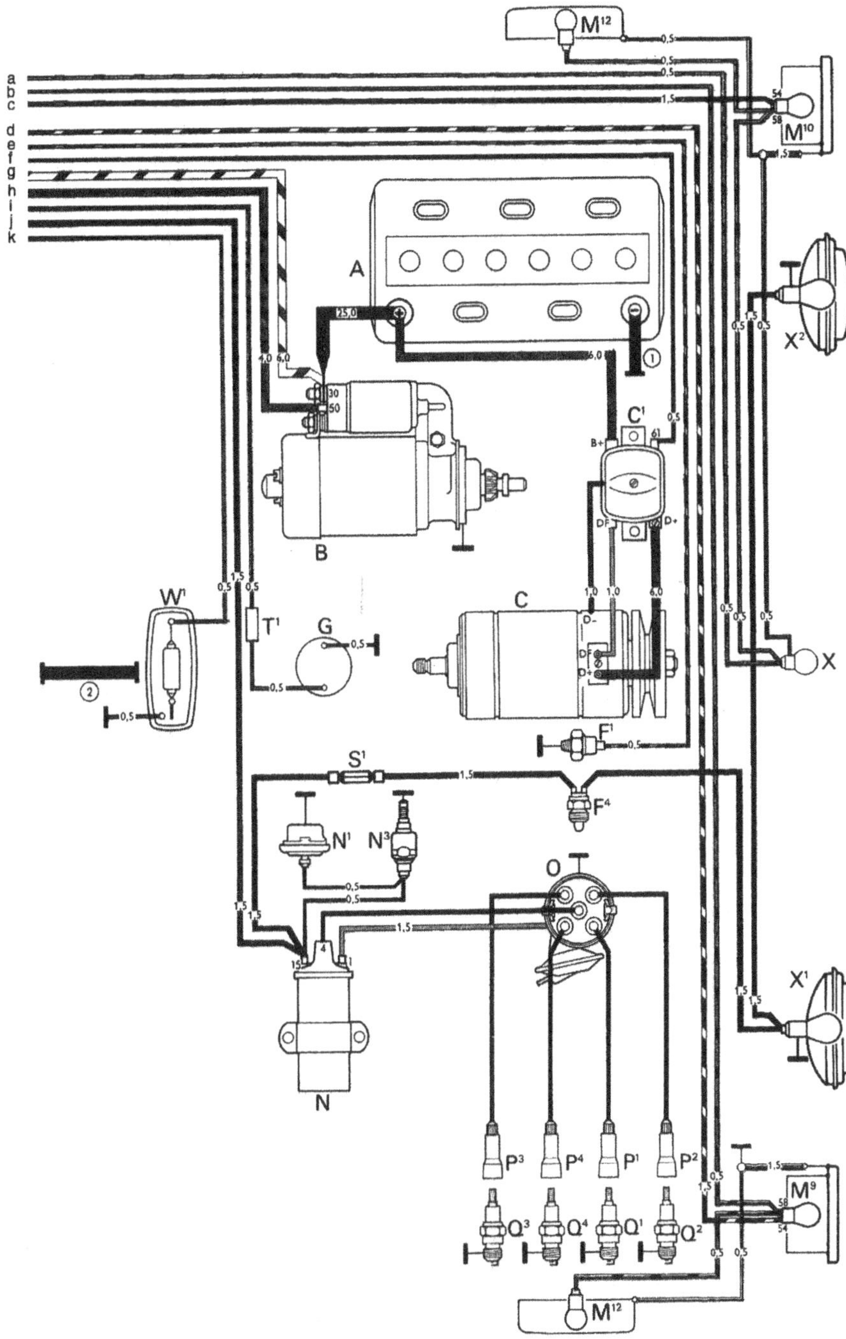

Q¹ – Spark plug, No. 1 cylinder
Q² – Spark plug, No. 2 cylinder
Q³ – Spark plug, No. 3 cylinder
Q⁴ – Spark plug, No. 4 cylinder
S – Fuse box
S¹ – Fuse for back-up light
T¹ – Cable connector, single
T⁵ – Cable connector, five connections
T⁶ – Cable connector, seren connections
V – Windshield wiper motor
W – Interior light, front
W¹ – Interior light, rear
X – License plate light
X¹ – Back-up light, left
X² – Back-up light, right
Y – Clock

① – Ground strap battery to frame
② – Ground strap transmission to frame
③ – Ground strap windshield wiper
④ – Ground strap horn button to steering coupling

E2. VW Type 2—from January 1971 (1971 Models)

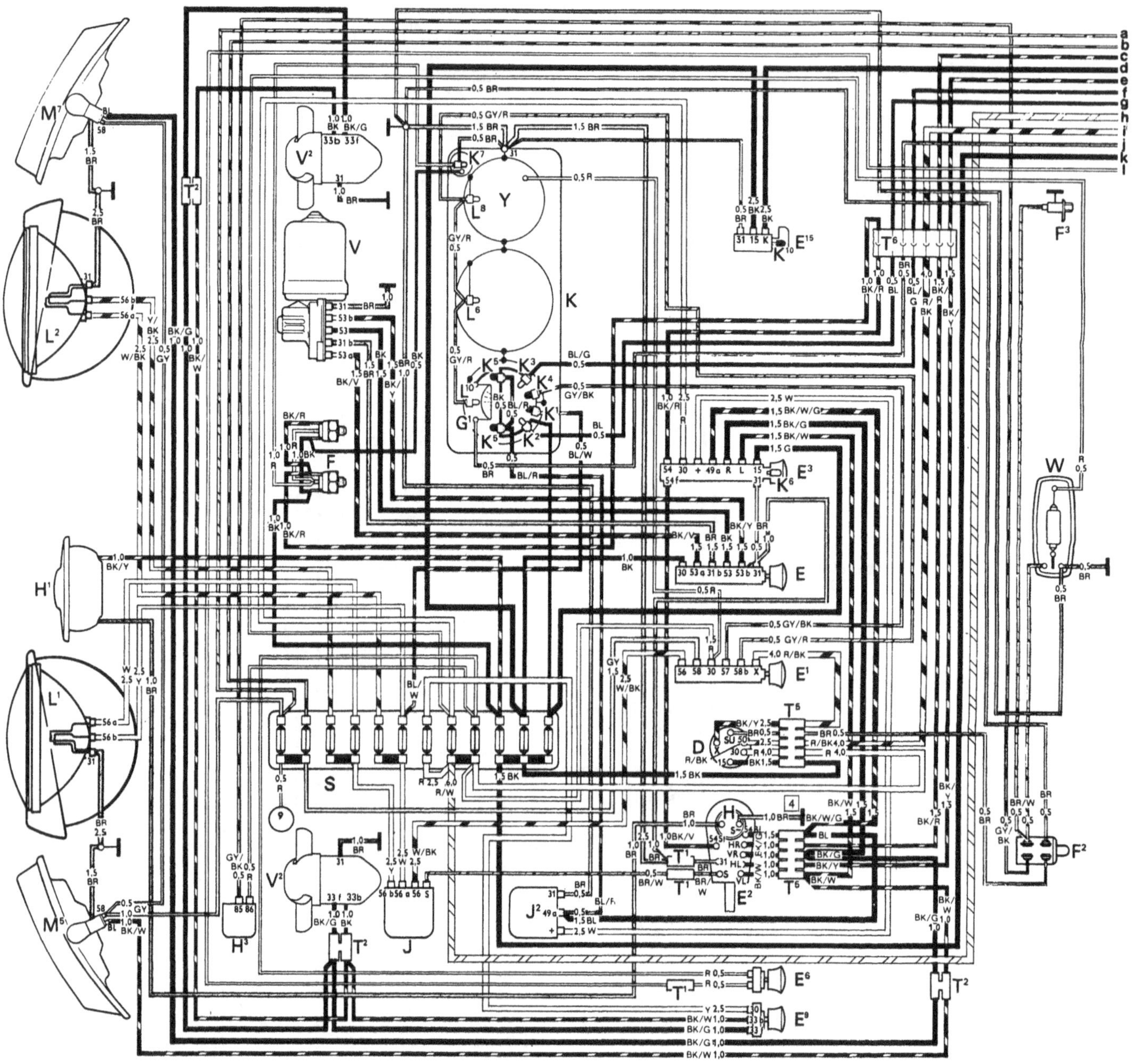

A – Battery
B – Starter
C – Generator
C[1] – Regulator
D – Ignition / starter switch
E – Windshield wiper switch
E[1] – Light switch
E[2] – Turn signal and headlight dimmer switch
E[3] – Emergency flasher switch
E[6] – Interior light switch, rear
E[9] – Switch for fan motors
E[15] – Rear window defogger switch
F – Brake light switch with warning switch
F[1] – Oil pressure switch
F[2] – Door contact switch, left, with contact for buzzer
F[3] – Door contact switch, right
F[4] – Back-up light switch
G – Fuel gauge sending unit
G[1] – Fuel gauge
H – Horn button
H[1] – Horn
H[3] – Ignition key warning buzzer
J – Dimmer relay
J[2] – Emergency flasher relay
K – Instrument panel insert
K[1] – High beam warning light
K[2] – Generator charging warning light
K[3] – Oil pressure warning light
K[4] – Parking light warning light
K[5] – Turn signal warning light
K[6] – Emergency flasher warning light
K[7] – Dual circuit brake system warning light
K[10] – Rear window defogger warning light
L[1] – Sealed beam unit, left headlight
L[2] – Sealed beam unit, right headlight
L[6] – Speedometer light
L[8] – Clock light
L[10] – Instrument panel light
M[5] – Turn signal and parking light, front, left
M[7] – Turn signal and parking light, front, right
M[9] – Tail-, brake-, turn signal light, left
M[10] – Tail-, brake-, turn signal light, right
M[12] – Sidemarker light, rear
N – Ignition coil
N[1] – Automatic choke
N[3] – Electro-magnetic pilot jet
O – Ignition distributor
P[1] – Spark plug connector, No. 1 cylinder
P[2] – Spark plug connector, No. 2 cylinder
P[3] – Spark plug connector, No. 3 cylinder
P[4] – Spark plug connector, No. 4 cylinder
Q[1] – Spark plug, No. 1 cylinder
Q[2] – Spark plug, No. 2 cylinder

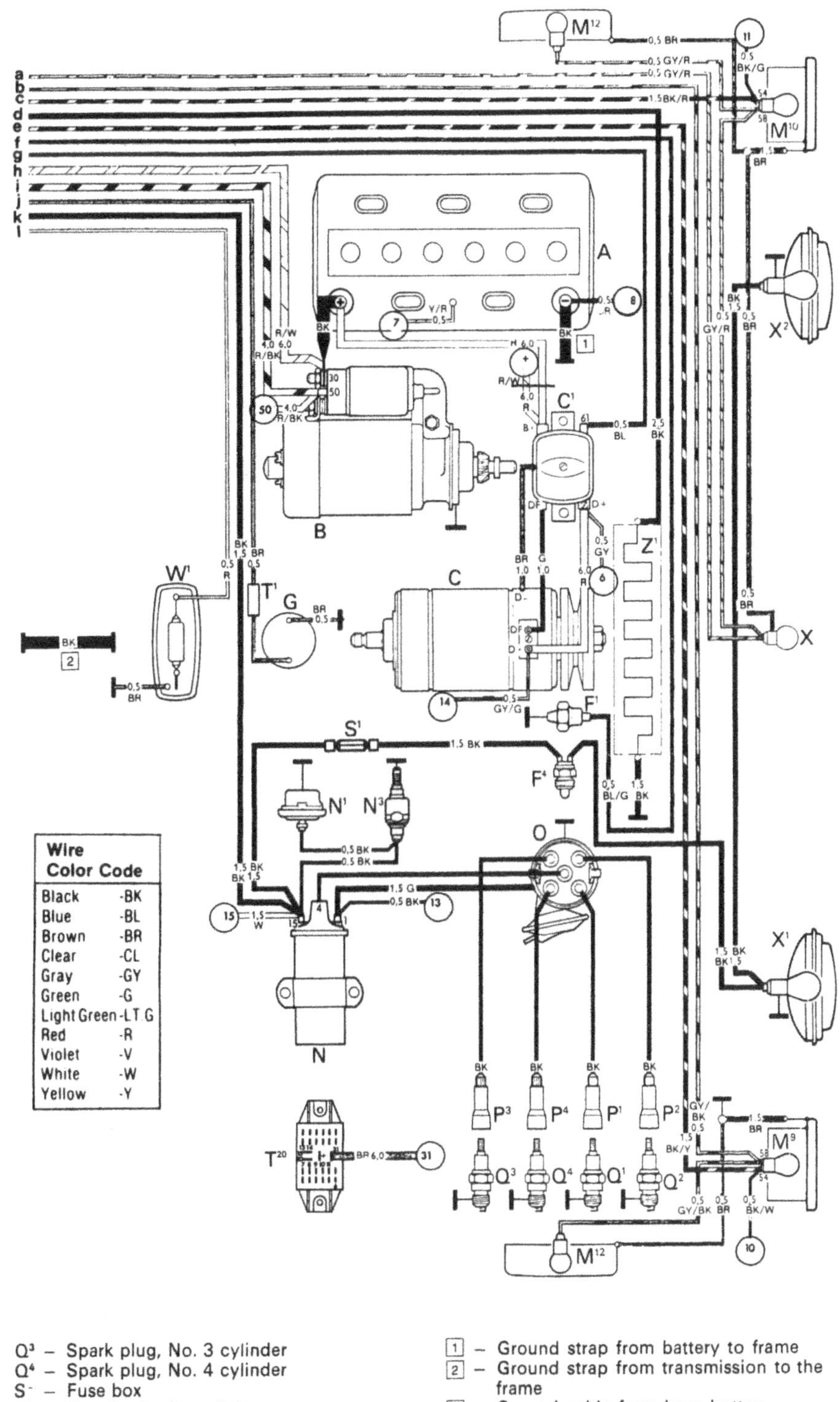

Wire Color Code	
Black	-BK
Blue	-BL
Brown	-BR
Clear	-CL
Gray	-GY
Green	-G
Light Green	-LT G
Red	-R
Violet	-V
White	-W
Yellow	-Y

4

Q^3 – Spark plug, No. 3 cylinder
Q^4 – Spark plug, No. 4 cylinder
S – Fuse box
S^1 – Fuse for back-up light
T^1 – Cable connector, single
T^2 – Cable connector, double
T^5 – Cable connector, five connections
T^6 – Cable connector, eight connections
T^{20} – Test network, central plug
V – Windshield wiper motor
Y^2 – Fan motor, front
W – Interior light, front
W^1 – Interior light, rear
X – License plate light
X^1 – Back-up light, left
X^2 – Back-up light, right
Y – Clock
Z^1 – Rear window defogger heating element

[1] – Ground strap from battery to frame
[2] – Ground strap from transmission to the frame
[4] – Ground cable from horn button to steering coupling

Test network (from January 1971 only)

The numbered circles are the connections in the test network which are wired to the central plug (T^{20}). The numbers correspond to the terminals in the central plug.

Note

Colors of the cables have changed:
cable 13 is grey/green
cable 14 is black

F. VW Type 2—from August 1971 (1972 Models)

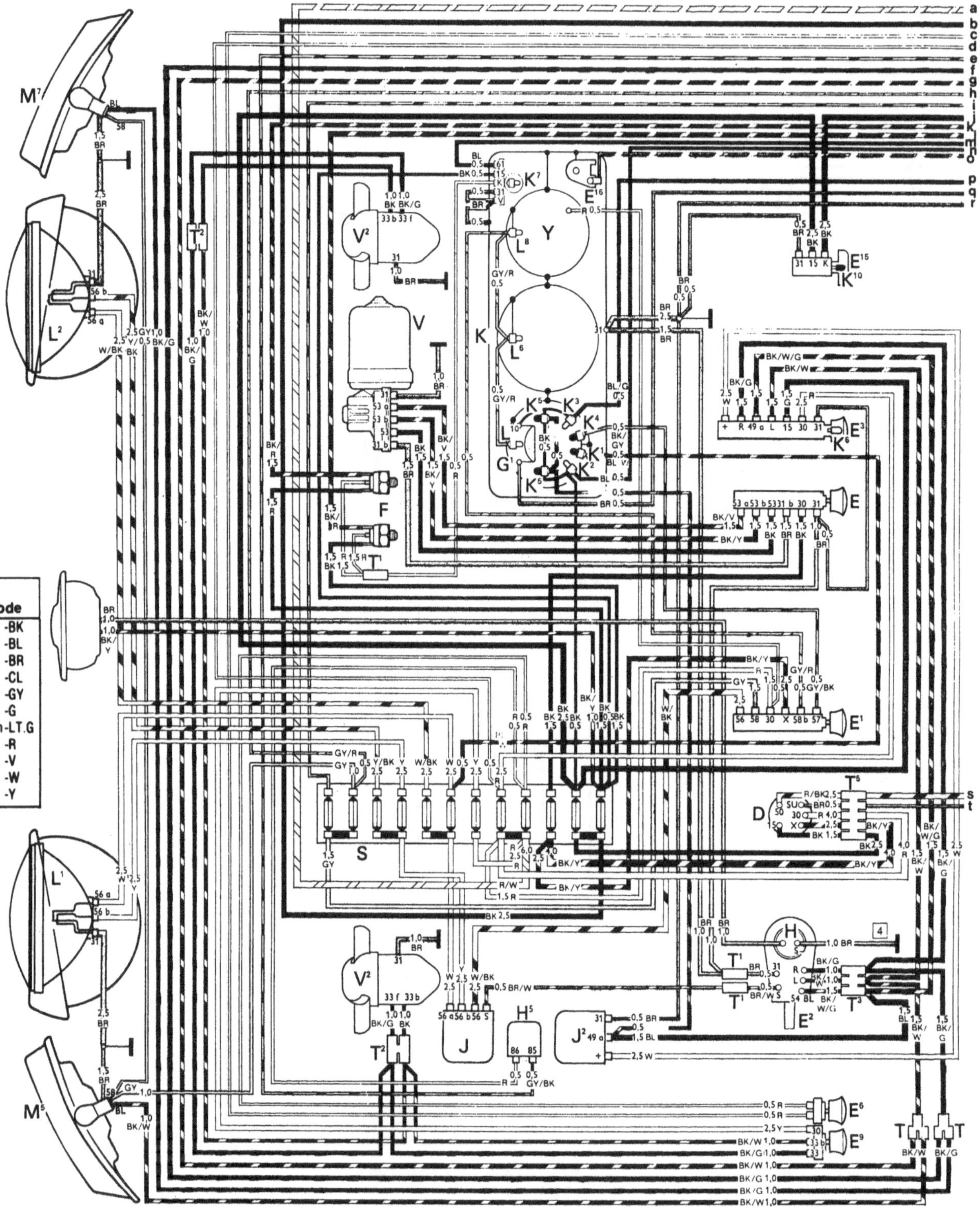

Wire Color Code

Black	-BK
Blue	-BL
Brown	-BR
Clear	-CL
Gray	-GY
Green	-G
Light Green	-LT.G
Red	-R
Violet	-V
White	-W
Yellow	-Y

A – Battery
B – Starter
C – Generator
C¹ – Regulator
D – Ignition / starter switch
E – Windshield wiper switch
E¹ – Light switch
E² – Turn signal and headlight dimmer switch
E³ – Emergency flasher switch
E⁶ – Interior light switch, rear
E⁹ – Switch for fan motors
E¹⁵ – Rear window defogger switch
E¹⁶ – Heater air blower switch
F – Brake light and dual / circuit warning light switch
F¹ – Oil pressure switch
F² – Door contact switch, left, with contact for buzzer
F³ – Door contact switch, right
F⁴ – Back-up light switch
G – Fuel gauge sending unit
G¹ – Fuel gauge
G⁴ – Ignition timing sensor
H – Horn button
H¹ – Horn
H³ – Ignition key warning buzzer
J – Dimmer relay
J² – Emergency flasher relay
J¹⁴ – Heater air blower relay
K – Instrument panel insert
K¹ – High beam warning light
K² – Generator charging warning light
K³ – Oil pressure warning light
K⁴ – Parking light warning light
K⁵ – Turn signal warning light
K⁶ – Emergency flasher warning light
K⁷ – Dual circuit brake system warning light
K¹⁰ – Rear window defogger warning light
L¹ – Sealed beam unit, left headlight
L² – Sealed beam unit, right headlight
L⁶ – Speedometer light
L⁸ – Clock light
L¹⁰ – Instrument panel light
M² – Tail- and brake light, right
M⁴ – Tail- and brake light, left
M⁵ – Turn signal and parking light, front, left
M⁶ – Turn signal, rear, left
M⁷ – Turn signal and parking light, front, right

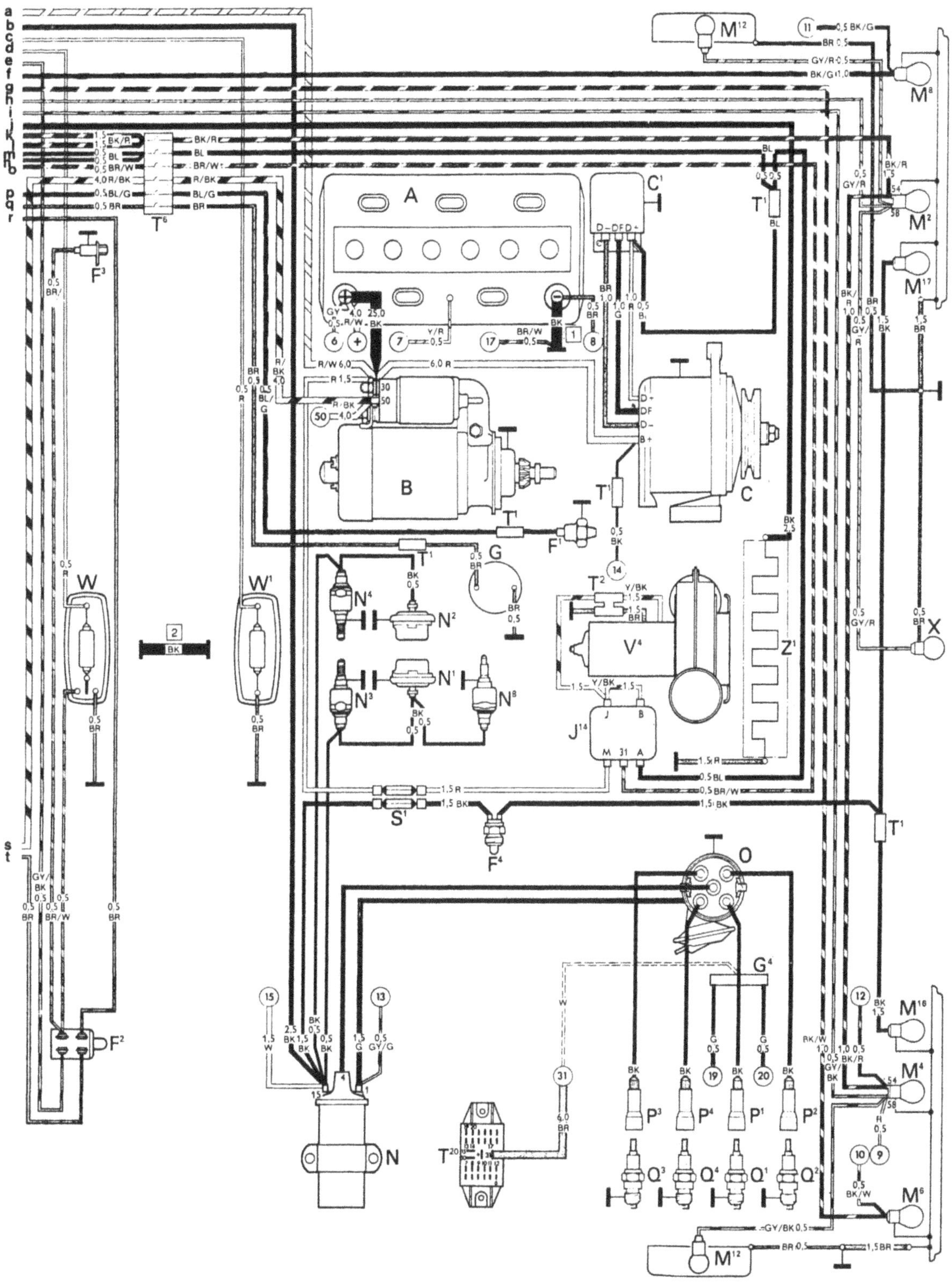

M⁸ – Turn signal, rear, right
M¹² – Sidemarker light, rear
M¹⁶ – Back-up light, left
M¹⁷ – Back-up light, right
N – Ignition coil
N¹ – Automatic choke, left carburetor
N² – Automatic choke, right carburetor
N³ – Electro-magnetic pilot jet, left carburetor
N⁴ – Electro-magnetic pilot jet, right carburetor
N⁸ – Cut-off valve for central idling system
O – Ignition distributor
P¹ – Spark plug connector, No. 1 cylinder
P² – Spark plug connector, No. 2 cylinder
P³ – Spark plug connector, No. 3 cylinder
P⁴ – Spark plug connector, No. 4 cylinder

Q¹ – Spark plug, No. 1 cylinder
Q² – Spark plug, No. 2 cylinder
Q³ – Spark plug, No. 3 cylinder
Q⁴ – Spark plug, No. 4 cylinder
S – Fuse box
S¹ – Fuse for back-up light
T – Cable adapter
T¹ – Cable connector, single
T² – Cable connector, double
T³ – Cable connector, triple
T⁶ – Cable connector, eight connections
T²⁰ – Test network, central plug
V – Windshield wiper motor
V² – Fan motor, front
V⁴ – Heater air blower
W – Interior light, front
W¹ – Interior light, rear

X – License plate light
Y – Clock
Z¹ – Rear window defogger heating element

[1] – Ground strap from battery to body
[2] – Ground strap from transmission to the frame
[4] – Ground cable from horn button to steering coupling

Test network

The numbered circles are the connections in the test network which are wired to the central plug (T²⁰). The numbers correspond to the terminals in the central plug.

G. VW Type 2—from August 1972 (1973 through 1975 Models)

page 1 of 4

	Description	current track
A	– Battery	6
B	– Starter	7, 8
C	– Alternator	1, 2, 3
C^{1}	– Regulator	1, 2, 3
D	– Ignition/starter switch	8, 15, 16
E	– Windshield wiper switch	17, 18
E^{1}	– Light switch	25, 26, 29
E^{2}	– Turn signal switch	44
E^{2}	– Headlight dimmer switch	22
E^{3}	– Emergency flasher switch	44, 46, 47, 48, 50
E^{6}	– Interior light switch, rear	14
E^{9}	– Fresh air fan motor switch	37
E^{15}	– Rear window defogger switch	19, 20
E^{16}	– Heater air blower switch	4
F	– Brake light switch	53, 54, 55, 56
F^{1}	– Oil pressure switch	39
F^{2}	– Door contact and buzzer alarm switch, left H 5	12, 13
F^{3}	– Door contact switch, right	10
F^{4}	– Backup light switch	65
G	– Fuel gauge sending unit	40
G^{1}	– Fuel gauge	40
G^{4}	– Ignition timing sensor	59
G^{7}	– TDC marker unit	57
H	– Horn button	51
H^{1}	– Horn	51
H^{5}	– Ignition key warning buzzer	12, 13
H^{6}	– Buzzer alarm contact in ignition / starter switch	13
J	– Dimmer relay	22, 23, 24
J^{2}	– Emergency flasher relay	44, 45
J^{6}	– Fuel gauge vibrator	40
J^{14}	– Heater air blower relay	4, 5
K^{1}	– High beam warning light	24
K^{2}	– Alternator charging warning light	42
K^{3}	– Oil pressure warning light	39
K^{5}	– Turn signal warning light	41
K^{6}	– Emergency flasher warning light	49
K^{7}	– Dual circuit brake warning light	54, 56
K^{10}	– Rear window defogger warning light	21
L^{1}	– Sealed beam unit, left headlight	23
L^{2}	– Sealed beam unit, right headlight	25
L^{10}	– Instrument panel light	27
L^{21}	– Light for heater lever illumination	27
M^{2}	– Tail / brake light, right	35, 53
M^{4}	– Tail / brake light, left	30, 54
M^{5}	– Turn signal and parking light front, left	32, 44
M^{6}	– Turn signal, rear, left	43
M^{7}	– Turn signal and parking light front, right	33, 46
M^{8}	– Turn signal, rear, right	47
M^{11}	– Sidemarker light front	32, 33
M^{12}	– Sidemarker light, rear	31, 34
M^{16}	– Backup light, left	66
M^{17}	– Backup light, right	65
N	– Ignition coil	60
N^{1}	– Automatic choke, left carburetor	62
N^{2}	– Automatic choke, right carburetor	64
N^{3}	– Electromagnetic pilot jet, left carburetor	61
N^{4}	– Electromagnetic pilot jet, right carburetor	63
N^{8}	– Cutoff valve for central idling system	62
O	– Distributor	58, 60

	Description	current track
P	– Spark plug connectors	59, 60
Q	– Spark plugs	59, 60
S^{1} to S^{12}	– Fuse box	11, 15, 18, 23, 25, 30, 33, 37, 44, 52
S^{21}	– Fuse for backup lights	65
S^{25}	– Fuse for heater air blower	6
T	– Cable adapter behind instrument panel	
T^{1}	– Wire connector, single a – in engine compartment, right b – in engine compartment c – beside the fuel tank, right side d – on the cluster of the instrument panel e – next to the tail / brake light, left	
T^{2}	– Wire connector, double point behind instrument panel	
T^{3a}	– Wire connector, three point behind instrument panel	
T^{3b}	– Wire connector, three point in engine compartment	
T^{8}	– Wire connector, eight point behind instrument panel	
T^{20}	– Test network, test socket	57
V	– Windshield wiper motor	16, 18
V^{2}	– Fan motor	37, 38
V^{4}	– Heater air blower	5
W	– Interior light, front	10
W^{1}	– Interior light, rear	14
X	– License plate light	36
Z^{1}	– Rear window defogger heating element	19

	Description	current track
[1]	– Ground strap from battery to body	6
[2]	– Ground strap from transmission to frame	1
[4]	– Ground strap via steering coupling	52
[10]	– Ground connection, dashboard	
[11]	– Ground connection, speedometer	

Test network

The numbered circles are the connections in the test network which are wired to the central plug (T^{20}). The numbers correspond to the terminals in the central plug.

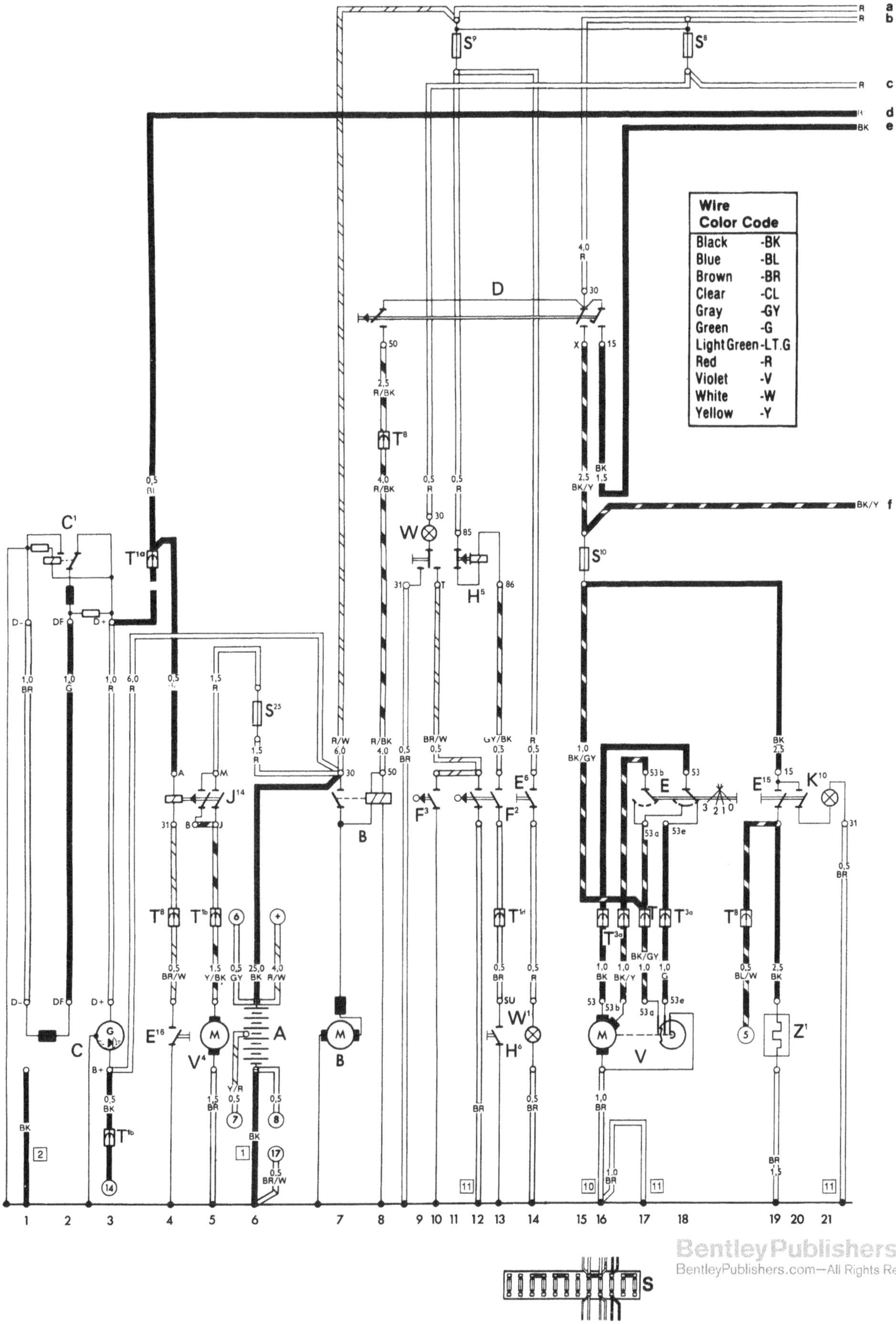

Wire Color Code	
Black	-BK
Blue	-BL
Brown	-BR
Clear	-CL
Gray	-GY
Green	-G
Light Green	-LT.G
Red	-R
Violet	-V
White	-W
Yellow	-Y

G. VW Type 2—from August 1972 (1973 through 1975 Models) page 3 of 4

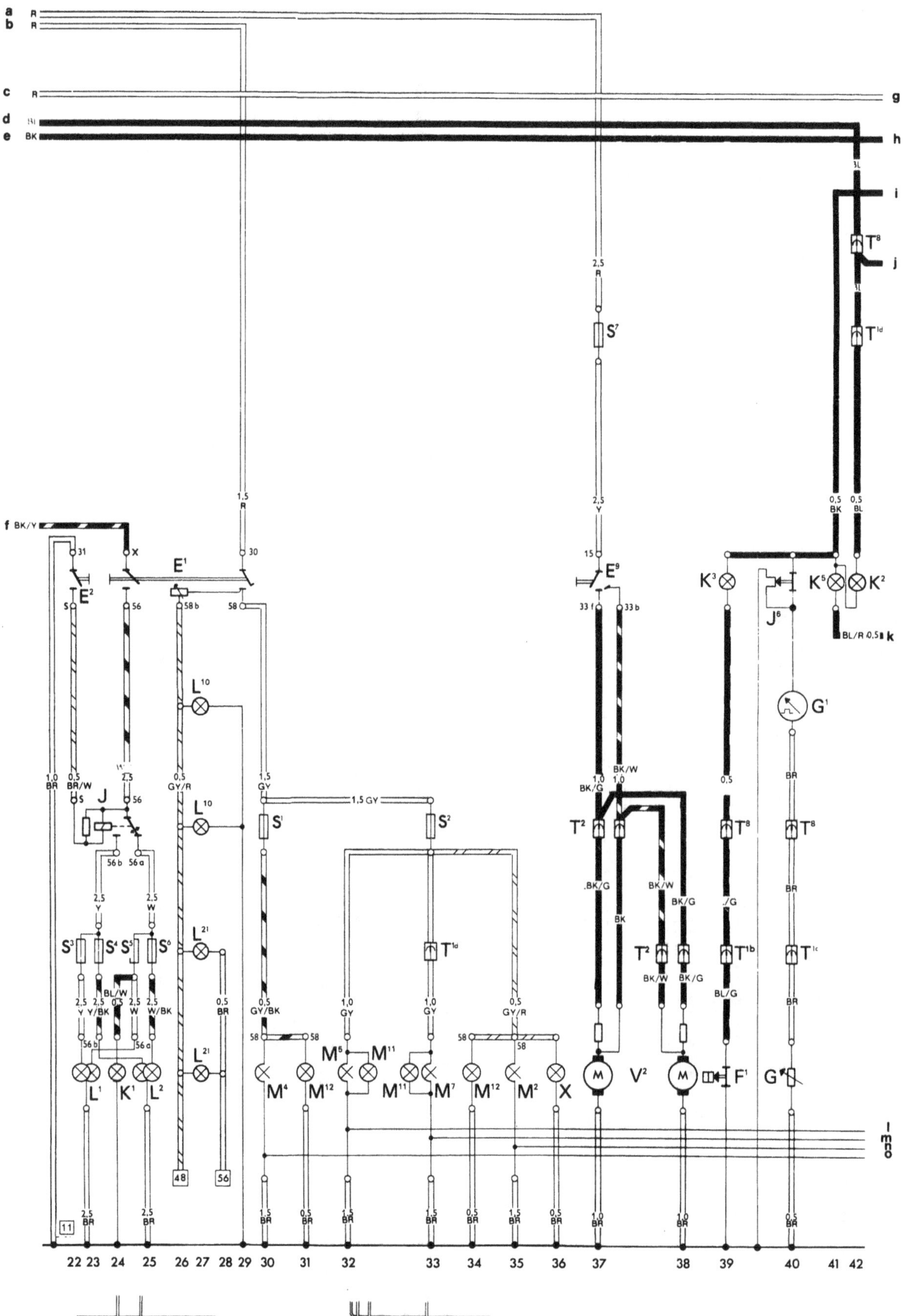

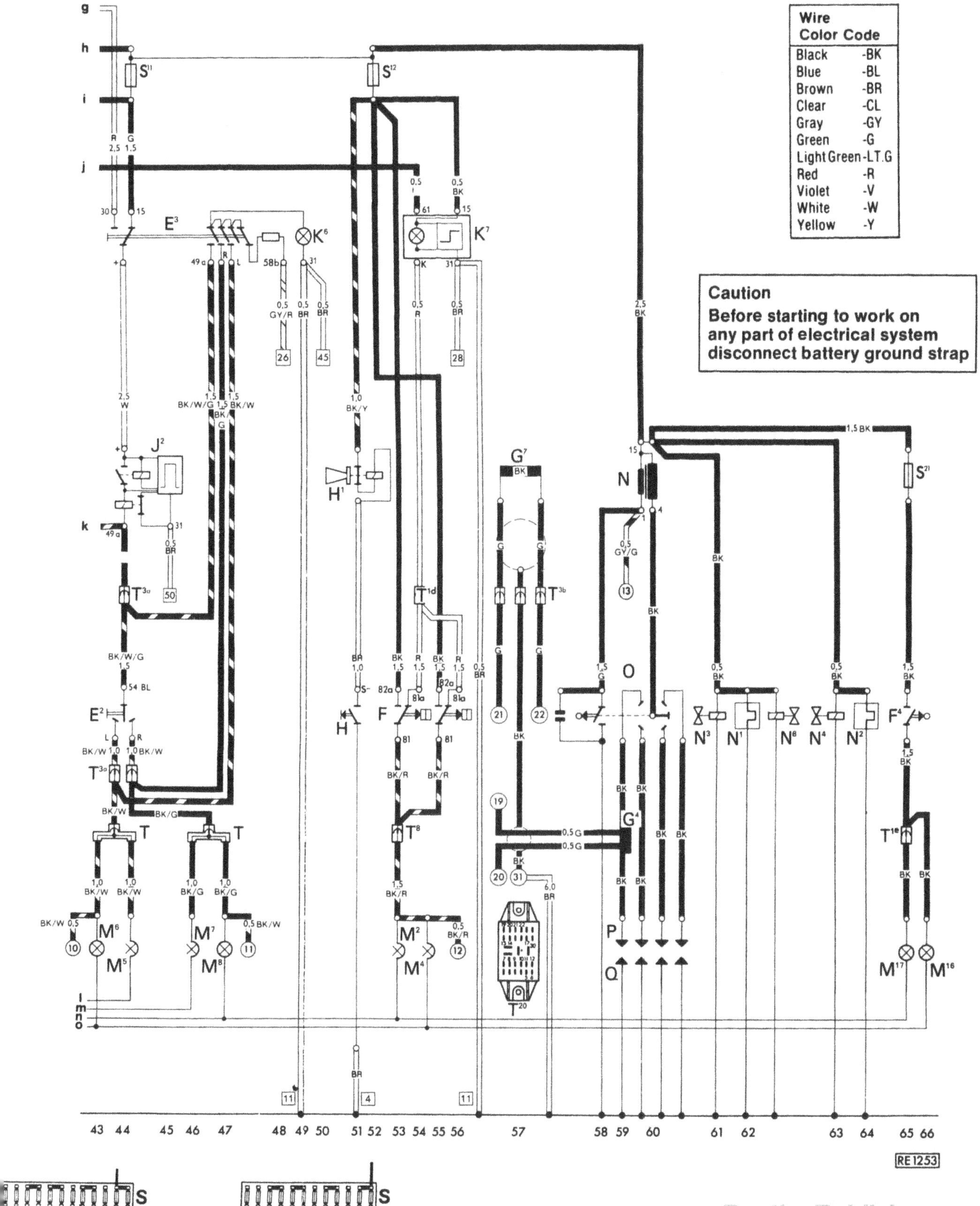

Wire Color Code	
Black	-BK
Blue	-BL
Brown	-BR
Clear	-CL
Gray	-GY
Green	-G
Light Green	-LT.G
Red	-R
Violet	-V
White	-W
Yellow	-Y

Caution

Before starting to work on any part of electrical system disconnect battery ground strap

Supplement to Wiring Diagram G

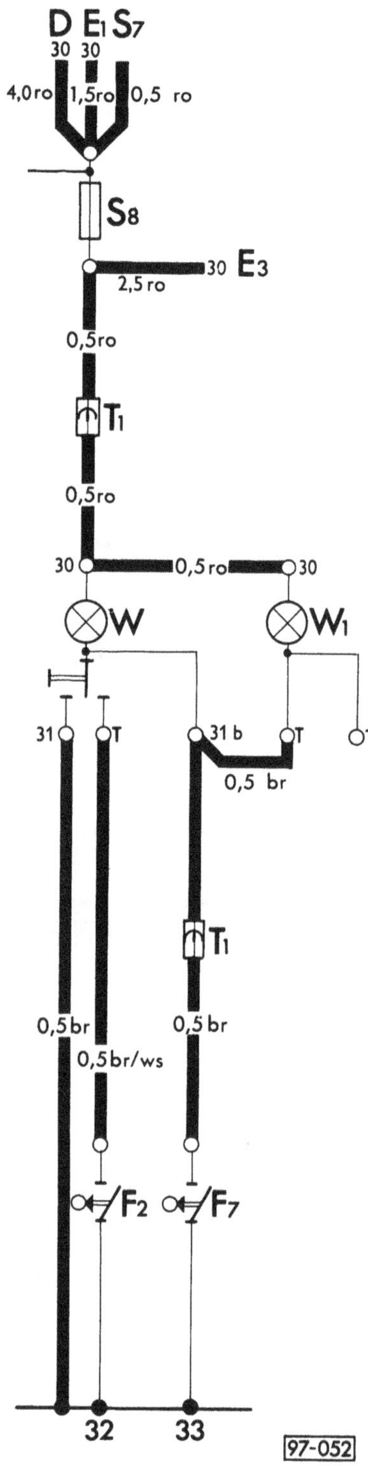

Interior Light Circuit Diagram

The 1974 current flow wiring diagram (Diagram G) remains unchanged for the 1975 model year except for the interior light circuit.

D - to ignition starter switch
E1 - to lighting switch
E3 - to emergency light switch
F2 - Door contact switch (driver's door)
F7 - Door contact switch (sliding door)
T1 - Wire connector—single
S7 - Fuse No. 7 (in fuse box)
S8 - Fuse (in fuse box)
W - Interior light, front
W1 - Interior light, rear

Color Codes

br - brown
ws - white
ro - red

Backup Lights, Starting Control, Shift Console Illumination circuit for VW Type 2 from August 1975 (1976 Models)

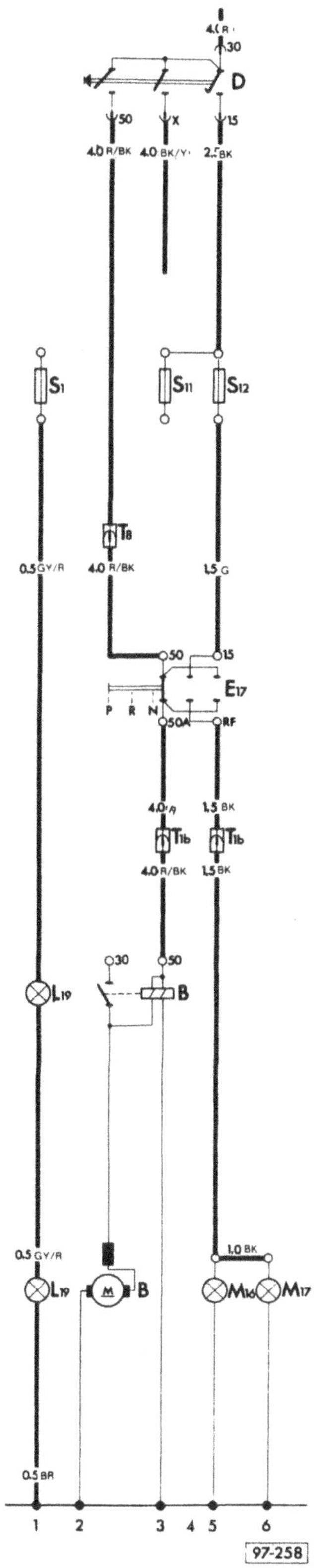

CAUTION

Before starting to work on any part of electrical system disconnect battery ground strap

Description		Current track
B	- Starter	2, 3
D	- Ignition/starter switch	2, 3, 5
E 17	- Starter cutout and backup lights switch	3, 4, 5
L 19	- Shift console light	1
M 16	- Backup light, left	5
M 17	- Backup light, right	6
S 1		
S 11	- Fuses in fuse box	
S 12		
T 1b	- Wire connector, single; behind dashboard	
T 8	- Wire connector, 8 point; behind dashboard	

Color code

br = brown
ge = yellow
gn = green
gr = grey
ro = red
sw = black

Wire Color Code	
Black	-BK
Blue	-BL
Brown	-BR
Clear	-CL
Gray	-GY
Green	-G
Light Green	-LT.G
Red	-R
Violet	-V
White	-W
Yellow	-Y

I.

VW Type 2—from August 1975 (1976 Models)
page 1 of 5

	Description	Current track
A	- Battery	7
B	- Starter	8, 9
C	- Alternator	1, 2, 3
C 1	- Regulator	1, 2, 3
D	- Ignition / starter switch	9, 10, 11, 12, 13, 14, 15, 16, 17
E	- Windshield wiper switch	18, 19
E 1	- Light switch	26,27,28,29,30,31
E 2	- Headlight dimmer switch	22
E 9	- Fan motor switch	40, 41
E 15	- Rear window defogger switch	20, 21
E 16	- Heater air blower switch	4
F 2	- Door contact and buzzer alarm switch, left	13, 14, 15
F 7	- Sliding door switch	10
H 5	- Ignition key warning buzzer	15
J	- Dimmer relay	22–28
J 14	- Heater air blower relay	4–6
K 1	- High beam warning light	25
K 10	- Rear window defogger warning light	21
L 1	- Sealed beam unit, left headlight	23
L 2	- Sealed beam unit, right headlight	28
L 10	- Instrument panel light	30
L 21	- Light for heater lever illumination	30
M 2	- Tail / brake light, right	38
M 4	- Tail / brake light, left	31
M 5	- Parking light, left	32
M 7	- Parking light, right	36
M 11	- Sidemarker light, front	34, 35
M 12	- Sidemarker light, rear	32, 37
S 1 to S 10	Fuse box	
S 25	- Fuse for heater air blower	7
T	- Cable adapter behind dashboard	18
T 1	- Wire connector, single	
	a - in engine compartment, right	4
	b - in engine compartment	6
T 2	- Wire connector, double behind dashboard	40, 41
T 30	- Wire connector, three connections behind dashboard	19, 17
T 8	- Wire connector, eight connections behind dashboard	47, 20
V	- Windshield wiper motor	17–19
V 2	- Fan motor	40, 41
V 4	- Heater air blower	6
W	- Interior light, front	12
W 1	- Interior light, rear	11
X	- License plate light	39
Z 1	- Rear window defogger heating element	20
[1]	- Ground strap from battery to body	7
[2]	- Ground strap from transmission to frame	1
[10]	- Ground connection, dashboard	17
[11]	- Ground connection, speedometer	13, 19, 21

	Description	Current tr
E 2	- Turn signal	52
E 3	- Emergency flasher switch	48–57
F	- Brake light switch	62–64
F 1	- Oil pressure switch	44
F 4	- Backup light switch	65
F 9	- Parking brake control light switch	54
F 24	- Elapsed CAT mileage odometer	42
F 25	- Throttle valve switch	82
F 26	- Thermo switch for cold starting valve	83, 84
F 27	- Elapsed EGR mileage odometer	43
G	- Fuel gauge sending unit	45
G 1	- Fuel gauge 45	45
G 6	- Fuel pump	72
G 7	- TDC marker unit	67
G 18	- Temperature sensor on cylinder head	77
G 19	- Intake air sensor	77–79
H	- Horn button	60
H 1	- Horn	60
J 2	- Emergency flasher relay	48–50
J 6	- Voltage stabilizer	45
J 21	- Control unit (electronic)	73–82
J 40	- Double relay	76–81
K 2	- Alternator charging warning light	47
K 3	- Engine oil pressure light	44
K 5	- Turn signal warning light	46
K 6	- Emergency flasher warning light	57
K 7	- Dual circuit brake warning light	63
K 21	- CAT converter warning light	42
K 22	- EGR warning light	43
M 5	- Turn signal, front left	52
M 6	- Turn signal, rear left	51
M 7	- Turn signal, front right	53
M 8	- Turn signal, rear right	55
M 9	- Brake light, left	64
M 10	- Brake light, right	62
M 16	- Backup light, left	66
M 17	- Backup light, right	65
N	- Ignition coil	71
N 17	- Cold starting valve	83
N 21	- Auxiliary air regulator	81
N 30	- Injector, cyl. 1	
N 31	- Injector, cyl. 2	
N 32	- Injector, cyl. 3	
N 33	- Injector, cyl. 4	
N 34	- Resistance for Injectors	73–76
O	- Ignition distributor	70–71
P	- Spark plug connectors	70–71
Q	- Spark plugs	70–71
S 11 to S 12	- Fuse box	48, 62
S 21	- Fuse for backup light	65
T 1	- Wire connector, single	42, 43
	b - engine compartment	
	c - fuel tank right side	45
	d - behind dashboard	64
	e - near tail light	65
T 3a	- Wire connector, three point, behind dashboard	52,53
T 3b	- Wire connector, three point, in engine compartment	67
T 8	- Wire connector, eight point, behind dashboard	44, 45, 47,
T 20	- Test network socket	68
[4]	- Ground strap, steering joint	60
[11]	- Ground connector, speedometer	

> **Caution**
>
> Before starting to work on any part of electrical system disconnect battery ground strap

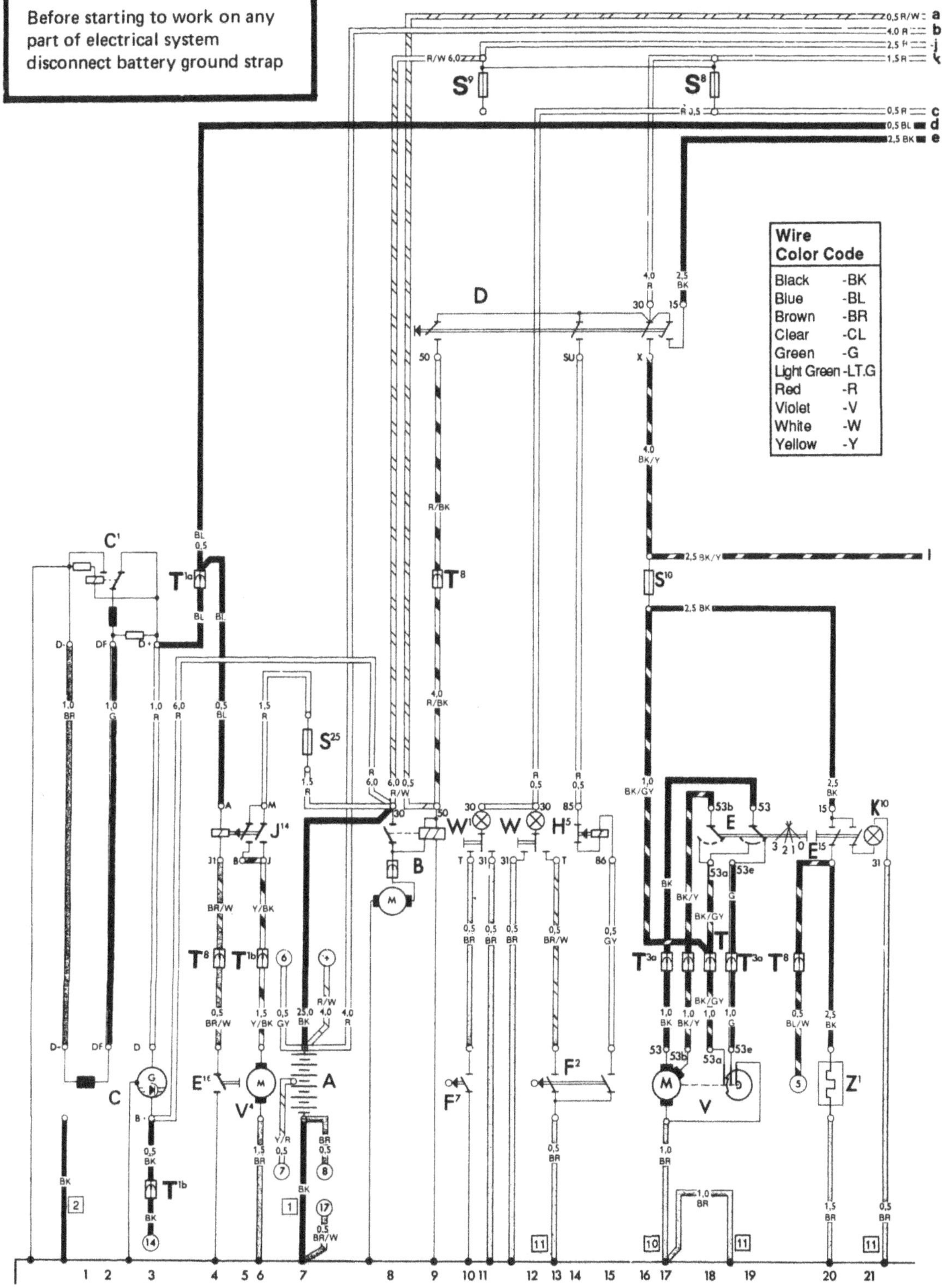

Wire Color Code	
Black	-BK
Blue	-BL
Brown	-BR
Clear	-CL
Green	-G
Light Green	-LT.G
Red	-R
Violet	-V
White	-W
Yellow	-Y

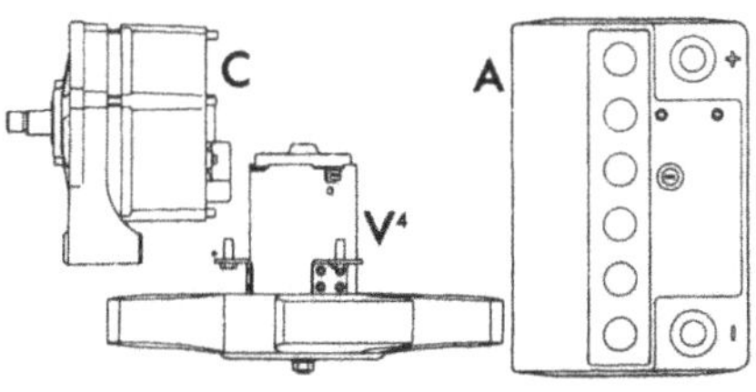

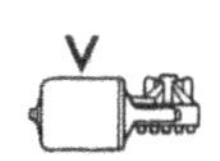

Test network

The numbered circles are the connections in the test network which are wired to the central plug (T[20]). The numbers correspond to the terminals in the central plug.

VW Type 2—from August 1975 (1976 Models)
page 3 of 5

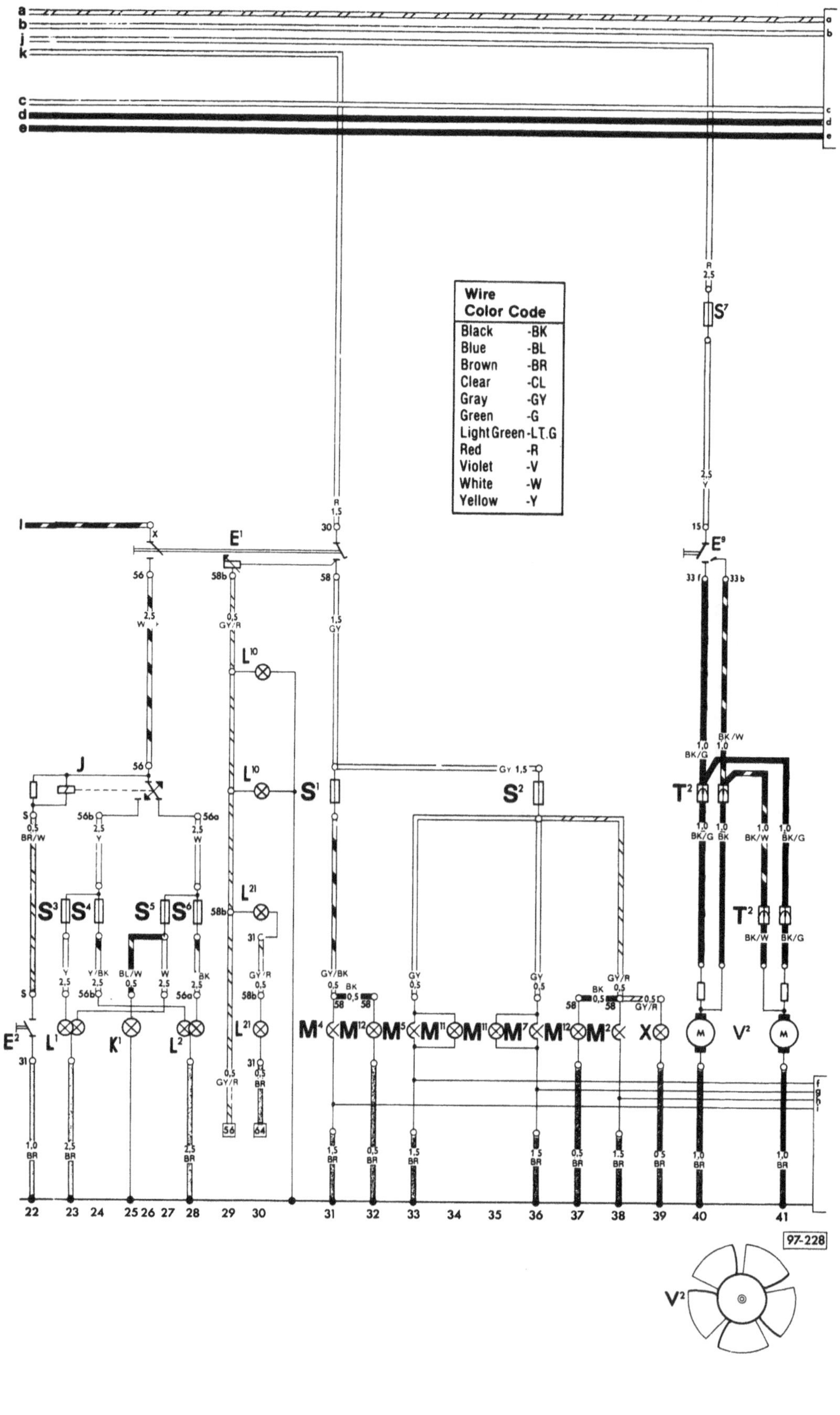

Wire Color Code	
Black	-BK
Blue	-BL
Brown	-BR
Clear	-CL
Gray	-GY
Green	-G
Light Green	-LT.G
Red	-R
Violet	-V
White	-W
Yellow	-Y

97-228

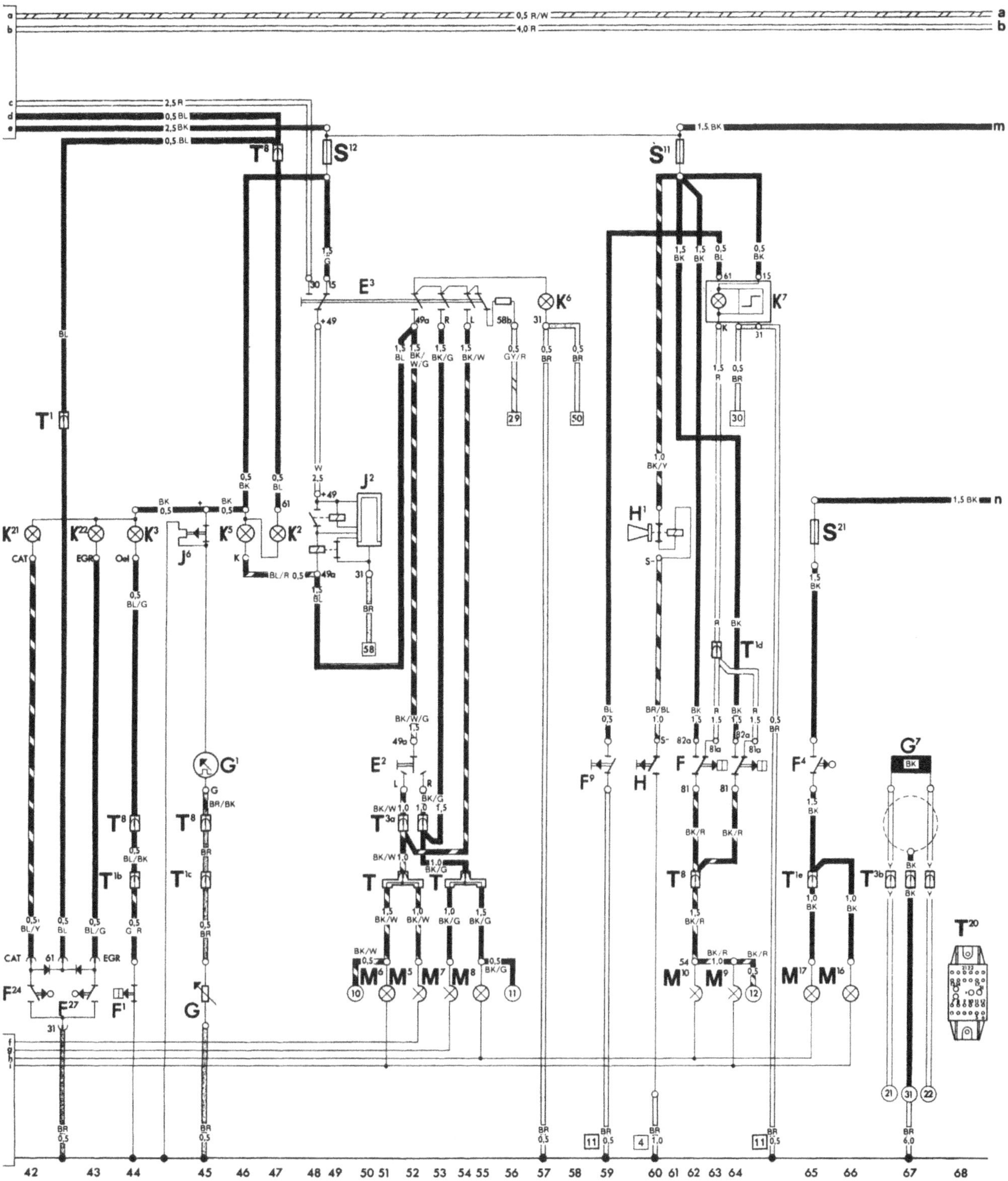

Caution—
Before starting to work on any part of electrical system, disconnect battery ground strap.

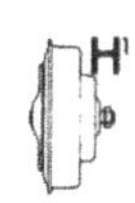

Test network
The numbered circles are the connections in the test network which are wired to the central plug (T[20]). The numbers correspond to the terminals in the central plug.

I.

VW Type 2—from August 1975 (1976 Models) page 5 of 5

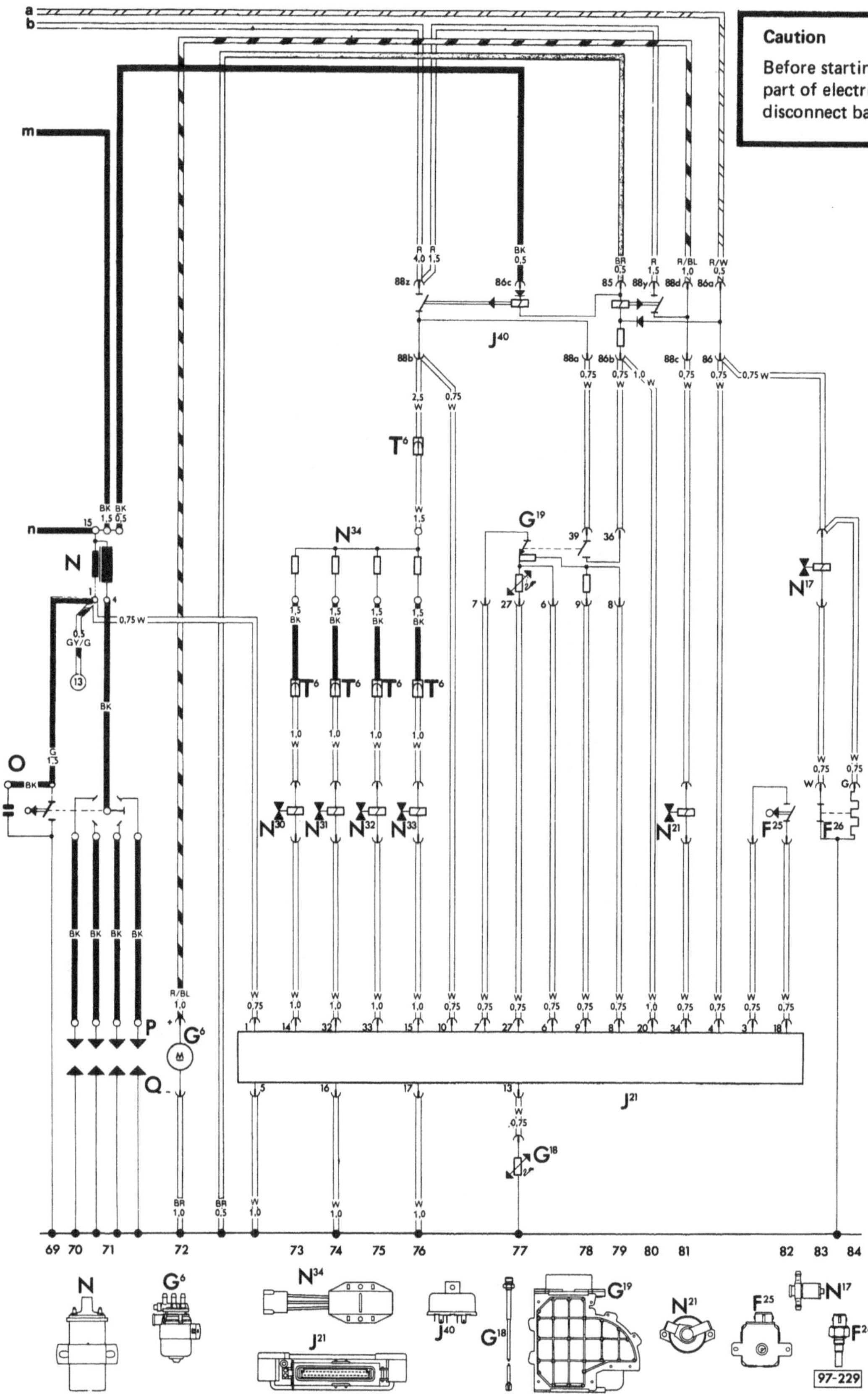

Caution

Before starting to work on any part of electrical system disconnect battery ground strap

Wire Color Code	
Black	-BK
Blue	-BL
Brown	-BR
Clear	-CL
Gray	-GY
Green	-G
Light Green	-LT.G
Red	-R
Violet	-V
White	-W
Yellow	-Y

VW Type 2—from August 1976 (1977 and 1978 Models) page 1 of 3

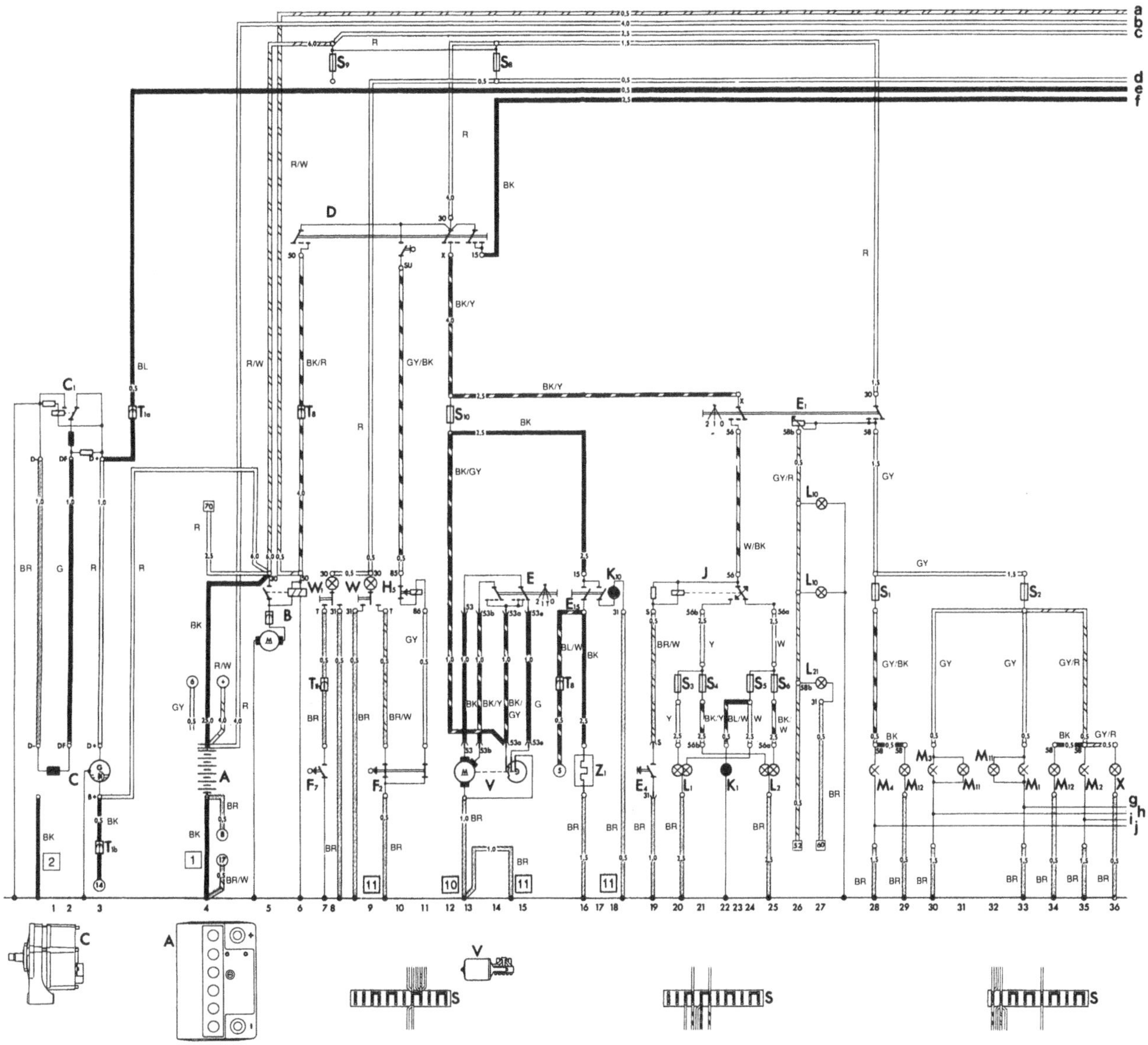

Description			Current track
A	-	Battery	4
B	-	Starter	5, 6
C	-	Alternator	1, 2, 3
C 1	-	Regulator	1, 2, 3
D	-	Ignition/starter switch	6–13
E	-	Windshield wiper switch	13–15
E 1	-	Light switch	22–28
E 4	-	Headlight dimmer switch	19
E 9	-	Fan motor switch	37
E 15	-	Rear window defogger switch	16–18
F 2	-	Door contact and buzzer alarm switch, left	10
F 7	-	Sliding door switch	7
H 5	-	Ignition key warning buzzer	11
J	-	Headlight dimmer relay	19–25
K 1	-	High beam warning light	22
K 10	-	Rear window defogger warning light	18
L 1	-	Sealed beam unit, left headlight	20
L 2	-	Sealed beam unit, right headlight	25
L 10	-	Instrument panel light	27
L 21	-	Light for heater lever illumination	27

M 1	-	Parking light, left	33
M 2	-	Taillight, right	35
M 3	-	Parking light, right	30
M 4	-	Taillight, left	28
M 11	-	Sidemarker light, front	31, 32
M 12	-	Sidemarker light, rear	29, 34
S 1 to S 10	-	Fuse box	
T 1	-	Wire connector, single a—in engine compartment, right b—in engine compartment, right h—next to interior light front	
T 2	-	Wire connector, double	
T 8	-	Wire connector, eight-point	13–15
V	-	Windshield wiper motor	37, 38
V 2	-	Fan motor	37, 38
W	-	Interior light, front	9
W 1	-	Interior light, rear	8
X	-	License plate light	36
Z 1	-	Rear window defogger heating element	16

J. VW Type 2—from August 1976 (1977 and 1978 Models) page 2 of 3

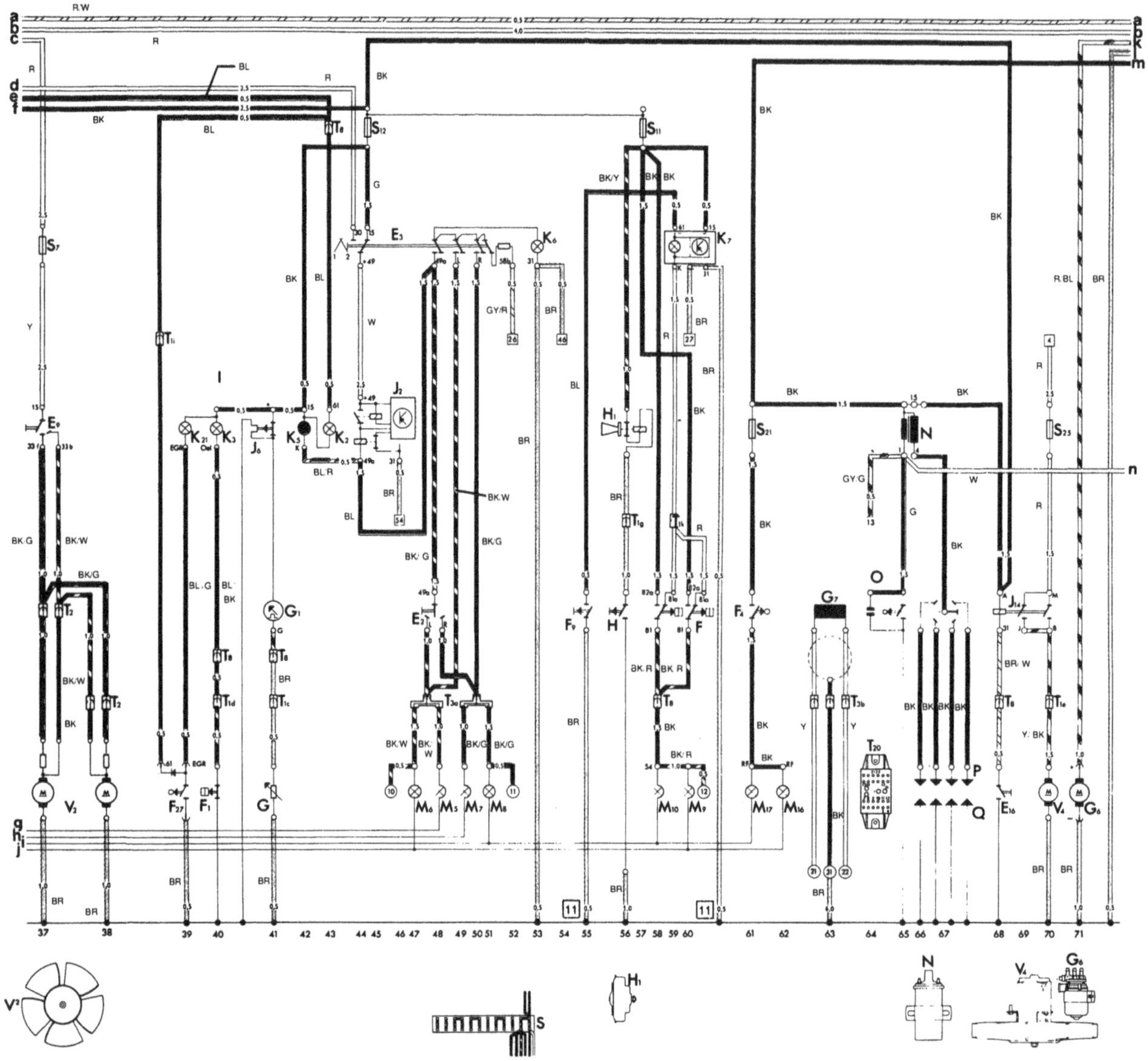

Wire Color Code	
Black	-BK
Blue	-BL
Brown	-BR
Clear	-CL
Green	-G
Light Green	-LT.G
Red	-R
Violet	-V
White	-W
Yellow	-Y

CAUTION ——

Disconnect the battery ground strap before starting to work on any part of the electrical system.

[1] - Ground strap from battery to body

[2] - Ground strap from transmission to frame

[10] - Ground connection, dashboard

[11] - Ground connection, speedometer

Test network

The numbered circles are the connections in the test network which are wired to the central plug (T^{20}). The numbers correspond to the terminals in the central plug.

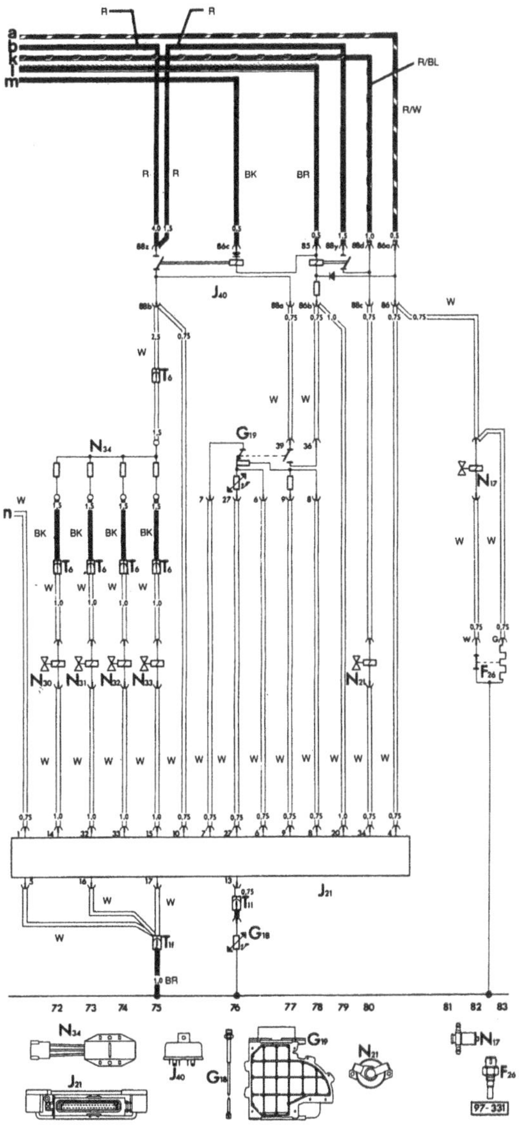

Wire Color Code	
Black	-BK
Blue	-BL
Brown	-BR
Clear	-CL
Green	-G
Light Green	-LT.G
Red	-R
Violet	-V
White	-W
Yellow	-Y

Description			Current track
E 2	–	Turn signal	48
E 3	–	Emergency flasher switch	44–52
E 16	–	Heater air blower switch	68
F	–	Brake light switch	57–60
F 1	–	Oil pressure switch	40
F 4	–	Back-up light switch	61
F 9	–	Parking brake control light switch	55
F 26	–	Thermo time switch for cold starting valve	82, 83
F 27	–	EGR elapsed mileage switch	39
G	–	Fuel gauge sending unit	41
G 1	–	Fuel gauge	41
G 6	–	Fuel pump	71
G 7	–	TDC marker unit	63
G 18	–	Temperature sensor on cylinder head	76
G 19	–	Intake air sensor	76–78
H	–	Horn button	56
H 1	–	Horn	56
J 2	–	Emergency flasher relay	44–47
J 6	–	Voltage stabilizer	41
J 14	–	Heater air blower relay	68–70
J 21	–	Control unit (electronic)	72–80
J 40	–	Double relay	75–80
K 2	–	Alternator charging warning light	43
K 3	–	Engine oil pressure light	40
K 5	–	Turn signal warning light	42
K 6	–	Emergency flasher warning light	53
K 7	–	Dual circuit brake warning light	59
K 21	–	CAT converter warning light	39
M 5	–	Turn signal, front left	48
M 6	–	Turn signal, rear left	47
M 7	–	Turn signal, front right	49
M 8	–	Turn signal, rear right	51
M 9	–	Brake light, left	60
M 10	–	Brake light, right	58
M 16	–	Back-up light, left	62
M 17	–	Back-up light, right	61
N	–	Ignition coil	65, 66
N 17	–	Cold starting valve	82
N 21	–	Auxiliary air regulator	80
N 30	–	Injector, cyl. 1	72
N 31	–	Injector, cyl. 2	73
N 32	–	Injector, cyl. 3	74
N 33	–	Injector, cyl. 4	75
N 34	–	Resistance for injectors	72–75
O	–	Ignition distributor	64–65
P	–	Spark plug connectors	66–67
Q	–	Spark plugs	66–67
S 11 to S 12	–	Fuse box	
S 21	–	Fuse for back-up light	61
S 25	–	Fuse for heater air blower	70
T 1	–	Wire connector, single	
c	–	fuel tank right side	
d	–	in engine compartment	
e	–	in engine compartment	
f	–	in engine compartment	
g	–	behind dashboard	
i	–	behind dashboard	
k	–	behind dashboard panel	
l	–	in engine compartment, center	
T 3a	–	Wire connector, three-point	
T 3b	–	Wire connector, three-point	
T 6	–	Wire connector, six-point	
T 8	–	Wire connector, eight-point	
T 20	–	Analysis system, analysis socket	
V 4	–	Heater air blower	70
11	–	**Ground connector, speedometer**	

4

K. VW Type 2—from August 1978 (1979 Models)
page 1 of 3

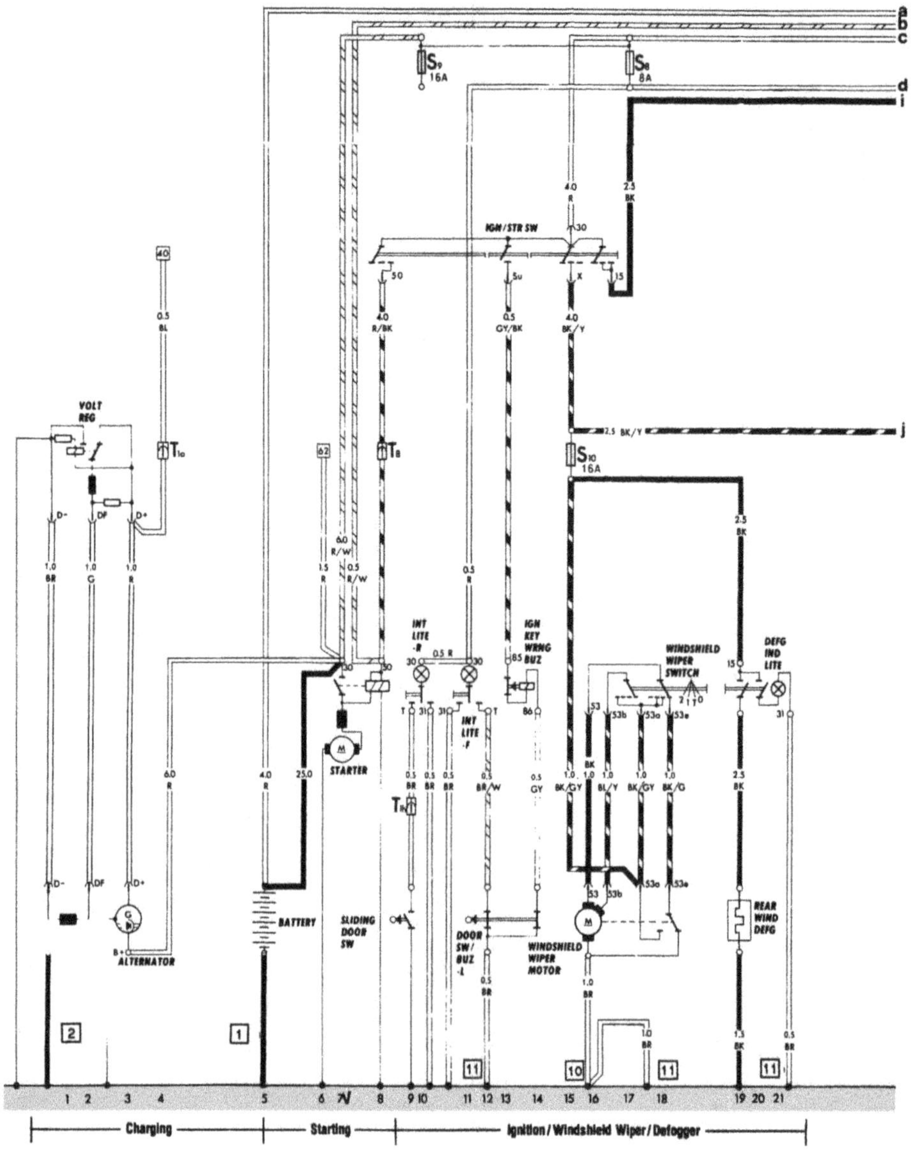

Description	Current track
Alternator	1–3
Alternator charging light	40
Battery	5
Directional signal indicator light	44
Door switch/buzzer (right)	15
Door switch/buzzer (left)	12–14
Fresh air fan ballast resistor	25
Fuel gauge	43
Fuel gauge sending unit	43
Headlights	23–28
Headlight dimmer relay	22–28
Headlight dimmer switch	22
Heater lever light	30
High beam indicator light	25
Ignition key warning buzzer	13–14
Ignition/starter switch	8–17
Instrument panel lights	30
Instrument panel dimmer	29
Interior light, front	11
Interior light, rear	10
License plate light	39
Light switch	25–31
Oil pressure switch	42
Oil pressure warning light	42
Oxygen sensor elapsed mileage switch	41
Oxygen sensor light	41
Parking light, left front	36
Parking light, right front	33
Rear window defogger	19
Rear window defogger indicator light	21
Rear window defogger switch	19–21
Side marker light, front	34, 35
Side marker light, rear	32, 37
Sliding door switch	9
Starter	6–8
Taillight, left	31
Taillight, right	38
Voltage regulator	1–3
Voltage stabilizer	43
Windshield wiper motor	16–18
Windshield wiper switch	16–18

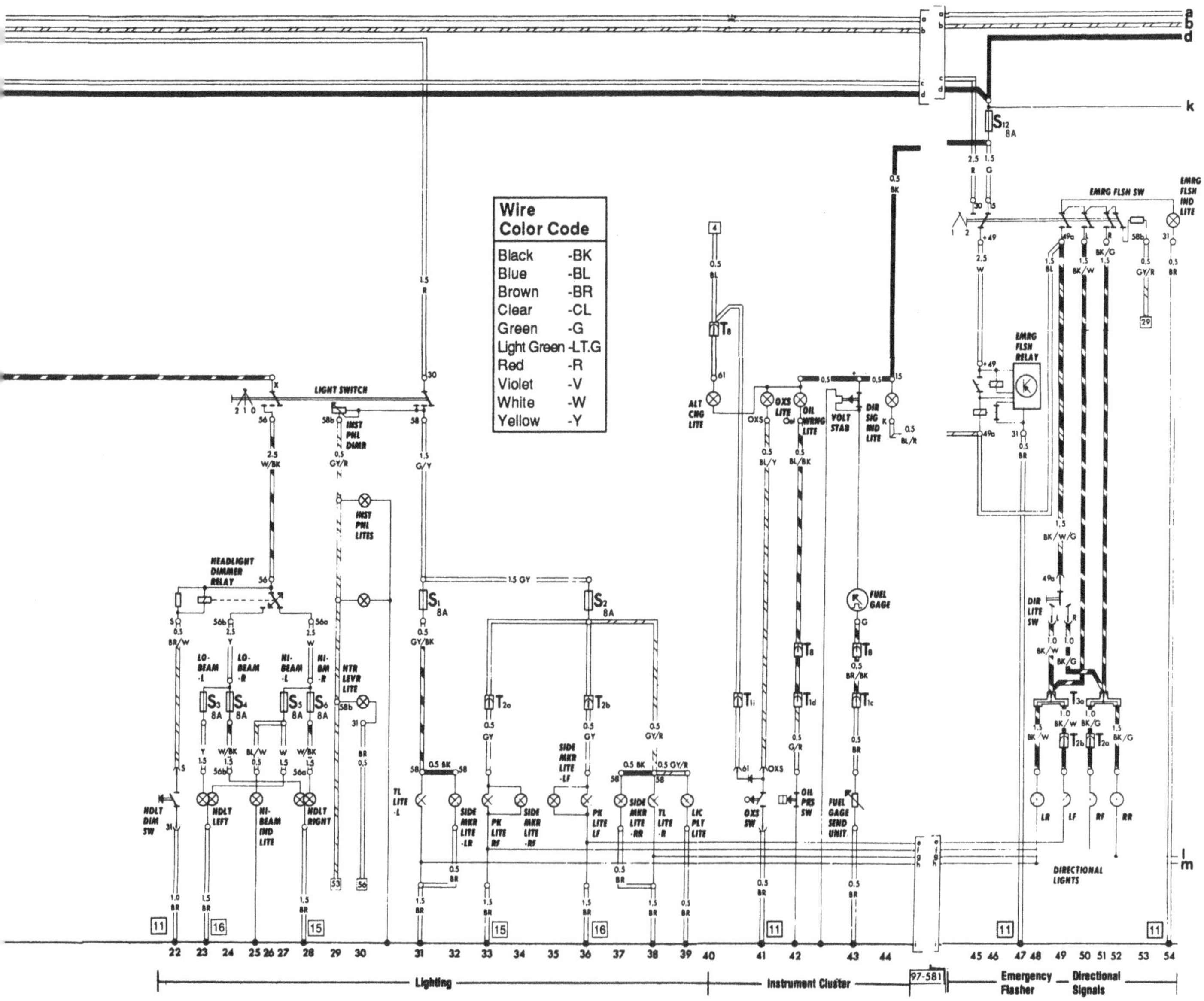

T 1a – Wire connector, single; in engine compartment, right
T 1c – Wire connector, single; fuel tank, right side
T 1d – Wire connector, single; in engine compartment
T 1h – Wire connector, single; next to interior light, front
T 1i – Wire connector, single; behind instrument panel
T 2a – Wire connector, double; rear of dashboard, right
T 2b – Wire connector, double; rear of dashboard, left
T 8 – Wire connector, 8-point
[1] – Ground connector, battery to body
[2] – Ground connector, transmission to body
[10] – Ground connector, dashboard
[11] – Ground connector, speedometer
[15] – Ground connector, headlight frame, right
[16] – Ground connector, headlight frame, left

CAUTION ——
Disconnect the battery ground strap before starting to work on any part of the electrical system.

Description	Current track
Auxiliary air regulator	77
Back-up lights	63–64
Back-up light switch	63
Brake lights	55–56
Brake light switch	55–56
Brake warning light	55–58
Cold-start valve	79
Coolant temperature sensor	75
Directional lights	48–52
Directional light switch	48–49
Distributor sending unit	68
Emergency flasher relay	78–81
Emergency flasher switch	45–54
Emergency flasher indicator light	54
Exhaust gas regulator	74
Fresh air fan	62
Fresh air fan switch	60
Fuel injection control unit	69–77
Fuel injection double relay	74–79
Fuel injectors	69–72

4

VW Type 2–from August 1978 (1979 Models)
page 3 of 3

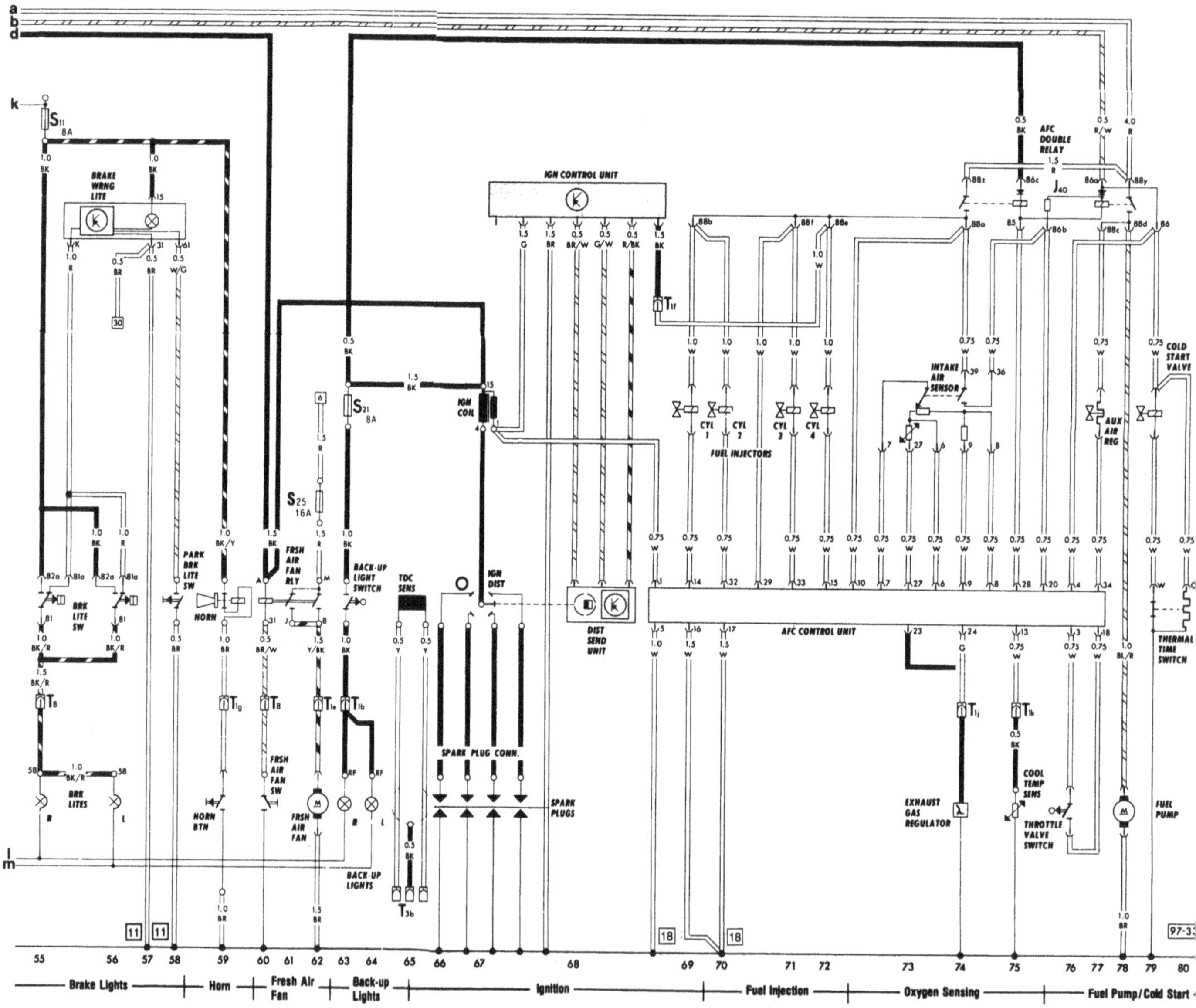

Fuel pump	78
Fuses S11, S12, on fuse/relay panel (under dash)	
Fuse S21 in engine compartment	63
Fuse S25 in engine compartment	62
Horn	59
Horn button	59
Ignition coil	67
Ignition control unit	67–69
Ignition distributor	67
Parking brake indicator light switch	58
Spark plug connectors	66–67
Spark plugs	66–67
Thermo-time switch	79–80
Throttle valve switch	76
Top dead center sensor	65

T 1b – Wire connector, single; in engine compartment, left
T 1e – Wire connector, single; in engine compartment
T 1f – Wire connector, single; in engine compartment
T 1g – Wire connector, single; behind instrument panel
T 1j – Wire connector, single, in engine compartment
T 1k – Wire connector, single; in engine compartment

T 2a – Wire connector, double; rear of dashboard, right
T 2b – Wire connector, double; rear of dashboard, left
T 3a – Wire connector, 3-point; rear of dashboard
T 3b – Wire connector, 3-point; in engine compartment
T 8 – Wire connector, 8-point
[11] – Ground connector, speedometer
[18] – Ground connector, engine

Color code

BK – Black
BR – Brown
CL – Clear
R – Red
Y – Yellow
G – Green
BL – Blue
V – Violet
GY – Gray
W – White

Section 5

ENGINE AND CLUTCH

Contents

5

TABLES

5

Engine and Clutch

Type 2 vehicles built during model years 1968 through 1971 are equipped with the VW 1600 engine. Its actual displacement is 1584 cc (96.6 cu. in.). The 1700 engine used on 1972 and 1973 models is a different design that shares no important parts with the earlier engine. The 1700 engine displaces 1679 cc (102.5 cu. in.). The 1800 engine, which was introduced on the 1974 models, is basically the same as the 1700 engine with the cylinder bore enlarged from 90 mm (3.543 in.) to 93 mm (3.661 in.). The larger bore diameter increased the displacement to 1795 cc (109.5 cu. in.). Beginning with the 1976 models, the same basic engine became a 2000 engine with a bore of 94 mm (3.701 in.) and a stroke of 71 mm (2.795 in.), giving it a displacement of 1970 cc (120.2 cu. in.).

The 1600 engine has a single carburetor. All 1700 engines have dual carburetors. On 1974 models, the 1800 engine is equipped with dual carburetors. The 1975 1800 engine and the 1976 2000 engine are equipped with an electronic fuel injection system. All four engines are four-cycle powerplants, rear mounted, and air-cooled. All have pushrod-operated overhead valves. As on other VW air-cooled engines, the four cylinders are divided into two horizontally-opposed pairs. The 1600 engines have DC generators and the 1700, 1800, and 2000 engines have alternators. All four engines have conventional coil and battery ignition systems.

Some VWs are equipped with a network of test wiring for the VW Computer Analysis system. On the engine there is a sensor in the high tension cable leading to the No. 1 cylinder for obtaining spark timing data, and a battery-testing wire attached to the starter solenoid. Another timing sensor is installed in the flywheel bellhousing. All sensor wires must remain properly attached if the VW Computer Analysis system is to work as designed. Never connect any device other than the test plug of the VW Computer Analysis system to the test network central socket in the engine compartment. Incorrect equipment could damage the plug connectors, the test sensors, or the components containing them.

The information in this section of the Manual is intended to serve as a guide to both car owners and professional mechanics. Some of the operations may require special equipment and experience that only a trained VW mechanic will normally have. If you lack the skills, tools, or a suitable workshop for servicing the engine, we suggest you leave such repairs to an Authorized VW Dealer or other qualified shop. We especially urge you to consult your Authorized VW Dealer before attempting any repairs on a car still covered by the new-car warranty.

1. GENERAL DESCRIPTION

Throughout this section of the Manual we will differentiate between the single-carburetor, dual-carburetor, and fuel injection engines only where there are significant design differences or different specifications. Anything not designated as applying to one engine or another applies to all.

Engine Mounting

Four threaded fasteners join the engine and transmission. The front (flywheel end) of the engine is supported by the transmission while the rear of the engine is supported by an engine bearer bolted to the vehicle frame.

Crankcase

The two-piece crankcase is cast in lightweight alloy and so precisely machined that the halves fit together without a gasket. If either half becomes defective, both must be replaced to ensure an exact fit.

Crankshaft and Bearings

The crankshaft rotates in four main bearings. Only one (No. 2) is of the split-shell type. The No. 1 bearing ring is flanged to take crankshaft end thrust. A seal at the clutch end of the crankshaft and a slinger at the pulley end prevent oil leakage.

Valve Train

The camshaft is gear driven off the crankshaft and runs in three split-shell bearings. A Woodruff key positions the crankshaft gear while the camshaft gear is riveted in place. Solid lifters, pushrods, and adjustable rocker arms make up the valve-operating linkage.

Cylinder Heads

Each pair of cylinders is covered by a deep-finned cast aluminum alloy cylinder head. The valve guides and seats are pressed in but they can be replaced only with special machinery. No cylinder head gaskets are used. However, oil seals are fitted at both ends of each pushrod tube.

Connecting Rods

The connecting rods are steel forgings. Split-shell bearings are used at the crankshaft end and lead-bronze coated steel bushings at the piston pin end.

Pistons and Cylinders

The pistons are of the three-ring type with full-floating pins secured by circlips. The cylinders are detachable and, with their matched pistons, fully interchangeable. The cylinders are cast iron and finned for cooling.

Flywheel and Drive Plate

The flywheel of single-carburetor engines is mounted on the crankshaft with four steel dowels and a gland nut. The nut contains the needle bearing that serves as a pilot bearing for the transmission's rear driveshaft. On dual-carburetor and fuel injection engines, the flywheel is mounted on the crankshaft with five self-locking bolts and the pilot bearing is a press-fit in the flywheel hub. Instead of a flywheel, engines used with automatic transmissions have a drive plate, to which the torque converter is bolted.

Cooling System

A fan forces cooling air over the finned cylinders and clyinder heads. On single-carburetor engines the fan is mounted on one end of the generator's armature shaft; on dual-carburetor and fuel injection engines the fan is mounted on the crankshaft. Thermostat-operated flaps regulate the volume of cooling air according to engine heat. Some air can be diverted through the heat exchangers. Thus warmed, it is used to heat the car's interior.

Emission Controls

All the engines covered by this Manual have closed PCV (positive crankcase ventilation) systems. Single-carburetor engines have a throttle valve positioner—supplemented, on 1971 models, by a throttle valve damper. Dual-carburetor engines have the throttle valve damper only. Fuel injection engines have a deceleration air enrichment valve. Exhaust gas recirculation is used on 1973 and later models. An air injection system for exhaust afterburning is used on 1973 and 1974 dual-carburetor engines. Some fuel injection engines have a catalytic converter for emission afterburning, and the 1979 and later California models have an oxygen sensor system. Emission control servicing and repair are covered in **FUEL SYSTEM** and in **FUEL INJECTION.**

Ignition System

The ignition distributor is driven by a gear on the crankshaft. Both the conventional ignition and the breakerless transistorized ignition (1979 and later California cars) are covered in this section of the Manual.

Lubrication System

A gear-type oil pump, driven by the camshaft, draws oil through a strainer in the bottom of the crankcase sump and then forces it through an oil cooler and into the engine's oil passages. The system is described in **9.7 Removing and Installing Oil Pressure Valves.**

Clutch

The engines used with manual transmissions have a dry, single-plate clutch fitted to the flywheel. The pressure plate assembly, which has a diaphragm spring, and the driven plate remain on the flywheel when the engine is removed from the car. The clutch release bearing remains with the clutch release shaft inside the transmission bellhousing.

All clutch adjustments and repairs are covered in this section of the Manual. Though the clutch pedal, clutch cable, and clutch release mechanism are not intimately associated with the engine, they are covered here along with the clutch parts that are attached directly to the engine.

2. MAINTENANCE

The following routine maintenance steps are covered briefly in **LUBRICATION AND MAINTENANCE**. Additional instructions can be found in this section under the headings listed after some of the items.

1. Changing engine oil, cleaning the oil strainer, and checking the oil level
2. Servicing and replacing the spark plugs. **19.5**
3. Checking the distributor and the ignition timing. **19.2** or **20.3.**
4. Checking the valve clearance. **11.9**
5. Checking compression
6. Servicing the air cleaner
7. Checking the exhaust system
8. Checking the clutch pedal freeplay. **15.5**
9. Checking the V-belt adjustment
10. Checking and adjusting the idle. (Described in detail in **FUEL SYSTEM** or **FUEL INJECTION**.)
11. Replacing catalytic converter. **5.5**.

3. REMOVING AND INSTALLING ENGINE

Aside from tune-up work and repair or replacement of engine accessories, virtually all engine repairs demand that you remove the engine from the car. The single-carburetor and dual-carburetor engines are covered in separate procedures.

To remove 1600 engine:

1. Disconnect the battery ground strap. Remove the air cleaner as described in **FUEL SYSTEM**. Then disconnect the wiring from the distributor, generator, oil pressure switch, carburetor, ignition coil, and starter solenoid. To simplify reconnection, attach identifying tags to each wire as you disconnect it.
2. Remove the screws indicated in Fig. 3-1. Then remove the rear engine compartment panel.

© 1974 VWoA—3646

Fig. 3-1. Screws (arrows) that hold the rear engine compartment panel.

3. Disconnect the accelerator cable from the throttle arm on the carburetor. On 1970 and 1971 models, disconnect the vacuum hoses for the throttle valve positioner.
4. On 1968 through 1970 models, remove the two upper engine mounting bolts; on 1971 models there is only one bolt, on the right.
5. Working under the vehicle, pull off the fuel hose and quickly plug it.

WARNING —
Do not smoke or work near heaters or other fire hazards. Have a fire extinguisher handy.

6. Disconnect the heater flap cables from the arms on the heat exchangers. Pull off the ducts that connect the heat exchangers with the car's interior. Pull the accelerator cable out of the guide tube.
7. Install a support bar as shown in Fig. 3-2.

© 1974 VWoA—3708

Fig. 3-2. Support bar VW 784 installed.

5

8. On 1971 models, remove the upper right-hand engine mounting bolt, which passes through the starter mounting flange. On all models, remove the two lower engine mounting nuts.
9. Place a floor jack with an engine adaptor beneath the engine. Raise the jack until the adaptor just supports the engine.
10. Remove the two M 10 bolts that hold the engine bearer on its bonded rubber mountings (Fig. 3-3).

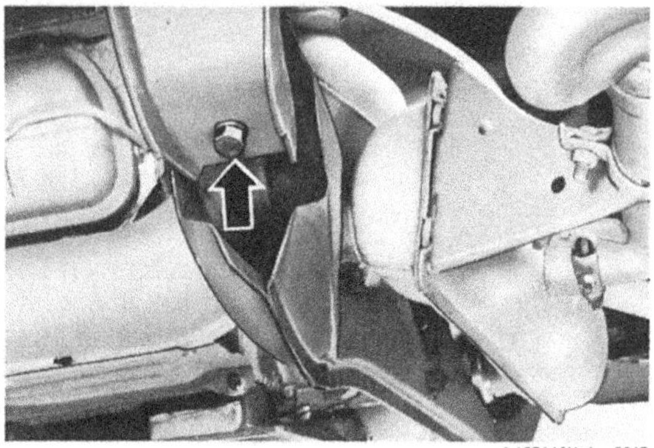

© 1974 VWoA—3647

Fig. 3-3. Mounting bolts for engine bearer.

11. Using the floor jack, pull the engine slightly to the rear until the transmission rear driveshaft is completely disengaged from the clutch. Then carefully lower the engine to floor level.

> ***CAUTION —***
> *If you intend to disassemble the engine, thoroughly clean its exterior following removal. Otherwise, dirt may accidentally enter the working parts during repair.*

Although installation is the reverse of removal, pay special attention to the following additional steps:

1. Check the clutch release bearing for wear. (See **15.4 Replacing Clutch Release Bearing**.) On both new and used release bearings, roughen the plastic facing with emery cloth. Then rub in molybdenum grease.
2. Wipe off the mating surfaces of the engine and transmission. Lightly lubricate the transmission rear driveshaft splines with molybdenum disulfide powder. Lubricate the starter drive bushing in the transmission case with multipurpose grease (Bosch starter only).
3. When installing the engine, be careful not to damage the transmission rear driveshaft or the clutch driven plate. Put the transmission in gear, set the parking brake, then hand-turn the crankshaft until the splines mesh.
4. Start the lower engine mounting studs in their holes, then press the engine into firm contact with the transmission flange. Install the upper mounting nuts and bolts. Torque all four mounting fasteners to 3.0 mkg (22 ft. lb.). Install new M-10 engine bearer bolts, then torque them to 4.0 mkg (29 ft. lb.).
5. Make certain that the rubber seal between the engine cover plates and the body is correctly positioned.
6. Make certain that the crankcase is filled with oil. Adjust the clutch pedal freeplay as described in **15.5 Adjusting Clutch Pedal Freeplay**; adjust the accelerator cable as described in **FUEL SYSTEM**; adjust ignition timing as described in **19.2 Distributor**; adjust the carburetor as described in **FUEL SYSTEM**.

To remove 1700, 1800, or 2000 engine:

1. Remove the air cleaner as described in **FUEL SYSTEM** or in **FUEL INJECTION**. Then disconnect the battery ground strap.
2. Disconnect the wiring from the distributor, the alternator regulator, the oil pressure switch, the transmission switch, the trigger for ignition timing and, on 1972 through 1974 cars, the carburetors. (To simplify reconnection, attach identifying tags to each wire as you disconnect it.)
3. Disconnect the wire for the backup lights at the in-line fuse holder near the ignition coil. Where applicable, disconnect the wire from the temperature sensor mounted in the upper right-hand side of the engine compartment.
4. On cars with fuel injection, disconnect the injection system wiring harness from the coil, the injectors, the intake air distributor, the crankcase, the temperature sensors, and the intake air distributor and its related parts.
5. Remove the ignition coil. Disconnect the hose from the activated charcoal filter canister. On dual-carburetor engines, remove the hose and left end piece from the top of the left carburetor (Fig. 3-4).

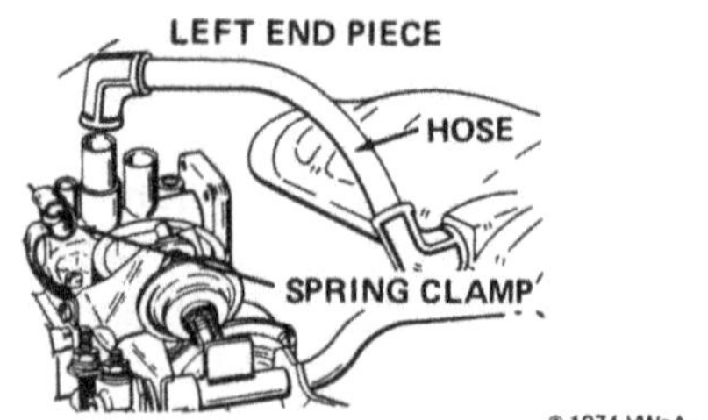

© 1974 VWoA—3722

Fig. 3-4. Hose removed from left-hand carburetor.

6. At the vacuum advance cutoff valve, located near the warm air fan, disconnect the wires and the two hoses. (Fuel injection engines and 1974 vehicles with manual transmissions have no vacuum advance cutoff valve.) Then remove the duct hoses that connect the warm air blower with the heat exchangers (Fig. 3-5).

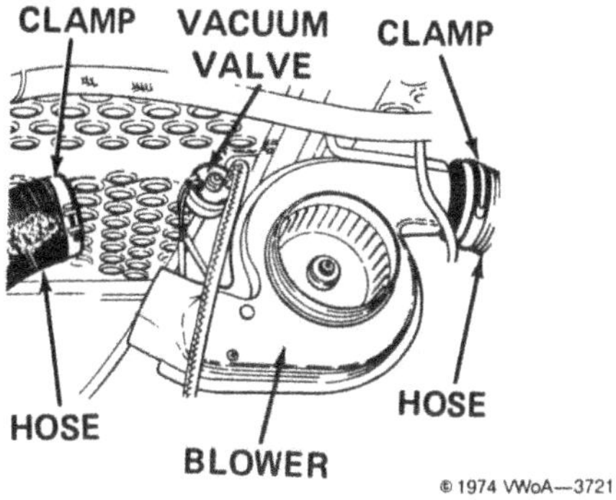

Fig. 3-5. Warm air blower and vacuum advance cutoff valve. There are also clamps at the lower ends of the duct hoses.

7. Disconnect the accelerator cable from the throttle operating crossrod or, on fuel injection engines, from the lever on the throttle valve shaft.
8. Take out the fillister head screws and remove the rear and side engine cover plates. You must also remove the oil filler neck and the oil dipstick before you can take off the right-hand rear cover plate.
9. On vehicles with automatic transmissions, pull out the ATF dipstick. Loosen the M 6 nuts that hold the ATF filler pipe, then turn the filler pipe as indicated in Fig. 3-6 and remove the pipe from the vehicle.

Fig. 3-6. ATF filler pipe removal. Loosen the nut at the lower arrow, then turn the pipe as indicated by the upper arrow.

10. Remove the gravel guard shown in Fig. 3-7.

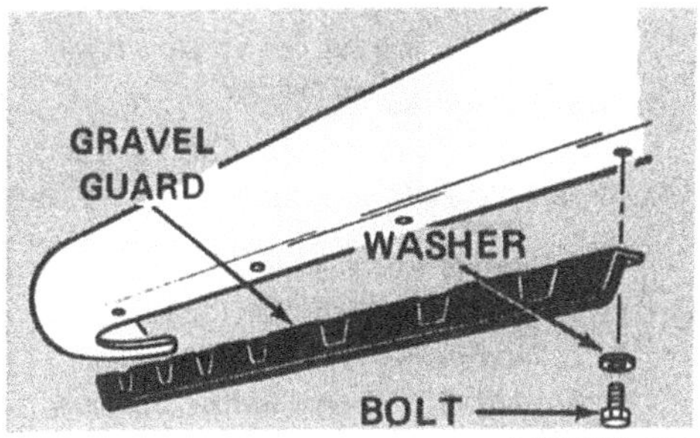

Fig. 3-7. Gravel guard being removed from rear bumper.

11. On vehicles with automatic transmissions, disconnect the vacuum line from the intake manifold balance tube or, on fuel injection engines, from the intake air distributor.
12. On vehicles with automatic transmissions, fully remove the three bolts that hold the torque converter to the drive plate. Turn the engine so that the bolts are accessible one at a time through the opening beneath a plastic plug in the upper left-hand part of the engine's mounting flange (inside the engine compartment).
13. Disconnect the wires and the battery cable from the starter solenoid.
14. Disconnect the heater flap cables from the arms on the heat exchangers or the heat control boxes. Pull off the ducts that connect the heat exchangers with the car's interior.
15. Remove the two upper engine-to-transmission mounting bolts.
16. Install a support bar under the transmission as shown in Fig. 3-8.

Fig. 3-8. Support bar installed under an automatic transmission. On both automatic and manual transmissions, support the end of the transmission that is farthest away from the engine.

5

17. Working under the vehicle, pull the accelerator cable out of its guide tube. Then disconnect the fuel lines from the pressure regulator of fuel injection engines or from the fuel pump of dual-carburetor engines. Quickly plug the line(s). (The fuel line connections mentioned are at the front, right-hand side of the engine.)

> ***WARNING —***
> *Do not smoke or work near heaters or other fire hazards. Have a fire extinguisher handy.*

18. Remove the two lower engine-to-transmission mounting bolts.
19. Place a floor jack with an engine adaptor beneath the engine. Raise the jack until the adaptor just supports the engine.
20. Remove the bolts that hold the engine bearer brackets on the frame (Fig. 3-9).

© 1974 VWoA—3273

Fig. 3-9. Bolts (arrows) that hold engine bearer to frame. There is a similar bracket at the opposite side of the vehicle.

21. Using the floor jack, pull the engine slightly to the rear until the transmission rear driveshaft is completely disengaged from the clutch or—on vehicles with automatic transmissions—the drive plate is freed from the torque converter.
22. Carefully lower the engine to floor level, making sure that none of the disconnected wires and hoses are caught on the engine.
23. On vehicles with automatic transmissions, install a retaining strap or stiff wire across the mouth of the transmission bellhousing in order to prevent the torque converter from falling off its support tube while the engine is out of the car.

> ***CAUTION —***
> *If you intend to disassemble the engine, thoroughly clean its exterior following removal. Otherwise, dirt may accidentally enter the working parts during repair.*

Although installation is the reverse of removal, pay special attention to the following additional steps:

1. Check the clutch release bearing for wear. (See **15.4 Replacing Clutch Release Bearing**.) On both new and used release bearings, roughen the plastic facing with emery cloth. Then rub in molybdenum grease.
2. Wipe off the mating surfaces of the engine and transmission. Lightly lubricate the transmission rear driveshaft splines with molybdenum disulfide powder. Lubricate the starter drive bushing in the transmission case with multipurpose grease (Bosch starter only).
3. When installing the engine, be careful not to damage the transmission rear driveshaft or the clutch driven plate. Put the transmission in gear, set the parking brake, then hand-turn the crankshaft until the splines mesh.
4. Press the engine into firm contact with the transmission flange, then install the engine-to-transmission mounting bolts. After you have loosely installed the bolts that hold the engine bearer brackets, torque all four engine-to-transmission mounting bolts to 3.0 mkg (22 ft. lb.).
5. Remove the transmission support bar. Shift the engine bearer brackets on their elongated bolt holes until the bearer is vertical and parallel to the engine fan housing as indicated in Fig. 3-10. Then torque the bearer bracket bolts to 2.0 mkg (14 ft. lb.).

© 1974 VWoA—3274

Fig. 3-10. Engine bearer alignment. The gap indicated by the arrows must be uniform along its entire length.

6. Make sure that the rubber seal between the engine cover plates and the body is correctly positioned.
7. On vehicles with automatic transmissions, torque the three bolts that hold the torque converter to the drive plate to 2.0 mkg (14 ft. lb.).
8. Make certain that the crankcase is filled with oil. Adjust the clutch pedal freeplay as described in **15.5 Adjusting Clutch Pedal Freeplay**; adjust the accelerator cable as described in conjunction with carburetor installation in **FUEL SYSTEM**; adjust ignition timing as described in **19.2 Distributor**; adjust the carburetor as described in **FUEL SYSTEM** or adjust the idle as described in **FUEL INJECTION**.

4. Replacing Outlet Pipes and Heater Control Flaps
(single-carburetor engine)

If necessary, you can pry up the cover plate tabs for access to the heater control flaps (Fig. 4-1).

© 1974 VWoA—1179

Fig. 4-1. Screwdriver being used to pry cover loose.

Check the heater flaps for free movement. Replace them if they are badly stuck or corroded. Install the cover plate, then crimp the tabs to hold it in place.

To replace the outlet pipes, use a 6-mm (¼-in.) drill to drill out the spot welds as shown in Fig. 4-2.

CAUTION—
If you lack welding skills and equipment for installing the new outlet pipes, we suggest that you leave such repairs to an Authorized VW Dealer or other qualified shop.

© 1974 VWoA—1180

Fig. 4-2. Spot welds being drilled out. Centerpunch the welds first.

After drilling, pry up the lip that holds the heat exchanger casing together (Fig. 4-3). Then bend the casing slightly apart and remove the outlet pipe.

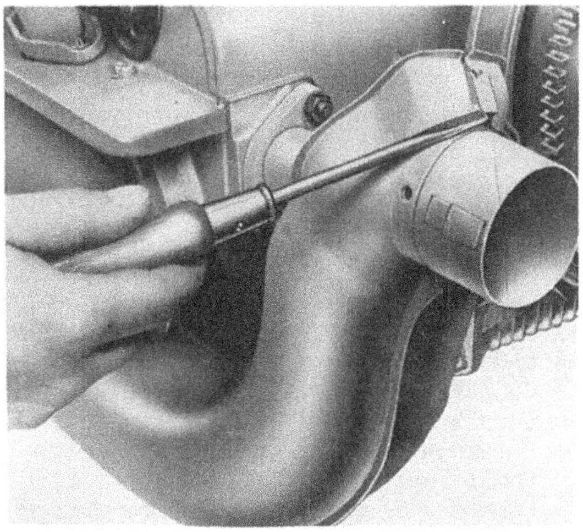

© 1974 VWoA—1181

Fig. 4-3. Heat exchanger casing halves being separated.

After removing the old outlet pipe, inspect the metal in the heat exchanger casing to make sure that it is sound. Keeping the heater flaps open, tack-weld the new outlet pipe to the heat exchanger. Then coat the pipe with cold zinc paint or a heat-resistant equivalent.

5. Exhaust System

If you intend to disassemble the engine following its removal, the exhaust system is the first thing you take off. However, you can also remove and install the muffler and the heat exchangers with the engine still in the car.

5.1 Removing and Installing Muffler
(single-carburetor engine)

The components of the exhaust system used on the single-carburetor engine are shown in Fig. 5-1. The main heat exchangers below the cylinder head covers seldom require replacement. You need not remove them when replacing the muffler. The small rear heat exchangers are a unit with the muffler.

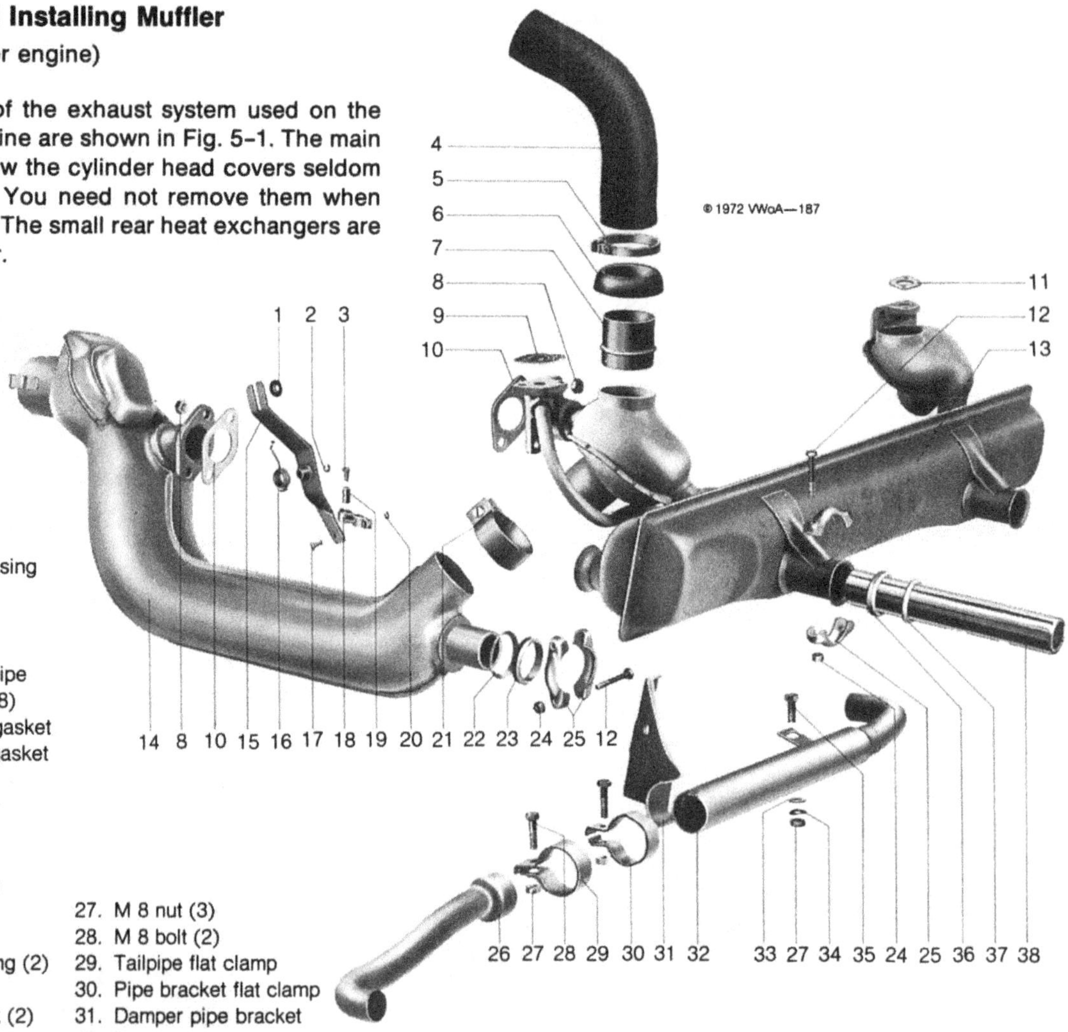

1. Clamp washer
2. E-clip (4)
3. Bolt (2)
4. Air duct from fan housing (2)
5. Hose clamp (4)
6. Air duct grommet (2)
7. Air duct connecting pipe
8. M 8 self-locking nut (8)
9. Left preheating pipe gasket
10. Exhaust pipe flange gasket (4)
11. Right preheating pipe gasket
12. M 6 bolt (6)
13. Muffler (1 outlet only)
14. Heat exchanger (2)
15. Heater flap lever (2)
16. Flap lever return spring (2)
17. Pin (2)
18. Cable connecting link (2)
19. Cable retaining pin (2)
20. E-clip (2)
21. Flat clamp (2)
22. Retaining ring (3)
23. Heat exchanger seal (2)
24. Self-locking nut (8)
25. Clamp (6)
26. Tailpipe
27. M 8 nut (3)
28. M 8 bolt (2)
29. Tailpipe flat clamp
30. Pipe bracket flat clamp
31. Damper pipe bracket
32. Damper pipe
33. Washer (3)
34. Lock washer (3)
35. M 8 bolt
36. Damper pipe seal
37. Retaining ring (2)
38. (This part not on Type 2 vehicles)

Fig. 5-1. Components of the exhaust system (1968 through 1971 models).

To remove muffler:

1. If the engine is in the car, loosen the two hose clamps. Then remove the air ducts and the air duct grommets. Remove the cover plates over the intake manifold preheater pipe connections, and unbolt the preheater pipes from the muffler. Then remove the crankshaft pulley cover plate, the rear cover plate, and the air duct connecting pipes.
2. Remove the screws and nuts from the flat clamps on the heat exchanger connections.
3. Remove the M 6 nuts and bolts, then remove the clamps that join the heat exchanger outlet pipes to the flanged pipes at the ends of the muffler.
4. Remove the four M 8 self-locking nuts that hold the muffler on the cylinder heads. Then, grasping the damper pipe, pull the muffler to the rear and off the vehicle.

Installation is the reverse of removal. Inspect the muffler for leaks and damage and make sure that the exhaust pipe flanges that bolt to the cylinder heads are not warped. If you install a new muffler, also install a new tailpipe and damper pipe. Use new gaskets at all points. Torque the M 8 self-locking nuts to 2.0 mkg (14 ft. lb.) and the M 6 nuts and bolts to 1.0 mkg (7 ft. lb.).

5.2 Removing and Installing Muffler
(dual-carburetor engine)

The components of the exhaust system used on the dual-carburetor engine are shown in Fig. 5-2. The main heat exchangers below the cylinder head covers seldom require replacement. You need not remove them when replacing the muffler.

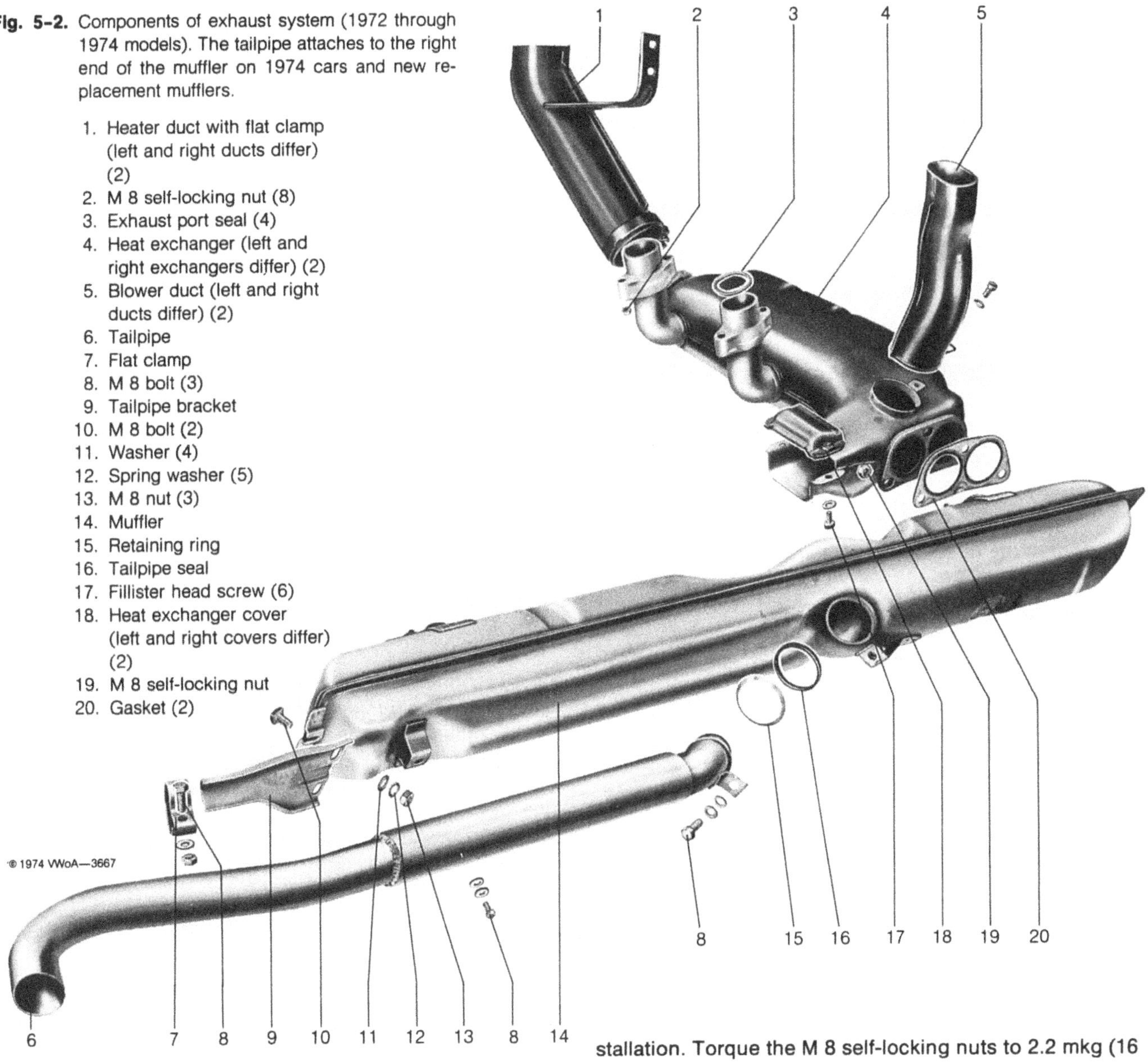

Fig. 5-2. Components of exhaust system (1972 through 1974 models). The tailpipe attaches to the right end of the muffler on 1974 cars and new replacement mufflers.

1. Heater duct with flat clamp (left and right ducts differ) (2)
2. M 8 self-locking nut (8)
3. Exhaust port seal (4)
4. Heat exchanger (left and right exchangers differ) (2)
5. Blower duct (left and right ducts differ) (2)
6. Tailpipe
7. Flat clamp
8. M 8 bolt (3)
9. Tailpipe bracket
10. M 8 bolt (2)
11. Washer (4)
12. Spring washer (5)
13. M 8 nut (3)
14. Muffler
15. Retaining ring
16. Tailpipe seal
17. Fillister head screw (6)
18. Heat exchanger cover (left and right covers differ) (2)
19. M 8 self-locking nut
20. Gasket (2)

To remove the muffler, remove the 6 self-locking nuts from the studs that hold the muffler on the two heat exchangers. On 1973 and later vehicles, remove the union nut and disconnect the exhaust gas recirculation system pipe from the left front side of the muffler. Then grasp the muffler and tail pipe and pull them as a unit to the rear and off of the heat exchangers. If only the tailpipe is faulty, you can remove it separately without taking off the muffler. However, if the muffler must be removed, it is best to remove it together with the tailpipe, then separate the two parts following removal.

Installation is the reverse of removal. If you install a new muffler, also install a new tailpipe. Replace the heat exchanger flange gaskets and the tailpipe seal during installation. Torque the M 8 self-locking nuts to 2.2 mkg (16 ft. lb.).

5.3 Removing and Installing Heat Exchanger
(single-carburetor engine)

The heat exchangers are located just below the cylinder heads on either side of the engine. You must remove the muffler before you can remove the heat exchangers.

To remove:

1. Remove the two fillister head screws that hold the heat exchanger casing to the lower warm air duct plate under the cylinders.
2. Disconnect the heater control cable from the heater flap lever.

3. Pull the flexible heated air duct off the front of the heat exchanger. Then remove the four M 8 self-locking nuts that hold the heat exchanger on the cylinder head.
4. Take a good grip on the heat exchanger (it is heavy); then push it forward and remove it.

Installation is the reverse of removal. Use new gaskets and torque the M 8 self-locking nuts to 2.0 mkg (14 ft. lb.).

5.4 Removing and Installing Heat Exchanger
(dual-carburetor engine)

The heat exchangers are located just below the cylinder head covers on either side of the engine. You must remove the muffler before you can remove the heat exchangers.

To remove:

1. To free the heat exchanger from the cooling air fan housing, remove the two fillister head screws that hold the heat exchanger cover on the heat exchanger.
2. Disconnect the flexible duct hoses from the tops of the blower ducts and the heater ducts.
3. Loosen the flat clamp that holds the heater duct on the front of the heat exchanger.
4. While supporting the heat exchanger to keep it from falling, remove the four M 8 self-locking nuts that hold the heat exchanger head pipes on the cylinder head. Remove the heat exchanger downward and to the rear.

Installation is the reverse of removal. Pry the old exhaust port seals out of the cylinder heads. Install new seals and then the heat exchanger. Torque the eight M 8 self-locking nuts to 2.0 mkg (14 ft. lb.).

5.5 Removing and Installing Muffler
(fuel injection engine)

A completely new and different exhaust system was introduced on the 1975 models. The main features of the new system are (1) improved ease of installation or removal; (2) improved heat output for the passenger compartment; (3) reduced exhaust emissions and, where applicable, easy replacement of the catalytic converter.

To remove muffler:

1. Working at the right-hand end of the muffler, remove the single bolt that holds the muffler support strap to the transverse exhaust pipe.
2. Working at the left-hand end of the muffler, remove the three nuts and bolts that join the muffler's mounting flange to the flange of the exhaust pipe or to the flange of the catalytic converter. Support the muffler as you remove the last bolt.

Installation is the reversal of removal. Use a new gasket between the muffler mounting flange and the exhaust pipe or catalytic converter flange.

Replacing Catalytic Converter

Because the muffler is easy to remove, it is best to remove first the muffler and then the catalytic converter, rather than to try to remove the catalytic converter without fully removing the muffler from the car. With the muffler removed, simply take out the three bolts and nuts that hold the catalytic converter to the exhaust pipe. Then remove the catalytic converter and its gasket.

Installation is the reverse of removal. Use new gaskets at both ends of the catalytic converter. See Fig. 5-3.

5.6 Removing and Installing Heat Exchanger
(fuel injection engine)

If you must remove both heat exchangers and the muffler, it is possible to leave the exhaust pipes bolted to the heat exchangers. After disconnecting the air ducts and the heater control cables, unbolt the EGR pipe and the support strap at the rear of each heat exchanger. Unbolt the front exhaust pipes from the exhaust manifolds (or unbolt the exhaust manifolds from the cylinder heads). It is then possible to lower the complete exhaust system from the engine as a unit.

> ***CAUTION —***
> *Support the exhaust system as you remove the mounting bolts. Otherwise, parts may be distorted or broken owing to the weight of the system.*

The procedure given below should be used if you intend to replace a heat exchanger only.

To remove heat exchanger:

1. Remove the muffler or remove the muffler and the catalytic converter as a unit. See **5.5 Removing and Installing Muffler**.
2. Loosen the flat clamp that holds the heat control box duct to the outlet on top the front of the heat exchanger. Then disengage the heat control box duct from the heat exchanger.
3. Disconnect the blower duct from the top of the rear end of the heat exchanger.

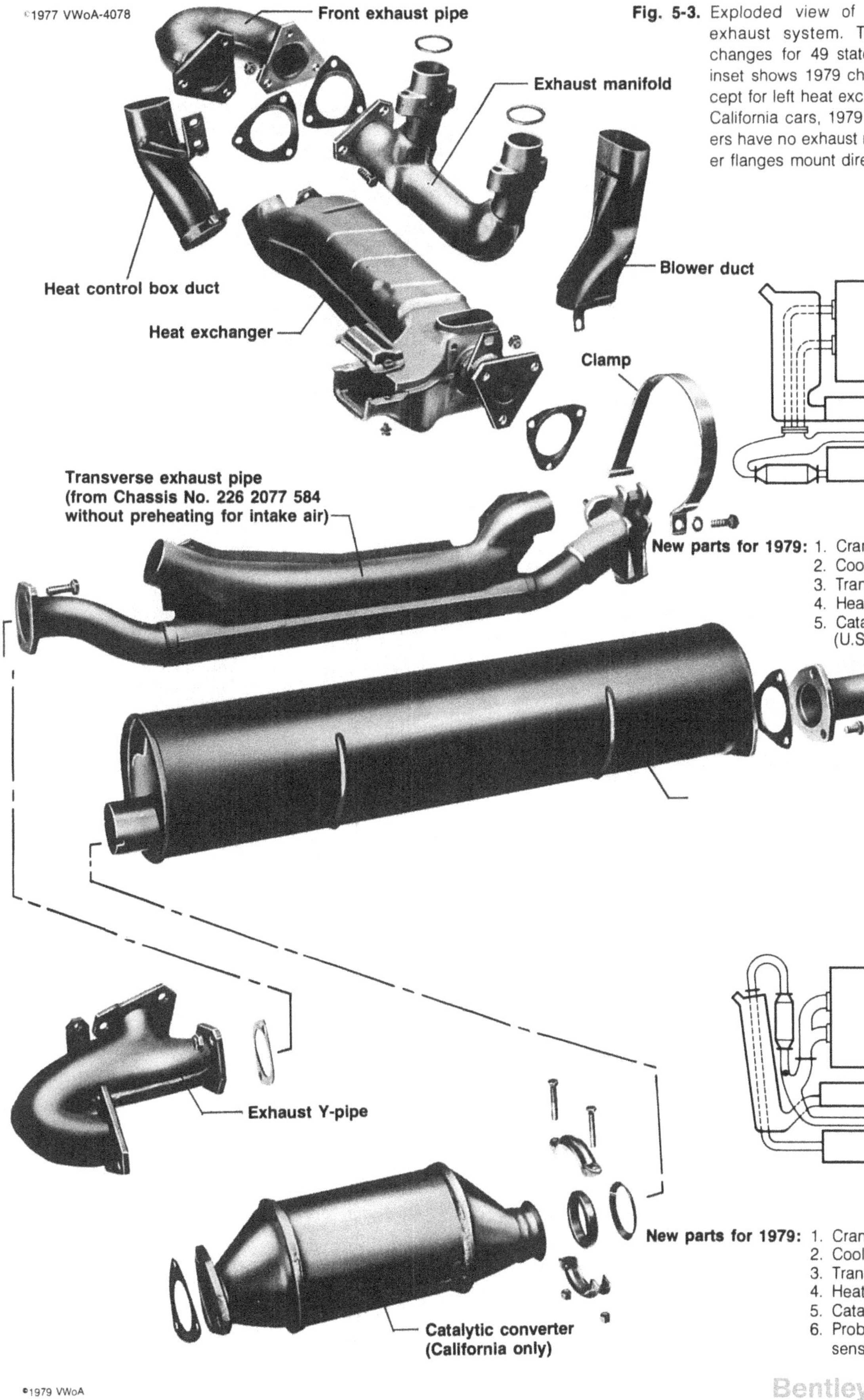

Fig. 5-3. Exploded view of fuel injection engine's exhaust system. Top inset shows 1979 changes for 49 states and Canada; bottom inset shows 1979 changes for California. Except for left heat exchanger of 1979 and later California cars, 1979 and later heat exchangers have no exhaust manifolds; heat exchanger flanges mount directly on cylinder head.

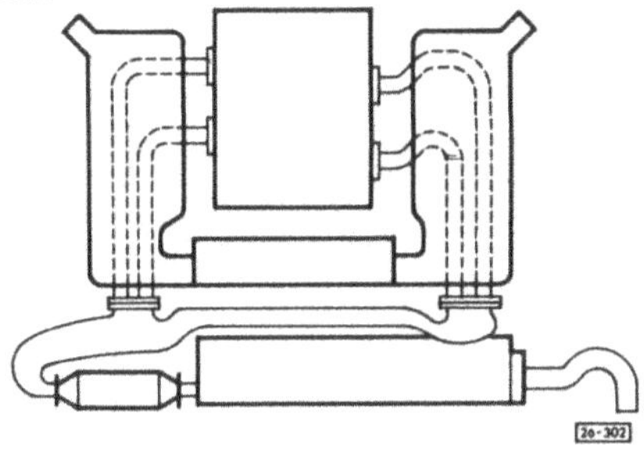

©1979 VWoA—4427

New parts for 1979:
1. Crankcase
2. Cooling fan
3. Transverse exhaust pipe
4. Heat exchangers
5. Catalytic converter (U.S. only)

5

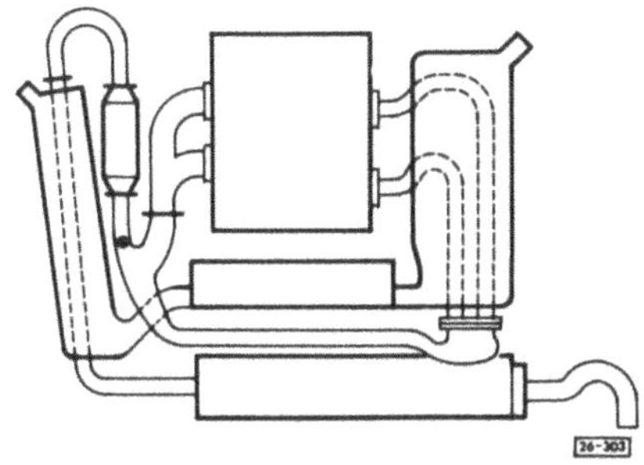

©1979 VWoA—4428

New parts for 1979:
1. Crankcase
2. Cooling fan
3. Transverse exhaust pipe
4. Heat exchangers
5. Catalytic converter
6. Probe receptacle for oxygen sensor

4. If you are removing the right-hand heat exchanger, remove the three bolts and nuts that hold the transverse exhaust pipe to the rear of the heat exchanger. Remove the two bolts and nuts that hold the left end of the transverse exhaust pipe to the exhaust Y-pipe. Then remove the transverse exhaust pipe and its gaskets.
5. If you are removing the left-hand heat exchanger, take out the two bolts that hold the EGR (exhaust gas recirculation) pipe to the left-side exhaust Y-pipe. Remove the three nuts and bolts that hold the Y-pipe to the heat exchanger and the two bolts that hold the Y-pipe to the transverse exhaust pipe. Then remove the Y-pipe and its gaskets.
6. Support the heat exchanger. Remove the bolt from the support strap that is at the rear of the heat exchanger. Remove the three bolts and nuts that hold the heat exchanger's front mounting flange to the front exhaust pipe. Then remove the heat exchanger and its gaskets.

Installation is the reverse of removal. Use new gaskets at all points.

5.7 Removing and Installing Exhaust Pipes

All of the exhaust pipes are held to the other parts of the exhaust system by nuts and bolts installed through the flat mounting flanges. The removal and installation of the exhaust pipes will be obvious from inspection of the vehicle. During installation of any exhaust pipe, always use new gaskets on the mounting flanges.

6. COOLING AIR SYSTEM AND INTAKE MANIFOLD

(single-carburetor engine)

With the engine in the car, you must partially remove the fan housing before you can remove the intake manifold, or before you can remove the generator as described in **ELECTRICAL SYSTEM**.

Partial Removal of Fan Housing

To remove the fan housing partially, remove the V-belt. Fully remove the air cleaner and the carburetor as described in **FUEL SYSTEM**. Pull the accelerator cable guide tube out of the fan housing. Then remove the strap that holds the generator to its support.

Disconnect all air ducts, hoses, and wires that might prevent the fan housing from being lifted upward. Disconnect the spark plug cables from the plugs and the high tension cable from the coil. On 1971 models, use a 10-mm wrench to remove the bolt that holds the oil cooler cover to the front of the fan housing and the bolt that holds the fan housing to the oil cooler flange. Then remove the oil cooler cover and air outlet duct.

Remove the screws that are at each end of the fan housing. Unbolt the thermostat as shown later in Fig. 6-3, then unscrew the thermostat from its rod. On 1971 models, pry off the spring clip and disconnect the connecting link from the left control flaps (Fig. 6-1). Raise the fan housing so that you can remove the generator or the intake manifold. Installation is the reverse of removal.

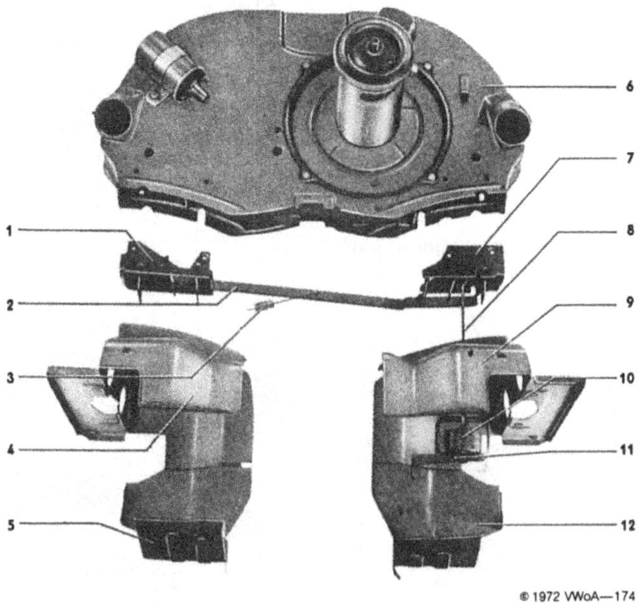

1. Left control flaps
2. Connecting link
3. Link return spring
4. Left cylinder cover plate
5. Left air duct lower part
6. Fan housing with fan
7. Right control flaps
8. Thermostat rod
9. Right cylinder cover plate
10. Thermostat
11. Thermostat bracket
12. Right air duct lower part

Fig. 6-1. Components of the cooling air system used on 1968 through 1971 models.

Removing and Installing Cooling Air System

For this procedure, it is necessary that the engine be removed. Otherwise, there will be insufficient working clearance around the engine.

To remove air ducts, fan housing, and control flaps:

1. If not removed previously, remove the air hoses that carry fresh air to the heat exchangers.
2. Remove the distributor cap and pull the spark plug connectors off the spark plugs.

3. If not removed previously, remove the crankshaft pulley upper cover plate, the two small cover plates over the intake manifold preheating pipes, and the rear cover plate.
4. Disconnect the vacuum hose(s) that go to the distributor. Disconnect the fuel line(s) from the fuel pump and carburetor. (Fuel pump and carburetor removal are covered separately in **FUEL SYSTEM**.)
5. Remove the generator pulley outer half as shown in Fig. 6-2, and then remove the fan belt.

© 1974 VWoA—1994

Fig. 6-2. Fan belt being removed. Screwdriver is held in a notch on the back of the pulley and braced against a generator through bolt. Use a 21-mm (13/16-in.) wrench to remove the nut.

6. Take out the fillister head screws that hold the air duct lower parts on the crankcase, on the cylinder cover plates, and on the heat exchangers (if the latter have not already been removed).
7. Unbolt the thermostat from its bracket (Fig. 6-3). Unscrew the thermostat from its rod. If you remove the thermostat bracket from the side of the crankcase, scribe its locations. Doing so will simplify thermostat adjustment.

© 1974 VWoA—1999

Fig. 6-3. Thermostat being unbolted from its bracket.

8. On 1971 models, use a 10-mm wrench to remove the bolt for the oil cooler cover and the bolt that holds the fan housing to the oil cooler flange. Disconnect the connecting link from the left control flaps. Then remove the oil cooler cover.
9. Loosen the bolt in the generator mounting strap. Then slide the strap forward and off the generator support stand.
10. Remove the fillister head screw at each side of the fan housing. Then lift the fan housing with the attached generator straight up until it is clear of the oil cooler.
11. If necessary, you can remove the right and left control flaps and the connecting link by unhooking the return spring and taking out the eight fillister head screws that hold the flap assembly in the lower part of the fan housing.

NOTE —
When installing the control flaps, make sure that the rubber stop is inserted in the right-hand flap housing.

Installing the fan housing is the reverse of removal. Following installation, check and adjust the fan belt tension as described in **LUBRICATION AND MAINTENANCE**. Check the thermostat adjustment as described in **6.1 Adjusting Thermostat.**

5

To remove intake manifold and cylinder cover plates:

1. First remove the fan housing. If the muffler has not already been removed, unbolt the intake manifold preheating pipes at their flanges.
2. Remove the four nuts that hold the intake manifold or, on 1971 models, the two cast aluminum intake pipes to the cylinder heads. (It is unnecessary to separate the intake manifold from its intake pipes by removing the hose clamps and the rubber sleeves.)
3. Lift off the intake manifold assembly.
4. Remove the two fillister head screws that hold each cylinder cover plate to the cylinder head. Then lift the plates off upward.

Installation is the reverse of removal. Use new gaskets between the cast aluminum intake pipes and the cylinder heads. On 1968 through 1970 vehicles replace the gaskets in the cylinder head intake ports. Torque the nuts to 2.0 mkg (14 ft. lb.). Torque the M 6 bolts for the preheating pipe flanges to 1.0 mkg (7 ft. lb.).

Lower Ducts and Cover Plates

Usually, it is not necessary to remove the cover plates unless the engine is being disassembled. The front cover plate cannot be removed until the engine has been removed from the vehicle. The pulley lower cover plate cannot be removed until after the crankshaft pulley has been removed from the engine. The pulley upper cover plate, which is easily removed, must be taken off before you can remove or install the fan belt.

Fig. 6-4 is an exploded view of the 1600 engine's cover plates and air duct lower parts. The right-hand air duct lower part must be removed for access to the thermostat. All of the components shown in Fig. 6-4 are held to the engine by fillister head screws. The removal of the various plates and ducts will be apparent from an inspection of the engine. During installation of the rear cover plate, it is essential that the plate correctly engage the rubber seal on the vehicle's body.

6.1 Adjusting Thermostat

The thermostat is located beneath the right-hand pair of cylinders. You should check the thermostat adjustment any time components of the cooling air system have been removed or replaced. The air duct lower part below the right-hand cylinders must be removed for access to the thermostat. The engine must be completely cold before accurate adjustments can be made.

Fig. 6-4. 1600 engine lower ducts and cover plates.

1. Accelerator cable tube rubber grommet
2. 38-mm rubber plug
3. Front cover plate
4. Air duct lower part (left)
5. Preheating pipe cover plate
6. Front half-round gasket
7. Rear half-round gasket
8. Air hose rubber grommet
9. Fillister head screw (typical)
10. Washer (typical)
11. Pulley lower cover plate
12. Rear cover plate
13. Pulley upper cover plate
14. Air duct lower part (right)
15. Air duct lower part (right front)

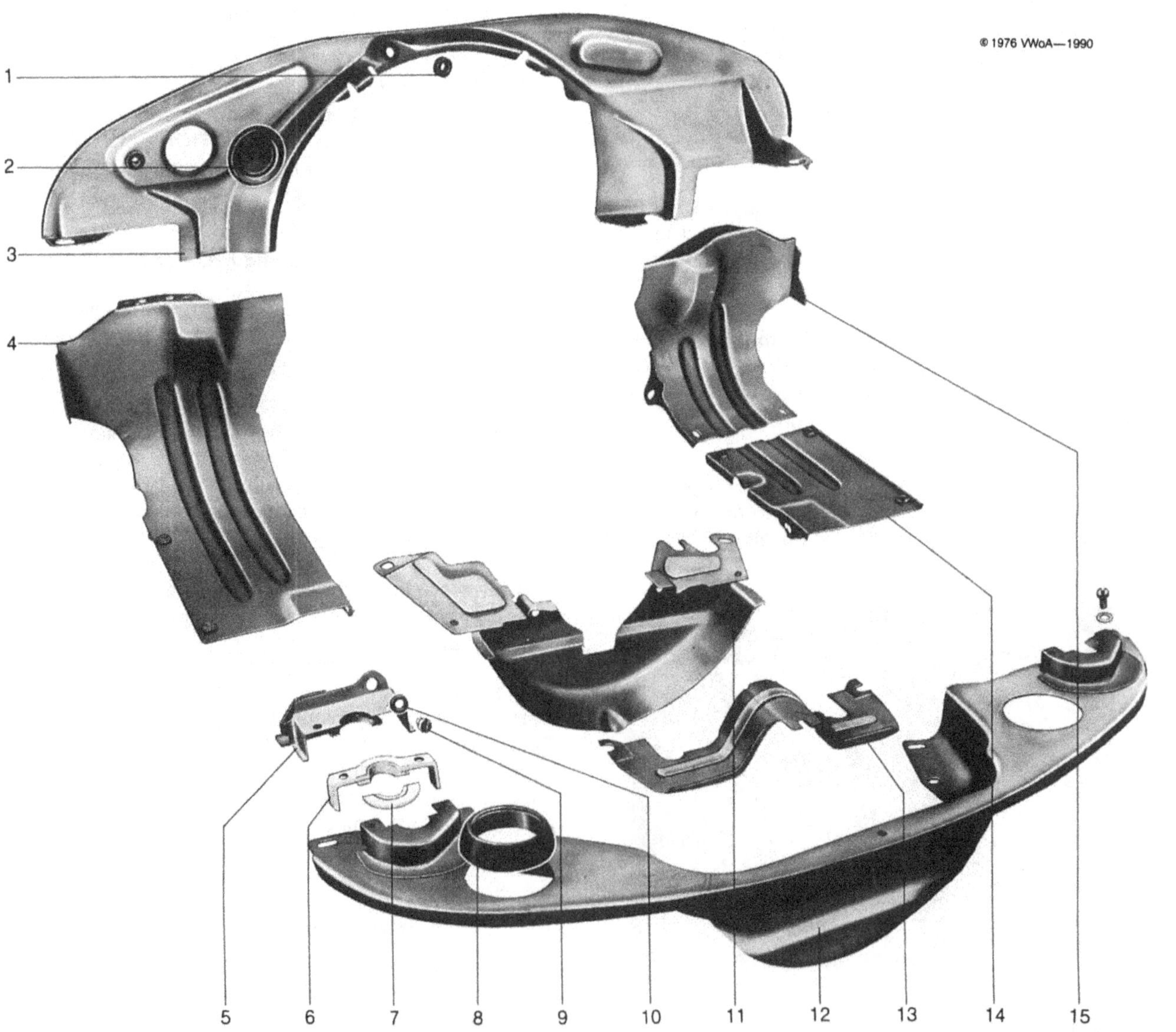

To check and adjust:

1. If a new thermostat is being installed, screw it onto the thermostat rod. If the adjustment is simply being checked, unbolt the thermostat from its bracket.
2. Raise the thermostat off its bracket until the control flaps are wide open. If the top of the thermostat does not just contact the upper part of the bracket, an adjustment is required.
3. To adjust, loosen the mounting nut that holds the thermostat bracket on the crankcase. Then move the bracket as indicated in Fig. 6-5 until the upper part of the bracket just contacts the top of the thermostat with the thermostat raised to open the control flaps fully.

© 1972 VWoA—175

Fig. 6-5. Bracket movement (double arrow) that permits thermostat to be adjusted.

4. Lock the bracket in position by tightening the nut.
5. Pull the thermostat down and bolt it on the lower part of the bracket.

6.2 Removing and Installing Crankshaft Pulley

To remove the crankshaft pulley, lock the flywheel with the flywheel holding fixture illustrated in **15.2 Removing and Installing Flywheel**. If the engine is installed, place the transmission in 4th gear and set the parking brake (manual transmission only). Then remove the M 20 bolt and washer from the center of the pulley. Use a puller, as shown in Fig. 6-6, to remove the pulley from the crankshaft.

CAUTION —
Do not attempt to pry the pulley off. You can easily damage the crankcase by doing so.

© 1974 VWoA—1996

Fig. 6-6. Crankshaft pulley being removed.

Check the pulley seating surface, the Woodruff key, the belt contact surface, and the crankshaft. Remove dirt and burrs; replace bent or cracked pulleys. Clean the oil return thread on the pulley hub and coat it with molybdenum grease. Start the pulley on the crankshaft, then draw it into place with the center bolt. Torque the bolt to 4.5 mkg (32 ft. lb.).

5

7. Cooling Air System and Intake Manifold

(dual-carburetor and fuel injection engines)

Whether the engines have dual carburetors or fuel injection, the cooling air systems of 1700, 1800 and 2000 engines are essentially identical. An extension shaft with pulley, used to drive the air injection pump of the exhaust afterburning system, is installed on the cooling air fan of 1973 and 1974 models only.

The various pressed steel air cooling ducts and cover plates are shown in Fig. 7-1 and Fig. 7-2 (on the following page). Though some of the plates shown in the illustrations may differ slightly from those on your engine, these two exploded views are useful in determining the correct locations for the plates and ducts during installation. The left rear air duct and the right rear air duct can be removed or installed only after removal of the fan housing, which is described under the next heading.

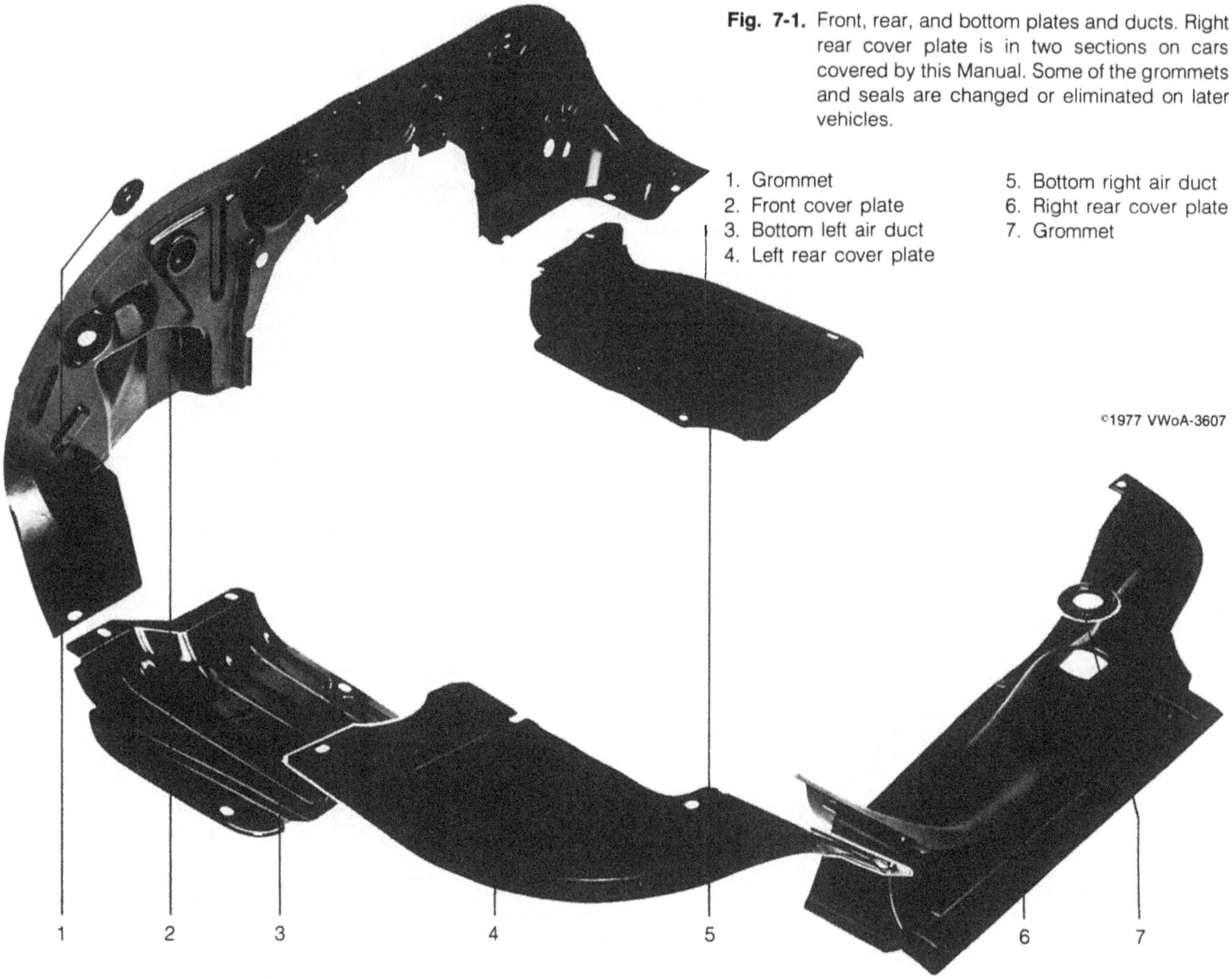

Fig. 7-1. Front, rear, and bottom plates and ducts. Right rear cover plate is in two sections on cars covered by this Manual. Some of the grommets and seals are changed or eliminated on later vehicles.

1. Grommet
2. Front cover plate
3. Bottom left air duct
4. Left rear cover plate
5. Bottom right air duct
6. Right rear cover plate
7. Grommet

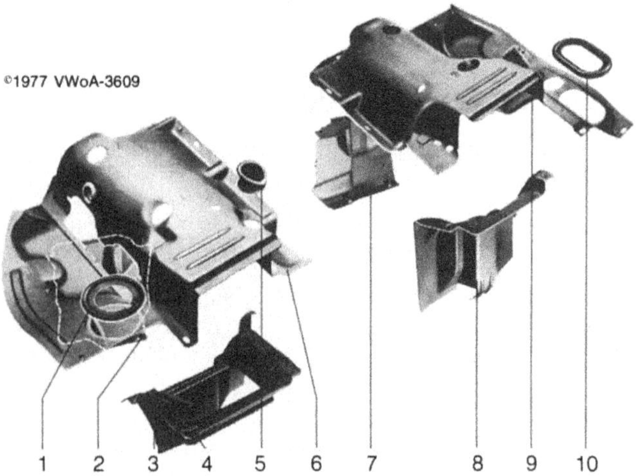

1. Grommet
2. (Not installed on cars covered by this Manual)
3. Left rear air duct
4. Oil cooler air seal
5. Seal
6. Left cylinder cover
7. Right front air duct
8. Right rear air duct
9. Right cylinder cover
10. Blower duct grommet

Fig. 7-2. Cylinder covers and air ducts.

7.1 Removing and Installing Fan Housing

It is unnecessary to remove the alternator before you remove the fan housing. The fan housing can be taken off with the alternator attached to it. Alternator removal, installation, and belt adjustment are described in **ELECTRICAL SYSTEM.** On 1973 and 1974 models, you must remove the air injection pump, the pump belt, and the extension shaft with pump drive pully. Fig. 7-3 gives an exploded view of the cooling air system used on the dual-carburetor and fuel injection engines. On 1972 models and 1975 and later models, there is a screen grille over the fan. The grille does not appear in the illustration.

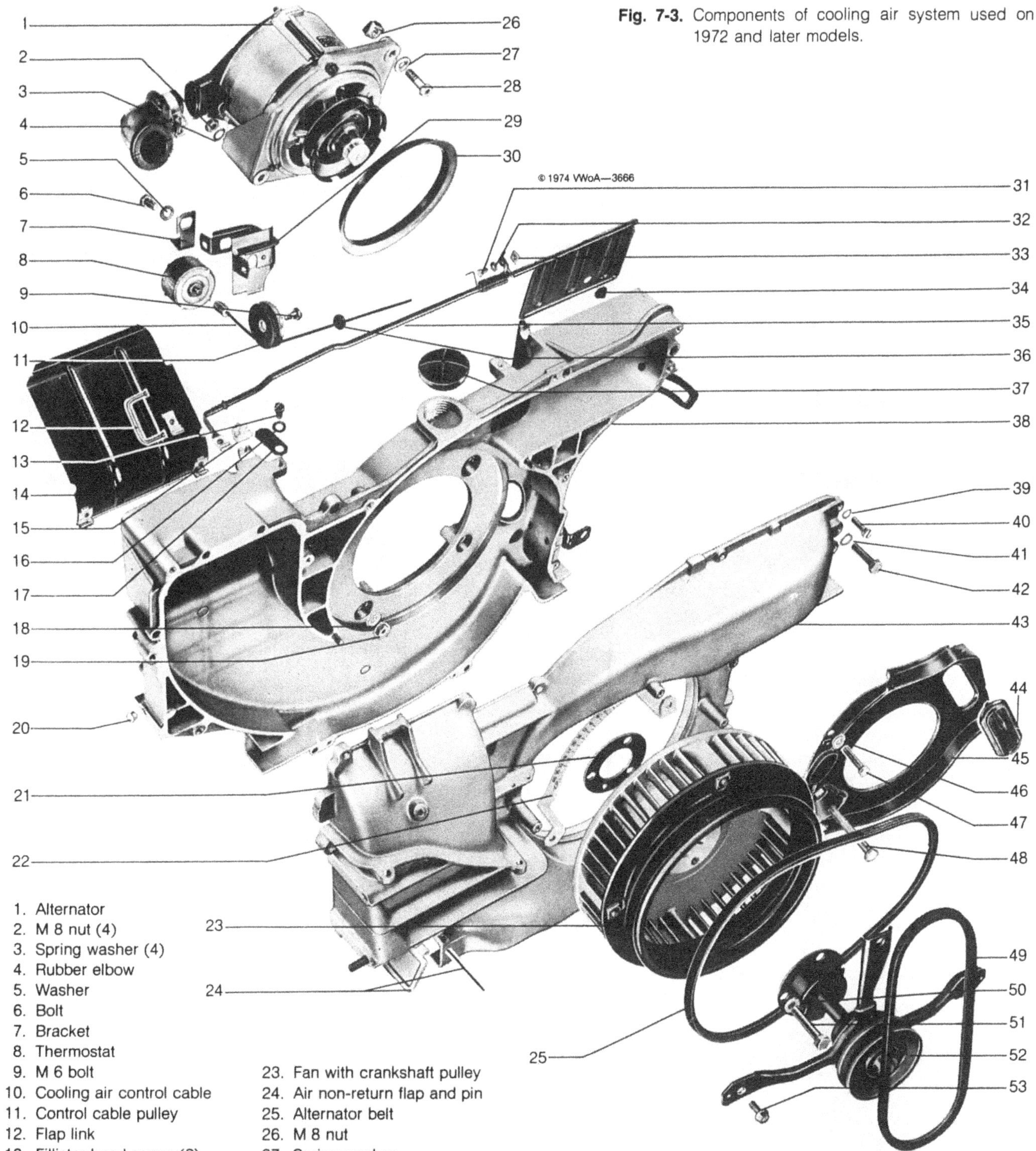

Fig. 7-3. Components of cooling air system used on 1972 and later models.

1. Alternator
2. M 8 nut (4)
3. Spring washer (4)
4. Rubber elbow
5. Washer
6. Bolt
7. Bracket
8. Thermostat
9. M 6 bolt
10. Cooling air control cable
11. Control cable pulley
12. Flap link
13. Fillister head screw (2)
14. Left control flap
15. Support (2)
16. Spring washer (2)
17. Shaft retaining spring (2)
18. Spring washer
19. M 8 nut
20. Plug
21. Fan spacer
22. Ignition timing scale
23. Fan with crankshaft pulley
24. Air non-return flap and pin
25. Alternator belt
26. M 8 nut
27. Spring washer
28. M 8 socket head screw
29. Thermostat mounting bracket
30. Alternator sealing ring
31. M 4 bolt
32. Washer
33. Square nut
34. Rubber flap stop
35. Right control flap with shaft
36. Sealing washer
37. Inspection hole cover
38. Front half of fan housing
39. Spring washer (7)
40. Fillister head screw (7)
41. Spring washer (3)
42. M 8 bolt (3)
43. Rear half of fan housing
44. Alternator cover plate insert
45. Alternator cover plate
46. Spring washer
47. Bolt
48. Alternator mounting bolt
49. Air injection pump belt
50. Spring washer (3)
51. M 8 bolt (3)
52. Extension shaft with pulley
53. M 6 screw (3)

5

To remove fan housing:

1. Loosen the bolt in the adjusting bracket. Then lower the air injection pump as indicated in Fig. 7-4. Take off the air pump drive belt.

Fig. 7-4. Pump belt adjustment (double arrow). To remove the belt, pull the pump pulley down.

2. Unbolt the air injection pump and the adjusting bracket and take them off the engine.
3. Except on engines without an air injection pump, unbolt the extension shaft with pulley from the fan housing and the fan. Then remove the extension shaft with pulley, the ignition timing scale, the fan with crankshaft pulley, and the alternator belt from the engine.
4. On engines without the air injection pump, carefully pry out the alternator cover plate insert. Using a 12-point driver, loosen the alternator belt adjusting screw. Slightly loosen the alternator mounting bolt. Then move the alternator as indicated in Fig. 7-5 and take off the alternator belt.

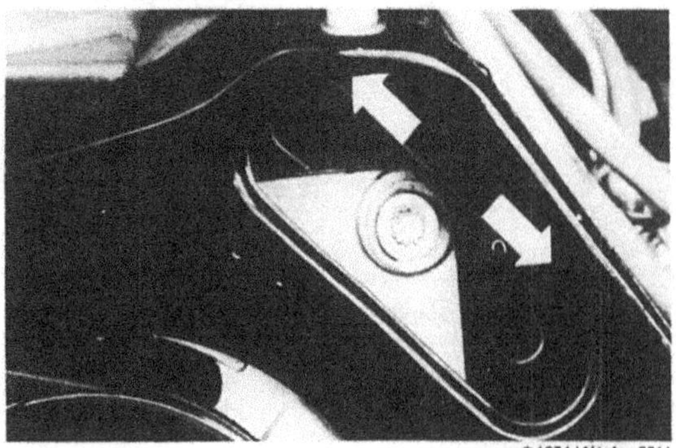

Fig. 7-5. Alternator belt adjustment. To remove the belt, move the pulley as indicated by the left arrow.

5. On engines without the air injection pump, unbolt the grille over the fan, then remove the grille together with the ignition timing scale. To remove the fan with crankshaft pulley, remove the three socket head screws indicated in Fig. 7-6.

Fig. 7-6. Socket head screws (arrows) that hold the fan with crankshaft pulley on engines that do not have an air pump.

6. Disconnect the cooling air control cable from the control flap shaft.
7. Pull the rubber elbow for the alternator out of the front half of the fan housing.
8. Remove the four nuts indicated in Fig. 7-7. Then pull the assembled fan housing to the rear and off the engine.
9. If necessary, remove the fillister head screws around the perimeter of the fan housing. Then separate the front half from the rear half.

Fig. 7-7. Nuts and studs that hold the fan housing on the crankcase (arrows).

Installation is the reverse of removal. Make sure that the rubber elbow for the alternator is correctly engaged in the front half of the fan housing. Torque the bolts that hold the fan with crankshaft pulley on its hub to 2.0 mkg (14 ft. lb.). Torque the bolts that hold the extension shaft with pulley to 2.0 mkg (14 ft. lb.). Adjust the alternator belt so that it yields 15 mm ($\frac{5}{8}$ in.) when firmly pressed at midpoint. Similarly, adjust the air pump belt so that it yields 5 to 8 mm ($\frac{3}{16}$ to $\frac{5}{16}$ in.). Adjust the cooling air control cable as described under the following heading.

7.2 Checking and Adjusting Thermostat and Cable

To adjust the cooling air control cable, loosen the screw and square nut that hold the cable on the control flap shaft. Hand-press the flaps to their fully closed position. Then tighten the screw and nut to hold the cable in that position.

You can check the thermostat by removing it from its mounting bracket and placing it in a pan of water. Heating the water to a temperature of 65° to 70°C (149° to 158°F) should cause dimension **a**, shown in Fig. 7-8, to be at least 46 mm ($1\frac{13}{16}$ in.). If it is not, replace the thermostat.

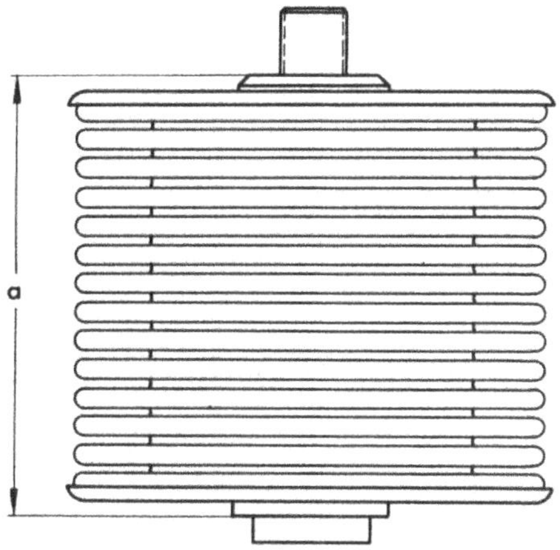

© 1974 VWoA—3616

Fig. 7-8. Thermostat length (dimension **a**).

8. DISTRIBUTOR AND DRIVE

The distributor shaft does not engage the crankshaft gear directly. Instead, a drive dog on the distributor shaft engages a separate distributor drive shaft. The distributor drive shaft has an integral gear that meshes with the crankshaft gear.

8.1 Removing and Installing Distributor

You should remove the distributor to protect it during engine repairs or before you attempt to give the distributor a thorough cleaning. Removal will also simplify breaker point installation when performing a tune-up.

To remove:

1. Remove the distributor cap.
2. Hand-turn the engine's crankshaft until the rotor tip points to the No. 1 cylinder mark on the distributor body (Fig. 8-1).

© 1974 VWoA—3669

Fig. 8-1. Distributor rotor aligned with No. 1 cylinder mark on distributor body. A dual-carburetor engine is shown, but single-carburetor engines and fuel injection engines have similar marks.

3. If not previously removed, disconnect all hoses and wiring from the distributor.
4. Remove the bolt that holds the distributor bracket to the crankcase.
5. Lift out the distributor and cover the opening in the engine to keep out dirt.

NOTE —
If it is necessary to remove the distributor bracket from the distributor, mark its position so that you can install the bracket at the original angle. Doing so will simplify starting and timing the engine after the distributor is installed.

5

To Install:

1. If necessary, hand-turn the crankshaft to bring the No. 1 cylinder to its firing point. On dual-carburetor and fuel injection engines, align the timing mark on the crankshaft pulley with the 0° mark on the ignition timing scale (Fig. 8-2). On single-carburetor engines, align the timing mark on the crankshaft pulley with the seam on the crankcase centerline (Fig. 8-3).

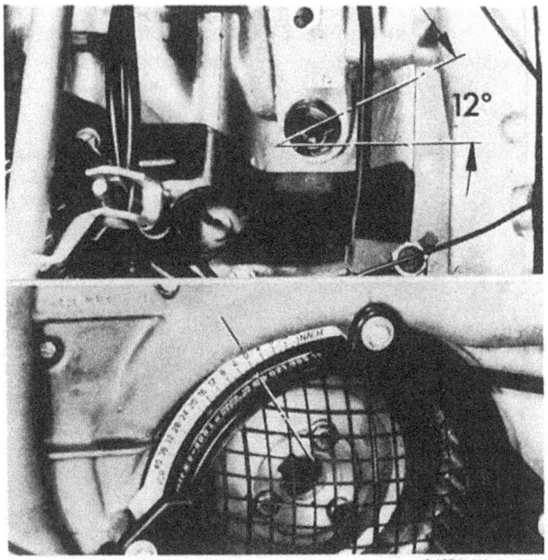

© 1974 VWoA—3671

Fig. 8-2. Dual-carburetor or fuel injection engine positioned for distributor installation. Arrows indicate the necessary 12° angle of the slot in the distributor drive shaft.

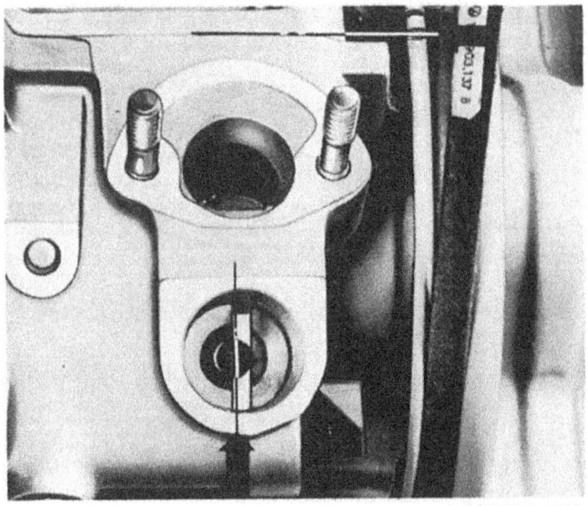

© 1974 VWoA—1989

Fig. 8-3. Single-carburetor engine positioned for distributor installation. Arrow at upper right indicates timing mark on the crankshaft pulley. Lower arrow indicates centerline of distributor mounting hole relative to the distributor drive shaft groove.

2. Insert the distributor with its rotor pointing to the No. 1 cylinder mark. Then install the bolt, hoses, and wiring. Check the ignition timing.

8.2 Removing and Installing Distributor Drive Shaft

You will simplify the installation of the distributor drive shaft if you leave the engine in position to fire the No. 1 cylinder while the shaft is out.

To remove:

1. Remove the distributor. Take the spacer spring out of the recess in the top of the drive shaft.
2. Using an expansion tool or magnet, withdraw the distributor drive shaft. Then use a magnet to remove the two shim washers that are under the shaft on single-carburetor engines or the single spacer washer that is under the shaft on dual-carburetor engines.

CAUTION —
Be careful not to drop the washers into the timing gear chamber. Getting them out could require complete engine disassembly.

To Install:

1. Inspect the drive shaft for worn gear teeth and the washer(s) for scoring or wear. Their individual thickness should be 0.60 mm (.024 in.) for single-carburetor engines or 1.30 mm (.051 in.) for dual-carburetor and fuel injection engines. Have the engine in position to fire the No. 1 cylinder.
2. Coat the washer(s) with grease so that they will stick in place until the distributor drive shaft can be installed, then insert a stiff wire or rod in the engine. Let the washers slide down the rod or wire and into position.
3. Insert the distributor drive shaft. It will turn slightly as it engages the crankshaft gear. Engage the teeth so that the shaft will be in the position shown previously in Fig. 8-2 or Fig. 8-3 when fully installed.
4. Install the distributor. Check the ignition timing.

NOTE —
On some single-carburetor rebuilt exchange engines, the drive shaft bore in the crankcase is machined out and the lower shim washer replaced by a 3 mm (.118 in.) spacer. When repairing these engines, always install the 0.60 mm (.024 in.) shim washer between the 3 mm (.118 in.) spacer and the drive shaft gear so that the spacer will be on the bottom.

9. LUBRICATION SYSTEM

Because of changes in the single-carburetor engine crankcase, the oil coolers used on 1970 and 1971 models are different from those used on earlier models. Beginning with the 1971 models, the single-carburetor engine's oil cooler is relocated on the top of the crankcase. On dual-carburetor and fuel injection engines, the oil cooler is mounted behind the No. 4 cylinder, as shown in Fig. 9-1.

1. Oil filter mounting flange
2. M 8 nut (2)
3. Spring washer (2)
4. Oil cooler seal (2)
5. Oil pressure switch
6. Breather seal
7. Oil breather
8. Oil pressure control valve plunger*
9. Oil pressure control valve spring*
10. Sealing ring*
11. Control valve plug*
12. Rubber bellows
13. Oil pump bearing plate
14. Oil pump driven gear
15. Oil pump drive gearshaft
16. Oil filler pipe
17. Oil dipstick
18. Oil filler clamp
19. Oil filler with cover
20. Oil cooler
21. Washer (3)
22. Spring washer (3)
23. M 6 nut (3)
24. Sealing ring
25. Plug
26. Oil filter
27. Oil strainer gasket (2)
28. Oil strainer
29. Oil strainer cover
30. Drain plug seal
31. Oil drain plug
32. Relief valve plug
33. Sealing ring
34. Oil pressure relief valve spring
35. Oil pressure relief valve plunger
36. Filter flange gasket
37. Oil filler pipe gasket
38. Oil pump gasket
39. Spring washer (2)
40. M 8 nut (2)
41. Oil pump O-ring
42. M 6 self-locking nut
43. Spring washer (4)
44. Oil pump housing
45. Spring washer (4)
46. M 8 nut (4)

* Discontinued after November 1975.

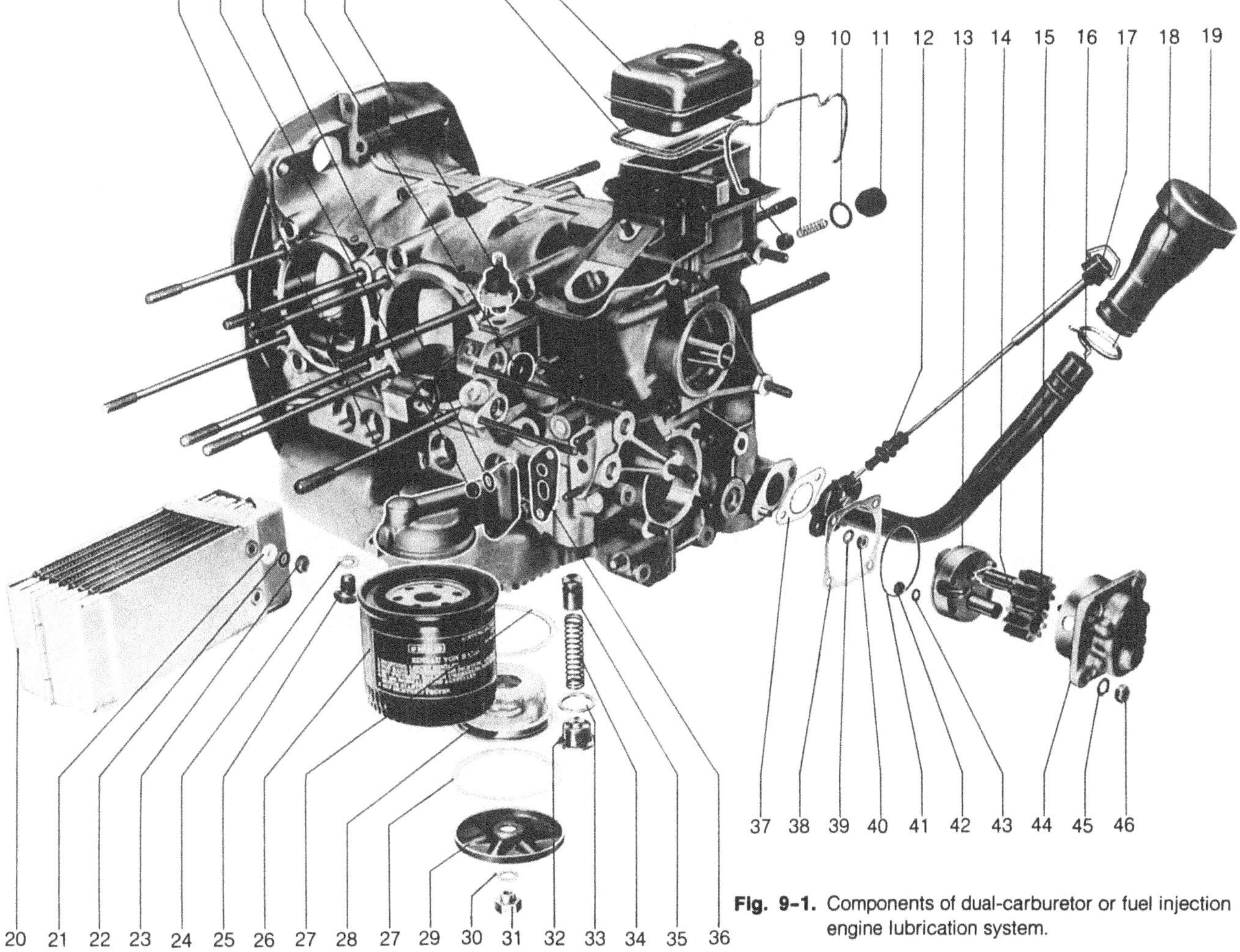

Fig. 9-1. Components of dual-carburetor or fuel injection engine lubrication system.

9.1 Removing and Installing Oil Cooler
(single-carburetor engine)

On 1968 through 1970 models, the oil cooler is mounted directly on the crankcase. On 1971 models, there is a light alloy intermediate flange bolted in the former oil cooler location. The oil cooler itself is mounted on the side of the intermediate flange. The oil cooler can either be removed from the flange by taking off the nuts or it can be removed complete with the flange, following a procedure similar to that used on the earlier engines. If the oil cooler is to be replaced, remove it from the intermediate flange. When making engine repairs, or when having the oil cooler pressure-tested by a VW Dealer, it is better to remove the cooler with the intermediate flange attached.

To remove the oil cooler (1968 through 1970 models) or the oil cooler in combination with the intermediate flange (1971 models), remove the one M 6 nut at the right-hand side of the oil cooler or flange, then remove the other two nuts from beneath the left-hand side as illustrated in Fig. 9-2.

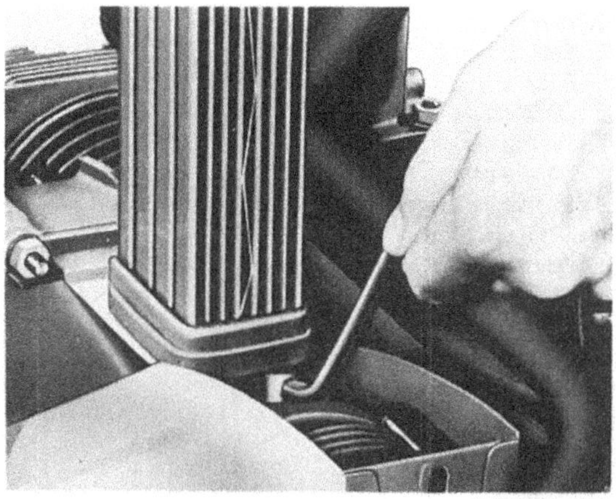

© 1974 VWoA—2002

Fig. 9-2. Removing one of two nuts under left-hand side of oil cooler (or intermediate flange).

During installation, use new rubber seals between the oil cooler and the crankcase (1968 through 1970 models) or between the intermediate flange and the crankcase (1971 models). If the later oil cooler is removed from the intermediate flange, use new rubber seals between the two parts during assembly. Torque the mounting nuts to 0.7 mkg (5 ft. lb. or 60 in. lb.).

9.2 Removing and Installing Oil Cooler

(dual-carburetor or fuel injection engine)

To remove the oil cooler from a dual-carburetor engine, remove the cooling air fan housing as described in **7.1 Removing and Installing Fan Housing**. Then remove the three M 6 nuts with washers that hold the oil cooler on the rear of the crankcase. Take out the bolts that hold the oil cooler support strap, then remove the support strap and oil cooler as a unit.

During installation, use new rubber seals between the oil cooler and the crankcase. Torque the mounting nuts to 0.7 mkg (5 ft. lb. or 60 in. lb.).

9.3 Testing and Replacing Oil Pressure Switch

If the oil pressure warning light does not come on when the ignition is on and the engine is not running, check the bulb and, if necessary, replace it. If the bulb is not at fault, remove the wire from the oil pressure switch and ground it on the crankcase with the ignition on. If the light does not come on, the wiring is faulty. If it does come on, there is trouble in the switch. Remove the switch as shown in Fig. 9-3 and inspect it for a clogged orifice. If cleaning does not restore the switch to serviceable condition, replace it.

© 1974 VWoA—2004

Fig. 9-3. Oil pressure switch being removed from single-carburetor engine. See Fig. 9-1 for the switch location on the dual-carburetor and fuel injection engines.

You can also test the pressure switch for accuracy by installing an oil pressure gauge between the switch and the crankcase as shown in Fig. 9-4. The warning light should go out at 2 to 6.4 psi (0.14 to 0.45 kg/cm²). On single-carburetor engines, the oil pressure at 2500 rpm should be about 42 psi (3.0 kg/cm²) with SAE 30 oil at a temperature of 70°C (158°F). On dual-carburetor and fuel injection engines, the oil pressure at 2500 rpm should be at least 28 psi (2.0 kg/cm²).

© 1974 VWoA—2005

Fig. 9-4. Oil pressure gauge installed between crankcase and oil pressure switch.

9.4 Removing and Installing Oil Pump
(single-carburetor engine)

The oil pump can be removed without draining the oil or separating the crankcase halves. Be careful not to let dirt enter the engine while the pump is out.

An exploded view of the oil pump is given in Fig. 9-5. A larger oil pump and a different camshaft with camshaft gear were introduced in May 1971. The larger pump can be used in earlier engines provided the camshaft and gear are also replaced. Never use the earlier pump with a late engine.

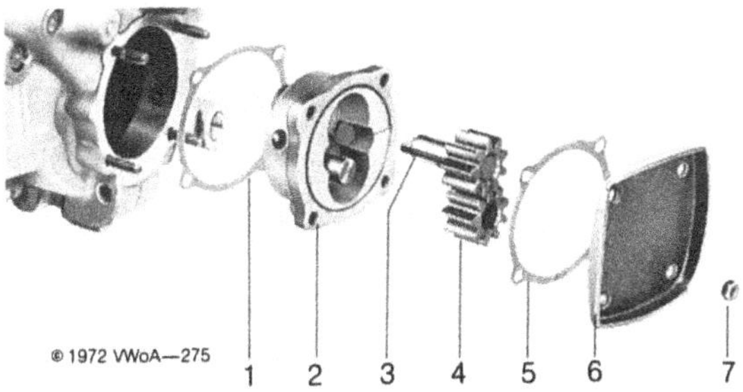

1. Housing gasket
2. Oil pump housing
3. Oil pump drive gearshaft
4. Oil pump driven gear
5. Cover gasket
6. Oil pump cover
7. Sealing nut

Fig. 9-5. Components of the oil pump used in the single-carburetor engine.

To remove:

1. Remove the engine. Remove the four sealing nuts that hold the oil pump cover, then take off the cover.
2. Remove the oil pump gears and cover gasket.
3. Attach a puller such as the one shown in Fig. 9-6, then use it to remove the oil pump housing.

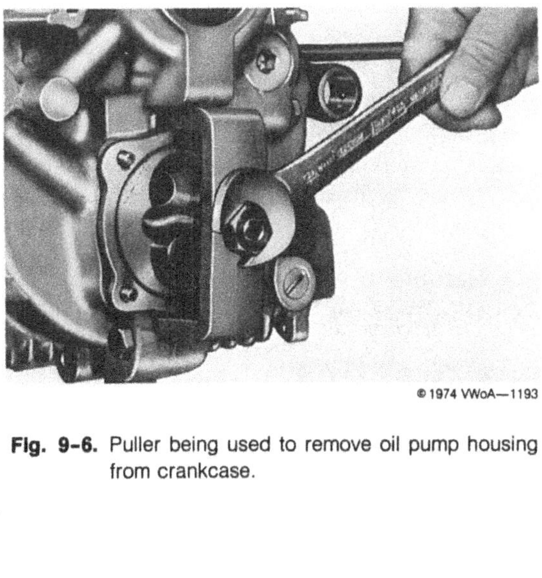

Fig. 9-6. Puller being used to remove oil pump housing from crankcase.

NOTE —
A different puller claw must be used for the larger pump introduced on 1971 cars.

4. Remove the housing gasket.

To install:

1. Thoroughly clean and inspect the pump as described in **9.6 Checking Oil Pump**. Replace worn parts.
2. While hand-turning the oil pump drive gearshaft to engage its drive dog in the camshaft, install the oil pump together with a new housing gasket.
3. Hand-turn the crankshaft through two revolutions so that the pump housing will align itself with the camshaft.
4. Using a new cover gasket, install the remaining pump parts on the studs.
5. Install new sealing nuts with their plastic facings toward the pump cover. Then torque them to 2.0 mkg (14 ft. lb.).

9.5 Removing and Installing Oil Pump
(dual-carburetor or fuel injection engine)

The oil pump can be removed without draining the oil or separating the crankcase halves. Be careful not to let dirt enter the engine while the pump is out.

To remove:

1. Remove the engine. Then remove the four M 8 nuts that hold the oil pump.
2. Using two levers as shown in Fig. 9-7, carefully pry the oil pump out of the crankcase.

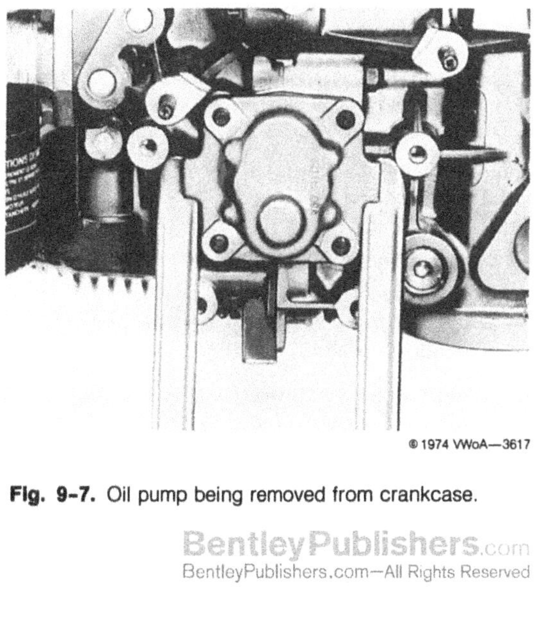

Fig. 9-7. Oil pump being removed from crankcase.

5

3. Use a puller such as the one shown in Fig. 9-8 to pull the oil pump bearing plate out of the oil pump housing.

© 1974 VWoA—3618

Fig. 9-8. Puller VW 803 being used to disassemble oil pump. Use a wrench to turn the bolt, which is threaded into a tapped hole in the bearing plate.

To install:

1. Thoroughly clean and inspect the pump as described in **9.6 Checking Oil Pump**. Replace worn parts.
2. While hand-turning the oil pump drive gearshaft to engage its drive dog in the camshaft, install the oil pump bearing plate together with a new O-ring.
3. Hand-turn the crankshaft through two revolutions so that the pump bearing plate will align itself with the camshaft.
4. Using a new gasket, install the remaining pump parts on the studs.
5. Install the four nuts and the spring washers. Then torque the nuts to 2.0 mkg (14 ft. lb.).

9.6 Checking Oil Pump

Check the oil pump thoroughly whenever you remove it. If it is faulty, it will not supply adequate oil pressure and must be replaced.

To check:

1. Inspect the housing for wear, especially in the gear-seating areas.
2. On single-carburetor engines, check the shaft for the driven gear. If it is loose, peen in the area indicated by the arrow in Fig. 9-9 or replace the housing.

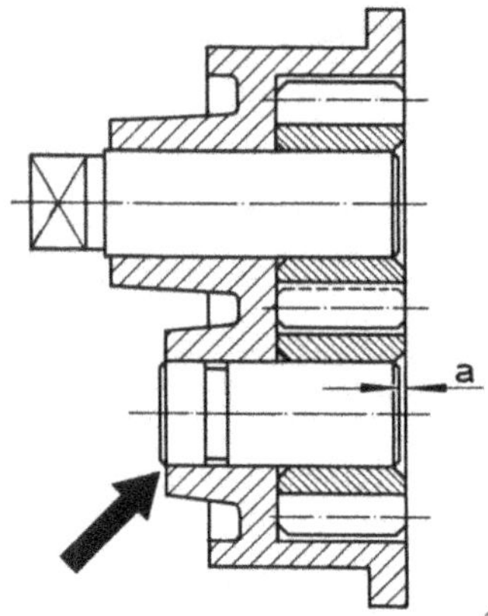

© 1974 VWoA—1194

Fig. 9-9. Shaft for driven gear (arrow). Dimension **a** must be 0.50 to 1.00 mm (.020 to .040 in.).

3. On all pumps, check the fit of the drive gearshafts.
4. Check the gears for wear. Backlash must not exceed 0.20 mm (.008 in.), and it should be even less.
5. On single-carburetor engines, measure the end play of the gears as shown in Fig. 9-10. It must not exceed 0.10 mm (.004 in.) without the gasket.

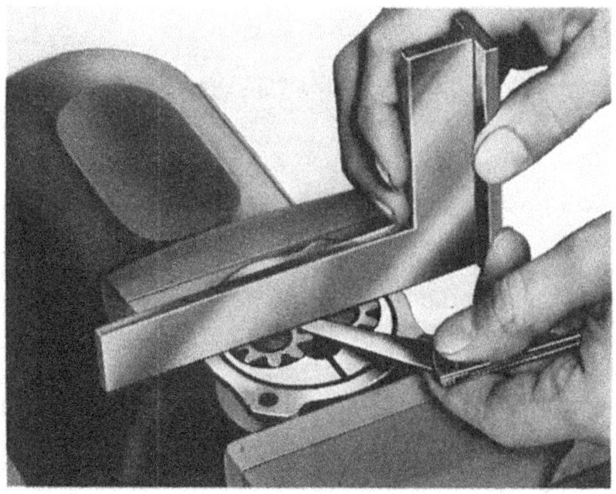

© 1974 VWoA—1195

Fig. 9-10. End play being measured with a feeler gauge and straightedge.

6. On dual-carburetor or fuel injection engines, replace the bearing plate if it is scored.

9.7 Removing and Installing Oil Pressure Valves

Unlike earlier VW engines, there are two oil pressure valves used in the engines of 1970 and later vehicles. The oil pressure valves are spring-loaded plungers that move in bores machined into the crankcase castings.

Fig. 9-11 is a schematic view of the oiling system used on 1970 and 1971 single-carburetor engines. On dual-carburetor and fuel injection engines, there is a filter located between the pump and the oil pressure switch.

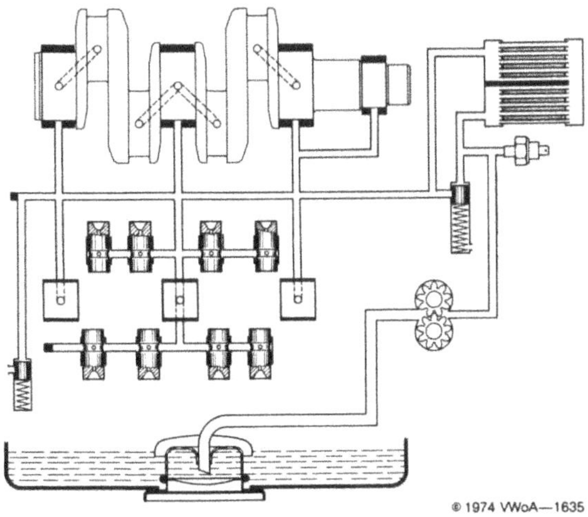

Fig. 9-11. Schematic view of oiling system. One oil pressure valve is at the right, below and to the left of the oil cooler; the other is at the lower left-hand side of the illustration.

The oil pressure valve at the right-hand side of the preceding illustration is known as the pressure relief valve. When the oil is cold and thick, the resulting increased oil pressure forces the valve down against spring tension. This allows oil from the pump to go directly to the engine bearings. In bypassing the oil cooler, warm-up is speeded and the oil cooler is protected against excessive pressures that might cause it to burst.

The oil pressure valve at the lower left-hand side of Fig. 9-11 is known as the oil pressure control valve. It is located at the extreme end of the main oil passage. When oil pressure rises above the point necessary to supply lubrication to the bearings, the valve's piston is forced down against the spring tension. This allows oil to be returned directly to the crankcase sump.

The oil pressure control valve ensures that oil pressure will remain constant at the bearings, regardless of engine speed. Thus, at high engine speeds, when pump output is greatest, the excess oil is allowed to escape. Delivered to the engine's working parts, the surplus oil might spray out onto the cylinder walls in excessive quantities and cause increased oil consumption. Avoiding excess oil pressure also reduces the power lost by the engine in driving the oil pump.

A cracked or otherwise leaking oil cooler should prompt the removal and inspection of both oil pressure valves. If either valve is stuck or sticking intermittently in its closed position, it can cause excessive oil pressure. Usually, the oil pressure valves become sticky only in neglected engines. If the oil is not changed according to the schedule given in **LUBRICATION AND MAINTENANCE**, foreign matter or corrosive agents may attack the valves and prevent them from moving freely.

Valve removal on single-carburetor engines is illustrated in Fig. 9-12; their removal on dual-carburetor and fuel injection engines is shown earlier in Fig. 9-1. Because the slotted plugs for the oil pressure valves are a tight fit in the crankcase, you should not try to remove them with an undersize screwdriver. Use a tool that will completely fill the slot. Otherwise, the alloy plug may be ruined or deformed.

Fig. 9-12. Oil pressure relief valve (**1**) and oil pressure control valve (**2**).

5

To remove oil pressure valve:

1. Using a properly fitting screwdriver, remove the slotted plug.
2. Remove the spring and plunger. If the plunger is stuck, remove it by screwing a 10-mm (3/8-in.) tap into it.

To install:

1. Check the plunger and bore for signs of seizure. Carefully remove rough spots and deposits, then install a new plunger if necessary. Oversize plungers are available for reconditioned bores in single-carburetor engines.
2. Examine the spring. It should not be deformed or damaged by abrasion or corrosion.

CAUTION —
If both valves are removed, be careful not to mix up the springs. The springs for the pressure relief valve and the pressure control valve are not interchangeable.

3. Check the spring tension and length. Then compare your findings with those given in **Table a.**

Table a. Spring Specifications

Model and part	Loaded length	Loaded tension
1968–1969 pressure relief valve spring	43.20 mm (1¹¹⁄₁₆ in.)	2.5–4.2 kg (5.5–9.2 lb.)
1970–1971 pressure relief spring	44.10 mm (1¾ in.)	5.6–7.3 kg (12.3–16.0 lb.)
1970 pressure control spring	20.20 mm (¹³⁄₁₆ in.)	2.9–3.6 kg (6.3–7.9 lb.)
1971 pressure control spring	20.20 mm (¹³⁄₁₆ in.)	3.1–3.8 kg (6.8–8.4 lb.)
1972 and later pressure relief spring	39.00 mm (1⁹⁄₁₆ in.)	6.8–8.8 kg (15.0–19.4 lb.)
1972 and later pressure control spring	26.00 mm (1¹⁄₁₆ in.)	1.7–2.0 kg (3.8–4.4 lb.)

4. Install the plunger and spring as indicated in Fig. 9-13.

CAUTION —
Do not allow the spring to scratch the wall of the bore as you install it. Burrs on the wall either could cause the plunger to jam or could limit its travel.

1. Plunger
2. Spring
3. Gasket
4. Plug

Fig. 9-13. Order in which oil pressure valve parts are installed in engine.

5. Using a new gasket, install the plug.

CAUTION —
Do not overtighten the plug. You could strip the threads in the crankcase casting. Also, do not apply sealing compound to the plug or gasket. Doing so could make future removal difficult or impossible.

10. Valve Gear and Cylinder Heads

Eight nuts hold each cylinder head on a pair of cylinders. These nuts thread onto long studs set in the crankcase casting. The rocker arm shaft(s) must be removed to gain access to the nuts holding the cylinder head. Fig. 10-1 gives an exploded view of the dual-carburetor or fuel injection engine cylinder head.

10.1 Removing and Installing Rocker Arm Shaft

Before removing the cylinder head cover, clean all road dirt from the cover and the cylinder head. To remove the cover, pry the heavy spring wire retainer clip outward and down. If you install a new gasket, do not use gasket cement between the cylinder head cover gasket and the cylinder head. Replace leaking gaskets.

To remove rocker arm shaft(s):

1. Remove the cylinder head cover.
2. Remove the rocker arm shaft retaining nuts. There are two nuts holding the shaft on single-carburetor engines. Four nuts hold two individual shafts on dual-carburetor and fuel injection engines.

NOTE —
Loosen the nuts a little at a time, working alternately to relieve spring tension evenly.

3. Lift off each rocker arm shaft assembly.

To install rocker arm shaft(s):

1. On single-carburetor engines only, install new silicone rubber seals on the rocker arm shaft support studs.
2. If previously removed, install the pushrods.

NOTE —
Roll the pushrods on a flat table to see that none are bent.

3. Carefully guiding the pushrods into the rocker arm sockets, install single-carburetor engine rocker arm

1. Cam follower (8)—hydraulic valve lifter on 1978 engine
2. Exhaust valve (4)
3. Intake valve (4)
4. Pushrod tube (8)
5. Exhaust valve guide (4)
6. Intake valve guide (4)
7. Cylinder head (2)
8. Washer (16)
9. M 10 nut (16)
10. Oil deflector ring (8)
11. Valve spring (8)
12. Spring retainer (8)
13. Valve keeper (16)
14. Spring washer (8)
15. M 7 nut (8)
16. Black sealing ring (8)
17. Pushrod (8)
18. White sealing ring (8)
19. Deflector plate (2)
20. Washer (4)
21. M 6 fillister head screw
22. Washer (2)
23. M 5 fillister head screw
24. Support (8)
25. Exhaust rocker arm (4)
26. Thrust washer—discontinued in early 1974, new type introduced in 1978 (8)
27. Spacer spring—spring washers with spacer bushing from 1978 (4)
28. Rocker shaft (4)
29. Intake rocker arm (4)
30. Adjusting screw (8)
31. M 8 nut (8)
32. Pushrod tube retaining wire (2)
33. Cylinder head cover gasket (2)
34. Cylinder head cover (2)

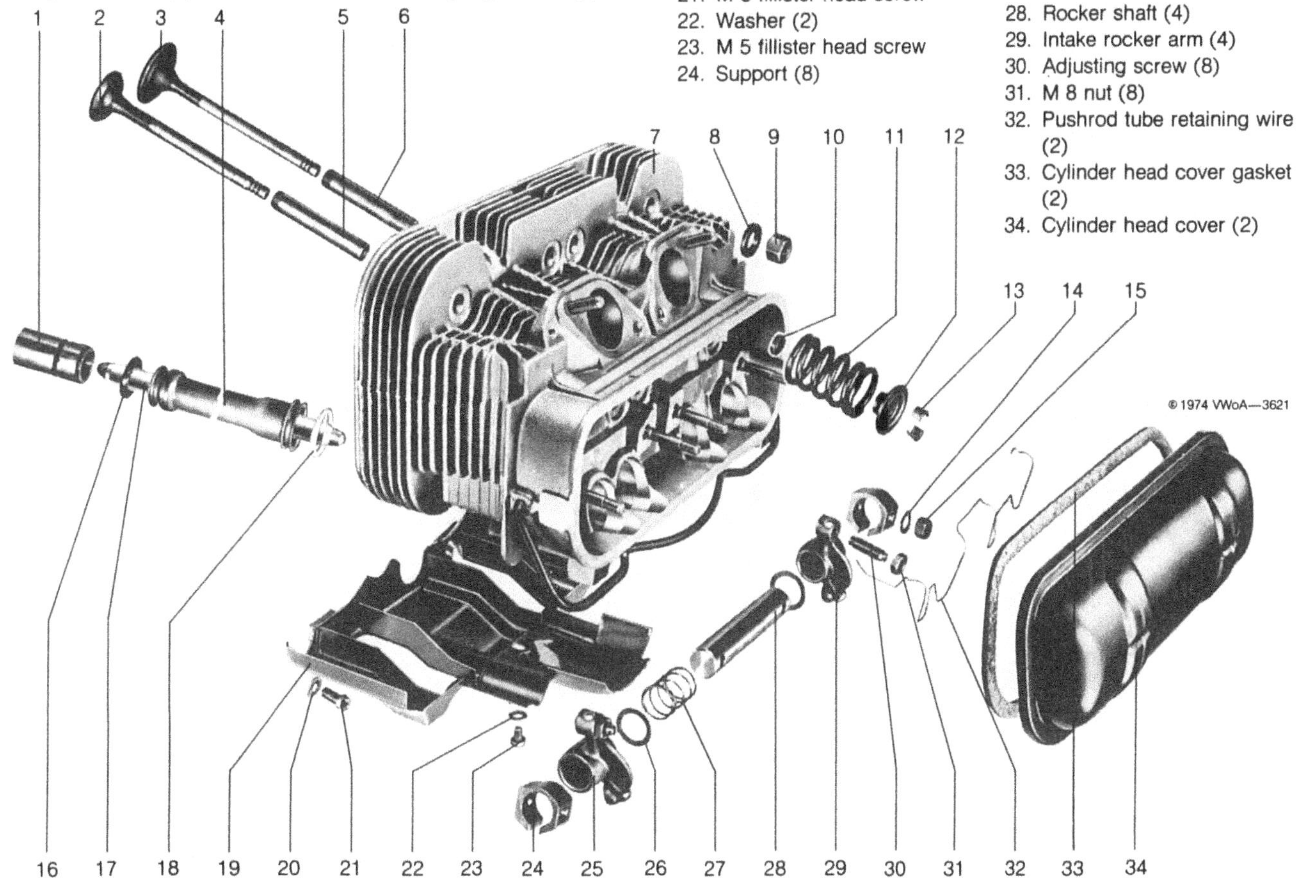

Fig. 10-1. Components of the dual-carburetor or fuel injection engine cylinder head.

shaft assemblies as indicated in Fig. 10-2; install dual-carburetor or fuel injection rocker arm shaft assemblies as indicated in Fig. 10-3.

Fig. 10-2. Single-carburetor rocker arm shaft installation. The slotted side of the support (left arrow) should point upward. The beveled end (right arrow) should be away from the cylinder head.

Fig. 10-3. Dual-carburetor and fuel injection engine rocker arm shaft installation. The slotted side of the support (arrows) should point downward. The beveled ends should be outward.

4. Install the nuts. On dual carburetor and fuel injection engines, torque the nuts to 1.4 mkg (10 ft. lb.); on single-carburetor engines torque the nuts to 2.5 mkg (18 ft. lb.).

5. On dual-carburetor and fuel injection engines only, install the pushrod tube retaining wire as shown in Fig. 10-4. Then install the cylinder head cover.

© 1974 VWoA—3623

Fig. 10-4. Pushrod tube retaining wire installation. The wire must bear against the ends of the pushrod tubes (lower arrows) and engage the slots in the rocker arm shaft supports (upper arrow).

10.2 Disassembling and Assembling Rocker Arm Shaft Assembly

If the rocker arms are worn or binding, the rocker arm shaft assembly may need to be taken apart in order to inspect the parts or to clean away hardened dirt accumulations. If the rocker arm shaft is galled, or if the rocker arm bores are worn out-of-round, either replace the worn parts individually or replace the entire rocker arm shaft assembly. If spacer springs without ground end faces, introduced in early 1974, are installed on an earlier dual-carburetor engine, do not install the thrust washers.

To disassemble:

1. On single-carburetor engines, remove the end clips. Then slide the rocker arms, supports, and washers off the shaft. On dual-carburetor and fuel injection engines, simply slide the rocker arms, supports, washers, and spacer spring off the shaft.

2. Inspect the shaft, the rocker arm bores, the adjusting screws, and the rocker arm sockets for wear. Replace worn parts.

3. Using Fig. 10-5 as a guide for single-carburetor engines or Fig. 10-1 as a guide for dual-carburetor and fuel injection engines, assemble the components.

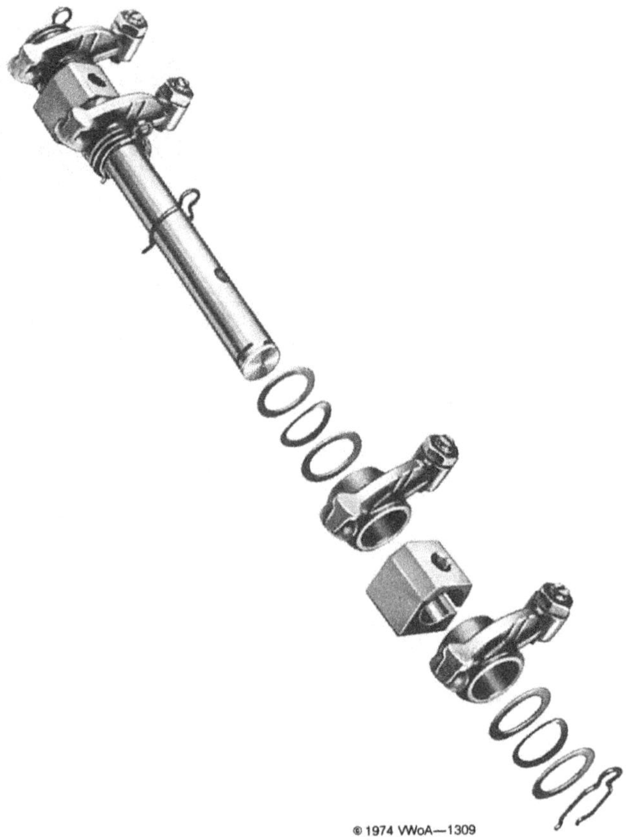

© 1974 VWoA—1309

Fig. 10-5. Exploded view of single-carburetor engine rocker arm shaft assembly. Replace worn or corroded washers and end clips.

10.3 Removing and Installing Cylinder Head and Pushrod Tubes

The engine must be out of the car before you can remove the cylinder heads. The pushrod tubes can be removed from dual-carburetor engines without removing the engine or taking off the cylinder heads. However, it is also possible, using the telescopic pushrod tubes described in the following procedure, to replace pushrod tubes on the single-carburetor engine without removing the engine or the heads.

To replace pushed tubes on single carburetor engine (cylinder heads installed):

1. Remove the cylinder head cover by prying the heavy spring wire retainer clip outward and down.
2. Remove the rocker arm shaft retaining nuts, loosening each nut a little at a time, working alternately to relieve spring tension evenly.
3. Lift off the rocker arm shaft assembly. Withdraw the pushrod(s) from the pushrod tube(s) that you intend to replace.
4. Using a large screwdriver, pry out the defective pushrod tube(s). Remove the old pushrod tube seals and wipe clean the recesses in the cylinder head and the crankcase.
5. Install new seals on the telescopic pushrod tube(s). Then squeeze together each telescopic tube and install it so that the small-diameter end is toward the cylinder head.
6. Install the pushrod(s). Install new silicone rubber seals on the rocker arm shaft support studs.
7. Carefully guiding the pushrods into the rocker arm sockets, install the rocker arm shaft assembly on the cylinder head. Tighten each nut a little at a time, working alternately in order to compress the valve springs evenly. Torque the nuts to 2.0 to 2.5 mkg (14 to 18 ft. lb.).
8. Adjust the valves as described in **10.8 Valve Clearance.** Then install the cylinder head cover.

> **NOTE ——**
> After running the engine for a short time, check for oil leaks around the pushrod tube(s) that you have replaced.

To remove cylinder head:

1. Remove the engine as described in **3. Removing and Installing Engine.**
2. Remove the cooling air system and the intake manifold(s) as described in **6. Cooling Air System and Intake Manifold (single-carburetor engine)** or **7. Cooling Air System and Intake Manifold (dual-carburetor and fuel injection engines).**
3. Remove the rocker arm assemblies as described in **10.1 Removing and Installing Rocker Arm Shaft.**
4. Withdraw the pushrods from the pushrod tubes. On 1700, 1800, and 2000 engines, pull the pushrod tubes out of the crankcase, shoving them through the floor of the cylinder head and out of the engine.
5. On 1700, 1800, and 2000 engines, remove the three screws that hold the deflector plate to the crankcase and to the bottom of the cylinder head. Remove the deflector plate.

6. Using a 15-mm (19/32-in.) socket wrench, remove the head nuts. Remove the washers, then pull the head off the cylinders.

NOTE ——

If the cylinders are to be left on the engine, use some type of retaining device to keep them from pulling free along with the heads.

To install head:

1. On dual-carburetor and fuel injection engines, remove the metal gasket rings from the peripheries of the combustion chambers. (Gasket rings may also be found in some single-carburetor engines.)
2. Check the head for cracks in the combustion chamber and valve ports. If there are cracks, replace the head.
3. Check the rocker arm shaft support studs and cylinder studs for tightness. Take any parts that have pulled studs to an Authorized VW Dealer or a qualified automotive machine shop for repair with special threaded inserts. See **16.4 Repairing Tapped Holes.**
4. Check the pushrod tubes for holes and replace any that is leaking. Measure the tubes for single-carburetor engines as indicated in Fig. 10-6. Dimension **a** must be 190 to 191 mm (7.48 to 7.52 in.). If it is not, stretch the tube bellows slightly.

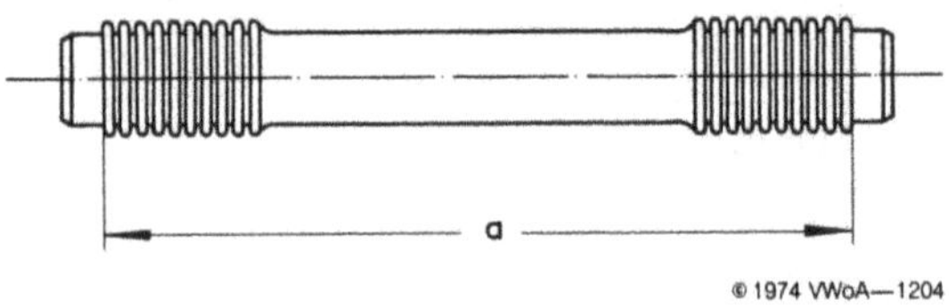

Fig. 10-6. Pushrod tube length (**a**). Check for cracks and pinholes.

5. On single-carburetor engines, install new seals on the pushrod tubes.
6. If the engine was originally so equipped, install new metal gasket rings in the peripheries of the combustion chambers.
7. On single-carburetor engines, slide the cylinder heads onto the cylinder studs. Install the pushrod tubes with their seams facing upward. Then push the head onto the cylinders to hold the tubes (Fig. 10-7). (Do not install the pushrod tubes of 1700, 1800, or 2000 engines until you have torqued the cylinder head nuts.)

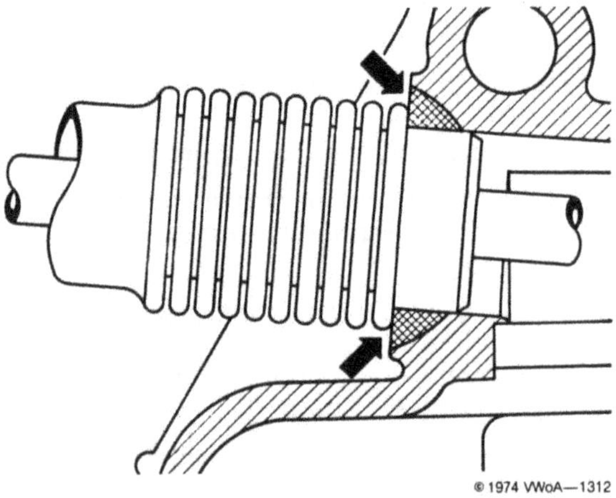

Fig. 10-7. Rubber seals correctly positioned (single-carburetor engine).

8. Loosely install the cylinder stud washers and nuts.
9. Following the sequence given in Fig. 10-8, torque the nuts on single-carburetor engines only to 1 mkg (7 ft. lb.).

NOTE —
This sequence is for initial tightening only. It is not the final sequence.

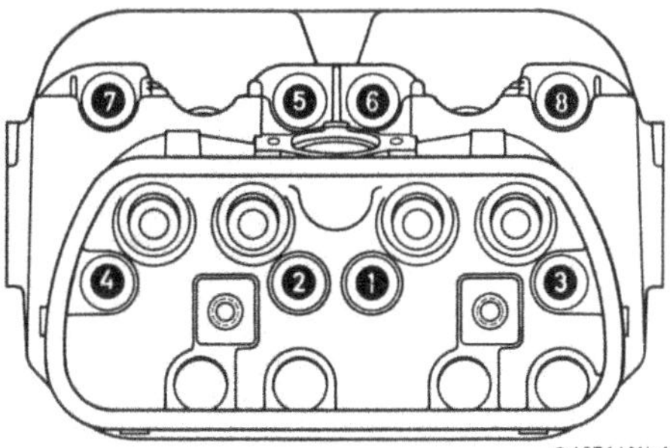

Fig. 10-8. Single-carburetor engine initial tightening sequence.

10. Following the sequence given in Fig. 10-9, torque the nuts on single-carburetor engines only to 3.0 to 3.2 mkg (22 to 23 ft. lb.).

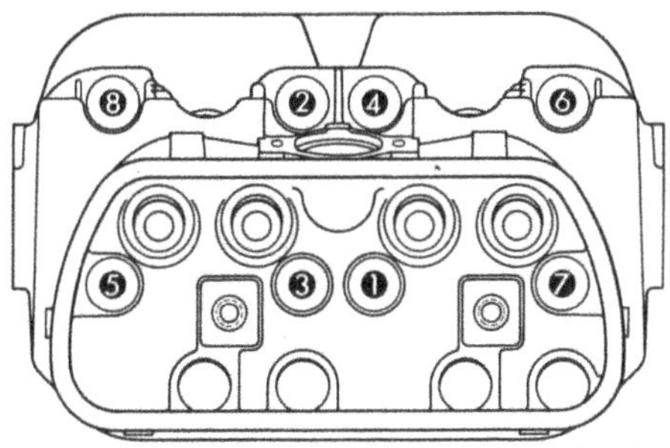

Fig. 10-9. Single-carburetor engine final torque sequence.

11. Following the sequence given in Fig. 10-10, uniformly torque the nuts on dual-carburetor and fuel injection engines to approximately 1.5 mkg (11 ft. lb.). Then go over the same sequence a second time, finally bringing the torque to 3.2 mkg (23 ft. lb.).

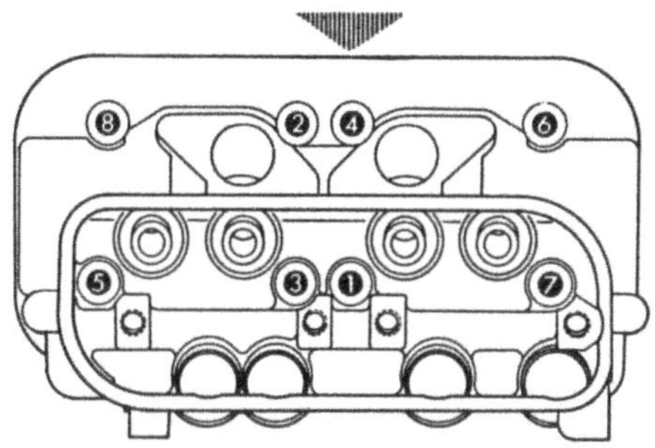

Fig. 10-10. Dual-carburetor and fuel injection engine tightening sequence.

12. On 1700, 1800, and 2000 engines, install the deflector plate beneath the cylinders and the cylinder head and attach it to the crankcase and to the head with the three screws.

NOTE —
The screw that holds the deflector plate to the cylinder head is smaller than the one that holds the deflector plate to the crankcase. You cannot install the deflector plate after the pushrod tubes are in place.

13. On 1700, 1800, and 2000 engines, install new seals on the pushrod tubes as shown previously in Fig. 10-1. Then slide the pushrod tubes into position through the tops of the cylinder heads.

14. On all engines, install the pushrods and the rocker arm shaft assemblies as described in **10.1 Removing and Installing Rocker Arm Shaft.** Fig. 10-4 under that heading shows the correct installation for the pushrod tube retaining wires on dual-carburetor and fuel injection engines.

5

11. VALVES

The components of a valve assembly for the VW engine are shown in Fig. 11-1.

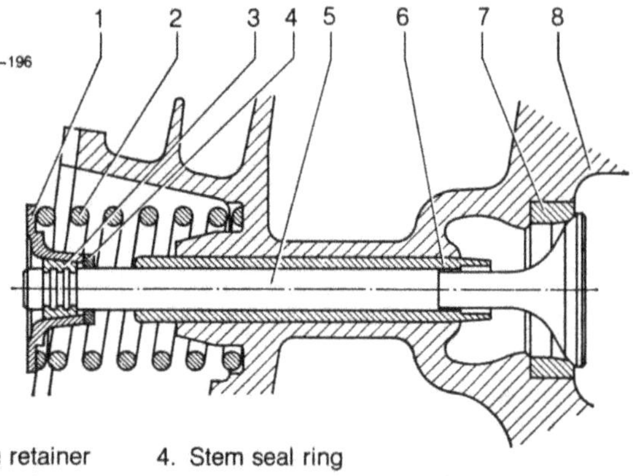

1. Spring retainer
2. Valve spring
3. Valve keeper
4. Stem seal ring
5. Valve
6. Valve guide
7. Valve seat insert
8. Cylinder head

Fig. 11-1. Cross section of valve assembly. The stem seal ring is not installed on 1973 and later models. The seal rings can be service-removed from 1972 dual-carburetor engines.

11.1 Removing and Installing Valves

A proper valve spring compressing tool is necessary to remove the valves from the cylinder head.

To remove valves:

1. Remove the cylinder head.
2. On single-carburetor engines, compress the valve spring as shown in Fig. 11-2.

Fig. 11-2. Valve spring compressing tool being used to push spring retainer down valve stem against spring tension (single-carburetor engine).

3. On dual-carburetor or fuel injection engines, compress the valve springs as shown in Fig. 11-3.

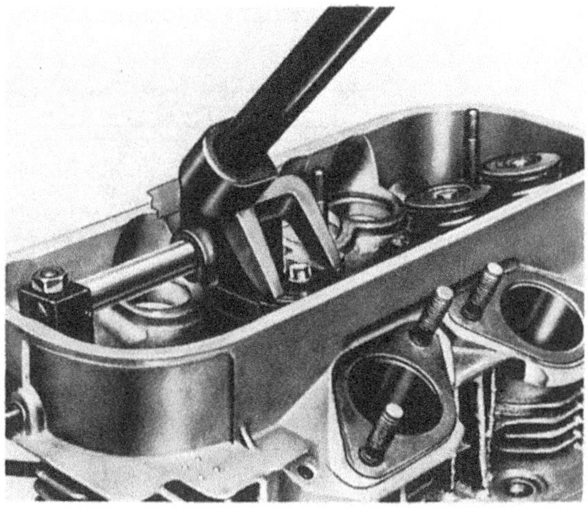

Fig. 11-3. Valve spring tool VW 311s being used to push spring retainer down valve stem against spring tension (dual-carburetor or fuel injection engine).

4. Remove the split keeper halves from the valve stem.
5. Release the compressing tool. Then take off the spring retainer, the valve spring, and (where fitted) the stem seal ring.
6. Remove the valve from the cylinder head.

> ***CAUTION—***
> *If the keeper grooves in the valve stem are burred, file them smooth before you remove the valve. If forced out, the burred stem will ruin the valve guide.*

To install:

1. Using a valve spring tester, check the valve spring tension.

> **NOTE—**
> Springs for a single-carburetor engine should require a load of 53.2 to 61.2 kg (117 to 135 lb.) to compress them to a loaded length of 31.0 mm (1$\frac{7}{32}$ in.). Springs for a dual-carburetor or fuel injection engine should require a load of 76.5 to 84.5 kg (168 to 186 lb.) to compress them to a loaded length of 29.0 mm (1$\frac{9}{64}$ in.).

> **NOTE—**
> If you do not have a valve spring tester, have the springs checked by an Authorized VW Dealer or a qualified automotive machine shop.

2. To check the keepers, oil them and fit them to the valve stem. Hold the keeper halves together while you turn the valve. The valve stem should rotate freely in the assembled keeper.
3. Check the valve seat and valve facings. If necessary, recondition (grind) them.

CAUTION—
If you lack the skills, tools, or a suitable workshop for reconditioning the valves, we suggest you leave such repairs to an Authorized VW Dealer or other qualified shop. We especially urge you to consult your Authorized VW Dealer before attempting repairs on a car still covered by the new-car warranty.

4. Check the valve guides as described in **11.3 Checking and Replacing Valve Guides.**

NOTE—
If a guide is within specifications but the valve can be moved from side to side in it, check the valve stem for wear. See **Table b.**

Table b. Valve Stem Diameters

Stem diameter	New valves mm (in.)	Wear limit mm (in.)
Intake valve all engines	7.94-7.95 (.3126-.3130)	7.90 (.3110)
Exhaust valve 1600 engines	7.91-7.92 (.3114-.3118)	7.88 (.3102)
1700, 1800, and 2000 engines	8.91-8.92 (.3508-.3512)	8.87 (.3492)

5. Install the valve and, on single-carburetor engines, the stem seal ring. Install the spring so that its closely-spaced coils are against the cylinder head.
6. Install the spring retainer. Then compress the spring with a valve spring compressing tool and install the keeper.

11.2 Removing and Installing Valve Springs
(head installed)

A worn stem seal ring, broken valve spring, damaged keeper, or spring retainer can be replaced without removing either the engine or the cylinder head. To do this, use the procedure that follows.

To remove spring:

1. Remove the spark plug, cylinder head cover, and rocker arm shaft.

2. Install an air hose adapter in the spark plug hole. Apply a constant air pressure of at least 85 psi (6 kg/cm^2) to the cylinder.
3. Attach the special valve spring compressor as shown in Fig. 11-4.
4. Compress the spring retainer and the spring. Then remove the keeper.

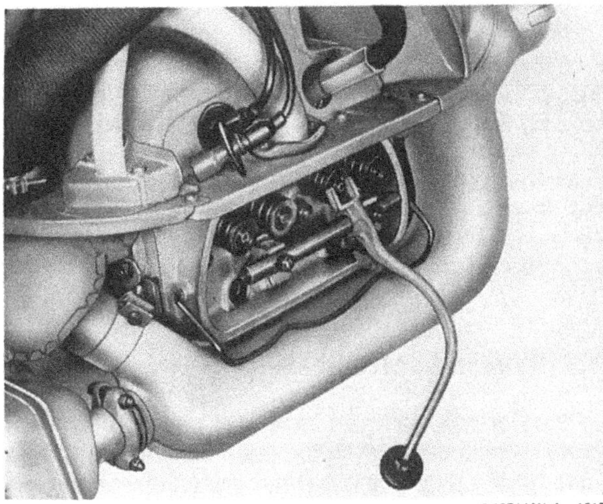

© 1974 VWoA—1217

Fig. 11-4. VW tool 653/2 being used to remove valve spring with cylinder head installed.

5

11.3 Checking and Replacing Valve Guides

If a new valve stem fits too tightly in the guide, the guide probably has "varnish" deposits inside it. You can correct this with either a guide hone or a brass bristle brush and carburetor cleaning solvent.

To check a guide for wear, insert a valve that is in good condition until the valve stem is flush with the end of the guide. On single-carburetor engines, it should not be possible to rock the valve head sideways more than 0.23 to 0.27 mm (.009 to .011 in.). The wear limit for this measurement is 0.80 mm (.031 in.). On dual-carburetor and fuel injection engines, the rock should be no more than 0.45 mm (.018 in.) with a wear limit of 0.90 mm (.035 in.).

Measure valve rock with a dial indicator. Mount the dial indicator on the cylinder head so that the gauge pin is against the valve head. If the guide permits excessive rocking motion even with a new valve, replace the guide.

The valve guides are a press fit in the cylinder head. They should be removed only with special equipment available at an Authorized VW Dealer or a qualified automotive machine shop. Guides with an oversize outside diameter are available. Following installation, the new guides must be reamed to the dimensions given in **20. Engine and Clutch Technical Data**.

11.4 Replacement of Valve Seats

It might appear that the valve seat inserts in the cylinder head can be replaced. Some specialty shops that rebuild VW engines for racing do make this modification. However, the job is impossible to carry out successfully without special equipment and is done only when valves of increased head diameter are to be used.

Such work is expensive and, done properly, costs more than a new replacement cylinder head. It is therefore not economically wise to replace valve seats when standard valve diameters are retained. Moreover, there is always the possibility of a replacement insert coming out. In salvaging an unserviceable cylinder head, an entire engine may be ruined. Therefore, we strongly recommend that you replace the entire cylinder head if the valve seats are no longer serviceable.

11.5 Refacing Valve Seats

Pitted or gas-cut valve seats can usually be reconditioned by refacing. A cylinder head may undergo this operation many times before it becomes unserviceable. Irreparable seat damage is usually the result of continued driving long after a compression test has indicated that the valves are leaking.

On single-carburetor engines only, the valve seat contact facing should be cut to 45° for both intake and exhaust valves. On dual-carburetor and fuel injection engines, cut the valve seat contact facing to 30° on intake valves and 45° on exhaust valves.

On single-carburetor engines, the width of the contact facing, dimension **a** in Fig. 11-5, must be kept to 1.30 to 1.60 mm (.051 to .063 in.) for intake valves, or 1.70 to 2.00 mm (.067 to .079 in.) for exhaust valves. On dual-carburetor and fuel injection engines, the width of the contact facing, dimension **a** in Fig. 11-5, must be kept to 1.80 to 2.20 mm (.071 to .086 in.) for intake valves or 2.00 to 2.50 mm (.079 to .098 in.) for exhaust valves.

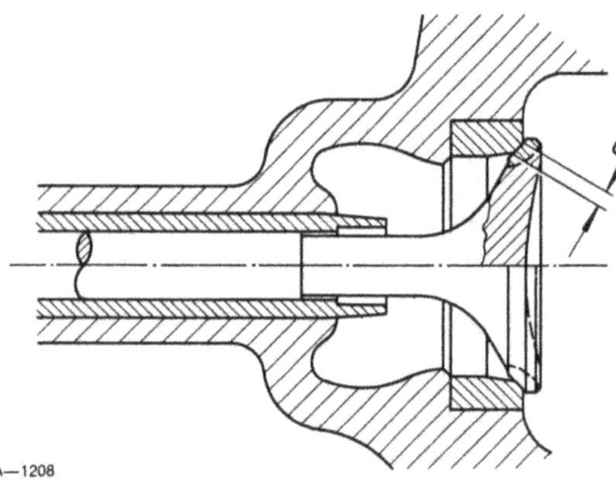

Fig. 11-5. Seat contact width (**a**).

The width of the contact facing is adjusted by cutting a 15° chamfer at the outer edge of the seat and a 75° chamfer at its inner edge. Seats cannot be refaced if too little material remains for a 15° chamfer to be cut without going beyond the boundary of the insert. If the 15° chamfer already extends to the edge of the insert, replace the cylinder head.

> ***CAUTION —***
> *Always replace worn valve guides before you reface the valve seats. If you install the new guides after refacing the seats, the new guide bores may not be concentric with the seats that were cut with the original valve guides installed.*

To obtain a true surface on valve seats, the refacing tools must be in good condition. Following the tool manufacturer's instructions will help you avoid chatter marks, out-of-round seats, or concentric scoring. Do not rely on lapping to correct careless work done earlier with a seat cutter or stone.

To reface:

1. On dual-carburetor and fuel injection engines only, cut the intake valve seat contact facing to a 30° angle, as shown in Fig. 11-6. Remove no more metal than is necessary to remove pitted or burned areas.

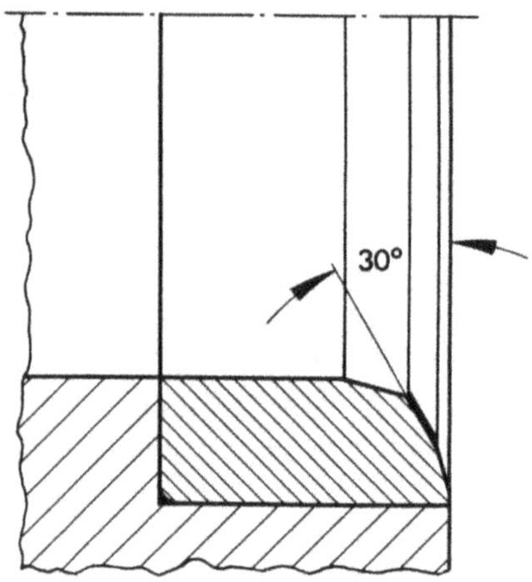

Fig. 11-6. 30° contact facing that is used on intake valve seats only on dual-carburetor and fuel injection engines. Work to obtain a glass-smooth finish.

2. On both intake and exhaust valve seats of single-carburetor engines and on the exhaust valve seats only of dual-carburetor and fuel injection engines, cut the valve seat contact facing to a 45° angle as shown in Fig. 11-7. Remove no more metal than is necessary to remove pitted or burned areas.

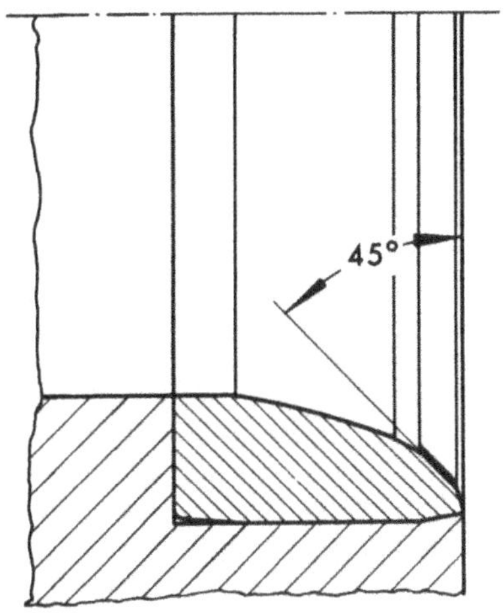

© 1974 VWoA—1209

Fig. 11-7. 45° contact facing used on exhaust valve seats of dual-carburetor and fuel injection engines and on both the intake and the exhaust valve seats of single-carburetor engines.

3. Using a 75° cutter, chamfer the valve seats to true and narrow the seats from their inner edge. See Fig. 11-8.

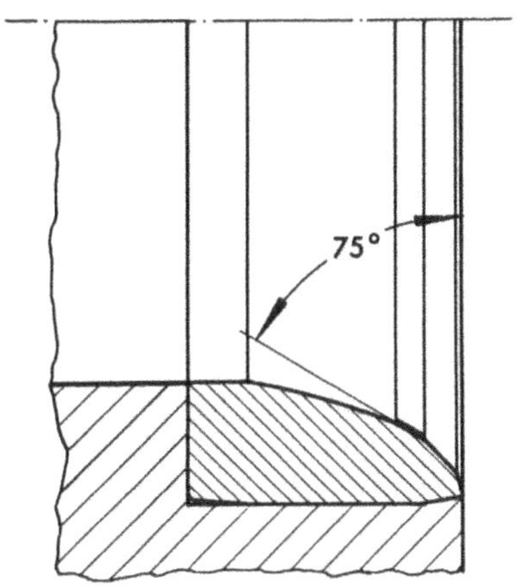

© 1974 VWoA—1210

Fig. 11-8. Lower edge of valve seat. Chamfer to 75°.

> ***CAUTION —***
> *Do not neglect narrowing the valve seats to specifications. Overly wide seats produced by using only a 45° cutter tend to trap carbon particles and other deposits.*

4. Cut a 15° chamfer, as shown in Fig. 11-9, to narrow the seat from its outer edge. The chamfer must not extend to the aluminum material of the cylinder head.

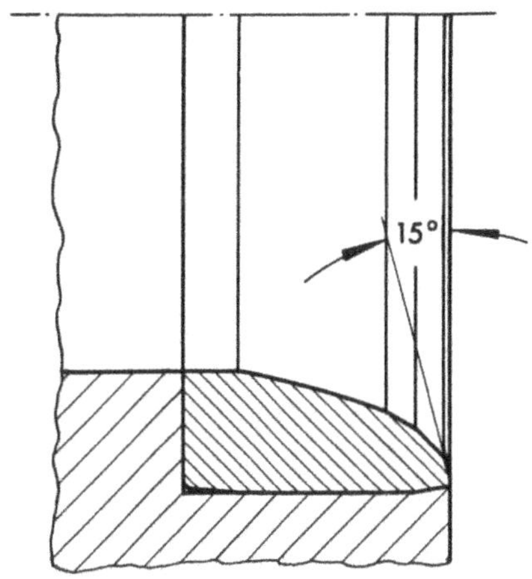

© 1974 VWoA—1211

Fig. 11-9. Outer edge of valve seat. Chamfer to 15°.

11.6 Refacing Valves

Reface worn or pitted valves on a grinding machine. On single-carburetor engines, grind the facings on the intake valves at an angle of 44°; on dual-carburetor and fuel injection engines, grind the intake valves at an angle of 29°30′. Doing so will produce a slight interference fit with the seats. Grind the exhaust valve facings to an angle of 45°. Both intake and exhaust valves should be lapped lightly into their seats following machine grinding. It is vital that a proper margin remain on the valve after it has been ground. This and other important dimensions are shown in Fig. 11-10 and listed in **Table c** or **Table d**.

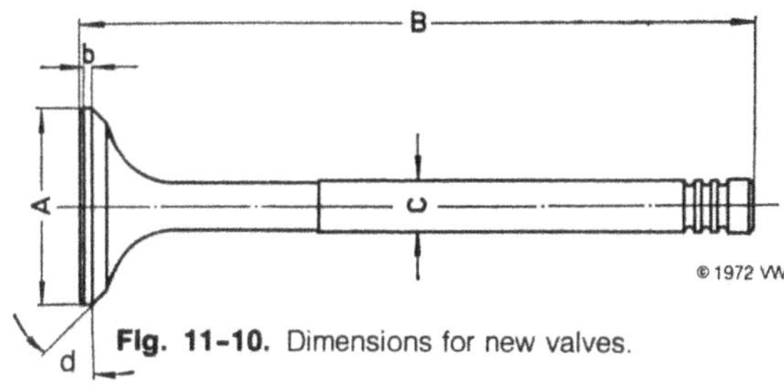

© 1972 VWoA—204

Fig. 11-10. Dimensions for new valves.

Table c. Single-carburetor Engine Valve Dimensions

Dimension **A**, valve head diameter	Intake: 35.60 mm (1.401 in.) Exhaust: 32.10 mm (1.263 in.)
Dimension **B**, valve length	112.00 mm (4.409 in.)
Dimension **C**, valve stem diameter	Intake: 7.94-7.95 mm (.3126-.3129 in.) Exhaust (through 1970): 7.91-7.92 mm (.3113-3118 in.) (1971 models): 7.92-7.94 mm (.3118-.3126 in.)
Dimension **b**, valve head margin	0.50 mm (.020 in.)
Dimension **d**, facing angle	Intake: 44° Exhaust: 45°

Table d. Dual-carburetor and Fuel Injection Engine Valve Dimensions

Dimension **A**, valve head diameter (1700 engine)	Intake: 39.30 mm (1.5472 in.) Exhaust: 33.00 mm (1.2992 in.)
Dimension **A**, valve head diameter (1800 dual-carburetor engine)	Intake: 41.00 mm (1.6142 in.) Exhaust: 34.00 mm (1.3386 in.)
Dimension **A**, valve head diameter (1800 fuel injection engine)	Intake: 39.30 mm (1.5472 in.) Exhaust: 33.00 mm (1.2992 in.)
Dimension **A**, valve head diameter (2000 fuel injection engine)	Intake: 37.50 mm (1.476 in.) Exhaust: 33.00 mm (1.2992 in.)
Dimension **B**, valve length	Intake: 117.00 mm (4.6063 in.) Exhaust: 117.00 (4.6063 in.)
Dimension **C**, valve stem diameter	Intake: 7.94-7.95 mm (.3126-.3129 in.) Exhaust: 8.91-8.92 mm (.3508-.3512 in.)
Dimension **b**, facing angle	0.50 mm (.020 in.)
Dimension **d**, facing angle	Intake: 29°30′ Exhaust: 45°

To check valve:

1. Using a motor-driven wire brush, remove the combustion chamber deposits from the valves.
2. Examine the seat contact area for pits, burns, and other signs of wear. If the damage is too extensive to be cured by refacing, replace the valve.

   NOTE —
   Because of the extreme conditions under which exhaust valves operate, many experienced mechanics routinely replace any exhaust valves that have been in service for 25,000 mi. (40,000 km) or more.

3. Discard any valve with damaged keeper grooves or with a stem that has been warped or galled by seizure.
4. Discard any valve that exceeds dimension **B** shown in Fig. 11-10. It has lost strength and is stretching.

After you have selected new valves or refaced used valves, check the valve fit in the refaced seats you have cut into the cylinder head.

To check valve seating:

1. Coat the face of the valve lightly with Prussian blue. Insert the coated valve into the valve seat. Applying light pressure, rotate the valve about a quarter turn.
2. Remove the valve and examine the contact pattern. If the seating is correct, the valve will leave an even coating of Prussian blue on the seat.
3. If areas of bare metal show on the valve seat, a more careful refacing of the valve or seat is required. Minor irregularities can be corrected by lapping.

If your check revealed an irregular contact pattern, inspect the seat first. A poorly fitting stone or cutter pilot may have caused you to cut the seat out-of-round. Another possible cause is a tapered valve guide that allows the cutter pilot to enter it at an angle. If the valve guide is tapered or out-of-round, replace it before you again attempt to reface the valve seat. If the valve is warped, replace it. Valve refacing will correct minor warping but if the remaining margin is noticeably irregular, replace the valve.

11.7 Lapping Valves

Using a tool such as that shown in Fig. 11-11, you can lap valves into their seats for a perfect seal. However, modern precision refacing machines that are in good condition will usually produce an excellent fit without significant lapping.

© 1974 VWoA—1216

Fig. 11-11. Valve being lapped with suction cup tool. Tool holds valve while you turn it against seat.

To lap valve:

1. Coat the seat with a small amount of valve grinding compound. Then insert the valve in the guide.
2. Using the suction cup tool, turn the valve clockwise and counterclockwise against the seat.

   NOTE —
   By lifting the valve off the seat every few turns and shifting its position slightly, you will avoid cutting concentric grooves into the seat.

3. Clean away every trace of grinding compound. Check the valve seating with Prussian blue.

11.8 Valve Clearance

On fuel injection engines and single-carburetor engines, the valve clearance is specified at 0.15 mm (.006 in.) for both intake and exhaust valves. The correct clearance for the intake valves of all dual-carburetor engines is also 0.15 mm (.006 in.).

On dual-carburetor engines with solid steel exhaust valves, the exhaust valve clearance should be 0.15 mm (.006 in.). On dual-carburetor engines with sodium-filled exhaust valves, the exhaust valve clearance should be 0.20 mm (.008 in.). The sodium-filled exhaust valves were installed in all 1974 dual-carburetor engines and in 1973 dual-carburetor engines from chassis No. 213 2210 554. If sodium-filled exhaust valves have been installed in an earlier dual-carburetor engine, adjust the exhaust valve clearance to 0.20 mm (.008 in.).

11.9 Adjusting Valves

For accurate adjustment, the engine must be completely cold (oil temperature no more than 50°C (122°F)) and the cylinder in position to fire. Adjust the valve clearance in cylinder sequence 1-2-3-4 (the firing order in reverse—starting with 1). Correct adjustment is essential for maximum efficiency and minimum wear.

To adjust:

1. Remove the cylinder head covers and the distributor cap. Hand-turn the engine until the distributor rotor is adjacent to the No. 1 cylinder notch in the top edge of the distributor body.

   NOTE —
   When the rotor is in this position, both valves for the No. 1 cylinder are closed, the piston is at approximately top dead center on its compression stroke, and the spark plug is ready to fire. No. 1 cylinder is at the right front of the engine, No. 2 behind it, No. 3 is at the left front of the engine, and No. 4 behind it.

2. Using a 0.15-mm (.006-in.) feeler gauge, check the clearance of both valves for the No. 1 cylinder. You should just be able to insert and withdraw the gauge with moderate drag.

   NOTE —
   Use a 0.20-mm (.008-in.) feeler gauge for the exhaust valves on 1974 1800 engines. Also use a 0.20-mm (.008-in.) feeler gauge for the exhaust valves of 1973 engines from chassis No. 213 2210 554 and earlier dual-carburetor engines that have been modified by installation of sodium-filled exhaust valves.

3. If the clearance is incorrect, loosen the locknut. Then turn the adjusting screw clockwise to reduce clearance or counterclockwise to increase it. (See Fig. 11-12.) When the clearance produces moderate drag on the feeler gauge, tighten the locknut while holding the adjusting screw stationary. Then recheck the clearance.

© 1974 VWoA—3627

Fig. 11-12. Clearance between rocker and valve stem tip being adjusted. Note feeler gauge and combination wrench and screwdriver being used. 5

NOTE —
If you lack experience in adjusting valves, a set of "go/no-go" feeler gauges may help you to determine clearances more accurately. Such gauges are stepped so that, when clearance is correct, the gauge will slide in about an inch and no further.

4. Turn the engine until the rotor moves counterclockwise 90° from the No. 1 cylinder firing position. This will correctly position the No. 2 cylinder for valve adjustments.

5. Check and adjust the No. 2 cylinder's valves as you did on the No. 1 cylinder. Turn the engine so that the rotor moves counterclockwise in 90° increments. Check and adjust the valves of the No. 3 and the No. 4 cylinders in the same way.

12. CYLINDERS

You can replace cylinders individually on the VW engine. However, you must replace the piston at the same time. Select a piston that fits the new cylinder bore and matches the weights of the other pistons in the engine. Weight and fitting dimensions are indicated by marks on the piston crown (see **13. Pistons and Rings**).

12.1 Removing and Installing Cylinders

The engine must be out of the car before you can remove the cylinders. Prior to disassembly, clean the engine to keep dirt from getting into the working parts.

To remove the cylinder, first remove all parts preparatory to cylinder head removal. Then remove the cylinder head. Remove the sheet metal air deflector plates below the cylinders. If more than one cylinder is to be removed, mark each cylinder so that you will be able to reinstall it in its original position. Then pull the cylinder off the crankcase and the piston.

To install cylinder:

1. Check the cylinder for wear. You must measure the piston diameter and the cylinder bore and compute the clearance. The procedure is described in **13.3 Piston Clearance.**

   NOTE —
   If the clearance exceeds 0.20 mm (.008 in.), install a new cylinder and matching piston. The clearance between new parts should be 0.04 to 0.06 mm (.0016 to .0024 in.).

2. Inspect the cylinder sealing surfaces shown in Fig. 12-1.

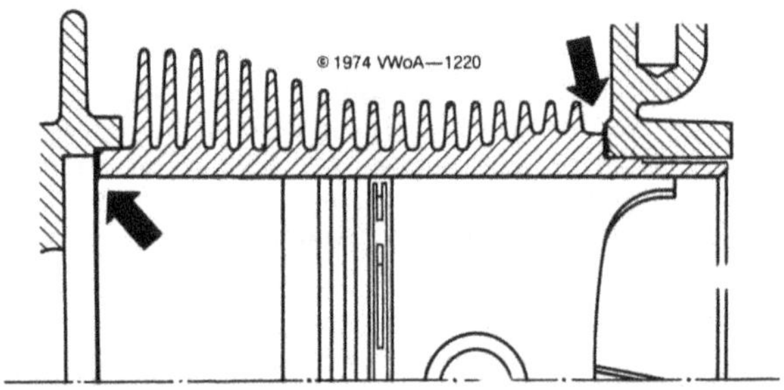

Fig. 12-1. Cylinder sealing surfaces. These areas must be perfectly clean and true prior to cylinder installation.

3. Stagger the ring gaps 90° apart so that the oil ring gap faces upward when the cylinder is installed.
4. Oil the piston, piston rings, and piston pin. Then install the paper sealing ring that fits between the cylinder and the crankcase.
5. Compress the piston rings with a piston ring compressor.

   NOTE —
   Use a piston ring compressor that can be taken apart. Otherwise, you will not be able to get it off the piston after the cylinder has been installed.

6. Oil the cylinder walls lightly and install the cylinder as shown in Fig. 12-2.
7. To complete the assembly, install the cylinder head and the other engine parts. Then adjust the valves.

Fig. 12-2. Cylinder being installed. Note design of ring compressor.

13. PISTONS AND RINGS

Piston size, weight, and installation position are marked on the crown as shown in Fig. 13-1.

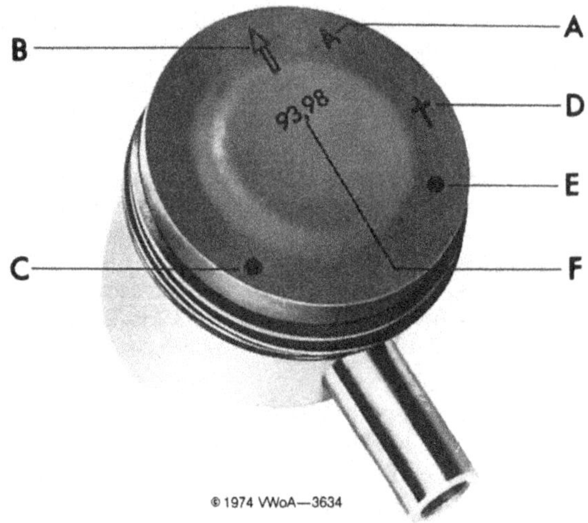

A. Index letter of piston part number
B. Arrow that must point toward flywheel
C. Paint spot indicating matching cylinder size (blue, pink, green)
D. Weight grading (+ or −)
E. Paint spot indicating weight grading (brown for − weight, gray for + weight)
F. Piston size in mm

Fig. 13-1. Piston markings used as fitting guide.

13.1 Removing and Installing Pistons

You can remove the pistons once you have taken off the cylinders. It is unnecessary either to remove the connecting rods from the crankshaft or to separate the crankcase halves.

To remove pistons:

1. Remove the cylinders.
2. Mark the pistons so that you will be able to reinstall them in their original positions.
3. Using suitable pliers, remove the piston pin circlips as shown in Fig. 13-2.

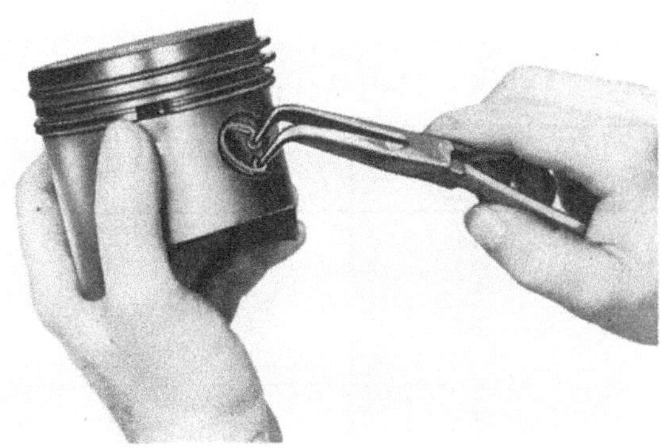

© 1974 VWoA—1227

Fig. 13-2. Piston pin circlip being removed or installed.

4. Push out the piston pin. Then take the piston off the connecting rod.
5. Using piston ring expanding pliers, remove the piston rings. Remove the top ring first; then work downward.

To install piston:

1. Clean the carbon and other deposits from the piston crown, skirt, and ring grooves.
2. Check the clearance as described in **13.3 Piston Clearance.**

NOTE —
If the clearance between the piston and the cylinder is excessive, install a new piston and cylinder combination. The new piston must be of the same weight grade as the original or within 10 g of the original piston weight.

3. Check the pin fit in the piston. It must be a light push fit at room temperature or with the piston heated to 75°C (167°F) in an oil bath. If not, replace both the piston and the pin.
4. Check the piston pin fit in the connecting rod. The wear limit is 0.04 mm (.0016 in.). New clearance should be 0.01 to 0.02 mm (.0004 to .0008 in.) for single-carburetor engines or 0.01 to 0.03 mm (.0004 to .0012 in.) for dual-carburetor and fuel injection engines.

NOTE —
Some pistons have oversize pin bores. Oversize pins, marked in green, are available for these. You should not, however, use oversize pins to reduce the clearance in worn pistons. If the clearance in the rod is excessive, fit new pins and new rod bushings and check them carefully for proper clearance.

5. Install one circlip in each piston in the side toward the flywheel, as shown by the arrow on the piston crown.
6. Install the piston on the rod so that the arrow on the piston crown points toward the flywheel. If necessary, heat the piston to 75°C (167°F) in an oil bath, then push in the piston pin and install the remaining circlip.

NOTE —
Install pistons for the No. 1 and No. 2 cylinders first, then No. 3 and No. 4.

7. Install the cylinders.

5

13.2 Piston Rings

New piston rings must always be fitted with reference to the markings on the piston crown. The various cylinder bore piston diameter grades each have similarly graded piston ring sets to match them.

It is highly important that the rings fit properly in the cylinder and piston ring grooves, and that they are correctly installed on the piston. If the piston ring end gap is insufficient, the ring will seize in the cylinder when the engine reaches operating temperature. Excessive end gap will reduce the piston ring's sealing capability.

To check ring gap:

1. Carefully hand-compress the ring and insert it in the bottom of the cylinder.
2. Using the matching piston as a pusher to prevent tilt, push the ring into the cylinder about 5 mm (¼ in.).

3. Measure the ring gap as shown in Fig. 13-3. The proper gaps are given in **Table e**.

© 1974 VWoA—1224

Fig. 13-3. Ring gap being measured with feeler gauge.

Table e. Piston Ring Gaps

Ring and Engine Model	New Ring mm (in.)	Wear Limit mm (in.)
Upper compression ring		
1600 engine	0.30-0.45 (.012-.018)	0.90 (.035)
1700, 1800, and 2000 engines	0.35-0.55 (.014-.022)	0.90 (.035)
Lower compression ring		
1600 engine	0.30-0.45 (.012-.018)	0.90 (.035)
1700, 1800, and 2000 engines	0.30-0.55 (.012-.022)	0.90 (.035)
Oil scraper ring		
1600 engine	0.25-0.40 (.010-.016)	0.95 (.037)
1700, 1800, and 2000 engines	0.25-0.40 (.010-.016)	0.95 (.037)

To check side clearance:

1. Using piston ring expanding pliers, install the piston rings in the piston grooves as shown in Fig. 13-4.

© 1974 VWoA—1225

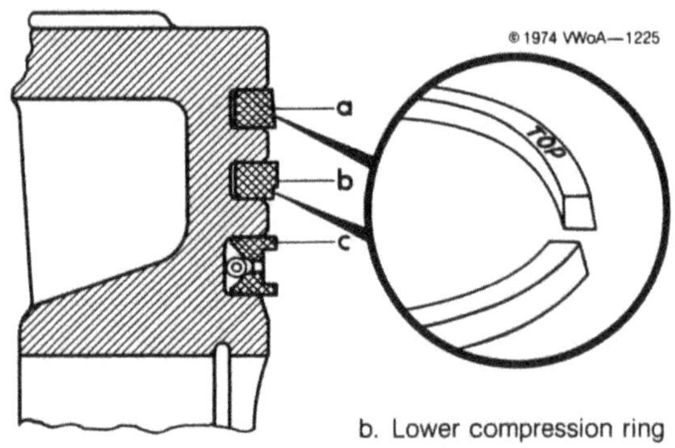

a. Upper compression ring b. Lower compression ring c. Oil scraper ring

Fig. 13-4. Proper ring installation. The word **TOP** must be toward the piston crown.

NOTE —
The rings must be installed as shown in Fig. 13-4. If installed otherwise, they will not produce a proper seal with the cylinder wall.

2. Check the ring side clearance as shown in Fig. 13-5. The proper clearances are given in **Table f**.

© 1974 VWoA—1226

Fig. 13-5. Ring side clearance being measured. Insert feeler gauge between ring and piston land.

Table f. Piston Ring Side Clearance

Ring and Engine Model	New Ring and Piston mm (in.)	Wear Limit mm (in.)
Upper compression ring		
1600 engine	0.07-0.10 (.003-.004)	0.12 (.005)
1700, 1800, and 2000 engines	0.06-0.09 (.002-.003)	0.12 (.005)
Lower compression ring		
1600 engine	0.05-0.07 (.002-.0027)	0.10 (.004)
1700, 1800, and 2000 engines	0.04-0.07 (.0015-.0027)	0.10 (.004)
Oil scraper ring		
1600 engine	0.03-0.05 (.001-.002)	0.10 (.004)
1700, 1800, and 2000 engines	0.02-0.05 (.0008-.002)	0.10 (.004)

13.3 Piston Clearance

Piston clearance must not exceed the prescribed range or the pistons will rock in their bores during engine operation. This would cause the piston rings to develop an elliptical wear pattern that would impair their sealing effectiveness.

Some clearance, however, is essential. If the clearance is inadequate, the aluminum pistons can expand and seize in their cylinders. To keep the necessary clearance from causing excessive rocking motion (and piston slap noises), the piston pin bores are offset slightly from the piston centerline. This offset, shown in Fig. 13-6, is the reason why the pistons must be installed so that the arrow on their crown points toward the flywheel.

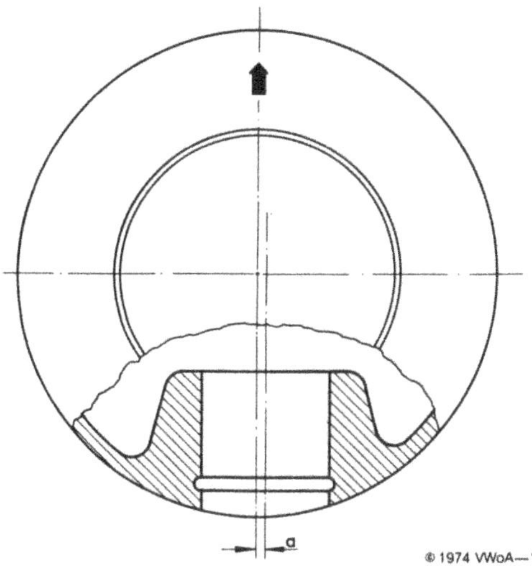

© 1974 VWoA—1228

Fig. 13-6. Piston pin offset. Arrow must point toward flywheel to position offset properly. Dimension **a** is 1.50 mm (.059 in.) on single-carburetor engines and 0.50 mm (.020 in.) on dual-carburetor and fuel injection engines.

The clearance between the piston and the cylinder wall is computed from measurements taken on the piston and in the cylinder. The clearance for new parts is specified at 0.04 to 0.06 mm (.0016 to .0024 in.) with a wear limit of 0.20 mm (.008 in.).

To check clearance:

1. Using a dial indicator device, measure the cylinder bore as shown in Fig. 13-7.

© 1974 VWoA—1230

Fig. 13-7. Cylinder bore being measured with special dial indicator.

NOTE —
When checking the cylinder bore, make your measurements at several points throughout the cylinder and at right angles to one another. If there are wide variations, the cylinder and piston may have to be replaced. To compute piston clearance, use the largest reading you obtain.

2. Write down the maximum cylinder diameter.
3. Measure the piston diameter as illustrated in Fig. 13-8.

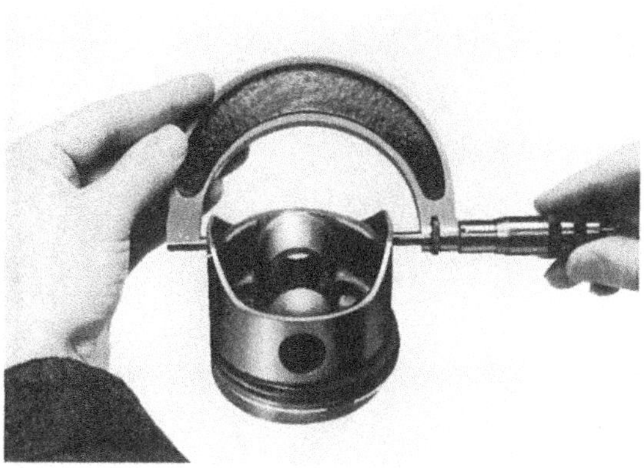

© 1974 VWoA—1231

Fig. 13-8. Piston diameter being measured. A micrometer is the best tool for this task.

4. To obtain the piston clearance, subtract the piston diameter from the cylinder diameter.

NOTE —
If wear proves to be almost entirely on the piston, the piston alone can be replaced with one of matching size. Only pistons of the same size and weight grade should be installed in the same engine.

14. TORQUE CONVERTER

(dual-carburetor and fuel injection engines only)

Vehicles equipped with automatic transmissions have a torque converter instead of a clutch, and a drive plate rather than a conventional flywheel. The torque converter is held to the drive plate by three bolts that are taken out when the engine is removed.

14.1 Removing and Installing Drive Plate

When the engine must be disassembled, it is best to remove the drive plate before separating the crankcase halves.

5

To remove:

1. Mark the position of the drive plate so that you can install it in the exact original position on the crankshaft.
2. Attach the drive plate holding fixture as shown in Fig. 14-1. Using a 12-point driver only, remove the five socket-head bolts that hold the drive plate on the crankshaft.
3. Remove the drive plate from the crankshaft.

© 1974 VWoA—3636

Fig. 14-1. Holding fixture installed on drive plate.

To install:

1. Check the converter mounting bolt holes for cracks and wear.
2. Check the drive plate mounting bolt holes for wear. Inspect the oil seal in the crankcase.
3. Install the rolled pin and the drive plate on the crankshaft so that the reference marks made earlier line up.
4. Attach the drive plate holding fixture. Then, working diagonally, torque the five socket head bolts to 9.0 mkg (65 ft. lb.).
5. If necessary, check and adjust the crankshaft end play as described in **17.5 Crankshaft End Play.**

15. CLUTCH AND FLYWHEEL

Vehicles covered by this Manual have one of two diaphragm spring clutches. However, some of the illustrations show the coil spring clutch used previously. The pressure plate assembly used on the late models is shown in Fig. 15-1. It can be used as a replacement for the earlier diaphragm spring pressure plate assembly shown in Fig. 15-2.

© 1974 VWoA—1282

Fig. 15-1. Late-type diaphragm spring clutch.

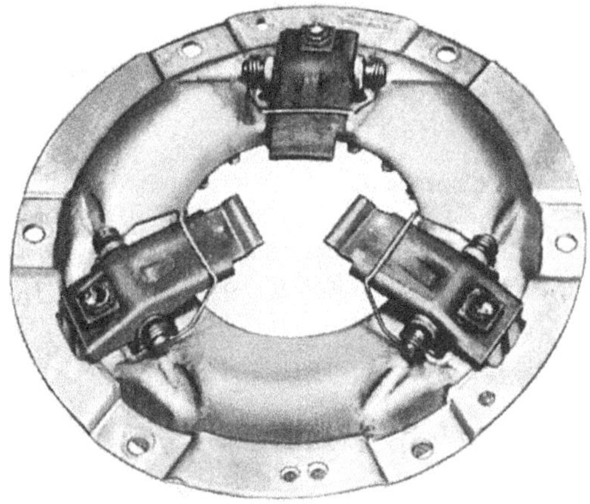

© 1974 VWoA—1282

Fig. 15-2. Diaphragm spring clutch used on early models.

15.1 Removing and Installing Clutch

The pressure plate assembly and driven plate remain on the flywheel when you remove the engine. The clutch release bearing and related parts stay in the transmission.

To remove clutch:

1. Remove the engine. Mark the position of the pressure plate so that you can install it in the exact original position on the flywheel.

2. If you have a tool such as the one shown in Fig. 15-3, use it to release the clutch. Gradually remove the bolts that hold the pressure plate to the flywheel. If you lack the special tool, loosen each bolt a quarter turn at a time. Work around the flywheel until the pressure is relieved, then remove the bolts completely.

© 1974 VWoA—1266

Fig. 15-3. Pressure plate being removed. Notice the locking dog being used to keep the flywheel from turning.

3. Remove the pressure plate assembly from the flywheel.

To Install:

1. Clean the friction surface on the flywheel and inspect it for wear, cracks, and grooves. Minor defects may be correctable by regrinding or machining at a specialty shop. If the flywheel is unserviceable, replace it.

2. Inspect the pressure plate for wear, cracks, and grooves. Alternate bright and dull areas indicate a warped plate. Shake the pressure plate assembly. The diaphragm spring should be under tension and should not rattle loosely. Also see that none of the release levers is out of line with the others. If the pressure plate assembly is in any way unserviceable, replace it.

3. Inspect the clutch release bearing. If it feels gritty when you turn it, or if it has been making noise, replace it. Never wash the bearing in solvent, since this will remove the factory-installed lubricant. If the release bearing is unserviceable, replace it as described in **15.4 Replacing Clutch Release Bearing.**

4. Use solvent to clean the pilot bearing. Check to see that the needles are not flattened by wear. Pack the bearing with 1 g (1/32 oz.) multipurpose grease (just enough to coat lightly all needles). Apply motor oil to the felt ring, then wipe away all excess lubricants.

5. Inspect the driven plate for wear. There should be at least 2.00 mm (about 1/16 in.) of friction material remaining above the rivet heads. Check the plate for runout as shown in Fig. 15-4. Runout should not exceed 0.50 mm (.020 in.).

© 1974 VWoA—1276

Fig. 15-4. Dial indicator being used to check driven plate for runout.

5

6. Inspect the splines in the driven plate and on the transmission rear driveshaft. They must not be broken or distorted. Lubricate the splines with molybdenum disulfide powder. Then see that the driven plate slides freely on the rear driveshaft without undue radial play. If the driven plate is in any way unserviceable, replace it.

7. Apply motor oil to the felt ring for the pilot bearing. Wipe away all excess lubricants.

8. Using a centering tool or a spare transmission rear driveshaft, install the driven plate against the flywheel.

9. Install the pressure plate assembly according to the reference marks made earlier.

   NOTE —
   On some replacement clutches, a white paint spot is used as a balance mark. This should be positioned 180° from the 5-mm (3/16-in.) countersunk hole or from the white paint balance mark on the flywheel.

10. Loosely install all the mounting bolts. Then, working diagonally, tighten the mounting bolts a turn or two at a time.

11. Working diagonally, torque the bolts to 2.5 mkg (18 ft. lb.).

> ***CAUTION —***
> *To simplify centering and mounting, replacement clutches are pretensioned by clips between the release levers and the clutch plate. Be sure to remove these clips following installation.*

15.2 Removing and Installing Flywheel

Remove the flywheel from the crankshaft before you separate the crankcase halves.

To remove:

1. Hold the flywheel with the locking dog illustrated earlier in Fig. 15-3. Mark the position of the flywheel so that you can install it in the exact original position on the crankshaft.
2. On single-carburetor engines, use a 36-mm ($1\frac{7}{16}$-in.) socket to remove the gland nut from the center of the flywheel.
3. On dual-carburetor and fuel injection engines, remove the five self-locking bolts that hold the flysheel on the crankshaft.
4. Pull the flywheel off the dowels or the rolled pin on the crankshaft.

To install:

1. Use solvent to clean the pilot bearing. Check to see that the needles are not flattened by wear. Pack the bearing with 1 g ($\frac{1}{32}$ oz.) multipurpose grease (just enough to coat lightly all needles). Apply motor oil to the felt ring, then wipe away all excess lubricants. Fig. 15-5 shows the pilot bearing for the single-carburetor engine.

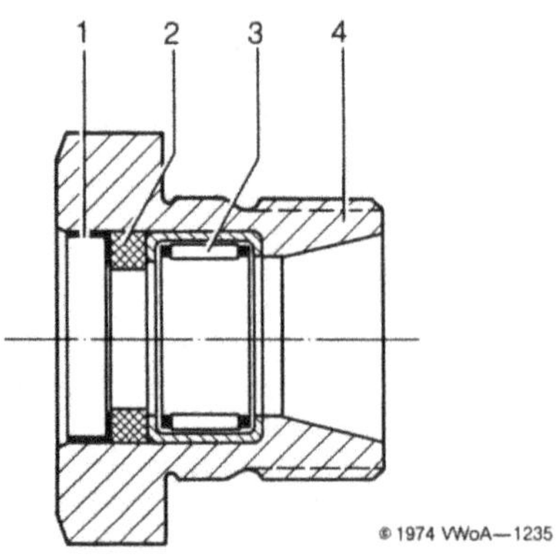

1. End ring
2. Felt ring
3. Needle bearing
4. Gland nut

Fig. 15-5. Pilot bearing in flywheel gland nut for single-carburetor engine.

2. Inspect the starter gear teeth for wear and damage.

> **NOTE —**
> Up to 2 mm ($\frac{1}{16}$ in.) can be removed from the clutch side of the flywheel to deburr the starter gear. Chamfer the teeth after you have machined the flywheel.

3. Inspect the dowel holes or, on dual-carburetor or fuel injection engines, the rolled pin hole in the crankshaft and flywheel. See that the dowels or the rolled pin fits tightly.
4. On single-carburetor engines, install the dowel pins and the flywheel on the crankshaft so that the reference marks made earlier line up. Then install the gland nut, but do not torque it.

> **NOTE —**
> Always use a new sealing ring (gasket) between the flywheel and the crankshaft. If the spring-type lock washer under the gland nut is deformed, replace it.

5. On dual-carburetor and fuel injection engines, install the rolled pin and the flywheel on the crankshaft so that the reference marks made earlier line up. Then install the five self-locking bolts together with a new bolt plate, but do not torque the bolts.
6. Check the crankshaft end play as described in **17.5 Crankshaft End Play**. Adjust if necessary.

> ***CAUTION —***
> *Crankshaft end play must be within the specified range. Incorrect end play will cause damage to the bearings or to other internal engine parts.*

7. Install the flywheel locking fixture. On single-carburetor engines, torque the gland nut to 35 mkg (253 ft. lb.). On dual-carburetor and fuel injection engines, torque the five self-locking bolts to 11 mkg (80 ft. lb.).
8. Lubricate the tip of the transmission rear driveshaft with a light coating of multipurpose grease. Install the clutch, and then the engine in the car.

15.3 Removing and Installing Crankshaft Oil Seals

If the oil seal in the flywheel end of the crankcase leaks, the clutch will be contaminated by engine oil. A leaking seal is easy to mistake for a leak in the crankcase joint, so check the oil seal before concluding that the crankcase has to be taken apart. Single-carburetor engines have an oil seal only at the flywheel end of the crankshaft; dual-carburetor and fuel injection engines have seals at both ends.

To remove the seal at the flywheel end, remove the flywheel. Inspect the oil seal lip and the contact surface on the flywheel shoulder. If the lip is worn or cracked, replace the seal. If the flywheel shoulder contact surface is deeply grooved, you will have to replace the flywheel in order to obtain a good seal. When you pry out the old oil seal, be careful not to scratch or gouge the magnesium alloy crankcase casting.

To Install:

1. If the outer edge of the crankcase is sharp, chamfer it as illustrated in Fig. 15-6. Clean away the metal shavings.

Fig. 15-6. Scraper being used to chamfer edge of seal recess in crankcase. This will prevent damage to the seal during installation.

2. Clean the oil seal recess to remove dirt, old sealer, and traces of oil. Then apply a thin film of sealing adhesive.
3. Start the seal into the chamfered recess by hand. The seal lip must point into the crankcase. Then press the new seal into position as shown in Fig. 15-7 or Fig. 15-8.

Fig. 15-7. Seal being installed in single-carburetor engine with tool VW 2046. A similar tool can be made with an M 28 × 1.5 bolt, a nut, washers, and a steel disk.

NOTE —
The installing tool centerbolt is screwed into the crankshaft. The guide piece is threaded and can be advanced along the bolt by using a wrench as shown in the illustration. The oil seal must be seated squarely in the crankcase recess.

Fig. 15-8. Seal installing tool VW 191 that is used on dual-carburetor engines. The inner part is held on the crankshaft by two of the flywheel bolts.

4. Remove the installing tool.
5. Using engine oil, lubricate the oil seal contact surface on the flywheel. Then install the flywheel on the engine.

CAUTION —
You should check the crankshaft end play during flywheel installation, although adjustment is seldom necessary if only the crankshaft oil seal is being replaced. By checking it whenever you make flywheel or related repairs, you guard against possible bearing failure or other engine damage.

6. Install the clutch. Then install the engine in the car.

NOTE —
The transmission shaft pilot bearing in the gland nut should be cleaned, inspected, and repacked as described in **15.2 Removing and Installing Flywheel** whenever the clutch and flywheel are being serviced.

To replace the oil seal at the fan end of a dual-carburetor or fuel injection engine's crankshaft, pull the fan hub off the crankshaft. To do this, use the special plate shown in Fig. 15-9 and the three fan mounting bolts.

Fig. 15-9. Fan hub being removed from dual-carburetor or fuel injection engine crankshaft. Install special plate VW 185 between the bolts and the crankcase, then gradually tighten the bolts.

Clean the oil seal recess to remove dirt, old sealer, and traces of oil. If necessary, chamfer the recess as shown earlier in Fig. 15-6. Then apply a thin film of sealing adhesive. Start the seal into the chamfered recess by hand so that the seal lip points into the crankcase. Use the tool shown in Fig. 15-10 to press in the seal. Remove the special tool, then lightly lubricate the fan hub and install it on the crankshaft.

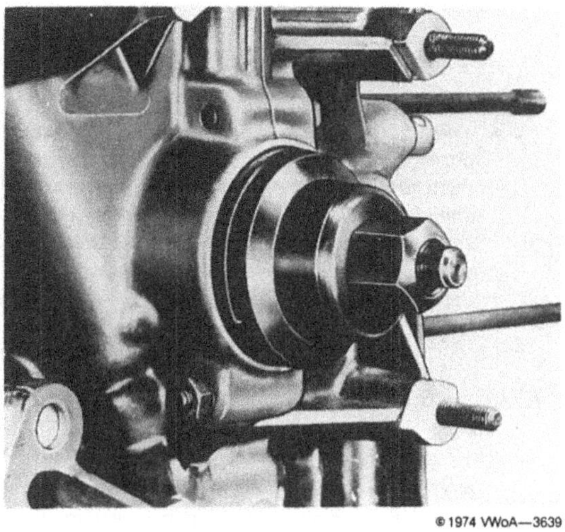

Fig. 15-10. Tool VW 190 in position to install fan end oil seal. This tool works the same way as the tool illustrated earlier in Fig. 15-7.

15.4 Replacing Clutch Release Bearing

The clutch release bearing is situated at the rear of the transmission. It is fastened to the clutch operating shaft by two retaining springs. On 1971 and later models, the clutch release bearing is supported by a central guide sleeve as shown in Fig. 15-11.

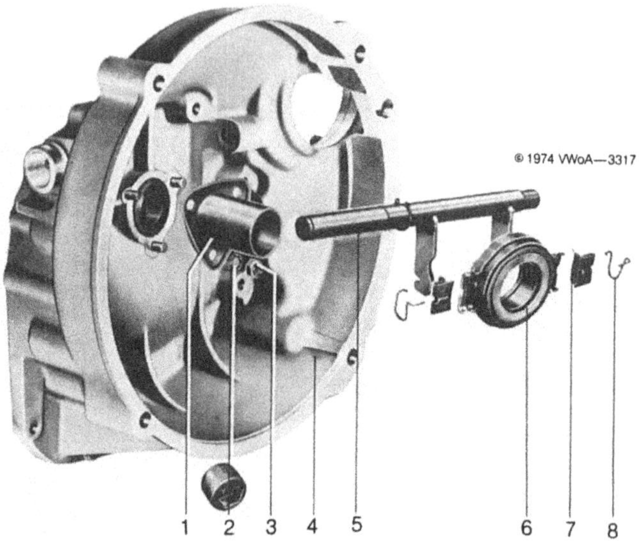

1. Release bearing guide sleeve
2. Lock washer (3)
3. M 7 nut (3)
4. Clutch housing (late type)
5. Clutch operating shaft
6. Clutch release bearing
7. Spring clip retainer (2)
8. Spring clip (2)

Fig. 15-11. Clutch release bearing and related parts. The sleeve, nuts, and lock washers are not installed on 1968 through 1970 vehicles and the clips that hold the bearing are different (see **TRANSMISSION AND REAR AXLE**).

The clutch release bearing is a maintenance-free item. However, if it is noisy, the plastic facing ring should be lubricated through a hole drilled in the transmission case, as shown in Fig. 15-12.

Fig. 15-12. Lubrication access. Drill 10-mm (0.4-in.) hole in transmission case as shown. Dimension **a** is 67 mm (2.6 in.); **b** is 10 mm (0.4 in.).

Use a spray can of molybdenum disulfide lubricant with a plastic hose attached to its nozzle. Insert the hose through the hole you have drilled in the transmission case and direct the spray against the plastic facing ring.

If the engine has been removed, simply roughen the surface of the plastic facing ring on the clutch release bearing with coarse emery cloth. Then rub in a small amount of molybdenum grease. Do this routinely whenever the engine is removed.

To remove the 1968- through 1970-model clutch release bearing, pry the retaining springs off with a screwdriver as shown in Fig. 15-13. On 1971 and later models, pry off the spring clip retainers and then the spring clips. Then remove the bearing, sliding it off the guide tube on 1971 and later vehicles.

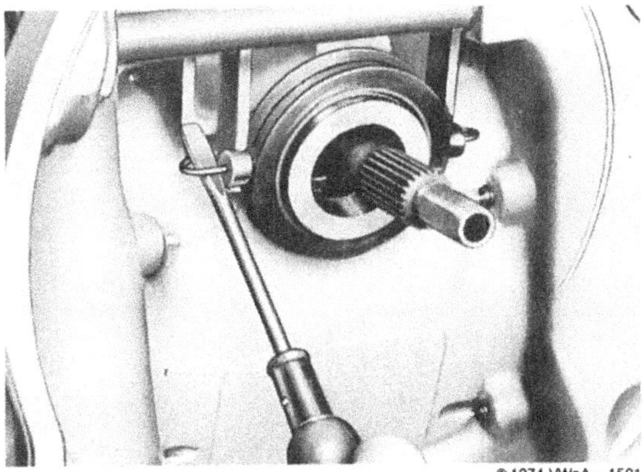

© 1974 VWoA—1501

Fig. 15-13. Clutch release bearing retaining springs being removed. Hold a rag over the springs to prevent them from flying off and becoming lost when tension is relieved.

To Install:

1. Wipe off the release bearing. Never wash the bearing in solvent since this will remove the factory-installed lubricant.
2. Hand-turn the bearing. It should not feel gritty or be difficult to turn. Make sure the plastic facing ring has not worn through. If necessary, replace the bearing.
3. Lightly lubricate the operating shaft bushings with multipurpose grease. On 1971 and later cars, lightly lubricate the guide sleeve with molybdenum grease.
4. Install the clutch release bearing and the retaining springs or spring clips and retainers. Make certain that the hooked ends of the springs engage behind the levers on the operating shaft. Replace loose-fitting retaining springs.
5. After you have installed the engine, check the clutch pedal freeplay. If necessary, adjust the freeplay.

15.5 Adjusting Clutch Pedal Freeplay

As the clutch linings wear, the clearance between the release bearing and the release levers is reduced. If this condition progresses until there is no clearance at all, clutch pressure will decrease and permit slippage that can lead to burned linings. When checking for the proper clearance, you should be able to depress the clutch pedal 10 to 20 mm (3/8 to 3/4 in.) on 1968 through early 1973 models—or 15 to 25 mm (5/8 to 1 in.) on vehicles built since February 1973—before encountering working resistance. If not, adjust the clutch cable to obtain this amount of clearance.

> **NOTE**
> A new clutch cable—20 mm longer—became standard with chassis No. 213 2129 107 manufactured in February 1973. If the threaded cable end is too short to permit adequate adjustment on earlier dual-carburetor engined vehicles, install the new cable (part No. 211 721 335J). Then adjust the freeplay to 15 to 20 mm (5/8 to 1 in.).

To adjust, raise the car on a lift or support it on safety stands. Grip the clutch cable with a pair of pliers to keep it from twisting, then adjust the wing nut shown in Fig. 15-14 until there is 1.0 to 2.0 mm (1/32 to 3/32 in.) clearance between the clutch lever and the wing nut. Recheck freeplay at the pedal to make sure that it is within specifications. Pump the pedal several times before you measure the distance.

5

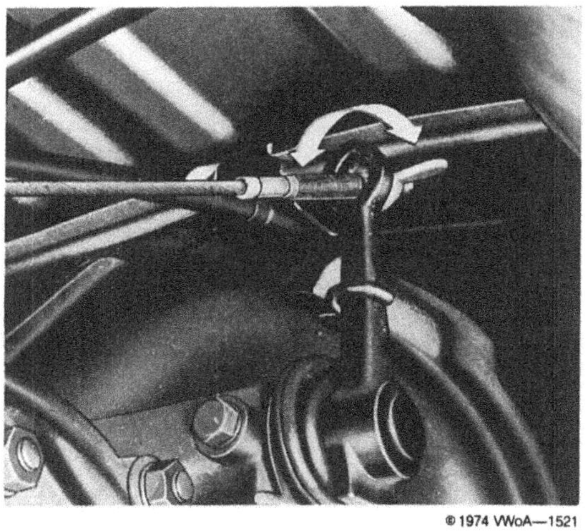

© 1974 VWoA—1521

Fig. 15-14. Clutch cable adjustment. Clean and lubricate cable end, clutch lever, and wing nut. Engage nut wings with recesses in clutch lever to prevent loosening.

Replacing Clutch Cable

You remove the clutch cable by working under the vehicle.

To remove:

1. Disconnect the cable from the clutch operating lever on the transmission. Then pull the rubber sleeve off the guide tube and off the rear of the cable.
2. Pull the flexible guide tube down in the middle and pull the guide tube and cable out of the bracket on the transmission.
3. Working at the front of the vehicle, remove the cover plate that is beneath the pedals.
4. Unbolt the clutch pedal from the pedal lever assembly. Then remove the pedal lever assembly from the frame by taking out the bolts indicated in Fig. 15-15.

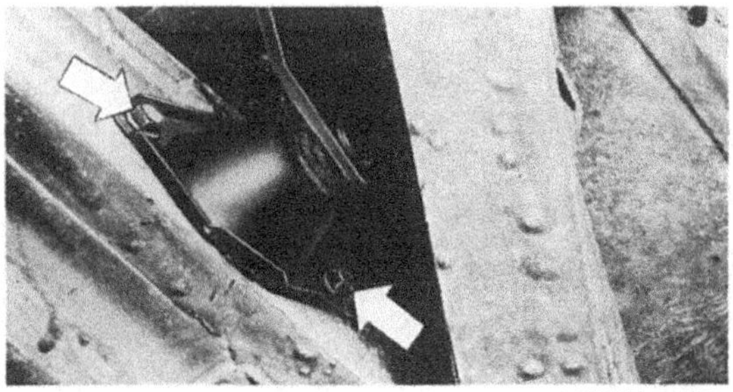

© 1974 VWoA—3684

Fig. 15-15. Bolts (arrows) that hold the pedal assembly on the vehicle frame.

5. Pull out the cable as shown in Fig. 15-16. Bend up the lockplate indicated by the arrow, then pull out the clutch cable clevis pin and disconnect the cable from the pedal lever assembly.

© 1974 VWoA—3685

Fig. 15-16. Clutch pedal lever assembly and clutch cable pulled out of frame. The arrow indicates the lockplate for the clevis pin.

6. Pull the clutch cable out toward the front.

Installation is the reverse of removal. Grease the entire replacement cable with molybdenum grease. Then insert it through the pedal lever assembly opening in the frame. Following installation, adjust clutch pedal freeplay as previously described.

When replacing a clutch cable, always check the condition of the rubber sleeve and the position of the flexible guide tube. Cracked or loose-fitting rubber sleeves must be replaced. The flexible guide tube should sag 25 to 45 mm (1 to 1¾ in.), as indicated at **B** in Fig. 15-17. The sag can be adjusted by adding or subtracting spacer washers at the point indicated by arrow **A**.

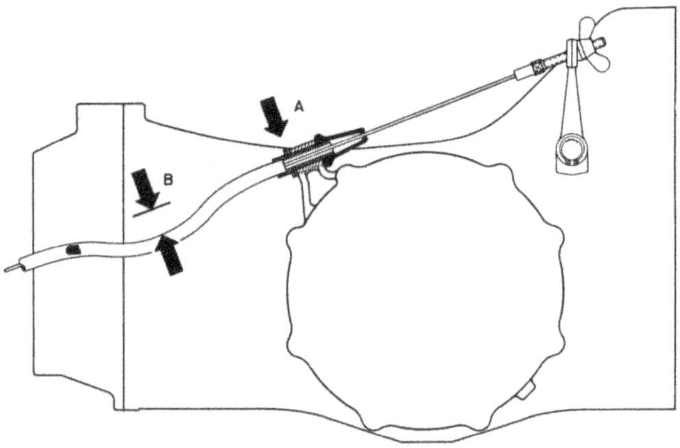

© 1974 VWoA—1839

Fig. 15-17. Flexible guide tube sag (between arrows at **B**). Spacer washers are at arrow **A**.

Excessive sag in the flexible guide tube will cause the cable to bind, creating noise, difficult clutch pedal operation, or even cable breakage. If the sag is inadequate, there will be insufficient preload on the cable. This can cause poor clutch pedal feel and accelerated wear.

In early 1972, a modified flexible guide tube was introduced. It has a hole in its front ferrule that permits grease to be forced into the guide tube with a standard lever-type hand-operated grease gun equipped with a tapered nozzle. Grease should be pumped into the guide tube only if the clutch cable creaks, jams, or is stiff in operation. The rubber boot over the flexible guide tube ferrule must be pushed forward against the frame crossmember to uncover the lubrication hole.

15.6 Troubleshooting Clutch

Road testing is an important part of troubleshooting the clutch because it lets you base your diagnosis on first-hand information.

For example, "lack of power" might turn out to be a slipping clutch and not inadequate engine output. Similarly, a complaint that "it's hard to shift gears" may mean a dragging clutch, not transmission trouble. Noises and clutch chatter are often described simply as "vibrations." It is important that such inaccurate descriptions do not result in troubleshooting errors or in a misinterpretation of the problem's real origin.

Troubleshooting Table

Table g lists the most common clutch complaints. They are grouped into four classes of symptoms. The actual causes, however, may vary. The numbers in bold type in the Remedy column refer to the headings in this section where the suggested repairs are described.

CAUTION —
If you lack the skills, tools, or a suitable workshop for clutch repairs, we suggest that you leave them to an Authorized VW Dealer or other qualified shop. We especially urge you to consult your Authorized VW Dealer before attempting repairs on a car still covered by the new-car warranty.

16. CRANKCASE

All disassembly procedures described up to this point must already have been carried out before you can disassemble the crankcase. After you have removed the various components, the crankcase of the single-carburetor

Table g. Clutch Troubleshooting

Problem	Probable Cause	Remedy
1. Clutch noise	a. Needles in pilot bearing worn	a. Replace pilot bearing; lubricate with 1 g ($\frac{1}{32}$ oz.) lithium grease. See **15.1**, **15.2**.
	b. Driven plate fouling pressure plate	b. Replace driven plate. See **15.1**.
	c. Release lever springs weak or tension uneven	c. Replace pressure plate assembly. See **15.1**.
	d. Release bearing defective	d. Fit new bearing. See **15.4**.
2. Clutch grabbing	a. Transmission mountings loose	a. Tighten mounting bolts and nuts. See **TRANSMISSION AND REAR AXLE**.
	b. Bend in cable guide too slight or excessive	b. Correct the bend to between 25 and 45 mm (1 and 1$\frac{3}{4}$ in.). See **15.5**.
	c. Pressure plate contacting unevenly	c. Replace pressure plate assembly. See **15.1**.
	d. Release plate not running true	d. Replace pressure plate assembly. See **15.1**.
	e. Spring segments deformed	e. Replace driven plate. See **15.1**.
3. Clutch dragging	a. Excessive pedal freeplay	a. Reduce pedal freeplay to lower limit of 10 mm ($\frac{3}{8}$ in.). See **15.5**.
	b. Excessive bend in cable guide	b. Reduce bend to 25 to 45 mm (1 to 1$\frac{3}{4}$ in.). See **15.5**.
	c. Driven plate not running true	c. Replace driven plate. See **15.1**.
	d. Spring segments excessively or unevenly set	d. Replace driven plate. See **15.1**.
	e. Plate linings broken	e. Replace driven plate. See **15.1**.
	f. Main drive shaft not running true with pilot bearing	f. It is sometimes sufficient to loosen the engine mounting bolts, move the engine slightly and retighten the bolts. Otherwise check pilot bearing. See **3**, **15.1**.
	g. Pilot bearing defective or insufficiently greased	g. Replace pilot bearing or lubricate needle bearing with 1 g ($\frac{1}{32}$ oz.) of lithium grease. See **15.1**, **15.2**.
	h. Splines on main drive shaft or clutch driven plate dirty or burred	h. Clean splines. Remove burr. See **15.1**.
	i. Sticky clutch linings (lining dust)	i. Replace driven plate. See **15.1**.
	j. Felt ring in pilot bearing too tight on main drive shaft	j. Replace the gland nut with one having a better fitting felt ring or install a new felt ring. See **15.2**.
	k. Stiffness in pedal cluster, clutch cable and operating shaft	k. Grease the parts thoroughly with universal grease. See **15.5**.
4. Clutch slipping	a. Insufficient pedal freeplay due to lining wear	a. Adjust freeplay at clutch pedal. Replace driven plate. See **15.5**, **15.1**.
	b. Oily linings	b. Replace driven plate. If necessary, replace engine or transmission oil seal. See **15.1**.

engine will appear as it does in Fig. 16-1. A similar view of the dual-carburetor and fuel injection engine crankcase appears in Fig. 16-2.

Fig. 16-1. Single-carburetor engine crankcase ready for disassembly.

1. M 8 Galvanized nut (4)
2. Galvanized lock washer (4)
3. Generator support
4. Oil deflector plate
5. Generator support gasket (2)
6. M 8 galvanized bolt
7. Galvanized lock washer (6)
8. M 8 galvanized nut (6)
9. M 8 bolt (2)
10. Lock washer (11)
11. M 8 nut (9)
12. Crankcase
13. Washer (6)
14. M 12 nut (6)
15. Crankshaft oil seal
16. Shim (3)
17. Steel dowel pin (4)
18. Rubber seal
19. Flywheel
20. Spring-type lock washer
21. Gland nut with bearing
22. Drive plate
23. Spring-type lock washer
24. Gland nut without bearing

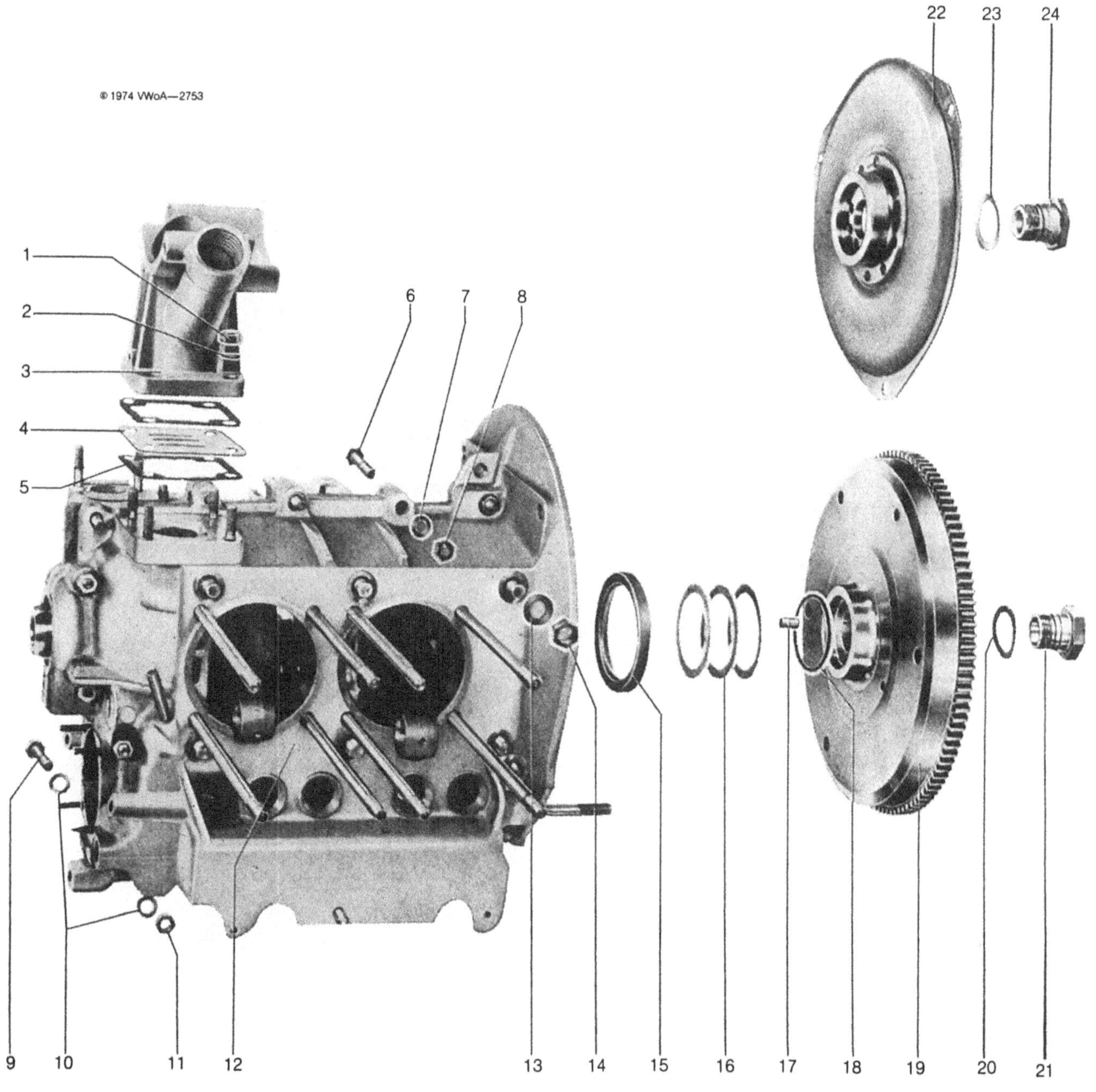

1. Fan hub retaining bolt
2. Washer
3. Fan hub
4. Fan hub sealing ring
5. Small crankshaft oil seal
6. Oil filler pipe bolt
7. Sealing ring
8. M 6 nut (2)
9. Spring washer (2)
10. Ignition cable holder
11. M 8 bolt (5)
12. Spring washer (12)
13. M 8 nut (12)
14. M 10 main bearing bolt (6)
15. Washer (12)
16. M 10 sealing nut (6)
17. Shim (3)
18. Large crankshaft seal
19. Flywheel sealing ring
20. Rolled pin
21. Pilot bearing (manual transmission only)
22. Flywheel (manual transmission only)
23. Felt ring for pilot bearing
24. Bolt plate
25. M 12 self-locking bolt (5)
26. M 8 self-locking bolt (4)
27. Support (2) (left and right differ)
28. M 8 nut (2)
29. M 8 self-locking nut (2)
30. Spring washer (2)
31. Bonded rubber mountings (2)
32. Woodruff key
33. Engine bearer
34. Oil pan cover
35. Cover sealing ring
36. M 6 bolt (2)
37. Seal washer (2)
38. Crankcase (two halves)
39. Drive plate (automatic transmission only)
40. Socket head screw (5)

Fig. 16-2. Dual-carburetor and fuel injection engine crankcase ready for disassembly.

5

16.1 Removing and Installing Oil Strainer
(single-carburetor engine only)

The oil strainer cover and oil strainer must be removed from the bottom of the crankcase before any other nuts or bolts are removed. To remove the oil strainer, remove the six cap nuts and seal washers. Then take off the cover, strainer, and the two oil strainer gaskets. Use new seal washers during installation. Full details on oil strainer servicing and installation are given in **LUBRICATION AND MAINTENANCE**.

16.2 Disassembling and Assembling Crankcase

The crankcase must be taken apart to replace the connecting rods, the connecting rod bearings, and the main bearings. It is also necessary to disassemble the crankcase in order to remove and install the crankshaft, camshaft, and camshaft bearings.

To disassemble:

1. On single-carburetor engines, remove the oil strainer and the cover plate below the flywheel pulley. (The pulley must first be removed as described in **6.2 Removing and Installing Crankshaft Pulley**.)
2. On dual-carburetor and fuel injection engines, remove the oil pan cover, the engine carrier and bonded rubber mountings, and the fan hub.
3. Remove the drive plate or the flywheel as described either in **14.1 Removing and Installing Drive Plate** or in **15.2 Removing and Installing Flywheel**.

4. On single-carburetor engines, remove the six nuts from the main bearing studs. On dual-carburetor and fuel injection engines, remove the six M 10 sealing nuts from the main bearing bolts, then take out the bolts.
5. On single-carburetor engines, remove the nine M 8 nuts and the one M 8 bolt and nut from the crankcase flange. On dual-carburetor and fuel injection engines, remove the five M 8 bolts and nuts that bind the crankcase flange together, then remove the oil filler pipe mounting bolt.
6. Using spring clips, clamp the cam followers in the right-hand crankcase half in place. Then lift off the right-hand crankcase half (Fig. 16-3).

© 1974 VWoA—1297

Fig. 16-3. Crankcase halves being separated. Note the spring clips holding the cam followers.

CAUTION —
Never insert any tools between the crankcase flanges to separate the halves. The slightest scratch can produce an oil leak. If the halves are stuck together with sealer, use a rubber hammer to loosen the right-hand half from the left.

To assemble:

1. Make sure that the mating surfaces of the crankcase halves are absolutely clean.
2. Clean any hardened sealer or dirt from the bolts and studs as such matter can cause inaccurate torque wrench readings.
3. Check the studs for tightness where they thread into the crankcase. Check the oil suction pipe for tightness at the points marked by the arrows in Fig. 16-4.

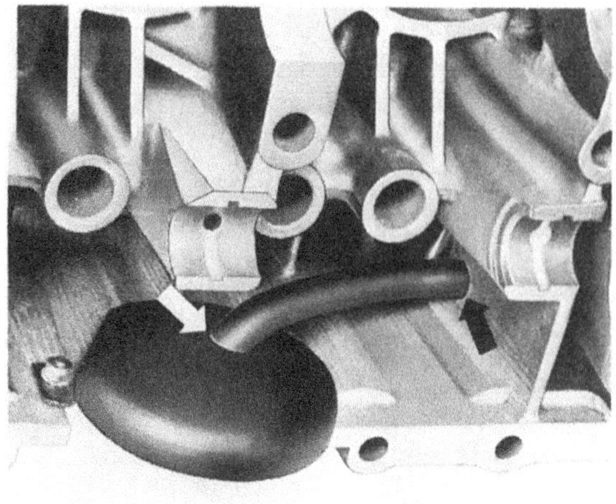

© 1974 VWoA—1295

Fig. 16-4. Checkpoints (arrows) for oil pipe tightness. If necessary, secure them with a peening tool.

4. Using sealing compound all around it, install the camshaft plug as shown in Fig. 16-5. Replace the plug if it is rough or deformed.

© 1974 VWoA—1296

Fig. 16-5. Camshaft end plug (arrow) installed in left half of crankcase.

5. Using spring clips, clamp the cam followers in the right-hand half of the crankcase.
6. Spread an even film of sealing compound over the mating surfaces of the crankcase halves.

CAUTION —
Do not let the sealing compound get into the oil passages for the crankshaft and camshaft bearings. It could block lubrication.

7. On single-carburetor engines, install new O-rings around the main bearing studs as indicated in Fig. 16-6.

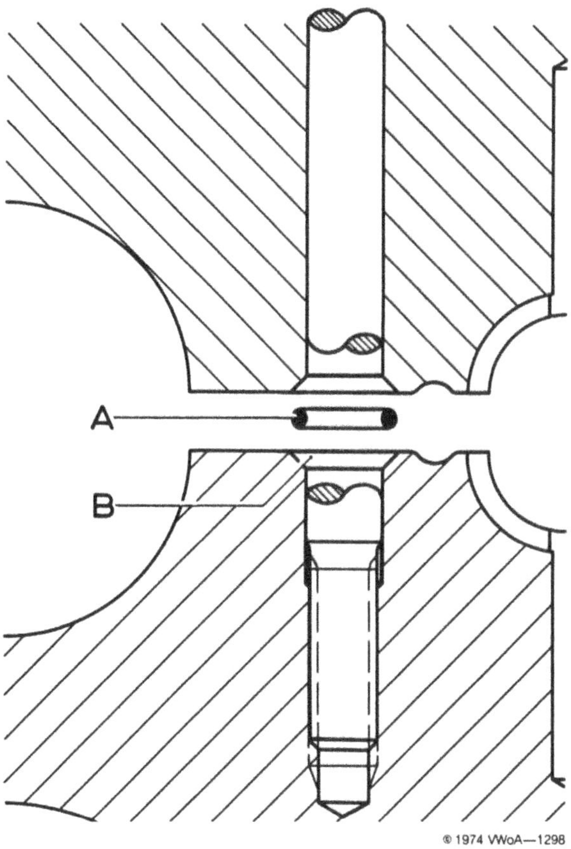

A. Rubber ring B. Chamfer

Fig. 16-6. O-ring installation. As the crankcase halves are joined, the silicone rubber O-rings will be compressed into the chamfers around the studs.

8. Join the crankcase halves. On single-carburetor engines, loosely install all the nuts.
9. On single-carburetor engines only, torque the M 8 nut near the lower main bearing stud adjacent to the flywheel to 2.0 mkg (14 ft. lb.).
10. On single-carburetor engines only, torque the six M 12 main bearing stud nuts.

NOTE —
If you are using sealing nuts on the main bearing studs, as are standard on some models, torque them to 2.5 mkg (18 ft. lb.). If you are using plain nuts, as originally installed on a majority of engines, torque them to 3.5 mkg (25 ft. lb.). If you service-install sealing nuts on an engine that originally had plain nuts, do not install the steel washers used under the original nonsealing nuts.

11. On dual-carburetor and fuel injection engines only, coat the main bearing bolt heads with sealing compound, then install the bolts in the crankcase together with their washers.

NOTE —
Install plastic dampers (part No. 021 101 107) on the shanks of the main bearing bolts whether or not the engine was originally so equipped.

12. On dual-carburetor and fuel injection engines only, coat the M 10 sealing nuts for the main bearing bolts with sealing compound. Install the nuts with their sealing rings outward, then torque them to 3.5 mkg (25 ft. lb.).
13. Hand-turn the crankshaft to check for free movement.
14. On single-carburetor engines only, torque the remaining M 8 nuts and the single M 8 bolt to 2.0 mkg (14 ft. lb.).
15. On dual-carburetor and fuel injection engines, coat the bolt heads and the nuts for the M 8 bolts with sealing compound. Then install the bolts and nuts in the crankcase flange together with their spring washers. Torque the nuts to 2.0 mkg (14 ft. lb.).
16. Remove the spring clips that hold the cam followers.
17. Check the crankshaft end play, as described in **17.5 Crankshaft End Play**. Then install a new crankshaft oil seal or oil seals.
18. Install the crankshaft pulley or the fan hub. Install the flywheel or drive plate.
19. On single-carburetor engines, install the crankcase oil strainer. Then torque the cap nuts to 0.7 mkg (5 ft. lb.). On dual-carburetor engines, install the oil pan cover. Torque the nuts to 1.3 mkg (9 ft. lb.).
20. Install the remaining engine parts to complete the assembly.

16.3 Removing and Installing Oil Filler and Breather

(single-carburetor engine)

Many repairs that require the crankcase halves to be separated do not demand the removal of the oil filler and breather. If the filler and breather do not require servicing, you can simply remove the generator stand from the crankcase and leave the oil filler and breather attached to it. However, if the oil filler and breather must be removed from the generator stand, it is best to do it before you unbolt the generator stand from the engine.

To remove the oil filler and breather, first take off the oil filler cap. Then, using the special tool VW 170 (Fig. 16-7), remove the gland nut that holds the oil filler and breather on the generator stand.

© 1974 VWoA—2003

Fig. 16-7. Special tool being inserted into oil filler to remove gland nut.

Installation of the oil filler and breather on the generator stand is the reverse of removal. Torque the gland nut to 5.5 mkg (40 ft. lb.).

If you remove the generator stand by taking off the four M 8 nuts that hold it on the crankcase, use two new gaskets during installation—one above the deflector plate and one below it. The deflector plate must be installed on the crankcase so that the louvers face downward with the slightly longer end of the central louver to the rear, as shown in Fig. 16-8.

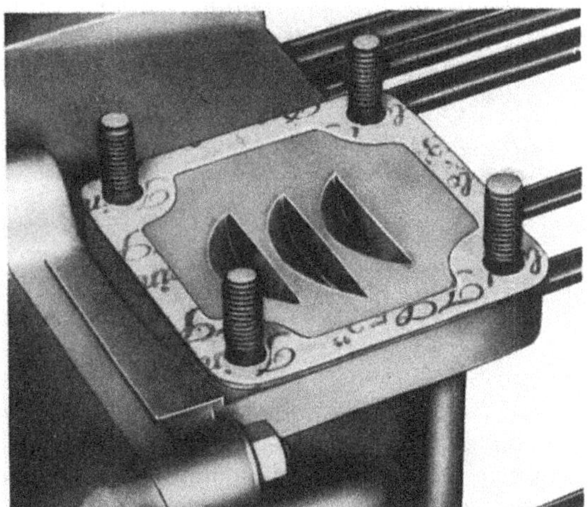

© 1974 VWoA—1237

Fig. 16-8. Oil deflector plate and gasket properly installed. The upper face of the deflector plate is marked **top**.

16.4 Repairing Tapped Holes

If the threads for the cylinder studs have been stripped, you can return the crankcase to service by installing steel inserts in the casting. The repair kit, containing 16 inserts plus the required tool set, may be uneconomical if you are repairing only one crankcase. If so, it is probably best to turn the job over to an Authorized VW Dealer or other qualified shop.

Sequence of operations:

1. Read the installation instructions supplied with the tool set.

   NOTE —
   The tool set contains a drill and tap jig, a drill pilot, a 33/64-in. drill, a special .5570-in. tap, and an insert driving tool.

2. After removing the studs, install the jig on the crankcase. Enlarge the stud holes to 13 mm (33/64 in.) following the instructions supplied with the tool set.
3. Using the jig as a guide, cut threads into the enlarged holes with the special .5570-in. tap.
4. Place each insert, tangs first, in the insert driving tool. Then thread them into the holes tapped in the crankcase. Screw in each insert until the driving tool is 1.00 mm (.040 in.) from the cylinder sealing surface on the crankcase.
5. Strike the insert driving tool with a hammer to drive the tangs ("Kee's") flush with the insert, thereby locking the insert in the crankcase casting.

17. CAMSHAFT AND CRANKSHAFT

After you have removed the right-hand half of the crankcase from the left-hand half, as described in **16.2 Disassembling and Assembling Crankcase**, simply lift the camshaft, crankshaft, bearings, and related parts out of the left-hand half of the crankcase. When servicing these internal engine parts, your major objective should be to carry out accurately the inspection and installation procedures given under the following headings.

17.1 Checking and Installing Camshaft and Bearings

Two camshafts and gears have been used on single-carburetor engines. On the early type, three rivets hold the gear onto the camshaft, whereas the later type has four rivets. The change was made to accommodate the larger oil pump introduced in May 1971.

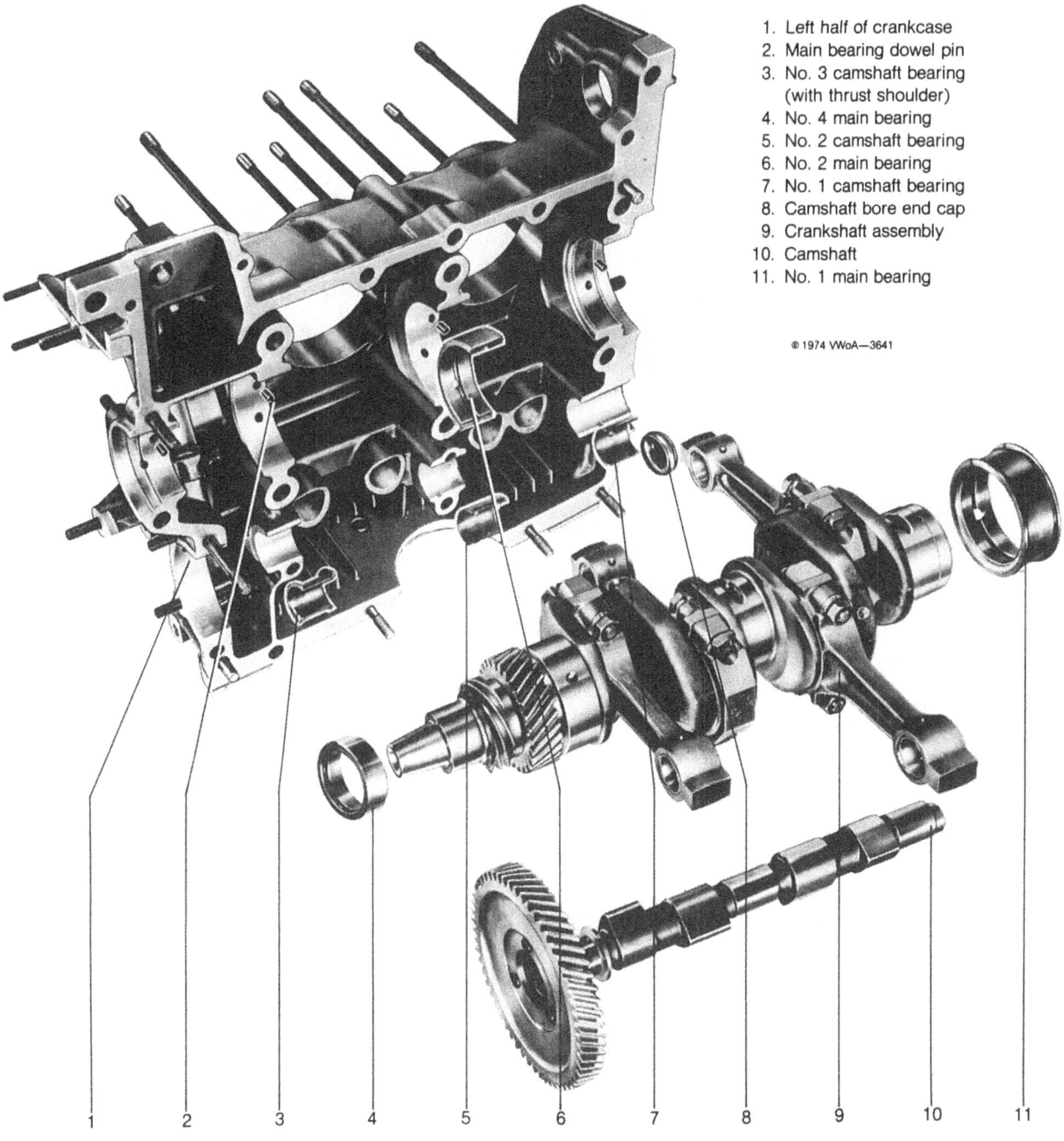

Fig. 17-1. Crankshaft assembly and camshaft removal. A dual-carburetor or fuel injection engine is shown, but the parts are similarly installed in earlier engines.

CAUTION —
The late-type single-carburetor engine oil pump, which has 26-mm (1.024-in.) gears, must be used with the four-rivet camshaft; the early-type pump, which has 21-mm (.827-in.) gears, must be used with the three-rivet camshaft. If the parts are mixed, the oil pump will either not fit or not supply lubrication.

The camshafts for dual-carburetor or fuel injection engines used with manual transmissions are different from those used with automatic transmissions. The camshafts for engines used with automatic transmissions have two narrow bands cast into the shaft on each side of the center bearing journal. Fig. 17-1 shows the camshaft, the crankshaft, and their bearings. Camshafts for hydraulic lifters have no cast bands at the center bearing.

To check and install camshaft:

1. Check the camshaft gear rivets for tightness.

5

2. Check the cam bearings and lobes for wear (Fig. 17-2).

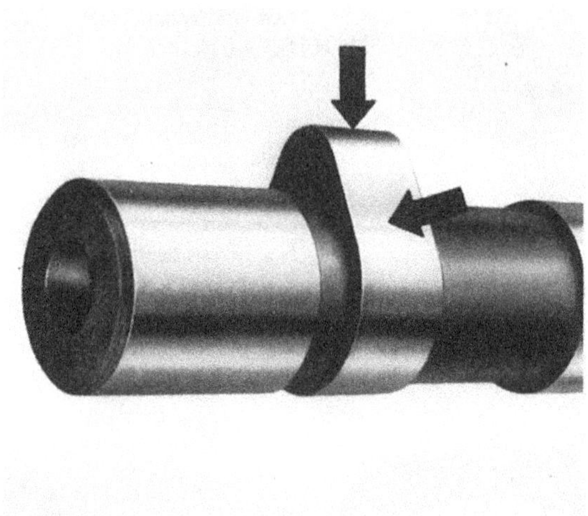

© 1974 VWoA—1242

Fig. 17-2. Cam lobes. Toe of cam (top arrow) should not be scored or worn unevenly. There should be no sign of wear at second arrow.

3. Check the camshaft for runout as in Fig. 17-3. New camshafts should not exceed 0.02 mm (.0008 in.); used camshafts should not exceed 0.04 mm (.0016 in.).

© 1974 VWoA—1243

Fig. 17-3. Camshaft runout being measured.

4. Examine the camshaft gear for wear and for a properly centered tooth contact pattern.
5. Clean the bearing saddles in the left-hand half of the crankcase. Check the oil holes for cleanliness.
6. Install the three camshaft bearing halves in their saddles, making sure that the oil holes in the bearing shells line up with those in the crankcase half.
7. Lightly coat the bearings with assembly lubricant (available from auto stores). Then install the camshaft as illustrated in Fig. 17-4.

© 1974 VWoA—3642

Fig. 17-4. Camshaft installation. The tooth marked with a small circle (arrow) goes between the two marked teeth on the crankshaft gear.

8. Check the axial play as shown in Fig. 17-5. The play permitted by the thrust bearing should be 0.04 to 0.13 mm (.0016 to .0051 in.) with a wear limit of 0.16 mm (.0063 in.).

© 1974 VWoA—1244

Fig. 17-5. Dial indicator being used to check axial play.

9. Check the backlash around the entire circumference of the gear. It should be between 0.00 and 0.05 mm (.000 and .002 in.). The cam gear size is correct if backlash is hardly noticeable and if the camshaft does not lift from its bearings when you turn the crankshaft backward.

NOTE—
Camshaft gears are marked −1, 0, +1, or +2, etc. on their inner face. These numbers indicate in 1/100 mm how much their pitch radius departs from the standard pitch radius denoted by 0. Do not confuse a zero on the inner face with the circular timing mark on the outer face.

10. Coat the upper halves of the bearing shells with assembly lubricant. Then install them atop the camshaft journals. Also coat the cam lobes and followers with assembly lubricant.

CAUTION—
If you are installing a new camshaft, install new cam followers or hydraulic lifters as well. The existing wear pattern of used followers or lifters may prevent a proper wear pattern from developing on the cam lobes.

11. Clean and install the right-hand half of the crankcase as described in **16.2 Disassembling and Assembling Crankcase.**
12. Install the remaining engine parts to complete the assembly.

17.2 Removing, Servicing, Installing and Adjusting Hydraulic Valve Lifters

The hydraulic valve lifters introduced on the 1978 models can be removed and installed without removing or disassembling the engine. Engines with hydraulic lifters can be identified by examining the pushrods, which are tubular steel (instead of tubular aluminum with steel tips).

To remove hydraulic lifters:

1. Remove the rocker arm shafts as described in **10.1 Removing and Installing Rocker Arm Shaft.**
2. Withdraw the pushrods and the pushrod tubes.
3. Using a magnetic tool, withdraw the hydraulic lifters from the crankcase, marking each lifter so that it can be reinstalled in the same crankcase bore.

NOTE—
Always place hydraulic lifters body-down on the workbench so that oil will not leak out and make bleeding necessary prior to reinstallation.

To clean, install, and adjust:

1. If metal particles have been found in the crankcase following failure of an engine component, disassemble the lifters for cleaning as shown in Fig. 17-6.

CAUTION—
Never repair faulty hydraulic lifters by installing parts from another lifter. Faulty lifters should be replaced as a unit. Worn-out lifters and dirty lifters will cause valve noises. However, intermittent valve noises are normal upon starting, during sudden acceleration, at high oil temperatures, and during high rpm operation.

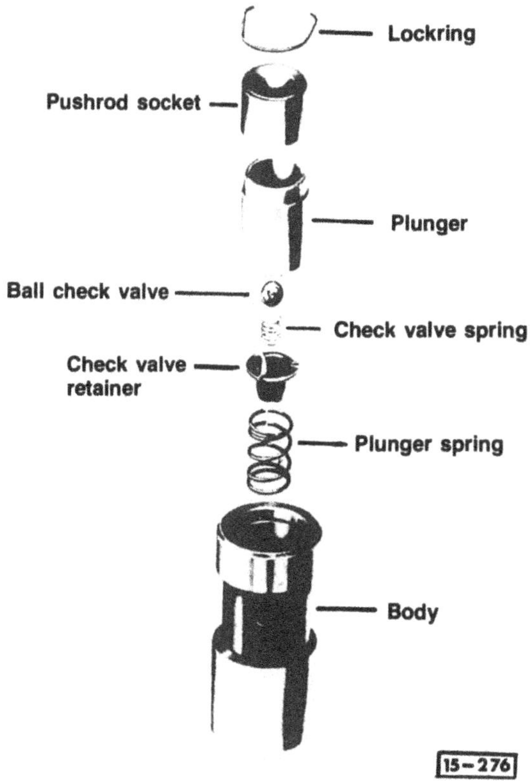

Fig. 17-6. Exploded view of hydraulic lifter. Remove lockring, then separate parts. Wash components in solvent; bleed lifters during reassembly.

2. If you are installing new hydraulic lifters, or hydraulic lifters that have leaked their oil, bleed the lifters as described later under **Bleeding Hydraulic Valve Lifters.**

NOTE—
If you are not sure whether the hydraulic lifters need to be bled, check their hydrostatic resistance to compression. To do this, apply firm pressure with your thumb at the point indicated in Fig. 17-7. A solid resistance to complete compressing of the spring should be felt. If not, bleed the lifter.

5

Fig. 17-7. Thumb pressure point (arrow) on face of lifter's pushrod socket.

3. Install the lifters in their original bores in the crankcase. Then install the pushrod tubes, the pushrods, and the rocker arm shafts.
4. Loosen the locknuts for the adjusting screws in the rocker arms (Fig. 17-8). Then back off each adjusting screw so that the unthreaded part of the screw, which contacts the valve stem, is flush with the surface of the rocker arm (no threads visible on the valve side of the rocker).

Fig. 17-8. Rocker arms, showing locknuts and adjusting screws at top. Arrow indicates spacer bushing that was introduced in 1978 along with hydraulic valve lifters.

5. Hand turn the crankshaft until the No. 1 cylinder is in firing position (distributor rotor pointed to the No. 1 firing mark on the distributor body).
6. With the No. 1 cylinder in position to fire, turn in the adjusting screws for both rocker arms of the No. 1 cylinder until the tips just touch the valve stems (zero clearance). Now turn each screw an additional two turns clockwise and tighten the locknuts.
7. Turn the crankshaft until the rotor moves counterclockwise 90° from the No. 1 cylinder position. This will correctly position the No. 2 cylinder for adjustment.
8. Repeat the adjustment described in step 6 on the rocker arms of the No. 2 cylinder. Turn the engine so that the rotor moves counterclockwise in 90° increments; adjust the No. 3 and No. 4 cylinders in the same way.

Bleeding Hydraulic Valve Lifters

The valve lifters should be filled with engine oil prior to their installation in the engine. This must be done correctly so that all air is bled from the lifter during filling.

To bleed lifter:

1. Fill a clean tin can with fresh engine oil.
2. Remove the lockring from the valve lifter. Then remove all of the internal parts (the pushrod socket, the plunger, the ball check valve, the check valve spring, the check valve retainer, and the plunger spring).
3. Place the valve lifter body in the can of oil so that the body is completely immersed.
4. Keeping the lifter body immersed in oil, insert the plunger spring, the check valve retainer, the check valve spring, the ball check valve, and the plunger. Use a scriber as indicated in Fig. 17-9 to open the ball check valve, thus allowing oil to flow through as you press down the plunger.

Fig. 17-9. Scriber being used (arrow) to press down ball check valve.

5. Insert the pushrod socket in the lifter. Then, keeping the lifter immersed, use an engine pushrod as shown in Fig. 17-10 to force down the pushrod

socket slowly. (Half of a discarded VW Beetle pushrod makes an excellent tool for this job.)

6. When the pushrod socket has been pressed down below the level of the lockring groove, install the lockring. Then remove the lifter from the oil. To prevent oil from leaking out, keep the lifter upright until it is to be installed in the engine.

Fig. 17-10. Pushrod socket being pressed down.

17.3 Removing and Installing Crankshaft

Crankshaft removal merely requires that you lift the crankshaft out after you have separated the crankcase halves. Installing the crankshaft and bearings, however, calls for careful work and clean working conditions.

To install crankshaft:

1. Inspect the crankshaft main bearing journals for scoring and wear. They should not be deeply grooved. The wear limit is 0.03 mm (.0012 in.). Maximum runout is also 0.02 mm (.0008 in.) measured at No. 2 and No. 4 journals with Nos. 1 and 3 on V-blocks.

NOTE —
Using a micrometer, check main journals 1, 2, and 3 for out-of-round. Their specified diameter for single-carburetor engines is 54.97 to 54.99 mm (2.1641 to 2.1649 in.) or, for dual-carburetor and fuel injection engines, 59.97 to 59.99 mm (2.3610 to 2.3618 in.). Then check the No. 4 journal; the specified diameter for all engines is 39.98 to 40.00 mm (1.5740 to 1.5748 in.). Replacement crankshafts are available in three undersizes of 0.25 mm (.010 in.) each.

2. Inspect the oil passages in the crankshaft and bearings to see that there are no sharp edges. Burrs can be removed with a half-round needle file.

NOTE —
Although foreign matter embedded in the bearing material can be removed with a sharp scraper, most experienced mechanics prefer to replace bearings that are in such condition, rather than to reuse them.

3. Check the main bearing dowels (Fig. 17-11) for tightness.

© 1974 VWoA—3643

Fig. 17-11. Main bearing dowels in bearing saddles (arrows). These dowels are found in the left-hand side of the crankcase only.

4. Liberally coat the bearings with assembly lubricant (available from auto supply stores). Then install Nos. 1, 3, and 4 on the crankshaft (see **17.4 Disassembling and Assembling Crankshaft** for instructions on replacing the No. 3 main bearing). Install the lower half of the split No. 2 bearing in the crankcase, carefully engaging the dowel with the hole in the bearing shell.

5. Turn the bearings on the crankshaft to position properly the oil holes and dowel holes. Then install the crankshaft assembly in the left-hand crankcase half, aligning the timing marks on the crankshaft and camshaft gears.

CAUTION —
Do not force the crankshaft into position; doing so could damage the bearings. Instead, lift the crankshaft slightly and hand-turn the bearings until their holes align with the dowels in the bearing saddles.

6. Install the right-hand half of the crankcase as described in **16.2 Disassembling and Assembling Crankcase.**

5

NOTE —
Main bearing clearance is partially determined by the preload (crush) imparted to the bearing shells as the crankcase halves are bolted together. Strict adherence to the prescribed torque figures is therefore of extreme importance. Bearing clearance is basically determined by the proper selection of bearings and by the use of a crankshaft within prescribed tolerances. However, you can measure individual clearances either with Plastigage or with a sensitive dial indicator. Clearances, as measured with the crankcase halves together, are given in **Table h.**

Table h. Bearing Clearance

	New	Wear limit
Crankshaft bearings 1 + 3 (1600 engine)	0.04–0.10 mm (.0016–.004 in.)	0.18 mm (.007 in.)
Crankshaft bearings 1 + 3 (1700, 1800, and 2000 engines)	0.05–0.10 mm (.002–.004 in.)	0.18 mm (.007 in.)
Crankshaft bearing 2 (all models)	0.03–0.09 mm (.001–.0035 in.)	0.17 mm (.0067 in.)
Crankshaft bearing 4 (all models)	0.05–0.10 mm (.002–.004 in.)	0.19 mm (.0075 in.)

7. Check the crankshaft end play as described in **17.5 Crankshaft End Play.**
8. Install the remaining engine parts to complete the assembly.

Plastigage, for measuring installed bearing clearances, is available at auto supply stores. However, it is not well suited for checking clearances of ring-type bearings, such as those used on journals 1, 3, and 4.

17.4 Disassembling and Assembling Crankshaft

When you remove the crankshaft from the crankcase, the connecting rods remain installed on the crankshaft. In addition to connecting rod removal, it is necessary to press off the distributor drive gear and crankshaft gear to get the No. 3 main bearing on or off its journal. The positions of the crankshaft assembly's components are shown in Fig. 17-12 and Fig. 17-13.

To disassemble:

1. Remove the nuts from the rod cap studs on each connecting rod. See **18.1 Removing Connecting Rods.**
2. Remove the connecting rods and the rod bearing shells.
3. Using circlip pliers, remove the distributor drive gear circlip.
4. Using a hydraulic press, press the crankshaft gear, spacer, and distributor drive gear off the crankshaft. Put a support plate under the crankshaft gear, then position the press ram against the end of the crankshaft.

Fig. 17-12. Exploded view of single-carburetor engine crankshaft assembly.

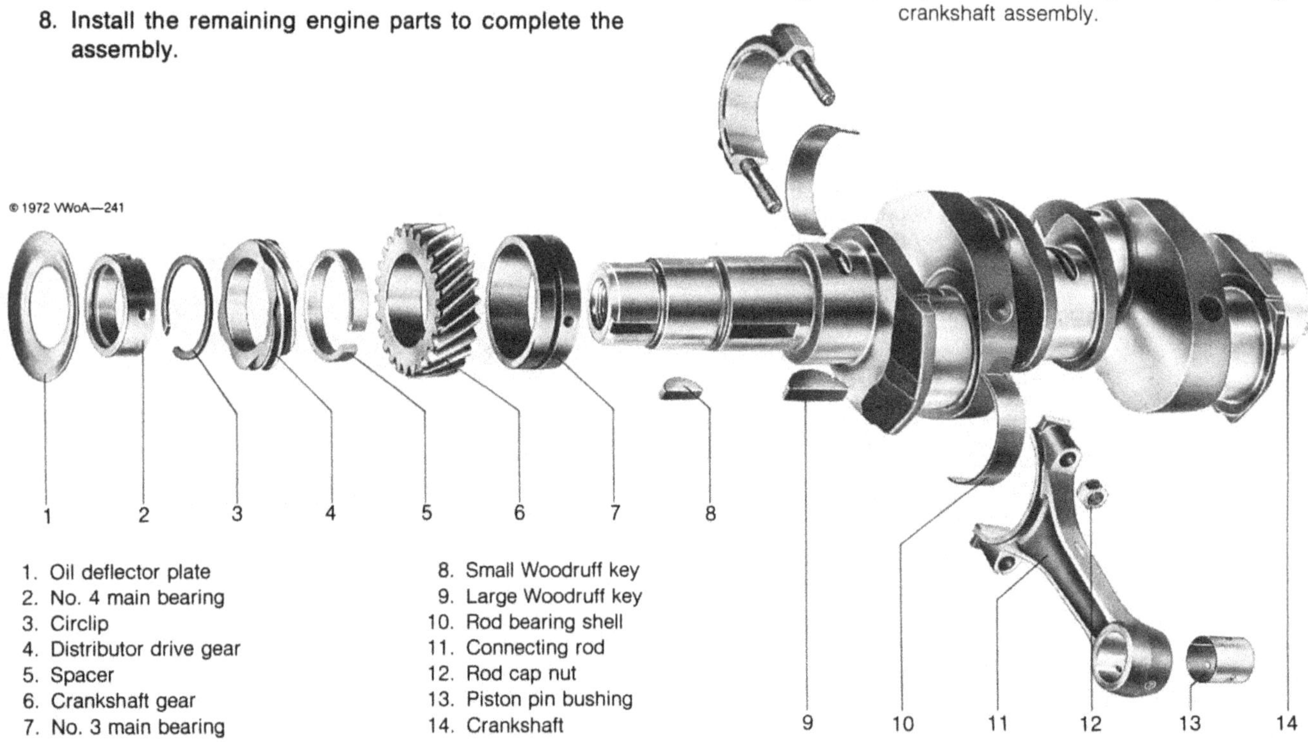

1. Oil deflector plate
2. No. 4 main bearing
3. Circlip
4. Distributor drive gear
5. Spacer
6. Crankshaft gear
7. No. 3 main bearing
8. Small Woodruff key
9. Large Woodruff key
10. Rod bearing shell
11. Connecting rod
12. Rod cap nut
13. Piston pin bushing
14. Crankshaft

Fig. 17-13. Exploded view of dual-carburetor or fuel injection engine crankshaft assembly.

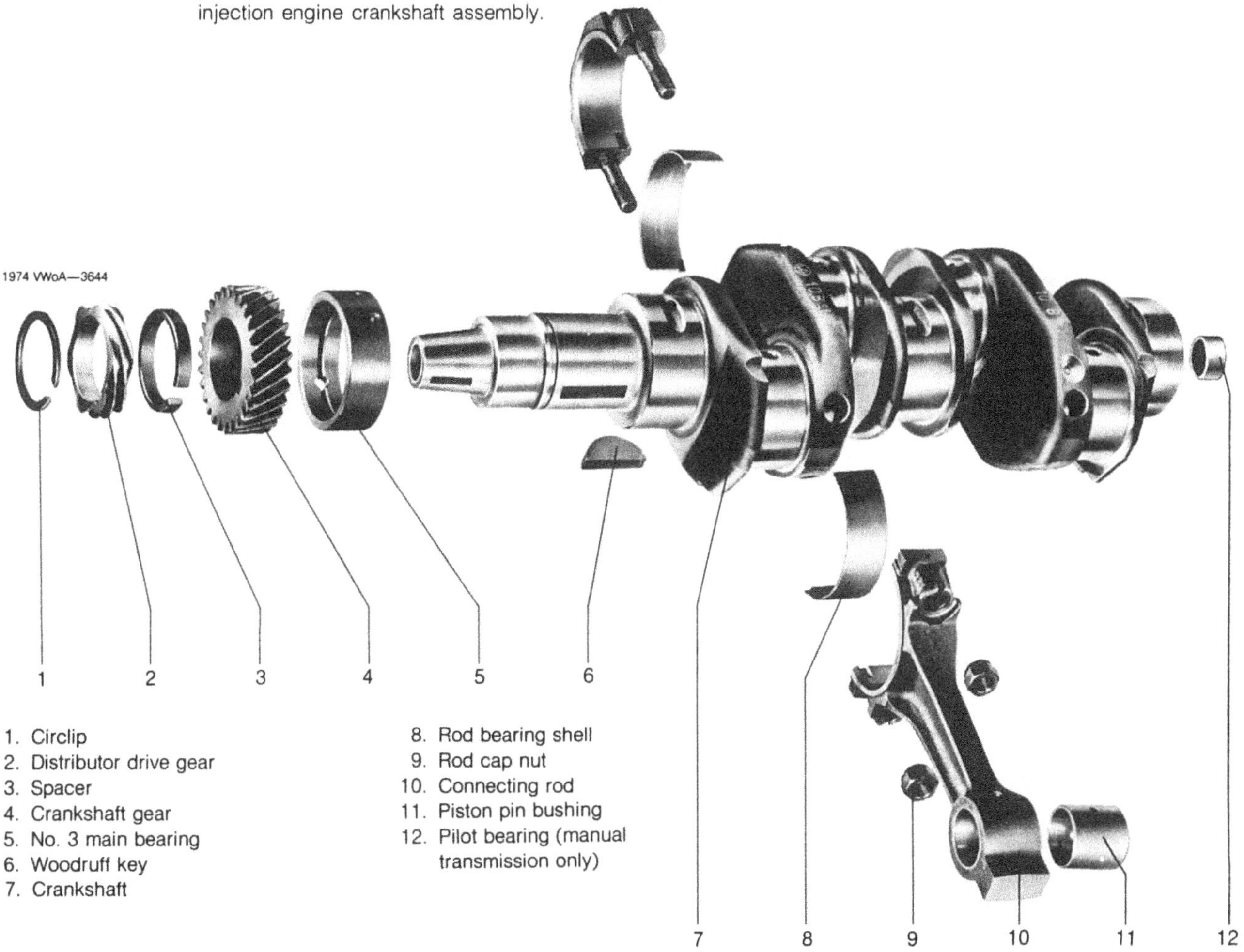

1. Circlip
2. Distributor drive gear
3. Spacer
4. Crankshaft gear
5. No. 3 main bearing
6. Woodruff key
7. Crankshaft
8. Rod bearing shell
9. Rod cap nut
10. Connecting rod
11. Piston pin bushing
12. Pilot bearing (manual transmission only)

5. Remove the No. 3 main bearing. Take out the Woodruff key(s).

To assemble:

1. Clean the crankshaft and blow out its oil passages with compressed air.
2. Check the gear teeth. Inspect the gear seating surfaces for seizure marks.

NOTE —
Light seizure marks can be removed, provided the press fit is not affected.

3. After coating the journal with assembly lubricant, install the No. 3 main bearing. Install the large Woodruff key in its recess.
4. Heat the crankshaft gear to about 80°C (176°F) in an oil bath. Then press the gear onto the crankshaft. The chamfer in the gear bore must face the No. 3 main bearing journal. The keyway must align with the Woodruff key on the crankshaft.
5. Slip on the spacer. On single-carburetor engines only, insert the small Woodruff key in its recess. Install the distributor drive gear as you did the crankshaft gear.
6. Install the circlip, being careful not to scratch the No. 4 bearing journal on the crankshaft.
7. Check to see that the gears fit tightly when cool.

17.5 Crankshaft End Play

Crankshaft end play is controlled by three shims between the flywheel and the No. 1 main bearing flange.

CAUTION —
Keep the crankshaft end play within the prescribed range. Excessive end play can cause early wear or failure of the main and connecting rod bearings, possibly damaging other engine parts.

5

The positions of the three shims are shown in Fig. 17-14.

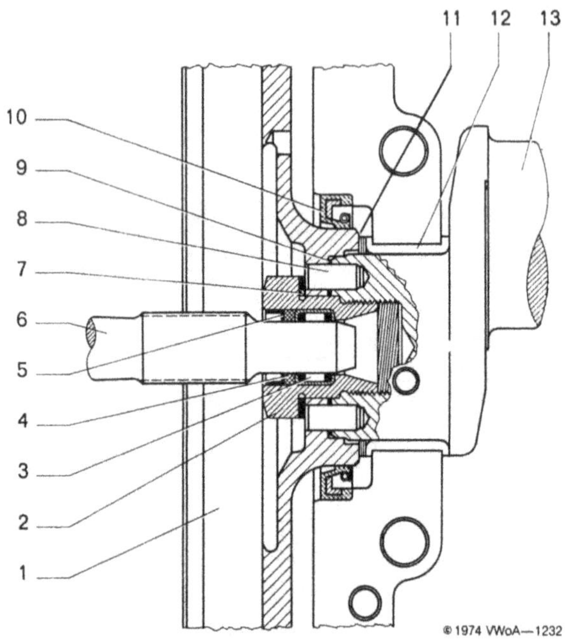

1. Flywheel
2. Gland nut
3. Needle bearing
4. Felt ring
5. Retaining ring
6. Rear driveshaft
7. Lock washer
8. Dowel pin
9. Rubber sealing ring
10. Crankshaft oil seal
11. Shims
12. Crankshaft bearing
13. Crankshaft

Fig. 17-14. Cross section of flywheel and crankshaft (single-carburetor engine).

To check end play:

1. Attach a dial indicator as shown in Fig. 17-15.

Fig. 17-15. Setup for measuring crankshaft end play.

2. Move the crankshaft in and out while watching the gauge. End play should be between 0.07 and 0.13 mm (.003 and .005 in.). The wear limit is 0.15 mm (.006 in.).
3. If necessary, make adjustments to bring end play within specifications.

To adjust end play:

1. Remove the flywheel and reinstall it with only two shims.

NOTE —
Shims come in five thicknesses: 0.24 mm (.0095 in.), 0.30 mm (.0118 in.), 0.32 mm (.0126 in.), 0.34 mm (.0134 in.), and 0.36 mm (.0142 in.). A sixth shim—0.38 mm (.0150 in.) thick—is available for the dual-carburetor and fuel injection engine only. Sizes are etched on new shims. You may have to check old shims with a micrometer.

2. Measure end play again.
3. Compute how thick the third shim must be in order to bring the end play within specifications.
4. Install the three-shim combination. Then install the oil seal and the flywheel.
5. Give the end play a final check.

17.6 Crankshaft Reconditioning

Crankshafts damaged by inadequate lubrication or crankshafts that have journals that are worn to a taper or that are out-of-round can sometimes be reconditioned by grinding. The crankshaft journals are reduced in diameter by 0.25, 0.50, or 0.75 mm (.010, .020, or .030 in.) to match the oversize bearing sets available.

Crankshaft grinding requires expensive precision machinery and is never done in repair shops. This job is left to specialty shops or engine reconditioning facilities. Factory reconditioned crankshafts are of the same quality as new shafts. It is always necessary, however, that new bearings of the proper dimensions be fitted to correct for the slight undersize of the reground crankshaft. Clean the oil passages thoroughly before you install a reground crankshaft.

18. CONNECTING RODS

Several connecting rod designs have been used in VW engines. Some rods shown in the illustrations may not look the same as those in your engine. However, the service and installation procedures are the same.

18.1 Removing Connecting Rods

The connecting rod caps have pressed-in studs that pass through holes in the connecting rod. Nuts are installed on the studs to join the connecting rod and cap.

CAUTION —
Do not drive the connecting rod studs out of the cap. Doing so could lead to connecting rod damage or failure during engine operation. If the studs are damaged, replace the entire connecting rod.

To remove connecting rod:

1. Remove the crankshaft assembly from the engine. Mount it on a crankshaft-holding fixture or in an old flywheel clamped in a vise.
2. Remove the connecting rod nuts.
3. Remove the connecting rod and cap complete with bearing shells from the crankshaft. Be careful not to scratch the crankshaft journals with the rod cap studs.

18.2 Checking and Installing Connecting Rods

Unless the connecting rods are in perfect condition, have uniform weight, and are perfectly fitted, there is always the possibility of rod failure and serious engine damage.

CAUTION —
If you lack the skills, tools, or special equipment needed to check, fit, or recondition connecting rods, we suggest you leave such repairs to an Authorized VW Dealer or other qualified automotive machine shop.

To check rods:

1. Examine the connecting rod and cap for obvious external damage or damage to the bearing bore.
2. Using steel blocks or a fixture such as that shown in **18.3 Reconditioning Connecting Rods**, check the connecting rod for bends and twists.
3. Check the weight of the connecting rod without the bearing shells but with the rod cap and nuts installed. The difference in weight between connecting rods in an engine must not exceed 10 g. You can remove metal from heavier rods at the points shown in Fig. 18-1.

NOTE —
Replacement rods come in two weights. Single-carburetor engine rods identified by a brown paint mark weigh from 580 to 588 g. Those marked with gray paint weigh from 592 to 600 g. Dual-carburetor and fuel injection engine rods identified by a white paint mark weigh from 746 to 752 g. Those marked with black paint weigh from 769 to 775 g.

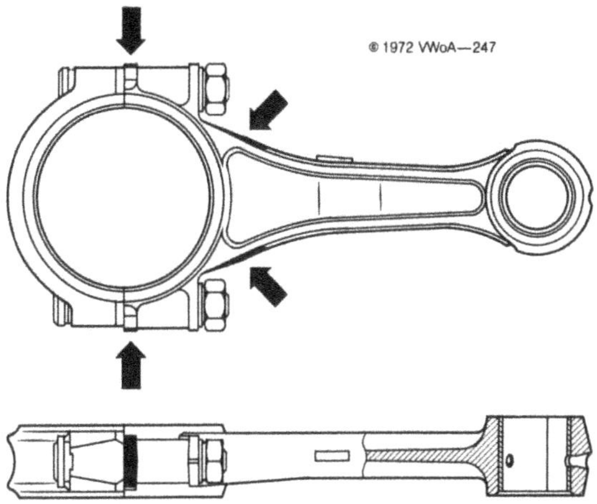

Fig. 18-1. Metal removal from overweight connecting rod. Arrows indicate where metal can be ground away.

CAUTION —
When grinding material from connecting rods to remove weight, be careful not to grind away the numbers on the side of the cap and rod. Nor should you grind off the projection forged into the side of the rod between the big and small ends. These are reference marks for installation.

4. Check the piston pin fit as shown in Fig. 18-2. Correct a loose fit by installing a new pin or a new bushing in the connecting rod. See **18.3 Reconditioning Connecting Rods** and **13. Pistons and Rings.**

Fig. 18-2. Piston pin fit being checked. The pin should be a push fit in a new bushing.

5

NOTE —
When you check the pin fit, be sure the parts are at room temperature. The pin and bushing in the connecting rod must be clean and unlubricated. A new pin will sometimes correct a loose fit, but you will get best results by installing a new bushing as well.

To install rods:

1. Install the bearing shells in the connecting rod and cap so that the tangs on the shells engage the notches in the rod bore.
2. Coat the bearing surfaces with assembly lubricant. Then install the connecting rod and cap on the crankshaft as shown in Fig. 18-3.

NOTE —
When the rod is properly installed, the numbers stamped into the connecting rod and cap will be on the same side.

Fig. 18-3. Connecting rod installations. Forged marks at arrows must face upward when crankshaft is positioned as it is in the engine.

3. Check the alignment of the rod and cap. Torque the nuts to 3.3 mkg (24 ft. lb.).

NOTE —
The bearing clearance should be 0.02 to 0.07 mm (.0008 to .0028 in.). This can be checked with Plastigage. The wear limit is 0.15 mm (.006 in.). A slight preload may develop between bearing halves as you torque the nuts. You can relieve this by lightly tapping both sides of the connecting rod with a mallet near the joint.

4. Check the axial play (side clearance) as illustrated in Fig. 18-4. It should be 0.10 to 0.40 mm (.004 to .016 in.). The wear limit is 0.70 mm (.027 in.).

Fig. 18-4. Side clearance being checked with feeler gauge.

NOTE —
Prior to the autumn of 1972, connecting rod nuts were of a special type that could be locked by peening their edges into recesses on the connecting rod. With the new nuts, no locking step is necessary and tightening torque remains unchanged. This modified procedure can be applied to all earlier engines.

CAUTION —
Do not reuse nuts that have been locked by peening. Always install new nuts of the later type to avoid having them come loose. The old-type nuts can be reused, however, if they have never been peened.

18.3 Reconditioning Connecting Rods

You can straighten slightly bent or twisted connecting rods if you have the proper equipment. You can also replace piston pin bushings.

To recondition rod:

1. Using a repair press and suitable driving tools, press out the bushing.

2. Place the connecting rod in a checking-and-straightening device such as that shown in Fig. 18-5.

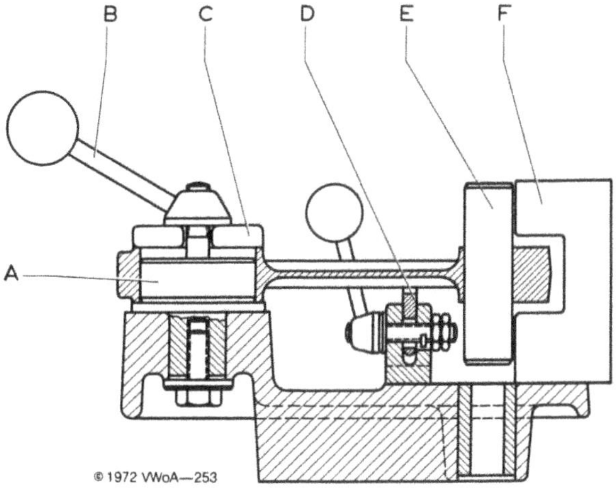

© 1972 VWoA—253

A. Eccentric mandrel
B. Locking lever
C. Washer
D. Support bar
E. Pin
F. Gauge

Fig. 18-5. Connecting rod reconditioning tool VW 214f/70. Rod is installed without piston pin bushing.

3. Turn the 7-mm eccentric mandrel to position it as shown in Fig. 18-6.

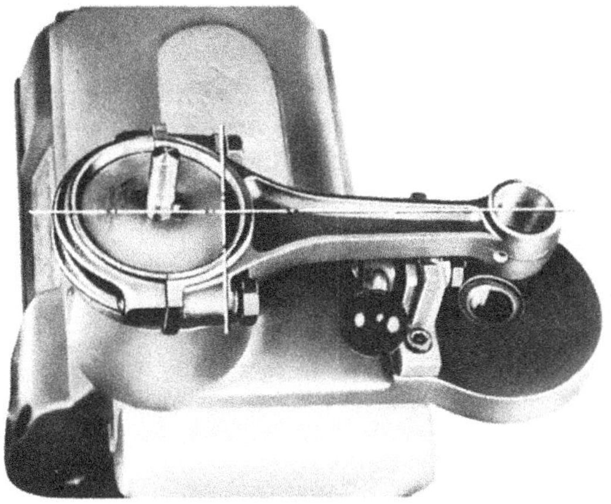

© 1972 VWoA—254

Fig. 18-6. Mandrel turned so that machined face is at a right angle to connecting rod centerline.

4. Install the washer over the mandrel. Tighten the locking lever until the connecting rod can just barely be turned on the mandrel.
5. Without tilting the rod, insert the pin in the connecting rod bushing. Move the gauge around the pin, as shown in Fig. 18-7, to check for bends and twists.

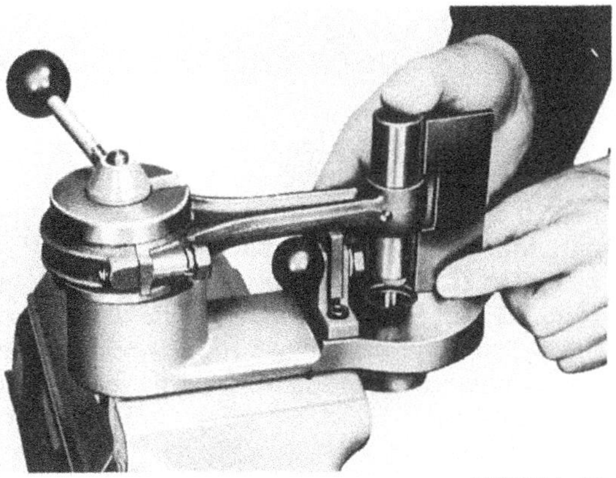

© 1972 VWoA—255

Fig. 18-7. Using gauge block to check for bends and twists. The piston pin should be perpendicular to the machined measuring surface of the device.

6. If you detect deviations, fully tighten the locking lever. Using the support bar, straighten the connecting rod.
7. Using a proper tool in a repair press, press in the new bushing. Then, using the hole in the rod eye as a guide, drill the 3.5-mm (9/64-in.) oil hole.
8. Using a pin-fitting machine or hand reamer (Fig. 18-8), fit the bushings to the pin. Clearance for single-carburetor engines is 0.01 to 0.02 mm (.0004 to .0008 in.). Clearance for dual-carburetor and fuel injection engines is 0.01 to 0.03 mm (.0004 to .0012 in.).

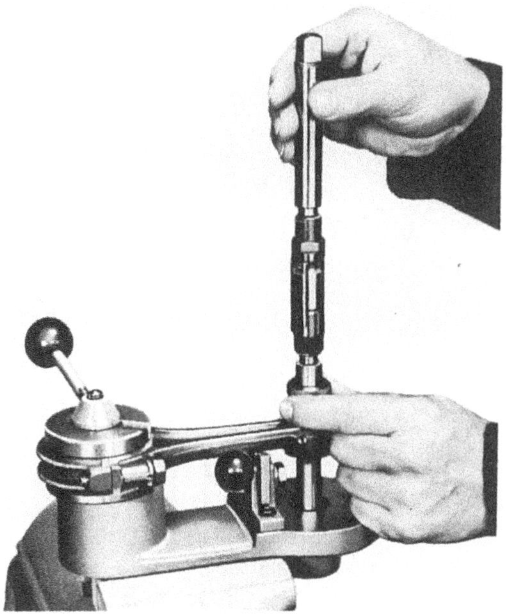

© 1972 VWoA—256

Fig. 18-8. New bushing being reamed. Afterward, using the piston pin, recheck rod alignment.

19. CONVENTIONAL IGNITION SYSTEM
(except 1979 and later California cars)

The standard VW ignition is a conventional coil and battery system. Its components and the wires that connect them are shown in Fig. 19-1. The primary circuit, carrying 12-volt battery current, is indicated by striped lines. The secondary circuit cables, carrying high voltage from the coil to the spark plugs, are shown by solid gray lines. The solid black lines represent the chassis of the car, which carries negative-polarity (ground) current.

19.1 Coil

By the process of electrical induction, the coil steps up the primary circuit's 12-volt current to as much as 20,000 volts in the secondary circuit. This high voltage is required to ionize the air in the spark plug gaps so that a spark can jump across them. Despite the painful electrical shock that spark current can produce, its low amperage (current flow) renders it harmless. When you test the spark, you can avoid getting shocks either by wearing rubber gloves or by holding the secondary cables with a spring-type clothespin. Cables with sound insulation, however, are safe to handle.

When removing and installing the coil, please be sure you connect the wire from the ignition switch to terminal 15 and the wire from the distributor to terminal 1. Reversing them will change the polarity of the secondary current from positive to negative. This reversed polarity limits the ability of the spark voltage to ionize the air in the spark plug gaps, which causes misfiring at high speeds or during hard acceleration.

Fig. 19-1. Ignition system shown schematically.

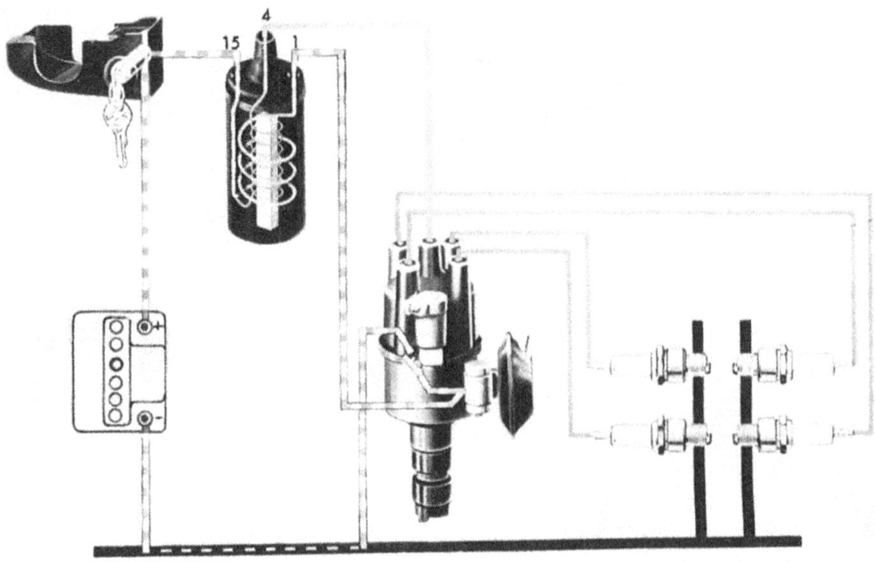

Keep the coil tower clean and dry and make sure that the rubber boot surrounding the high tension cable is tight-fitting and waterproof. Otherwise, spark current may begin arcing from terminal 4 to ground. Such a short can prevent the engine from running well or even from running at all. The arcing may also burn irreparable carbon tracks into the coil tower.

Ignition coils seldom produce trouble. However, if all the other ignition components are sound and the car still misses or is hard to start, check the coil. First test the battery as described in **ELECTRICAL SYSTEM** to make sure it is not run down. Then disconnect the high tension cable from the center terminal of the distributor cap and hold it about 10 mm (3/8 in.) from the engine's crankcase. Have someone run the starter while you observe the spark produced. If the spark is weak and yellowish, fires only when the cable is moved close to the crankcase, or fails to fire at all, the coil is probably weak and should be replaced.

Since faulty wiring between the ignition switch and coil, distributor faults, or a weak condenser can produce similar symptoms, it is best to have the coil tested before investing in a new one. If the coil proves satisfactory, use a voltmeter to check the voltage between terminals 15 and 1. If the voltage is less than 9.6 volts, check for poor distributor point contact, high resistance in the ignition switch, or high resistance in the wire that connects the switch to coil terminal 15.

19.2 Distributor

Servicing the distributor includes checking, replacing, and adjusting the breaker points; checking and possibly replacing the rotor and cap; checking the spark advance mechanism; and adjusting the ignition's timing. The condenser should not be replaced routinely but only when electrical testing proves it defective. The breaker points should be checked anr replaced at the intervals specified in **LUBRICATION AND MAINTENANCE**. Look for the following conditions when you check the breaker points:

1. Moderate pitting and build-up with bright point surfaces can be considered normal wear.
2. Gray point surfaces indicate that the point gap is too narrow, that the spring tension is inadequate, or both. Normal tension is 400 to 600 g (14⅛ to 21⅛ oz.).
3. Blued points indicate a defective coil or condenser.
4. Points fouled by grease, oil, or dirt will display yellow or black markings.

To replace breaker points:

1. Take off the distributor cap and rotor.
2. Remove the screw that holds the fixed contact to the distributor's advance plate, then pull the point wire, shown in Fig. 19-2, off the connector.
3. Lift the breaker point assembly out of the distributor.

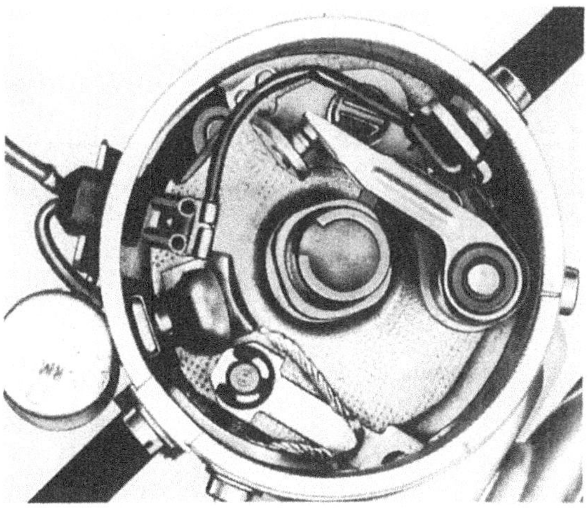

© 1974 VWoA—1167

Fig. 19-2. Breaker points installed in the distributor. The point assembly's wire can be pulled off the connector at the left side of the distributor housing.

To install:

1. Wipe clean the inside of the distributor.
2. Install the new points, making sure the contacts meet squarely. If necessary, align them by bending the fixed contact.
3. Apply one drop of engine oil to the breaker point pivot and a small amount of multipurpose grease to the distributor cam and breaker point rubbing block. Wipe off any excess lubricants so that they will not be thrown onto the point contacts.

Point Adjustments

The breaker points must be adjusted after you have installed them. You can adjust the points either with a feeler gauge or with a dwell meter.

To adjust with feeler gauge:

1. Remove the distributor cap and rotor.
2. Gripping the generator pulley or fan belt, hand-turn the crankshaft until the breaker point rubbing block is on a high point on the distributor cam (points wide open).

NOTE —
If you are doing a complete tune-up, adjust the points while the spark plugs are out. This will make it easier to turn the engine to the correct position.

3. Insert a 0.40-mm (.016-in.) feeler gauge in the point gap and determine whether the points are gapped correctly, too close together, or too far apart. The tolerance range is 0.40 to 0.50 mm (.016 to .020 in.). 5
4. If necessary, loosen the fixed contact mounting screw. Then, using the tip of a screwdriver as shown in Fig. 19-3, move the fixed contact one way or the other until you get the specified breaker point gap.

© 1974 VWoA—1291

Fig. 19-3. Breaker points being adjusted. Screwdriver tip goes between projections on advance plate, with one edge engaged in the notch on the fixed contact mount. Note feeler gauge inserted in gap.

5. Tighten the fixed contact mounting screw, then re-check the gap.
6. Install the distributor rotor and cap.
7. Adjust the timing.

To adjust with dwell meter:

1. If the points have not already been gapped with a feeler gauge, do so to make sure that the engine will start. The adjustment can be approximate.
2. Connect the dwell meter. The black clip should be attached to ground on any bare metal part and the colored clip attached to terminal 1 at the coil.
3. Switch the dwell meter to the tachometer position. Start the engine and run it at 1000 rpm.
4. Move the switch to measure dwell angle. Read the dwell.

   NOTE —
   Adjust new breaker points to a dwell angle of 44° to 50°. Used points can be retained without adjustment so long as they are within 42° to 58° of dwell. Despite wear, further adjustment is unnecessary if the dwell falls within this range.

5. Run the engine at about 2000 rpm. The dwell should not vary more than ±1° from the first reading. Larger deviations indicate a worn distributor shaft.
6. If necessary, remove the distributor cap and rotor and adjust the dwell. To do this, loosen the fixed contact's mounting screw. Then, using the tip of a screwdriver, move the fixed contact either to widen the gap and decrease dwell or to narrow the gap and increase dwell.
7. Fully tighten the fixed contact's mounting screw. You can run the engine with the starter to check the dwell during adjustment but always make the final check with the engine running.
8. Install the distributor rotor and cap. Then adjust the timing.

Adjusting Timing

You should adjust the ignition timing whenever the distributor has been removed and reinstalled or after installing or adjusting the breaker points. A change of only 0.10 mm (.004 in.) in the point gap will alter ignition timing about 3°.

To adjust timing:

1. Following the instrument manufacturer's instructions, install a tachometer and a stroboscopic timing light.
2. On 1968 through 1970 models, disconnect and plug the vacuum hose to the distributor. On other models, leave the hose or hoses attached.
3. Let the engine warm up until the oil temperature is between 30° and 70°C (86° and 158°F).
4. Adjust the idle to the rpm specified in **Table i**.

   NOTE —
   Except on 1968 and 1969 models, the idle must be adjusted by turning the bypass air screw and not by adjusting the throttle valve opening. See **FUEL SYSTEM** or **FUEL INJECTION**.

5. Aim the timing light at the crankshaft pulley. On single-carburetor engines, the timing mark should fall relative to the seam in the crankcase centerline as shown in **Table i**. On dual-carburetor engines the mark should fall relative to the ignition timing scale as shown in the same table. Alternatively, use the setting given on the engine's emission decal.

Table i. Ignition Timing

Model	Ignition setting	Marking
1968 through 1970 (Engine code letter B)	0° before TDC Vacuum hose disconnected and plugged	0° 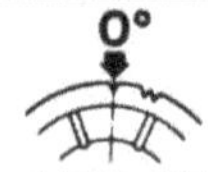 at 800–900 rpm
1971 (Engine code letter AE)	5° after TDC Vacuum hose(s) on	5°  AFTER at 800–900 rpm
1972 (Engine code letter CB)	5° after TDC Vacuum hose(s) on	5°  AFTER at 850–950 rpm
1973 and 1974 dual-carburetor models with manual transmission (Engine code letter CB or AW)	10° after TDC Vacuum hose(s) on	10°  AFTER at 850–950 rpm
1973 through 1975 dual-carburetor and fuel injection models with automatic transmissions (Engine code letter CD, AW, or ED)	5° after TDC Vacuum hose(s) on	5°  AFTER at 900–1000 rpm
1975 fuel injection models with manual transmissions (Engine code letter ED)	5° after TDC Vacuum hose(s) on	5°  AFTER at 850–950 rpm
1976 and later fuel injection models (Engine code letter GD) **NOTE** — See Topic **20.3** for engine code letter GE.	7.5° before TDC Vacuum hose on	7.5°  BEFORE at 850–950 rpm (manual) or 900–1000 rpm (automatic)

©1979 VWoA

NOTE —
On 1971 through 1975 engines with a double vacuum chamber (two hoses to the distributor), the throttle valve(s) must close adequately for accurate timing adjustments. To check the 1971 single-carburetor engine, disconnect the green (retard) hose from the distributor with the engine idling. The timing mark should move 15 to 18 mm ($^{19}/_{32}$ to $^{23}/_{32}$ in.) to the left. If it does not, the throttle valve is not closing adequately and the carburetor must be adjusted.

6. If necessary, adjust the timing by loosening the through bolt in the distributor clamp enough so that you can hand-turn the distributor housing as indicated in Fig. 19-4.

© 1974 VWoA—3602

Fig. 19-4. Distributor being turned (curved double arrow). The single arrow indicates the through bolt.

7. Start the engine. Then hand-turn the distributor clockwise or counterclockwise in the crankcase until the timing marks, illuminated by the timing light, are correctly positioned.

WARNING —
Be careful of the fan belt and pulleys. Do not wear a necktie or necklace. Tie back long hair and do not wear loose-fitting clothing. Serious injury will result if you are caught in the moving machinery.

8. While holding the distributor in the correct position, tighten the through bolt to lock the distributor in place.

Checking Vacuum Advance Cutoff

Except for 1974 models with manual transmissions, all dual-carburetor engines are equipped with an automatic vacuum advance cutoff. The cutoff valve is electrically operated by a switch on the transmission and a temperature switch. The system is shown schematically in Fig. 19-5.

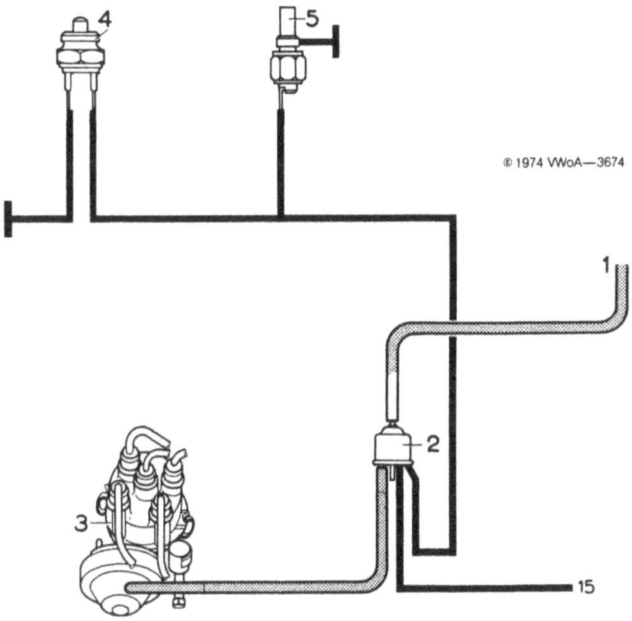

1. Advance vacuum hose to carburetor
2. Two-way cutoff valve
3. Distributor
4. Transmission switch
5. Temperature switch

Fig. 19-5. Vacuum advance cutoff system.

Except in 4th gear, the vacuum advance is cut off whenever the air temperature is above 12°C (54°F). This means that the vacuum advance is only effective in 4th gear and in all other gears at temperatures below 12°C (54°F).

To check cutoff:

1. Following the instrument manufacturer's instructions, install a tachometer and a stroboscopic timing light.
2. Run the engine at 2000 rpm and direct the flashes of the timing light on the timing mark on the crankshaft pulley.
3. Have someone depress the clutch pedal and engage 4th gear—keeping the clutch pedal depressed so that the vehicle does not move. At the moment when 4th gear is engaged, the timing mark should move about 30 mm ($1^{3}/_{16}$ in.) to the left. If not, there is trouble in the vacuum advance cutoff system.

19.3 Disassembling and Assembling Distributor

The distributor is shown disassembled in Fig. 19-6. The individual components shown are those available as replacement parts. If the distributor body assembly must be cleaned, disassemble the distributor as shown. Wash the distributor body assembly thoroughly in clean solvent. Then blow the distributor body assembly dry with compressed air.

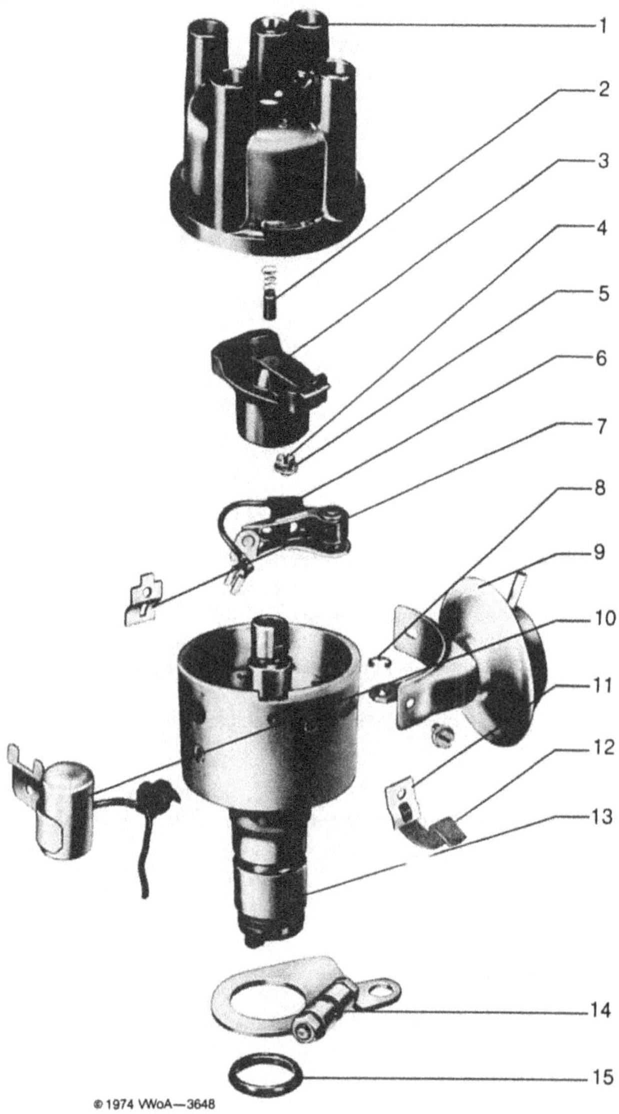

1. Distributor cap
2. Carbon brush and spring
3. Distributor rotor
4. Screw (2)
5. Washer (1)
6. Breaker point assembly
7. Retaining clip mount with cap locating lug
8. E-clip for pull rod
9. Vacuum chamber unit
10. Condenser
11. Retaining clip mount
12. Retaining clip (2)
13. Distributor body assembly
14. Distributor clamp
15. Rubber sealing ring

Fig. 19-6. Distributor disassembled.

CAUTION —
Be careful not to direct a strong blast of air into the housing at close range. Doing so could damage the calibrated springs in the centrifugal advance mechanism.

Thoroughly lubricate the shaft and bearings with engine oil by applying the oil to the space between the drive dog and the distributor body. Hand-turn the shaft with the distributor body inverted until the oil has worked its way all along the shaft. The shaft should turn smoothly without binding. Apply one drop of motor oil only to each pivot point in the centrifugal advance mechanism and to the breaker plate bearing. If necessary, you can remove the breaker plate to gain access to the centrifugal advance.

Checking Spark Advance Mechanism

The distributors used on 1968 through 1970 models have a vacuum advance mechanism only. The 1971 and later models have both vacuum and centrifugal advance mechanisms, with vacuum retard at idle. Special timing lights are available that incorporate a built-in meter you can use to measure spark advance. The instructions supplied tell you how to use the device to check the centrifugal advance mechanism. You can also check the vacuum advance mechanism by installing a vacuum gauge with a T-fitting in the vacuum hose(s) between the carburetor and the distributor.

Table j lists the rpm and vacuum levels where critical changes in the spark advance curves take place. Comparing actual distributor operation to the data in the table will help you locate dirty, binding advance mechanisms or mechanical faults in the distributor. The distributor number will be found stamped on the distributor itself.

If testing reveals irregularities in the spark advance curve, first clean and lubricate the moving parts of the distributor. If the vacuum advance still fails to conform to specifications, install a new vacuum chamber unit. However, on the early distributor with a vacuum advance only, the advance curve can be adjusted by turning the eccentric pin onto which the breaker plate return spring hooks. Any discrepancies revealed in retesting the centrifugal advance curve indicate worn internal parts. In such cases, the distributor body and internal parts should be replaced as a unit.

Even if you do not have the equipment for checking the spark advance curve, you can still quick-check the spark advance mechanisms. To check the vacuum advance, hand-turn the breaker plate counterclockwise. It should move without grittiness and spring back solidly to its original position when released. To check the diaphragm for leaks, turn the breaker plate as far as it will go counterclockwise, take off the vacuum hose(s), and cover the hose connection(s) with your finger(s). Then release

Table j. Spark Advance Curves with Distributor Installed

Centrifugal timing range							
Distributor	**Begin**					**End**	
	rpm	**rpm**	**degrees**	**rpm**	**degrees**	**rpm**	**degrees**
113 905 205 M	—	—	—	—	—	—	—
113 905 205 T	—	—	—	—	—	—	—
111 905 205 AA	—	—	—	—	—	—	—
211 905 205 Q	1000–1250	1500	6–12°	—	—	3800	22–25°
021 905 205 E	1050–1250	—	—	2000	13–16°	3600	23–26°
021 905 205 J	1000–1200	1600	9–14°	—	—	3400	21–25°
021 905 205 F	1000–1200	1700	14–19°	2400	15–19°	3400	21–25°
021 905 205 N	1000–1200	1600	9–14°	—	—	3400	21–25°
021 905 205 S	1000–1200	1600	9–14°	—	—	3400	21–25°
021 905 205 AB 022 905 205 AB	1000–1225	—	—	2000	11–17°	3600	21–25°
021 905 205 AC 022 905 205 AC	1000–1250	—	—	2000	11–17°	3600	21–25°
021 905 205 P 022 905 205 S	1000–1250	1600	8–13°	—	—	3400	20.5–24.5°

Vacuum timing range						
Distributor	**Begin**			**End**		**Timing direction**
	mm Hg (in. Hg)	**mm Hg (in. Hg)**	**degrees**	**mm Hg (in. Hg)**	**degrees**	
113 905 205 M	3–7 (.12–.27)	30 (1.2)	8.5–9.5°	74 (2.9)	16–17.5°	advance
113 905 205 T	3–7 (.12–27)	30 (1.2)	8.5–9.5°	74 (2.9)	16–17.5°	advance
111 905 205 AA	3–7 (.12–.27)	30 (1.2)	8.5–9.5°	74 (2.9)	16–17.5°	advance
211 905 205 Q	110–160 (4.3–6.3)	—	—	150–170 (5.9–6.7)	2–5°	advance
	80–130 (3.2–5.1)	—	—	160–230 (6.3–9.1)	11–13°	retard
021 905 205 E	80–110 (3.2–4.3)	—	—	200 (7.9)	8–12°	advance
	80–110 (3.2–4.3)	—	—	150–170 (5.9–6.7)	11–15°	retard
021 905 205 J	100–140 (3.9–5.5)	—	—	180 (7.1)	7–12°	advance
	70–120 (2.8–4.7)	—	—	170–210 (6.7–8.2)	16–18°	retard
021 905 205 F	80–110 (3.2–4.3)	—	—	200–210 (7.9–8.2)	8–12°	advance
	80–110 (3.2–4.3)	—	—	140–170 (5.5–6.7)	11–14°	retard
021 905 205 N	100–140 (3.9–5.5)	—	—	180 (7.1)	7–12°	advance
	70–120 (2.8–4.7)	—	—	170–210 (6.7–8.2)	16–18°	retard
021 905 205 S	90–120 (3.5–4.7)	—	—	200 (7.9)	9–12°	advance
	45–90 (1.8–3.5)	—	—	142 (5.6)	11–13°	retard
021 905 205 AB 022 905 205 AB	70–130 (2.8–5.1)	—	—	180–200 (7.1–7.9)	8–11°	advance
	50–120 (2.0–4.7)	—	—	160–200 (6.3–7.9)	11–13°	retard
021 905 205 AC 022 905 205 AC	70–130 (2.8–5.1)	—	—	140–160 (5.5–6.3)	3–7°	advance
	50–120 (2.0–4.7)	—	—	160–210 (6.3–8.2)	11–13°	retard
021 905 205 P 022 905 205 S	80–130 (3.2–5.1)	—	—	200 (7.9)	8.5–11°	advance

the breaker plate. The vacuum in the chamber should keep the breaker plate from returning fully to its original position until you uncover the hose connection(s). If it returns fully, there is a leak in the diaphragm or the chamber.

To quick-check the centrifugal advance, hand-turn the distributor rotor clockwise. When you release it, the rotor should return automatically to its original position. If it does not, either the mechanism is dirty or the distributor body and internal parts are faulty and should be replaced as a unit.

19.4 Secondary Circuit

The secondary circuit of the ignition system consists of the distributor rotor, the distributor cap, the spark plug cables, and the high tension cable that links the coil and the distributor cap. Secondary circuit resistance, for radio suppression and for plug gap erosion reduction, is built into the distributor rotor and spark plug connectors. This resistance allows the use of metallic conductor ignition cables.

You can check the rotor resistor with an ohmmeter as shown in Fig. 19-7. You can test individual plug connectors the same way. Rotor resistance should not exceed 10,000 ohms; connector resistance should not exceed 5000 to 10,000 ohms.

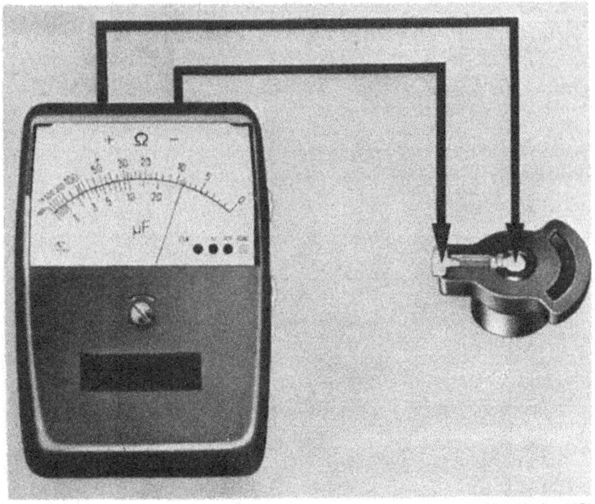

© 1974 VWoA—1171

Fig. 19-7. Rotor resistance being checked.

Keep the distributor rotor and cap clean and dry to prevent spark flashover, which could cause irreparable carbon tracks. Also keep the ignition cables clean and replace them when the insulation shows signs of cracking or deterioration. The rubber boots at the coil tower, distributor cap towers, and plug connectors must be sound and tight-fitting if the secondary circuit is to remain waterproof.

19.5 Spark Plugs

Spark plugs must be of the correct heat range and physical dimensions. The plug reach (length of the threaded portion of the shell) should be 12 mm (.472 in.) for single-carburetor engines or ¾ in. for dual-carburetor engines. The threaded diameter for both engines is 14 mm. Suitable spark plugs are listed in **20. Engine and Clutch Technical Data.**

Replace the spark plugs at the mileages specified in **LUBRICATION AND MAINTENANCE**. After these distances, the electrodes will have worn to a degree that will inhibit ionization of the spark plug gap. Adjust the gaps of new and, if necessary, used spark plugs as shown in Fig. 19-8. The color of the deposits on a used plug's firing tip may vary, indicating the following conditions:

1. Gray or light tan deposits indicate good combustion, proper fuel mixtures, and consistently good spark plug performance.
2. Light gray or chalky-white deposits indicate too lean a fuel/air mixture or an overheated spark plug.
3. Soft, fluffy, black deposits indicate too rich a fuel/air mixture or a plug that is misfiring.
4. A spark plug fouled by oil indicates severe oil leakage past the piston rings or valve guides, or a spark plug that is no longer firing.

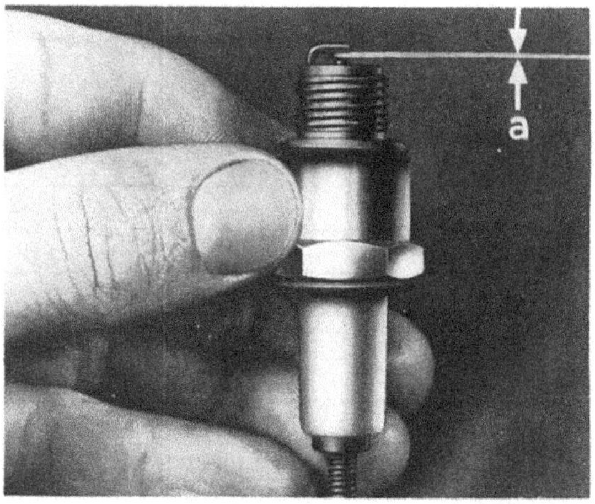

© 1974 VWoA—1172

Fig. 19-8. Spark plug gap being set. Dimension **a** should be 0.60 mm (.024 in.).

You can clean spark plugs with a sharp stick or plastic rod or in a sandblast machine. Never sandblast an oil-fouled plug until after the oily deposit has been removed with solvent. After sandblasting, always file the spark plug electrodes to the same sharp, square profiles of a new plug. This is also an advisable step when regapping used plugs that have not been sandblasted. Never use a wire brush to clean spark plugs. The bristles will leave electrically conductive "pencil marks" on the insulator nose.

You can safely clean the spark plug threads with a thread-cutting die. The spark plug hole in the cylinder head can be cleaned with a thread chaser made especially for the task. Do not clean the plug threads with a power-driven wire brush. This can deform the threads so that they will damage the cylinder head.

Another important precaution during spark plug installation is to coat the threads lightly with a good quality anti-seize compound. This coating will prevent corrosion damage to the soft aluminum threads in the cylinder head. If nothing else is available, lubricate the spark plug threads with multipurpose grease or even with motor oil. The important thing is never to install the plugs dry.

Anti-seize compounds usually contain graphite, which must not come into contact with plug electrodes or insulator. The graphite's metallic composition could short out the spark plug. Therefore, coat the threads with your fingertip rather than dipping the plug into the compound.

Start the plug in the cylinder head with your fingertips or by hand-holding the spark plug socket. This will give improved feel and help prevent accidental cross-threading. Turn the spark plug in until you feel the gasket contact the cylinder head. Then install a torque wrench on the spark plug socket and torque the plug to 3.5 (25 ft. lb.) in single-carburetor engines or to 3.0 mkg (22 ft. lb.) in dual-carburetor engines.

20. BREAKERLESS TRANSISTORIZED IGNITION SYSTEM

(1979 and later California cars)

Fig. 20-1 is a schematic diagram of the breakerless transistorized ignition system used on 1979 and later California vehicles. The distributor in this system has no breaker points. Instead, it is equipped with a Hall electromagnetic sending unit. The signals generated by this unit are sensed by the transistorized ignition control unit and amplified. The 12-volt primary current for the special ignition coil is switched on and off electronically by the switching transistor in the ignition control unit. Because the primary current is switched electronically instead of mechanically, there are no breaker points to be adjusted, replaced, or given attention during routine maintenance.

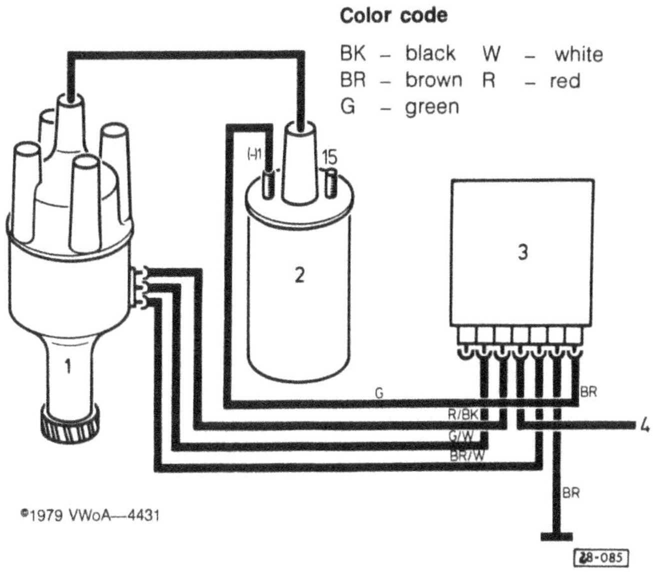

1. Distributor with Hall sending unit
2. Special ignition coil
3. Ignition control unit
4. Wire to ignition switch (terminal 15)

Fig. 20-1. Breakerless transistorized ignition system shown schematically.

To prevent damage to the breakerless transistorized ignition system, certain cautions should be observed in troubleshooting, maintenance, and repair work.

5

CAUTION —

- *Do not disconnect and connect ignition system wires or ignition test and analysis equipment while the ignition is switched on. This can damage the solid-state components of the ignition control unit.*
- *When operating the starter without starting the engine (as you would in making a compression test), disconnect the high tension wire from the distributor cap and ground it to keep the engine from starting. Cranking the engine with the primary wires disconnected can damage the ignition system.*
- *Do not replace the special ignition coil with a standard ignition coil. Doing this will create an electrical imbalance that will damage the system and cause poor running.*
- *Do not connect a condenser to terminal 1, the coil's negative (–) terminal. This can cause misfiring and damage to the system.*
- *Do not connect a quick-charge battery charger for boost starting for longer than 1 minute. Long operation with excessive current will damage electronic components of the ignition.*

20.1 Special Coil

The basic description of coil operation, installation, and maintenance given in **19.1 Coil** applies equally to the special coil installed on vehicles with breakerless transistorized ignitions. However, under no circumstances should a coil other than VW Part No. 211 905 115 C be used with the breakerless transistorized system.

Though the kinds of ohmmeters commonly used in automotive electrical testing lack the sensitivity required, you can test the special coil with sensitive ohmmeters such as those used in radio and TV repair work. If the measured resistances are outside the ranges given in Fig. 20-2 and Fig. 20-3, the special coil should be replaced with a new special coil (Part No. 211 905 115 C).

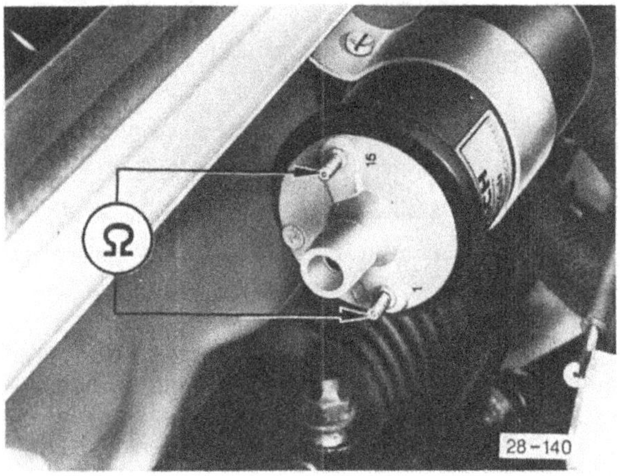

©1979 VWoA—4432

Fig. 20-2. Primary coil resistance being checked with ohmmeter (shown schematically) connected to terminals 1 and 15. The correct resistance is 0.55 to 0.75 ohms.

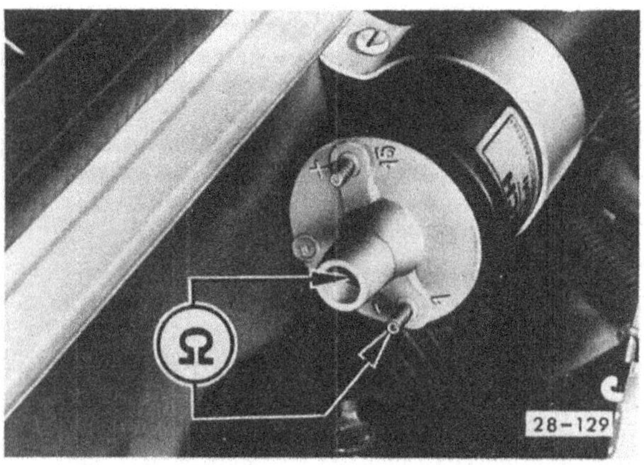

©1979 VWoA—4433

Fig. 20-3. Secondary coil resistance being checked with ohmmeter (shown schematically) connected to terminals 1 and 4. The correct resistance is 3000 to 5000 ohms.

20.2 Ignition Control Unit

The ignition control unit is a transistorized device with two basic functions. First, it receives and amplifies the triggering pulses generated by the distributor's Hall sending unit. Second, it switches on and off the 12-volt primary current by means of a switching transistor that is actuated by the amplified signals from the distributor.

To test ignition control unit:

1. Make sure that the ignition is turned off. Then disconnect the high tension cable from the center of the distributor cap and ground it on some clean, unpainted, metal part of the engine.
2. Disconnect the 3-point plug for the control unit wires from the distributor. Then attach a voltmeter's positive lead to terminal 1 (–) at the coil, without disconnecting either primary wire from the coil. Ground the voltmeter's negative lead on some clean, unpainted, metal part of the engine.
3. Switch on the ignition. The voltmeter should indicate at least 12 volts at the coil.

> ***CAUTION —***
>
> *If the voltage is considerably below 12 volts, switch off the ignition immediately to avoid additional damage to the system. Then replace the ignition control unit. If there is no voltage at all, test the special coil's primary resistance and also make sure that battery voltage is reaching coil terminal 15(+).*

4. With the ignition off, use a jumper wire to ground the central (green/white) wire's terminal at the plug that you disconnected from the distributor. Keep the voltmeter attached to coil terminal 1 (–).
5. Switch on the ignition. The voltmeter should indicate at least 12 volts at terminal 1 (–). Then, keeping the ignition turned on for this test only, disconnect the green/white wire from ground. The voltage at terminal 1 (–) should drop briefly to approximately 6 volts. Replace the ignition control unit if the voltage is incorrect in either of these tests.

20.3 Breakerless Distributor

The breakerless distributor introduced on the 1978 California models is similar in construction to the distributors with breaker points. The main difference is the installation of the Hall sending unit in place of the breaker points used in conventional distributors.

The Hall sending unit cannot be tested with common workshop equipment. So if the sending unit is suspected of a malfunction, replace it. Then check and, if necessary, adjust the ignition timing as described under the next heading.

Adjusting Timing

You should adjust the ignition timing whenever the distributor has been removed and reinstalled, or after installing a replacement Hall sending unit. Incorrect timing can cause decreased fuel economy and excessive exhaust emissions.

To adjust timing:

1. Connect a tachometer as indicated in Fig. 20-4. Then, following the instrument manufacturer's instructions, install a stroboscopic timing light.

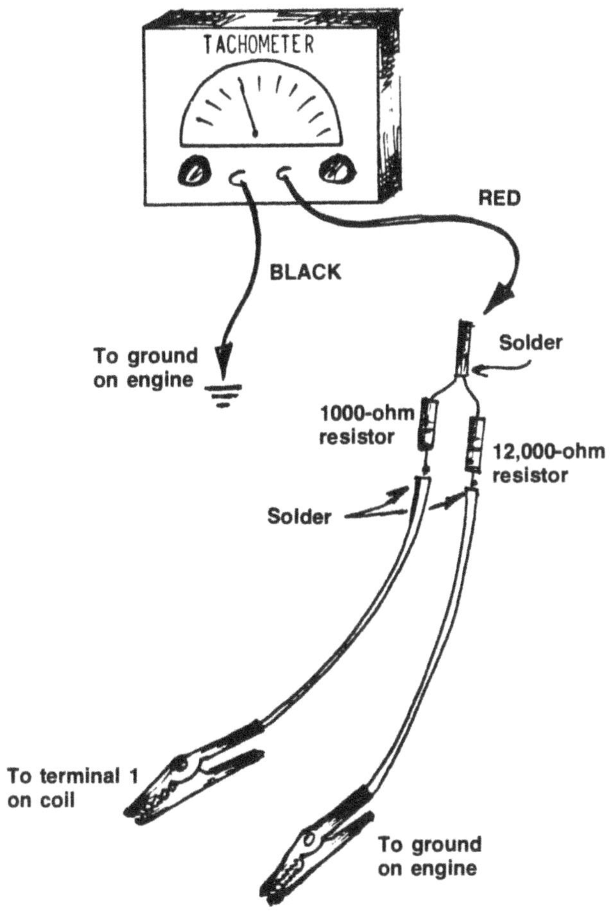

Fig. 20-4. Tachometer correctly attached to breakerless distributor. 1000-ohm and 12,000-ohm resistors can be obtained from radio supply store. Use 3-mm² (or 12-gauge) wire, and insulate resistors and connections with electrical tape.

2. Start the engine and allow it to warm up until the oil temperature is between 30° and 70°C (86° and 158°F). If necessary, adjust the idle to 800 to 950 rpm for vehicles with manual transmissions or to 850 to 1000 rpm for vehicles with automatic transmissions. See **FUEL INJECTION** for the correct idle speed adjustment procedure.

3. Without disconnecting the vacuum hoses from the ignition distributor, aim the timing light at the crankshaft pulley. The correct ignition timing is 5° ATDC (after top dead center), as indicated in Fig. 20-5.

©1979 VWoA—4430

Fig. 20-5. Timing marks as they should appear when illuminated by stroboscopic timing light with engine running. Arrow indicates V-notch on periphery of crankshaft pulley aligned with 5° dot on scale.

5

4. If necessary, adjust the timing by loosening the through bolt in the distributor clamp enough so that you can hand-turn the distributor housing as indicated in Fig. 20-6.

© 1974 VWoA—3602

Fig. 20-6. Distributor being turned (curved double arrow). The single arrow indicates the through bolt.

5. With the engine running, hand-turn the distributor clockwise or counterclockwise in the crankcase until the timing marks, illuminated by the timing light, are correctly positioned. Then, while holding the distributor in the correct position, tighten the through bolt to lock the distributor in place.

Checking Spark Advance Mechanism

The breakerless distributor has both a centrifugal spark advance mechanism and a vacuum spark control that incorporates a vacuum retard and a vacuum advance. **Table k** lists the rpm and vacuum levels where critical changes in the spark advance curves take place. Comparing actual distributor operation to the data in the table will help you locate dirty, binding advance mechanisms or mechanical faults in the distributor. The distributor number will be found stamped on the distributor itself.

If testing reveals irregularities in the spark advance curve, first clean and lubricate the moving parts of the distributor. If the vacuum advance still fails to conform to specifications, install a new vacuum chamber unit. Any discrepancies revealed in retesting the centrifugal advance curve indicate worn internal parts. In these cases, the distributor body and internal parts should be replaced as a unit.

Even if you do not have the equipment for checking the spark advance curve, you can still quick-check the spark advance mechanisms. To check the vacuum advance, hand-turn the sending unit plate counterclockwise. It should move without grittiness and spring back solidly to its original position when released. To check the diaphragm for leaks, turn the sending unit plate as far as it will go counterclockwise, take off the vacuum hoses, and cover the hose connections with your fingers. Then release the plate. The vacuum in the chamber should keep the sending unit plate from returning fully to its original position until you uncover the hose connections. If it returns fully, there is a leak in the diaphragm or the chamber.

To quick-check the centrifugal advance, hand-turn the distributor rotor clockwise. When you release it, the rotor should return automatically to its original position. If it does not, either the mechanism is dirty or the distributor body and internal parts are faulty and should be replaced as a unit.

20.4. Spark Plugs and Secondary Circuit

The spark plugs and the secondary circuit used with the breakerless transistorized ignition are unchanged from those of vehicles with conventional ignition systems. Please consult **19.4 Secondary Circuit** or **19.5 Spark Plugs.**

Table k. Spark Advance Curves with Breakerless Distributor Installed

Centrifugal timing range							
Distributor	**Begin**					**End**	
	rpm	**rpm**	**degrees**	**rpm**	**degrees**	**rpm**	**degrees**
039 905 205 C	1050–1250	—	—	2800	12–16°	3600	21–25°

	Vacuum timing range				
Distributor	**Begin**		**End**		**Timing direction**
	mm Hg (In. Hg)	**mm Hg (In. Hg)**	**degrees**	**mm Hg (In. Hg)**	**degrees**	
039 905 205 C	60–120 (2.4–4.7)	—	—	240 (9.4)	11–13°	advance
	60–120 (2.4–4.7)	—	—	240 (9.4)	11–13°	retard

21. ENGINE AND CLUTCH TECHNICAL DATA

I. Basic Tune-up Specifications (fuel injection engine)

Idle speed (through 1978)	Manual transmission: 900 rpm ± 50 Automatic transmission: 950 rpm ± 50
Idle speed (1979 and later)	Manual transmission: 875 rpm ± 75 Automatic transmission: 925 rpm ± 75
Oil capacity	Oil change with filter change: 7.4 U.S. pints (6.1 Imperial pints, 3.5 liters) Oil change without filter change: 6.3 U.S. pints (5.3 Imperial pints, 3.0 liters)
Ignition	12 volt
Ignition timing (1975 cars and 1979 and later Calif. cars through vin 2262077583)	5° after TDC with vacuum hoses connected
Ignition timing (1976 and later — except 1979 and later Calif. cars from vin 2262077584)	7.5° before TDC with vacuum hoses connected
Firing order	1-4-3-2
Cylinder location	No. 1: front cyl. on right side of car No. 2: rear cyl. on right side of car No. 3: front cyl. on left side of car No. 4: rear cyl. on left side of car
Point gap (where applicable)	0.40 to 0.50 mm (.016 to .020 in.)
Dwell angle (where applicable)	New points: 44° to 50° Used points: 42° to 58°
Spark plug gap	0.70 mm (.028 in.)
Spark plugs	Champion N-288 Bosch W 145 M2 or W 8CO Beru 145/14/3L or plugs with similar specifications from other manufacturers
CO volume % at idle	1975, 1976 (through vin 2262077583) 0.2% to 2.0% 1976 (from vin 2262077584) 0.9% to 1.1% 1978, 1979 (except Calif.) 0.5% to 1.5% 1979 Calif. (with oxygen sensor disconnected) 0.2% to 1.2%

II. Basic Tune-up Specifications (single-carburetor engine)

Idle speed	850 rpm ± 50
Oil capacity	5.3 U.S. pints (4.4 Imperial pints, 2.5 liters)
Ignition	12 volt
Ignition timing	Through July 1970: 0° before TDC with vacuum hose disconnected and plugged From August 1970: 5° after TDC with vacuum hose(s) connected
Firing order	1-4-3-2
Cylinder location	No. 1: front cyl. on right side of car No. 2: rear cyl. on right side of car No. 3: front cyl. on left side of car No. 4: rear cyl. on left side of car
Point gap	0.40 to 0.50 mm (.016 to .020 in.)
Dwell angle	New points: 44° to 50° Used points: 42° to 58°
Spark plug gap	0.60 mm (.024 in.)
Spark plugs	Normal service: Champion L-88-A Bosch W 145 T1 Beru 145/14 High speed service: Champion L-85 or L-86 Bosch W175 T1 Beru 175/14 or plugs with similar specifications from other manufacturers
CO volume % at idle	34 PICT-3 carburetor: 3% ± 1%

5

III. Basic Tune-up Specifications (dual-carburetor engine)

Idle speed	Manual transmission: 850 rpm ± 50 Automatic transmission 950 rpm ± 50
Oil capacity	Oil change with filter change: 7.4 U.S. pints (6.1 Imperial pints, 3.5 liters) Oil change without filter change: 6.3 U.S. pints (5.3 Imperial pints, 3.0 liters)
Ignition	12 volt
Ignition timing	From August 1971 through July 1972: 5° after TDC with vacuum hoses connected Manual transmission from August 1972: 10° after TDC with vacuum hoses connected Automatic transmission from August 1972: 5° after TDC with vacuum hoses connected
Firing order	1-4-3-2
Cylinder location	No. 1: front cyl. on right side of car No. 2: rear cyl. on right side of car No. 3: front cyl. on left side of car No. 4: rear cyl. on left side of car
Point gap	0.40 to 0.50 mm (.016 to .020 in.)
Dwell angle	New points: 44° to 50° Used points: 42° to 58°
Spark plug gap	0.60 mm (.024 in.)
Spark plugs	Champion N-88 Bosch W 145 T2 Beru 145/14/3 or plugs with similar specifications from other manufacturers
CO volume % at idle	3% ± 1% through July 1972 2% ± 1% from August 1972 through December 1972 1% ± 0.5% from January 1973

IV. Tolerances, Wear Limits, and Settings (single-carburetor engine)

Designation	New parts on installation mm (in.)	Wear limit mm (in.)
A. Crankcase bores		
1. Bores for main bearings		
a) Bearings 1, 2, and 3 .... diameter	65.00-65.02 (2.5590-2.5598)	65.03 (2.5602)
b) Bearing 4 .... diameter	50.00-50.03 (1.9685-1.9696)	50.04 (1.9700)
2. Bore for oil seal/flywheel end .... diameter	90.00-90.05 (3.5433-3.5452)	—
3. Bores for camshaft bearings .... diameter	27.50-27.52 (1.0826-1.0834)	—
4. Bore for oil pump housing .... diameter	70.00-70.03 (2.7559-2.7570)	—
5. Bores for cam followers .... diameter	19.00-19.02 (.7480-.7488)	19.05 (.7500)
B. Crankshaft		
1. Journal dimensions		
a) Main journals 1, 2, and 3 .... diameter	54.97-54.99 (2.1641-2.1649)	—
b) Main journal 4 .... diameter	39.98-40.00 (1.5740-1.5748)	—
c) Connecting rod journals .... diameter	54.98-55.00 (2.1645-2.1653)	—
d) Three undersizes of 0.25 mm (.010 in.) each		
2. Crankshaft at Nos. 2 and 4 main journals (Nos. 1 and 3 journals on V-blocks) .... runout	—	0.02 (.0008)
3. Crankshaft .... imbalance	max. 12 cmg	—
4. Main bearing journal .... out of round	—	0.03 (.0012)
5. Connecting rod journal .... out of round	—	0.03 (.0012)
6. Crankshaft/main journals (taking housing preload into account):		
a) Bearings 1 and 3 .... radial play	0.04-0.10 (.0016-.004)	0.18 (.007)
b) Steel bearing 2 .... radial play	0.03-0.09 (.001-.003)	0.17 (.0067)
c) Bearing 4 .... radial play	0.05-0.10 (.002-.004)	0.19 (.0075)
7. Crankshaft/main journal 1 .... end play	0.07-0.13 (.0028-.005)	0.15 (.006)
8. Connecting rod journal/connecting rod .... radial play	0.02-0.07 (.0008-.0027)	0.15 (.006)
end play	0.10-0.40 (.004-.016)	0.70 (.027)
C. Connecting rods		
1. Weight difference between connecting rods in one engine	Max. 5 g	Max. 10 g
Weight of replacement connecting rods		
− Weight (white) brown/white	580-588 g	—
+ Weight (black) gray/black	592-600 g	—
2. Piston pin .... diameter	21.996-22.00 (.8660-.8661)	—
3. Piston pin bush .... diameter	22.009-22.017 (.8664-.8667)	—
4. Piston pin/pin bush .... radial play	0.01-0.02 (.0004-.0008)	0.04 (.0016)
D. Camshaft		
1. Bearings 1, 2, and 3 .... inside diameter	24.99-25.00 (.9838-.9842)	—
2. Measured at center bearing (bearings 1 and 3 on V-blocks) .... runout	0.02 (.0008)	0.04 (.0016)
3. Camshaft/camshaft bearings (taking housing preload into account) .... radial play	0.02-0.05 (.0008-.002)	0.12 (.0047)
Thrust bearing .... end play	0.04-0.13 (.0016-.0051)	0.16 (.0063)
4. Camshaft gear .... backlash	0.00-0.05 (.000-.002)	—
5. Cam follower .... diameter	18.96-18.98 (.7464-.7472)	18.93 (.7452)
6. Bore/cam follower .... radial play	0.02-0.06 (.0008-.0024)	0.12 (.0047)
7. Pushrod .... runout	max. 0.30 (.012)	—
E. Lubrication		
1. Oil pressure (for SAE 30 only) at an oil temperature of 70°C (158°F) and 2500 rpm	approx. 3 kg/cm² (42 psi)	2 kg/cm² (28 psi)
2. Spring for pressure relief valve through July 1969. Length loaded 43.20 (1 11/16 in.) .... load	2.5-4.2 kg (5.5-9.2 lb.)	—
3. Spring for pressure relief valve from August 1969. Length loaded 44.10 mm (1¾ in.) .... load	5.6-7.3 kg (12.3-16.0 lb.)	—
4. Spring for oil pressure control valve from August 1969 through July 1970. Length loaded 20.20 mm (13/16 in.) .... load	2.9-3.6 kg (6.3-7.9 lb.)	—
5. Spring for oil pressure control valve from Aug. 1970. Length loaded 20.20 mm (13/16 in.) .... load	3.1-3.8 kg (6.8-8.4 lb.)	—
6. Oil pressure switch open at .... pressure	0.15-0.45 kg/cm² (2.1-6.4 psi)	—
F. Flywheel		
1. Flywheel (measured at center of friction surface) .... lateral runout	max. 0.30 (.012)	—
imbalance	max. 20 cmg	—

IV. Tolerances, Wear Limits, and Settings (single-carburetor engine) (continued)

Designation	New Parts on Installation mm (in.)	Wear Limit mm (in.)
2. Shoulder for oil seal . . . . . . outside dia.	69.9-70.1 (2.752-2.756)	—
3. Drive plate, torque converter . . . . . . imbalance	max. 5 cmg	—
G. Cylinders and pistons		
Two oversizes of 0.5 mm (.020 in.) each		
1. Cylinders . . . . . . out of round	max. 0.01 (.0004)	—
2. Cylinder/piston . . . . . . clearance	0.04-0.06 (.0016-.0023)	0.20 (.008)
3. a) Upper piston ring . . . . . . side clearance	0.07-0.10 (.0027-.0039)	0.12 (.0047)
b) Lower piston ring . . . . . . side clearance	0.05-0.07 (.0020-.0027)	0.10 (.004)
4. Oil scraper ring . . . . . . side clearance	0.03-0.05 (.0012-.0020)	0.10 (.004)
5. a) Upper piston ring . . . . . . gap	0.30-0.45 (.012-.018)	0.90 (.035)
b) Lower piston ring . . . . . . gap	0.30-0.45 (.012-.018)	0.90 (.035)
6. Oil scraper ring . . . . . . gap	0.25-0.40 (.010-.016)	0.95 (.037)
7. Piston weight		
− Weight (brown)	398-410 g	—
+ Weight (gray)	406-418 g	—
− Weight (brown)	402-412 g	—
+ Weight (gray)	410-420 g	—
8. Weight difference between pistons in one engine	max. 5 g	max. 10 g
H. Cylinder head and valves		
1. Cylinder seating depth in cylinder head	13.45-13.55 (.5295-.5334)	—
a) Rocker arm . . . . . . inside dia.	18.00-18.02 (.7086-.7094)	18.04 (.7102)
b) Rocker shaft . . . . . . diameter	17.97-17.99 (.7074-.7082)	17.95 (.7067)
2. Valve springs		
Length loaded 31.0 mm (1⁷⁄₃₂ in.) . . . . . . load	53.2-61.2 kg (117-135 lb.)	—
3. Valve seats		
a) Intake . . . . . . width	1.30-1.60 (.051-.063)	—
b) Exhaust . . . . . . width	1.70-2.00 (.067-.078)	—
c) Intake . . . . . . seat angle	45°	—
d) Exhaust . . . . . . seat angle	45°	—
e) Outer correction angle	15°	—
f) Inner correction angle	75°	—
4. Valve guides		
Intake . . . . . . inside dia.	8.00-8.02 (.3150-.3157)	8.06 (.3173)
Exhaust . . . . . . inside dia.	8.00-8.02 (.3150-.3157)	8.06 (.3173)
5. Valve stem		
Intake . . . . . . diameter	7.94-7.95 (.3126-.3129)	
Exhaust (through 1970) . . . . . . diameter	7.91-7.92 (.3114-.3118)	7.90 (.3110)
Exhaust (from 1971) . . . . . . diameter	7.92-7.94 (.3118-.3126)	7.88 (.3102)
out of round	0.01 (.0004)	—
6. Valve head		
Intake . . . . . . diameter	35.6 (1.40)	—
Exhaust . . . . . . diameter	32.1 (1.26)	—
7. Valve facing		
Intake . . . . . . angle	44°	—
Exhaust . . . . . . angle	45°	—
8. Valve guide/valve stem		
Intake and exhaust . . . . . . rock	0.23-0.27 (.009-.010)	0.80 (.031)
9. Valve clearance (cold)		
Intake . . . . . . setting	0.15 (.006)	—
Exhaust . . . . . . setting	0.15 (.006)	—
10. Compression pressure	8.0-10.0 kg/cm² (114-142 psi)	7 kg/cm² (100 psi)
Pressure difference between cylinders	max. 2 kg/cm² (28 psi)	—
I. Cooling		
1. Thermostat . . . . . . opening temperature	65-70°C (149-158°F)	—
2. Fan/belt pulley . . . . . . out of balance	max. 5 cmg	—
J. Clutch		
1. Total pressure	380-420 kg (837-925 lb.)	—
2. Complete clutch . . . . . . imbalance	max. 15 cmg	—
3. Pressure plate . . . . . . runout	—	0.10 (.004)
4. Clutch plate [measured at 195 mm (7.677 in.) dia.] . . . . . . lateral runout	max. 0.50 (.020)	—
5. Play at clutch pedal	10-20 (⅜ to ¾)	—

5

V. Tolerances, Wear Limits, and Settings (dual-carburetor and fuel injection engines)

Designation	New part mm (in.)	Wear limit mm (in.)
A. Crankcase bores		
1. Bores for main bearings		
a. Bearings 1, 2, and 3 ... diameter	70.00-70.02 (2.7559-2.7567)	70.03 (2.7570)
b. Bearing 4 ... diameter	50.00-50.03 (1.9685-1.9696)	50.04 (1.9700)
2. Bore for oil seal/flywheel end ... diameter	95.00-95.05 (3.7401-3.7421)	—
3. Bore for oil seal/fan end ... diameter	62.00-62.05 (2.4409-2.4429)	—
4. Bores for camshaft bearings ... diameter	27.50-27.52 (1.0826-1.0834)	—
5. Bore for oil pump housing ... diameter	70.00-70.03 (2.7559-2.7570)	—
6. Bores for cam followers ... diameter	24.00-24.02 (.9448-.9456)	24.05 (.9468)
B. Crankshaft		
3 undersizes of 0.25 mm (.010 in.) each		
1. Crankshaft		
a. Main journals 1, 2, and 3 ... diameter	59.97-59.99 (2.3610-2.3618)	—
b. Main journal 4 ... diameter	39.98-40.00 (1.5740-1.5748)	—
c. Connecting rod journals (1972-1975) ... diameter	54.98-55.00 (2.1645-2.1653)	—
d. Connecting rod journals (1976) ... diameter	49.98-50.00 (1.9677-1.9685)	
2. Crankshaft at No. 2 and No. 4 main journals (No. 1 and No. 3 journals on V blocks) ... runout	—	0.02 (.0008)
3. Crankshaft ... imbalance	max. 12 cmg	—
4. Main bearing journal ... out-of-round	—	0.03 (.0012)
5. Connecting rod journal ... out-of-round	—	0.03 (.0012)
6. Crankshaft/main journals (taking housing preload into account):		
a. Bearings 1 and 3 ... radial play	0.05-0.10 (.002-.004)	0.18 (.007)
b. Steel bearing 2 ... radial play	0.03-0.09 (.0012-.0035)	0.17 (.0067)
c. Bearing 4 ... radial play	0.05-0.10 (.0019-.004)	0.19 (.0074)
7. Crankshaft/main journal 1 ... end play	0.07-0.13 (.0027-.005)	0.15 (.006)
8. Connecting rod journal/connecting rod ... radial play	0.02-0.07 (.0008-.0027)	0.15 (.006)
end play	0.10 -0.40 (.004-.016)	0.70 (.0275)
C. Connecting rods		
1. Weight difference between connecting rods in one engine	max. 6 g	max. 10 g (during repair)
Weight of replacement connecting rods		
− Weight (white) brown/white	746-752 g	—
+ Weight (black) gray/black	769-775 g	—
2. Piston pin ... diameter	23.996-24.00 (.9445-.9448)	—
3. Piston pin bush ... diameter	24.015-24.024 (.9454-.9457)	—
4. Piston pin/bush ... radial play	0.01-0.03 (.0004-.0012)	0.04 (.0016)
D. Camshaft		
1. Bearings 1, 2, and 3 ... diameter	24.99-25.00 (.9839-.9843)	—
2. Measured at center bearing (bearings 1 and 3 on V blocks) ... runout	0.02 (.0008)	0.04 (.0016)
3. Camshaft/camshaft bearings (taking housing preload into account) ... radial play	0.02-0.05 (.0008-.002)	0.12 (.0047)
Thrust bearing ... end play	0.04-0.13 (.0016-.005)	0.16 (.0063)
4. Camshaft gear ... backlash	0.00-0.05 (.000-.002)	—
5. Cam follower ... diameter	23.96-23.98 (.9433-.9440)	23.93 (.9421)
6. Bore/cam follower ... radial play	0.02-0.06 (.0008-.0023)	0.12 (.0047)
7. Push rod ... runout	max. 0.3 (.012)	—
E. Lubrication system		
1. Oil pressure (for SAE 30 only) at an oil temperature of 70°C (158°F) and 2500 rpm ...	approx. 42 psi (3 kg/cm²)	28 psi (2 kg/cm²)
2. Spring for pressure relief valve Length loaded 44.1 mm (1.73 in.) ... load	—	—
3. Spring for oil pressure relief valve Length loaded: 39.0 mm (1.54 in.) ... load	6.8-8.8 kg (15-19⅜ lbs.)	—
4. Spring for oil pressure control valve Length loaded 20.2 mm (.795 in.) ... load	—	—
5. Spring for oil pressure control valve Length loaded: 26.0 mm (1.02 in.) ... load	1.7-2.0 kg (3¾-4⅜ lbs.)	—
6. Oil pressure switch open at ... pressure	2.1-6.4 psi (.15-.45 kg/cm²)	—
F. Flywheel		
1. Flywheel (measured at center of friction surface) .. lateral runout	max. 0.40 (.016)	—
imbalance	20 cmg	—
2. Shoulder for oil seal ... outside diameter	74.9-75.1 (2.948-2.956)	—
3. Drive plate ... imbalance	max. 5 cmg	—

V. Tolerances, Wear Limits, and Settings (dual-carburetor and fuel injection engines) (continued)

Designation	New part mm (in.)	Wear limit mm (in.)
G. Pistons and cylinders		
2 oversizes of 0.5 mm (.020 in.) each		
1. Cylinders ... out-of-round	max. 0.01 (.0004)	—
2. Cylinder/piston ... clearance	0.04-0.06 (.0016-.0023)	0.20 (.008)
3. a. Upper piston ring ... side clearance	0.06-0.09 (.0023-.0035)	0.12 (.0047)
b. Lower piston ring ... side clearance	0.04-0.07 (.0016-.0027)	0.10 (.004)
4. Oil scraper ring ... side clearance	0.02-0.05 (.0008-.0020)	0.10 (.004)
5. a. Upper piston ring ... gap	0.35-0.55 (.014-.022)	0.90 (.035)
b. Lower piston ring ... gap	0.30-0.55 (.012-.022)	0.90 (.035)
6. Oil scraper ring ... gap	0.25-0.40 (.010-.016)	0.95 (.037)
7. Piston weight		
– Weight (brown)	496-504 g 456-464 g	—
+ Weight (gray)	504-512 g 464-472 g	—
8. Weight difference between pistons in one engine	max. 4 g	max. 10 g
H. Cylinder head and valves		
1. Cylinder seating depth in cylinder head	5.4-6.5 (.2126-.2559)	—
a. Rocker arm ... inside diameter	20.00-20.02 (.7874-.7882)	20.04 (.7890)
b. Rocker shaft ... diameter	19.95-19.97 (.7854-.7862)	19.93 (.7846)
2. Valve springs		
Length loaded 29.0 mm (1.14 in.) ... load	76.5-84.5 kg (168-186 lb.)	—
3. Valve seats		
a. Intake ... width new	1.8-2.2 (.070-.086)	—
b. Exhaust ... width new	2.0-2.5 (.078-.098)	—
c. Intake ... repair width	1.3-1.6 (.051-.063)	—
d. Exhaust ... repair width	1.7-2.0 (.067-.079)	—
e. Intake ... seat angle	30°	—
f. Exhaust ... seat angle	45°	—
g. Outer seat edge ... correction angle	15°	—
h. Inner seat edge ... correction angle	75°	—
4. Valve guides:		
a. Intake ... inside diameter	8.00-8.02 (.3150-.3157)	8.06 (.3173)
b. Exhaust ... inside diameter	8.98-8.99 (.3535-.3539)	9.06 (.3566)
5. Valve stem		
a. Intake ... diameter	7.94-7.95 (.3126-.3129)	7.90 (.3110)
b. Exhaust ... diameter	8.91-8.92 (.3507-.3511)	8.87 (.3492)
6. Valve head out-of-round	0.01 (.0004)	—
a. Intake (1700 dual-carburetor engine) ... diameter	39.30 (1.5472)	—
Intake (1800 dual-carburetor engine) ... diameter	41.00 (1.6142)	—
Intake (1800 fuel-injection engine) ... diameter	39.30 (1.5472)	—
Intake (2000 fuel-injection engine) ... diameter	37.50 (1.4764)	—
b. Exhaust (1700 dual-carburetor engine) ... diameter	33.00 (1.2992)	—
Exhaust (1800 dual-carburetor engine) ... diameter	34.00 (1.3386)	—
Exhaust (1800 fuel-injection engine) ... diameter	33.00 (1.2992)	—
Exhaust (2000 fuel-injection engine) ... diameter	33.00 (1.2992)	—
7. Valve guide/valve stem:		
Intake ... rock	0.45 (.018)	0.90 (.035)
Exhaust ... rock	0.35 (.014)	1.20 (.047)
8. 1972 through 1977 valve clearance (cold)		
a. Intake (all engines) ... setting	0.15 (.006)	—
b. Exhaust (1700 dual-carburetor engines through chassis No. 213 2210 553 and all fuel-injection engines) ... setting	0.15 (.006)	—
Exhaust (1800 engines, 1700 engines from chassis No. 213 2210 554, and earlier 1700 engines that have been modified by the installation of sodium-filled exhaust valves) ... setting	0.20 (.008)	—
9. 1978 valve clearance ... setting	controlled automatically by hydraulic valve lifters	—
10. Compression		
a. 1700 engine ... pressure	100-135 psi (7.0-9.5 kg/cm²)	85 psi (6.0 kg/cm²)
b. 1800 and 2000 engines ... pressure	85-135 psi (6.0-9.5 kg/cm²)	71 psi (5.0 kg/cm²)
c. Pressure difference between cylinders ...	max. 28 psi (2 kg/cm²)	—
I. Cooling		
1. Thermostat (through Sept. 1972) ... opening temperature	65-70°C (149-158°F)	—
Thermostat (from Oct. 1972) ... opening temperature	85-90°C (185-194°F)	—
2. Fan/belt pulley ... imbalance	max. 5 cmg	—

(continued)

V. Tolerances, Wear Limits, and Settings (dual-carburetor and fuel injection engines) (continued)

Designation	New part mm (in.)	Wear limit mm (in.)
J. Clutch		
1. Total pressure 1700	380-440 kg (837-970 lbs.)	—
1800	420-485 kg (924-1067 lbs.)	—
2000	460-520 kg (1014-1146 lbs.)	—
2. Complete clutch . . . . . imbalance	max. 15 cmg	—
3. Pressure plate . . . . . runout	—	0.10 (.004)
4. Clutch plate . . . . . lateral runout	max. 0.5 (.020)	—
5. Clutch pedal		
a. Through chassis No. 213 2129 106 . . . . . freeplay	10-20 (3/8-3/4)	—
b. Through chassis No. 213 2129 107 . . . . . freeplay	15-25 (5/8-1)	—

VI. Tightening Torques (single-carburetor 1600 engine)

Location	Designation	Thread	Quality grade	mkg	ft. lb.
Connecting rods	connecting rod nut	M 9 x 1	8 G	3.3	24
Crankcase halves	nut	M 8	5 S	2.0	14
Crankcase halves	nut	M 12 x 1.5	35 S 20 K	3.5	25
Cylinder head	nut	M 10	35 S 20 KV	3.2	23
Rocker shaft to cylinder head	nut	M 8	8 G	2.5	18
Heat exchangers on cylinder head	self-locking nut	M 8	—	2.0	14
Muffler clamps	bolts	M 6	—	1.0	7
Intake manifold to cylinder head	nut	M 6	—	2.0	14
Intake manifold to cylinder head	nut	M 8	—	2.0	14
Preheating pipe flange bolts	bolt	M 6	—	1.0	7
Oil pump to crankcase	nut	M 8	—	2.0	14
Oil drain plug	plug	M 14 x 1.5	9 S 20 K	3.5	25
Oil strainer to crankcase	nut	M 6	—	0.7	5 (60 in. lb.)
Oil cooler to crankcase	nut	M 6	—	0.7	5 (60 in. lb.)
Oil filler and breather to generator stand	gland nut	—	—	5.5	40
Flywheel to crankshaft	gland nut	M 28 x 1.5	45 S 20 KN	35.0	253
Clutch to flywheel	bolt	M 8 x 1.5	8 G	2.5	18
Spark plugs	spark plug	M 14 x 1.25	—	3.5	25
Engine to transmission	nut	M 10	—	3.0	22
Engine bearer to crankcase	self-locking bolt	M 8	8 G	2.5	18
Engine bearer to crankcase	self-locking bolt	M 10	—	4.0	29
Engine bearer to engine bearer mounting	bolt	M 10	—	4.0	29
Generator pulley	nut	M 12 x 1.5	—	6.0	43
Special nut for fan	nut	M 12 x 1.5	9 S 20 K	6.0	43
Crankshaft pulley	bolt	M 20 x 1.5	90 S 20 K	4.5	32
Engine bearer mounting to body	self-locking nut	M 8	8 G	2.5	18

VII. Tightening Torques (dual-carburetor and fuel injection 1700, 1800, and 2000 engines)

Location	Designation	Thread	Quality grade	mkg	ft. lb.
Connecting rods	connecting rod nut	M 9 x 1	41 Cr 4	3.3	24
Crankcase halves	nut	M 8	6 G	2.0	14
Crankcase halves	sealing nut	M 10 x 1.25	35 S 20 KV	3.5	25
Cylinder head	nut	M 10	35 S 20 KV	3.2	23
Rocker shaft to cylinder head	nut	M 7	6 G	1.4	10
Heat exchangers to cylinder head	self-locking nut	M 8	—	2.2	16
Muffler to heat exchangers	self-locking nut	M 8	—	2.0	14
Oil drain plug	plug	M 12 x 1.5	9 S 20 K	2.2	16
Oil pan cover to crankcase	sealing nut	M 8	9 S 20 K	1.3	9
Drive plate to crankshaft	socket head screw	M 12 x 1.5	10 K	9	65
Hub to crankshaft	self-locking bolt	M 8	SAE 1022	3.2	23
Fan to hub	socket head screw	M 8	—	2.0	14
Extension shaft with pulley to fan	bolt	M 8	—	2.0	14
Spark plugs	spark plug	M 14 x 1.25	—	3.0	22
Engine to transmission	nut	M 10	—	3.0	22
Engine bearer support to crankcase	self-locking bolt	M 8	—	3.0	22
Bonded rubber mounting to support/engine bearer	self-locking nut	M 8	—	2.0	14
Engine bearer to engine bearer mounting	self-locking nut	M 8	—	2.0	14
Engine bearer mounting to body	self-locking nut	M 8	—	2.0	14
Oil pump to crankcase	nut	M 8	—	2.0	14
Oil cooler to crankcase	nut	M 6	—	0.7	5 (60 in. lb.)
Flywheel to crankshaft	self-locking bolt	M 12 x 1.5	22 Cr 4	11.0	80
Clutch to flywheel	bolt	M 8 x 20	8 G	2.5	18
Torque converter to driveplate	bolt	M 8	8 G	2.0	14

VIII. General Engine Data

Engine code letters	From August 1967 through July 1970: B From August 1970 through July 1971: AE Manual transmission from August 1971 through July 1973: CB Automatic transmission from August 1972 through July 1973: CD Manual transmission from August 1973 through July 1974: AW Automatic transmission from August 1973 through July 1974: AW Fuel injection engines from August 1974: ED Fuel injection engines from August 1975: (except 1979 and later Calif.) GD Calif. fuel injection from August 1977: GE
Number of cylinders	4
Cylinder layout	*Horizontally* opposed pairs
Bore mm (in.)	Code letters B, AE: 85.50 (3.365) Code letters CB, CD: 90.00 (3.540) Code letters AW, ED: 93.00 (3.660) Code letter GD, GE: 94.00 (3.701)
Stroke mm (in.)	Code letters B, AE: 69.00 (2.720) Code letters CB, CD, AW, ED: 66.00 (2.600) Code letter GD, GE: 71.00 (2.795)
Piston displacement cc (c.i.)	Code letters B, AE: 1584 (96.7) Code letters CB, CD: 1679 (102.5) Code letters AW, ED: 1795 (109.5) Code letter GD, GE: 1970 (120.2)
Compression ratio	Code letters B, AE: 7.5 (7.7 through Engine No. B 5 039 389) Code letters CB, CD, AW, ED, GD, GE: 7.3

VIII. General Engine Data (continued)

Fuel requirement	91 octane RON—lead-free only for cars with catalytic converters
Horsepower (DIN)	Code letter B: 47 @ 4000 rpm
(SAE)	Code letter B: 57 @ 4400 rpm
(DIN)	Code letter AE: 50 @ 4000 rpm
(SAE)	Code letter AE: 60 @ 4400 rpm
(DIN)	Code letter CB: 66 @ 4800 rpm
(SAE)	Code letter CB: 63 @ 4800 rpm
(DIN)	Code letter CD: 62 @ 4000 rpm
(SAE)	Code letter CD: 59 @ 4200 rpm
(DIN)	Code letter AW: 68 @ 4200 rpm
(SAE)	Code letter AW: 65 @ 4200 rpm
(DIN)	Code letter ED: 70 @ 4800 rpm
(SAE)	Code letter ED: 67 @ 4400 rpm
(DIN)	Code letter GD or GE: 70 @ 4200 rpm
(SAE)	Code letter GD or GE: 67 @ 4200 rpm
Torque (DIN)	Code letter B: 10.6 mkg @ 2200 rpm
(SAE)	Code letter B: 81.7 ft. lb. @ 3000 rpm
(DIN)	Code letter AE: 10.8 mkg @ 2800 rpm
(SAE)	Code letter AE: 81.7 ft. lb. @ 3000 rpm
(DIN)	Code letter CB: 11.6 mkg @ 3200 rpm
(SAE)	Code letter CB: 81.0 ft. lb. @ 3200 rpm
(DIN)	Code letter CD: 11.8 mkg @ 3200 rpm
(SAE)	Code letter CD: 83.0 ft. lb. @ 3200 rpm
(DIN)	Code letter AW: 13.2 mkg @ 3000 rpm
(SAE)	Code letter AW: 92.4 ft. lb. @ 3000 rpm
(DIN)	Code letter ED: 13.0 mkg @ 2400 rpm
(SAE)	Code letter ED: 90.0 ft. lb. @ 2400 rpm
(DIN)	Code letter GD or GE: 14.3 mkg @ 2800 rpm
(SAE)	Code letter GD or GE: 101.0 ft. lb. @ 3000 rpm

5

TRANSMISSION AND REAR AXLE

Contents

6

2 Transmission and Rear Axle

TABLES

Transmission and Rear Axle

The four-speed manual transmission is standard equipment on all Type 2 vehicles and is fully synchronized in all forward gears. Both the transmission and the final drive are housed in a single lightweight alloy case. The clutch is housed in a separate lightweight alloy clutch housing that bolts to the rear of the transmission case. The manual transmission shares very few parts with the fully automatic transmission and operates on entirely different principles. See **AUTOMATIC TRANSMISSION** for automatic transmission repair instructions.

The information in this section of the Manual should be of interest to nearly every VW Type 2 owner, even though much of the repair data may prove practical only to the professional mechanic. A general knowledge of transmission operation and repair can often be helpful when troubleshooting. It can also lead to better understanding between you and your VW Service Advisor or mechanic.

6

Many of the repair procedures described in this section require tools and skills that only a trained VW mechanic or transmission specialist is likely to have. So if you do not feel completely confident in these areas, we recommend that such repairs be left to an Authorized VW Dealer or other qualified repair shop. We especially urge you to consult an Authorized VW Dealer before attempting any repairs on a car still covered by the new-car warranty.

Though the home mechanic often lacks the tools and experience necessary to actually repair his transmission, he may be perfectly capable of removing and installing the unit. If so, he should deliver the transmission to an Authorized VW Dealer or other qualified repair shop after giving it a thorough exterior cleaning but without doing any disassembly work. A partially disassembled transmission in a box or a basket is a mechanic's nightmare, so the car owner is not likely to be greeted with sympathy after trying to tackle a job over his head and then giving up. Also, several critical measurements must be made on some transmissions before disassembly.

Cleanliness and a careful approach are imperative when repairing the transmission. Clean and lay out the parts. If necessary, mark them to show their proper assembly order. Also make sure that you have the necessary tools—particularly for procedures given only with metric specifications. Work that is designated by the lack of U.S. equivalents can only be carried out with metric tools and instruments.

1. General Description

A cutaway drawing of the Type 2 manual transmission is given in Fig. 1-1. Though its basic design has gone unchanged since its introduction on the 1968 models, there have been a considerable number of detail improvements and modifications.

Fig. 1-1. Cutaway views of the manual transmission.

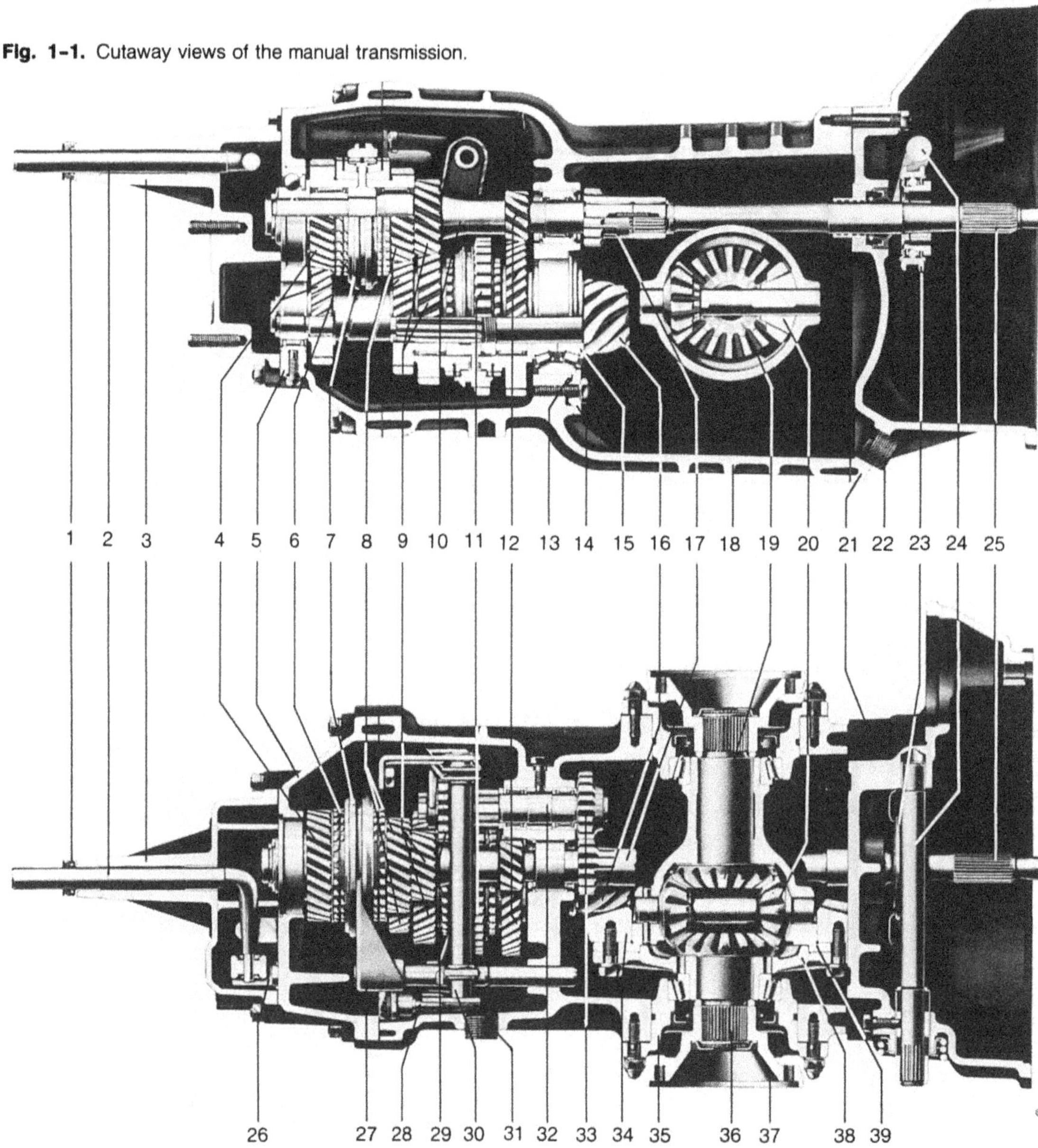

1. Oil seal
2. Inner shift lever
3. Gearshift housing
4. 4th speed gears
5. Gear carrier
6. 4th gear synchronizer ring
7. 3rd/4th gear clutch gear assembly
8. 3rd speed gears
9. 2nd speed gears
10. Mainshaft
11. 1st/2nd gear clutch gear assembly with integral reverse gear
12. 1st speed gears
13. Circlip (1968 only)
14. Bearing retainer (1968 only)
15. Double tapered-roller bearing
16. Drive pinion
17. Reverse drivegear
18. Transmission case
19. Differential gearshaft
20. Differential pinion
21. Clutch housing
22. Magnetic drain plug
23. Clutch release bearing
24. Clutch operating shaft
25. Rear driveshaft
26. Selector link
27. 3rd/4th gear shift fork
28. Reverse gear lever
29. Reverse sliding gear
30. Relay shaft
31. Oil filler plug
32. Reverse shaft
33. Reverse driven gear
34. Ring gear
35. Side cover (early models only)
36. Sealing cap
37. Axle flange
38. Differential housing cover
39. Differential housing

Transmission Case

The transmission case, which contains the transmission and final drive gearsets, is a tunnel-type die casting of magnesium alloy. A cast-in partition divides the transmission housing from the final drive housing and provides a mounting for the rear bearings of the mainshaft, reverse shaft, and pinion shaft. A die cast magnesium clutch housing is mounted on studs at the rear of the case. A gear carrier and shift housing of the same material are similarly mounted at the front.

The flanged shafts that carry driving torque to the rear wheels extend from two openings at either side of the final drive portion of the transmission case. Cover plates, mounted on studs in the transmission case casting (early models) or adjusting rings (since early 1969) form a support for the differential bearings.

Transmission Gears and Synchronizers

The transmission gears are of the constant-mesh type with balk ring synchronizers. The 3rd and 4th gear synchronizers are on the mainshaft; the 1st and 2nd gear synchronizers are on the pinion shaft.

Final Drive

The ring and pinion gearset has a hypoid offset that elevates the axis of the differential assembly 10-mm above that of the drive pinion. The differential gearset, which consists of the two differential gearshafts and the two differential pinions, allows the rear wheels to turn at different speeds, as is necessary when making turns (the outside wheel must travel farther than the inside wheel in the same amount of time).

Rear Axle and Suspension

Each rear wheel is suspended on a diagonal arm and is independently sprung by means of torsion bars, trailing spring plates, and hydraulically dampened shock absorbers. The wheels are driven by double-jointed axles, which have two constant velocity universal joints on each driveshaft.

2. MAINTENANCE

There are four items that require routine inspection or maintenance at a prescribed mileage. These are listed here and described briefly in **LUBRICATION AND MAINTENANCE**.

1. Checking the constant velocity joint screws.
2. Checking the axle boots.
3. Checking and correcting the transmission oil level
4. Lubricating the rear wheel bearings.

3. REAR AXLE REPAIRS

(engine and transmission in car)

The following repair jobs do not require engine or transmission removal. Shock absorber replacement is covered in **10. Rear Suspension**, along with other repairs related to the suspension.

3.1 Removing and Installing Driveshafts

The driveshafts used on vehicles with manual transmissions are not the same length as either of the two driveshafts used with the automatic transmission. When you buy replacement parts, dimension **b** indicated in Fig. 3-1 must be 476 mm (about 18¾ in.)—over half an inch longer or shorter than the shafts for any other VW vehicle.

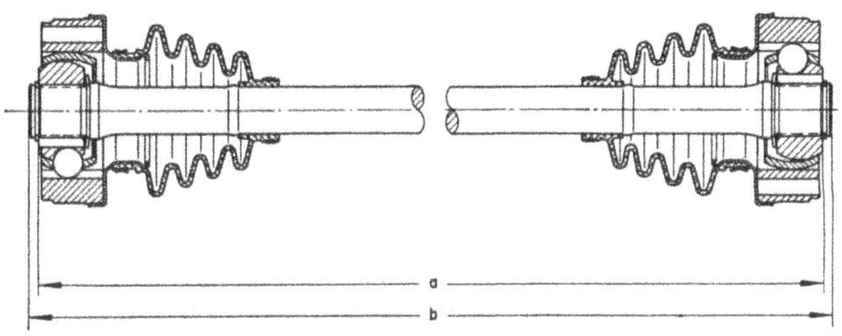

© 1974 VWoA—894

Fig. 3-1. Driveshaft dimensions. Dimension **a** does not apply to Type 2 replacement axles.

To remove a driveshaft, apply the parking brake. Remove first the socket head screws at the transmission end of the driveshaft (Fig. 3-2), then the socket head screws at the wheel end. Pull the driveshaft off the flanges.

6

© 1974 VWoA—3261

Fig. 3-2. Socket head screws that hold the constant velocity joint on the flange at the transmission.

Inspect the constant velocity joint boots after the driveshaft has been removed. Replace them if they are cracked or torn. Move the joints by hand. If they are stiff or feel gritty, remove and disassemble the joints so that they can be inspected and, if necessary, replaced or cleaned and lubricated.

Driveshaft installation is the reverse of removal. Install the socket head screws, new lockwashers, the spacers, and the protective cap as shown in Fig. 3-3. Working diagonally, torque the socket head screws to 3.5 mkg (25 ft. lb.).

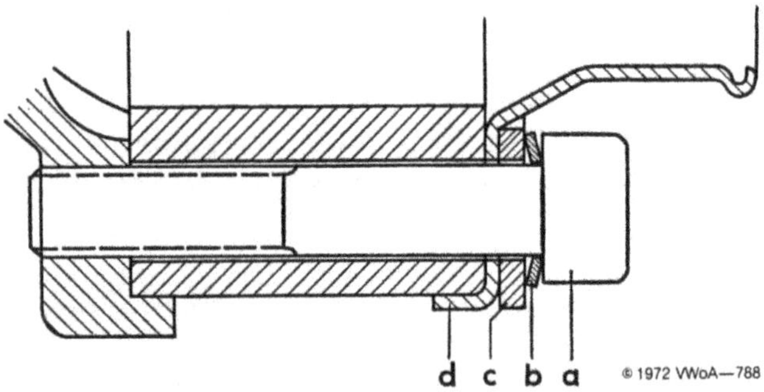

a. Socket head screw
b. Lock washer
c. Spacer
d. Protective cap

Fig. 3-3. Components used in mounting a constant velocity joint on its flange.

3.2 Removing and Installing Constant Velocity Joints

A driveshaft with one constant velocity joint removed is shown in Fig. 3-4. Boots with integral protective caps have been installed since the middle of 1972 and are the only type available as replacement parts. The new boot also eliminates the need for the pinch clamp and hose clamp.

1. Driveshaft
2. Boot with protective cap
3. Pinch clamp (early models) (2)
4. Boot (2)
5. Socket head screw (12)
6. Lock washer (12)
7. Hose clamp (early models) (2)
8. Protective cap (2)
9. Spacer (6)
10. Dished washer (discontinued on 1975 and later cars)
11. Constant velocity joints (2)
12. Circlip
13. End cap (early models) (2)

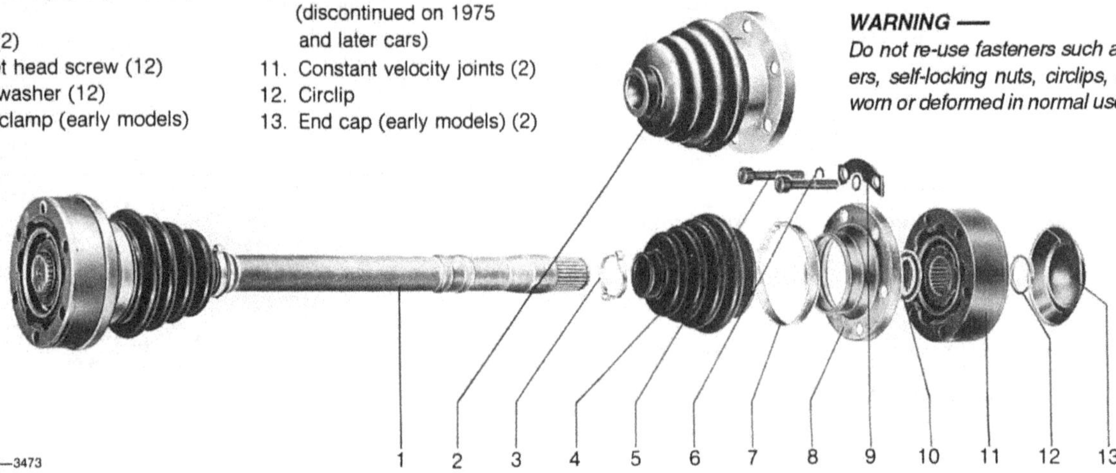

Fig. 3-4. Constant velocity joint removal.

To remove joint:

1. Remove the driveshaft.
2. On early joints, remove the hose clamp. Then pull the boot off the protective cap and turn it inside out on the driveshaft.
3. Carefully drive the protective cap off the joint as shown in Fig. 3-5. On late joints, pull the cap back so that the boot is turned inside out on the driveshaft.

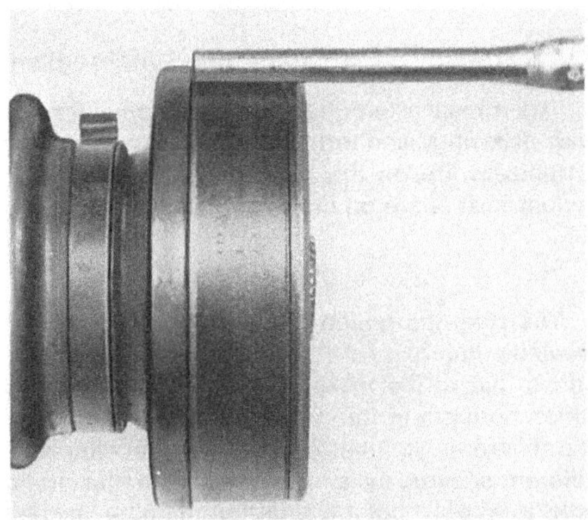

Fig. 3-5. Protective cap being driven off joint.

4. Using a small screwdriver, carefully pry the end cap off the joint (where fitted).

WARNING —
Do not re-use fasteners such as nuts, bolts, w ers, self-locking nuts, circlips, cotter pins tha worn or deformed in normal use. Always replac

5. Remove the circlip from the groove in the driveshaft.

> ***CAUTION —***
> *After removing the protective cap, do not tilt the ball hub more than 20° in the joint outer ring. If you do, the balls may fall out.*

6. Slide the joint outer ring as far as possible toward the outer end of the driveshaft. Then, with the ball hub supported, press the driveshaft out of the ball hub as shown in Fig. 3-6.

© 1974 VWoA—3476

Fig. 3-6. Driveshaft being pressed out of constant velocity joint. Note that only the ball hub is supported so that no stress is placed on the balls or outer ring.

7. If there is a dished washer on the driveshaft, remove and discard the dished washer.
8. If necessary, remove the pinch clamp (where fitted), then remove the protective cap and boot from the driveshaft.

To install:

1. Examine all parts. Replace any that are worn or damaged.
2. When installing the rubber boot, either use a plastic guide sleeve or cover the driveshaft splines with tape to protect the boot's sealing surface as you slide it onto the driveshaft. Use a new pinch clamp if an early-type boot is being reinstalled.
3. If the protective cap is a separate part from the boot, install a new hose clamp and then the protective cap but do not assemble the two until the joint is in place.
4. Do not install a dished washer. With the opposite end of the driveshaft against the beam under the press table, press the constant velocity joint onto the driveshaft as shown in Fig. 3-7.

© 1974 VWoA—3479

Fig. 3-7. Constant velocity joint being pressed onto the driveshaft with a sleeve-type driving tool. Though shown here, the dished washer used on 1968 through 1974 driveshafts should not be reinstalled.

5. Raise the press ram. Install the circlip on the driveshaft. Then use the press and driving tool to press the circlip onto the driveshaft until the circlip snaps into its groove.
6. Using pliers, squeeze the circlip around its circumference until the circlip is completely seated in the groove.
7. Repack the joint with 90 g (3.2 oz.) of molybdenum grease. Pack two-thirds of the grease between the joint and protective cap. Use the remaining third to pack the open front of the joint.
8. If necessary, position both clamps and tighten them.

> **NOTE —**
> When installing the hose clamp, be sure to position its projections between the screw holes in the protective cap. Otherwise, the projections will interfere with the installation of the socket head screws.

9. Hand-squeeze the boot tightly so that grease will be forced into the joint from the rear.
10. Press in the end cap (where fitted). Then install the driveshaft. Torque the socket head screws to 3.5 mkg (25 ft. lb.).

3.3 Servicing Constant Velocity Joints

It is not possible to repair the constant velocity joints. The components of each joint are factory-matched and cannot be replaced individually. The joints can, however, be disassembled for cleaning, inspection, and repacking with molybdenum grease.

To disassemble joint:

1. Remove the constant velocity joint from the driveshaft as described in **3.2 Removing and Installing Constant Velocity Joints.**
2. Using hand pressure, push the ball hub and ball cage out of the outer ring as shown in Fig. 3-8.

© 1974 VWoA—1494

Fig. 3-8. Ball cage and ball hub being separated from outer ring. Position the ball hub and ball cage perpendicular to the outer ring, then push in the direction indicated by the arrow.

3. Lift the six steel balls out of the ball cage, taking care not to drop them.

CAUTION —
The ball cage, the outer ring, and the balls themselves are selected for matching tolerances. When disassembling more than one joint, be sure not to intermix the parts. If you do, the joints may seize, make noise, or wear rapidly.

4. Rotate the ball hub into the position shown in Fig. 3-9. The groove in the ball hub must be in line with the outer edge of the ball cage. The ball hub now has sufficient clearance to be tipped out of the ball cage.

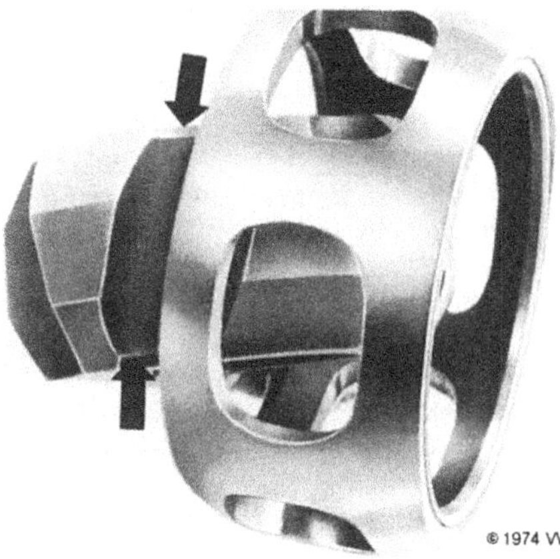

© 1974 VWoA—1495

Fig. 3-9. Ball hub being removed from or installed in the ball cage. Arrows indicate the alignment of the ball hub groove with the ball cage edge.

To assemble joint:

1. Clean and inspect all parts. If any are worn or damaged, replace the entire joint.
2. Thoroughly coat all parts with molybdenum grease. Then install the ball hub in the ball cage as illustrated in Fig. 3-9.
3. Install the balls in the ball cage and ball hub. If necessary, use additional molybdenum grease as an adhesive to hold the balls in place.
4. Insert the ball hub together with the balls and ball cage into the outer ring. The chamfer on the splines and the larger diameter portion of the outer ring must be on the same side of the joint.

NOTE —
Insert the ball and hub assembly in the outer ring at a 90° angle, as shown in Fig. 3-10. The narrow ball hub grooves **b** and the wide outer ring grooves **a** must be positioned as illustrated.

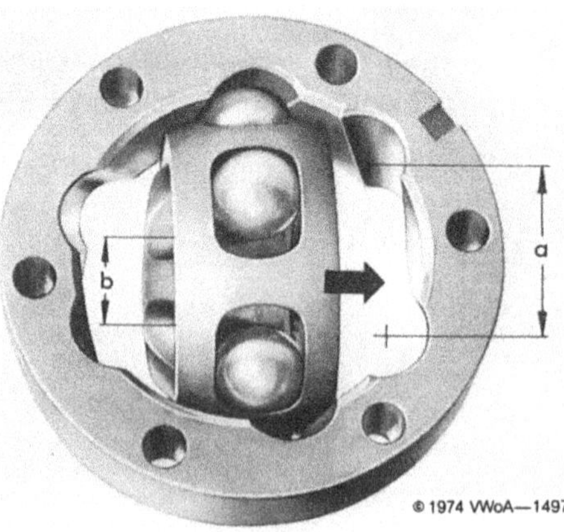

© 1974 VWoA—1497

Fig. 3-10. Ball and hub assembly positioned in the outer ring. Arrow indicates the direction in which the ball and hub assembly must be turned to complete the installation.

5. Hold the ball and hub assembly steady and push it in the direction indicated in Fig. 3-11 (left arrow). This will align the balls with their respective grooves (right arrows) in the outer ring.

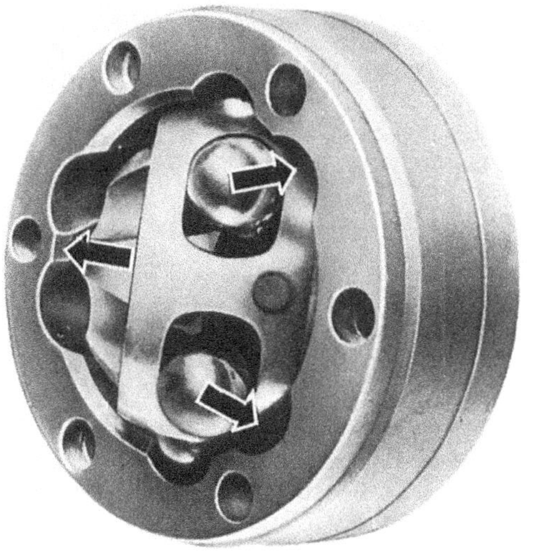

© 1974 VWoA—1498

Fig. 3-11. Balls being aligned with the grooves of the outer ring. Apply finger pressure in the direction indicated by the left arrow.

6. When the balls are aligned with their grooves in the outer ring, firmly press the ball cage, as indicated in Fig. 3-12, until it swings fully into place. Heavy pressure should not be required.

© 1974 VWoA—1499

Fig. 3-12. Balls being engaged in the grooves of the outer ring. Apply hand pressure at the point indicated by the arrow.

CAUTION —
Excessive force will result in improper joint assembly. Double-check the previous steps if the joint does not go together readily. An improperly assembed joint will lock solidly and render the unit unserviceable.

7. Check the operation of the joint. The ball hub should be able to turn smoothly throughout the entire range of travel.

8. Using the procedures given in **3.2 Removing and Installing Constant Velocity Joints,** pack the joint with molybdenum grease and install it on the driveshaft.

3.4 Removing, Installing, and Adjusting Gearshift Lever and Shift Rods

The gearshift lever and shift rods are virtually the same on all models covered by this Manual. Minor changes were made to the gearshift lever bracket and stop plate from chassis No. 211 2041 707. These changes were to simplify installation and adjustment and do not affect the operation of the gearshift lever.

To remove lever and rods:

1. Working inside the vehicle, remove the two bolts with washers that hold the gearshift lever assembly on the vehicle's floor panel. Then lift out the gearshift lever assembly.
2. Working under the vehicle, remove the cover plate that is beneath the pedal cluster.
3. Cut the locking wire on the square head setscrew for the coupling that joins the front and rear shift rods. Then remove the screw (Fig. 3-13).

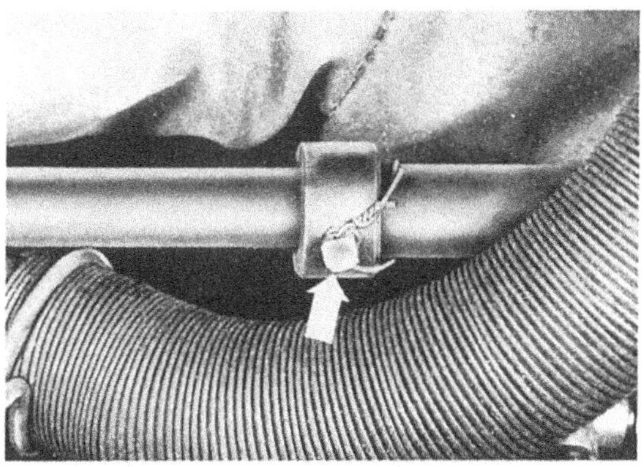

© 1974 VWoA—3506

Fig. 3-13. Square head setscrew (arrow) for the coupling that joins the front and rear shift rods. Note the locking wire that is installed around the front shift rod and through the hole in the head of the setscrew.

6

4. Pull apart the front and rear shift rods. Disengage the front shift rod from the guide pin at its forward end and remove the front rod and coupling from the vehicle. (See Fig. 3-14.)

5. You must remove the transmission as described in **4. Removing and Installing Engine and Transmission** before you can remove the rear shift rod. To disconnect the rear shift rod from the transmission, remove the square head setscrew indicated in Fig. 3-15.

Fig. 3-14. Components of the gearshift lever and shift rods. Parts **6a** and **9a** are used from chassis No. 211 2041 707. The new bracket has an added guide for the stop plate; the new stop plate has added locating lugs.

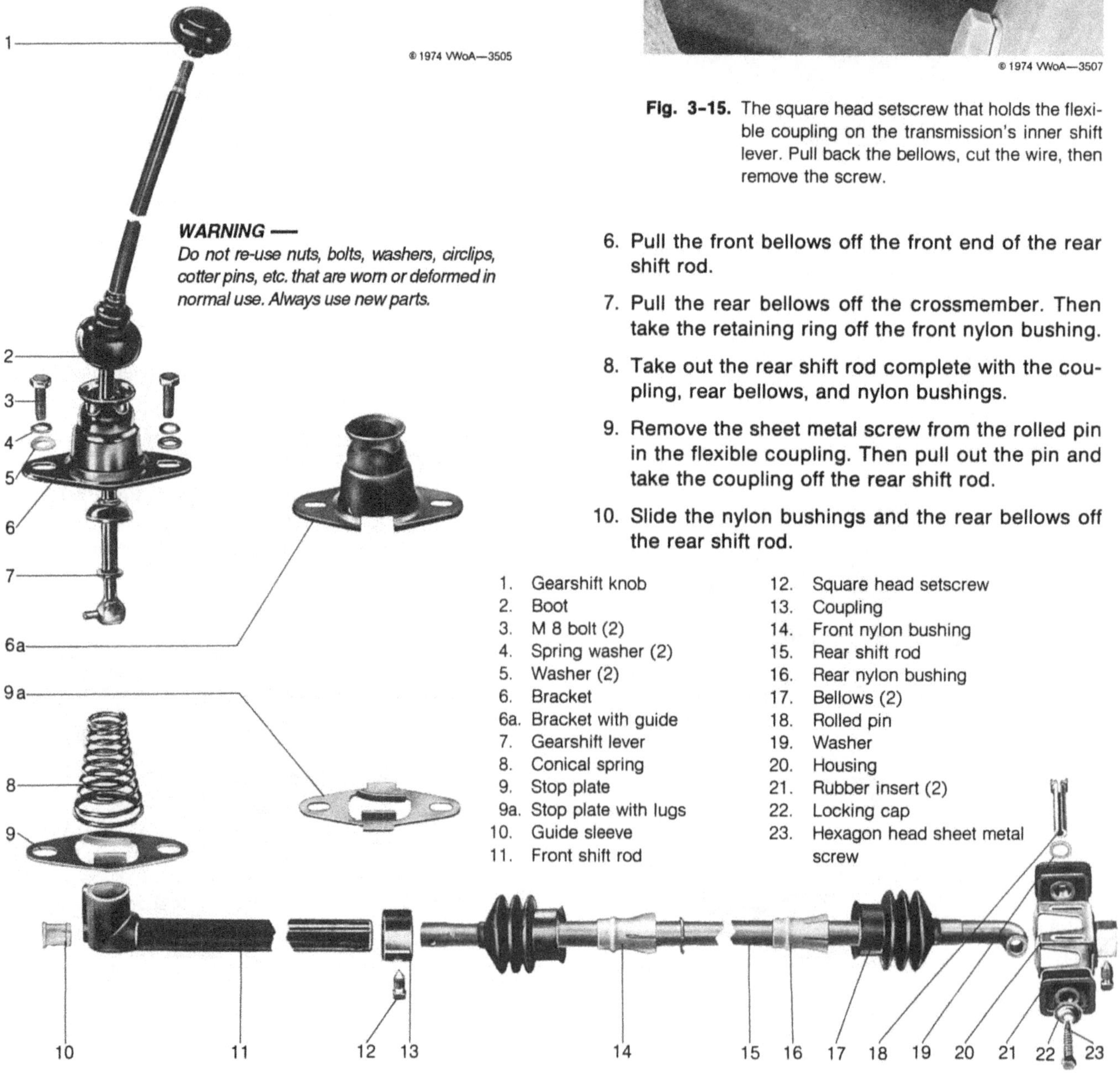

WARNING —
Do not re-use nuts, bolts, washers, circlips, cotter pins, etc. that are worn or deformed in normal use. Always use new parts.

Fig. 3-15. The square head setscrew that holds the flexible coupling on the transmission's inner shift lever. Pull back the bellows, cut the wire, then remove the screw.

6. Pull the front bellows off the front end of the rear shift rod.

7. Pull the rear bellows off the crossmember. Then take the retaining ring off the front nylon bushing.

8. Take out the rear shift rod complete with the coupling, rear bellows, and nylon bushings.

9. Remove the sheet metal screw from the rolled pin in the flexible coupling. Then pull out the pin and take the coupling off the rear shift rod.

10. Slide the nylon bushings and the rear bellows off the rear shift rod.

1. Gearshift knob
2. Boot
3. M 8 bolt (2)
4. Spring washer (2)
5. Washer (2)
6. Bracket
6a. Bracket with guide
7. Gearshift lever
8. Conical spring
9. Stop plate
9a. Stop plate with lugs
10. Guide sleeve
11. Front shift rod
12. Square head setscrew
13. Coupling
14. Front nylon bushing
15. Rear shift rod
16. Rear nylon bushing
17. Bellows (2)
18. Rolled pin
19. Washer
20. Housing
21. Rubber insert (2)
22. Locking cap
23. Hexagon head sheet metal screw

To install and adjust:

1. Clean and inspect all parts. If any are worn or damaged, replace them.
2. Install the rear bellows and the nylon bushings on the rear shift rod.

   NOTE —
   The larger ends of both bushings point to the rear of the vehicle. The retaining ring should be installed on the front bushing.

3. Assemble the flexible coupling and install it on the rear shift rod.

   NOTE —
   On 1972 models only, the flexible coupling is installed on a separate extension rod that is bolted to the end of the rear shift rod.

4. Install the front bellows. Then install the rear shift rod in the vehicle.
5. Install the transmission. Then torque the square head setscrew to 1.5 mkg (11 ft. lb.) and secure it with a new locking wire. Pull the bellows over the transmission's inner shift lever forward to cover the rear of the flexible coupling.
6. Coat the sliding parts of both the front and rear shift rods with multipurpose grease. Then engage the front part of the front shift rod on the guide pin on the frame as indicated in Fig. 3-16.

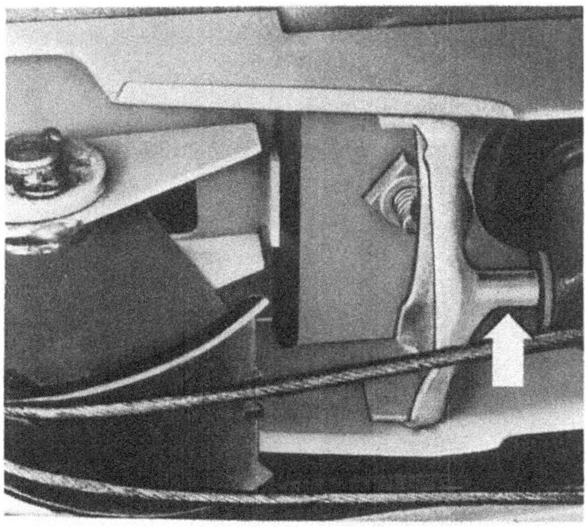

© 1974 VWoA—3508

Fig. 3-16. Guide pin (arrow) engaged in the plastic guide sleeve in the front shift rod.

7. Assemble the front shift rod, the coupling, and the rear shift rod. Torque the square head setscrew to 1.5 mkg (11 ft. lb.), then secure it with a new locking wire.
8. Coat the gearshift lever ball, the gearshift lever bracket, and the socket in the front shift rod with multipurpose grease. Then assemble the gearshift lever, conical spring, and stop plate.
9. Loosely install the gearshift lever assembly on the floor panel. Engage 2nd gear. Then slide the gearshift lever bracket on its elongated holes to position the lever as shown in Fig. 3-17.

© 1974 VWoA—3510

Fig. 3-17. Lower part of gearshift positioned at 90° to the floor with 2nd gear engaged.

10. Using the tip of a screwdriver, position the stop plate against the gearshift lever as shown in Fig. 3-18. Then torque the M 8 bolts to 2.0 mkg (14 ft. lb.).
11. Engage each of the four forward gears and reverse to see that the stop plate is adjusted properly.

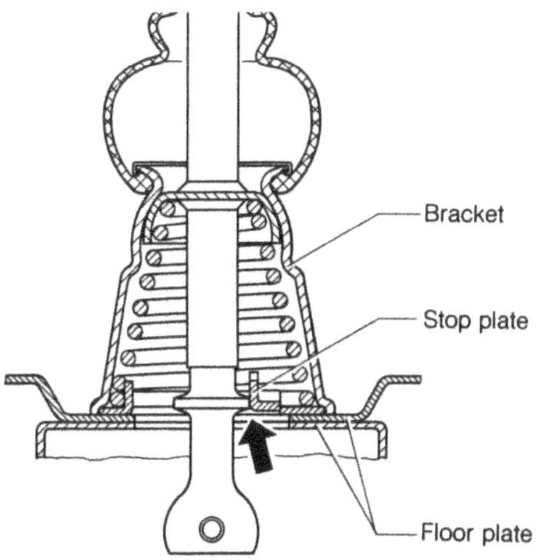

© 1974 VWoA—3511

Fig. 3-18. Stop plate positioned against the gearshift lever (arrow).

6

4. Removing and Installing Engine and Transmission

It is best if you remove the engine and transmission as a unit and then separate them following removal. However, several transmission repairs, covered in **5.2 Disassembling and Assembling Clutch Housing**, can be made with the transmission in the vehicle and the engine removed.

4.1 Removing and Installing Dual-carburetor Engine and Transmission

The transmission and the 1700/1800 engine installed in 1972 and later vehicles are mounted differently from the engines and transmissions of earlier models. The late-model transmission mountings are shown in Fig. 4-1.

To remove:

1. Disconnect the battery ground strap from the negative pole of the battery.
2. Disconnect the electric wires from the voltage regulator, ignition coil, carburetors, oil pressure switch, and the starter solenoid. On 1975 and later vehicles, disconnect the fuel injection wiring from the coil, throttle valve switch, EGR (exhaust gas recirculation) valve, intake air sensor, thermo-time switch, cold-start valve, auxiliary air regulator, temperature sensor II, and the injectors.
3. Disconnect all hoses that connect the engine with components mounted on the vehicle body.

NOTE —
As you disconnect each wire or hose, attach a tag to help you return it to its original position during installation. Position the disconnected hoses and wires so that they will not become caught or tangled as the engine and transmission are removed.

4. Disconnect the accelerator cable from the engine. Pull the cable out of its guide tube forward.

Fig. 4-1. Transmission mounts, 1700/1800 engine.

1. M 10 nut
2. Bellows
3. Transmission carrier bolt
4. Spring washer
5. Sealing ring
6. Sleeve
7. Transmission carrier
8. Spring washer
9. M 10 bolt
10. Bonded rubber bushing
11. Rubber stop
12. Washer
13. Limiting stop
14. M 10 bolt
15. Washer
16. Support
17. Bonded rubber mounting
18. Transmission

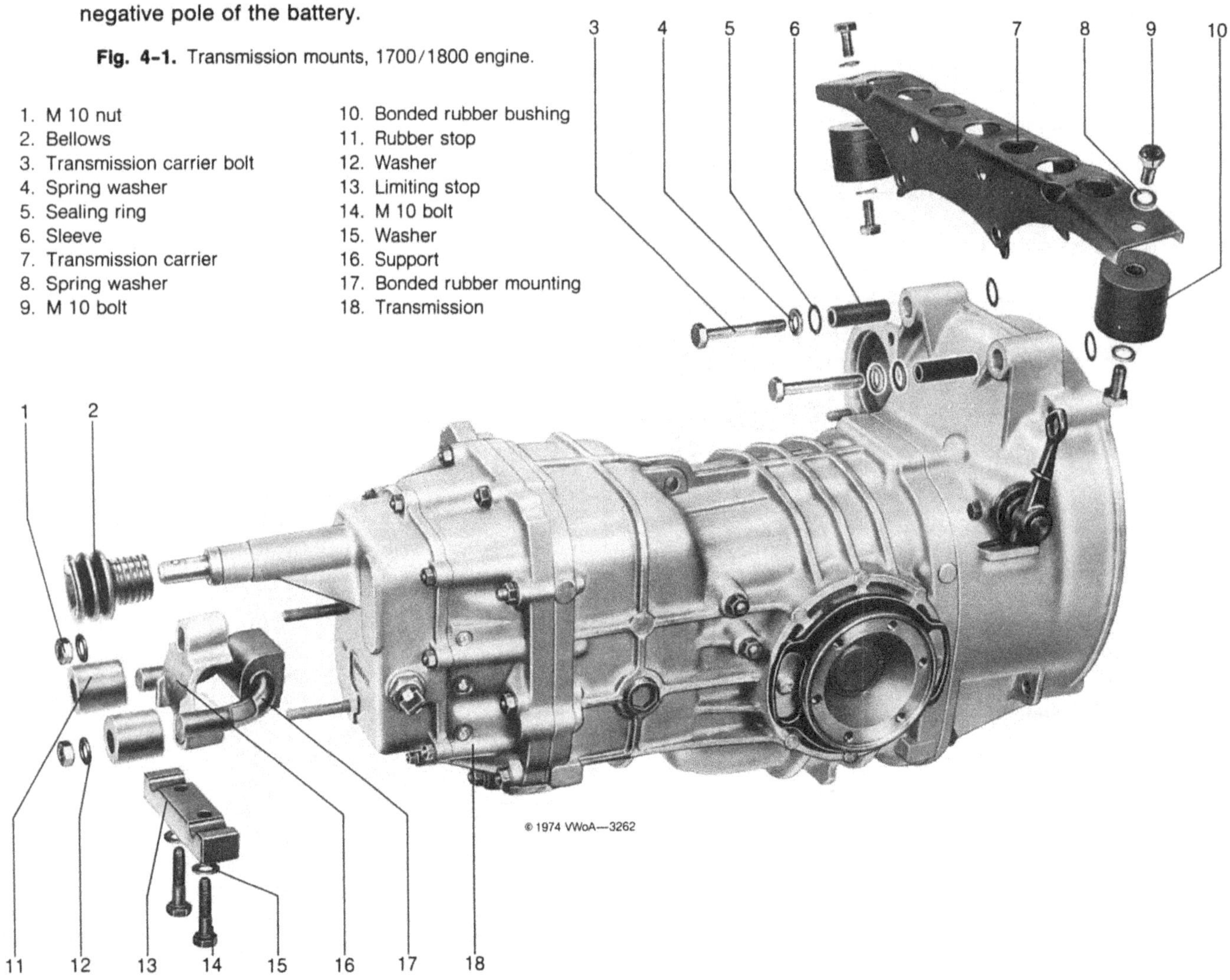

5. Remove the wing nut from the end of the clutch cable where it is connected to the clutch lever. Then remove the two nuts that hold the bracket indicated in Fig. 4-2 and take the clutch cable and its guide tube off the transmission.

© 1974 VWoA—3263

Fig. 4-2. Angled support bracket for clutch cable guide tube. Mounting nuts have been removed.

6. Disconnect the driveshafts from the transmission as described in **3.1 Removing and Installing Driveshafts**, then suspend them from the frame on wire hooks. Cover the joints to keep out dirt. If the vehicle must be moved, either completely remove the driveshafts or hang them so that they are free to turn.

7. Install a transmission support bar (Fig. 4-3).

© 1974 VWoA—3264

Fig. 4-3. Support bar installed with its contact plate centered under the transmission. Connect the chains to the frame sidemembers.

8. Turn the T-handle on the support bar until the contact plate raises the transmission enough so that you can remove the carrier bolts (Fig. 4-4). Then remove the support bar from the vehicle.

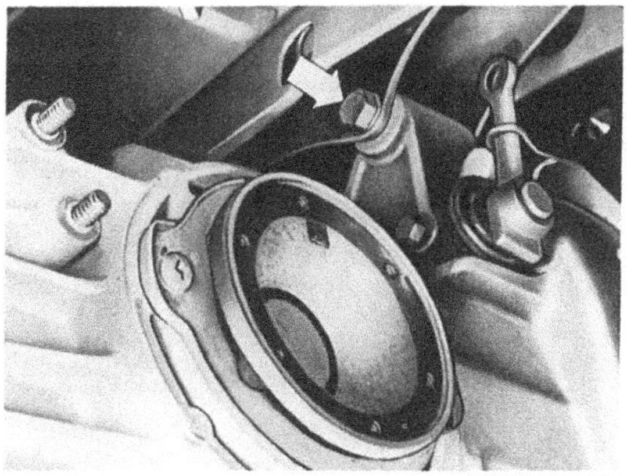

© 1974 VWoA—3265

Fig. 4-4. Transmission carrier bolt (arrow). A similar bolt is at the opposite side of the transmission.

9. Working at the front of the transmission, pull back the rubber bellows that covers the inner shift lever where it projects from the transmission.

10. With the gearshift lever in 3rd gear, cut the locking wire and remove the square head setscrew that holds the rear shift rod's flexible coupling on the transmission inner shift lever. To disconnect the coupling from the inner shift rod, move the gearshift lever into its 2nd gear position.

11. Position a floor jack under the vehicle together with an engine adaptor and a transmission adaptor. Raise the jack until the engine and transmission are slightly lifted.

12. Remove the two bolts that hold the limiting stop for the front transmission mount (Fig. 4-5).

6

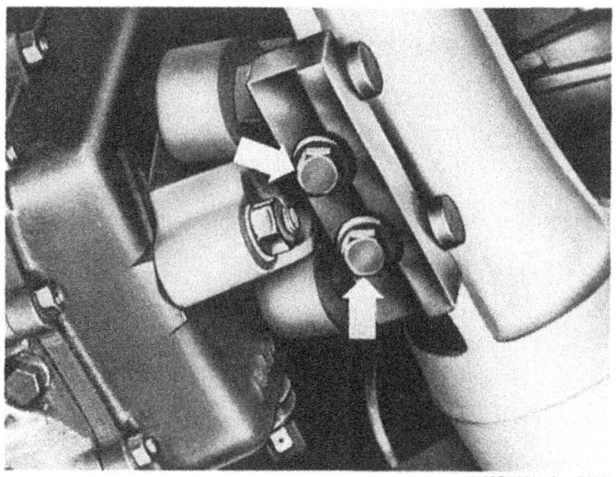

© 1974 VWoA—3267

Fig. 4-5. Bolts (arrows) that hold the limiting stop.

13. Remove the nuts and bolts that hold the engine bearer on its mountings (Fig. 4-6).

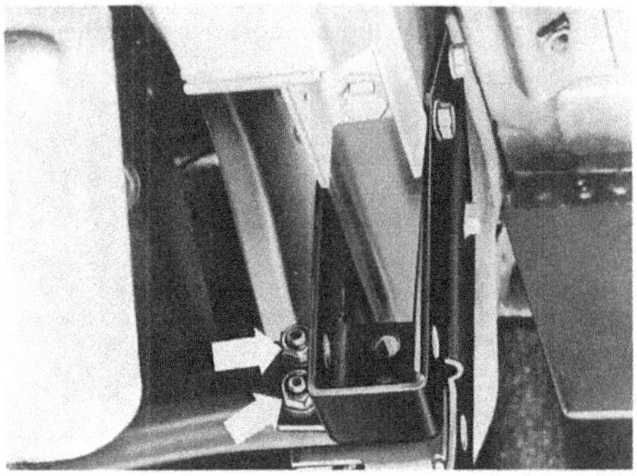

© 1974 VWoA—3268

Fig. 4-6. Nuts and bolts that hold the engine bearer. There are two similar bolts at the opposite side of the engine.

14. Slightly lower the floor jack together with the engine/transmission assembly. Then, watching to see that no hoses or wires are caught, pull the assembly to the rear, lower it somewhat more, and remove it toward the rear of the vehicle.

After transmission repairs have been completed, inspect and service the clutch before installing the engine on the transmission. See that the Teflon facing on the clutch release bearing is not worn through and that the bearing turns smoothly and without noise.

> ***CAUTION —***
> *Wipe the clutch release bearing clean, do not wash it in solvent. Doing so would destroy the factory-installed lubricant, which cannot be replaced.*

Lightly lubricate the clutch release bearing guide sleeve with molybdenum grease. Apply Moly-Kote® to the Teflon facing on the clutch release bearing. Lubricate the rear driveshaft splines with molybdenum disulphide powder and lubricate the starter drive bushing and the pilot bearing in the crankshaft with multipurpose grease.

To install:

1. Make sure that the sleeves and the sealing rings for the transmission carrier bolts are correctly positioned in the transmission's clutch housing. If necessary, refer back to Fig. 4-1.
2. Using the floor jack with adaptors, lift the engine/transmission assembly into the vehicle.
3. Install the engine bearer bolts with the new self-locking nuts, but do not tighten them.
4. Install the limiting stop for the front transmission mount. Torque the bolts to 4.5 mkg (33 ft. lb.).
5. Using a transmission support bar or a jack, lift the transmission at the clutch housing until the transmission carrier bolts can be screwed in.
6. Torque the transmission carrier bolts to 3.5 mkg (25 ft. lb.). Torque the engine bearer bolts to 2.5 mkg (18 ft. lb.). Then reconnect the shift rod to the transmission.
7. Remove the jack or support bar. If necessary, adjust the engine and transmission position as described following the procedure given here.
8. Install the driveshafts as described in **3.1 Removing and Installing Driveshafts**. Use new lock washers with their convex side toward the screw head and torque the socket head screws to 3.5 mkg (25 ft. lb.).
9. Install the angled support bracket for the clutch cable guide tube and torque the nuts to 2.0 mkg (14 ft. lb.).
10. Install the wing nut on the clutch cable. Through chassis No. 213 2129 106, adjust the clutch pedal freeplay to 10 to 20 mm (3/8 to 3/4 in.); from chassis No. 213 2129 107, adjust freeplay to 15 to 25 mm (5/8 to 1 in.). See **ENGINE AND CLUTCH**.
11. Install and adjust the accelerator cable. Reconnect all hoses and wires, then connect the battery ground strap.

Adjusting Position of Engine and Transmission

If a new transmission is being installed, or if excessive driveline noise is being transmitted to the interior of the vehicle, the position of the engine and transmission should be checked and, if necessary, adjusted.

To adjust limiting stop:

1. Loosen the nuts indicated in Fig. 4-7.

© 1974 VWoA—3272

Fig. 4-7. Two nuts that hold the front mount support on the transmission gearshift housing.

2. Taking hold of the heat exchangers on the engine, rock the engine/transmission assembly about its longitudinal axis.
3. Torque the two nuts for the front mount support to 2.5 mkg (18 ft. lb.).

To adjust engine bearer:

1. Loosen the bolts that hold the engine bearer mounting brackets on the frame sidemembers (Fig. 4-8).

Fig. 4-8. Bolts (arrows) that hold the engine bearer mounting brackets on the frame sidemembers.

2. Align the engine bearer so that it is vertical and its rear edge parallel to the fan housing as shown in Fig. 4-9.
3. Once aligned, torque the engine bearer bracket mounting bolts to 3.5 mkg (25 ft. lb.).

Fig. 4-9. Gap between bearer and fan housing (arrows). The two parts must be evenly spaced and parallel.

Removing and Installing Transmission Carrier

If necessary, the transmission carrier can be removed from the frame member after the transmission has been removed.

To remove:

1. Remove the bolts indicated in Fig. 4-10.

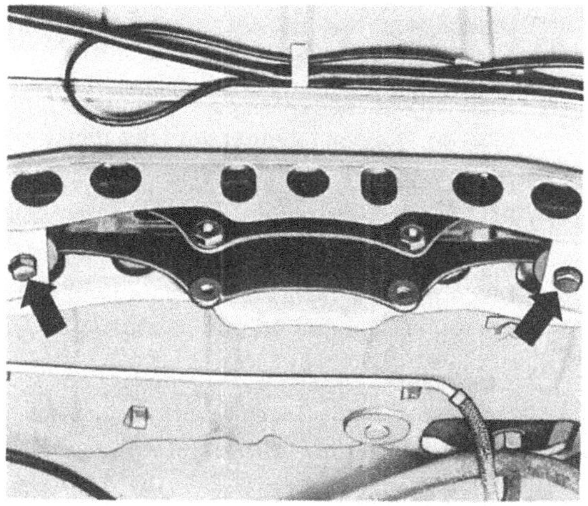

Fig. 4-10. Bolts that hold transmission carrier to frame crossmember.

2. Push the transmission carrier toward the left (arrow **A** in Fig. 4-11), then tilt it downward as at arrow **B** and take it out of the crossmember.
3. When you install the transmission carrier, make sure that the captive nuts for the transmission carrier bolts are toward the rear of the vehicle.

6

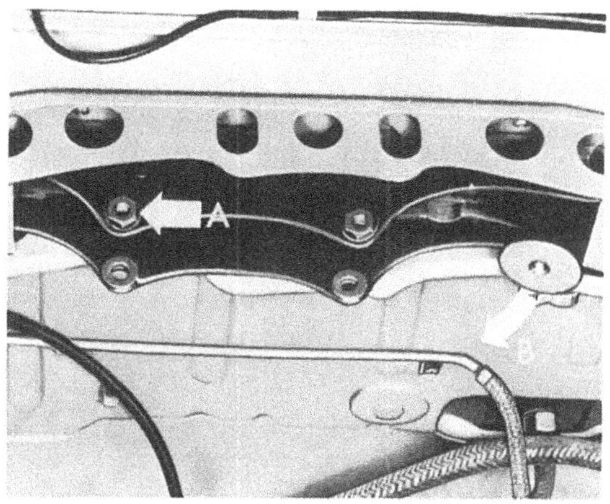

Fig. 4-11. Transmission carrier removal. Push carrier left, as indicated by arrow **A**. Then pull the carrier downward and out as shown by arrow **B**.

4.2 Removing and Installing Single-carburetor Engine and Transmission

The single-carburetor engine and transmission are mounted at two points: the front of the transmission and—by means of the engine bearer—at the rear of the engine crankcase.

To remove:

1. Disconnect the battery ground strap from the negative pole of the battery.
2. Remove the rear cross panel. Then take off the oil bath air cleaner together with the intake elbow.
3. Disconnect the electric wires from the generator, ignition coil, carburetor, oil pressure switch, and the starter solenoid.
4. Disconnect all hoses that connect the engine with components mounted on the vehicle body.

NOTE —
As you disconnect each wire or hose, attach a tag to help you return it to its original position during installation. Position the disconnected hoses and wires so that they will not become caught or tangled as the engine and transmission are removed.

5. Disconnect the throttle cable from the carburetor. Then pull the cable out of its guide tube forward.
6. Disconnect the heater flap cables and remove the heater duct hoses.
7. Remove the wing nut from the end of the clutch cable where it is connected to the clutch lever. Then remove the two nuts that hold the bracket indicated in Fig. 4-12 and take the clutch cable and its guide tube off the transmission.

© 1974 VWoA—3257

Fig. 4-12. Angled support bracket for clutch cable guide tube (right arrow). Upper mounting nut has been removed. Also disconnect the ground strap indicated by the left arrow.

8. Disconnect the driveshafts from the transmission as described in **3.1 Removing and Installing Driveshafts**, then suspend them from the frame on wire hooks. Cover the joints to keep out dirt. If the vehicle must be moved, either completely remove the driveshafts or hang them so that they are free to turn.
9. Working at the front of the transmission, pull back the rubber bellows that covers the inner shift lever where it projects from the transmission.
10. With the gearshift lever in 3rd gear, cut the locking wire and remove the square head setscrew that holds the rear shift rod's flexible coupling on the transmission inner shift lever. To disconnect the coupling from the inner shift rod, move the gearshift lever into its 2nd gear position.
11. Position a floor jack under the vehicle together with an engine adaptor and a transmission adaptor. Raise the jack until the engine and transmission are slightly lifted.
12. Remove the two bolts indicated in Fig. 4-13.

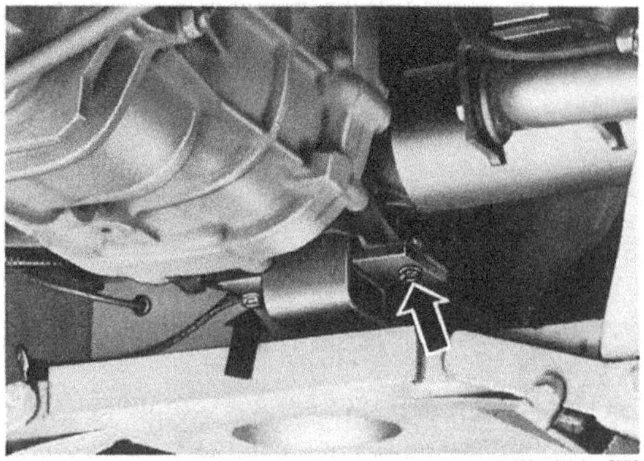

© 1974 VWoA—3258

Fig. 4-13. Bolts (arrows) that mount the front of the transmission.

13. Remove the bolts from both ends of the engine bearer.
14. Using the jack, lift the engine/transmission assembly slightly—if necessary—then pull it to the rear. Lower the assembly and then take it out toward the rear of the vehicle.

After transmission repairs have been completed, inspect and service the clutch before installing the engine on the transmission.

CAUTION —
Wipe the clutch release bearing clean, do not wash it in solvent. Doing so would destroy the factory-installed lubricant, which cannot be replaced.

See that the Teflon facing on the clutch release bearing is not worn through and that the bearing turns smoothly and without noise. From chassis No. 211 2021 717, lightly lubricate the clutch release bearing guide sleeve with molybdenum grease. Apply Moly-Kote® to the Teflon facing. Lubricate the rear driveshaft splines with molybdenum disulphide powder and lubricate the starter drive bushing and the pilot bearing in the flywheel gland nut with multipurpose grease.

Installation is the reverse of removal. Install—but do not tighten—the bolts in the front transmission mount and in the engine bearer. Torque first the transmission mount bolts to 3.5 mkg (25 ft. lb.). Then, using the new self-locking nuts, torque the engine bearer bolts to 2.5 mkg (18 ft. lb.).

5. CLUTCH HOUSING

The components of both the early-type and late-type clutch housings are shown in Fig. 5-1.

5.1 Removing and Installing Clutch Housing

The clutch housing must be removed for access to the differential and other parts of the final drive. To remove the clutch housing from the transmission case, remove the ten nuts from the studs. On early transmissions, loosen all 12 cap nuts on both final drive side covers. Tap the clutch housing lightly with a rubber mallet to free it from the transmission case.

CAUTION—
Do not insert tools between the clutch housing and the transmission case in order to pry them apart. This will damage the magnesium castings and cause leaks.

Installation is the reverse of removal. Use a new gasket and torque the nuts to 2.0 mkg (14 ft. lb.). On early transmissions, torque the cap nuts for the final drive side covers to 2.0 mkg (14 ft. lb.).

Fig. 5-1. Components of clutch housing.

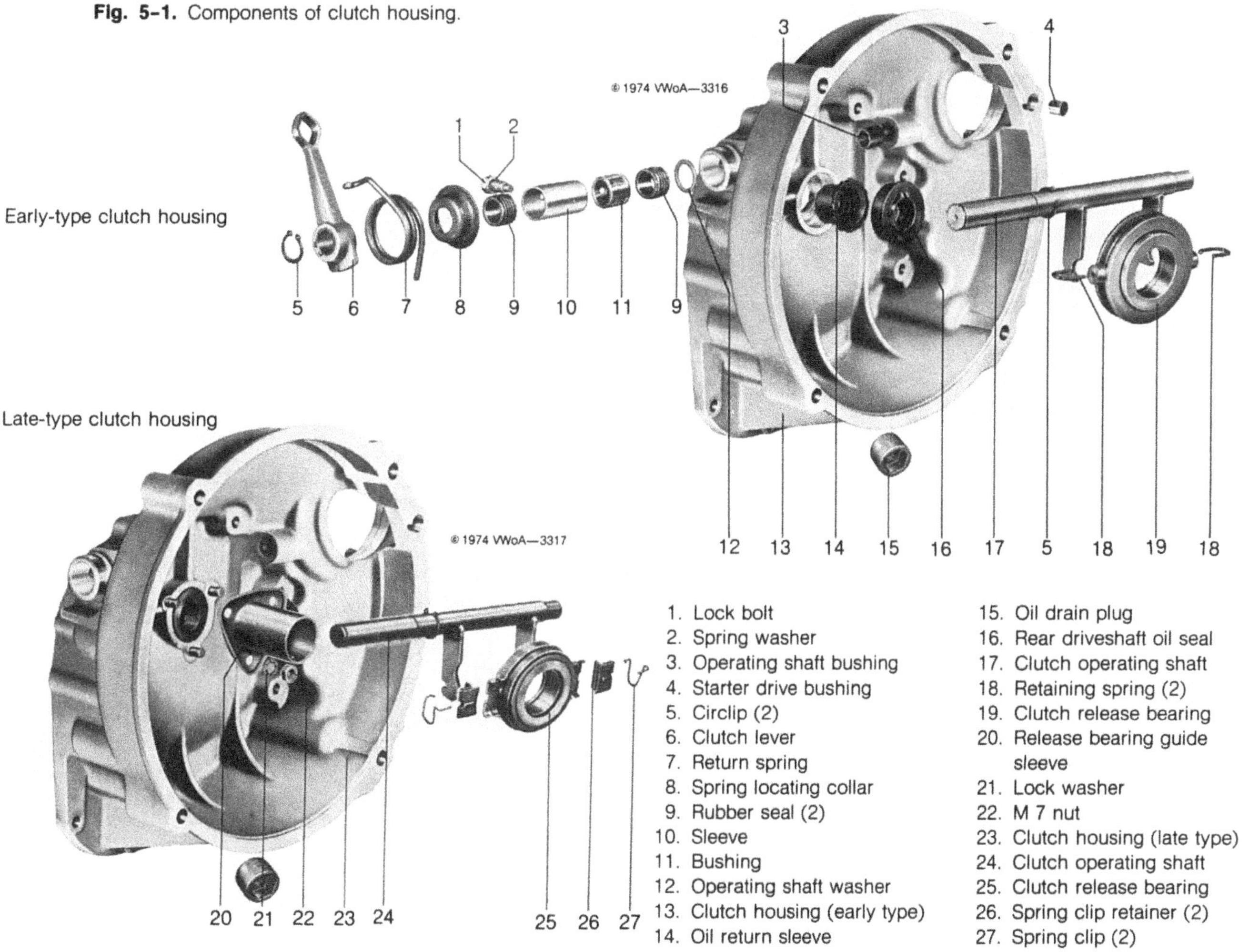

1. Lock bolt
2. Spring washer
3. Operating shaft bushing
4. Starter drive bushing
5. Circlip (2)
6. Clutch lever
7. Return spring
8. Spring locating collar
9. Rubber seal (2)
10. Sleeve
11. Bushing
12. Operating shaft washer
13. Clutch housing (early type)
14. Oil return sleeve
15. Oil drain plug
16. Rear driveshaft oil seal
17. Clutch operating shaft
18. Retaining spring (2)
19. Clutch release bearing
20. Release bearing guide sleeve
21. Lock washer
22. M 7 nut
23. Clutch housing (late type)
24. Clutch operating shaft
25. Clutch release bearing
26. Spring clip retainer (2)
27. Spring clip (2)

6

5.2 Disassembling and Assembling Clutch Housing

Although the following repairs are illustrated with a clutch housing that has been removed, the repairs can also be carried out with the clutch housing installed. The repairs may also be carried out with the transmission installed and the engine removed.

Replacing Starter Drive Bushing

The bushing for the starter drive pinion is a light press fit in the clutch housing. To remove it, take off the starter, then use a bushing puller as shown in Fig. 5-2 to extract the old bushing from the housing. Install the new bushing with a bushing driver that fits the inside of the bushing properly. Make certain that the bushing is started correctly. If it enters at an angle, both the bushing and clutch housing may be damaged. Lightly lubricate the bushing with multipurpose grease before you install the starter.

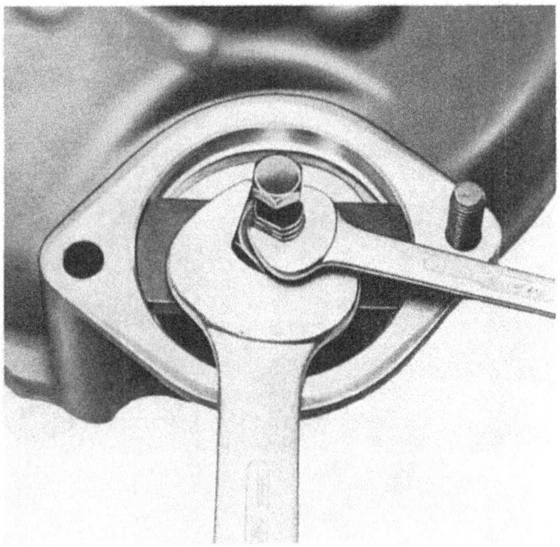

© 1974 VWoA—3319

Fig. 5-2. Starter drive bushing being removed from the clutch housing with puller VW 228b.

Replacing Rear Driveshaft Oil Seal

The rear driveshaft oil seal is located behind the clutch release bearing in the center of the clutch housing. It can be replaced without removing the rear driveshaft.

To remove seal:

1. Pry off the retaining springs or spring clips and remove the clutch release bearing from the arms on the clutch operating shaft.
2. If there is a clutch release bearing guide sleeve, remove the three nuts with washers and remove the sleeves.
3. Carefully pry out the damaged oil seal, as shown in Fig. 5-3.

© 1974 VWoA—3318

Fig. 5-3. A hook-type tool being used to remove the rear driveshaft oil seal. Be careful not to damage the seal recess in the housing or the oil return sleeve that is behind the seal.

To install:

1. Lightly coat the outer surface of the new seal with sealing compound. Lubricate the rear driveshaft and seal lip with transmission oil.

> **NOTE** —
> Make certain that the spring is correctly positioned around the seal lip and that the seal lip is installed toward the transmission.

2. Slide the oil seal onto the rear driveshaft (if installed). Then drive it in as shown in Fig. 5-4.

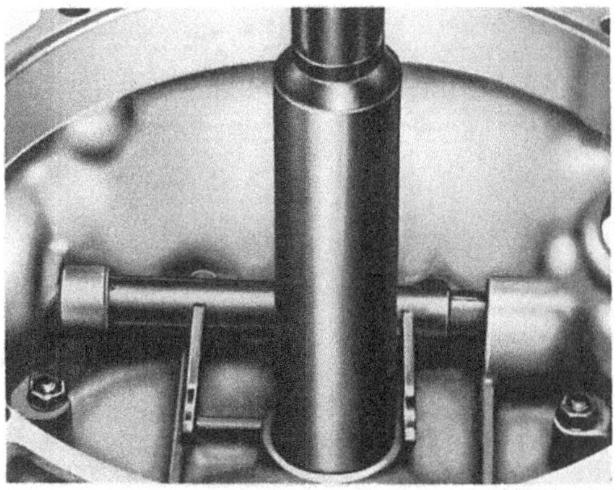

© 1974 VWoA—3322

Fig. 5-4. A sleeve-type bushing driver in place over the oil seal. Use a mallet to drive the seal in until it is flush with the clutch housing.

3. If the clutch housing is so equipped, install the clutch release bearing guide sleeve. Torque the nuts to 1.5 mkg (11 ft. lb.).
4. Service the clutch as described in **4. Removing and Installing Engine and Transmission**. Then install the clutch release bearing.

Replacing Oil Return Sleeve

The oil return sleeve is not subject to wear and does not normally need to be replaced. It must, however, be replaced if damaged.

To remove:

1. Remove the clutch housing. Then remove the oil seal as previously described.
2. Press the damaged oil return sleeve out toward the engine side of the housing.

To install:

1. Using a flat scraper, remove all burrs from the clutch housing where it will be contacted by the new oil return sleeve.
2. Align the new oil return sleeve so that the claws around its edge do not engage in the old holes. (Because of the holes left by prior sleeve installation, the oil return sleeve can be replaced one time only.)
3. Using a driver such as VW 295a, press the new bushing into position in the clutch housing. Use a pressure of 4 tons to ensure that the claws engage properly (Fig. 5-5).

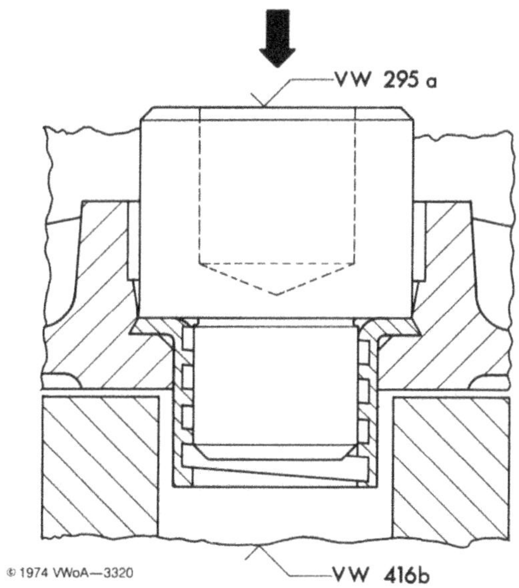

Fig. 5-5. Oil return sleeve being pressed in (arrow). Note that driver VW 295a has caused the claws to engage the casting. Support VW 416b must be used to support the housing as shown.

4. Use a machinist's square to check the position of the sleeve. If it is more than 0.20 mm (.008 in.) out of line with the clutch housing, replace the clutch housing.

Removing and Installing Clutch Operating Shaft

The names and positions of the parts for the clutch operating shaft are shown earlier in Fig. 5-1.

To remove:

1. Pry off the retaining springs on spring clips and remove the clutch release bearing from the arms on the clutch operating shaft.
2. If there is a clutch release bearing guide sleeve, remove the three nuts with washers and remove the sleeve.
3. Remove the circlip from the end of the operating shaft. Then pry off the clutch lever and remove the return spring and spring locating collar.
4. Remove the lock bolt for the clutch operating shaft sleeve and bushing.
5. Slide the clutch operating shaft outward so that the sleeve—together with the bushing and rubber seals—is pressed out of bore in the clutch housing.
6. Remove the sleeve, rubber seals, and flat washer from the operating shaft. Then remove the operating shaft from the housing inward.
7. To remove the bushing from the sleeve, use a screwdriver to compress the bushing at the slot. Then pull the bushing out of the sleeve.

6

Installation is the reverse of removal. Always replace bushings that have been removed from the sleeve. Inspect the rubber seals also and, if necessary, replace them.

Inspect the inner operating shaft bushing that is pressed into the boss inside the clutch housing. If it is worn, remove it with an extractor, then drive in a new bushing. Lightly lubricate both bushings with multipurpose grease before installing the operating shaft.

Examine the return spring. Replace it if it is badly corroded or weak. Before installing the return spring and clutch lever, make sure that the hole in the outer bushing and its sleeve are aligned with the lock bolt hole. Then install the lock bolt to a torque of 1.5 mkg (11 ft. lb.).

Before you install the engine, service the clutch as described in **4. Removing and Installing Engine and Transmission.** Additional information on servicing the clutch is given in **ENGINE AND CLUTCH**. The troubleshooting table given there is particularly useful in pinpointing the causes of clutch malfunctions.

6. DIFFERENTIAL

Fig. 6-1 gives exploded views that show the differential and its related parts in both early and late transmissions. Study the procedures for removal and disassembly carefully before undertaking any actual work.

Fig. 6-1. Exploded views showing differential removal.

Late-type transmission case

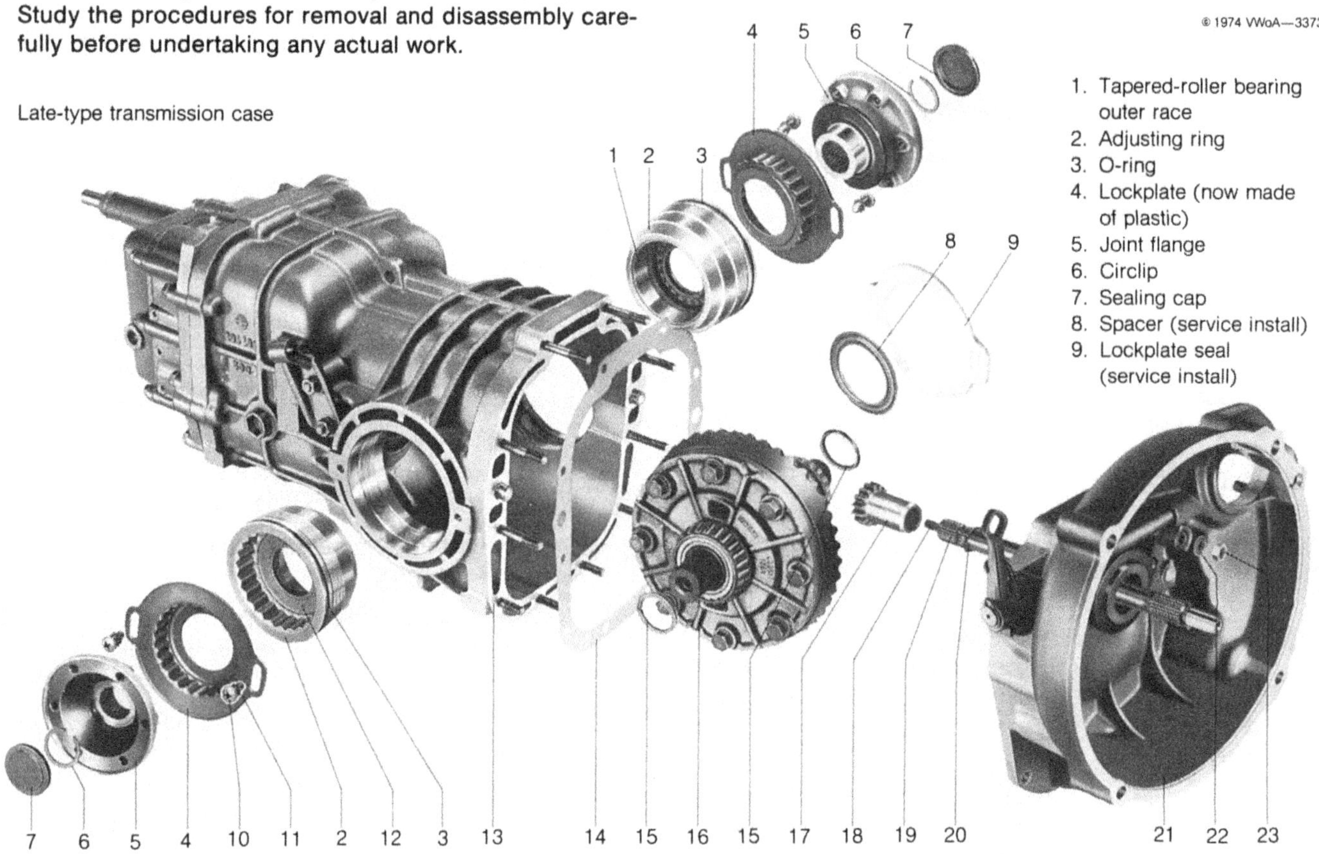

© 1974 VWoA—3373

1. Tapered-roller bearing outer race
2. Adjusting ring
3. O-ring
4. Lockplate (now made of plastic)
5. Joint flange
6. Circlip
7. Sealing cap
8. Spacer (service install)
9. Lockplate seal (service install)

Early-type transmission case

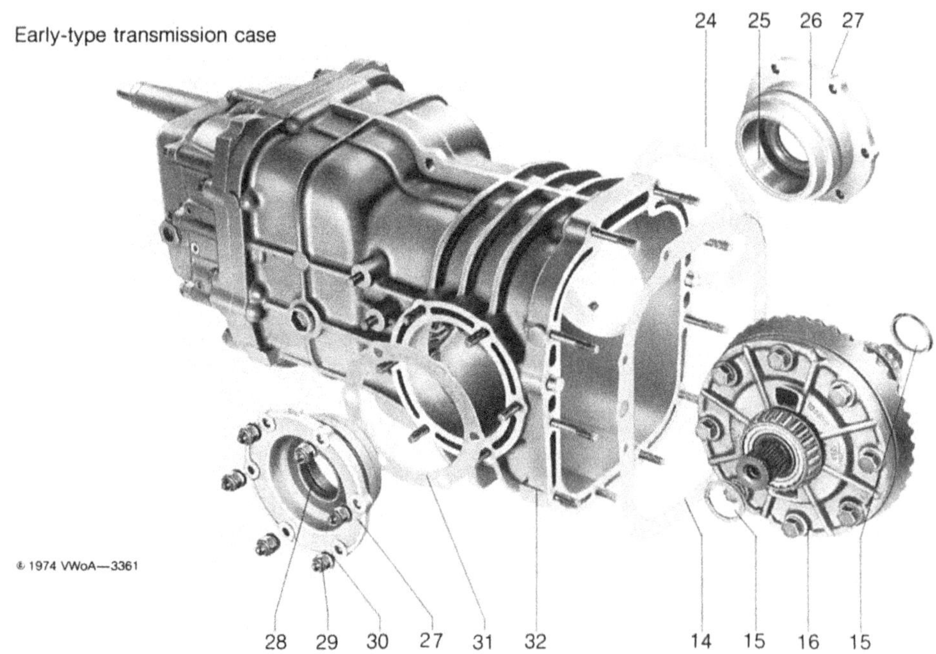

© 1974 VWoA—3361

10. Phillips head screw
11. Lock washer (or metal plate)
12. Oil seal
13. Transmission case (from August 1968)
14. Gasket
15. Thrust ring
16. Differential
17. Reverse drivegear
18. M 7 stud
19. Rear driveshaft
20. Circlip
21. Clutch housing
22. Spring washer
23. M 8 nut
24. Shim(s) S_2
25. Tapered-roller bearing outer race
26. O-ring
27. Final drive side cover
28. Oil seal
29. M 8 cap nut
30. Spring washer
31. Shim(s) S_1
32. Transmission case (through July 1968)

6.1 Removing and Installing Differential

You can replace the constant velocity joint flanges, the oil seals, and O-rings without removing the differential. The differential must be removed before you can take the transmission gears out of the case or repair the differential itself. If you intend to replace any part of the final drive, read **9. Adjusting Final Drive** to determine whether precision adjustments must be made following repair.

> ***CAUTION—***
> *If you lack the skills, tools, or a clean workshop for transmission repairs, we suggest you leave such repairs to an Authorized VW Dealer or other qualified shop. We especially urge you to consult your Authorized VW Dealer before attempting repairs on a car still covered by the new-car warranty.*

To remove differential:

1. Remove the engine and starter from the transmission. On early transmissions, loosen the cap nuts on the final drive side covers.
2. Remove the ten M 8 nuts, then remove the clutch housing as described in **5.1 Removing and Installing Clutch Housing**.
3. Remove the circlip from the rear driveshaft. Pull the reverse drivegear rearward, then remove the driveshaft by unscrewing it from the M 7 stud.
4. Using a screwdriver, pierce the plastic sealing caps in the centers of the joint flanges. Then pry out the sealing caps.
5. Remove the circlips from the centers of both joint flanges. Pry the flanges off the differential gearshafts as shown in Fig. 6-2.

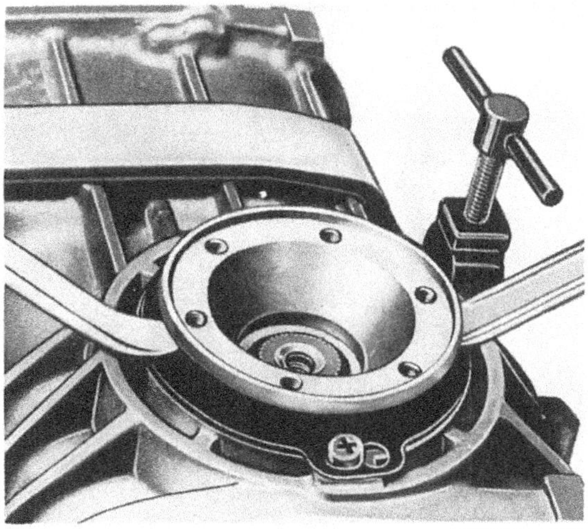

Fig. 6-2. Levers being used to pry off a joint flange.

6. On early transmissions, remove the six cap nuts from each final drive side cover. Lift the side covers off as shown in Fig. 6-3. Then, using string or wire, attach the $\mathbf{S_1}$ or $\mathbf{S_2}$ shims to their side covers and mark the covers to identify which side of the transmission they belong to.

Fig. 6-3. A slide hammer and puller being used to remove a side cover from an early transmission.

7. On late transmissions, remove the Phillips head screws. Then remove the lockplates.
8. On late transmissions, use a depth micrometer to measure the depth to which the adjusting rings are screwed into the case (or mark their position). Record the depth, then remove the rings as shown in Fig. 6-4.

6

Fig. 6-4. Tool VW 231/15 being used to remove an adjusting ring from a late transmission.

9. Keeping the teeth of the ring gear in mesh with those of the drive pinion, angle the ring gear end of the differential toward the rear of the transmission case, then pull the differential out of the case to the rear.

To Install:

1. Inspect the gears and bearings; replace any that are worn or damaged as described in **6.2 Disassembling and Assembling Differential** or in **6.3 Removing and Installing Tapered-roller Bearings.**

   NOTE —
   Consult **9. Adjusting Final Drive** to determine whether precision adjustments are required following replacement.

2. Inspect the oil seals. If their lips are worn or cracked, replace them as described in **6.3 Removing and Installing Tapered-roller Bearings.**

   NOTE —
   The final drive adjustments described in **9. Adjusting Final Drive** must be made without the oil seals installed. Install the oil seals only if adjustments are unnecessary.

3. Insert the differential assembly into the transmission case. The ring gear goes on the left side and must be in mesh with the drive pinion.

4. Install new O-rings on the final drive side covers (early transmissions) or on the adjusting rings (late transmissions).

5. On early transmissions, install the side covers on their original sides and with the original number and thickness(es) of S_1 or S_2 shims. Do not torque the cap nuts until the clutch housing has been installed.

   NOTE —
   To protect the bearings, hand-support the differential while installing the left side cover. Place the transmission left side down, then carefully lower the differential into the left side cover and keep it centered while installing the right side cover.

6. On late transmissions, apply Moly-Kote® to the threads on the adjusting rings. Install one adjusting ring to its original depth (Fig. 6-5).

7. On late transmissions, place the side of the transmission that has the adjusting ring installed downward. Carefully lower the differential into the adjusting ring. Then install the other adjusting ring to its original depth (Fig. 6-5).

8. On all late transmissions, service install new spacers and lockplate seals. Then install the lockplates, washers or metal plates, and Phillips head screws.

© 1974 VWoA—3382

Fig. 6-5. Adjusting ring being installed. Note the dial indicator being used with a bridge base to find the original screw-in depth. Though slower, a depth micrometer can also be used.

9. Install the clutch housing. Torque the ten M 8 nuts to 2.0 mkg (14 ft. lb.). On early transmissions, torque the cap nuts for the final drive side covers to 2.0 mkg (14 ft. lb.).

10. Install the thrust rings, joint flanges, and new circlips. If necessary, lift the differential gearshaft with an M 10 bolt and cross bar as shown in Fig. 6-6. Then, using pliers, squeeze the circlip into its groove.

11. Using a sleeve-type driver, drive new plastic sealing caps into the joint flange centers.

© 1974 VWoA—3383

Fig. 6-6. An M 10 bolt threaded into the differential gearshaft to pull the gearshaft up into the flange while the circlip is being installed.

6.2 Disassembling and Assembling Differential

An exploded view of the differential is given in Fig. 6-7. If you intend to replace any part of the differential, read **9. Adjusting Final Drive** to determine whether precision adjustments must be carried out following repair.

1. Tapered-roller bearing inner race (2)
2. Ring gear
3. Differential housing
4. Shaft retaining pin
5. Large thrust washer (2)
6. Long differential gearshaft
7. Spacer sleeve
8. Differential pinion (2)
9. Small thrust washer (2)
10. Differential pinion shaft
11. Short differential gearshaft
12. Differential housing cover
13. Washer (8)
14. M 9 bolt (8)

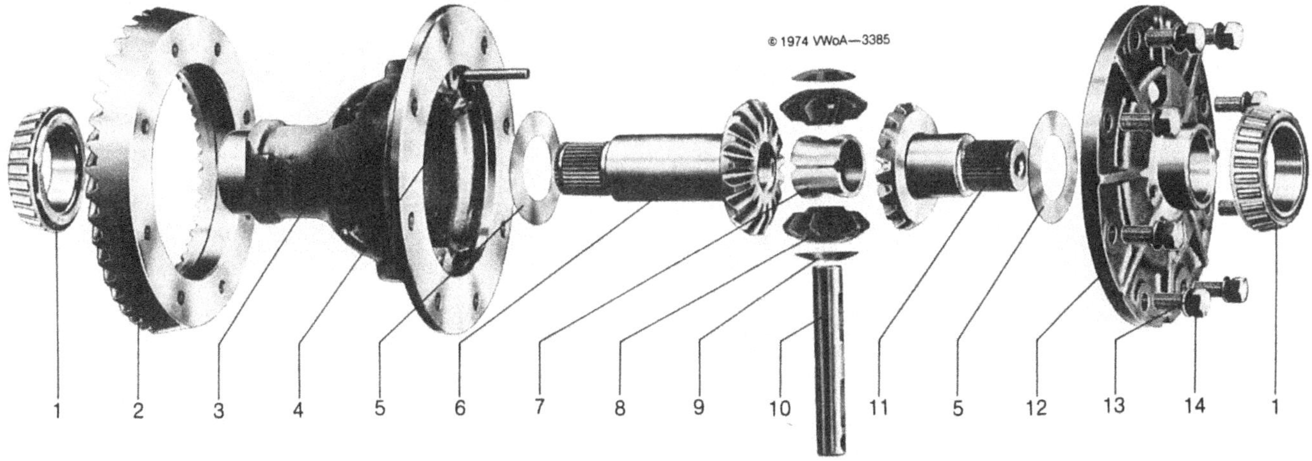

Fig. 6-7. Exploded view of differential.

CAUTION —
When removing the ring gear, be careful not to let it fall against the vise. Doing so could damage the ring gear teeth. A pad of rags placed beneath the ring gear is a worthwhile precaution.

To disassemble:

1. Remove the eight bolts, then drive off the ring gear as shown in Fig. 6-8.

Fig. 6-8. A drift being used to drive off the ring gear. Work alternately on both sides of the differential cover, going all around the ring of bolt holes until the gear is free.

2. Using a slide hammer as shown in Fig. 6-9, lift the cover from the differential housing.

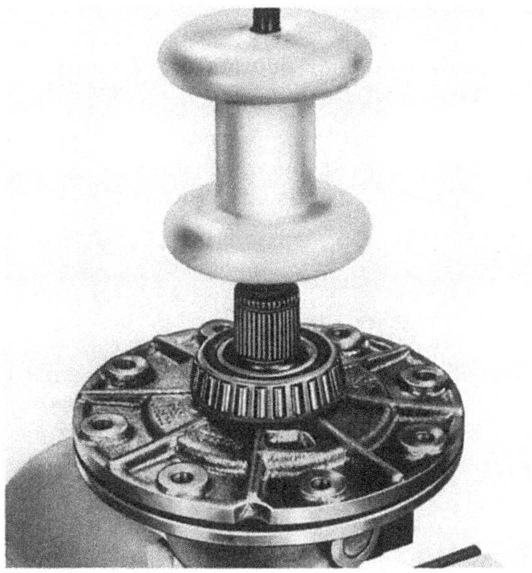

Fig. 6-9. Differential housing cover being removed by lifting the short gearshaft with a slide hammer.

6

3. Remove the short differential gearshaft and large thrust washer from the cover.

NOTE —
If necessary, the tapered-roller bearing inner race can be removed at this time as described in **6.3 Removing and Installing Tapered-roller Bearings.**

4. Using a drift, drive out the shaft retaining pin. Then drive the differential pinion shaft out of the differential housing (Fig. 6-10) and remove the spacer sleeve, thrust washers, and pinions.

© 1974 VWoA—3391

Fig. 6-10. Pinion shaft removal. Drive out first the retaining pin (top arrow) and then the pinion shaft (bottom arrow) in the directions that the arrows indicate.

To assemble:

1. Inspect the gear teeth for burrs or excessive wear. Replace as necessary. Examine all thrust surfaces on the differential housing, cover, ring gear, spacer sleeve, and thrust washers.
2. If the tapered-roller bearing inner races have been removed, install them as described in **6.3 Removing and Installing Tapered-roller Bearings.**
3. Install the long differential gearshaft and its large thrust washer in the differential housing.

NOTE —
If you notice gear-tooth damage (usually the result of insufficient backlash) on models not equipped with a spacer sleeve, the sleeve should be service installed as described in the procedure for adjusting gearshaft axial play. The axial play must also be adjusted if the gearshafts, thrust washers, differential housing, or cover are replaced.

4. Install the differential pinions, spacer sleeve, small thrust washers, and pinion shaft in the differential housing.
5. Install a new pinion shaft retaining pin. Then peen the housing over the pin as shown in Fig. 6-11.

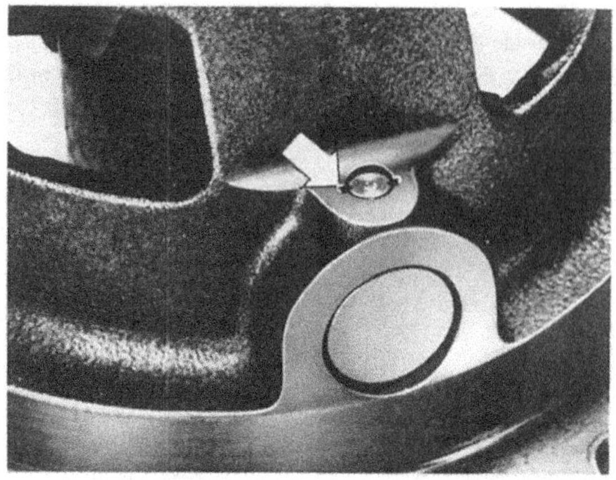

© 1974 VWoA—3392

Fig. 6-11. Metal of differential housing peened over the retaining pin (arrow).

6. Using an oilstone, remove all burrs and pressure marks from the mating surfaces of the housing, the housing cover, and the ring gear.
7. Install the short differential gearshaft and its large thrust washer in the differential housing cover. Then install the cover on the differential housing without the bolts.
8. Heat the ring gear to about 212°F (100°C) in a pan of oil placed in a larger pan of boiling water.
9. Install two centering pins in the housing bolt holes, then install the ring gear as shown in Fig. 6-12. Working diagonally, torque the bolts to 5.0 mkg (36 ft. lb.).

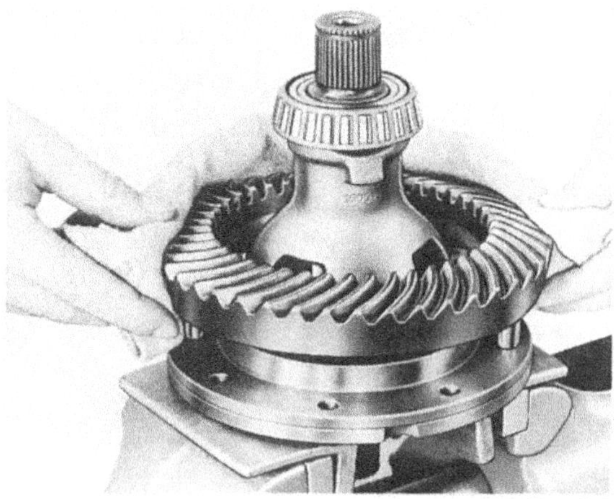

© 1974 VWoA—3395

Fig. 6-12. Heated ring gear being installed on the differential housing. Note the centering pins.

To adjust gearshaft axial play:

1. Install the short differential gearshaft with both large thrust washers in the differential cover, then clamp the gear tightly against the cover as shown in Fig. 6-13.

© 1974 VWoA—3396

Fig. 6-13. Differential gearshaft, cover, and thrust washers clamped together. Note clamping device.

2. Install the long differential gearshaft in the differential housing.
3. Measure the shortest available spacer sleeve—part No. 002 517 241—as shown in Fig. 6-14. Record the dimension.

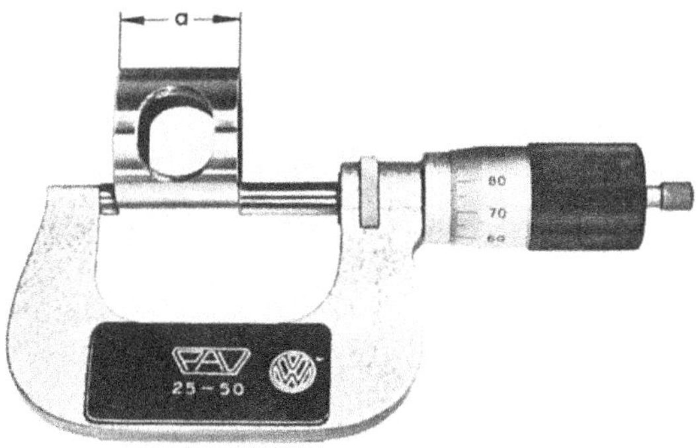

© 1972 VWoA—754

Fig. 6-14. Spacer sleeve length being measured with a micrometer to determine dimension **a**.

4. Install the spacer sleeve atop the machined face on the short gearshaft. Then install the differential housing and long gearshaft on top of the cover.
5. Install four M 8 bolts with nuts and washers and tighten them to draw the housing and cover completely together.
6. Install a measuring bar on the long differential gearshaft and attach a dial indicator as shown in Fig. 6-15. The gauge pin must contact the differential housing neck.

© 1974 VWoA—3397

Fig. 6-15. Gearshaft axial play being measured with a dial indicator.

7. Press down the long gearshaft and zero the dial indicator with a 1-mm (or .050 in.) preload. Ascertain the gearshaft axial play by moving the long gearshaft up and down.
8. Add the measured play to the known spacer sleeve length to obtain the X-range. Select the proper sleeve length as given in **Table a**.

6

Table a. Differential Spacer Sleeve Lengths

x range mm (in.)	Sleeve length (nominal) mm (in.)	Part No.
31.84-31.92 (1.253-1.256)	31.84 (1.253)	002 517 241
31.93-32.01 (1.257-1.260)	31.93 (1.257)	002 517 242
32.02-32.10 (1.261-1.263)	32.02 (1.261)	002 517 243
32.11-32.20 (1.264-1.268)	32.11 (1.264)	002 517 244

9. Disassemble the differential and install the correct spacer sleeve.
10. Assemble the differential. Then recheck the gearshaft axial play. It should be from 0.00 to 0.14 mm (.000 to .0055 in.).

If you expect to service differentials in the future, save the short spacer sleeve you used in making measurements. Scribe the length on its side and keep the sleeve as a permanent measuring tool.

6.3 Removing and Installing Tapered-roller Bearings

The tapered-roller bearing inner races are a press fit on the differential housing and differential housing cover. The bearing race on the housing can be removed and installed without removing the cover or ring gear. The outer races are a press fit in the final drive side covers (early transmissions) or the adjusting rings (late transmissions).

> **NOTE —**
> If you replace any of the double-tapered roller bearing races, you must carry out the precision adjustments described in **9. Adjusting Final Drive** following installation of the new bearings.

To remove inner races:

1. Support the differential as shown in Fig. 6-16. Then press the differential housing out of the inner bearing race.

> ***CAUTION —***
> *Do not let the differential fall suddenly. Doing so could damage the gearshaft or the bearing at the cover end.*

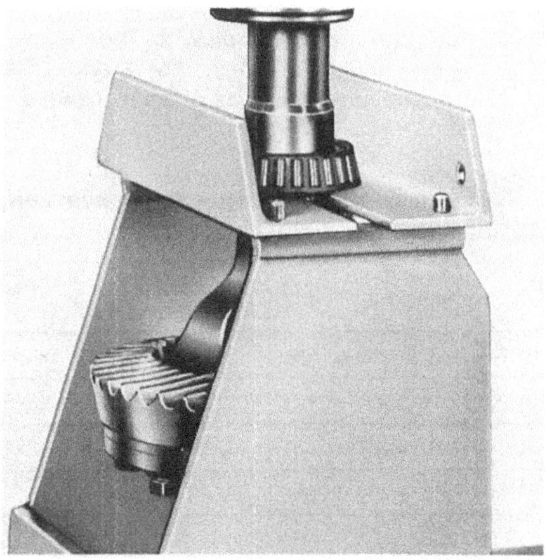

© 1974 VWoA—3390

Fig. 6-16. Inner bearing race being pressed off the differential housing.

2. To remove the bearing inner race from the differential housing cover, first remove the cover.
3. Using the special tool shown in Fig. 6-17, press off the bearing race. Alternately, you can drive off the bearing race with a drift.

© 1974 VWoA—3389

Fig. 6-17. Inner bearing race being pressed off the differential housing cover.

To install inner races:

1. Heat the inner races to about 212°F (100°C) in a pan of oil placed in a larger pan of boiling water.
2. Using a pressure of 3 tons, press the inner race onto the differential housing or cover with the setup shown in Fig. 6-18.

© 1974 VWoA—3393

Fig. 6-18. Inner race being pressed onto differential housing.

To remove outer races:

1. Press the oil seal out of the final drive side cover (early transmissions) or the adjusting ring (late transmissions) as shown in Fig. 6-19.

© 1974 VWoA—3377

Fig. 6-19. Oil seal being pressed out. The inner side of the side cover or adjusting ring should be up.

2. Using a late Type 2 adjusting ring as a support, press the tapered-roller bearing outer race out of the final drive side cover (early transmissions—as shown) or the adjusting ring. (See Fig. 6-20.)

© 1974 VWoA—3366

Fig. 6-20. Outer race being pressed out of final drive side cover (use same tools for adjusting ring).

3. Using the tools shown in Fig. 6-21, press the new outer bearing race into the side cover (early transmissions) or adjusting ring (late transmissions—as shown).

NOTE —
Before pressing in the outer races, use an oilstone to remove any burrs or pressure marks from the outer race and the bore in the side cover or adjusting ring.

© 1974 VWoA—3380

Fig. 6-21. Outer race being pressed into adjusting ring. The driving tool is of a diameter that will apply force only to the outermost edge of the outer race.

NOTE —
No lubricant should be applied to the outer race before it is pressed into the final drive side cover or the adjusting ring.

4. Moisten the seals with transmission oil. Then press the oil seal into the final drive side cover (early transmissions) or the adjusting ring (late transmissions) as shown in Fig. 6-22.

NOTE —
Unless the original bearings are being reinstalled, do not press in the oil seals until the final drive has been adjusted as described in **9. Adjusting Final Drive.**

6

© 1974 VWoA—3381

Fig. 6-22. Oil seal being pressed in from the outer end of the adjusting ring. On early transmissions, place a support tube under the center of the final drive side cover while the seal is being pressed in.

7. Transmission

Fig. 7-1 gives exploded views of both the early and late transmissions. The principal difference, insofar as the transmission gears are concerned, is in the method of retaining the pinion bearing.

The double tapered-roller pinion bearing for the later transmission has threads on its outer surface. It is retained in the transmission by a threaded retaining nut, whereas the double tapered-roller bearing in the early transmission is held by a large circlip and three bolt-on retainers.

Through chassis No. 219 020134

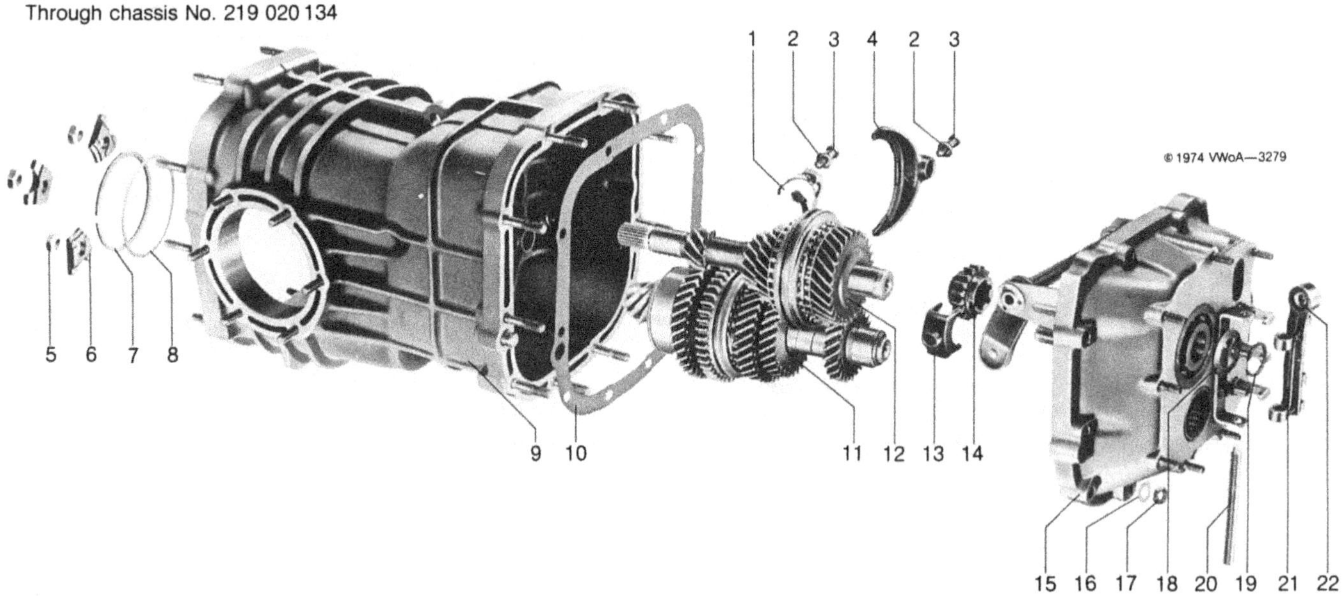

From chassis No. 219 020 135

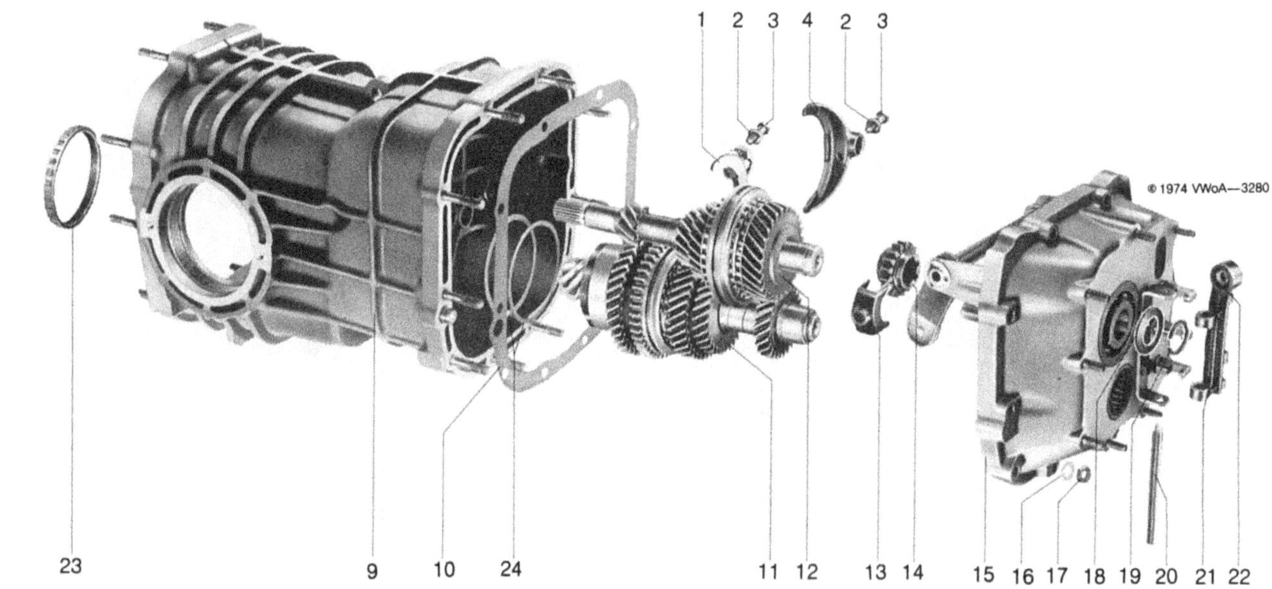

1. 1st/2nd gear shift fork
2. Spring washer (2)
3. Bolt (2)
4. 3rd/4th gear shift fork
5. Nut (3)
6. Retainer (3)
7. Circlip
8. **S₃** shim (early location)
9. Transmission case
10. Gear carrier gasket
11. Drive pinion
12. Mainshaft
13. Reverse gear shift fork
14. Reverse sliding gear
15. Gear carrier
16. Spring washer
17. Nut (9)
18. Dished washer
19. Circlip
20. Selector link shaft
21. Selector link
22. Selector link ball joint
23. Retaining nut
24. **S₃** shim (late location)

Fig. 7-1. Transmission gearset removal.

7.1 Disassembling and Assembling Shift Housing

The shift housing is a separate casting at the extreme front end of the transmission. It contains the inner shift lever which engages the ball joint in the selector link. The shift housing is removed from the transmission gear carrier by taking the nuts off eight studs.

To disassemble shift housing:

1. Using water pump pliers, twist out the bushing with oil seal. See exploded view in Fig. 7-2.

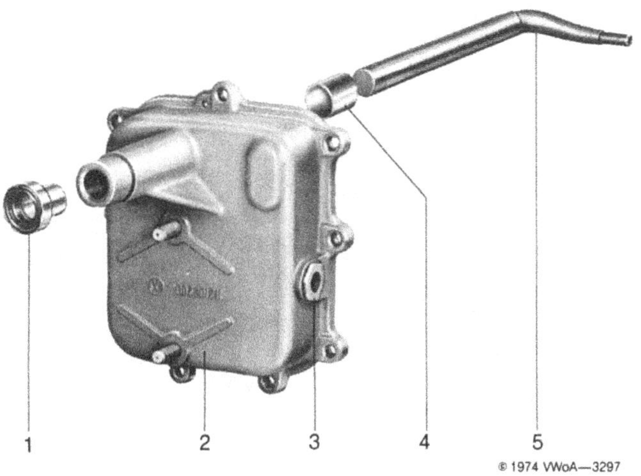

1. Bushing with oil seal
2. Shift housing
3. Plug with washer
4. Guide bushing
5. Inner shift lever

Fig. 7-2. Exploded view of shift housing.

2. Using a drift, drive out the guide bushing.
3. Unscrew the back-up light switch with washer or the plug with washer.

To assemble shift housing:

1. Press in the guide bushing.
2. Press in the bushing with oil seal, being careful not to damage the oil seal lip.

NOTE —
The bushings are finished to size and do not normally need reaming. However, if the inner shift lever operates stiffly, ream the bushings to 15.03 to 15.05 mm (.591 to .593 in.) with a 15-mm reamer. Be careful not to ream the oil seal lip.

3. Screw in the back-up light switch or plug together with its washer.

7.2 Removing and Installing Transmission Gear Train

When you disassemble the early-type transmission installed through chassis No. 219 020 134, it is always necessary to loosen the cap nuts for the final drive side covers before removing the clutch housing.

To remove gear train:

1. Remove the differential as described in **6.1 Removing and Installing Differential.**
2. Remove the shift housing.
3. On early transmissions, remove the three self-locking nuts shown in Fig. 7-3 and take off the retainers; on late transmissions, remove the retaining nut as shown in Fig. 7-4.

Fig. 7-3. Double tapered-roller pinion bearing held by retainers inside differential housing. This method applies to early models only.

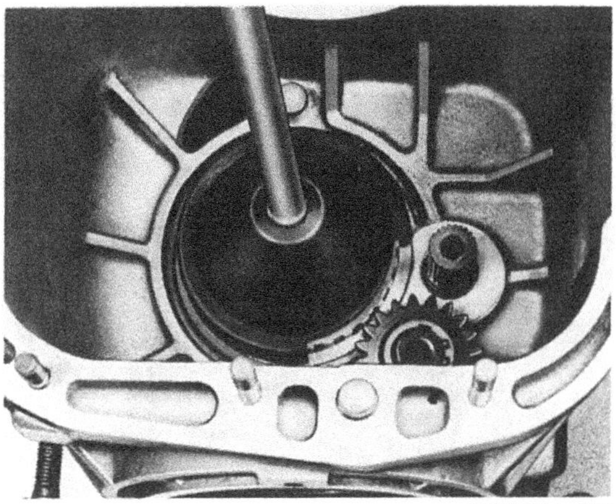

Fig. 7-4. Special tool VW 381/14 being used to remove the retaining ring for late-type pinion bearing.

6

4. Remove the selector link shaft and selector link. Then unbolt the bracket for the selector link from the face of the gear carrier.

5. On early transmissions, use a rubber mallet to lightly tap the end of the drive pinion where it projects from the gear carrier. This will expose the circlip on the double tapered-roller bearing inside the differential housing.

6. On early transmissions, remove the large circlip as shown in Fig. 7-5. Then remove the S_3 shim(s) and make note of the number and thickness(es).

© 1974 VWoA—3283

Fig. 7-5. Circlip pliers being used to remove the large circlip that surrounds the double tapered-roller pinion bearing on early transmissions.

7. Remove the nine nuts and spring washers that hold the gear carrier on the transmission case.

8. Press the gear train out of the case by applying leverage to the face of the pinion gear (Fig. 7-6).

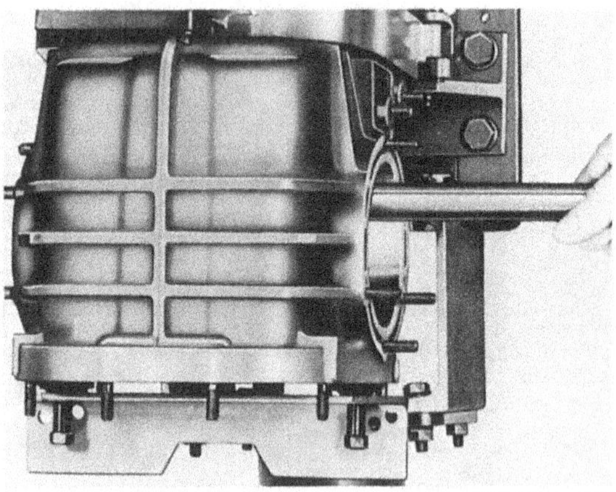

© 1974 VWoA—3285

Fig. 7-6. Gear train being pressed out. Notice the piece of angle iron installed across the transmission case. This provides support for removal tool VW 296, being used to lever out the gear train.

9. Working on the gear carrier and gear train, loosen the union nut on the clamp sleeve for the reverse lever support. Then turn the relay shaft just far enough so that you can remove the reverse sliding gear together with its shift fork.

10. Slide the shift forks for 3rd/4th and 1st/2nd gears off the shift rods.

11. Remove the circlip from the mainshaft as shown in Fig. 7-7.

WARNING —
The dished washer beneath the circlip is under considerable tension and can cause injury if it snaps out suddenly.

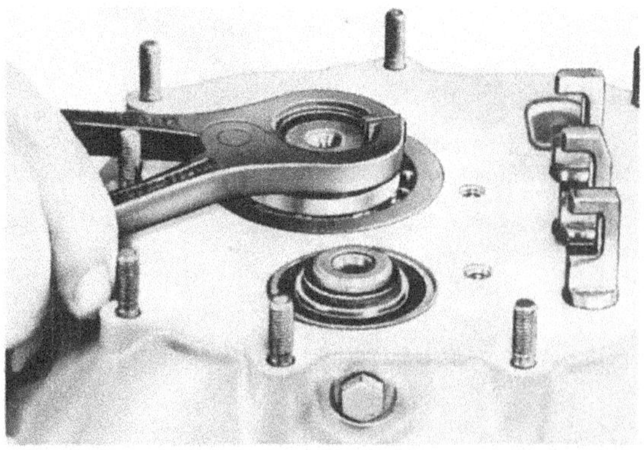

© 1974 VWoA—3286

Fig. 7-7. Circlip being removed from the mainshaft.

12. Press out the mainshaft and drive pinion by applying pressure to the mainshaft (Fig. 7-8). Carefully guide the gears to avoid damaging the teeth.

© 1974 VWoA—3287

Fig. 7-8. Gear train being pressed out of gear carrier.

To install:

1. Inspect the bearings, selector shafts, and other parts of the gear carrier. If necessary, replace worn or damaged parts as described in **7.3 Disassembling and Assembling Gear Carrier.**
2. Inspect the shift forks, dished washer, and reverse gear for wear and damage and, if necessary, replace parts that are no longer serviceable.
3. Place the mainshaft and drive pinion in mesh and support them in the position shown in Fig. 7-9. Then, using a sleeve-type driver that will apply pressure only to the mainshaft bearing inner race, press the gear carrier down onto the gear train.

CAUTION —
Make certain that the drive pinion enters its bearing properly. Otherwise the gear train or bearings may be severely damaged.

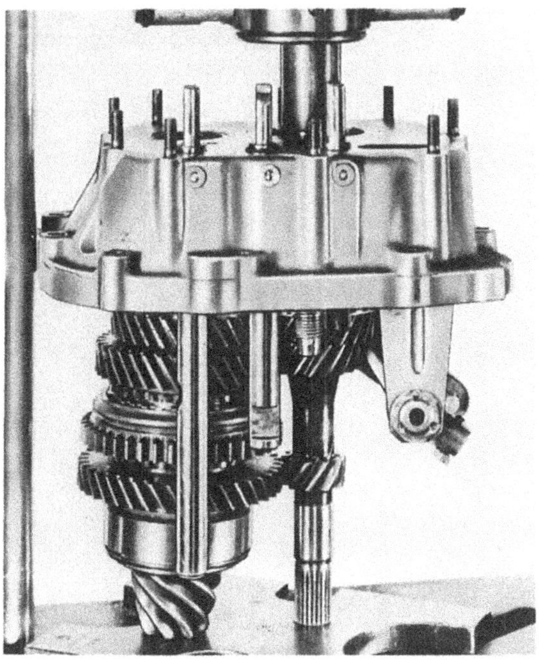

Fig. 7-9. Gear carrier being pressed down over gear train. The sleeve-type driving tool is placed over the bearing for the mainshaft.

4. Without removing the gear train from the press, install the dished washer and a new circlip on the mainshaft.
5. Using the same press tool, press the circlip down against the tension of the dished washer until the circlip snaps into the groove in the mainshaft.
6. Raise the press tool. Using waterpump pliers, squeeze the circlip all around until it bottoms in its groove.

7. Install the gear train in the adjusting jig. Then install and adjust the shift forks as described in **7.4 Adjusting Shift Forks**. Torque the shift fork setscrews to 2.5 mkg (18 ft. lb.) and the union nut for the reverse lever support to 3.0 mkg (22 ft. lb.).
8. On late-type transmissions, install the original number and thickness(es) of $\mathbf{S_3}$ shim(s) on the pinion bearing.

NOTE —
If the drive pinion or the double tapered-roller pinion bearing have been replaced, you must make precision adjustments as described in **9. Adjusting Final Drive.**

9. Place a new gear carrier gasket over the studs on the transmission case. Then insert the gear train into the case, carefully aligning the 12-sided shoulder on the pinion bearing with the recesses in the case (late transmissions only).
10. On early transmissions, use a rubber mallet to tap the end of the drive pinion where it projects from the gear carrier. This will drive the pinion bearing to the rear so that the $\mathbf{S_3}$ shims and circlip can be installed.
11. On early transmissions, install the original number and thickness(es) of $\mathbf{S_3}$ shim(s). Then install a new large circlip. Install the retainers and new self-locking nuts. Torque to 4.5 mkg (32 ft. lb.), back the nuts off slightly, then torque to 3.0 mkg (22 ft. lb.).
12. On late transmissions, install a new retaining nut. Torque to 22 mkg (160 ft. lb.), back the nut off slightly, then torque to 22 mkg (160 ft. lb.). Lock the nut by peening it as shown in Fig. 7-10.

6

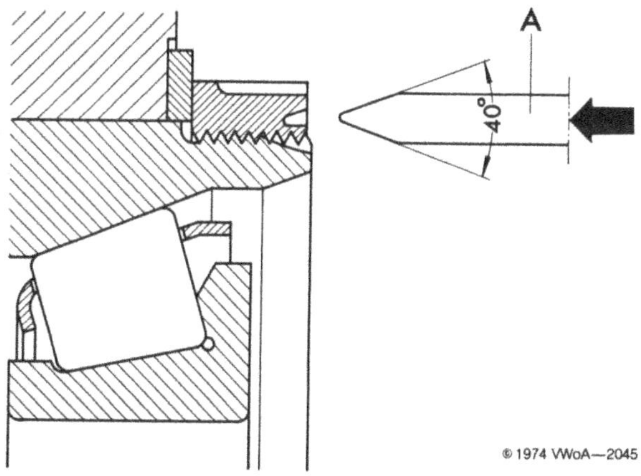

Fig. 7-10. Peening tool **A** being driven, as shown by arrow, to lock the retaining nut. On 1969 through 1971 models, grind a recess in the bearing so that the nut can be peened into it. This type of nut has been installed on all 1972 and later vehicles and is the only replacement part available.

13. Working diagonally, torque the 9 nuts that hold the gear carrier to 2.0 mkg (14 ft. lb.).
14. Install the selector link, bracket, then install the selector link and selector link shaft. Torque the bolts for the bracket to 2.5 mkg (18 ft. lb.).
15. Install the shift housing, making sure that the inner shift lever engages the ball joint in the selector link. Then torque the nuts to 1.5 mkg (11 ft. lb.).

7.3 Disassembling and Assembling Gear Carrier

The gear carrier can be disassembled after it has been removed from the transmission case. The selector shafts should be removed only if they are damaged or binding. See Fig. 7-11 for an exploded view of the gear carrier.

To disassemble:

1. Remove the selector link shaft and the selector link. Then remove the two special bolts and take the selector link bracket off the gear carrier.
2. Remove the lock bolt for the drive pinion needle bearing.
3. Use one side of press tool VW 433 to press out the mainshaft bearing (Fig. 7-12) and the other side to press out the pinion needle bearing. Alternately, use any press tool that will apply pressure only to the outer bearing race.

Fig. 7-12. Mainshaft bearing being pressed out.

4. Using an open end wrench, remove the clamp sleeve from the transmission case together with the reverse lever support and union nut.

Fig. 7-11. Exploded view of the gear carrier.

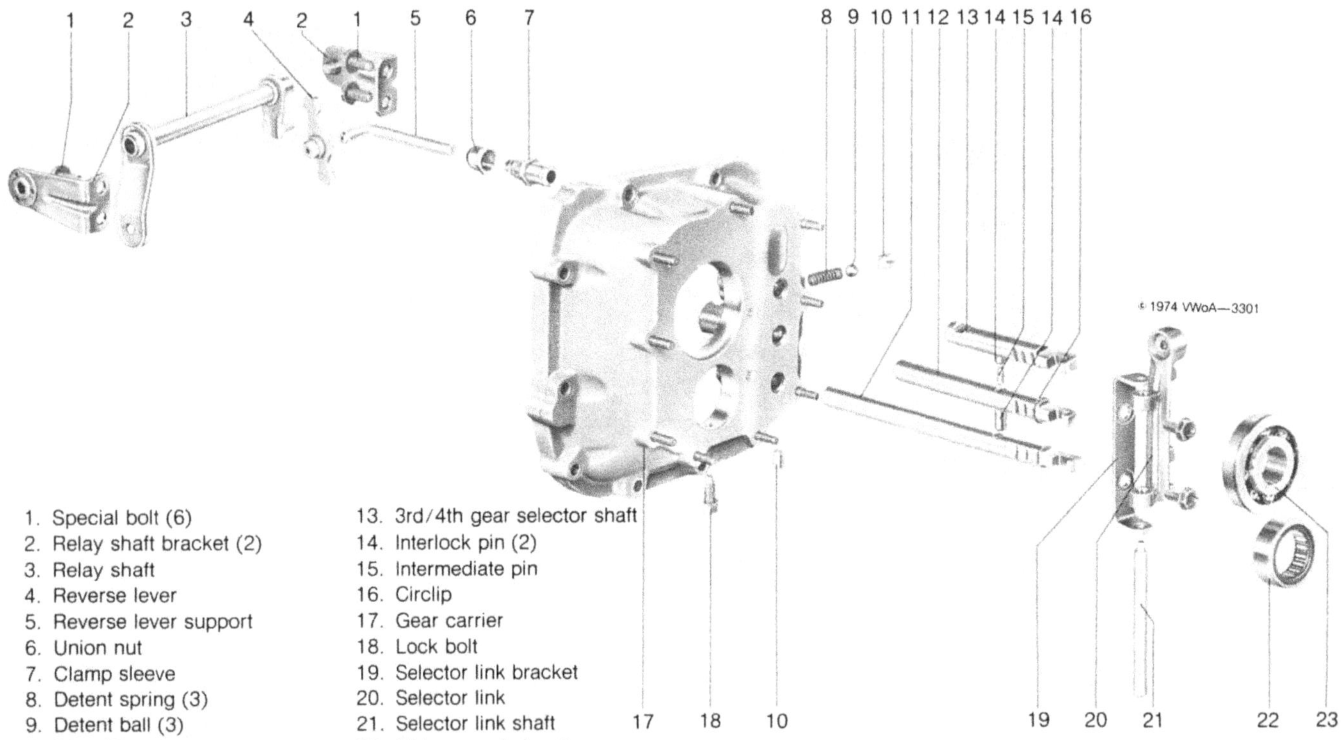

1. Special bolt (6)
2. Relay shaft bracket (2)
3. Relay shaft
4. Reverse lever
5. Reverse lever support
6. Union nut
7. Clamp sleeve
8. Detent spring (3)
9. Detent ball (3)
10. Plug (3)
11. 1st/2nd gear selector shaft
12. Reverse selector shaft
13. 3rd/4th gear selector shaft
14. Interlock pin (2)
15. Intermediate pin
16. Circlip
17. Gear carrier
18. Lock bolt
19. Selector link bracket
20. Selector link
21. Selector link shaft
22. Pinion needle bearing
23. Mainshaft bearing

5. Remove the four special bolts as shown in Fig. 7-13, then take off the relay shaft brackets together with the relay shaft.

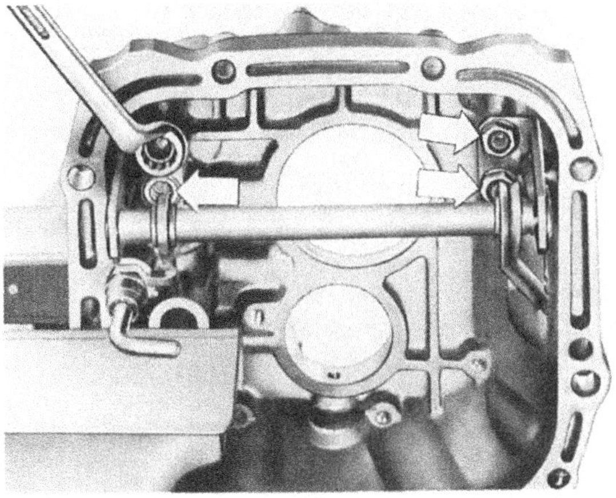

© 1974 VWoA—3305

Fig. 7-13. Special bolts (arrows) being removed.

6. Only drill out the plugs if it is necessary to replace the detent springs. Install a self-tapping bolt and pull out the plugs as indicated in Fig. 7-14.

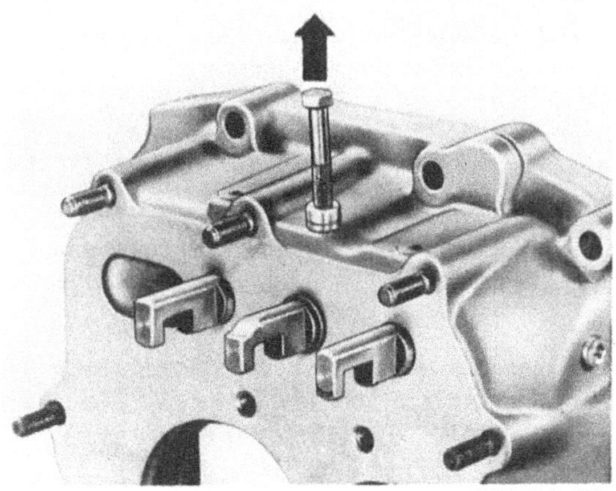

© 1974 VWoA—3304

Fig. 7-14. Plug being pulled out (arrow) with the help of a self-tapping bolt.

7. Only if it is necessary to replace a selector shaft, remove the circlip and then remove the selector shaft from the gear carrier.

NOTE —
If gearshifting is difficult, the selector shafts and detents should be checked as shown in Fig. 7-15 before assuming that the trouble is in the selector shafts. It should take 15 to 20 kg (33 to 44 lb.) to overcome the detent springs.

© 1974 VWoA—3307

Fig. 7-15. Detents being checked with a spring scale. If shaft moves too stiffly, check for bending or dirt in the bore; if it moves too easily, check the spring length. The springs should be 23 to 25 mm ($^{29}/_{32}$ to $^{63}/_{64}$ in.) long.

To assemble:

1. Inspect all parts. Replace any that are worn or damaged.
2. If removed previously, install the detent balls and detent springs, then drive in new plugs.
3. If removed previously, install the selector shafts using new circlips. Make certain that the interlock pins and intermediate pin are positioned properly.
4. Check the operation of the interlock mechanism. It should not be possible to engage two gears at the same time.
5. Using a suitable driving tool, press in the drive pinion needle bearing. Install the lock bolt and torque it to 1.5 mkg (11 ft. lb.).
6. Using a press tool that will apply pressure only to the outer race, press in the mainshaft bearing until its shoulder contacts the outer surface of the gear carrier.
7. Install the clamp sleeve together with the reverse lever support and union nut. Torque the sleeve to 4.5 mkg (32 ft. lb.).
8. Loosely install the reverse gear relay shaft together with its brackets.
9. Install the selector link bracket on the gear carrier and torque the special bolts to 2.5 mkg (18 ft. lb.).
10. Install the selector link and selector link shaft.
11. Adjust the play between the reverse sliding gear and its shift fork as described in **7.4 Adjusting Shift Forks.**

6

7.4 Adjusting Shift Forks

The shift forks can only be adjusted properly with the adjusting appliance VW 294b illustrated here. If the gear carrier has been disassembled, or the reverse relay shaft and its brackets removed for any reason, adjust the play between the reverse sliding gear and its shift fork before installing the gear train in the gear carrier and adjusting the 1st/2nd and 3rd/4th gear shift forks.

To adjust reverse sliding gear shift fork play:

1. Install the gear carrier together with the reverse sliding gear and its shift fork in the adjusting appliance VW 294b.
2. Loosen the special bolts on the reverse relay shaft brackets.
3. Slide the bracket on the shift-fork side of the relay shaft onto the shaft until the shift fork firmly contacts the reverse sliding gear.
4. To obtain running clearance, slide the bracket slightly sideways as indicated in Fig. 7-16. Then torque the special bolts that hold the bracket to 2.0 mkg (14 ft. lb.).

© 1974 VWoA—3310

Fig. 7-16. Right-hand bracket adjustment. Slide the bracket on its elongated bolt holes, as indicated by the left arrow, to obtain a slight running clearance between the shift fork and reverse sliding gear (right arrow).

5. Press the other bracket against the relay shaft until there is no end play. Torque the special bolts indicated in Fig. 7-17 to 2.0 mkg (14 ft. lb.), then remove the gear carrier from the adjusting appliance.

© 1974 VWoA—3311

Fig. 7-17. Left-hand bracket adjustment. After tightening bolts (top arrow), clearance should be equal at both ends (lower arrows) of reverse lever. If it is not, loosen union nut, rotate reverse lever support to obtain equal clearance, then torque union nut to 3.0 mkg (22 ft. lb.).

To adjust 1st/2nd and 3rd/4th gear shift forks:

1. If the final drive requires precision adjustments, make them before adjusting the shift forks.
2. Install the gear train in the gear carrier. Then install both in the adjusting appliance.

NOTE —
The gear carrier gasket must be in place when you adjust the transmission installed through chassis No. 219 020 134. Install the late gear carrier in the appliance without a gasket but with the S_3 shim(s) on the pinion bearing.

3. On early transmissions, install the correct S_3 shim(s) and the large circlip. Then install the clamping flange shown in Fig. 7-18.

© 1974 VWoA—3291

Fig. 7-18. Clamping flange installed over pinion bearing on early transmission. Tighten bolts uniformly.

4. On late transmissions, install the special retaining nut (supplied with the adjusting appliance) on the pinion bearing threads and hand-tighten only (Fig. 7-19).

© 1974 VWoA—3290

Fig. 7-19. Special retaining nut used on the pinion bearing of the late transmission.

5. Loosen the union nut on the reverse lever support. Move the support and lever aside, then install the 1st/2nd and 3rd/4th gear shift forks on the selector shafts, but do not tighten the clamp bolts.
6. Reconnect the reverse lever as shown in Fig. 7-20.

© 1974 VWoA—3292

Fig. 7-20. Reverse lever and support being moved (arrows) to reconnect lever. Slightly tighten union nut.

7. Pull out the lower selector shaft (for the 1st/2nd gear shift fork) until it engages the detent for 2nd gear.
8. Slide the operating sleeve (Fig. 7-21) together with the shift fork in the same direction until the operating sleeve slides over the teeth on the synchronizing ring and presses firmly against the 2nd gear.

© 1974 VWoA—3293

Fig. 7-21. Operating sleeve being moved. 2nd gear is at the right end of the double-ended arrow.

9. Center the shift fork in the groove in the operating sleeve and torque the clamp bolt to 2.5 mkg (18 ft. lb.).
10. While turning the transmission, select 2nd gear, neutral, and 1st gear. The selector fork must not rub against the sides of the groove in any of these positions. If it does, readjust it.
11. Push in the upper selector shaft (for the 3rd/4th gear shift fork) until it engages the detent for 3rd gear.
12. Adjust the shift fork using the same procedure employed in adjusting the 1st/2nd gear shift fork. Then torque the clamp bolt to 2.5 mkg (18 ft. lb.).

CAUTION —
The mainshaft ball bearing in the gear carrier must be fully seated when you adjust the 3rd/4th gear shift fork. Otherwise, your adjustment will be inaccurate.

13. Adjust the union nut on the reverse lever support until the support is just free to slide in and out of its clamp sleeve.
14. Push in the center selector shaft (for reverse gear) until its circlip contacts the face of the gear carrier. If necessary, tighten the reverse lever support union nut just enough to keep the reverse lever in this position.
15. Without touching its shift fork, press the reverse sliding gear toward the gear carrier until there is about 0.50 to 1.00 mm (.020 to .039 in.) clearance between the reverse sliding gear and the 2nd gear on the mainshaft (Fig. 7-22). Then, with the sliding gear in this position, torque the union nut to 3.0 mkg (22 ft. lb.).

© 1974 VWoA—3294

Fig. 7-22. Clearance between reverse sliding gear and 2nd gear on the mainshaft. Dimension **a** must be 0.50 to 1.00 mm (.020 to .039 in.).

NOTE —
There should be no axial play in the reverse sliding gear shift fork or relay shaft when you adjust the sliding gear. If there is, carry out the procedure given earlier for adjusting reverse sliding gear shift fork play.

16. Select reverse gear and check the engagement of the teeth on the reverse sliding gear with those on the 1st/2nd gear operating sleeve. If necessary, correct the position of the sliding gear.

CAUTION —
To prevent contact with 2nd gear, correct the sliding gear's position only as shown in Fig. 7-23.

© 1974 VWoA—3295

Fig. 7-23. Sliding gear engagement correction. Move sliding gear away from 2nd gear (arrow) by moving the shift lever support. Then torque the union nut to 3.0 mkg (22 ft. lb.).

17. Disengage reverse and engage 2nd gear. Make sure there is clearance between the reverse sliding gear and the teeth on the 1st/2nd gear operating sleeve (Fig. 7-24). If not, make adjustments to the sliding gear by moving the shift lever support. Then torque the union nut to 3.0 mkg (22 ft. lb.).

© 1974 VWoA—3296

Fig. 7-24. Point where clearance must exist (arrow) between the sliding gear and the operating sleeve.

18. Check the interlock mechanism. You should be able to move one selector shaft into a gear only when the other two are in neutral.

7.5 Disassembling and Assembling Mainshaft

Fig. 7-25 gives an exploded view of the mainshaft used through 1971. The new mainshaft, introduced on the 1972 models, is shown in Fig. 7-26.

1. Circlip
2. Dished washer
3. 4th gear thrust washer
4. 4th gear
5. 4th gear needle bearing
6. 4th gear needle bearing inner race
7. 4th gear synchronizing ring
8. Clutch gear assembly
9. 3rd gear synchronizing ring
10. 3rd gear
11. 3rd gear needle bearing
12. Woodruff key
13. Mainshaft

Fig. 7-25. Exploded view of early mainshaft.

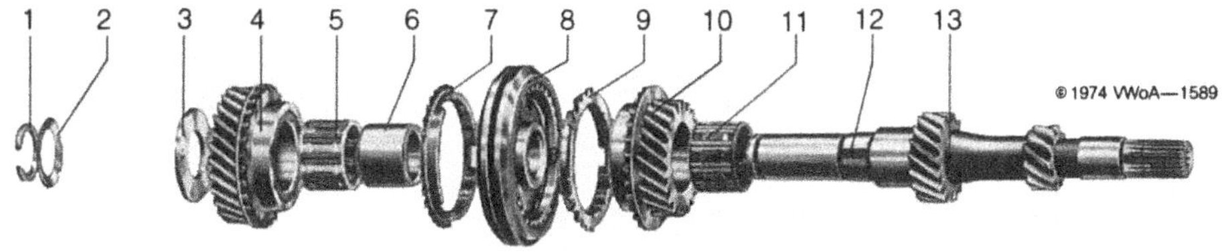

Fig. 7-26. Exploded view of late mainshaft. The 4th gear needle bearing inner race is integral with the mainshaft. The synchronizer hub is located by splines rather than by a Woodruff key. The axial location of the synchronizer hub is controlled by circlips (arrows) rather than by the 4th gear needle bearing inner race as on the early-type mainshaft.

To disassemble:

1. Remove the mainshaft from the gear carrier. Then remove the 4th gear thrust washer, 4th gear, its needle bearing, and synchronizing ring.
2. On early mainshafts, press off the inner race, the clutch gear assembly, and 3rd gear, as shown in Fig. 7-27.
3. On late-type mainshafts, remove the circlip. Then slide off the clutch gear assembly.
4. On late-type mainshafts, remove the second circlip. Then slide off 3rd gear. Open up the split needle bearing cage just enough to lift it over the splines on the mainshaft.
5. If necessary, remove the spring rings from the clutch gear assembly and separate the synchronizer hub from its operating sleeve.

NOTE ——
Individual parts are available for the clutch gear assembly. However, if the assembly is in serviceable condition, disassemble it only if cleaning is necessary. The spring rings for the 3rd/4th clutch gear assembly are 74 mm ± 1 mm (2.913 in. ± .040 in.) in diameter while those for the 1st/2nd clutch gear are larger—78 mm ± 1 mm (3.071 in. ± .040 in.) in diameter.

6

Fig. 7-27. 4th gear needle bearing inner race being pressed off together with synchronizer hub and 3rd gear.

To assemble:

1. Inspect all parts. Replace any that are worn or damaged.

   NOTE —
   Especially check the synchronizing rings. Modified synchronizing rings became available in 1972. These prevent the "howling noise" sometimes caused by the earlier type synchronizing rings when the transmission is cold. You may wish to install these rings in earlier cars even if the original parts are not badly worn.

2. Hand-press the synchronizing rings into the gears as illustrated in Fig. 7-28. Measure clearance **a** with a feeler gauge. The wear limit is 0.60 mm (.023 in.). With new parts, dimension **a** should be between 1.00 and 1.9 mm (.040 and .075 in.).

Fig. 7-28. Clearance measurement **a**.

3. After aligning the marks indicated in Fig. 7-29, assemble the clutch gear assembly if it was taken apart.

   NOTE —
   Late-model and replacement synchronizer hubs and operating sleeves are not marked for alignment. Mesh the teeth in various positions until a free-sliding fit is found.

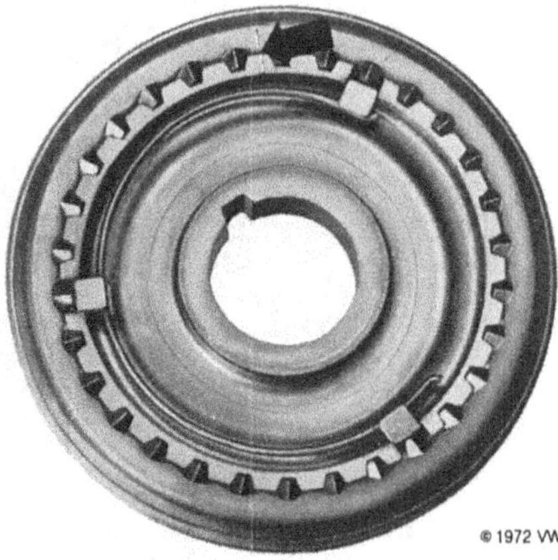

Fig. 7-29. Alignment marks (arrow).

NOTE —
The spring rings on opposite sides of the clutch gear assembly must be installed 120° apart as indicated in Fig. 7-30.

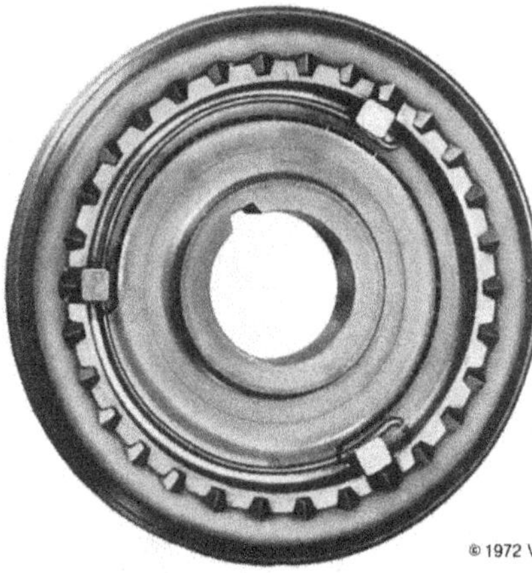

Fig. 7-30. Positions for the spring rings. Make sure the spring ring ends are fully hooked over the synchronizer keys.

The remainder of assembly is the reverse of disassembly with the following additional steps:

1. When installing the clutch gear assembly, the side of the operating sleeve containing the 1-mm-deep groove goes toward 4th gear, and the side of the clutch gear hub having the wide chamfer on the teeth goes toward 3rd gear.

2. The needle bearing inner race (where fitted) must be heated to approximately 212°F (100°C) in a pan of oil placed in a larger pan of boiling water before it is pressed onto the mainshaft as shown in Fig. 7-31.

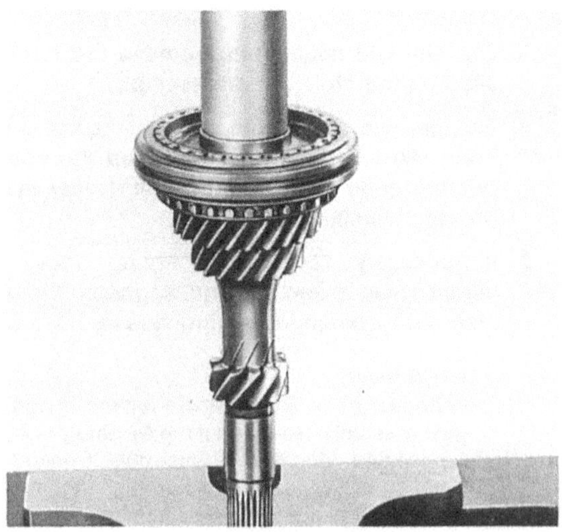

Fig. 7-31. Needle bearing inner race being pressed onto the mainshaft.

7.6 Disassembling and Assembling Drive Pinion

Fig. 7-32 gives an exploded view of the drive pinion. Familiarize yourself with the part names. They are used frequently in the text.

© 1974 VWoA—1596

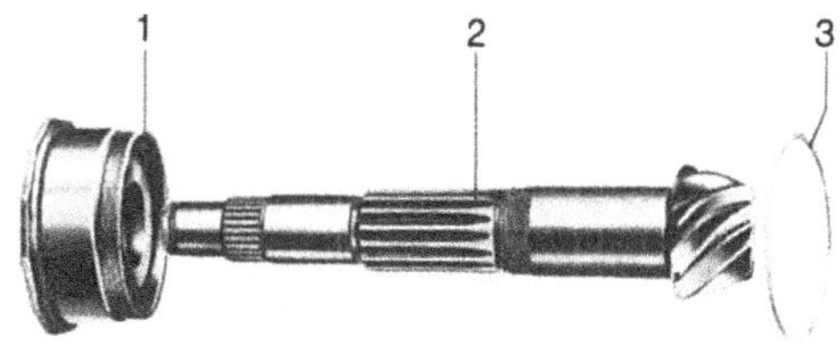

Fig. 7-32. Exploded view of the drive pinion.

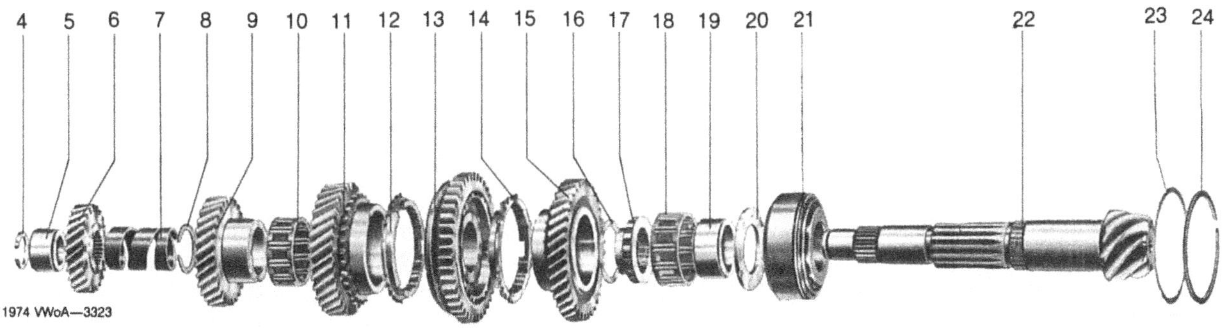

© 1974 VWoA—3323

1. Double tapered-roller pinion bearing (late)
2. Pinion (late)
3. S_3 shim
4. Small circlip
5. Pinion bearing inner race
6. 4th gear
7. Spacer spring
8. Medium-size circlip
9. 3rd gear
10. 2nd gear needle bearing
11. 2nd gear
12. 2nd gear synchronizing ring
13. Clutch gear assembly
14. 1st gear synchronizing ring
15. 1st gear
16. Shim for round nut
17. Round nut
18. 1st gear needle bearing
19. Inner race for 1st gear needle bearing
20. 1st gear thrust washer
21. Double tapered-roller pinion bearing (early)
22. Drive pinion (early)
23. S_3 shim
24. Large circlip

NOTE —
Read **9. Adjusting Final Drive** before deciding to replace the drive pinion or the double tapered-roller bearing.

To disassemble:

1. Hold 4th gear down with the repair press to keep the gear from flying off and to make removing the circlip easier. Then remove the small circlip.
2. If the needle bearing inner race is tight, press the drive pinion out of the inner race with 4th gear supported.
3. Lift off the spacer spring and then remove the medium-size circlip.
4. Remove 3rd gear, 2nd gear, the needle cage, the synchronizing rings and clutch gear assembly, the shim for the round nut, and the needle cage for 1st gear.
5. Mount the drive pinion in special appliance VW 293 as shown in Fig. 7-33. Then unscrew the round nut by turning the drive pinion counterclockwise at the splined end.

   NOTE —
   Make certain that the drive pinion is securely fastened because it will take considerable leverage to break the round nut free.

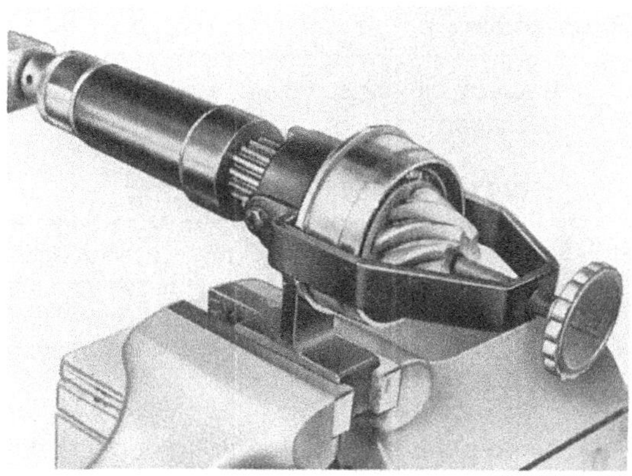

© 1972 VWoA—856

Fig. 7-33. Drive pinion mounted in special appliance. A special splined socket wrench is required that will fit over the splines on the drive pinion. Otherwise there is no way to attach a torque wrench.

6. If necessary, remove the spring rings from the clutch gear assembly and separate the synchronizer hub from its operating sleeve.

   NOTE —
   Individual parts are available for the clutch gear assembly. However, if the assembly is in serviceable condition, disassemble it only if cleaning is necessary.

6

7. Press the double tapered-roller bearing, 1st gear thrust washer, and the inner race for the 1st gear needle bearing off the drive pinion together as shown in Fig. 7-34.

© 1974 VWoA—1599

Fig. 7-34. Drive pinion being pressed out of double tapered-roller bearing and other parts. Make sure that only the inner race of the bearing is supported.

To assemble:

1. Inspect all parts. Replace any that are worn or damaged.

NOTE —
Especially check the synchronizing rings. A modified synchronizing ring became available in 1972. It helps 1st gear to synchronize in both directions. You may wish to install this ring in earlier vehicles even if the original part is not badly worn.

2. Hand-press the synchronizing ring into the gear as illustrated in Fig. 7-35. Measure clearance **a** with a feeler gauge. With new parts, dimension **a** should be between 1.10 and 1.80 mm (.043 and .071 in.). The wear limit is 0.60 mm (.023 in.).

© 1974 VWoA—1591

Fig. 7-35. Clearance measurement **a**.

3. Reassemble the clutch gear assembly (if it was taken apart earlier). See **7.5 Disassembling and Assembling Mainshaft** for details.

4. Heat the double tapered-roller bearing to about 212°F (100°C) in a pan of oil placed in a larger pan of boiling water. Then install the bearing on the drive pinion. When it has cooled to room temperature, seat the drive pinion in the bearing using a pressure of 3 tons.

5. Similarly heat the inner race for the 1st gear needle bearing and press it onto the drive pinion together with the 1st gear thrust washer.

6. Using the special appliance illustrated earlier in Fig. 7-32, install the round nut. Torque it to 20.0 mkg (145 ft. lb.).

7. Install the drive pinion in the transmission case (see **9. Adjusting Final Drive**). Using a special torque gauge, check the turning torque of the double tapered-roller bearing. Turning torque should be 6 to 21 cmkg (5.2 to 18.3 in. lb.) for a new bearing or 3 to 7 cmkg (2.6 to 6.1 in. lb.) for a bearing that has been in service for 30 mi. (50 km) or more.

NOTE —
Lubricate the bearing with hypoid oil only. See **9.1 Adjusting Drive Pinion** for complete details.

8. Using a blunt cold chisel (Fig. 7-36), peen the locking shoulder of the round nut into the drive pinion splines at three places 120° apart. Make sure you do not crack or burr the locking shoulder. Replace round nuts that have a cracked locking shoulder.

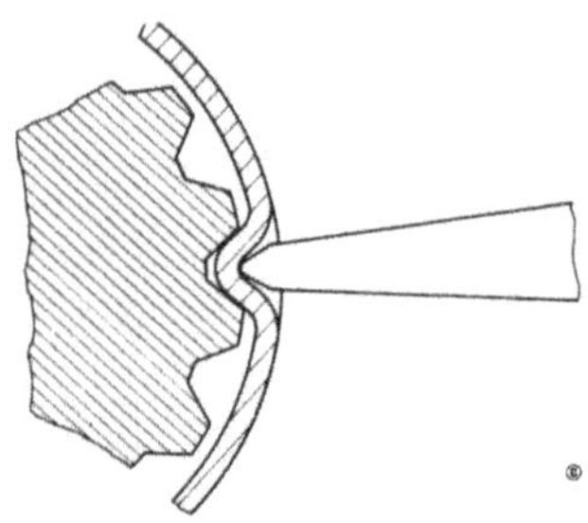

© 1974 VWoA—2069

Fig. 7-36. A blunt cold chisel being used to peen the round nut locking shoulder into the splines.

9. Inspect the 1st/2nd gear synchronizer hub of the clutch gear assembly. If there is no groove at the point indicated in Fig. 7-37, install the shim for the round nut. If the hub has the groove, do not install a shim. (The illustrated hub is the standard replacement part, Part No. 113 311 242 C.)

10. Install the 1st gear needle bearing, 1st gear, its synchronizing ring, and the clutch gear assembly.

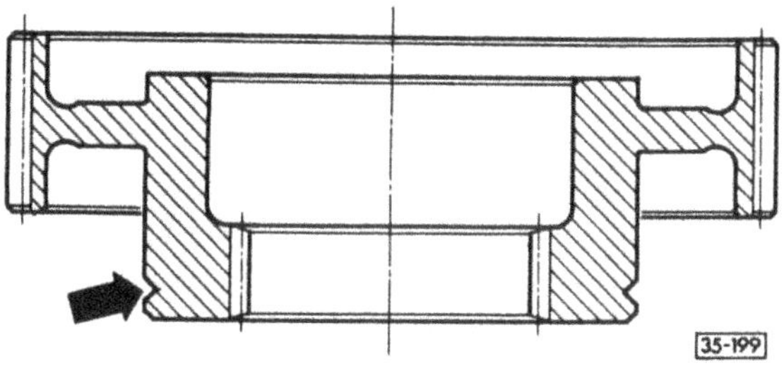

Fig. 7-37. Thicker 1st/2nd gear synchronizer hub with identification groove (arrow).

11. Only if you have installed the shim for the round nut, check the 1st gear axial play (Fig. 7-38). To do this, hold down the clutch gear. Then insert feeler gauges between the 1st gear and 1st gear thrust washer. The clearance between the gear and the thrust washer should be 0.10 to 0.25 mm (.004 to .010 in.). If not, select a thicker or thinner shim for the round nut.

© 1974 VWoA—1603

Fig. 7-38. Measuring 1st gear axial play. Shims for the round nut are available in thicknesses from 0.55 to 0.85 mm in increments of 0.05 mm.

12. Assemble the drive pinion up to 3rd gear. Install the medium-size circlip, then check axial play as illustrated in Fig. 7-39. Clearance should be between 0.10 and 0.25 mm (.004 and .010 in.), preferably the lower limit.

© 1972 VWoA—696

Fig. 7-39. Axial play for 3rd gear being measured with a feeler gauge. To adjust this clearance, circlips are available in thicknesses from 1.45 to 2.20 mm in increments of 0.15 mm (see **11. Transmission and Rear Axle Technical Data**).

13. Install the spacer spring. Heat the pinion needle bearing inner race to about 212°F (100°C) as described earlier. Then press the inner race onto the drive pinion together with 4th gear.
14. While holding down 4th gear against spacer spring tension, install the small circlip.

8. REVERSE GEARS AND MAINSHAFT BEARING

The reverse driven gear, reverse gear shaft, its bearings, and the needle bearing for the mainshaft are shown in Fig. 8-1. These parts remain in the transmission case after the gear train and the final drive have been removed. From February 1972, the reverse gear and shaft were modified as shown in Fig. 8-2.

© 1974 VWoA—3312

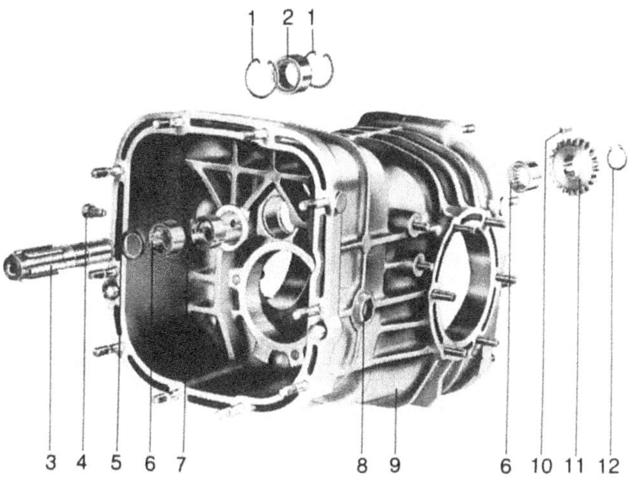

1. Needle bearing lockring
2. Mainshaft needle bearing
3. Reverse gear shaft (through January 1972)
4. Lock bolt (through January 1972)
5. Reverse gear shaft thrust washer
6. Reverse gear shaft needle bearing
7. Spacer sleeve
8. Oil filler plug
9. Transmission case
10. Woodruff key (through January 1972)
11. Reverse driven gear (through January 1972)
12. Circlip

Fig. 8-1. Parts mounted in transmission case.

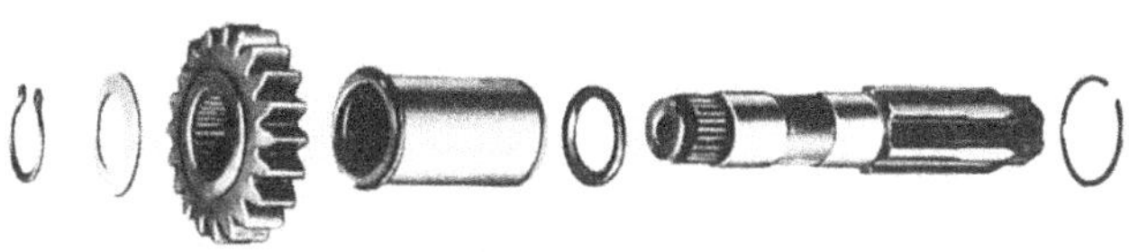

© 1974 VWoA—1551

Fig. 8-2. Reverse gear shaft from February 1972. Note the one-piece bearing with lockrings.

6

A further modification to the reverse driven gear and reverse gear shaft was made beginning with the 1974 models. The strengthened parts are shown in Fig. 8-3.

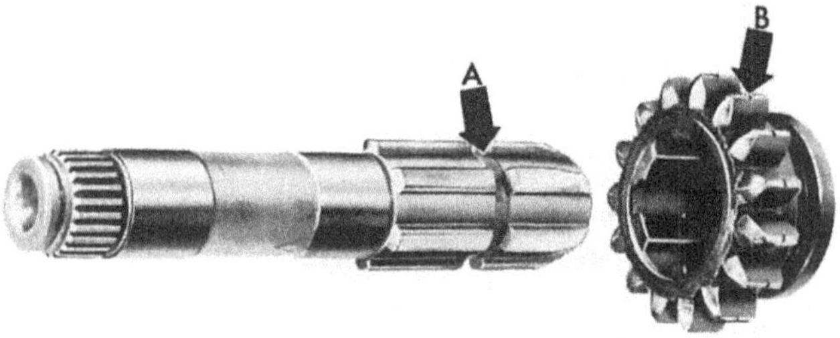

Fig. 8-3. Gear and shaft used in 1974 transmissions. The grooves at arrows **A** and **B** identify the new parts, which can be service installed as a set in vehicles built since November 1972.

To dissassemble transmission case:

1. Remove the circlip from the reverse gear shaft. Then pull off the reverse driven gear and remove the Woodruff key (where fitted).
2. Remove the reverse gear shaft by pushing it out into the transmission housing.
3. Remove both lockrings for the mainshaft needle bearing. Drive out the bearing with a suitable bushing driver (Fig. 8-4).

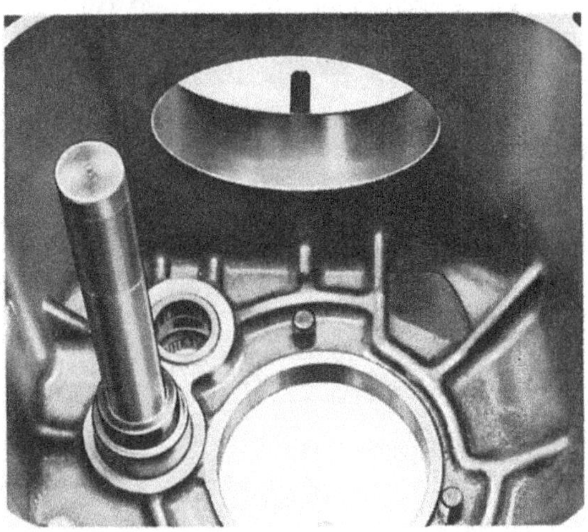

Fig. 8-4. Mainshaft needle bearing being driven out.

4. On transmissions built through January 1972, remove the lock bolt. On later transmissions, remove the lockring from the final drive end of the reverse gear shaft needle bearing. Then, using a suitable bushing driver, drive out the bearing from the final drive end.

To assemble:

1. Inspect all parts. Replace any that are worn or damaged.
2. Using the same bushing driver used in removal drive in the mainshaft needle bearing and install two new lockrings.
3. On transmissions built before February 1972, insert the spacer sleeve for the reverse shaft needle bearings so that the hole in its side aligns with the lock bolt hole in the transmission case. Then install the lock bolt to a torque of 1.5 mkg (11 ft. lb.).
4. On transmissions built before February 1972, drive one needle bearing race in on either side of the spacer sleeve until both contact the spacer sleeve. The side of the bearings with lettering must be toward the driving tool as shown in Fig. 8-5.

Fig. 8-5. Correct relationship of bearing to driving tool.

5. On transmissions built from February 1972, install one lockring on the reverse shaft needle bearing assembly. Drive the bearing in until the lockring contacts the transmission case, then install the other lockring.
6. Using the reverse of the removal procedure, install the reverse gear shaft, reverse driven gear, and related parts.

9. ADJUSTING FINAL DRIVE

Careful adjustment of the mesh between the ring gear and pinion is essential to ensure long rear axle service and quiet operation. The adjustments described here are necessary to the life of the gearset, the differential bearings, and drive pinion bearing. The final drive gearset requires adjustments only when parts directly affecting the adjustment have been replaced or when careless disassembly has resulted in the loss of the original shims or the adjusting ring screw-in depth measurements.

Table b lists what adjustments must be made when certain parts are replaced. In all cases, adjust the shift forks following pinion adjustment and not before.

Table b. Necessary Adjustments for Replaced Parts

Part replaced	Parts to be adjusted		
	Pinion	**Shift forks**	**Ring gear**
Transmission case	X	X	X
Final drive covers			X
Differential tapered-roller bearings			X
Circlip on pinion bearing	X	X	
Pinion double tapered-roller bearing	X	X	
Ring gear and pinion	X	X	X
Differential housing			X
Differential housing cover			X

Final drive adjustments are made by adjusting bearing turning torque and the positions of the ring and pinion gears. Through chassis No. 219 020 134, the S_1 and S_2 shims indicated in Fig. 9-1 control ring gear position. The S_3 shim determines pinion gear position.

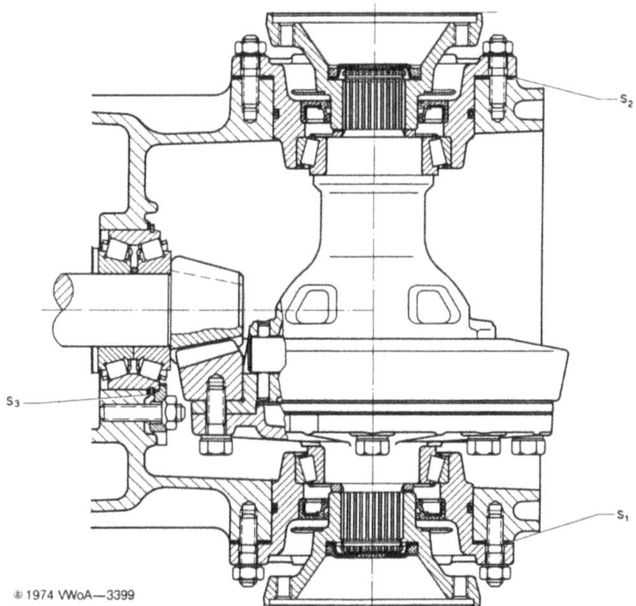

© 1974 VWoA—3399

Fig. 9-1. Early final drive adjustments. S_3 shim(s) control pinion position; S_1 shim(s) on ring gear side and S_2 shim(s) on opposite side control ring gear position.

From chassis No. 219 020 135, the S_3 shim is relocated and the bearing adjusting rings determine ring gear position—there are no shims (Fig. 9-2). The symbols used in final drive adjustment are defined in **Table c.**

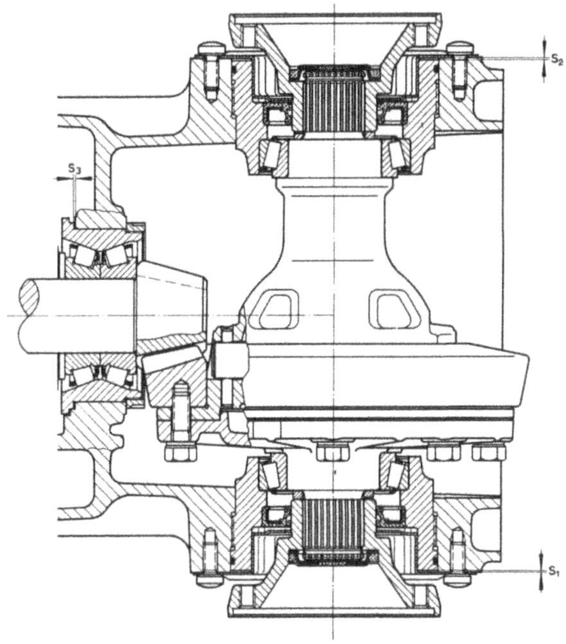

© 1974 VWoA—3400

Fig. 9-2. Late final drive adjustment.

Table c. Symbols Used in Final Drive Adjustment

Symbol	Description	Dimension
S_1	Screw-in depth of adjusting ring (ring gear side)	measured in 1/100 mm
S_2	Screw-in depth of adjusting ring (opposite side)	—
Sv_o mean	Average of several backlash measurements	measured in 1/100 mm
M.S.	Measuring shim VW 381/10	1.30 mm
w	Correction factor for individual gearset	—
h	Ring gear lift from no-play mesh position with its matching pinion	measured in 1/100 mm
R_o	Length of master gauge used in factory testing machine	63.00 mm
r	Deviation from R_o	marked on gearset in 1/100 mm (25 = 0.25)
e	Difference between mandrel and measuring pin	measured in mm range: 0.85–1.30
S_3	Thickness(es) of shim(s) that determine axial placement of the pinion	measured in 1/100 mm

6

The ring gear and pinion are run on special testing machines during transmission manufacture to check the tooth contact pattern and silent running under both drive and coast conditions. When the optimum relationship of the two gears is found, they are installed in the transmission using S_3, S_1, and S_2 adjustments that will duplicate the gearset's position in the testing machine. The purpose of all subsequent adjustments is to restore the gearset to this position following repair.

9.1 Adjusting Drive Pinion

If the pinion must be adjusted, do so before adjusting either the shift forks or the ring gear. Dimension **R**, given in Fig. 9-3, is the actual distance from the pinion face to the ring gear centerline. It differs from the length of the master gauge R_o that is used in the factory testing machine, by dimension **r**. Through April 1972, the deviation **r** is recorded on the outer face of the ring gear.

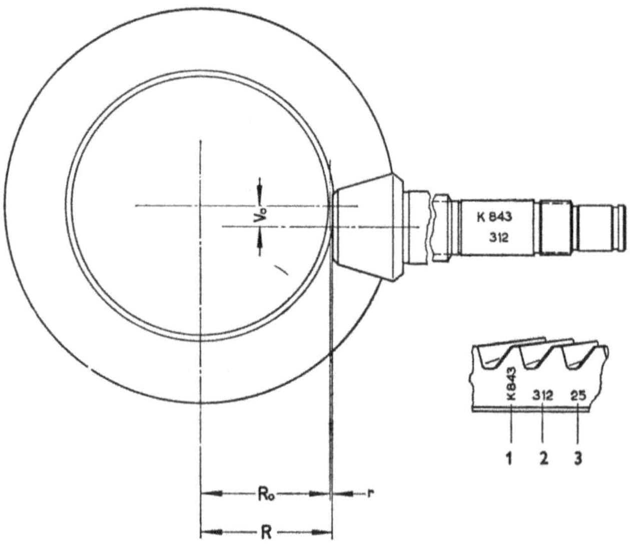

1. Manufacturer (Klingelnberg) and tooth ratio (8/43)
2. Matching number to show that the ring gear and pinion are a matched set (gearset 312)
3. Deviation **r** (given in 1/100 mm)

Fig. 9-3. Pinion adjustment dimensions and gearset marking used through April 1972. Dimension V_o is the 10-mm hypoid offset built in during manufacture and is not adjustable.

Gearsets manufactured from approximately April 1972 are adjusted to a flexible dimension called P_o, as shown in Fig. 9-4. The deviation **r** and the gearset matching number no longer appear on these gearsets. If final drive adjustments will be required (see **Table b** given earlier), measurements must be made before you disassemble one of these gearsets so that the pinion can be adjusted to its original position following repairs.

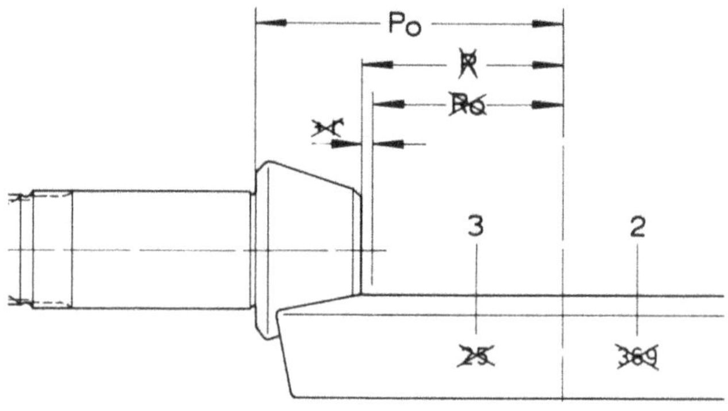

P_o. New setting dimension
X. Discontinued setting dimensions
2. Discontinued matching number on both ring gear and pinion
3. Discontinued deviation **r** stamped on ring gear

Fig. 9-4. Current pinion setting dimension. The manufacturer's initial and tooth ratio still appear.

NOTE —
Replacement gearsets are still marked with deviation **r** and the matching number. If a new ring gear and pinion are to be installed, you do not need to make measurements before disassembling a late model transmission.

The transmissions installed through chassis No. 219 020 134 require several different tools and somewhat different adjusting procedures from the transmissions installed on later models. Therefore, the adjustment of early and late transmissions will be covered separately.

CAUTION —
If you lack the skills, special measuring tools, or a clean workshop for adjusting the final drive, we suggest you leave such repairs to an Authorized VW dealer or other qualified and properly equipped shop. We especially urge you to consult your Authorized VW Dealer before attempting repairs on a vehicle covered by the new-car warranty.

To adjust pinion (through chassis No. 219 020 134):

1. Assemble the drive pinion up to the needle bearing for 1st gear. Tighten the round nut to 20 mkg (144 ft. lb.) but do not lock it.
2. Install the partially assembled drive pinion in the transmission case without any S_3 shim(s). Install the large circlip.
3. Install the three pinion bearing retainers and torque the special nuts to 4.0 mkg (29 ft. lb.), back them off, then retorque to 3.0 mkg (21 ft. lb.).

4. Lubricate the double tapered-roller bearing with hypoid oil, then install a torque gauge as shown in Fig. 9-5.

© 1974 VWoA—3402

Fig. 9-5. Gauge for measuring pinion turning torque.

5. Using the handle of the torque gauge, spin the pinion rapidly 15 or 20 turns in each direction. Take torque readings while continuing to spin the pinion. Turning torque should be 6 to 21 cmkg (5.2 to 18.3 in. lb.) for a new bearing or 3 to 7 cmkg (2.6 to 6.1 in. lb.) for a bearing that has been in service for 30 mi (50 km) or more.

NOTE —
Used bearings must have no visible axial play. Use only hypoid oil as a lubricant. The test results will be inaccurate if the turning torque is checked with the bearings dry or lubricated with another kind of oil.

6. Install the left (ring gear side) final drive cover without an oil seal or O-ring and with the special 1.30-mm-thick measuring washer VW 381/10 in place of the **S_1** shim(s). Torque the nuts to 2.0 mkg (14 ft. lb.).

NOTE —
On some very early transmissions, final drive covers were used which had convex ribs between the stud holes. When adjusting one of these transmissions, a later-type cover must be temporarily installed.

7. Install mandrel VW 381/2. The cylindrical measuring surface must be on the reverse gear shaft side of the transmission case (Fig. 9-6).

NOTE —
All surfaces must be perfectly clean in order to ensure that the measurements will be accurate.

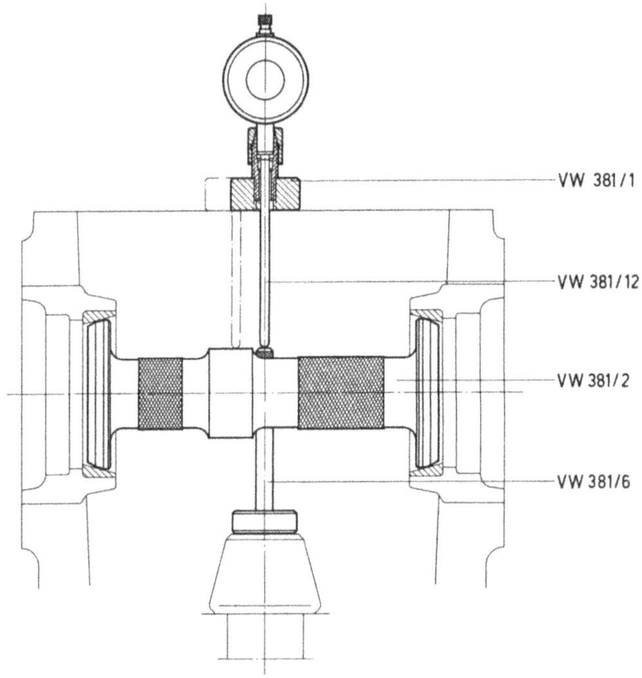

© 1974 VWoA—3404

Fig. 9-6. Mandrel VW 381/2 in position. The cylindrical measuring surface has a 40-mm diameter. Other special measuring tools are also shown. Two measuring positions are shown for the dial indicator.

8. Install the right final drive cover without an oil seal or O-ring and with special measuring washer VW 381/10 in place of the **S_2** shim(s). Install VW 381/2 over the cover as shown in Fig. 9-7. Then torque the nuts to 2.0 mkg (14 ft. lb.).

© 1974 VWoA—3405

Fig. 9-7. Mandrel and drive pinion setting tool in place.

9. Using the twin-handled spindle on the setting tool, press in the right-hand differential bearing outer

race until the mandrel can just barely be hand-turned.

10. Install setting tool VW 381/6 (83.00 mm long) on the drive pinion face with its pin positioned as illustrated earlier in Fig. 9-6.

11. Install a dial indicator that has a 3-mm range on the measuring bridge. Position the dial indicator with its gauge pin against the setting pin VW 381/6, as shown in Fig. 9-8. Then zero the gauge with no preload.

© 1974 VWoA—3406

Fig. 9-8. Dial indicator zeroed against setting pin on drive pinion face.

12. Shift the position of the dial indicator so that the gauge pin is against the highest point on the cylindrical measuring surface on the mandrel.

13. Note the maximum reading on the dial indicator. Then subtract deviation **r** (stamped on the ring gear) from the reading on the gauge. The remainder is the thickness of the S_3 shim(s) that must be installed.

14. Remove the measuring tools. Remove the large circlip. Then install the proper number and thickness(es) of S_3 shim(s).

NOTE —
Table d lists the shims available as replacement parts. If the S_3 measurement you obtain exceeds a standard shim dimension, use the next thicker shim. Shims can be used singly or in combination as prescribed in **Table e**. Measure the shims carefully at several points with a micrometer and check them for burrs or damage. Use only shims that are in perfect condition.

15. After installing the S_3 shim(s) and large circlip, recheck the measurement with the same tool setup used earlier. The dial indicator reading should not deviate from the **r** dimension marked on the ring gear by more than 0.04 mm.

Table d. Available S_3 Shim Sizes through Chassis 219 020 134

Shim No.	Part No.	Thickness
1	002 311 391	0.15 mm (.0059 in.)
2	002 311 392	0.20 mm (.0079 in.)
3	002 311 393	0.30 mm (.0118 in.)
4	002 311 394	0.40 mm (.0157 in.)
5	002 311 395	0.50 mm (.0197 in.)
6	002 311 396	0.60 mm (.0236 in.)

Table e. S_3 Shim Combinations

Calculated Shim Thickness Required (mm)	Shim Thickness (mm)	Shim Number(s)
0.28–0.32	0.30	3
0.33–0.37	0.35	1 + 2
0.38–0.42	0.40	4
0.43–0.47	0.45	1 + 3
0.48–0.52	0.50	5
0.53–0.57	0.55	1 + 4
0.58–0.62	0.60	6
0.63–9.67	0.65	1 + 5
0.68–0.72	0.70	3 + 4
0.73–0.77	0.75	1 + 3 + 4
0.78–0.82	0.80	3 + 5
0.83–0.87	0.85	1 + 3 + 5
0.88–0.92	0.90	3 + 6
0.93–0.97	0.95	1 + 3 + 6

To adjust drive pinion (from chassis No. 219 020 135):

1. Assemble the drive pinion up to the 1st gear needle bearing. Tighten the round nut to 20 mkg (144 ft. lb.) but do not lock it.

2. Install the partially assembled drive pinion in the transmission case without any S_3 shim(s). Install the retaining nut to a torque of 22 mkg (160 ft. lb.), back it off, then retorque it to the same figure.

3. Lubricate the double tapered-roller bearing with hypoid oil, then install a torque gauge as shown in Fig. 9-9.

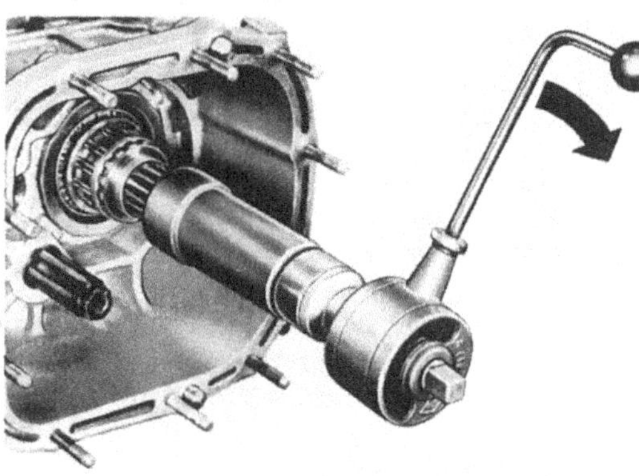

© 1974 VWoA—3408

Fig. 9-9. Gauge for measuring pinion turning torque.

4. Using the handle of the torque gauge, spin the pinion rapidly 15 or 20 turns in each direction. Take torque readings while continuing to spin the pinion. Turning torque should be 6 to 21 cmkg (5.2 to 18.3 in. lb.) for a new bearing or 3 to 7 cmkg (2.6 to 6.1 in. lb.) for a bearing that has been in service for 30 mi. (50 km) or more.

NOTE —
Used bearings must have no visible axial play. Use only hypoid oil as a lubricant. The test results will be inaccurate if the turning torque is checked with the bearings dry or lubricated with another kind of oil.

5. Screw in the left (ring gear side) adjusting ring until its outer face is 1.00 mm below the surface of the transmission case.

NOTE —
On late transmissions that do not have the deviation **r** marked on the ring gear, the following measurement procedure must also be carried out before you remove the drive pinion. Record the measurement obtained (dimension $\mathbf{P_o}$) for use during subsequent assembly and adjustment.

6. Install Mandrel VW 381/2. The cylindrical measuring surface must be on the reverse gearshaft side of the transmission case (Fig. 9-10).

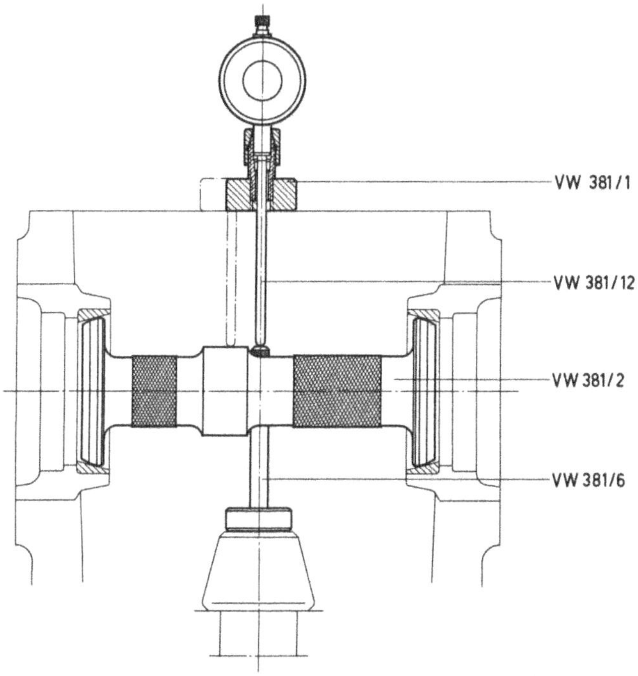

Fig. 9-10. Mandrel VW 381/2 in position. The cylindrical measuring surface has a 40-mm diameter. Other special measuring tools are also shown. Two measuring positions are shown for the dial indicator.

NOTE —
All surfaces must be perfectly clean in order to ensure that the measurements will be accurate.

7. Install the right adjusting ring without an oil seal or O-ring. Tighten the adjusting ring until the mandrel can just barely be hand-turned.

8. Install setting tool VW 381/6 (83.00 mm long) on the drive pinion face with its pin positioned as illustrated earlier in Fig. 9-10.

9. Install a dial indicator that has a 3-mm range on the measuring bridge. Position the dial indicator with its gauge pin against the setting pin VW 381/6, as shown in Fig. 9-11. Then zero the gauge with no preload.

Fig. 9-11. Dial indicator zeroed against setting pin on drive pinion face.

10. Shift the position of the dial indicator so that the gauge pin is against the highest point on the cylindrical measuring surface on the mandrel.

11. Note the maximum reading on the dial indicator. Then subtract deviation **r** (stamped on the ring gear) from the reading on the gauge. The remainder is the thickness of the $\mathbf{S_3}$ shim(s) that must be installed.

NOTE —
If you are measuring a late, unmarked gearset prior to transmission disassembly, the measurement made here (with the original $\mathbf{S_3}$ shims installed) will be dimension $\mathbf{P_o}$. When adjusting one of these unmarked gearsets, the difference between the gauge reading (with no $\mathbf{S_3}$ shim(s) installed) and dimension $\mathbf{P_o}$ will give you the thickness of the $\mathbf{S_3}$ shim(s) that must be installed to return the pinion to its factory-installed position.

6

12. Remove the measuring tools. Remove the retaining nut and take the drive pinion out of the transmission case. Then install the proper number and thickness(es) of S_3 shim(s).

NOTE —
Table f lists the shims available as replacement parts. If the S_3 shim measurement you obtain exceeds a standard shim dimension, use the next thicker shim. Shims can be used singly or in combination as prescribed in **Table g**. Measure the shims carefully at several points with a micrometer and check them for burrs or damage. Use only shims that are in perfect condition.

13. After installing the S_3 shim(s) and retaining nut, recheck the measurement with the same tool setup used earlier. The dial indicator reading should not deviate from the **r** dimension marked on the ring gear, or from the dimension P_0 measured prior to transmission disassembly, by more than 0.04 mm.

Table f. Available S_3 Shim Sizes from Chassis 219 020 135

Shim No.	Part No.	Thickness
1	001 311 391	0.15 mm (.0059 in.)
2	001 311 392	0.20 mm (.0079 in.)
3	001 311 393	0.30 mm (.0118 in.)
4	001 311 394	0.40 mm (.0157 in.)
5	001 311 395	0.50 mm (.0197 in.)
6	001 311 396	0.60 mm (.0236 in.)

Table g. S_3 Shim Combinations

Calculated Shim Thickness Required (mm)	Shim Thickness (mm)	Shim Number(s)
0.28-0.32	0.30	3
0.33-0.37	0.35	1 + 2
0.38-0.42	0.40	4
0.43-0.47	0.45	1 + 3
0.48-0.52	0.50	5
0.53-0.57	0.55	1 + 4
0.58-0.62	0.60	6
0.63-0.67	0.65	1 + 5
0.68-0.72	0.70	3 + 4
0.73-0.77	0.75	1 + 3 + 4
0.78-0.82	0.80	3 + 5
0.83-0.87	0.85	1 + 3 + 5
0.88-0.92	0.90	3 + 6
0.93-0.97	0.95	1 + 3 + 6

Alternate Adjusting Procedure with Universal Measuring Bar VW 385

Most VW Dealers use Universal Measuring Bar VW 385 to adjust the final drive. This measuring bar is a more modern tool than the mandrel described previously.

To adjust with VW 385:

1. Check the double tapered-roller pinion bearing turning torque as described in the previously given adjustment procedures.
2. On late transmissions, screw in the right-side adjusting ring until its outer surface is flush with the transmission case.
3. On early transmissions, install the right-side final drive cover and torque the nuts to 2.0 mkg (14 ft. lb.).
4. Place the magnetic measuring plate VW 385/17 on the drive pinion face as shown in Fig. 9-12.

Fig. 9-12. Measuring plate in place on pinion face.

5. Adjust the setting ring on the universal measuring bar. Dimension **a** shown in Fig. 9-13 is 75 mm.

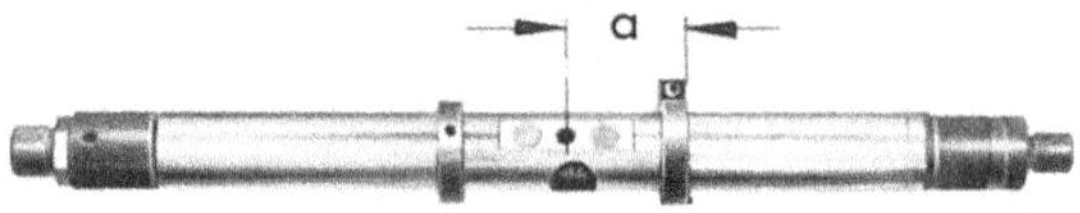

Fig. 9-13. Adjusting setting ring.

6. Slide the centering discs (VW 385/3) onto the bar until they contact the setting rings. Then attach measuring pin VW 385/14 with extension VW 385/16 to the gauge pin hole in the center of the bar (Fig. 9-14). Install the dial indicator.

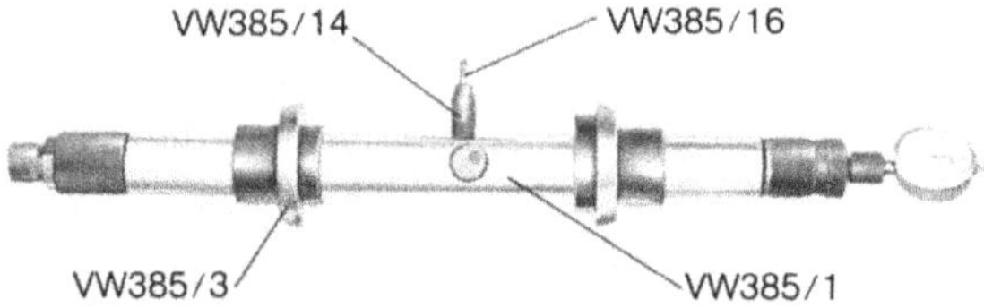

Fig. 9-14. Measuring bar ready for use.

Fig. 9-15. Arrangement of measuring tools inside the final drive housing.

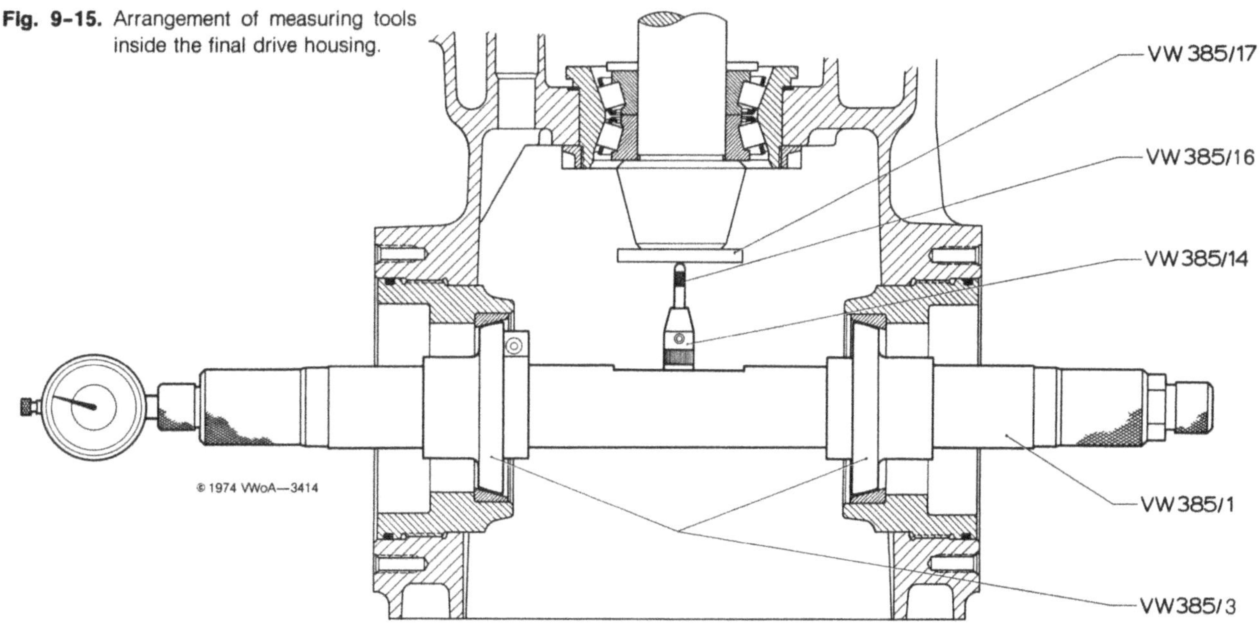

7. On early transmissions, place the universal measuring bar in the transmission case and install the left-side final drive cover. Torque the nuts to 2.0 mkg (14 ft. lb.).
8. On later transmissions, place the universal measuring bar in the transmission case and screw in the left-side adjusting ring until its outer surface is flush with the transmission case.
9. Loosen the second setting ring and move the centering disc outward until the measuring bar can just barely be hand-turned. Then tighten the screw in the setting ring. See Fig. 9-15.
10. Using setting block VW 385/11, zero the dial indicator. The U-shaped setting block zeros the indicator at R_o (63.00 mm). See Fig. 9-16.

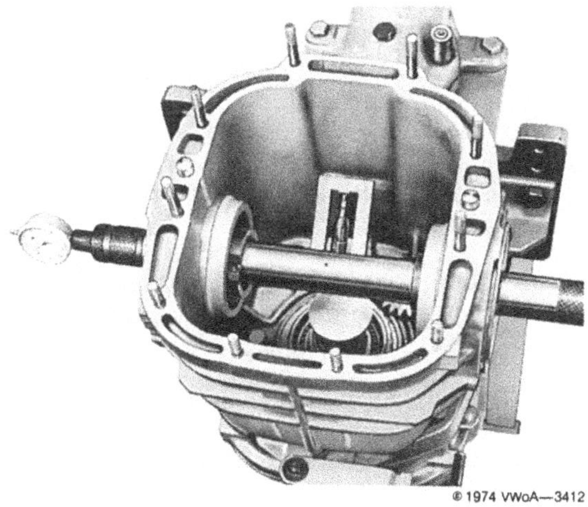

Fig. 9-16. Setting block in place so that dial indicator can be zeroed at R_0.

11. Hand-turn the universal measuring bar until the measuring pin extension rests against the measuring plate on the pinion face.
12. Rotate the bar back and forth over center. The maximum dial indicator reading should be observed and written down. This is measurement **e.**
13. Read the deviation **r** marked on the ring gear. Subtract deviation **r** from measurement **e**. The remainder is the thickness of the S_3 shim(s) that must be installed.

6

If you are measuring a late, unmarked gearset prior to transmission disassembly, the measurement made here—with the original S_3 shim(s) installed—will be dimension P_o. When adjusting one of these unmarked gearsets, the difference between the gauge reading **e**—with no S_3 shim(s) installed—and dimension P_o will give you the thickness of the S_3 shim(s) that must be installed to return the pinion to its factory-installed position.

Using tables **d, e,** and **f** given in the earlier adjustment procedure, select and install the necessary shim(s). Then recheck the pinion position using the universal measuring bar. The gauge reading should not deviate from **r** (or P_o) by more than 0.04 mm.

9.2 Adjusting Ring Gear

The differential bearing preload and ring gear depth-of-mesh must be adjusted following pinion adjustment. Separate procedures are given for the early-type and late-type transmissions.

To adjust differential bearings (through chassis No. 219 020 134):

1. Remove the oil seals and O-rings from the final drive side covers if they were not removed previously. Make sure the outer races for the tapered-roller bearings are pressed all the way into the covers.
2. Install the left (ring gear side) final drive side cover with measuring shim VW 381/10 in place of the $\mathbf{S_1}$ shim(s). Working diagonally, torque the special nuts to 2.0 mkg (14 ft. lb.).
3. Install the right final drive cover with measuring shim VW 381/10 in place of the $\mathbf{S_2}$ shim(s). Install the setting appliance VW 381/3 together with thrust piece VW 381/4 as shown in Fig. 9-17. Then, working diagonally, torque the special nuts to 2.0 mkg (14 ft. lb.).

© 1974 VWoA—3417

Fig. 9-17. Setting appliance installed. Notice that the dial indicator (to be installed in Step 6) must be positioned toward the bottom of the transmission.

4. By turning the twin-handled spindle clockwise on the setting appliance, advance the thrust piece until it contacts the tapered-roller bearing outer race without moving the race.
5. Install a dial indicator that has a 3-mm range, as shown earlier in Fig. 9-16, together with a 52-mm extension. Make sure that the gauge pin contacts the thrust piece inside the setting appliance, then zero the gauge with a 1-mm preload.
6. Position the transmission with the differential at the top. Then install the spacer bridge (VW 381/8) and the torque gauge as shown in Fig. 9-18.

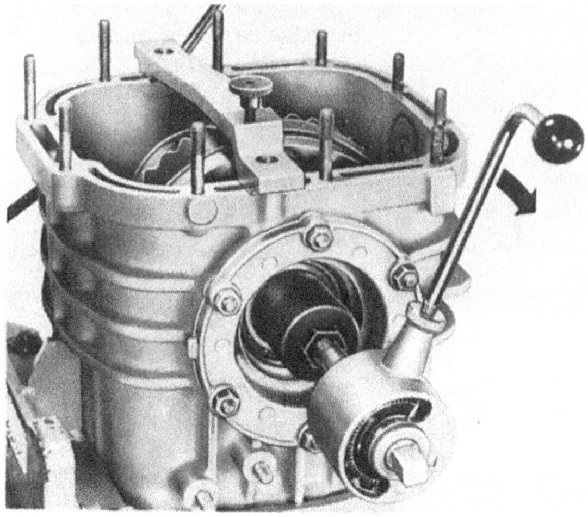

© 1974 VWoA—3419

Fig. 9-18. Spacer bridge installed over studs and torque gauge installed on differential.

7. Using the handle of the torque gauge, spin the differential 15 or 20 turns in each direction while applying hypoid oil to the tapered-roller bearings.

NOTE —
Use only hypoid oil as a lubricant. The test results will be inaccurate if the turning torque is checked with the bearings dry or lubricated with another kind of oil.

8. While spinning the differential rapidly, slowly increase the bearing preload by screwing in the spindle on the setting appliance. Continue slowly until the preload indicated by the torque gauge reaches 30 to 35 cmkg (26 to 30 in. lb.) for new bearings or 3 to 7 cmkg (2.6 to 6.1 in. lb.) for bearings that have been in service for 30 mi. (50 km) or more.

NOTE —
If the specified torque is accidentally exceeded, the right side cover must be removed, the bearing outer race pressed back to its original position, and the adjustment procedure repeated.

9. Record the dial indicator reading you obtained with the correct bearing preload. This is value **S** which must be between 0.50 and 1.60 mm.
10. Remove the setting appliance, spacer bridge, and torque gauge. Then remove the side covers and remove the differential from the transmission case.

NOTE —
Mark the side covers so that you will not accidentally reinstall them on the wrong sides.

To adjust backlash (through chassis No. 219 020 134):

1. Using the correct number and thickness(es) of S_3 shim(s), as determined during pinion adjustment, install the transmission gear train in the transmission case.
2. Install the differential. Install the final drive covers on their correct sides with measuring shims VW 381/10 in place of the S_1 and S_2 shims. Install the spacer bridge. On the right side, reinstall the setting appliance together with the thrust piece, spindle, and dial indicator. Torque the nuts to 2.0 mkg (14 ft. lb.).
3. The dial indicator should read the value **S** obtained earlier. If not, use the twin-handled spindle to press in the tapered-roller bearing outer race while you turn the differential via the main shaft, 4th gear, and the drive pinion.
4. With the correct value **S** obtained, lock the drive pinion by installing a clamping bar on the gear carrier as shown in Fig. 9-19.

© 1974 VWoA—3420

Fig. 9-19. Pinion locked by clamping bar VW 381/11.

5. Press the plastic bracket VW 381/7 on two adjacent ring gear bolts as far as it will go (Fig. 9-20).

© 1974 VWoA—3422

Fig. 9-20. Bracket in place on ring gear bolts.

6. Install the dial indicator in a holder and mount it on the spacer bridge as shown in Fig. 9-21.

© 1974 VWoA—3423

Fig. 9-21. Dial indicator installed. The gauge pin rests against the bracket on the ring gear bolts.

7. Loosen the clamping bar on the drive pinion. Then turn the ring gear via the mainshaft until the bracket contacts the dial indicator's guide pin as shown in Fig. 9-22. Turn the ring gear further until the dial indicator shows a 1.5-mm preload.

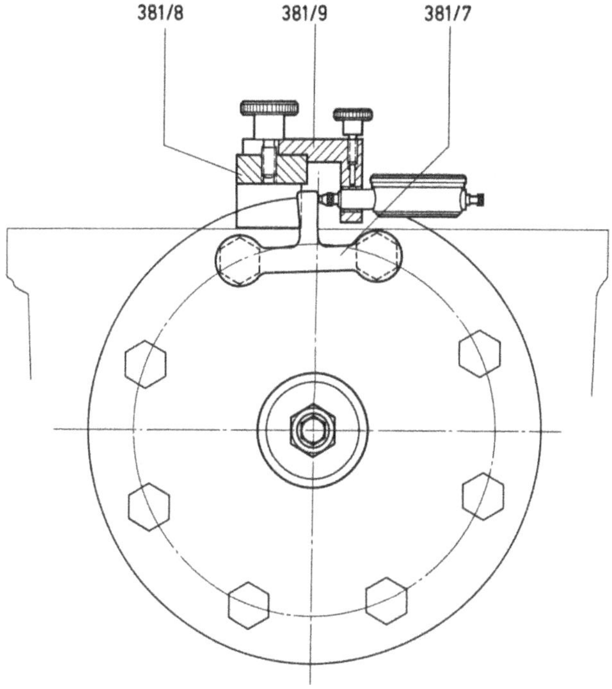

© 1974 VWoA—3424

Fig. 9-22. Relative positions of bracket on ring gear bolts and dial indicator gauge pin. It is important that the point where these parts meet is at the vertical axis of the ring gear (broken lines).

6

8. Tighten the clamping bar on the drive pinion. Hand-turn the ring gear away from the dial indicator until it is stopped by the locked drive pinion. Then zero the dial indicator.

9. From the zeroed position, hand-turn the ring gear toward the dial indicator until it is stopped by the locked drive pinion. Write down the reading. This is the backlash, sometimes abbreviated as $\mathbf{Sv_o}$.

10. Repeat the measurement procedures at three other points 90° apart around the ring gear. Add the four measurements, then divide the sum by four. The quotient is the $\mathbf{Sv_o}$ average.

CAUTION —
Backlash should be between 0.52 and 0.56 mm and the difference between readings must be smaller than 0.06 mm; if it is greater, something is wrong with the gearset or its installation. Left in this condition, the final drive will be noisy and wear rapidly.

11. Compute the $\mathbf{S_1/S_2}$ correction factor. To do this, multiply the $\mathbf{Sv_o}$ average by the correction factor **w** that is listed for your particular gearset in the following chart. Then subtract lift **h** as listed in the same chart. The remainder is the $\mathbf{S_1/S_2}$ correction factor.

Gearset	Correction factor "w"	Lift "h"
K 843	1.00	0.20
K 741	0.90	0.18

NOTE —
For example, if the $\mathbf{Sv_o}$ average is found to be 0.54 for a K 843 gearset, you would multiply 0.54 by 1.00. From the product (0.54) you would subtract 0.20. The $\mathbf{S_1/S_2}$ correction factor would therefore be 0.34.

12. Compute the nominal thickness of the $\mathbf{S_1}$ shim. To do this, subtract the $\mathbf{S_1/S_2}$ correction factor computed in Step 11 from the thickness of the measuring shim VW 381/10 which is 1.30 mm.

NOTE —
For example, if the $\mathbf{S_1/S_2}$ correction factor is 0.34, you would subtract it from 1.30 to get a remainder of 0.96. So 0.96 is the nominal thickness of the $\mathbf{S_1}$ shim required.

13. Compute the nominal thickness of the $\mathbf{S_2}$ shim. To do this, add the $\mathbf{S_1/S_2}$ correction factor computed in Step 11 to the thickness of the measuring shim VW 381/10 which is 1.30 mm. Then subtract the value **S** obtained with the dial indicator during the adjustment of differential bearing preload.

NOTE —
For example, if the $\mathbf{S_1/S_2}$ correction factor is 0.34, you would add it to 1.30 to obtain a sum of 1.64. If value **S** was earlier found (in Step 9 of the differential bearing adjustment procedure) to be 0.59, you would subtract it from the sum of the correction factor and the measuring shim to get a remainder of 1.05—the nominal thickness of the $\mathbf{S_2}$ shim required.

14. Using the nominal $\mathbf{S_1}$ and $\mathbf{S_2}$ shim thicknesses you have computed, find the actual $\mathbf{S_1}$ and $\mathbf{S_2}$ shim thicknesses in **Table h**. **Table i** gives the thickness of each shim and its part number.

Table h. Actual S_1 and S_2 Shim Thicknesses

Calculated thickness (mm) S_1/S_2 nominal	Shim thickness (mm) to be used S_1/S_2 actual	Shim numbers
0.23–0.27	0.25	4
0.28–0.32	0.30	1 + 3
0.33–0.37	0.35	2 + 3
0.38–0.42	0.40	5
0.43–0.47	0.45	3 + 4
0.48–0.52	0.50	6
0.53–0.57	0.55	2 + 5
0.58–0.62	0.60	7
0.63–0.67	0.65	2 + 6
0.68–0.72	0.70	1 + 7
0.73–0.77	0.75	2 + 7
0.78–0.82	0.80	8
0.83–0.87	0.85	4 + 7
0.88–0.92	0.90	1 + 8
0.93–0.97	0.95	2 + 8
0.98–1.02	1.00	9
1.03–1.07	1.05	4 + 8
1.08–1.12	1.10	1 + 9
1.13–1.17	1.15	2 + 9
1.18–1.22	1.20	7 + 7
1.23–1.27	1.25	4 + 9

Table i. Shim Thicknesses and Part Numbers

Shim No.	Thickness d (mm)	Part No.
1	0.10	002 517 201 B
2	0.15	002 517 202 B
3	0.20	002 517 203 B
4	0.25	002 517 204 B
5	0.40	002 517 205 B
6	0.50	002 517 206 B
7	0.60	002 517 207 B
8	0.80	002 517 208 B
9	1.00	002 517 209 B

15. Install the $\mathbf{S_1}$ and $\mathbf{S_2}$ shims under their respective side covers. Remeasure the $\mathbf{Sv_o}$ average. It should be from 0.15 to 0.25 mm and one reading must not vary from another by more than 0.05 mm.

16. Remove the adjusting and measuring equipment. Remove the side covers and press in the oil seals and install the O-rings. Then install the differential as described in **6.1 Removing and Installing Differential.**

To adjust differential bearings (from chassis No. 219 020 135):

1. Remove the oil seals from the adjusting rings if they were not removed previously. Make sure the outer races for the tapered-roller bearings are pressed all the way into the adjusting rings.
2. Screw in the adjusting ring on the ring gear side until it is approximately 0.10 to 0.20 mm (.004 to .008 in.) below the measuring surface on the case.
3. Install the differential with ring gear. Then screw in the other adjusting ring until the differential is firmly supported, but without bearing preload.
4. Position the transmission so that the differential is on top. Then install spacer bridge VW 381/8 and a torque gauge as shown in Fig. 9-23.

© 1974 VWoA—3428

Fig. 9-23. Spacer bridge and torque gauge in position for measuring tapered-roller bearing preload.

5. Using the handle of the torque gauge, spin the differential 15 or 20 turns in each direction while applying hypoid oil to the tapered-roller bearings.

NOTE —
Use only hypoid oil as a lubricant. The test results will be inaccurate if the turning torque is checked with the bearings dry or lubricated with another kind of oil.

6. While spinning the differential rapidly, slowly increase the bearing preload by screwing in the adjusting ring that is opposite the torque gauge. Continue until the preload indicated by the torque gauge reaches 30 to 35 cmkg (26 to 30 in. lb.) for new bearings or 3 to 7 cmkg (2.6 to 6.1 in. lb.) for bearings that have been in service for 30 mi. (50 km) or more.
7. Measure depth S_1 and S_2 as illustrated in Fig. 9-24 and write down the readings.

© 1974 VWoA—3429

Fig. 9-24. A dial indicator with bridge being used to measure the depth of an adjusting ring. A depth micrometer is also suitable for this job.

8. Using the proper number and thicknesses of S_3 shims, as computed during pinion adjustment, install the transmission gears. Torque and lock the pinion retaining nut as described in **7.2 Removing and Installing Transmission Gear Train.**

To adjust backlash (from chassis No. 219 020 135):

6

1. Press the plastic bracket VW 381/7 on two adjacent ring gear bolts as far as it will go (Fig. 9-25). Then install a dial indicator in a holder and mount it on the spacer bridge as shown. The gauge pin must contact the plastic bracket on the ring gear bolts.

© 1974 VWoA—3423

Fig. 9-25. Plastic bracket on ring gear bolts in contact with gauge pin of dial indicator.

2. Turn the ring gear via the mainshaft until the bracket contacts the dial indicator gauge pin as shown in Fig. 9-26. Turn the ring gear further until the dial indicator shows a 1.5-mm preload.

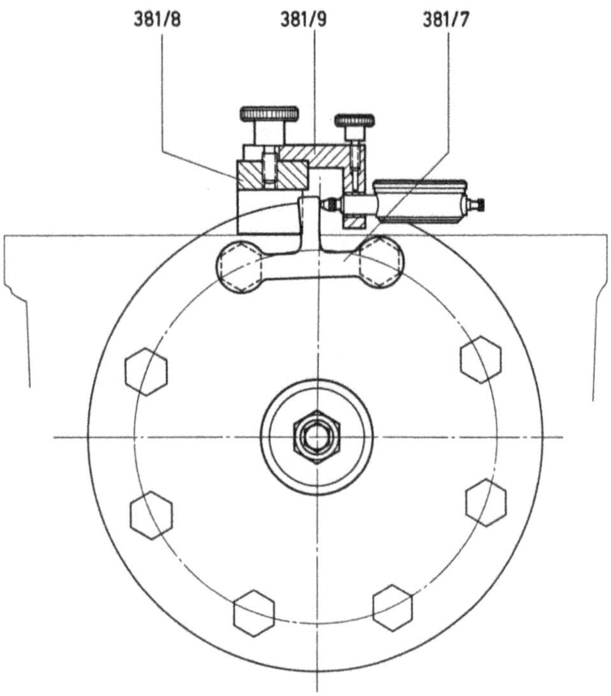

Fig. 9-26. Relative positions of bracket on ring gear bolts and dial indicator pin. It is important that the point where these parts meet is at the vertical axis of the ring gear (broken lines).

3. Clamp the drive pinion as shown in Fig. 9-27 to keep it from turning.

Fig. 9-27. Pinion locked by clamping bar VW 381/11.

4. Hand-turn the ring gear away from the dial indicator until it is stopped by the locked drive pinion. Then zero the dial indicator.

5. From the zeroed position, hand-turn the ring gear toward the dial indicator until it is stopped by the locked drive pinion. Write down the reading. This is the backlash, sometimes abbreviated **Sv_o**.

6. Repeat the measurement procedures at three other points 90° apart around the ring gear.

7. Add the four measurements, then divide the sum by four. The quotient is the **Sv_o** average.

CAUTION —
Backlash should be between 0.52 to 0.56 mm and the difference between readings must be smaller than 0.06 mm; if it is greater, something is wrong with the gearset or its installation. Left in this condition, the final drive will be noisy and wear rapidly.

8. Compute the **S_1** correction factor. To do this, multiply the **Sv_o** average by the correction factor **w** that is listed for your particular gearset in the following chart. Then subtract lift **h** as listed in the same chart. The remainder is the **S_1** correction factor.

Gearset	Correction factor "w"	Lift "h"
K 843 K 741	1.00 0.90	0.20 0.18

NOTE —
For example, if the **Sv_o** average is found to be 0.54 for a K 843 gearset, you would multiply 0.54 by 1.00. From the product (0.54) you would subtract 0.20. The **S_1** correction factor would therefore be 0.34.

9. Unscrew the adjusting ring opposite the ring gear by the distance of the **S_1** correction factor computed in Step 8. Screw in the adjusting ring on the ring gear side by exactly the same amount. Stay within a tolerance of ± 0.01 mm (.0004 in.) at both sides.

10. Recheck backlash. The **Sv_o** average should be from 0.15 to 0.25 mm and individual readings should not vary by more than 0.05 mm (.002 in.).

11. Install new oil seals and the O-rings on the adjusting rings, if they were not installed following the adjustment of the differential bearings.

12. Coat the outer surface of the adjusting rings with anti-rust preservative, then install new plastic lockplates. Torque the Phillips head screws to 1 mkg (7 ft. lb.).

10. REAR SUSPENSION

Each rear wheel is independently suspended by means of a trailing arm called the spring plate and also by a diagonal arm that ensures proper lateral location for the wheel. The spring plate links the rear wheel to the torsion bar which provides springing and ensures proper longitudinal location of the wheel. The inherent self-damping qualities of the torsion bar are supplemented by a hydraulic shock absorber.

Rubber buffers mounted on the frame limit the upward travel of the suspension. The lower limit is determined by a stop built into the torsion bar housing's frame casting. These parts and their relationship can be seen in the exploded view of the rear suspension that is given in Fig. 10-1.

The rear suspension system is fully adjustable to allow accurate rear axle alignment. However, adjustments are seldom required unless the vehicle has received physical damage or has been subjected to many miles of hard service.

10.1 Shock Absorbers

Maker sure you install only shock absorbers intended for the rear of the vehicle. Mismatched shock absorbers will impair handling and ride. It is not necessary, however, to replace both rear shock absorbers if only one is defective. Also, rear axle shock absorbers of different manufacture can be combined as long as their damping characteristics are identical.

If the vehicle is to be subjected to heavy loads, rough roads, or extremely high temperatures, it may be advisable to install heavy-duty shock absorbers in the interest of longer service—despite their effect on the riding qualities of the car.

> **CAUTION—**
> *Install heavy-duty shock absorbers on all four wheels at the same time. Otherwise, handling will be adversely affected.*

> **WARNING —**
> *Do not re-use any fasteners that are worn or deformed in normal use. Many fasteners are designed to be used only once and become unreliable and may fail when used a second time. This includes, but is not limited to, nuts, bolts, washers, self-locking nuts or bolts, circlips, cotter pins. Always replace these fasteners with new parts.*

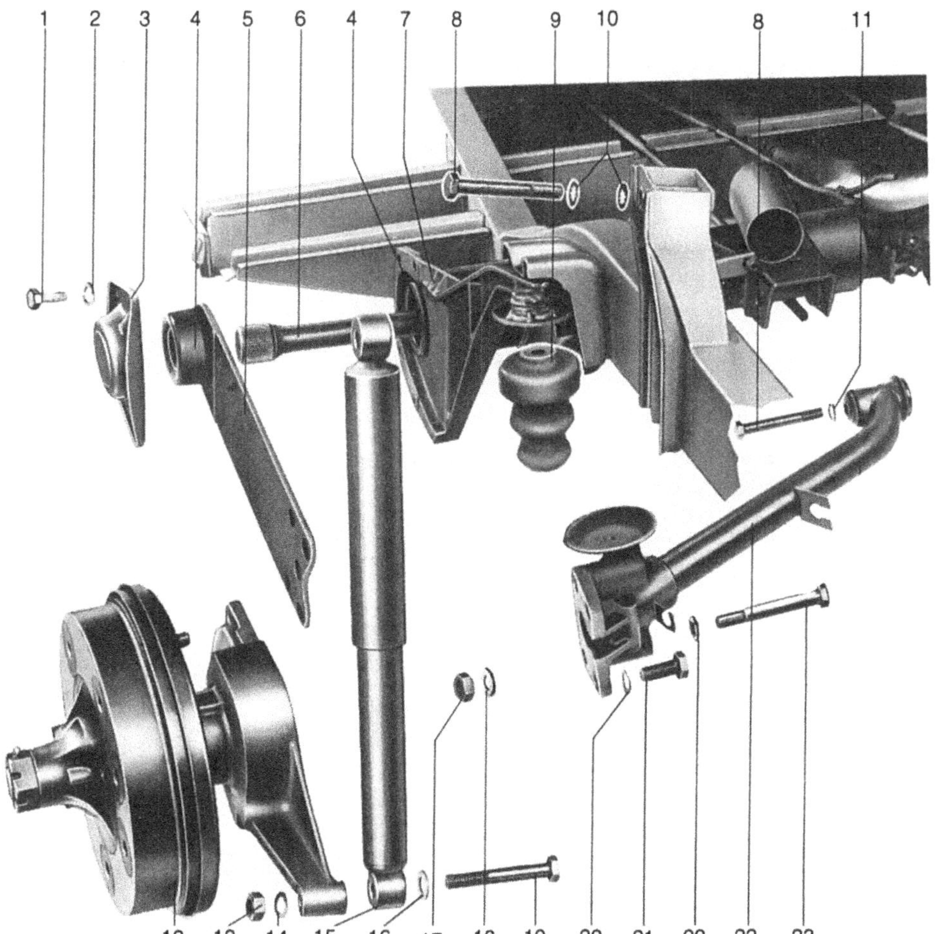

Fig. 10-1. Exploded view of rear suspension components.

© 1974 VWoA—3483

1. M 12 bolt
2. Spring washer
3. Spring plate bearing cover
4. Inner and outer bushings
5. Spring plate
6. Torsion bar
7. Frame
8. M 12 bolt
9. Rubber buffer
10. Lock washer
11. Spring washer
12. Brake assembly and bearing housing
13. M 12 nut
14. Spring washer
15. Shock absorber
16. Washer
17. M 14 nut
18. Washer
19. M 14 bolt
20. Spring washer
21. M 14 bolt
22. Diagonal arm
23. M 12 bolt

Checking Shock Absorbers

You can quick-check the shock absorbers by grasping the rear bumper and rocking the car vigorously. When you let go, the vehicle should rebound only once and then settle into its normal attitude. If the car continues to rock or bob, the shock absorbers are worn. Excessive rear-end bobbing on the highway also signals defective shock absorbers. Badly worn shock absorbers often make knocking noises when the vehicle is driven.

You can hand-check a shock absorber by extending and compressing it while holding it in its installed position. It should operate smoothly and with uniform resistance throughout its entire stroke. If possible, compare the used shock absorber with a new one. New shock absorbers that have been in storage may have to be pumped several times before they reach full efficiency.

An adequate supply of fluid is placed in the shock absorbers during manufacture to compensate for small leaks. Minor traces of fluid are acceptable if the shock absorber still functions efficiently.

Replacing Shock Absorbers

The shock absorbers are hydraulic double-acting units designed to dampen excessive suspension rebound. They are mounted on the rear wheel bearing housing and frame by two bolts with nuts and washers. Shock absorbers cannot be repaired and should be replaced if faulty.

> **NOTE —**
> If necessary, the rubber bushings in the ends of an otherwise serviceable shock absorber can be replaced. The procedure is described in conjunction with replacing shock absorbers in **FRONT AXLE.**

To replace:

1. Raise the vehicle and remove the rear wheel.
2. Remove the M 12 nut and bolt that hold the upper end of the shock absorber to the frame.
3. Remove the M 12 nut and bolt from the lower shock absorber mounting and remove the shock absorber from the vehicle.
4. Install the lower end of the new shock absorber on the rear wheel bearing housing.
5. Keeping the shock absorber in a vertical position, torque the lower mounting nut and bolt to 6.0 mkg (43 ft. lb.).
6. Extend the shock absorber until the upper end is in line with the hole in the frame. Install the bolt, nut, and washer. Torque the bolt to 6.0 mkg (43 ft. lb.).
7. Install the road wheel and lower the vehicle.

10.2 Removing and Installing Rear Suspension Components, Including Bearing Housing

Though the removal of the rear wheel bearing housing, spring plate, torsion bar, and diagonal arm are given as one procedure, you can remove the rear wheel bearing housing separately. The spring plate can also be removed individually, or together with the torsion bar.

To remove:

1. Remove the cotter pin, then loosen the large castellated nut that holds the rear brake drum or rear wheel hub on the axle shaft.

> ***WARNING —***
> *Loosen axle shaft nuts with the car on the ground. The leverage needed for this job is enough to topple a car off the lift.*

2. Unless only the spring plate and torsion bar are being removed, detach the driveshaft from the axle shaft as described in **3.1 Removing and Installing Driveshafts**. Suspend the driveshaft from the frame.
3. Remove the lower shock absorber mounting bolt.
4. Remove the castellated nut, then remove the brake drum and wheel hub. Remove the brake assembly as described in **BRAKES AND WHEELS.**

> **NOTE —**
> If you are only going to disassemble the rear wheel bearings, do not disconnect the brake line and parking brake cable. Instead, suspend the brake assembly from the frame. If only the spring plate is being removed, suspend the brake assembly together with the rear wheel bearing housing.

5. Using a cold chisel, mark the spring plate, diagonal arm, and bearing housing both top and bottom as shown in Fig. 10-2.

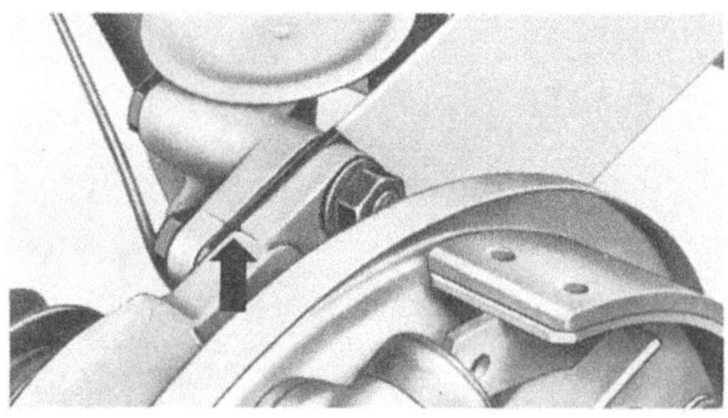

© 1974 VWoA—3484

Fig. 10-2. Alignment mark on spring plate, diagonal arm, and bearing housing.

6. Remove the four nuts and bolts that hold the diagonal arm and the bearing housing on the spring plate. Then take the bearing housing off to the rear.
7. If necessary, remove the diagonal arm from the vehicle by dismounting the brake hose from its bracket on the diagonal arm, then removing the bolt shown in Fig. 10-3.

© 1974 VWoA—3486

Fig. 10-3. Diagonal arm being unbolted from the frame.

8. Remove the four bolts that hold the spring plate bearing cover, then remove the cover.
9. Using a lever, as shown in Fig. 10-4, lift the spring plate off its lower stop. Once lifted, pry it outward *off* the stop. Then slowly release the leverage until the torsion bar is unloaded.
10. Slide the spring plate off the torsion bar splines.

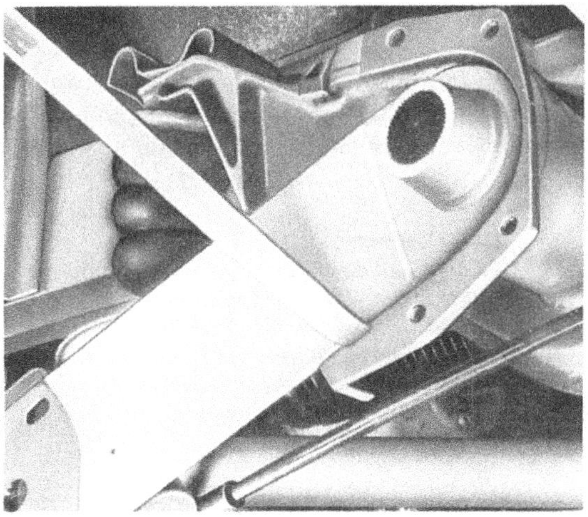

© 1974 VWoA—3487

Fig. 10-4. Spring plate being pried up and off its lower stop.

11. Being careful not to scratch the protective paint on the torsion bar, pull the torsion bar out of the tube.

NOTE —
If the torsion bar is broken, remove the torsion bar on the opposite side of the vehicle, then drive the broken piece out of the tube with a long drift. Alternately, flare the end of a piece of tubing until it just fits over the end of the broken bar. Drive the tubing onto the broken piece, then pull the broken piece out with the tubing.

To install:

1. Inspect the rubber bushings for the spring plate hub and replace them if they are worn or damaged.
2. Make sure the torsion bar and spring plate splines are not damaged. Chipped or scratched areas in the protective paint on the torsion bar must be touched up with paint in order to avoid fatigue fractures due to corrosion.
3. Apply multipurpose grease to the inner torsion bar splines, then insert the torsion bar.

CAUTION —
Torsion bars are prestressed in their working direction. If both rear torsion bars have been removed, you must be careful not to interchange them from one side of the vehicle to the other. The left bar has an **L** *on its outer face; the right bar has an* **R**. *If the bars are installed on the wrong side, they will break when placed in service.*

4. Coat the inner and outer rubber bushings with talcum powder only. Install the inner rubber bushing.

CAUTION —
Do not coat the rubber bushings with graphite. It will permit the spring plate hub to turn in the bushings, resulting in rapid wear. Do not allow multipurpose grease or other petroleum-based lubricants to contaminate the bushings since these substances are harmful to rubber.

5. Install the spring plate and adjust the torsion bar as described in **10.3 Adjusting Torsion Bars.**

NOTE —
If you install a new spring plate, the rear wheel alignment must be checked on an axle alignment stand and, if necessary, adjusted.

6. Install the outer rubber bushing.

6

7. Attach the spring plate bearing cover with two bolts—using extra-long bolts, if necessary, to pull the cover tightly against the partially installed spring plate and bushing.
8. Using a tensioning device such as the one shown in Fig. 10-5, lift the spring plate above the level of its lower stop.

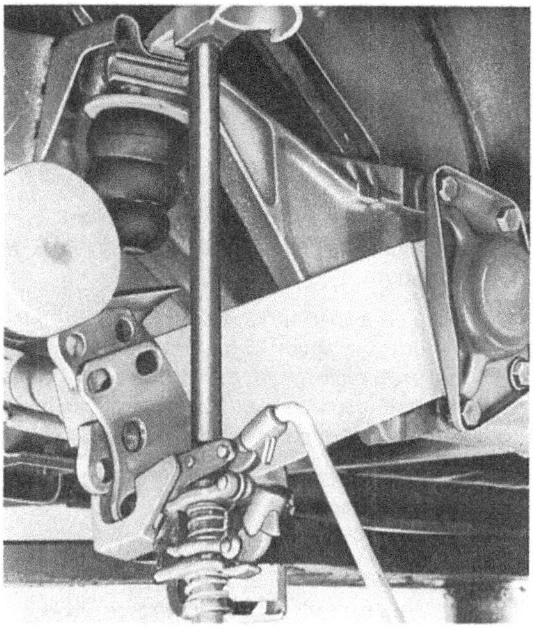

© 1974 VWoA—3489

Fig. 10-5. Spring plate being lifted onto its stop.

9. To press the spring plate fully onto the torsion bar splines, tighten the two bolts in the spring plate bearing cover.
10. Install the remaining bearing cover bolts. If necessary, replace the extra-long bolts with the correct bolts. Then torque the bolts to 3.5 mkg (25 ft. lb.).
11. Install the diagonal arm on the frame. To prevent tension in the bonded rubber bushing, hold the diagonal arm in its extended (unladen) position. Then torque the bolt to 8.0 mkg (58 ft. lb.).

NOTE —
If the bonded rubber bushing is worn or damaged, you must replace the entire diagonal arm since the bushing is vulcanized to the metal and cannot be replaced separately. If a new diagonal arm is installed for any reason, or if the rear wheel bearing has been replaced, the rear wheel alignment must be checked on an axle alignment stand and, if necessary, adjusted.

12. With careful reference to the alignment marks made during removal, install the bearing housing on the spring plate and diagonal arm.
13. Torque the four bolts to 13 mkg (94 ft. lb.). Then again check the alignment marks.
14. Lightly grease the constant velocity joints. Make sure that the joint face and the axle shaft flange are both grease-free, then install the joint on the axle shaft using new lock washers. Torque the socket head screws to 3.5 mkg (25 ft. lb.).
15. Install the brake assembly, if necessary. Loosely install the brake drum or wheel hub and the road wheel.

WARNING —
Torque the axle shaft nuts with the car on the ground. The leverage needed for this job is enough to topple a car off the lift.

16. Install the lower end of the shock absorber. Torque the nut and bolt to 6.0 mkg (43 ft. lb.).
17. If necessary, bleed and adjust the brakes as described in **BRAKES AND WHEELS.**
18. Lower the vehicle, then torque the castellated nut on the axle shaft to 35 mkg (253 ft. lb.). Torque the wheel bolts or nuts to 13 mkg (94 ft. lb.).

10.3 Adjusting Torsion Bars

The torsion bar setting is determined by the unloaded position of the spring plate—that is, with the spring plate off its lower stop. The specifications are expressed as degrees of angle between the spring plate and the longitudinal axis of the vehicle.

There are 44 splines at the inner end of each torsion bar and 48 splines at the outer end. This provides a vernier adjustment for setting the spring plate angle. By rotating the bar—one spline (8° 10′) at its inner end and one spline in the opposite direction at its outer end (7° 30′)—the spring plate angle is changed in increments of 0° 40′ in whichever direction the outer spline is rotated.

CAUTION —
On high-mileage vehicles, always adjust both torsion bars. The bars tend to settle with extended use and, while the side being serviced may be properly set to specifications, the opposite side, being lower, would cause the vehicle to set at an angle.

To adjust:

1. Using a bubble protractor placed against a frame rail (Fig. 10-6), determine the deviation of the vehicle from the horizontal.

© 1974 VWoA—3490

Fig. 10-6. Bubble protractor being used to determine the number of degrees that the vehicle's frame rail departs from the horizontal.

2. With the spring plate bearing cover removed and the spring plate pried off its lower stop (torsion bar unloaded), place the bubble protractor on the spring plate as shown in Fig. 10-7.

© 1974 VWoA—3491

Fig. 10-7. Bubble protractor in position atop spring plate. Lift the spring plate to remove all play from the splines while measuring the spring plate's deviation from the horizontal.

3. If the front of the vehicle is lower than the rear, add the deviation measurement made at the frame rail to that made at the spring plate; if the rear of the vehicle is lower, subtract the deviation measurement made at the frame rail from that made at the spring plate. The result is the spring plate angle.
4. Compare the spring plate angle with the specified angles listed in **Table III** in **11. Transmission and Rear Axle Technical Data**. If it deviates from the specified angle by more than 40′, adjust the torsion bar.
5. Depending on the direction of deviation from specifications, move the torsion bar one spline forward and the spring plate one spline back on the torsion bar to obtain an angle that is no more than 50′ greater than that specified.

10.4 Rear Wheel Alignment

Rear wheel alignment work should be carried out under the same conditions described for front wheel checks in **FRONT AXLE.** However, the rear can be checked properly only with optical wheel alignment equipment. The rear wheel toe is adjusted by loosening the four bolts that bind the bearing housing, spring plate, and diagonal arm together, then shifting the position of the bearing housing in the elongated holes in the spring plate. However, rear wheel camber must be adjusted before toe adjustments are made.

Rear wheel camber is determined primarily by the adjustment of the torsion bars, as just described in **10.3 Adjusting Torsion Bars.** However, minor camber changes can be effected by loosening the four bolts that bind the bearing housing, spring plate, and diagonal arm together, then rotating the diagonal arm one way or the other with a pipe wrench. The range of adjustment by this method is quite small and should be considered as supplementary to careful torsion bar adjustment. After installing a new torsion bar, do not adjust the alignment until the bar has been in service for at least 300 mi. (500 km).

The camber should be $-50' \pm 30'$ with a maximum permissible difference of 30′ between sides. With camber correctly adjusted, rear wheel toe-out should be set at $+10' \pm 20'$. This is the total toe-out and is the sum of the angles measured for each rear wheel on the optical alignment stand. The toe angle for one wheel must be kept within 10′ of the angle for the other wheel. The special tool shown in Fig. 10-8 is very helpful in moving the bearing housing on the spring plate during toe adjustments. However, tension should be removed from the tool while measurements are being made.

6

© 1974 VWoA—3493

Fig. 10-8. Special tool being used to move bearing housing on spring plate.

10.5 Disassembling and Assembling Rear Wheel Bearing Housing

Fig. 10-9 is an exploded view of the rear wheel bearing housing and brake assembly as installed on 1968 through 1970 models. Fig. 10-10 is a similar view of the rear bearing housing and brake assembly used on 1971 and later models. New seals should always be obtained for use during assembly. Also, the outer roller bearing race is destroyed when it is driven out of the housing. If you intend to remove it, obtain a replacement roller bearing set in advance.

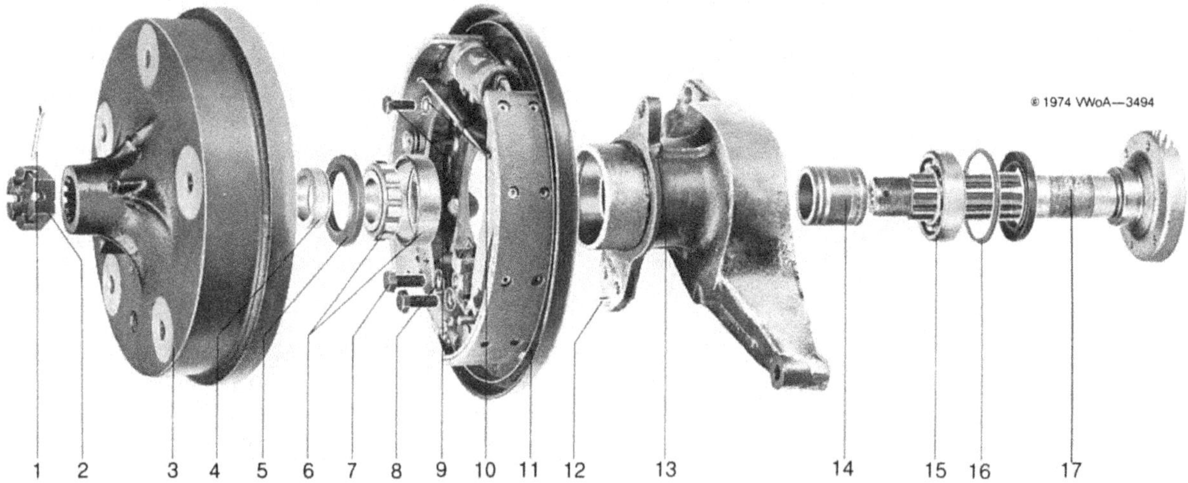

1. Cotter pin
2. Castellated nut
3. Brake drum
4. Spacer ring
5. Oil seal (2)
6. Roller bearing assembly
7. M 10 bolt (2)
8. Spring washer (2)
9. M 8 bolt
10. Lock washer
11. Brake backing plate with complete brake assembly
12. Dowel pin
13. Bearing housing
14. Spacer sleeve
15. Ball bearing
16. Circlip (2)
17. Axle shaft

Fig. 10-9. Exploded view of rear wheel bearings, bearing housing, and related parts (1968 through 1970 models).

WARNING —
Do not re-use fasteners that are worn or deformed in normal use. Many are designed to be used only once and may fail when used a second time. This includes nuts, bolts, washers, self-locking nuts or bolts, circlips, cotter pins. Always use new parts.

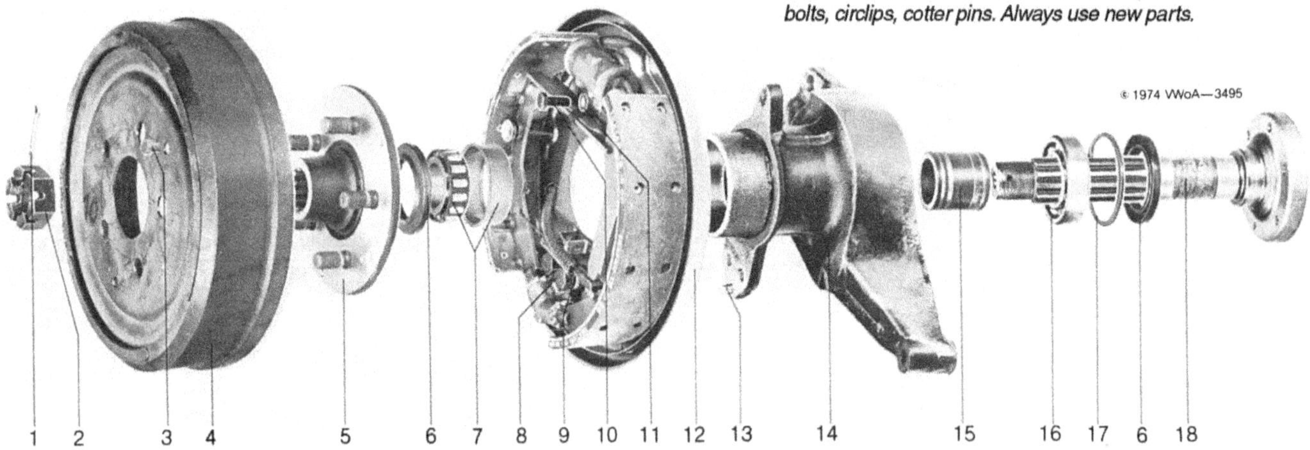

1. Cotter pin
2. Castellated nut
3. M 7 bolt (2)
4. Brake drum
5. Wheel hub
6. Oil seal (2)
7. Roller bearing assembly
8. M 10 bolt (2)
9. Spring washer
10. M 8 bolt
11. Lock washer
12. Brake backing plate with complete brake assembly
13. Dowel pin
14. Bearing housing
15. Spacer sleeve
16. Ball bearing
17. Circlip (2)
18. Axle shaft

Fig. 10-10. Exploded view of rear wheel bearing, bearing housing, and related parts (1971 and later models).

To disassemble:

1. Remove the bearing housing, with or without brake assembly, from the diagonal arm and spring plate. See **10.2 Removing and Installing Rear Suspension Components, Including Bearing Housing.**
2. To support the bearing housing during disassembly, clamp its spring plate flange in a vise.
3. If necessary, remove the castellated nut, the brake drum (or brake drum and rear wheel hub), and the brake backing plate assembly.
4. Using a puller as shown in Fig. 10-11, press the axle shaft out of the bearings and bearing housing.

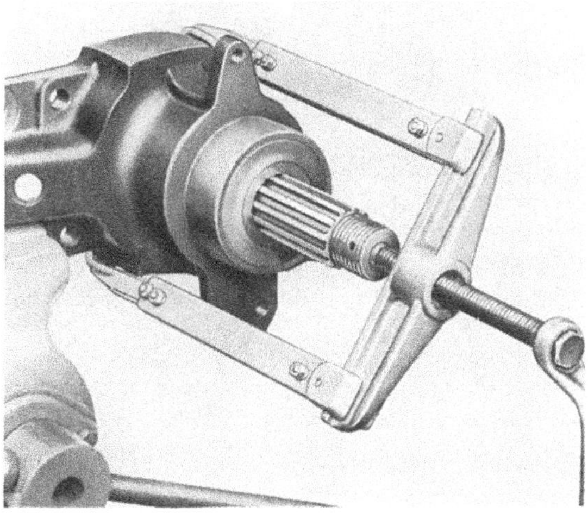

© 1974 VWoA—3497

Fig. 10-11. A puller being used to press out the axle shaft.

5. Pry out the oil seals at both ends of the bearing housing. Then remove the circlips (Fig. 10-12).

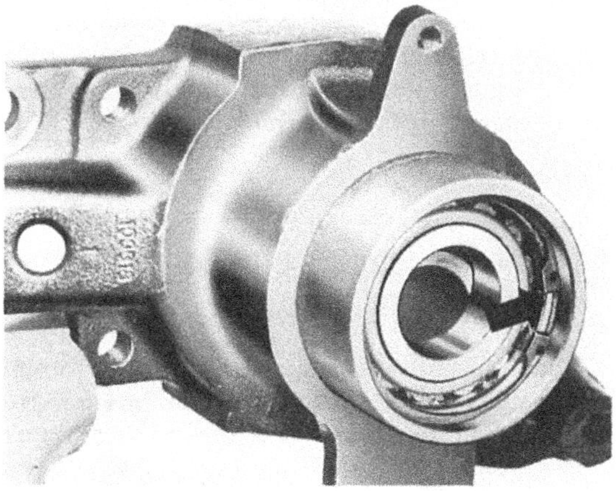

© 1974 VWoA—3503

Fig. 10-12. Circlip for bearing. There is an identical circlip at the opposite end of the housing.

6. Take out the roller bearing inner race and the spacer sleeve.
7. Using a drift that will contact only the ball bearing outer race, drive out the ball bearing as shown in Fig. 10-13.

© 1974 VWoA—3499

Fig. 10-13. Ball bearing being driven out.

8. Only if the roller bearing assembly is damaged, drive out the roller bearing outer race. Leave the outer race in the housing if you only intend to pack the bearing housing with new grease.

To assemble:

1. Clean and inspect all parts. Replace any that are worn or damaged. Pack the ball bearing with multipurpose grease, then install it as shown in Fig. 10-14.

© 1974 VWoA—3500

Fig. 10-14. Ball bearing being pressed into housing.

6

2. Install the circlip over the ball bearing. Then, using a suitable driving tool, drive in the inner oil seal.
3. If necessary, press in a new roller bearing outer race as shown in Fig. 10-15.

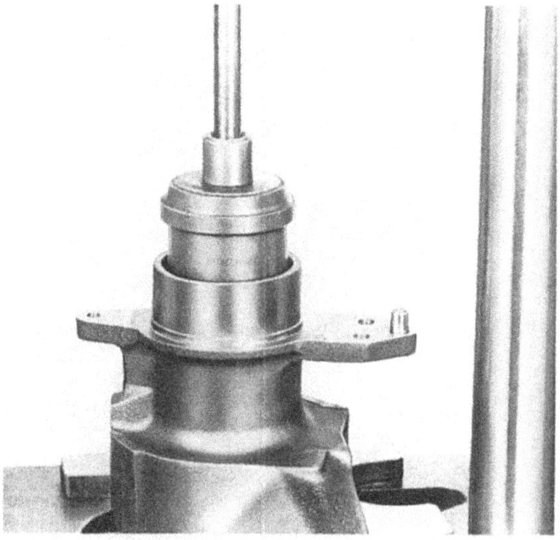

© 1974 VWoA—3502

Fig. 10-15. Roller bearing outer race being pressed into housing. If the outer race has no shoulder, have the side with the manufacturer's name to the outside.

4. Fill the housing with multipurpose grease until the spacer sleeve can just be inserted. Then temporarily remove the spacer sleeve from the bearing housing.

NOTE —
Spacer sleeves with an outside diameter of 42 mm (1.654 in.) and spacer sleeves with an outside diameter of 46 mm (1.811 in.) have both been installed during manufacture. However, only the 46-mm sleeve is available as a replacement part. If one sleeve must be replaced, also replace the sleeve at the opposite rear wheel.

5. Drive in the axle shaft until its shoulder enters the oil seal and contacts the ball bearing.
6. Install the spacer sleeve over the axle shaft.
7. Pack the roller bearing inner race with multipurpose grease. Then install the inner race over the axle shaft, with the rounded side of the cage inwards, until the rollers contact the outer race.

CAUTION —
If it is necessary to press the inner race down over the shaft, keep the sheet metal cage square with the shaft. Otherwise, the cage may be damaged.

8. Install the circlip over the roller bearing. Then, using a suitable driving tool, drive in the other oil seal.
9. Before installing the brake assembly, apply a ring of sealing compound around the bearing housing as indicated in Fig. 10-16.

CAUTION —
Do not apply excessive sealing compound as it may prevent the brake backing plate from seating squarely on the bearing housing.

© 1974 VWoA—3504

Fig. 10-16. Sealing compound (arrow) applied around bearing housing.

10. After installing the brake assembly, apply another ring of sealing compound around the inner surface of the brake backing plate where it abuts against the bearing housing.
11. Install the bearing housing on the spring plate and diagonal arm and install the brake drum or wheel hub and brake drum as described in **10.2 Removing and Installing Rear Suspension Components, Including Bearing Housing.**

11. TRANSMISSION AND REAR AXLE TECHNICAL DATA

The tables of technical data that appear on the following pages contain all the specifications needed for rebuilding the transmission and rear axle. For convenience, these specifications have also been included in the text wherever they are applicable. Technical data for the constant velocity joints, and the driveshafts that connect the rear axle to the rear wheels, are given in the Technical Data section of **AUTOMATIC TRANSMISSION.**

I. Tightening Torques

Location	Designation	mkg	ft. lb.
Gears/housing	retaining nut	22.0	160
Round nut/pinion	round nut	20.0	145
Union nut/sleeve	union nut	3.0	22
Bracket/reverse shifter shaft on gear carrier	bolt	2.5	18
Support/rocker lever on gear carrier	bolt	2.5	18
Selector shaft/fork	bolt	2.5	18
Locking screw with dog point	bolt	1.5	11
Clamp sleeve on gear carrier	clamp sleeve	4.5	32
Shift housing to gear carrier	nut	1.5	11
Nuts for gear carrier, side cover, transmission, and clutch housing	nut	2.0	14
Ring gear to differential housing	bolt	5.0	36
Lockplate/adjusting ring	Phillips head screw	1.0	7
Diagonal arm to rear wheel bearing housing	bolt	13.0	94
Diagonal arm to frame	bolt	8.0	58
Shock absorber to frame and rear wheel bearing housing	bolt	6.0	43
Drive shaft/flange	socket head screws	3.5	25
Cover/spring plate bushing	bolt	4.5	32
Backing plate to housing	M 8 bolt	2.5	18
	M 10 bolt	3.5	25
Rear wheel hub to rear wheel shaft	castellated nut	35	253
Gearshift lever bracket to floor	bolt	2.0	14
Transmission to transmission carrier	bolt	3.5	25
Engine bearer to bearer brackets	bolt	2.5	18
Clutch cable guide tube bracket to transmission case	nut	2.0	14
Front transmission mount support to transmission	nut	2.5	18
Engine bearer bracket to frame	bolt	3.5	25
Front transmission mount to mount support	bolt	3.5	25
Clutch release bearing guide sleeve to bellhousing	nut	1.5	11
Pinion bearing retainer to transmission case	special nut	3.0	22
Limiting stop on cross tube	bolt	4.5	33
Relay shaft bracket on gear carrier	special bolt	2.0	14

II. Transmission Markings

Transmission	Code letter	Final drive no. of teeth	Engine displacement	Introduction date
Manual	CA	8:43	1600/1700	prior to August 1967
Manual	CN	7:34	1800	August 1973

6

III. Torsion Bar Adjustment (spring plates unloaded)

Model	Transmission	Installed from chassis No.	Torsion bar length mm (in.)	Torsion bar diameter mm (in.)	Setting
Delivery Van Kombi Campmobile Pickup Truck	Manual	218 000 002	610 (24 1/64)	28.10 (1.106)	21° 10′ + 50′
Station Wagon	Manual	218 000 002	610 (24 1/64)	26.20 (1.031)	23° + 50′
Delivery Van Kombi Campmobile	Manual and Automatic (Campmobile only)	212 2000 001	610 (24 1/64)	28.90 (1.138)	20° + 50′
Station Wagon	Manual and Automatic	212 2000 001	610 (24 1/64)	26.90 (1.059)	23° + 50′

IV. Tolerances, Wear Limits, and Settings

Designation	New part mm (in.)	Wear limit mm (in.)
A. Gears, drive pinion		
1. 1st gear .... end play (try to keep to lower limit)	0.10–0.25 (.004–.010)	—
2. 3rd gear .... end play	0.10–0.25 (.004–.010)	—
3. 4th gear .... end play	0.10–0.25 (.004–.010)	—
4. Synchromesh units clearance **a** between coupling teeth and synchronizer ring		
1st/2nd gears ..... clearance	1.0–1.9 (.040–.075)	0.60 (.023)
3rd/4th gears...... clearance	1.1–1.8 (.043–.071)	0.60 (.023)
5. Shift fork/operating sleeves for 1st/2nd and 3rd/4th gears........ end play	0.10–0.30 (.004–.012)	—
operating sleeve with wide slot .... end play	0.50–0.85 (.020–.033)	—
6. Preload of pinion tapered-roller bearing		
Turning torque........ new	6–21 cmkg (5.2–18.3 in. lb.)	—
used more than 30 mi. (50 km)	3–7 cmkg (2.6–6.1 in. lb.)	—
B. Drive shaft		
1. Drive shaft, front (surface for 3rd gear needle bearing) .... runout	max. 0.02 (.0008)	—
C. Final drive		
1. Play at differential gears with differential housing bolted together .... axial	0.00–0.14 (.000–.0055)	—
Play between differential housing and cover/gearshaft........ radial, new	0.025–0.06 (.001–.002)	0.12 (.005)
old	0.03–0.08 (.001–.003)	0.12 (.005)
2. Rear axle shafts		
a. Flange/differential gears (measured across the convex faces)...... clearance	0.03–0.10 (.001–.004)	0.20 (.008)
b. Measured at bearing seat shaft between centers........ runout	max. 0.05 (.002)	—
D. Gearbox and gearshift housing		
1. Preload of final drive covers on tapered-roller bearings		
Turning torque........ new	30–35 cmkg (26–30 in. lb.)	—
used more than 30 mi (50 km)	3–7 cmkg (2.6–6.1 in. lb.)	—
2. Plastic packing/transmission case/axle tube/tube retainer .... clearance	0.00–0.20 (.000–.008)	—
3. Shift rod........ effort required to overcome detent	15–20 kg (33–44 lb.)	—
4. Gearshift housing bushings........ inside diameter	15.05–15.03 (.592–.591)	15.25 (.600)
5. Inner shift lever .... diameter	15.00–14.96 (.590–.588)	14.75 (.580)
6. Starter bushing........ inside diameter	12.55–12.57 (.494–.495)	12.65 (.498)
7. Starter shaft/bushing .... radial clearance	0.09–0.14 (.0035–.0055)	0.25 (.010)
8. Main drive and pinion shafts (measured between centers)........ runout	0.5 (.020)	—

V. Ratios

Gear	Tooth ratio	Numerical ratio
1st gear	34/9	3.78
2nd gear	35/17	2.06
3rd gear	29/23	1.26
4th gear (through chassis No. 213 2068 547)	23/28	0.82
4th gear (from chassis No. 213 2068 548)	27/24	0.89
Reverse (through chassis No. 211 2276 560)	20/14 x 43/17	3.61
Reverse (from chassis No. 211 2276 561 through chassis No. 213 2068 547)	20/14 x 40/15	3.80
Reverse (from chassis No. 213 2068 548)	17/12 x 40.15	3.79
Final drive (through chassis No. 213 2300 001)	8/43	5.375
Final drive (from chassis No. 214 2000 001)	7/34	4.857
Final drive (2.0L engine)	7/32	4.57

VI. 3rd-gear Circlip Sizes

Shim S mm (in.)	Part No.	Color
1.45 (.0571)	113 311 381	plain
1.60 (.0630)	113 311 382	black
1.75 (.0689)	113 311 383	blue
1.90 (.0748)	113 311 384	brown
2.05 (.0807)	113 311 385	gray
2.20 (.0866)	113 311 386	copper

WARNING —
Do not re-use fasteners that are worn or deformed in normal use. Many are designed to be used only once and may fail when used a second time. This includes nuts, bolts, washers, self-locking nuts or bolts, circlips, cotter pins. Always use new parts.

Section 7

AUTOMATIC TRANSMISSION

Contents

TABLES

Automatic Transmission

The VW automatic transmission covered in this section of the Manual is optional on 1973 through 1975 models. Though the final drive of the 1976 transmission is the same as the final drive of the 1973 through 1975 automatic transmissions, the transmission itself is different on the 1976 models. The 1976 automatic transmission is basically the same as the automatic transmission used in the VW Rabbit. Unfortunately, insufficient data is available from the factory for the inclusion of the 1976 automatic transmission in this printing of the Manual.

This section should interest almost every VW owner, although many of the repair procedures will be of practical value only to the professional mechanic. Some operations require equipment and experience that only a trained mechanic is likely to have. So if you lack skills, tools, or a clean workshop, we suggest that you leave automatic transmission repairs to an Authorized VW Dealer or other qualified shop.

The nonprofessional may, however, be able to remove the engine and transmission which can help to reduce service time. In such cases, we recommend that the transmission be thoroughly cleaned on the outside and then taken to the shop as is—partial disassembly will not make repairs easier and may indeed complicate them. We especially urge you to consult an Authorized VW Dealer before attempting repairs on a car still covered by the new-car warranty.

Cleanliness and a careful approach are imperative when repairing the transmission. Familiarizing yourself with procedures, notes, cautions, and warnings before beginning work is good insurance. Clean and lay out the parts. If necessary, mark them to show their proper assembly order. Also make sure that you have the necessary tools—particularly for procedures given with metric specifications only. Specifications that lack U.S. equivalents require that the related work be carried out only with metric tools and instruments.

The Owner's Manual supplied with new Type 2 vehicles illustrates the operation of the selector lever and describes normal performance in each of the driving ranges. Please familiarize yourself with this information. It will help you drive your car safely and economically without placing undue strain on the machinery. It can also be a useful standard of comparison for road testing and troubleshooting—especially for mechanics who are not familiar with the VW transmission.

1. GENERAL DESCRIPTION

The automatic transmission is housed in a case assembled from two main castings. At the front of the assembly is a cast aluminum transmission case containing the automatic transmission fluid (ATF) pump, the hydraulic controls, and the planetary gear system. Attached at the rear of this case by four steel studs is a final drive housing also cast in light alloy. The bellhousing for the torque converter is an integral part of the final drive housing.

The flanged shafts that carry driving torque to the rear wheels extend from two openings at either side of the final drive housing. Adjusting rings, threaded into the case, form a support for the differential bearings. A pressed steel pan on the bottom of the final drive housing can be removed for access to the differential.

Torque Converter

The torque converter is a large doughnut-shaped assembly located between the engine and the automatic transmission. The converter not only receives engine output and passes it on to the transmission, but also multiplies engine torque at low vehicle speeds and serves as a fluid coupling between the engine and the transmission. The converter housing spins with the engine's crankshaft. Curved vanes inside the housing set up a flow of ATF that drives another vaned wheel called the turbine. The turbine drives a hollow shaft that transmits power to the transmission.

ATF Pump

ATF must be circulating under pressure before the automatic transmission can function. The ATF pump that creates this pressure is located at the extreme front of the transmission case. A long pump driveshaft that passes through the center of the hollow turbine shaft drives the ATF pump.

The pump driveshaft is splined directly to the converter housing. Therefore, the pump circulates ATF whenever the engine is running, regardless of selector lever position. Since the circulating ATF is also the transmission's only lubricant, it is important to remember that the ATF does not circulate when the engine is not running and the car is being towed.

CAUTION —
Never tow a car with automatic transmission faster than 30 mph (48 kph) or farther than 30 miles (48 kilometers). Bearings can be damaged by lack of lubrication. If you must tow the car farther, lift the rear wheels or remove the driveshafts that connect the rear wheels to the transmission.

Planetary Gears

A torque converter alone cannot supply the torque multiplication needed for all driving conditions. The output of the torque converter is therefore routed into a planetary gearset. The planetary gearset is located at the rear of the transmission case, just ahead of the final drive housing.

The planetary gear system used in the VW automatic transmission operates on the same principles as similar gearsets found in other automatic transmissions, though it differs in numerous construction details. The planetary gear system has one large sun gear (51 teeth), one small sun gear (30 teeth), three small planet pinions (16 teeth each), three large planet pinions (35 teeth each), and one large annulus (ring) gear. The planet pinions are all mounted on the planet carrier which is coupled to the final drive pinion (transmission output shaft). The annulus has a one-way roller clutch to provide free wheeling when the driver's foot is removed from the accelerator with the selector lever at **D** and the transmission in 1st gear.

Clutches

Two hydraulically operated multiple disk clutches control the delivery of turbine output to the planetary gear system. The clutch at the front of the transmission is called the direct and reverse clutch because it transfers power to the small sun gear of the planetary gearset only when the transmission is in direct (3rd gear) or in reverse. The other clutch, located between the direct and reverse clutch and the planetary gearset, is called the forward clutch because it transfers power to the large sun gear in all forward gears.

Brake Bands

Two hydraulically operated brake bands are used to hold various parts of the planetary gear system stationary, thereby obtaining reverse and 2nd gears. One brake band operates on the outer surface of the planetary gearset's annulus (ring) gear. It is called the 1st and reverse brake band because its purpose is to provide reverse operation and to keep the transmission from freewheeling in 1st gear when the selector lever is in **1**. This brake band does not engage when the transmission is in 1st gear and the selector lever is at **D**.

The other brake band locks the drum that houses the direct and reverse clutch and thereby prevents free rotation of the small sun gear in the planetary gearset. It is called the 2nd gear brake band because it is applied in 2nd gear with the selector lever at **D** or at **2**, or during 2nd gear kickdown. Both brake bands are fitted with adjusting screws. However, adjustments can be performed only after the transmission has been removed from the vehicle.

Hydraulic Controls

The hydraulic control system directs and regulates hydraulic pressure from the ATF pump, thereby controlling shifting of the planetary gearset. Shifts are produced by applying ATF pressure to the ring-shaped clutch pistons and the two piston-type brake band servos in the bottom of the transmission case. Hydraulic pressure is directed to the proper clutch or brake band servo by a number of spring-loaded control valves inside cylinders machined into the valve body. See Fig. 1-1.

Fig. 1-1. Cutaway view of automatic transmission. Final drive is different on vehicles covered by this Manual.

© 1974 VWoA—967

Three primary control devices regulate the movement of the control valves. One of these devices, the manual valve, is connected to the selector lever by a flexible cable. Moving the lever changes the setting of the valve to produce the necessary application of hydraulic pressure for the drive range selected. A second primary control device, the primary throttle pressure valve, which operates on engine vacuum, makes the transmission responsive to variations in engine speed and load. A third primary control device, the governor, which is gear-driven off the final drive pinion and controls ATF pressure relative to its rotational speed, makes the transmission responsive to variations in vehicle speeds.

NOTE —
The final drive housing shown in Fig. 1-1 is not the one used on Type 2 vehicles.

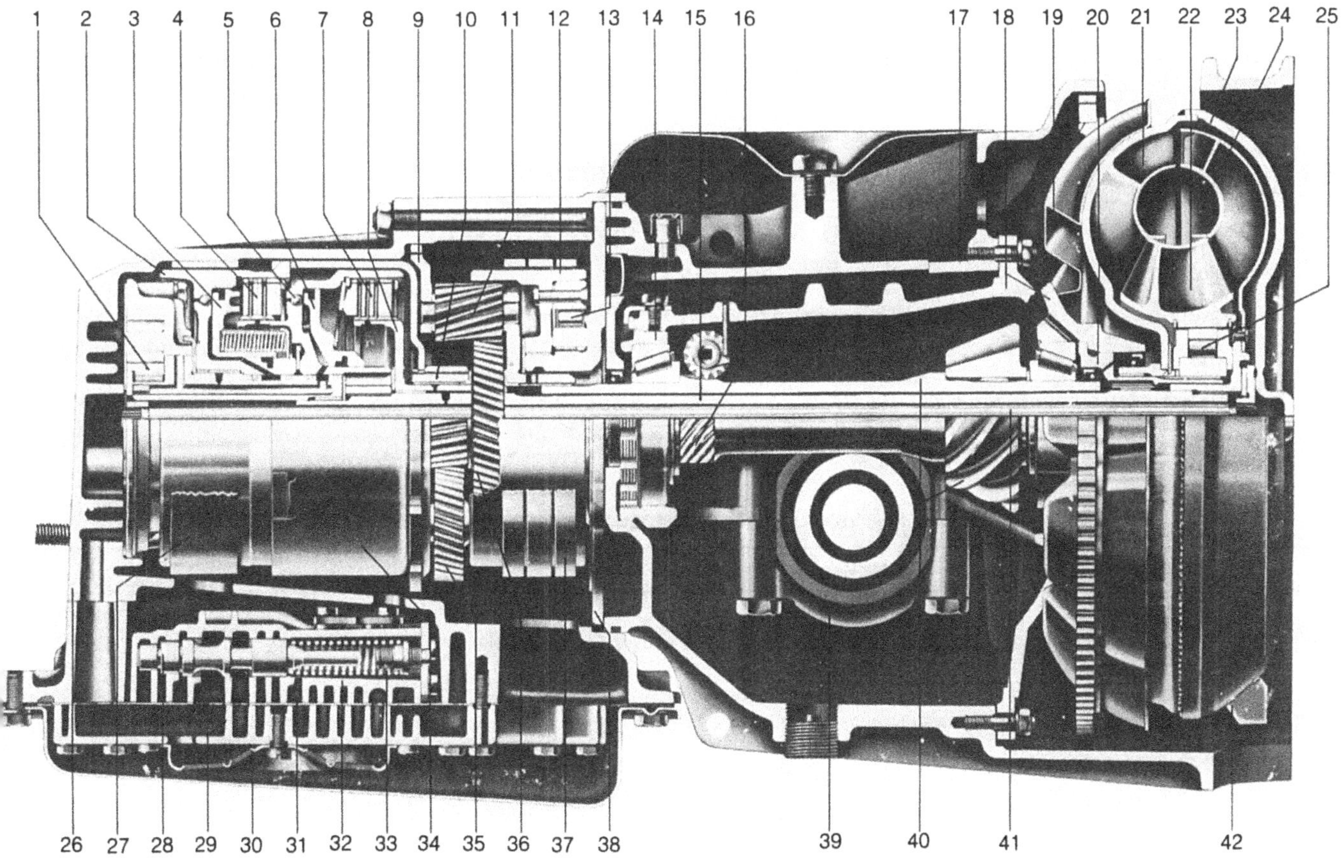

1. Pump
2. Clutch drum
3. Piston for direct and reverse clutch
4. Direct and reverse clutch
5. Forward clutch drum with ball valve
6. Piston for forward clutch
7. Forward clutch
8. Forward clutch hub
9. Planetary gear carrier
10. Small sun gear
11. Small planet pinion
12. Annulus or ring gear
13. 1st gear one-way clutch
14. Adjusting ring for pinion bearing
15. Turbine shaft
16. Governor drive
17. Final drive housing
18. Bearing web
19. Cooling fins
20. One-way clutch support
21. Impeller
22. Stator
23. Converter housing
24. Turbine
25. One-way clutch
26. Transmission case
27. 2nd gear brake band
28. Control valve
29. Transfer plate
30. ATF strainer
31. Separator plate
32. Valve body
33. Spring for valve
34. Driving shell
35. Large planet pinion
36. Large sun gear
37. 1st and reverse brake band
38. Bearing flange
39. Differential
40. Pinion with shaft
41. Pump shaft
42. Connecting lug

Final Drive

The final drive consists of a hypoid drive pinion and ring gear with a differential gearset. The differential gearset, which consists of the two differential gearshafts and the two differential pinions, allows the rear wheels to turn at different speeds, as is necessary when making turns (the outside wheel must travel farther than the inside wheel in the same amount of time).

2. MAINTENANCE

Many automatic transmission malfunctions can be traced to dirty ATF, too little or too much ATF, or other improper maintenance and lubrication conditions. Instructions and suggestions for automatic transmission care are found in **LUBRICATION AND MAINTENANCE**. The following operations are covered there:

1. Checking ATF level
2. Checking transmission pan screws
3. Checking constant velocity joint screws
4. Checking constant velocity joint seals
5. Checking kickdown operation
6. Changing ATF; cleaning the sump and strainer
7. Filling transmission
8. Changing final drive hypoid oil
9. Lubricating rear wheel bearing.

3. TROUBLESHOOTING

Before diagnosing automatic transmission troubles, review the history of the unit. Such a review may offer important clues to present difficulties. The following should be checked before making repairs or adjustments:

1. Be sure that the engine is tuned up and running right.
2. Inspect the transmission for external damage, loose or missing screws, and obvious leaks. Check the final drive hypoid oil for ATF contamination.
3. Check ATF level. Rub some ATF between your fingers and sniff it to detect the burned odor that means burned friction linings. If the ATF is dirty, it may be clogging the automatic controls.
4. Check the adjustment and operation of the kickdown switch and the adjustment of the selector lever cable. See if the vacuum unit on the primary throttle pressure valve is bent.

3.1 Road Testing

Drive the car in all transmission ranges and under as many road conditions as possible. Note the shift points both up and down. They should take place quickly, without interrupting the power flow. Listen for engine racing between gears, a possible indication of slipping clutches or brake bands. **Table a** may suggest remedies for defects you observe. The numbers in bold type in the Remedy column refer to numbered headings in **AUTOMATIC TRANSMISSION.**

Table a. Automatic Transmission Troubleshooting

Problem	Probable Cause	Remedy
1. No drive in any selector lever position	a. Automatic transmission fluid level low	a. Check and correct ATF level. Repair any leaks.**
	b. Manual valve not hooked to selector lever	b. Replace selector lever cable or attachments. See **13.1**.
	c. Pump or pickup screen clogged	c. Clean pump and pickup.**
	d. Defect in transmission pump or pump drive	*d. Repair or replace pump. If drive is broken check pump gears for free movement. See **10.6**.
	e. Broken shaft or planetary gear set	*e. Replace broken parts. See **10**.
2. ATF dark colored and smells burned	This may accompany or signal the start of trouble caused by burned friction linings on the brake bands or clutches	Drain contaminated ATF. Remove as much ATF as possible from converter. Replace with fresh ATF. See **8.5**.**
3. No drive in forward gears	Forward clutch defective	*Repair clutch. See **10.2**, **10.8**.
4. No drive in **R** and no engine braking with lever at **1**	1st and reverse band or servo defective	*Check and repair 1st and reverse band and servo. See **7**, **10.2**, **10.4**.
5. Car will not move off when lever is at **2** or **D**	1st gear one-way clutch in annulus defective	*Repláce annulus gear and one-way clutch. See **10.2**, **10.9**.
6. No drive in 2nd gear when lever is at **2** or **D**	2nd gear band or servo defective	*Repair band or servo. See **7**, **10.2**, **10.4**.
7. Transmission stays in 1st gear with lever at **2** or **D**	a. Governor dirty or defective	a. Clean or repair governor. Fit new lock washer if old one is missing. Remove pan to retrieve old washer. See **6**.
	b. Valve body assembly dirty	b. Remove oil pan. Clean valve body. See **5**.

Table a. Automatic Transmission Troubleshooting (continued)

Problem	Probable Cause	Remedy
8. No drive in 3rd gear or reverse	Direct and reverse clutch defective	*Repair clutch. See **10.2**, **10.7**.
9. Erratic power transmission, noisy reverse (accelerator may have to be depressed several times before car moves)	a. ATF level too low or high b. Selector lever out of adjustment c. Oil strainer dirty d. Primary throttle pressure valve sticking	a. Check and correct ATF level. Repair any leaks.** b. Adjust cable. See **13.2**. c. Remove oil pan, clean strainer.** d. Check valve, replace if necessary. See **3.3**, **4**.
10. Engine surges on upshifts. Shift time too long	a. ATF level too low or high b. Primary throttle pressure valve misadjusted c. Direct and reverse clutch defective	a. Check and correct ATF level. Repair any leaks.** b. Adjust primary throttle pressure. See **3.3**, **4**. *c. Repair clutch. See **10.2**, **10.7**.
11. Shifts take place at too low speeds	a. Primary throttle pressure valve misadjusted b. Governor or governor drive defective c. Valve body assembly dirty	a. Adjust primary throttle pressure. See **3.3**, **4**. b. Inspect and repair governor. See **6**. c. Remove oil pan. Clean valve body. See **5**.
12. Shifts take place at too high speeds	a. Primary throttle pressure valve misadjusted b. Vacuum hose leaky c. Kickdown switch lever bent d. Kickdown solenoid switch defective e. Valve body assembly dirty f. ATF pressure low due to internal transmission leaks	a. Adjust primary throttle pressure. See **3.3**, **4**. b. Replace hose. See **4**. c. Repair or replace kickdown switch.** d. Replace solenoid. See **5.2**. e. Remove oil pan. Clean valve body. See **5**. f. Disassemble transmission and replace all seals and gaskets. See **10**, **7**, **6**, **8.3**.
13. Transmission does not shift into 3rd gear with lever at **D**	a. Governor or governor drive defective b. Valve body assembly dirty c. Direct and reverse clutch defective	a. Inspect and repair governor. See **6**. b. Remove oil pan. Clean valve body. See **5**. *c. Repair clutch. See **10.2**, **10.7**.
14. Heavy jerk when selecting a drive range from neutral	a. Engine idle too fast b. Primary throttle pressure valve misadjusted c. Vacuum hose leaky d. Primary throttle pressure vacuum unit leaking	a. Adjust idle speed. See **FUEL SYSTEM**. b. Adjust primary throttle pressure. See **3.3**, **4**. c. Replace hose. See **4**. d. Replace primary throttle pressure vacuum chamber. See **4**.
15. Kickdown will not function	a. Kickdown switch defective or misadjusted b. Kickdown solenoid switch defective	a. Replace kickdown switch.** b. Replace solenoid. See **5.2**.
16. Poor acceleration. Top speed low despite good engine output	a. ATF level too low or high b. Torque converter one-way clutch defective c. Forward clutch defective d. Direct and reverse clutch defective e. 2nd gear brake band or servo defective	a. Check and correct ATF level. Repair any leaks.** b. Replace torque converter. See **8.1**. *c. Repair clutch. See **10.2**, **10.8**. *d. Repair clutch. See **10.2**, **10.7**. *e. Repair brake band or servo. See **7**, **10.2**, **10.4**.
17. Screeching noise when moving off or accelerating	a. Torque converter one-way clutch defective b. 1st gear one-way clutch in annulus defective	a. Replace torque converter. See **8.1**. *b. Replace annulus gear and one-way clutch. See **10.2**, **10.9**.
18. Scraping, grinding noise from converter. Fluid silver-colored	Thrust washer in converter worn	Replace annulus gear and one-way clutch. See **10.2**, **10.9**.
19. High ATF consumption without external leak	a. Leaking vacuum chamber on primary throttle pressure valve (exhaust will be smoky) b. Oil seals for pinion or governor shafts leaking (oil may be leaking from transmission breather)	a. Replace primary throttle pressure vacuum chamber. See **4**. *b. Replace seals. See **7**, **11.1**, **11.2**.
20. Parking lock will not hold vehicle	a. Selector lever out of adjustment b. Parking lock linkage defective	a. Adjust cable. See **13.2**. b. Repair linkage. See **10.11**.
21. Heavy leakage of ATF. Transmission case and under side of car oily	Leaking converter oil seal. Welded seam in converter may be leaking	Replace oil seal, or seal and converter if necessary. See **8**, **11.3**.

***Transmission must be out of the car and disassembled for this repair**
****See LUBRICATION AND MAINTENANCE**

7

When troubleshooting the automatic transmission, try to pin down the main component involved: converter, planetary gear system, or hydraulic controls. If **Table a** has failed to pinpoint the malfunction adequately, the following tests should help to isolate the problem.

3.2 Stall Speed Testing

This test provides a quick check of the torque converter operation, but should be performed only if the car accelerates poorly or fails to reach the specified maximum speed. An electronic tachometer is required.

CAUTION —
Never extend this test beyond the time it takes to read the gauges. Doing so may overheat the transmission and damage the oil seals.

To test:

1. Connect the tachometer according to the instrument manufacturer's instructions.
2. Start the engine.
3. Set the parking brake and depress the foot brake firmly to hold the vehicle stationary.
4. Shift the selector lever to position **D** and floor the accelerator pedal. Instead of revving up, the engine will run at a reduced rpm, known as stall speed.

If the rpm at stall is about 400 rpm below the specified 1900 to 2000 rpm stall speed—and the engine is in a proper state of tune—something is wrong with the torque converter. If the rpm at stall is above 2000 rpm, something is wrong in the forward clutch or the 1st gear one-way clutch located in the annulus gear of the planetary gearset. The test can also be made with the selector lever at **R**. If the reverse stall speed is too high, it indicates slippage in either the direct and reverse clutch or the 1st gear brake band.

3.3 Testing Hydraulic Control Circuit

A stall speed test is valuable mainly for isolating problems in the converter and the planetary gear system. Troubleshooting the hydraulic control system requires pressure testing. Although the pressure tests described here do not include tests for every valve in the hydraulic control system, the tests are adequate for determining whether or not the trouble is in the hydraulic controls. The actual source of the trouble is not significant since you will have to remove the valve body to correct the trouble, regardless of where it lies. If there are physical defects—even minor ones—the entire valve body and governor must be replaced. In many cases, however, a thorough cleaning will be all that is required.

Pressure Testing

Only two of the six pressure test connections need be used. One is for primary throttle pressure and the other for main pressure. By attaching pressure gauges at these points, using long hoses, it is possible to determine whether there are internal leaks, wear, clogged ATF passages, or sticking valves.

Most automatic transmission malfunctions can be isolated without pressure tests. However, such tests are often valuable for confirming the need for certain repairs or isolating one of two possible causes for the same malfunction. A gauge reading from 0 to 140 psi (0 to 10 kg/cm^2) is needed to measure primary throttle pressure and a gauge with a 0 to 350 psi (0 to 25 kg/cm^2) range is needed to measure main pressure. Fig. 3-1 shows the test connection points on the transmission case.

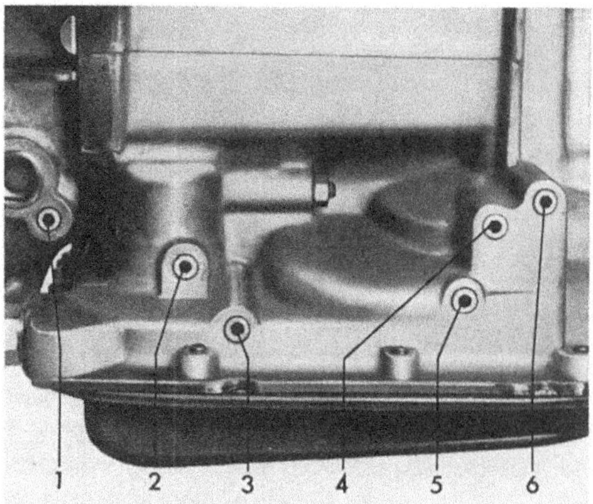

© 1974 VWoA—1079

1. Governor pressure
2. Main pressure
3. Primary throttle pressure
4. Main pressure. Release side of 2nd gear brake band servo piston
5. Secondary throttle pressure
6. Main pressure. Apply side of 2nd gear brake band servo piston

Fig. 3-1. Pressure test connections. The main pressure and primary throttle pressure connections are the most important points to know.

CAUTION —
If you lack the skills, tools, or a suitable workshop for automatic transmission work, we suggest you leave such repairs to an Authorized VW Dealer or other qualified shop. We especially urge you to consult your Authorized VW Dealer before attempting repairs on a car still covered by the new-car warranty.

To test pressures:

1. Connect the 0 to 350 psi (0 to 25 kg/cm^2) gauge to the main pressure test point.
2. Connect the 0 to 140 psi (0 to 10 kg/cm^2) gauge to the primary throttle pressure test point.
3. Route the long gauge hoses through an open window into the car.

> **CAUTION —**
> *The car must be driven during this test. Be sure the hoses do not drag on the pavement or rub the wheels of the car. This could damage the testing equipment.*

4. Remove and plug the vacuum line. See Fig. 3-2.

© 1974 VWoA—1078

Fig. 3-2. Disconnected vacuum hose. Pull the hose off the vacuum unit and seal it with an 8-mm (3/8-in.) punch. The test connection is behind the punch. The hex key in the vacuum unit's hose connection is in position to adjust the primary throttle pressure.

5. Move the selector lever to **N**, start the engine, and measure both pressures at a fast (1000-rpm) idle. The throttle pressure should be 45.5 psi (3.20 kg/cm^2). The main pressure should be 92.0 psi (6.50 kg/cm^2).
6. If the throttle pressure is not within specifications, correct it by turning the adjusting screw inside the vacuum chamber connection. Replace vacuum units that will not adjust to specifications.

> **NOTE —**
> Be careful not to bend the vacuum unit. It must be perfectly straight to work properly.

7. Reconnect the vacuum hose and again measure the pressures at a fast idle. The throttle pressure should be 5 to 6 psi (0.35 to 0.42 kg/cm^2), and the main pressure 47 to 50 psi (3.30 to 3.50 kg/cm^2).
8. Move the selector lever to **D**.
9. Check both pressures at full throttle while you hold the car stationary with the foot and parking brakes. The throttle pressure should be 31 to 44 psi (2.20 to 3.10 kg/cm^2), and the main pressure should be 92 psi (6.50 kg/cm^2).

> **CAUTION —**
> *Never continue a full-throttle pressure test longer than it takes to read the gauges. Doing so may overheat the transmission and damage the oil seals.*

10. Move the selector lever to **R**. Check the main pressure while you hold the car stationary with the foot and parking brakes. The pressure should be 142 to 156 psi (10.00 to 11.00 kg/cm^2) at a fast idle and 256 to 327 psi (18.00 to 23.00 kg/cm^2) at full throttle.
11. Take the car on the road and measure the main pressure at speeds above 19 mph (30 kph). Main pressure should be 92 psi (6.50 kg/cm^2) at full throttle.
12. After removing the test hose connections, torque the pressure connection plugs to 1.0 mkg (7 ft. lb.). Then run the engine and check the pressure connection plugs for leaks.

Pressures lower than those specified mean one or more of the following: (1) a worn pump; (2) internal ATF leaks past seals, gaskets, and metal mating surfaces; (3) sticking pressure regulating valves. High pressures always indicate sticking valves or a bent primary throttle pressure valve vacuum unit.

4. REPLACING PRIMARY THROTTLE PRESSURE VACUUM UNIT

7

The primary throttle pressure vacuum unit screws into the transmission case. A hex below the vacuum chamber provides a grip for the wrench. When installing a new unit lubricate the threads lightly with a good anti-seize compound. Torque to 2.5 mkg (18 ft. lb.). Make certain that the vacuum hose to the engine is in good condition. Replace it if it is cracked or fits loosely. Do not overtighten the hose clamps.

> **NOTE —**
> When replacing the primary throttle pressure vacuum unit or installing a new or exchange transmission, check the primary throttle pressure to bring it within the specifications given under Step 5 of the pressure testing procedures. A hex key is inserted in the vacuum connection to make any adjustment required as shown in Fig. 3-2.

5. SERVICING VALVE BODY AND VALVE BODY ASSEMBLY

Servicing the valve body assembly normally involves only removal and cleaning. However, it may also be necessary to remove the assembly from the transmission in order to replace a faulty kickdown solenoid.

5.1 Removing and Installing Valve Body Assembly

The valve body assembly can be removed with the engine and transmission in the car. The assembly must also be removed for cleaning during more extensive repairs with the transmission out of the car. The location of the valve body assembly together with related parts in the transmission case can be seen in Fig. 5-1.

1. Manual valve
2. Kickdown solenoid
3. ATF strainer
4. Transfer plate
5. Valve body
6. Primary throttle pressure valve vacuum unit

Fig. 5-1. Valve body and related parts. The transmission pan has been removed.

To remove:

1. Remove the transmission pan screws. Then take off the transmission pan and the gasket.
2. Disconnect the solenoid electrical wire from its terminal on the transmission case.
3. Fourteen bolts and a screw hold the valve body together and fasten it to the transmission. These fasteners are shown in Fig. 5-2. Take out the 14 bolts, but for the moment leave the screw in place.

Fig. 5-2. Fasteners for valve body assembly. Arrow **1** points to screw which is the fifteenth and last fastener to be removed. Arrow **2** points to the electrical connection for the kickdown solenoid. Arrow **3** indicates the manual valve.

4. Remove the screw. Take off the valve body assembly, but do not separate it into its three main components as shown in Fig. 5-3 or you may lose the ball valves.

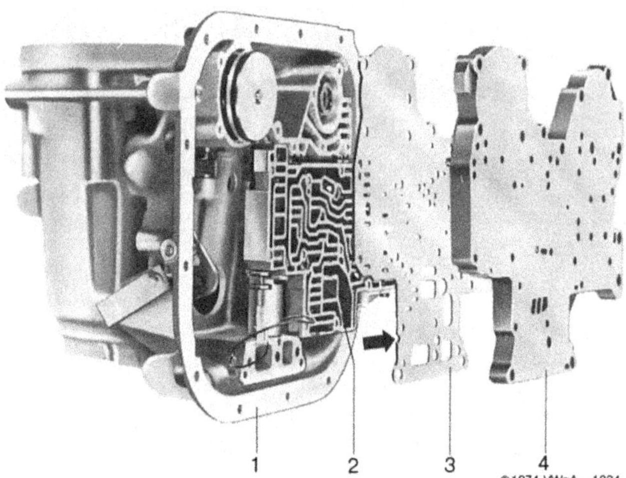

1. Transmission case
2. Valve body
3. Separator plate
4. Transfer plate

Fig. 5-3. Valve body assembly apart. After removal, disassemble only on a clean workbench.

CAUTION —
If the valve body assembly is removed while the transmission is in the car, the servo piston for the 1st gear band might fall out. Be ready to catch it.

To disassemble valve body assembly:

1. Place the assembly on a clean workbench with the valve body down. Then remove the 15 Phillips head screws that hold the three components together.
2. Lift off the separator plate and transfer plate as a unit so that the ball valves in the valve body remain in the valve body.
3. Place the transfer plate and separator plate on the workbench with the transfer plate down. Then lift off the separator plate.

Assembly is the reverse of disassembly. The locations for the ball valves are shown in Fig. 5-4 and Fig. 5-5.

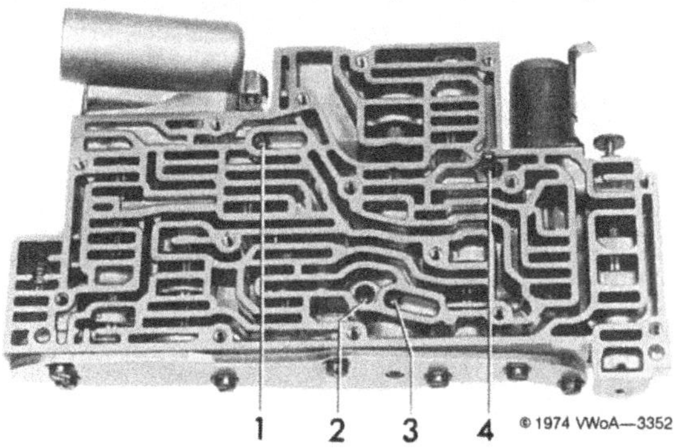

Fig. 5-4. Ball valves in valve body. Ball for direct and reverse clutch valve is at **1**, ball for 1st gear valve is at **2**, ball for 1st and reverse brake band valve is at **3**, and ball and spring for forward pressure relief valve is at **4**.

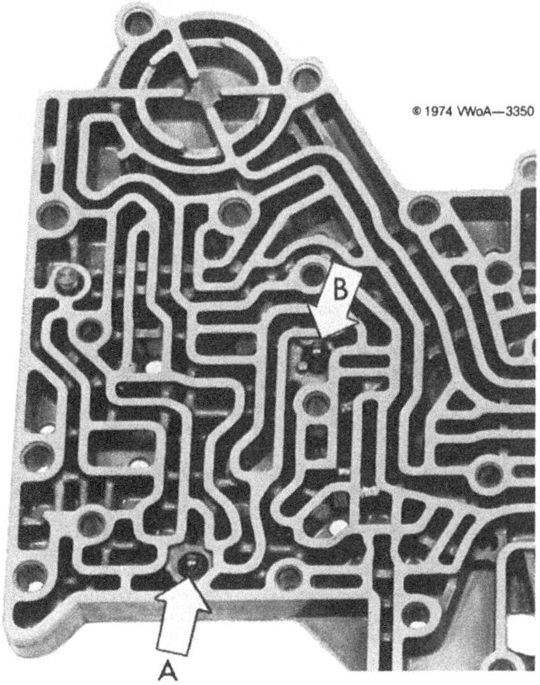

Fig. 5-5. Ball valves in transfer plate. Ball and spring for 2nd/3rd gear valve is at **A**; ball and spring for 3rd/2nd gear valve is at **B**.

To install valve body assembly:

1. Attach the valve body assembly to the transmission case with the screw indicated by arrow **1** in Fig. 5-2. The manual valve (arrow **3**) must be engaged in the operating lever.
2. Install the 14 bolts and washers finger-tight.
3. Working diagonally, torque all bolts to 0.4 mkg (2.8 ft. lb.) and the screw to 0.35 mkg (2.5 ft. lb.).
4. Connect the wire to the kickdown solenoid.
5. Install a new pan gasket and then the transmission oil pan. Working diagonally, torque the pan screws to 1 mkg (7 ft. lb.).
6. Wait 5 minutes for the new gasket to compress, then retorque the screws to 1 mkg (7 ft. lb.).
7. Repeat procedure 6 until the screws remain at 1 mkg (7 ft. lb.).

> ***CAUTION —***
> *Never tighten the transmission pan screws to more than 1 mkg (7 ft. lb.) in an attempt to cure a leaking gasket. Overtightening will deform the pan and make it impossible to get a new seal. Always install a new gasket to correct leaks.*

5.2 Removing and Installing Kickdown Solenoid

The kickdown solenoid is an electromechanical device that moves the kickdown valve in the hydraulic control system. It is operated by an electrical switch under the accelerator. If the transmission will not kick down, use a voltmeter to check the wire leading to the transmission. If current is reaching the transmission, the solenoid may be at fault.

Test the solenoid for continuity by removing the wire from the outside of the transmission. Attach a battery-powered test light to the terminal on the transmission case and to the case itself. If the test light does not come on, the solenoid is faulty or disconnected from the terminal inside the transmission case. Remove the transmission pan in order to inspect the wire's connection.

To replace solenoid:

1. Take off the transmission pan and remove the valve body assembly.
2. Remove the two screws that hold the solenoid to the valve body.

> **NOTE —**
> The solenoid mounting screws are inaccessible while the valve body is in place.

7

5.3 Disassembling and Assembling Valve Body

As a rule, the valve body is disassembled only for cleaning. Unless the fluid is very dirty or contaminated by large solid particles, it usually is sufficient to immerse the complete assembly in cleaning fluid and dry it with compressed air. However, be careful not to hold the air jet so close that it moves the valves violently. This could damage the springs.

Fig. 5-6 is an exploded view of the valve body. Because many of the parts look alike, it is easy to mix them up. This is especially true of the springs. Unless you keep the springs separated and marked for identification, you will have to measure each spring with a micrometer prior to reassembly in order to find its correct place. See **14. AUTOMATIC TRANSMISSION TECHNICAL DATA** for spring dimensions. To avoid the need for such measurements, use a compartmental storage tray. Such a tray will also keep the springs from getting bent or stretched, which would upset their precisely calibrated tensions.

Fig. 5-6. Valve body. The kickdown valve, throttle pressure limiting valve, and converter pressure valve are physically identical. Used valves, however, must not be interchanged.

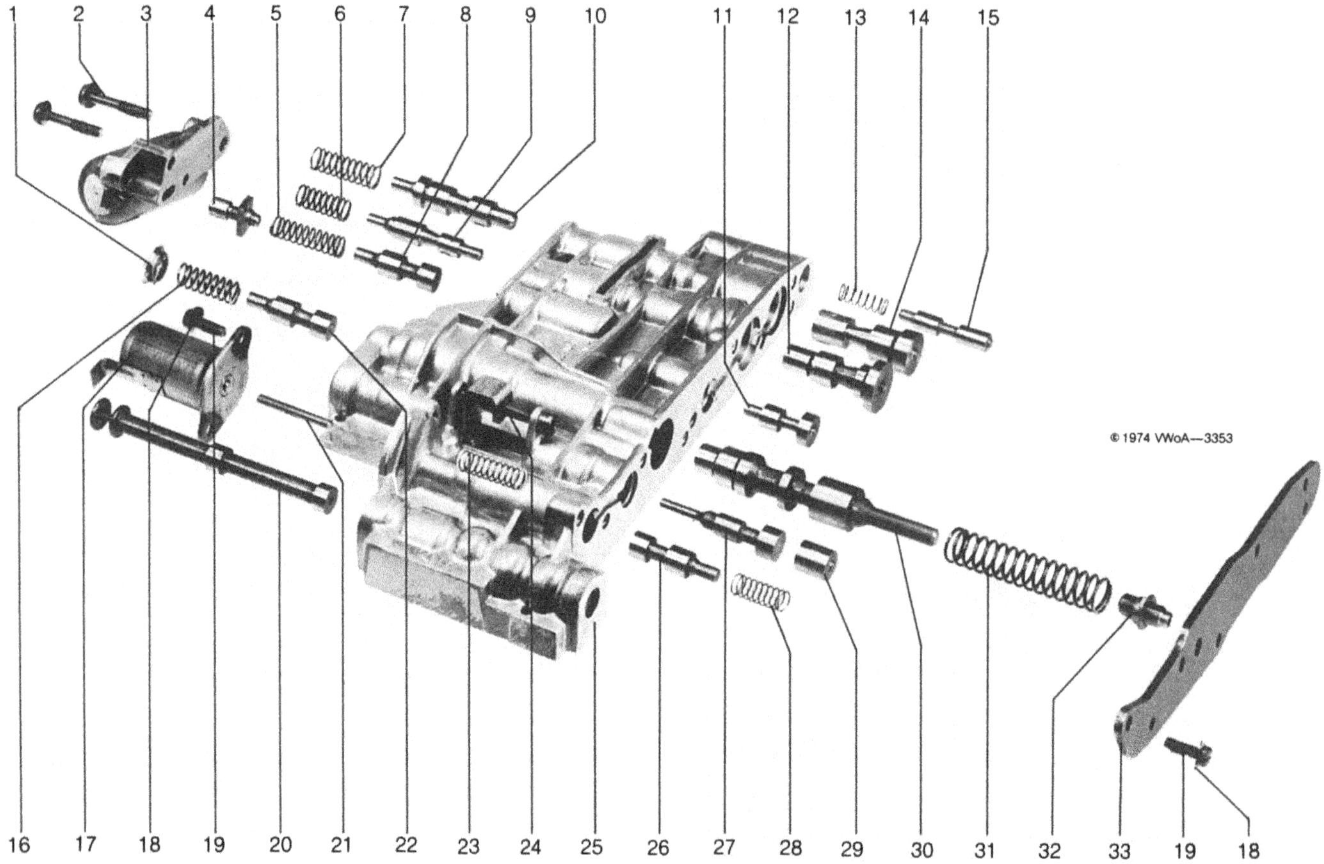

1. Converter pressure valve spring cup
2. M 5 screw
3. 2nd/3rd gear accumulator
4. Adjusting screw with spring seat
5. Throttle pressure limiting valve spring
6. 1st/2nd gear shift valve spring
7. 2nd/3rd gear shift valve spring
8. Throttle pressure limiting valve
9. 1st/2nd gear shift valve
10. 2nd/3rd gear shift valve
11. 2nd/3rd gear control valve
12. 1st/2nd gear governor plug
13. 3rd/2nd gear part throttle valve spring
14. 2nd/3rd gear governor plug
15. 3rd/2nd gear part throttle valve
16. Converter pressure valve spring
17. Kickdown solenoid
18. Spring washer
19. M 5 screw
20. Manual valve
21. Kickdown solenoid plunger
22. Converter pressure valve
23. Secondary throttle pressure valve spring
24. Secondary throttle pressure adjusting screw
25. Valve body
26. Kickdown valve
27. Secondary throttle pressure valve
28. Kickdown valve spring
29. 1st gear plug for secondary throttle pressure valve
30. Main regulating valve
31. Main regulating valve spring
32. Adjusting screw with spring seat
33. Rear endplate

If a storage tray (Fig. 5-7) for the valve body components is not available, you can make one by drilling holes in a thick board and numbering them.

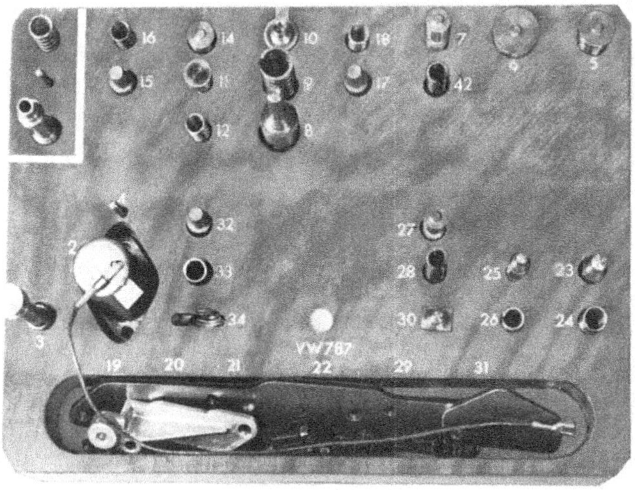

Fig. 5-7. Storage tray for valve body parts. Number holes to conform with part numbers in Fig. 5-6.

To disassemble:

1. Remove the valve body assembly from the transmission. Then disassemble it into its three main components as described in **5.1 Removing and Installing Valve Body Assembly.**
2. Remove the rear endplate.

> ***CAUTION —***
> *Do not, under any circumstances, alter the setting of the main pressure regulating valve adjusting screw that is under the endplate. The screw can be adjusted properly only at the factory. The adjusting screws for the throttle pressure limiting and secondary throttle pressure valves must also be left alone for the same reason.*

3. One at a time, remove each valve and spring and place it in the storage tray. Use a brass rod to carefully press out sticking or tight-fitting valves.
4. Remove the solenoid. Then take out the plunger and the manual valve. Remove the 2nd/3rd gear accumulator.
5. Remove the remaining valves and springs.
6. Wash all parts in clean kerosene, then dry them with compressed air.

Never use water to clean the valves and valve body, and never dry the parts with fluffy rags or by rubbing them against your clothing. Even a microscopic piece of lint or a small patch of rust can cause a valve to stick in its bore.

Assembly is basically the reverse of disassembly. Clean the workbench before you start to work. Lubricate each valve with ATF as you reinstall it, then make certain that the valve moves freely of its own weight. Used valves that have worn to fit individual bores must be returned to their original locations. When you install the endplate, be careful not to overtighten the screws. Doing so could easily strip the threads in the valve body.

Join together the three main components of the valve body assembly. Then install the valve body assembly in the transmission. Make certain that moving the selector lever moves the manual valve properly, and that when the valve is fully forward, it contacts the lug on the solenoid.

6. SERVICING GOVERNOR

The governor for the hydraulic control system is located beneath a round, black pressed steel cover just ahead of and slightly above the left driveshaft on the transmission. The cover is held in place by a spring wire clip.

6.1 Removing and Installing Governor

The governor can be removed with the engine and transmission installed in the car. The governor is usually removed for cleaning or for replacing worn parts. However, if a new valve body is being installed, a matching governor is supplied with the new valve body and must be installed whether the old governor is serviceable or not.

To remove:

1. Release the clip and take off the cover. Pull the governor out of the transmission case.
2. Inspect the thrust plate and the drive end of the shaft for wear and scoring.

> **NOTE —**
> Because replacing the entire governor could possibly change the governor pressure, new governor shafts are available separately to replace those that are worn or damaged.

Governor installation is the reverse of removal. Check the seal at the innermost end of the governor shaft bore in the final drive housing—especially if the hypoid oil is contaminated by ATF. Pry out faulty seals. Install the new seal with the lip outward, toward the governor body.

Make sure that a new governor is the proper replacement part for the vehicle. Turn the governor as you install it so that the square drive at the end of the shaft engages the drive gear. If the O-ring for the cover is broken or deformed, replace the O-ring.

6.2 Disassembling and Assembling Governor

Disassemble the governor only if it contains debris from burned clutch or brake band linings. Otherwise, just dip it in solvent and dry it with compressed air.

To disassemble the governor, remove the two M 5 × 40 screws and take off the thrust plate and housing. Then take out the transfer plate and the balance weight. The weight has been matched to the governor, so do not exchange the original weight with a weight from another governor. You can remove the weight, valve, spring, and dished washer from the pin by prying off the E-clips. See Fig. 6-1.

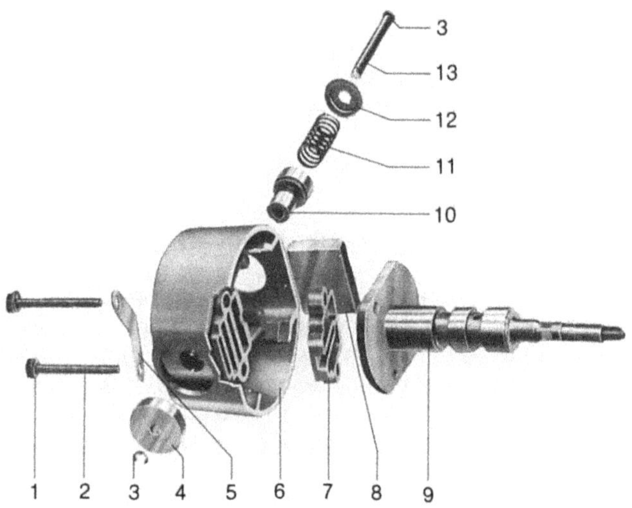

1. Spring washer (2)
2. M 5 screw (2)
3. E-clip (2)
4. Centrifugal weight
5. Thrust plate
6. Housing
7. Transfer plate
8. Balance weight
9. Governor shaft
10. Valve
11. Spring
12. Dished washer
13. Pin

Fig. 6-1. Centrifugal governor disassembled.

Before you assemble the governor, wash all parts in kerosene and dry them with compressed air. Lubricate the parts with ATF as you install them. The drillings in the transfer plate must taper down toward the weight. Make sure the angle in the thrust plate is at the center of the housing so that the cover will bear against it.

7. REMOVING AND INSTALLING SERVO PISTONS AND 1ST/2ND GEAR ACCUMULATOR

The servo pistons and accumulator can be removed and installed with the transmission in the car. This job is usually done to replace the piston seals and the O-ring. Fig. 7-1 shows the relative positions of the parts in the transmission case.

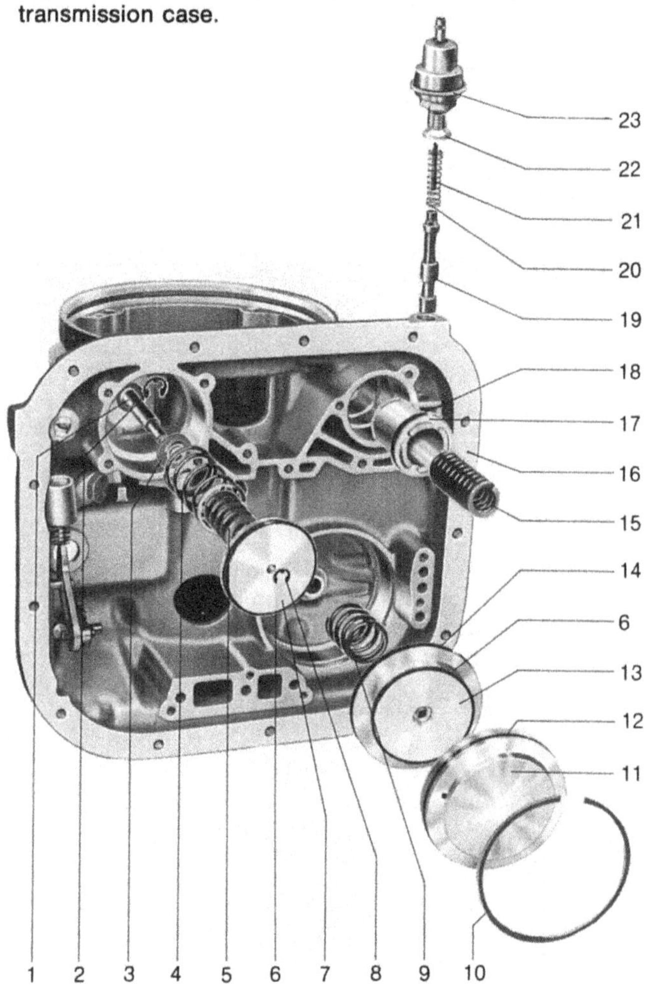

1. 1st gear band piston rod
2. E-clip
3. Dished washer
4. 1st gear band return spring
5. 1st gear accumulator spring
6. Piston seal (2)
7. 1st gear band piston
8. E-clip
9. 2nd gear band return spring
10. Sealing cover circlip
11. Sealing cover
12. O-ring
13. Piston with rod
14. 2nd gear band piston seal
15. Accumulator spring
16. Case
17. Accumulator piston
18. Accumulator piston seal
19. Primary throttle pressure valve
20. Primary throttle pressure valve spring
21. Thrust pin
22. Aluminum ring
23. Vacuum unit

Fig. 7-1. Exploded view of servo assemblies.

To remove:

1. Remove the transmission pan and the valve body assembly.
2. Take out the 1st gear band piston, then remove its seal.
3. Take the 9-mm and 6-mm E-clips off the piston rod and disassemble the piston and springs. Check the parts for wear.
4. Remove the sealing cover circlip. Take out the sealing cover for the 2nd gear band piston. Take out the piston and remove its two seals.
5. Replace the O-ring on the sealing cover.
6. Take out the accumulator spring and the accumulator piston. Remove the piston seal.

To install:

1. Dip the new seals in ATF and install them on their respective pistons.

   NOTE —
   When installing the seals on the pistons, be sure to position them with the seal lips toward the pressure side of the pistons. The lip of the large seal on the 2nd gear brake band servo piston should point upward, the lip of the small seal downward. The lip of the accumulator piston seal should point upward and the 1st and reverse servo piston seal should point downward.

2. Lubricate the 2nd gear brake band servo piston with ATF. Then insert it into the sealing cover using a twisting motion.
3. Insert the cover together with the piston and band return spring into the transmission case. Then install the circlip.
4. Slide the 1st and reverse piston and spring onto the piston rod. Lock them on the rod with the E-clips.
5. Lubricate the accumulator piston and the 1st and reverse brake band servo piston with ATF. Using a twisting motion, install the pistons together with their springs.
6. Install the valve body assembly as described in **5.1 Removing and Installing Valve Body Assembly**.
7. Install a new pan gasket and then the transmission pan.

When you install the pan, follow a diagonal pattern as you torque the pan screws to 1 mkg (7 ft. lb.). Wait 5 minutes for the new gasket to compress, then retorque the screws to 1 mkg (7 ft. lb.). Do this several times until the screws remain at the specified torque.

8. TORQUE CONVERTER

Up to this point we have covered service operations that can be carried out with the engine and transmission in the car—although they may also be carried out with the transmission removed. However, servicing the torque converter demands that the engine be removed from the car.

8.1 Removing and Installing Torque Converter

The torque converter usually requires removal only when you must replace the oil seal, replace the bushing, or clean the converter after a transmission failure has contaminated the ATF. Since the converter is a welded assembly, you must replace it as a unit if it is leaky or noisy, if it has a defective starter ring gear, or if a stall speed test shows the unit to be defective.

To remove:

1. Take the engine out of the car as described in **ENGINE AND CLUTCH**. Remove the securing bracket that you installed during engine removal.
2. Grasp the converter with both hands. Remove it by pulling it with a twisting motion off its support tube on the final drive carrier.

   CAUTION —
   Do not rock or tilt the converter when removing or installing it. This could damage the oil seal, the one-way stator clutch, or other parts in the hub.

Installation is the reverse of removal. Before you install the converter, inspect it thoroughly as described in **8.2 Inspecting Converter**. If the oil seal seat on the hub is rough, worn, or pitted, you should replace the torque converter. Otherwise the seal will wear out in a very short time. Slowly turn the converter clockwise and counterclockwise as you install it so that the turbine and pump shaft splines can engage.

8.2 Inspecting Converter

Inspect the converter seal inside the support tube on the final drive carrier. Replace the seal if necessary. Check the converter hub for signs of scoring from the oil seal. If the scoring is deep, replace the converter. Check for broken welds on the starter ring gear and air deflector plate. Remove any burrs from the ring gear. Insert the turbine shaft and turn the turbine to see that it spins freely. Check the condition of the torque converter bushing. It must be within the specifications given in **8.4 Replacing Converter Bushing.**

8.3 Replacing Converter Seal

The converter seal is located inside the converter support tube on the final drive housing. If the converter seal is worn or damaged, pry it out. Then, using an appropriate seal-installing tool, carefully drive in a new seal. The seal is soft, easily torn, and weakened by many common solvents. Replace the seal if it has been exposed to cleaning solvents or to gasoline or if it has been torn.

8.4 Replacing Converter Bushing

A leaking converter seal is usually caused by a worn bushing in the converter hub, so check the bushing every time you replace a seal. The bushing inside diameter must not exceed 34.25 mm (1.348 in.). Maximum out-of-round is 0.03 mm (.001 in.).

To replace the converter bushing, use an extractor and a slide hammer to remove the old bushing, as shown in Fig. 8-1. Drive in the new bushing with a properly fitting bushing driver. The special VW driver is made so that it just barely comes free after the bushing is pressed in. The bushing inside diameter should be between 34.03 and 34.10 mm (1.340 and 1.343 in.) following installation.

© 1974 VWoA—1092

Fig. 8-1. Extracting converter bushing. Use clean tools to keep dirt out of converter.

8.5 Cleaning Torque Converter

When charred material from a burned clutch disk or other pollutants have entered the ATF, remove the residual fluid from the converter with the home-made siphon shown in Fig. 8-2.

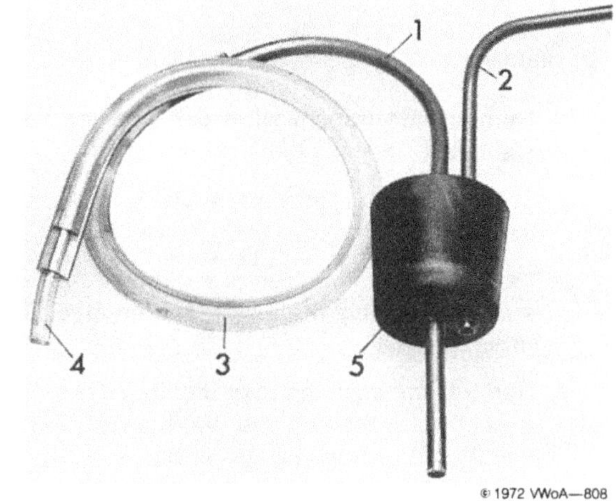

© 1972 VWoA—808

1. Steel or copper tube 3/16 × 8 in. (4 × 200 mm)
2. Steel or copper tube ⅛ × 6 in. (3 × 150 mm)
3. PVC hose ¼ × 14 in. (6 × 350 mm)
4. PVC hose ⅛ × 1¼ in. (3 × 30 mm)
5. Conical rubber plug 1½ in. (35 mm) diam.

Fig. 8-2. Siphon parts available at auto stores. Do not use tubing diameters larger than those specified.

To drain the converter, install the siphon as shown in Fig. 8-3. Push the siphon line pipe through the rubber plug until it contacts the converter bottom, then place the siphon hose over the oil receptacle. Blow into the short tube to start the siphon. Let the converter drain overnight or for about eight hours.

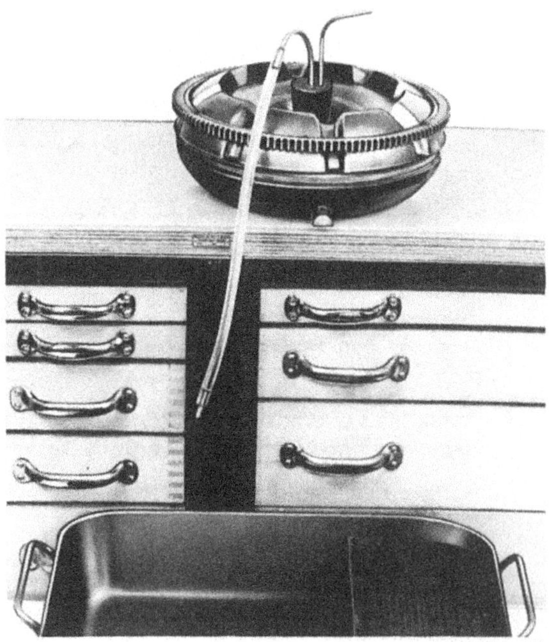

© 1974 VWoA—995

Fig. 8-3. Siphon and oil receptacle in position.

9. REMOVING AND INSTALLING TRANSMISSION

The automatic transmission and its mountings are shown in Fig. 9-1. The transmission carrier remains bolted to the frame when you remove the automatic transmission. The same carrier is used both for the manual transmission and for the automatic transmission. If you must remove the carrier, follow the procedure given in **TRANSMISSION AND REAR AXLE**. However, the carrier seldom requires removal unless there has been collision damage.

The automatic transmission and the engine can be removed individually or as a unit. The most efficient procedure is to remove them as a unit, then separate the engine from the transmission after both are out of the vehicle. However, depending on the capacity of your floor jack, it may be necessary to remove the engine as described in **ENGINE** and to then remove the transmission separately.

Whether you remove the engine and transmission separately or as a unit, all the necessary removal points for both the engine and the transmission are indicated in

Fig. 9-1. Automatic transmission and its mountings.

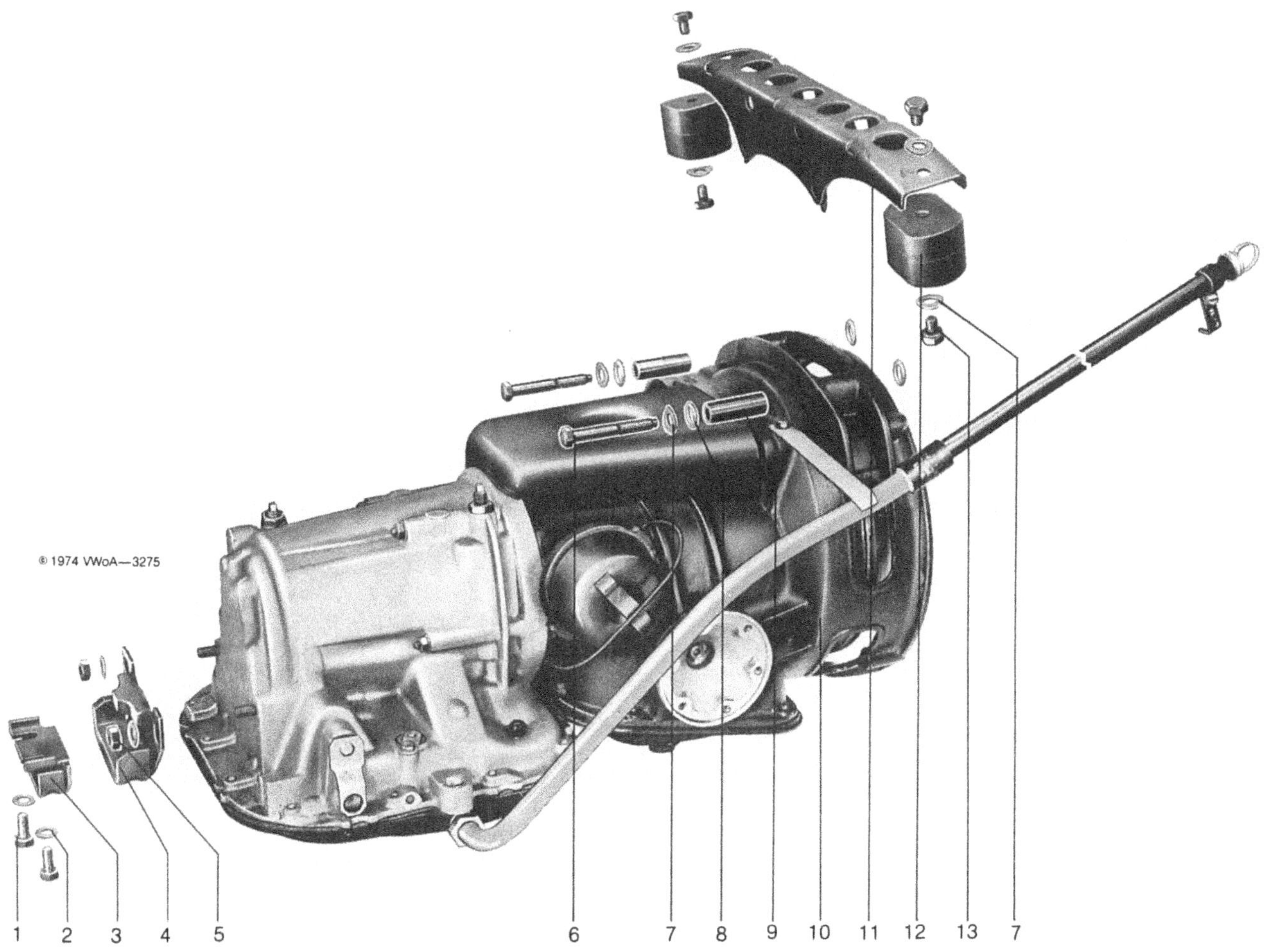

1. M 10 bolt (2)
2. Washer (4)
3. Limiting stop
4. Front bonded rubber mounting
5. M 10 nut (2)
6. Long M 10 bolt (2)
7. Spring washer (6)
8. Sealing ring (4)
9. Sleeve (2)
10. Transmission
11. Transmission carrier
12. Bonded rubber mounting (2)
13. Short M 10 bolt (4)

7

Fig. 9-2. We will refer to this illustration throughout the removal procedure.

To remove:

1. Disconnect the ground strap from the battery.

Fig. 9-2. Dual-carburetor engine and automatic transmission.

2. Remove the hose clamps at **2a** and **2b** in Fig. 9-2. Then take off the hoses for the fuel tank ventilation system and the ducts for the fresh air fan.

3. Remove the wires connected to the engine. These are on the regulator (**3a** in Fig. 9-2), the oil pressure switch (**3b** in Fig. 9-2), and the ignition coil (**3c** in Fig. 9-2). On fuel injection engines, disconnect the wires from all fuel injection components.

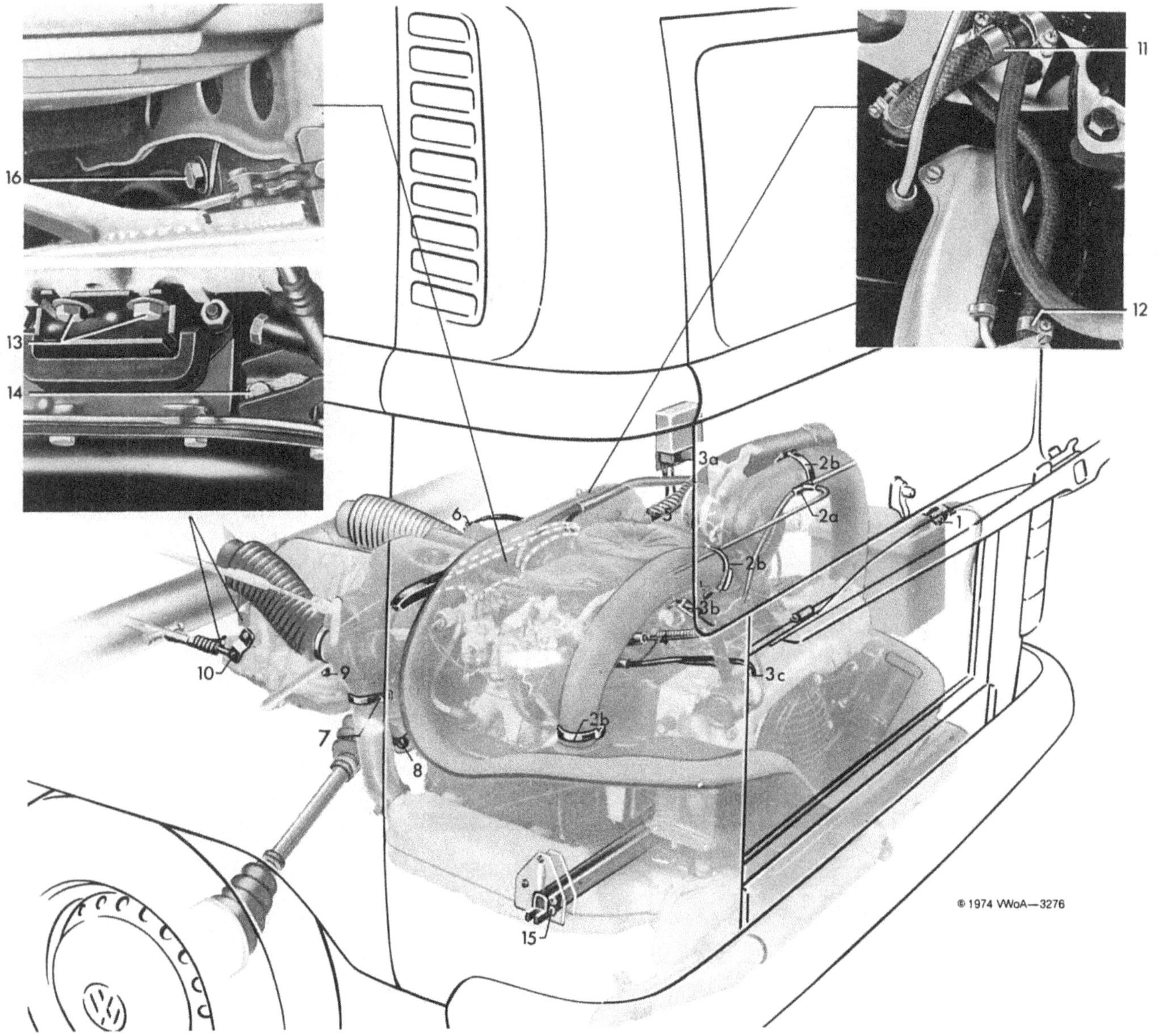

1. Battery ground strap
2a. Hose clamps at T-piece for fuel tank ventilation system (2)
2b. Hose clamps for fresh air fan ducts (2)
3a. Wires on regulator (2)
3b. Wire on oil pressure switch
3c. Wires on ignition coil
4. Accelerator cable
5. Duct for oil bath air cleaner
6. Wires on starter solenoid
7. Heat exchanger duct clamps (2)
8. Driveshaft inner constant velocity joints (2)
9. Wire to kickdown switch
10. Selector lever cable
11. Vacuum hose for vacuum powered brake servo
12. Fuel hose
13. Limiting stop
14. Bracket for selector lever cable
15. Nuts on engine bearer (4)
16. Bolts on transmission carrier (2)

4. Disconnect the accelerator cable (**4** in Fig. 9-2).
5. Remove the duct from the air cleaner (**5** in Fig. 9-2). On fuel injection engines, remove the entire air cleaner.
6. Disconnect the wires from the starter solenoid (**6** in Fig. 9-2).
7. Loosen the clamps on both the left and right heat exchangers (**7** in Fig. 9-2). Then remove the heat exchanger ducts.
8. Disconnect the driveshafts (**8** in Fig. 9-2) from the transmission as described in **TRANSMISSION AND REAR AXLE**. Using home-made wire hooks, suspend the disconnected driveshaft ends from the frame.
9. Disconnect the kickdown switch wire from the side of the transmission (**9** in Fig. 9-2).
10. Loosen the clamp nut on the selector cable lever (**10** in Fig. 9-2), then disconnect the cable.
11. Loosen the clamp that secures the vacuum hose for the vacuum powered brake servo (**11** in Fig. 9-2). Then slide the vacuum hose off the T-piece.
12. Loosen the clamp for the fuel hose (**12** in Fig. 9-2), then remove the fuel hose and plug it. On fuel injection engines, disconnect and plug both fuel hoses for the pressure regulator.

> ***WARNING —***
> *Some gasoline will escape from the hose(s). To prevent fire, do not smoke or work near heaters or other fire hazards. Have a fire extinguisher handy.*

13. Remove the two M 10 bolts that hold the front transmission mount limiting strap on the frame. The bolts are shown at **13** in Fig. 9-2.
14. Remove the bolt that holds the selector lever cable bracket (**14** in Fig. 9-2). This bolt also holds the transmission/frame ground strap.
15. Remove the four nuts and bolts on the engine bearer (**15** in Fig. 9-2).
16. Loosen, but do not remove, the two long M 10 bolts that hold the transmission on the transmission carrier (**16** in Fig. 9-2).
17. Place a floor jack with an engine adapter and a transmission adapter under the engine/transmission assembly. Slightly raise the engine/transmission assembly. Then fully remove the two long M 10 bolts from the transmission carrier.
18. Using a floor jack, slightly lower the engine/transmission assembly, pull it to the rear, then fully lower it to floor level.
19. Working through the openings beneath the plastic plug in the left-hand part of the engine's mounting flange, remove the three bolts that hold the torque converter on the engine's drive plate. Then take off the nuts and bolts that hold the transmission on the engine and separate the engine from the transmission. See **ENGINE AND CLUTCH** for full details and illustrated procedures.

Installation is the reverse of removal. Torque the engine/transmission mounting nuts and bolts to 3.0 mkg (22 ft. lb.) and the 12-point bolts that hold the converter on the drive plate to 2.5 mkg (18 ft. lb.). Torque the long M 10 bolts in the transmission carrier to 3.5 mkg (25 ft. lb.), and the bolts for the front transmission mount limiting stop to 4.5 mkg (33 ft. lb.). Following installation, adjust the accelerator cable as described in **FUEL SYSTEM**. Adjust the selector lever cable as described in **13.2 Adjusting Selector Lever Cable**.

10. REPAIRING AUTOMATIC TRANSMISSION

Thoroughly clean the outside of the transmission before disassembly so that dirt will not enter the hydraulic controls or mechanical parts. Study the repair procedures on the following pages. If they require equipment you do not have, the transmission should be turned over to a specialist before any disassembly.

> ***CAUTION —***
> *If you lack the skills, tools, or a suitable workshop for transmission work, we suggest you leave such repairs to an Authorized VW Dealer or other qualified shop. We especially urge you to consult an Authorized VW Dealer before attempting repairs on a car still covered by the new-car warranty.*

10.1 Separating Transmission Case From Final Drive Housing

A transmission stand is a great help when you must take the automatic transmission apart. Do not disassemble the unit on the shop floor as dirt and debris may get into the working parts.

To disassemble:

1. After cleaning the outside of the transmission and draining the ATF, remove the converter and withdraw the pump shaft.

> **NOTE —**
> You need not drain the hypoid oil unless you intend to disassemble the final drive.

2. Remove the four M 8 nuts from the steel studs that hold the final drive housing to the transmission case. Separate the main parts as shown in Fig. 10-1.

Fig. 10-1. Main parts of the transmission separated. The final drive shown here is not for a Type 2 vehicle.

1. M 8 nut (4)
2. Spring washer (4)
3. Oil filler tube
4. Oil dipstick
5. Capscrew
6. Spring washer
7. Air deflector plate
8. Ground strap
9. Selector lever cable bracket
10. M 6 bolt (2)
11. Spring washer (2)
12. O-ring
13. Connecting piece with O-ring
14. Transmission assembly (complete)
15. Gasket
16. Vacuum unit
17. O-ring
18. Breather pipe cover (2)
19. Final drive (complete)
20. Turbine shaft
21. Pump shaft
22. Torque converter

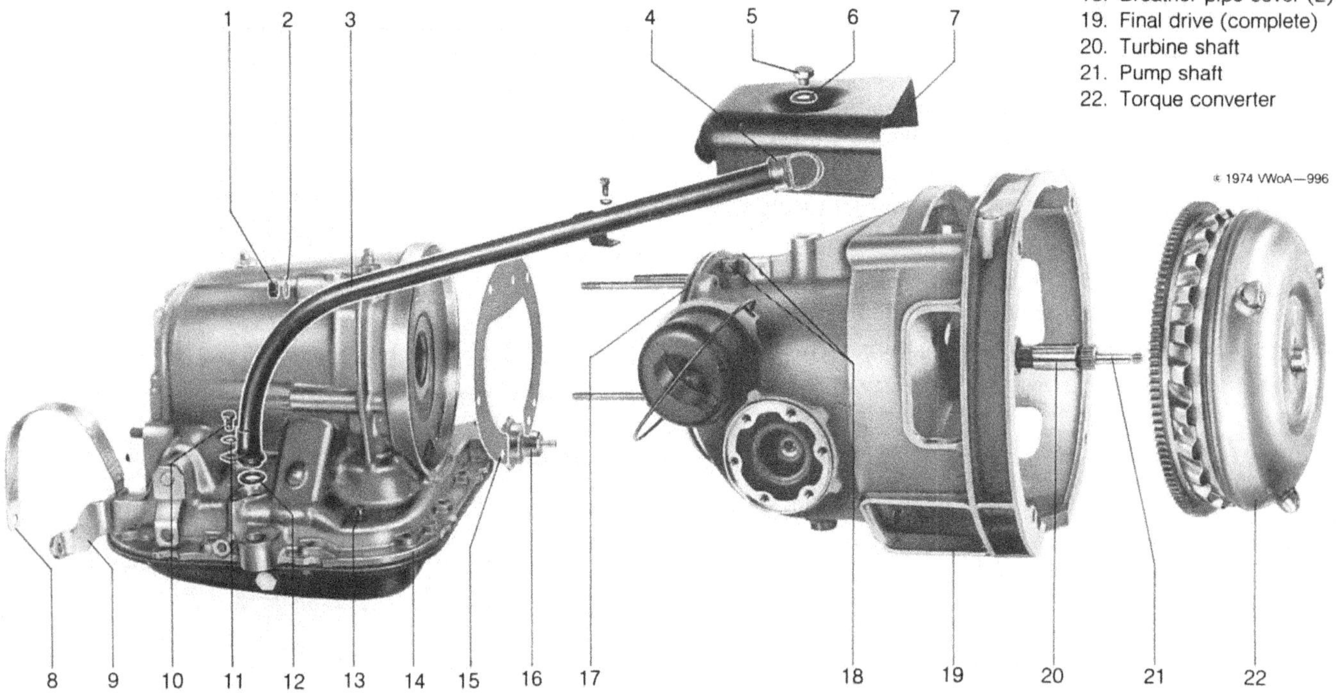

3. Cover the converter hub opening and the final drive housing to prevent dirt from entering. Take the transmission itself to the workbench.

Assembly is covered in **10.5 Installing Final Drive on Automatic Transmission**.

> **NOTE —**
> A special stand is used in Authorized VW Dealers' shops to hold the transmission while the transmission case and final drive housing are separated and to hold the transmission for subsequent repair.

10.2 Disassembling and Assembling Automatic Transmission

Study Fig. 10-3 carefully so that you become familiar with the names of the various parts. These names will be referred to frequently on the following pages. The illustration also shows the positions of the parts inside the transmission case.

To disassemble:

1. Mount the transmission on a stand, as in Fig. 10-2, or secure it to the workbench.

Fig. 10-2. Transmission case mounted on a repair stand. The bearing flange shown here is an early type not used in Type 2 transmissions.

© 1974 VWoA—997

1. Transmission case
2. Adjusting screw locknut
3. 2nd gear band adjusting screw
4. 1st gear band adjusting screw
5. Spring washer (5)
6. M 6 bolt
7. ATF pump
8. 2nd gear band
9. Thrust washer 1
10. Direct and reverse clutch
11. Thrust washer 2
12. Thrust washer 3
13. Forward clutch
14. Thrust washer 4
15. Clutch hub
16. Driving shell
17. Small sun gear
18. Thrust washer 5
19. Planetary gearset
20. Shim(s)
21. Thrust washer 6
22. 1st gear band
23. Support fork
24. Annulus gear with one-way clutch
25. Gasket for bearing flange
26. Bearing flange
27. M 6 screw (2)

Fig. 10-3. Disassembled planetary gear system. The bearing flange is screwed to the transmission case.

© 1974 VWoA—3335

Fig. 10-4. Slide hammer being used to pull bearing flange.

7

2. Remove the countersunk bearing flange screw.
3. Using a slide hammer and a puller hook, as shown in Fig. 10-4, pull out the bearing flange.
4. **Loosen the brake band adjusting screw. Lift out the 1st gear band, the annulus gear with one-way clutch, and the planetary gearset.**

NOTE —
Record the number and thicknesses of the shims you find in the gearset.

5. Remove the driving shell, the small sun gear, the clutch hub, and the forward clutch.
6. Take out the direct and reverse clutch and the 2nd gear brake band. Remove the pump bolts and lift out the ATF pump. (If you are also removing the transmission pan, keep its bolts separate from the bolts you will find in the valve body assembly.)

To assemble:

1. Insert the pump in the case so that the single lug and the part number face upward (see Fig. 10-5). Install the five pump bolts together with the spring washers finger-tight. Working diagonally, torque the bolts to 0.4 mkg (2.8 ft. lb.).

© 1974 VWoA—1001

Fig. 10-5. ATF pump installed. Lower arrow indicates the lug which must face upward. Test the installation by turning the pump as indicated by the curved arrow.

2. Turn the pump with the pump shaft. The pump should turn smoothly and easily.
3. Slide the thrust washer **1** for the direct and reverse clutch over the pump housing and engage the washer on the lug.
4. Install the 2nd gear brake band. Turn the adjusting screw until it enters the recess in the band.
5. Install the direct and reverse clutch, pressing it down in the case until the clutch drum bears on the thrust washer.
6. Using grease as an adhesive, stick thrust washers **2** and **3** to the inside of the forward clutch (see Fig. 10-6). Then install the forward clutch in the case.

© 1974 VWoA—1002

Fig. 10-6. Thrust washer **3** (arrow **A**) and thrust washer **2** (arrow **B**) in forward clutch.

7. Using a screwdriver, align the internal splines of the forward clutch plates to receive the splines on the clutch hub.
8. Using grease as an adhesive, stick thrust washer **4** on the forward clutch hub. Then insert the clutch hub into the clutch plate splines.
9. Install the driving shell and the small sun gear. Engage the driving shell's skirt notches with the lugs on the direct and reverse clutch.
10. Insert thrust washer **5** between the large planet pinions so that the projecting shoulder faces the small sun gear, as in Fig. 10-7.

© 1974 VWoA—1003

Fig. 10-7. Thrust washer **5** for large sun gear. Shoulder (arrow) should face the small sun gear.

11. Install the planetary gearset. All the parts are correctly installed if the planet carrier and parking lock pawl are aligned axially, as in Fig. 10-8.

© 1974 VWoA—1004

Fig. 10-8. Transmission parking lock. Arrow indicates the pawl, which snaps into the carrier notches.

12. Measure the axial play as described in **10.3 Adjusting Axial Play**. If necessary, adjust it.
13. If axial play is correct, slip the 1st gear brake band with support fork over the annulus gear. Then turn the adjusting screw until it enters the fork.
14. Install the bearing flange and the gasket with the countersunk flat head screw.

10.3 Adjusting Axial Play

The axial play of the planetary gearset and clutches can be from 0.45 to 1.05 mm (.018 to .041 in.). Measure the play after you have assembled the transmission. If necessary, adjust the play to bring it within the specified range.

To measure and adjust:

1. Assemble the transmission without the shims and the thrust washer. Do not install the 1st gear brake band. Use a new gasket under the bearing flange.
2. With the transmission horizontal, place a dial indicator mounted in a measuring bridge against the face of the bearing flange. The gauge pin should contact the planet carrier of the planetary gearset.
3. Using suitable pliers, push the planet carrier in and out as indicated by the double arrow in Fig. 10-9.

© 1974 VWoA—3337

Fig. 10-9. Axial play being measured. Move the planet carrier as indicated by the double arrow.

4. Compare the measured play with the axial play ranges given in **Table b**. Then assemble the transmission with the proper number and thickness(es) of shim(s) and the thrust washer.
5. Recheck the axial play to make sure that it is between 0.45 and 1.05 mm (.018 and .041 in.).

Table b. Axial Play

Measured play range	Number and thickness of shims
1.95–2.25 mm (.077–.089 in.)	Thrust washer only
2.25–2.65 mm (.089–.104 in.)	one 0.4 mm (.0157 in.)
2.65–3.05 mm (.104–.120 in.)	two 0.4 mm (.0157 in.)
3.05–3.45 mm (.120–.136 in.)	one 1.2 mm (.0472 in.)
3.45–3.85 mm (.136–.152 in.)	one 0.4 mm (.0157 in.) and one 1.2 mm (.0472 in.)

10.4 Adjusting Brake Bands

For these adjustments, the transmission must be in a horizontal position. If it is not, any adjustments will be inaccurate.

To adjust:

1. Center the 2nd gear brake band by torquing the adjusting screw to 1.0 mkg (86.8 in. lb.).
2. Loosen the adjusting screw. Then retorque it to 0.5 mkg (43 in. lb.).
3. From this setting, back the screw off from 1¾ to 2 turns. Then, while holding the adjusting screw to prevent it from turning, torque the adjusting screw locknut to 2.0 mkg (14 ft. lb.).

7

4. After adjusting the 2nd gear brake band, use the same procedure for adjusting the 1st and reverse gear band, but back this screw off from 3¼ to 3½ turns (see Fig. 10-10).

© 1974 VWoA—1007

Fig. 10-10. Torquing 1st and reverse brake band adjusting screw. Arrow indicates 2nd gear brake band screw.

10.5 Installing Final Drive on Automatic Transmission

For this procedure the transmission should be on a repair stand and in a vertical position. When moving or carrying the transmission, do not use the filler tube as a handle. You might bend it or cause it to leak.

To assemble:

1. Inspect the chamfer on the transmission case. Remove burrs, dirt, or rust.
2. Dip the O-ring in ATF.
3. Place a new paper gasket on the transmission case sealing surface and carefully set the final drive housing on it. When you install the housing, be careful not to crush or otherwise damage the O-ring.
4. Working diagonally, torque the four attaching nuts to 2.0 mkg (14 ft. lb.).
5. Insert the pump and the turbine shafts.
6. Set the converter in place over the end of the turbine shaft. Turn the converter clockwise and counterclockwise until it engages the splines of the turbine shaft.

10.6 Disassembling and Assembling ATF Pump

Whenever you remove the ATF pump, carefully inspect the housing, both gears, and the cover plate. Replace the pump if these parts are worn or damaged. The drive plate and the piston rings (Fig. 10-11) can be replaced individually.

To disassemble:

1. Remove the two M 4 screws, then remove the cover plate.
2. Remove the 11-mm ball and ball spring, the inner gear, outer gear, and drive plate.
3. Using needle nose pliers, remove the piston rings (Fig. 10-12).

1. M 4 screw (2)
2. Cover plate
3. Drive plate
4. Inner gear
5. Outer gear
6. 11-mm ball
7. Ball spring
8. Pump housing
9. Piston rings (4)

Fig. 10-11. ATF pump.

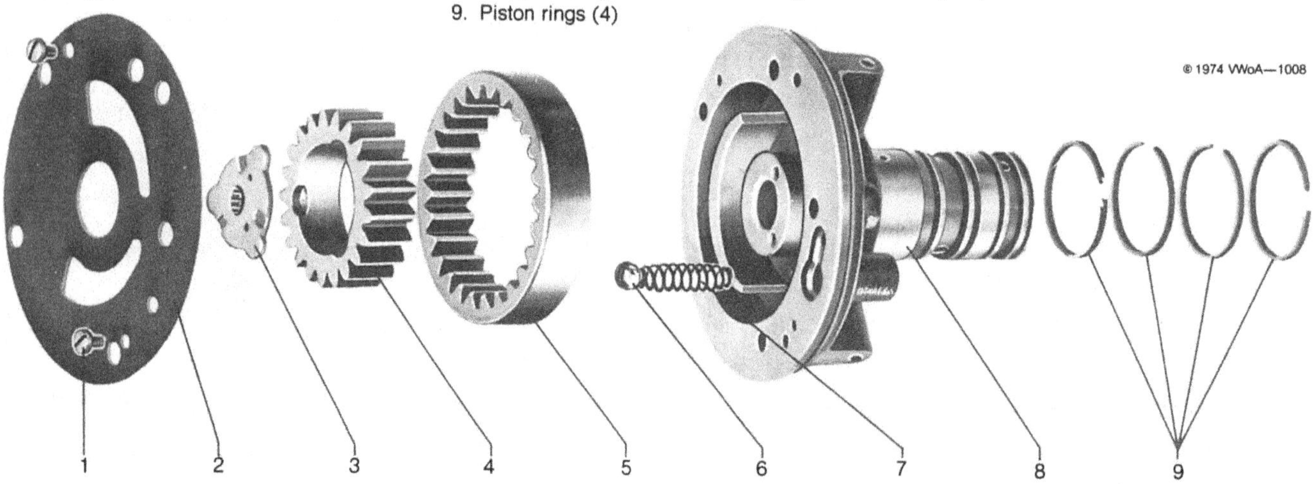

© 1974 VWoA—1009

Fig. 10-12. Clutch piston rings being removed from pump.

To assemble:

1. Clean all parts thoroughly. Blow out the fluid passages with compressed air.
2. After lubricating all parts thoroughly with ATF, install the gears and the drive plate. Be sure that the drive plate is in the position shown in Fig. 10-11, with the long side of the hub toward the pump body.
3. Insert the ball valve spring and the 11-mm ball.
4. Screw on the cover plate.

10.7 Disassembling and Assembling Direct and Reverse Clutch

The direct and reverse clutch and the 2nd gear brake band that operates against the clutch drum are shown in an exploded view in Fig. 10-13. Familiarize yourself with the names of the parts, as they are used frequently in the disassembly and repair procedures.

To disassemble:

1. Using a screwdriver, pry out the large circlip.
2. Remove the pressure plate, the three lined plates, and the three steel plates.
3. Put the clutch on a press and force down the spring plate until you can pry out the small circlip. Then raise the press ram and remove the spring plate.
4. With a twisting movement, pull the clutch piston with return springs out of the clutch drum. Remove the piston seal and the clutch drum seal.
5. Using an extractor tool and the repair press, withdraw the clutch drum bushing as shown in Fig. 10-14.

© 1974 VWoA—1012

Fig. 10-14. Removing clutch drum bushing. Phantom drawing (dashed line) shows extractor.

7

Fig. 10-13. Direct and reverse clutch. The pressure plate and steel plates alternate with the three lined plates.

1. 2nd gear brake band
2. Clutch drum bushing
3. Adjusting screw with large dog point
4. Clutch drum
5. Clutch drum seal
6. Clutch piston seal
7. Clutch piston
8. Return spring (18)
9. Spring plate
10. Small circlip
11. Steel plate (3)
12. Lined plate (3)
13. Pressure plate
14. Large circlip

© 1974 VWoA—3340

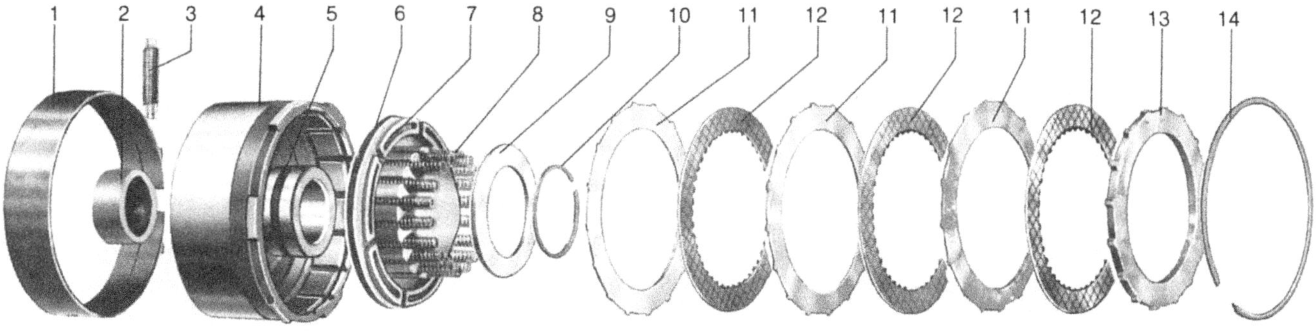

© 1974 VWoA

To check clutch:

1. Check for wear or damage on the friction surfaces of the piston and the clutch drum and in the grooves that the steel clutch plates ride in.
2. Check the ball valve for freedom of movement. Make sure that the drilling is clear.
3. Inspect the steel plates. If any plate is scored or grooved, replace it.
4. Check the lined plates. Replace any plate that is worn, damaged, or burned.
5. Check the 2nd gear brake band. Replace it if it is worn, damaged, or burned.

To assemble:

1. Install new seals on the clutch drum and the piston. The seal lips should point into the drum toward the source of hydraulic pressure.
2. Place a stiff plastic sheet in the clutch drum as shown in Fig. 10-15. Lubricate the seals with ATF, then insert the piston into the drum with a twisting motion.

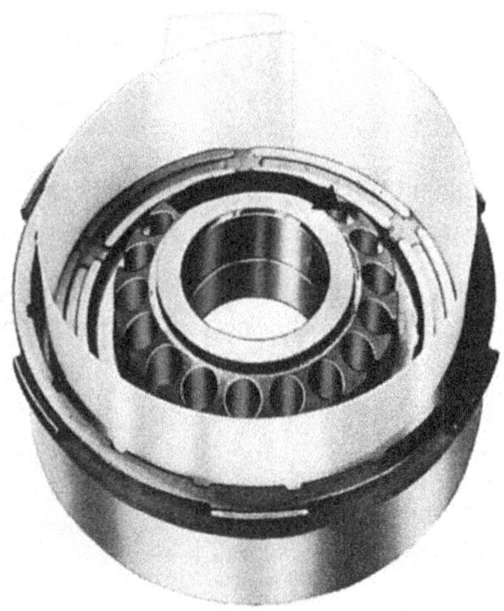

© 1974 VWoA—3341

Fig. 10-15. Plastic sheet used to prevent damage to the seals during piston installation. Insert piston with twisting motion (arrow).

3. Insert the 18 return springs and the spring plate. Press down the spring plate with the repair press until you can snap the small circlip into its groove.
4. Install the waffled clutch lined plates (see Fig. 10-16) and the steel plates. The sequence is important. Check your work by referring to Fig. 10-13. Be sure to soak the new lined plates in ATF for at least an hour before you install them.

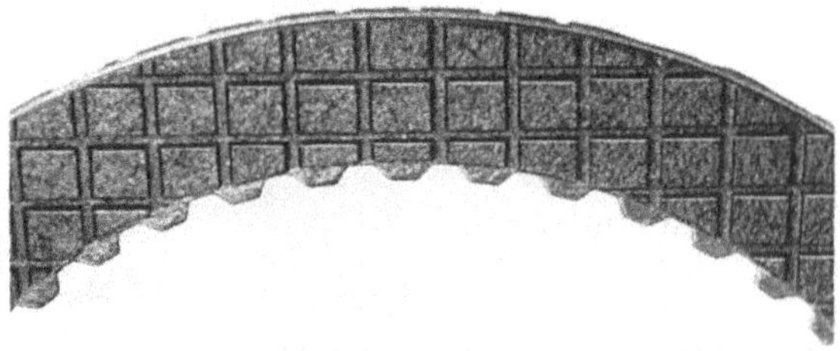

© 1974 VWoA—1013

Fig. 10-16. Direct and reverse clutch lined plate. The waffled surface of the lining distinguishes these plates from those of the forward clutch.

5. Install the pressure plate and the large circlip. The tolerance range for the thickness of the pressure plate is 6.15 to 6.30 mm (.242 to .248 in.).
6. Check the axial play as shown in Fig. 10-17. There should be 1.70 to 2.20 mm (.067 to .086 in.) of axial play.

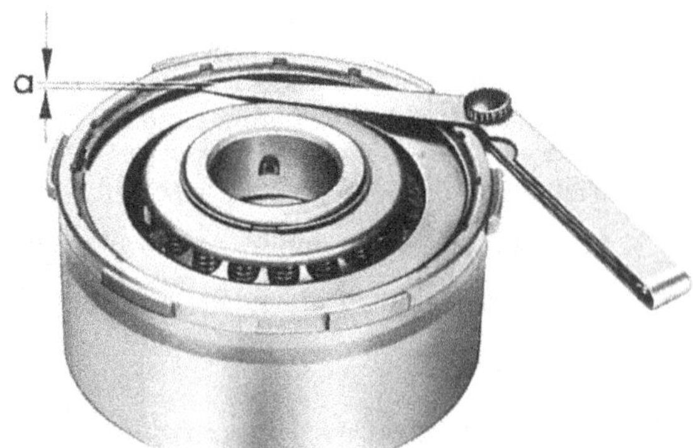

© 1974 VWoA—3342

Fig. 10-17. A feeler gauge being used to measure the axial play of the clutch plates in the clutch drum.

7. If axial play is outside the prescribed range, install a thicker or thinner large circlip. Circlips are available in thicknesses of 1.5 mm, 1.7 mm, 2.0 mm, 2.3 mm, 2.5 mm, and 2.7 mm (.059 in., .067 in., .079 in., .090 in., .098 in., and .106 in.).
8. When axial play is within the correct range, use the repair press and an appropriate bushing driver to press in a new clutch drum bushing.

10.8 Disassembling and Assembling Forward Clutch

The forward clutch transmits torque from the converter turbine to the direct and reverse clutch and the large sun gear. Fig. 10-18 gives an exploded view.

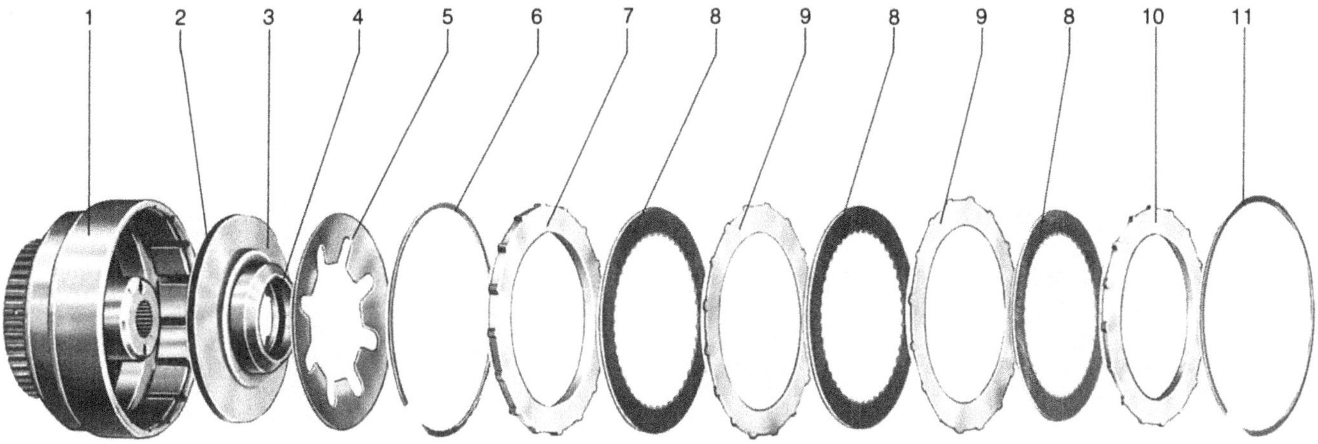

© 1974 VWoA—1015

1. Clutch drum
2. Large piston seal
3. Clutch piston
4. Small piston seal
5. Diaphragm spring
6. Spring circlip (2 mm thick)
7. Pressure plate
8. Lined plate (3)
9. Steel plate (2)
10. End plate
11. Outer circlip

Fig. 10-18. Forward clutch disassembled.

To disassemble:

1. Using a screwdriver, pry out the outer circlip. Then remove the clutch plates.
2. Pry out the spring circlip. Remove the diaphragm spring.
3. Pull the piston out of the clutch drum. Then remove the small and large piston seals.

To check clutch:

1. Shake the drum. If the drilling (Fig. 10-19) is clear, you should hear the ball valve rattle.

© 1974 VWoA—1016

Fig. 10-19. Ball valve (arrow) in clutch drum.

2. Check the diaphragm spring. When the piston is installed, the spring should reach at least to the lower edge of the circlip groove.
3. Inspect the steel plates. Replace any plate that is scored or grooved.
4. Look for signs of burning and wear on the lined plates. Replace damaged plates.

To assemble:

1. Install new piston seals. The seal lips should point into the drum toward the source of hydraulic pressure.
2. Place a stiff plastic sheet in the clutch drum as illustrated in Fig. 10-15. Lubricate the seals with ATF, then insert the piston into the drum with a twisting motion.
3. Install the diaphragm spring so that the curved side is toward the piston. To retain the diaphragm, use only a spring circlip 2 mm thick with a lug.

NOTE —
The diaphragm spring should be under some tension when the spring circlip is installed and it should not be easy to snap the circlip into its groove. If inserting the circlip does not put the diaphragm spring under tension, replace the diaphragm spring.

4. Install the pressure plate so that the chamfered side is toward the diaphragm spring.

7

5. Install lined plates alternately with steel plates as shown earlier in Fig. 10-18. New lined plates (Fig. 10-20) should be soaked for at least an hour before installing. Used plates can simply be lubricated with ATF.

Fig. 10-20. Forward clutch lined plate. The concentrically grooved lining distinguishes these plates from the lined plates of the direct and reverse clutch.

6. Install the end plate and outer circlip. The specified thickness of the end plate is 5.85 to 6.10 mm (.230 to .240 in.).
7. Using a feeler gauge, as shown in Fig. 10-21, check the clearance between the end plate and the outer circlip.

Fig. 10-21. Measuring forward clutch end play. Use a feeler gauge to measure clearance **a** between the end plate and the large outer circlip.

8. If the clearance measured in Step 7 is not between 0.80 and 1.20 mm (.031 and .047 in.), select a circlip to give this fit. **Table c** lists the circlips that are available.

Table c. Circlip Thickness

Thickness	Part No.
1.50 mm (.0590 in.)	003 323 157 D
1.70 mm (.0669 in.)	003 323 157
2.00 mm (.0787 in.)	003 323 157 A
2.30 mm (.0905 in.)	003 323 157 C
2.50 mm (.0984 in.)	003 323 157 B
2.70 mm (.1063 in.)	003 323 157 E

10.9 Disassembling and Assembling Annulus Gear

The annulus gear assembly includes the annulus gear, the one-way clutch, and the 1st gear brake band. Fig. 10-22 gives an exploded view of the assembly. The annulus engages the small pinions of the planetary gearset.

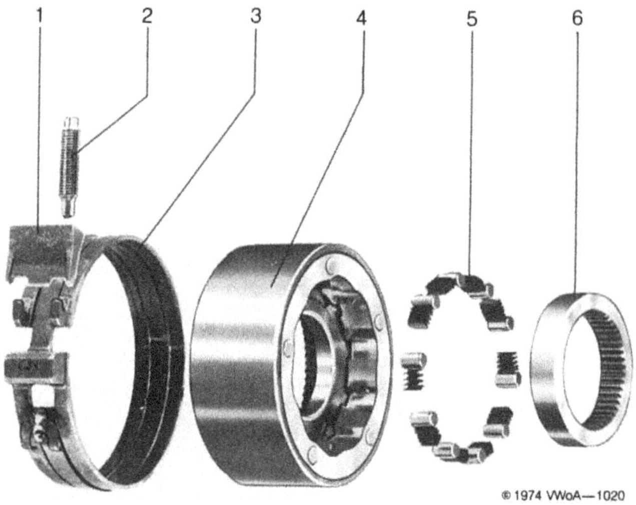

1. Support fork
2. Adjusting screw
3. 1st gear brake band
4. Annulus gear
5. Spring spacers (10) and rollers (10)
6. One-way clutch inner ring

Fig. 10-22. Annulus gear assembly. The adjusting screw for the 1st gear brake band passes through the transmission case into a recess in the support fork on the end of the band. The spring spacers and rollers are part of the one-way clutch assembly.

To disassemble the annulus gear, pull out the inner ring. Then lift out the 10 rollers and 10 spring spacers. Inspect the parts. The outer surface of the annulus should not be scored and the annulus gear teeth should not be noticeably worn. The rollers should not be flattened and the spring spacers should be crack-free. The inner and outer rings of the one-way clutch should not show signs of wear from contact with the rollers. Unless the one-way clutch is in perfect condition, it will not hold the annulus gear against the rotation of the turbine shaft.

To assemble:

1. Insert the inner ring of the one-way clutch in the annulus.
2. Insert the 10 rollers into the space between the one-way clutch inner ring and the annulus gear assembly.
3. Place the spring spacers between the lugs and the rollers. As you look down on the gear, the installation sequence should be: lug, spring, roller, lug, spring, roller, . . . and so on.
4. Install the one-way clutch on the bearing flange so that the splined part of the flange meshes with the inner ring.
5. To check the locking effect and the direction of rotation, try to turn the clutch both clockwise and counterclockwise (Fig. 10-23).

© 1974 VWoA—1022

Fig. 10-23. One-way clutch. When installed on the bearing flange, the clutch turns in direction **A** but holds in direction **B**.

6. Inspect the 1st gear brake band. Check the lining for burning and excessive wear. Remove any embedded bits of metal. Check the operating parts for wear. Replace faulty brake bands.

10.10 Planetary Gears

The planetary gearset shown in Fig. 10-24 should not be disassembled. The planetary pinions run on needle bearings on peened-in shafts that can be properly installed only at the factory. Worn or damaged gearsets must be replaced as a unit since individual parts are not supplied.

© 1974 VWoA—1021

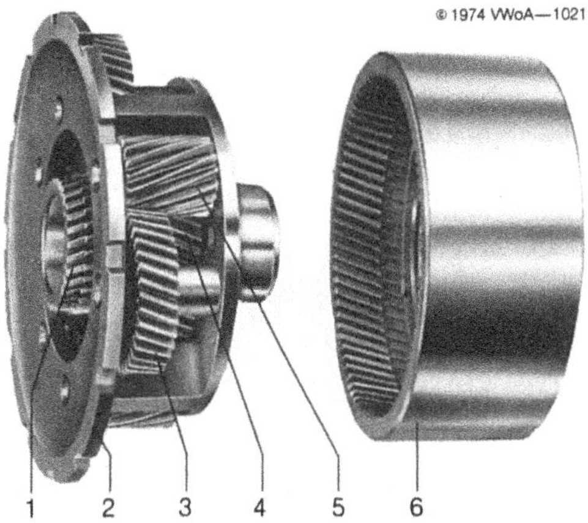

1. Small sun gear
2. Planet carrier
3. Large planet pinion
4. Large sun gear
5. Small planet pinion
6. Annulus (ring) gear

Fig. 10-24. Planetary gear system with annulus removed. The annulus engages the small pinions only.

To check the planetary gearset, inspect all gear teeth and thrust surfaces, and check for excessive backlash in meshing gears. Check for excessive radial play of the planetary gears. Replace worn gearsets. Check the internal splines and the flange of the parking lock for wear. Remove any burrs that you find on the flange.

10.11 Parking Lock

Disassembly of the parking lock requires removal of the transmission pan and the valve body. The position of the parking lock relative to that of the planetary gearset is shown in Fig. 10-25.

© 1974 VWoA—980

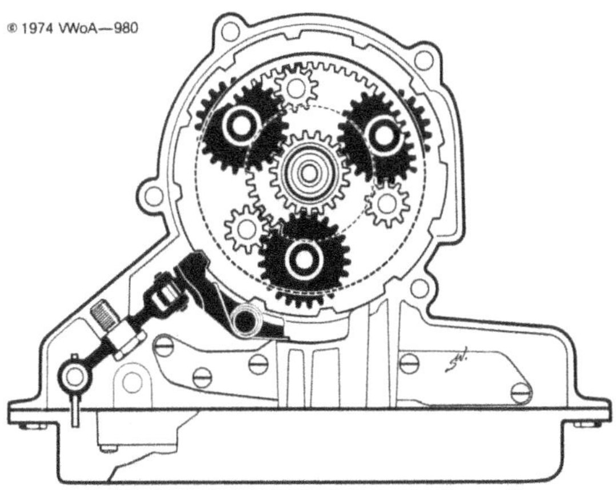

Fig. 10-25. Parking lock in engaged position. The notches around the edge of the planet carrier are shaped to hold the pawl firmly when the vehicle is stationary, but to force the pawl from engagement if the selector lever is accidentally moved to **P** while the car is moving.

7

To disassemble:

1. Remove the transmission pan and the valve body (Fig. 10-26). Unscrew the threaded pin for the operating lever. Remove the E-clip and then the lever.

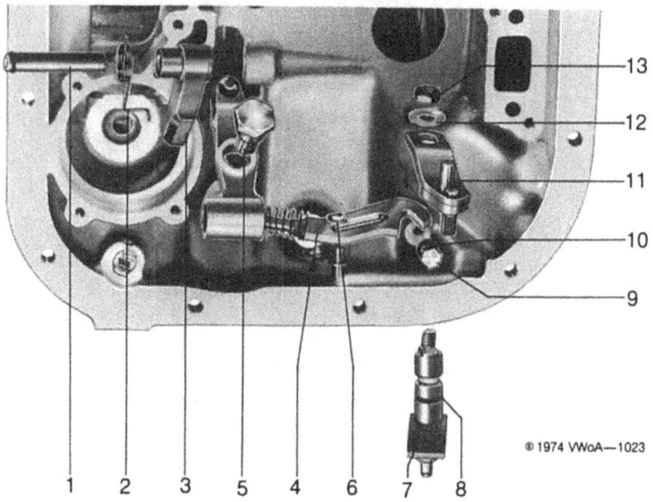

1. Pawl pin
2. Return spring
3. Pawl
4. Operating lever
5. Threaded pin
6. E-clip (2)
7. Cable lever and shaft
8. O-ring
9. Retaining bolt
10. Spring washer
11. Manual valve lever
12. Washer
13. Self-locking nut

Fig. 10-26. Parking lock assembly (viewed from below).

2. Drive out the pawl pin. Remove the pawl and the return spring.
3. Remove the self-locking nut from the manual valve lever, then remove the lever.
4. Remove the retaining bolt and then the cable lever and shaft.

To assemble:

1. Using a new O-ring, install the cable lever and shaft. Install the manual valve lever with the washer and the self-locking nut. Torque the nut to 0.6 mkg (4.3 ft. lb.). Install the retaining bolt together with its spring washer.
2. Insert the pawl together with the return spring. Holding the pawl away from the planet carrier, drive in the pawl pin.
3. Install the operating lever and then the E-clip. Torque the threaded pin to 0.6 mkg (4.3 ft. lb.).
4. Check the operation of the parking lock. If it works properly, install the valve body and the transmission pan as described in **5.1 Removing and Installing Valve Body Assembly**.

11. FINAL DRIVE

Fig. 11-1 illustrates the disassembly of the final drive. The factory position for the S_3 shim is shown. For repair adjustments, there are larger-diameter S_3 shims available which can be installed between the rear tapered-roller bearing outer race and the final drive housing.

CAUTION —
Before you decide to replace any part of the final drive, carefully read ***12. Adjusting Final Drive****. For some replacements, you must make precision measurements before you disassemble the final drive. Failure to do so can make proper reassembly impossible.*

11.1 Removing and Installing Differential

You can replace the flanged shafts, the oil seals, and the O-rings for the differential bearing adjusting rings without removing the differential. If you intend to replace the ring gear, an adjusting ring, a differential bearing, or any part related to the drive pinion, read **12. Adjusting Final Drive** to determine whether precision adjustments must be made before or after repair.

CAUTION —
If you lack the skills, tools, or a clean workshop for final drive repairs, we suggest you leave such repairs to an Authorized VW Dealer or other qualified shop. We especially urge you to consult your Authorized VW Dealer before attempting repairs on a car still covered by the new-car warranty.

To remove differential:

1. Remove the socket head screws from the centers of the flanged shafts. If necessary, use a slide hammer. Pull out the flanged shafts.

   NOTE —
   If only the oil seals require replacement, you can do it now. Pry out the faulty seals. Then, using a suitable seal driver, drive in the new seals.

2. Carefully and permanently mark the position of each differential bearing adjusting ring in the final drive housing. Using a depth micrometer, measure the screw-in depth of each adjusting ring. Write down the measurements for use during installation.
3. Remove the 10 M 6 bolts together with their spring washers. Then remove the final drive pan and its gasket.
4. Remove the M 4 bolts together with the lockplates for the differential bearing adjusting rings.

Fig. 11-1. Exploded view of final drive and related parts.

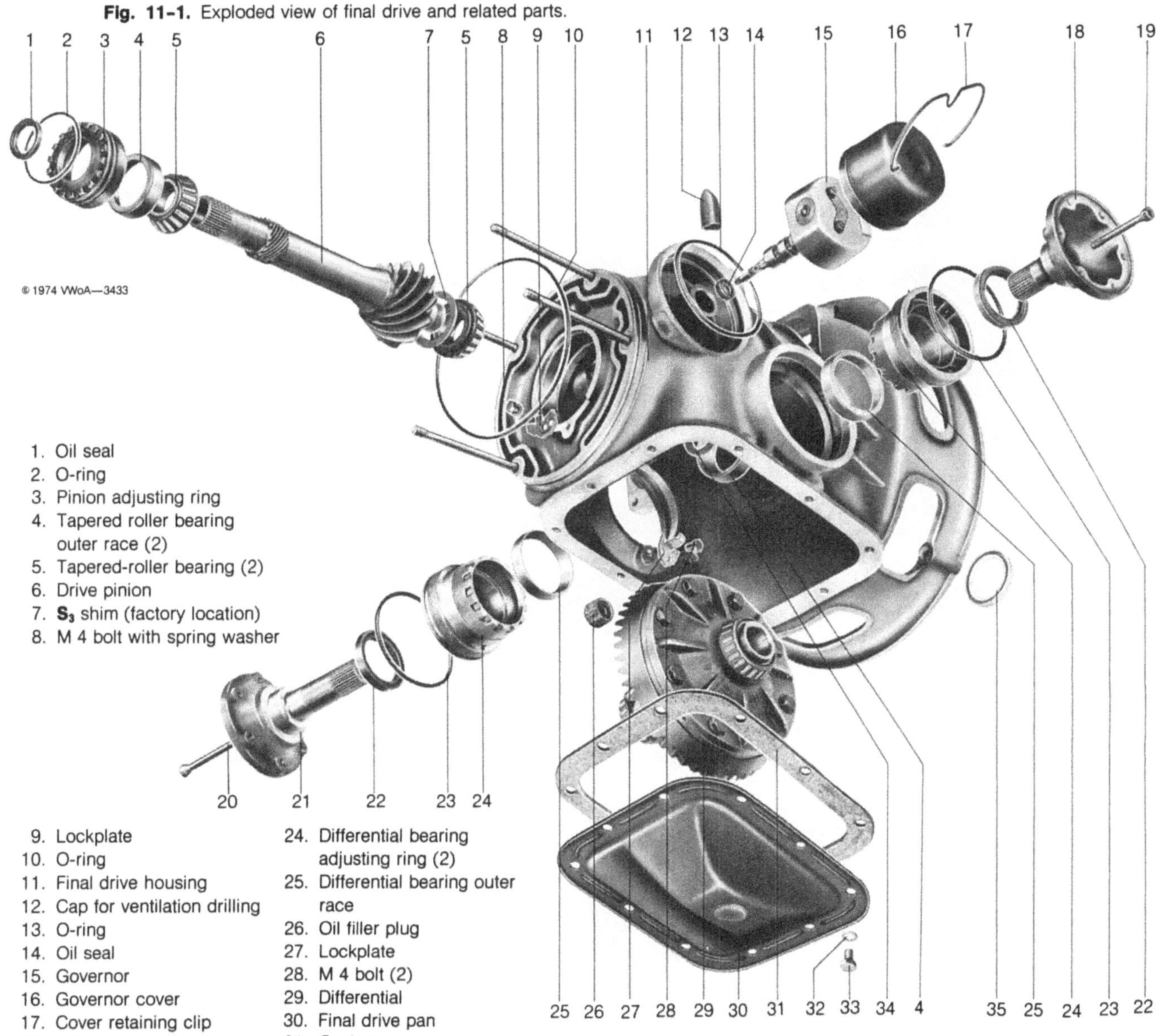

© 1974 VWoA—3433

1. Oil seal
2. O-ring
3. Pinion adjusting ring
4. Tapered roller bearing outer race (2)
5. Tapered-roller bearing (2)
6. Drive pinion
7. **S_3** shim (factory location)
8. M 4 bolt with spring washer
9. Lockplate
10. O-ring
11. Final drive housing
12. Cap for ventilation drilling
13. O-ring
14. Oil seal
15. Governor
16. Governor cover
17. Cover retaining clip
18. Left flanged shaft
19. Left socket head screw
20. Right socket head screw
21. Right flanged shaft
22. Oil seal (2)
23. O-ring (2)
24. Differential bearing adjusting ring (2)
25. Differential bearing outer race
26. Oil filler plug
27. Lockplate
28. M 4 bolt (2)
29. Differential
30. Final drive pan
31. Gasket
32. Spring washer (10)
33. M 6 bolt (10)
34. Oil seal
35. Torque converter seal

7

NOTE —
If only the O-rings require replacement, you can do it now. One at a time, unscrew each adjusting ring just far enough so that you can remove the old O-ring and install a new one. Then screw in the adjusting ring to its original depth and position before installing the lockplate.

5. While supporting the differential so that it does not suddenly fall, fully remove both adjusting rings as shown in Fig. 11-2.

© 1974 VWoA—3434

Fig. 11-2. Special wrench being used to remove the adjusting rings.

6. Move the differential toward the side of the housing that is opposite the ring gear. Then tilt the ring gear upward and remove the differential from the final drive housing.

To install:

1. Inspect the gears and bearings. Replace worn or damaged parts as described in **11.2 Disassembling and Assembling Differential.**

> **NOTE —**
> Consult **12. Adjusting Final Drive** to determine whether precision adjustments are required following replacement.

2. Inspect the oil seals. Replace them if their lips are worn or cracked.

> ***CAUTION —***
> *The original fit of oil seals can never be duplicated once the seals have been removed. Therefore, never reinstall used seals.*

3. Insert the differential assembly into the transmission case. The ring gear goes on the left side and must be in mesh with the drive pinion.
4. Install new O-rings on the final drive covers. Apply Molykote® to the threads, then install one adjusting ring to its original depth and position.
5. Hand-support the differential in the installed adjusting ring. Then install the other adjusting ring to its original depth and position.
6. Install the lockplates for the adjusting rings.
7. Install the flanged shafts. Torque the socket head screws to 3.5 mkg (25 ft. lb.).
8. Install the final drive pan together with a new gasket. Torque the 10 M 6 bolts to 1 mkg (7 ft. lb.). Wait five minutes for the gasket to compress, then retighten the bolts to the same torque.

11.2 Disassembling and Assembling Differential

An exploded view of the differential is given in Fig. 11-3. If you intend to replace any part of the differential, read **12. Adjusting Final Drive** to determine whether precision adjustments must be carried out following repair. If the final drive is not correctly adjusted following certain repairs, the ring and pinion gearset will be noisy and will wear rapidly.

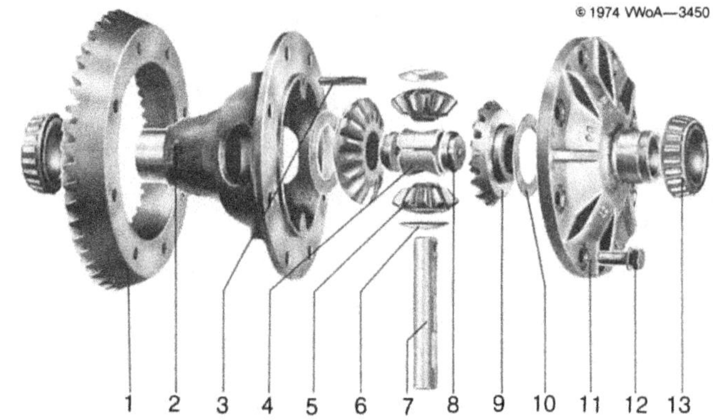

1. Ring gear
2. Differential housing
3. Shaft retaining pin
4. Spacer sleeve
5. Differential pinion (2)
6. Small thrust washer (2)
7. Differential pinion shaft
8. Nut for flanged shaft socket head bolt (2)
9. Differential sidegear (2)
10. Large thrust washer (2)
11. Differential housing cover
12. M 9 bolt with spring washer (8)
13. Tapered-roller bearing inner race (2)

Fig. 11-3. Exploded view of differential.

To disassemble:

1. Loosen the eight bolts, then drive off the ring gear as shown in Fig. 11-4.

> ***CAUTION —***
> *When removing the ring gear, be careful not to let it fall against the vise. Doing so could damage the ring gear teeth. A pad of rags placed beneath the ring gear is a worthwhile precaution.*

© 1974 VWoA—3451

Fig. 11-4. Rubber mallet being used against bolt heads to drive off ring gear.

2. If the tapered-roller bearing inner races are to be replaced, press them off the differential housing and the differential housing cover as shown in Fig. 11-5.

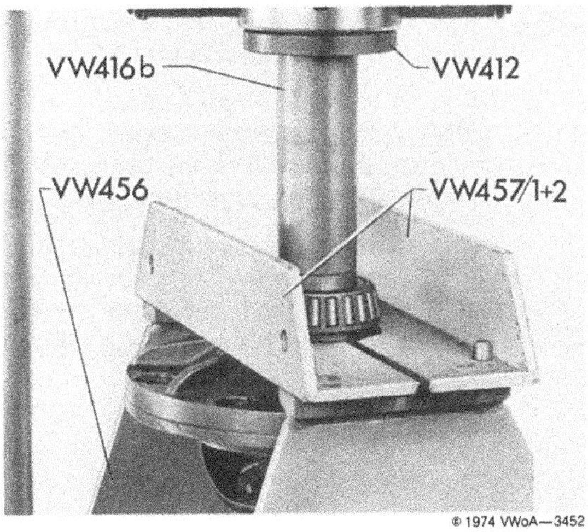

© 1974 VWoA—3452

Fig. 11-5. Press tool setup for removing bearing inner races.

3. Using a screwdriver, separate the differential housing cover from the differential housing as shown in Fig. 11-6.

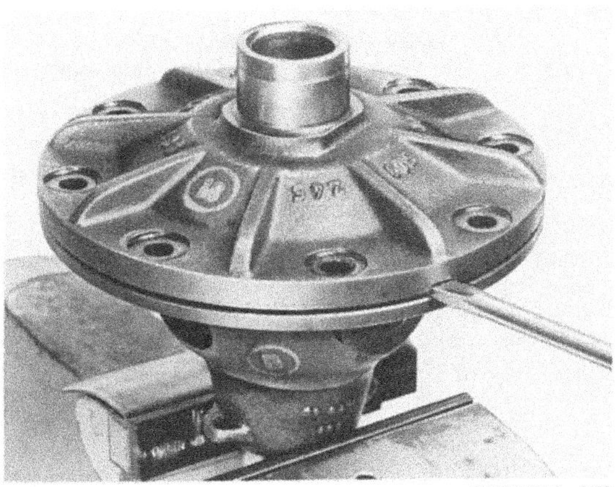

© 1974 VWoA—3453

Fig. 11-6. Cover being pried off differential housing. There is another notch for the screwdriver at the opposite side of the differential.

4. Using a drift, drive out the shaft retaining pin. Then drive the differential pinion shaft out of the differential housing and remove the spacer sleeve, pinions, sidegears, and other internal parts.

To assemble:

1. Examine all thrust surfaces on the differential housing, cover, ring gear, spacer sleeve, and thrust washers. Replace worn parts. Inspect the gear teeth for burrs or excessive wear and replace as necessary.

2. If the housing, cover, a side gear, or the spacer sleeves are replaced, determine the correct length for the spacer sleeve as described in Steps 3 through 10. If these parts are not being replaced, continue with Step 11.

NOTE —
If you notice gear-tooth damage (usually the result of insufficient backlash), it is probable that the original spacer sleeve is too long. Experience has shown that the spacer sleeve is unnecessary. You need not reinstall the spacer sleeve as described in the next eight steps unless you choose to do so.

3. To determine spacer sleeve length, place the sidegear and one large thrust washer in the differential housing.
4. Measure the shortest spacer sleeve available—part No. 004 517 241—as shown in Fig. 11-7. Then write down the measured length.

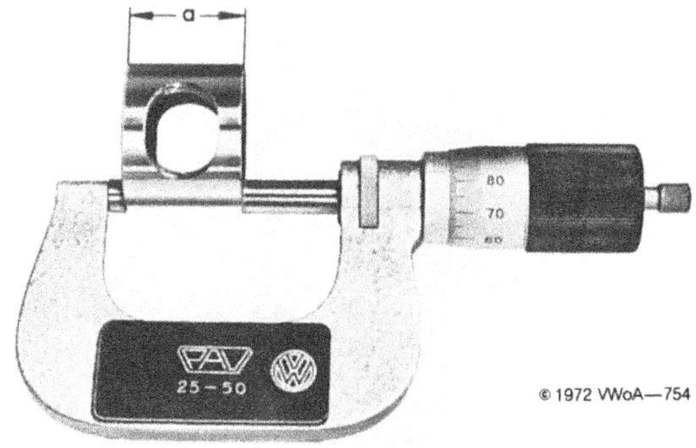

© 1972 VWoA—754

Fig. 11-7. Smallest spacer sleeve being measured to determine dimension **a**.

5. Place the measured spacer sleeve in the differential housing atop the sidegear (Fig. 11-8).

7

© 1974 VWoA—3458

Fig. 11-8. Spacer sleeve in place on sidegear.

6. Place the second sidegear and the other large thrust washer atop the spacer sleeve. Then install the differential cover with two of the eight cover bolts.
7. Install a dial indicator on the differential housing cover. Position the gauge pin against the special locking sleeve VW 521/4, which must be attached to the uppermost sidegear. Then zero the dial indicator.
8. Move the locking sleeve up and down as indicated by the double arrow in Fig. 11-9. Note the play range shown on the dial indicator.

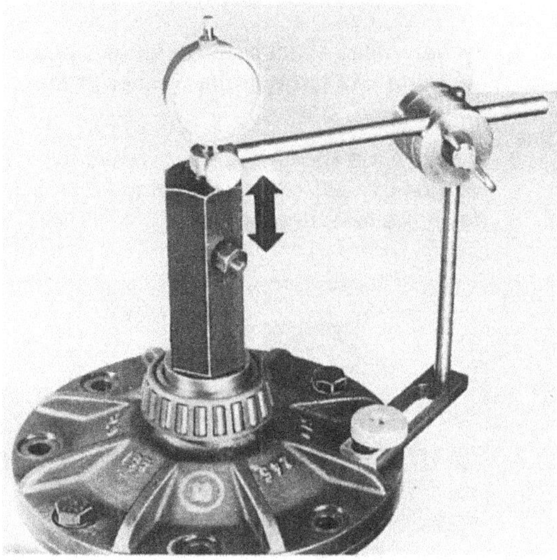

Fig. 11-9. Axial play of differential sidegear being checked with a dial indicator.

9. Add the measured play to the known spacer sleeve length to obtain the X-range. Select the proper sleeve length as given in **Table d**.

Table d. Differential Spacer Sleeves

X range mm (in.)	Sleeve length mm (in.)	Part No. of sleeve
28.82-28.91 (1.135-1.138)	28.82 (1.135)	004 517 241
28.91-28.99 (1.138-1.141)	28.91 (1.138)	004 517 242
29.00-29.03 (1.142-1.143)	29.00 (1.142)	004 517 243
29.09-29.18 (1.145-1.149)	29.09 (1.145)	004 517 244

10. Remove the differential cover and install the correct spacer sleeve. Reinstall the cover and check the sidegear axial play as before. It should be from 0.00 to 0.14 mm (.000 to .0055 in.).
11. Install the differential sidegears, large thrust washers, differential pinions, spacer sleeve, nuts for socket head screws, and small thrust washers in the differential housing.
12. Install the differential pinion shaft. Then drive in a new pinion shaft retaining pin.
13. Using an oilstone, remove all burrs and pressure marks from the mating surfaces of the housing, the housing cover, and the ring gear. Then install the differential housing cover without the bolts.
14. Heat the ring gear to about 100°C (212°F) in a pan of oil placed in a larger pan of boiling water. Then install the ring gear as shown in Fig. 11-10. Torque the eight bolts to 5.0 mkg (36 ft. lb.).

Fig. 11-10. Ring gear being installed on differential. Use a rag to prevent burning your fingers.

15. If the tapered-roller bearing inner races have been removed, heat them in oil also. Then install the bearing races as shown in Fig. 11-11.

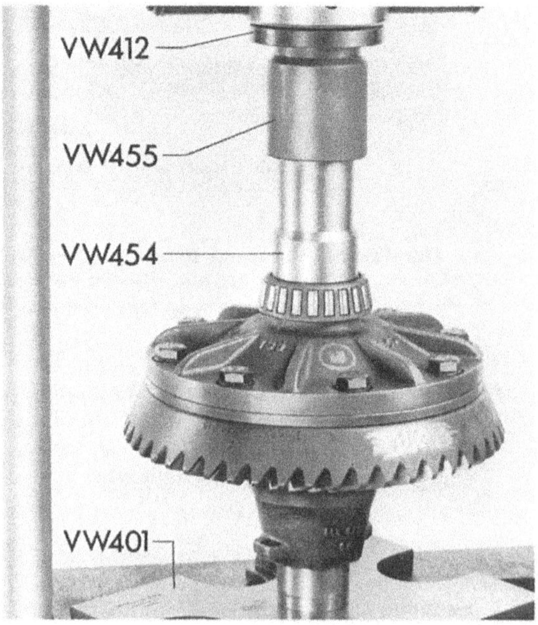

Fig. 11-11. Tapered-roller bearing inner race being pressed onto differential. The press tool must not contact the roller cage.

A new tapered-roller bearing outer race must be installed at the same time as the new inner race. Press out the old outer race, or drive it out of the adjusting ring with a drift. Then press in the new race as shown in Fig. 11-12.

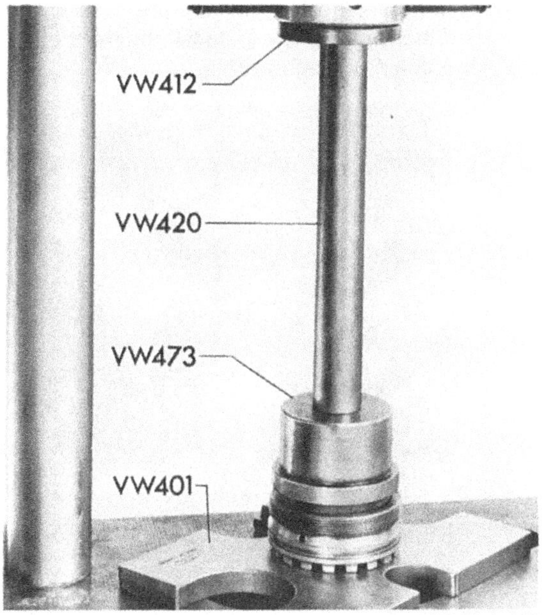

© 1974 VWoA—3446

Fig. 11-12. Tapered-roller bearing outer race being pressed into adjusting ring.

11.3 Removing and Installing Drive Pinion

The drive pinion is a matched pair with the ring gear, so the ring and pinion must be replaced as a unit.

> ***CAUTION —***
> *If you intend to replace the final drive housing, the pinion tapered-roller bearings, or the pinion adjusting ring, you must make precision measurements before you remove the drive pinion. See* ***12. Adjusting Final Drive****. If a new ring and pinion gearset is to be installed, measurements are unnecessary before disassembly since replacement gearsets have measuring numbers stamped on them.*

To remove drive pinion:

1. Remove the differential as described in **11.1 Removing and Installing Differential**.
2. Carefully and permanently mark the position of the pinion adjusting ring in the final drive housing. Using a depth micrometer, measure the screw-in depth of the adjusting ring. Write down the measurement for use during installation.
3. Remove the lockplate for the pinion adjusting ring. Then remove the adjusting ring as shown in Fig. 11-13.

© 1974 VWoA—3435

Fig. 11-13. Special wrench being used to unscrew the pinion adjusting ring. Support the drive pinion so that it does not fall suddenly.

4. Withdraw the drive pinion from the final drive housing.
5. If necessary, use a hook-type puller to remove the oil seal from the torque converter end of the final drive housing.
6. Using a puller (Fig. 11-14), remove the tapered-roller bearing outer race. If you find the S_3 shim(s) here, record the number and thickness(es).

7

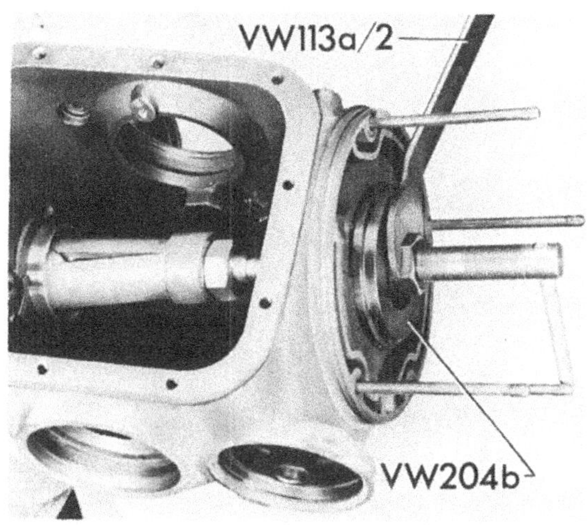

© 1974 VWoA—3437

Fig. 11-14. Expansion-type puller being used to extract the tapered-roller bearing outer race.

7. Using the expansion-type puller in combination with the repair press, remove the tapered-roller bearing outer race from the pinion adjusting ring. Do not drive out the outer race with a drift and hammer, since doing so may damage the threads on the adjusting ring. (See Fig. 11-15).

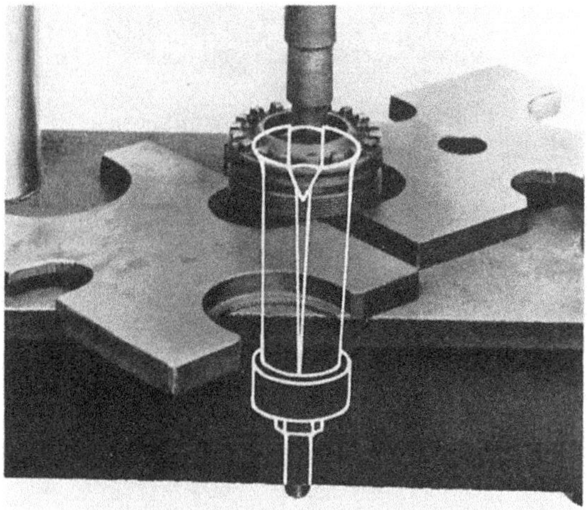

© 1974 VWoA—3439

Fig. 11-15. Expansion-type puller being used together with repair press to remove tapered-roller bearing outer race from pinion adjusting ring.

8. Press off the tapered-roller bearing inner races (Fig. 11-16). If you find the S_3 shim(s) here, record the number and thickness(es).

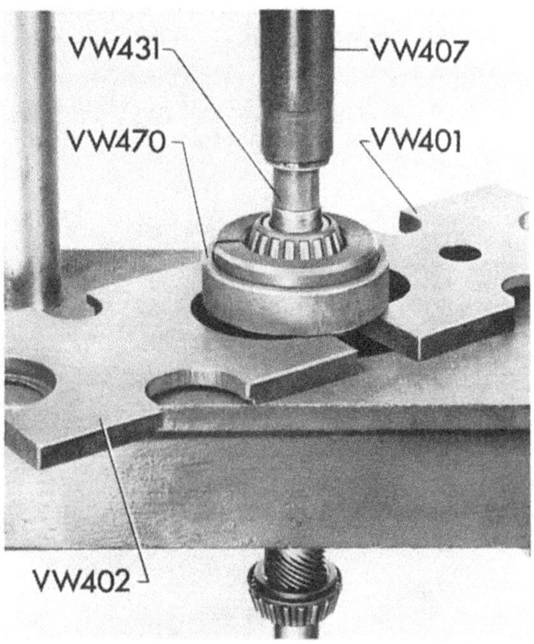

© 1974 VWoA—3440

Fig. 11-16. Tapered-roller bearing inner race being pressed off drive pinion. The split-ring collar must not contact the roller cage.

To install:

1. If necessary, install the S_3 shim(s) in the repair location (see **12. Adjusting Final Drive**). Then use a tapered-face tool (Fig. 11-17) to drive in the rear tapered-roller bearing outer race. Use a similar tool and the repair press to install the front outer race in the pinion adjusting ring.

© 1974 VWoA—3448

Fig. 11-17. Tapered-roller pinion bearing outer race being driven into final drive housing.

2. Unless you must adjust the final drive, install the pinion oil seal in the final drive housing. Otherwise, install it after making adjustments.

3. If you need not adjust the final drive, install the factory's original S_3 shim(s). Press the tapered-roller bearings onto the pinion (Fig. 11-18).

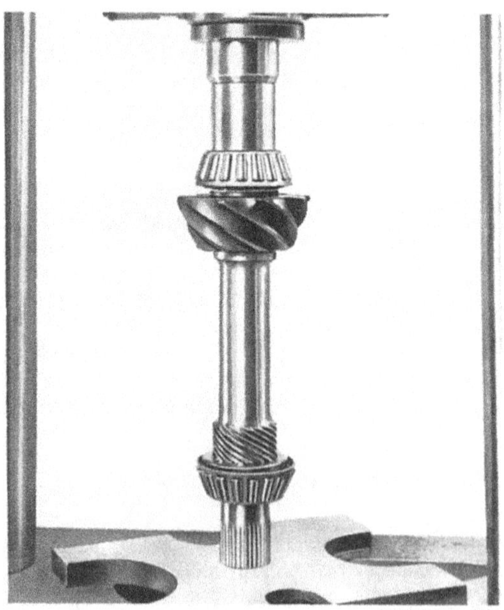

© 1974 VWoA—1055

Fig. 11-18. Tapered-roller bearing inner race being installed on drive pinion. The press tool must not contact the roller cage.

4. Install the drive pinion in the final drive housing. Use a new O-ring on the pinion adjusting ring.
5. Install the pinion adjusting ring to its original depth and position, then install the adjusting ring lockplate.
6. Install the differential as described in **11.1 Removing and Installing Differential**.

12. ADJUSTING FINAL DRIVE

The adjustments described here are necessary to the life of the gearset, the differential bearings, and the drive pinion bearings. They are also necessary to ensure quiet operation. The final drive requires adjustment only when parts directly affecting the adjustment have been replaced or when careless disassembly has resulted in the loss of the original shims or of the adjusting ring screw-in depth measurements. **Table e** lists what adjustments must be made when certain parts are replaced.

Table e. Required Final Drive Adjustments

Part replaced	Adjust pinion	Adjust ring gear (differential bearings)
Final drive housing	X	X
Differential tapered roller bearings		X
Adjusting rings for differential		X
Differential housing		X
Differential housing cover		X
Pinion tapered-roller bearings	X	
Adjusting ring for pinion	X	
Pinion and ring gear. Installation of new gearset requires replacement of tapered-roller bearings	X	X

There are two factors involved in making final drive adjustments: bearing turning torque and the backlash between the ring and pinion gears. The backlash is adjusted by repositioning the ring gear axially. The position of the pinion is determined by the thickness(es) of the $\mathbf{S_3}$ shim(s). The position of the ring gear is determined by the screw-in depth of the differential bearing adjusting rings. The screw-in depths are called $\mathbf{S_1}$ and $\mathbf{S_2}$. These and other standard symbols used in making final drive adjustments are defined in **Table f**.

Table f. Standard Symbols

Symbol	Description	Dimension
$\mathbf{S_1}$	Screw-in depth of adjusting ring (ring gear end)	
$\mathbf{S_2}$	Screw-in depth of adjusting ring (opposite end)	
$\mathbf{S_3}$	Shim between tapered roller bearing and pinion or between bearing outer race and final drive housing	See table for thickness
$\mathbf{R_o}$	Length of master gauge used in factory testing machine	40.55 mm (1.5965 in.)
R	Position of pinion in relation to centerline of ring gear at quietest running point (nominal dimension)	$\mathbf{R} = \mathbf{R_o} + \mathbf{r}$
r	Deviation from $\mathbf{R_o}$, marked on gear set	
$\mathbf{Sv_o}$	Backlash	0.15–0.25 mm (.0059–.0098 in.)
$\mathbf{V_o}$	Hypoid offset	42.50 mm (1.6732 in.)
G 933	Gear set: **G** = Gleason; .933 = 9/33 number of teeth	3.67:1 ratio
D/2	Half diameter of setting mandrel	**D**/2 = 10.00 mm (.3937 in.)
$\mathbf{E_o}$	Length of setting pin VW 380/3; $\mathbf{E_o} = \mathbf{R_o} + \mathbf{D}/2$ mm	$\mathbf{E_o}$ = 50.55 mm (1.9902 in.)
e	Difference between mandrel and setting pin	Measured in 1/100 mm

The ring gear and pinion are run on special testing machines during transmission manufacture to check the tooth contact pattern and silent running under both drive and coast conditions. When the optimum relationship of the two gears is found, they are installed in the final drive housing using $\mathbf{S_3}$, $\mathbf{S_1}$, and $\mathbf{S_2}$ adjustments that will duplicate the gearset's position in the testing machines. The purpose of all subsequent adjustments is to restore the gearset to this position following repair.

12.1 Adjusting Drive Pinion

If the pinion must be adjusted, do so before adjusting the ring gear. Dimension **R**, given in Fig. 12-1, is the actual distance from the pinion face to the ring gear centerline. It differs from the length of the master gauge $\mathbf{R_o}$, that is used in the factory testing machine, by deviation **r**. Deviation **r** is recorded on the outer face of the ring gear only on replacement gearsets.

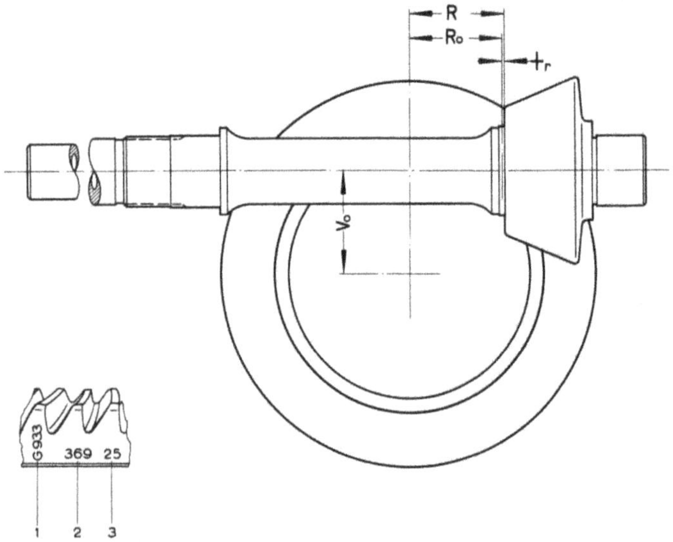

1. Manufacturer (Gleason) and tooth ratio (9/33)
2. Matching number to show that the ring gear and pinion are a matched set (gearset 369)
3. Deviation **r** (given in 1/100 mm)

Fig. 12-1. Pinion adjustment dimensions and gearset marking found on replacement gearsets only. Dimension $\mathbf{V_o}$ is the hypoid offset which is built in during manufacture and is not adjustable.

Because of modified procedures used to test gearsets in production, the original gearset in the transmission is adjusted to a flexible dimension called $\mathbf{P_o}$, as shown in Fig. 12-2. The deviation **r** and the gearset matching number are discontinued and do not appear on these gearsets. If drive pinion adjustments will be required (see **Table e** given earlier), measurements must be made before you disassemble one of these gearsets so that the pinion can be adjusted to its original position following repairs.

CAUTION —
If you lack the skills, special measuring tools, or a clean workshop for adjusting the final drive, we suggest you leave such repairs to an Authorized VW Dealer or other qualified and properly-equipped shop. We especially urge you to consult your Authorized VW Dealer before attempting repairs on a vehicle covered by the new-car warranty.

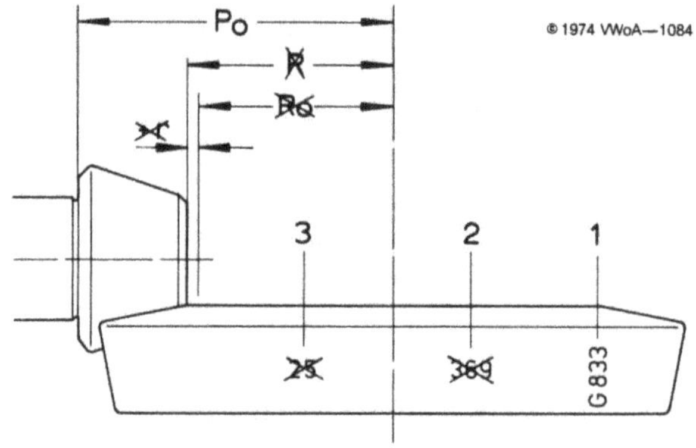

$\mathbf{P_o}$. New setting dimension
X. Discontinued setting dimensions
1. Manufacturer (Gleason) and tooth ratio
2. Discontinued matching number on both ring gear and pinion
3. Discontinued deviation **r** stamped on ring gear

Fig. 12-2. Production pinion setting dimension (a manual transmission gearset is shown).

To adjust pinion:

1. Install the pinion tapered-roller bearing outer race in the torque converter end of the final drive housing without any $\mathbf{S_3}$ shims. Do not install the oil seal.

   NOTE —
   If you are measuring an unmarked gearset prior to pinion removal, begin with Step 5 of this procedure rather than with Step 1.

2. If there is a shim between the pinion gear and the tapered-roller bearing (the production location for the $\mathbf{S_3}$ shim), press off the bearing race, remove the shim, then press on the bearing.
3. Install the drive pinion. Screw in the pinion adjusting ring until the drive pinion has no axial play.
4. Lubricate the tapered-roller pinion bearings with hypoid oil. Install a torque gauge (Fig. 12-3).

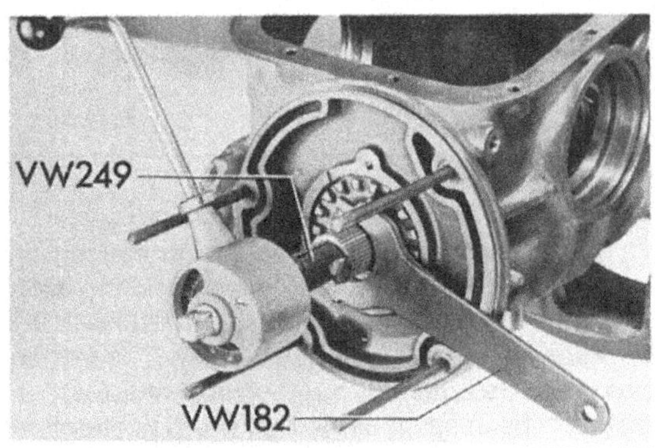

Fig. 12-3. Gauge for measuring pinion turning torque. Notice the special wrench that will be used to turn the pinion bearing adjusting ring.

NOTE —
Use only hypoid oil as a lubricant. The test results will be inaccurate if the turning torque is checked with the bearings dry or lubricated with another kind of oil.

5. Using the handle of the torque gauge, spin the pinion rapidly 15 or 20 turns in each direction. While continuing to spin the pinion, tighten or loosen the pinion adjusting ring until the turning torque reading is 14 to 20 cmkg (12 to 18 in. lb.) for new bearings or 2 cmkg (1.7 in. lb.) for a bearing that has been in service for 30 mi. (50 km) or more.

6. To determine the thickness(es) of the S_3 shim(s), remove the torque gauge. Then install setting gauge VW 385/22 and spring support VW 385/19 against the drive pinion as shown in Fig. 12-4.

© 1974 VWoA—3461

Fig. 12-4. Setting gauge VW 385/22 in place on pinion face. This gauge compensates for hypoid offset.

7. Screw in the left-side differential bearing adjusting ring until its outer surface is flush with the final drive case.

8. Adjust the setting ring on the universal measuring bar. Dimension **a** in Fig. 12-5 is about 57 mm.

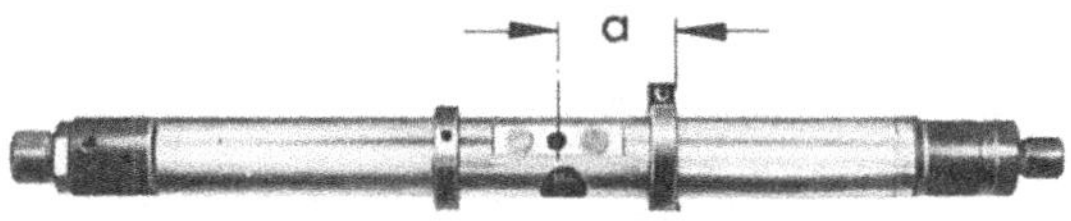

© 1974 VWoA—2560

Fig. 12-5. Adjusting setting ring.

9. Install the centering disks (VW 385/2) onto the bar until they contact the setting rings. Then attach measuring pin VW 385/14 with extension VW 385/20 to the gauge pin hole in the center of the bar (Fig. 12-6). Install the dial indicator.

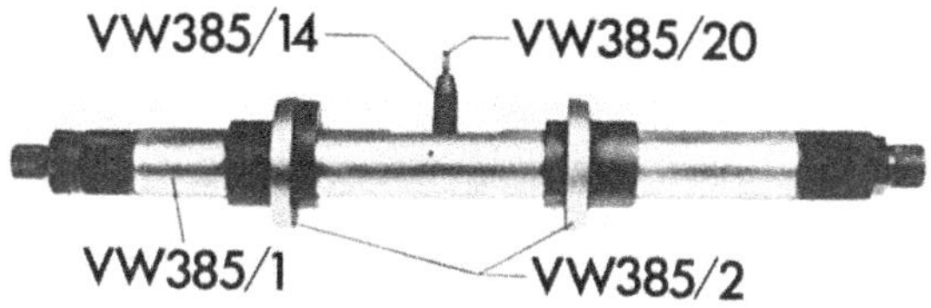

© 1974 VWoA—3462

Fig. 12-6. Measuring bar ready for use.

10. Place the universal measuring bar in the final drive housing and screw in the right-side adjusting ring until its outer surface is flush with the transmission case.

11. Loosen the second setting ring and move the centering disk outward until the measuring bar can just barely be hand-turned. Then tighten the screw in the setting ring. See Fig. 12-7.

Fig. 12-7. Arrangement of measuring tools inside the final drive housing.

© 1974 VWoA—3465

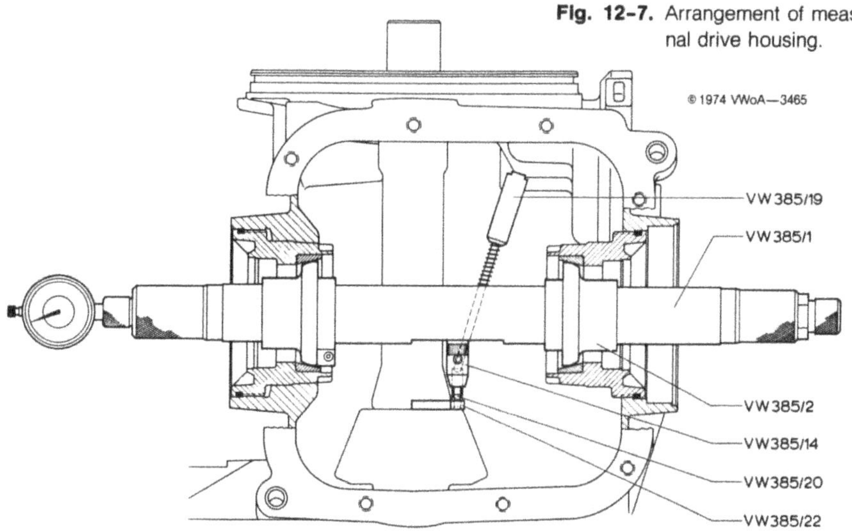

7

12. Using setting block VW 385/21, zero the dial indicator. The U-shaped setting block zeros the indicator at R_0 (52.60 mm). See Fig. 12-8.

© 1974 VWoA—3464

Fig. 12-8. Setting block in place so that dial indicator can be zeroed at R_0.

13. Hand-turn the universal measuring bar until the measuring pin extension rests against the setting gauge on the pinion face.
14. Rotate the bar back and forth over center. The maximum dial indicator reading should be observed and written down. This reading is measurement **e**.
15. Read the deviation **r** marked on the ring gear. Subtract deviation **r** from measurement **e**. The remainder is the thickness(es) of the S_3 shim(s) that must be installed.

If you are measuring an unmarked gearset prior to pinion removal, the measurement made here—with the original S_3 shim(s) installed—will be P_0. When adjusting one of the unmarked gearsets, the difference between the gauge reading **e**—with no S_3 shim(s) installed—and dimension P_0 will give you the thickness(es) of the S_3 shim(s) that must be installed to return the pinion to its factory-installed position.

Use **Table g** to select the necessary shim(s). After you have installed the shim(s), recheck the pinion position using the universal measuring bar. If you have installed the correct shim(s), the dial indicator will show a reading within ± 0.04 mm of dimension P_0 or deviation **r** marked on replacement gearsets. If the adjustment is correct, again adjust the pinion bearing turning torque as described in Steps 3 and 4 of the pinion adjustment procedure. Then install the pinion adjusting ring lockplate.

Table g. S_3 Shims

S_3 Measured Range mm (in.)	Shim Thickness mm (in.)	Part No. through chassis No. 213 2194 781	Part No. from chassis No. 213 2194 782
0.98–1.02 (.0385–.0400)	1.00 (.0395)	003 519 141	—
1.03–1.07 (.0405–.0420)	1.05 (.0415)	003 519 142	—
1.08–1.12 (.0425–.0440)	1.10 (.0435)	003 519 143	082 519 141 AE
1.13–1.17 (.0445–.0460)	1.15 (.0455)	003 519 144	082 519 141 AG
1.18–1.22 (.0465–.0480)	1.20 (.0470)	003 519 145	082 519 141 AJ
1.23–1.27 (.0485–.0500)	1.25 (.0490)	003 519 146	082 519 141 AL
1.28–1.32 (.0505–.0520)	1.30 (.0510)	003 519 147	082 519 141 AN
1.33–1.37 (.0525–.0540)	1.35 (.0530)	003 519 148	082 519 141 AQ
1.38–1.42 (.0545–.0560)	1.40 (.0550)	003 519 149	082 519 141 AS
1.43–1.47 (.0565–.0580)	1.45 (.0570)	003 519 150	082 519 141 BA
1.48–1.52 (.0580–.0600)	1.50 (.0590)	003 519 151	082 519 141 BC
1.53–1.57 (.0600–.0620)	1.55 (.0610)	003 519 152	082 519 141 BE
1.58–1.62 (.0620–.0635)	1.60 (.0630)	003 519 153	082 519 141 BG
1.63–1.67 (.0640–.0655)	1.65 (.0650)	003 519 154	082 519 141 BJ
1.68–1.72 (.0660–.0675)	1.70 (.0670)	003 519 155	082 519 141 BL

12.2 Adjusting Ring Gear

The differential bearing preload and ring gear depth-of-mesh backlash must be adjusted following pinion adjustment. The correct S_1 and S_2 adjustments—the screw-in depths of the differential bearing adjusting rings—are not measured directly. Instead, they are arrived at by adjusting the differential bearings to the correct turning torque and then adjusting the backlash between the ring gear and the drive pinion.

To adjust:

1. Install the differential. Install the differential bearing adjusting rings so that the ring gear and drive pinion are in mesh, but with some backlash between the two gears.

2. Install a torque gauge on the drive pinion as shown in Fig. 12-9.

© 1974 VWoA—3466

Fig. 12-9. Torque gauge in position for measuring total final drive turning torque. The special wrench is in position to turn the right-side differential bearing adjusting ring.

3. Using the handle of the torque gauge, spin the final drive gears 15 or 20 turns in each direction while applying hypoid oil to the tapered-roller bearings.

NOTE —
Use only hypoid oil as a lubricant. The test results will be inaccurate if the turning torque is checked with the bearings dry or lubricated with another kind of oil.

4. While spinning the final drive gears rapidly, slowly increase the bearing preload by screwing in the right-side differential bearing adjusting ring. Increase the preload until the total turning torque reaches the specifications given in **Table h**.

Table h. Total Turning Torque

Actual pinion torque cmkg (in. lb.)		Total torque cmkg (in. lb.)	
new bearings	**used bearings run 30 mi. (48 km)**	**new bearings**	**used bearings run 30 mi. (48 km)**
14 (12)		16–18 (14–16)	
15 (13)		17–19 (15–17)	
16 (14)		18–20 (16–18)	
17 (15)	2 (1.7)	19–21 (17–19)	3–5 (2.6–4.3)
18 (16)		20–22 (18–20)	
19 (17)		21–23 (19–21)	
20 (18)		22–24 (20–22)	

5. Remove the torque gauge. Screw out the right-side differential bearing adjusting ring, and screw in the adjusting ring at the opposite side by exactly the same amount until you can feel a backlash of approximately 0.20 mm between the ring gear and the pinion.

NOTE —
The adjusting rings must be turned in opposite directions by exactly the same amount to maintain the total turning torque that was obtained earlier.

6. Adjust the length of measuring bar VW 388 to precisely 80.00 mm as shown in Fig. 12-10.

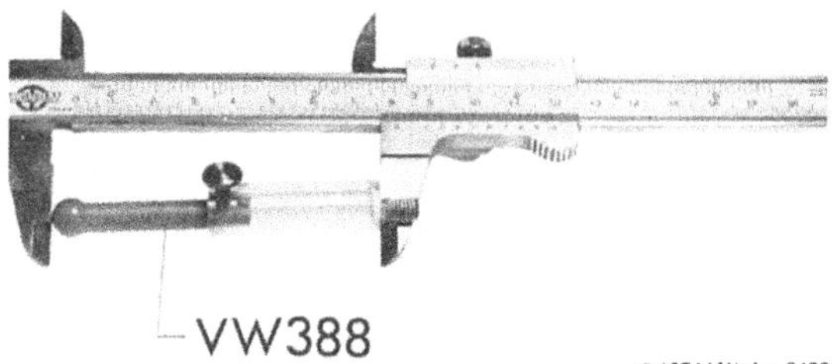

© 1974 VWoA—3468

Fig. 12-10. Vernier caliper being used to adjust measuring bar VW 388 to as near 80.00 mm as possible.

7. Using locking sleeve parts VW 521/4 and VW 521/7, install measuring bar VW 388 on the ring gear side of the differential. Install a dial indicator so that its gauge pin contacts the measuring bar. See Fig. 12-11.

7

© 1974 VWoA—3469

Fig. 12-11. Measuring bar and dial indicator installed on final drive. The mounting for the dial indicator must be rigidly mounted on the final drive housing.

8. Turn the ring gear via the drive pinion until the measuring bar contacts the dial indicator gauge pin as shown in Fig. 12-12. Turn the ring gear further until the dial indicator shows a 1.00-mm preload.

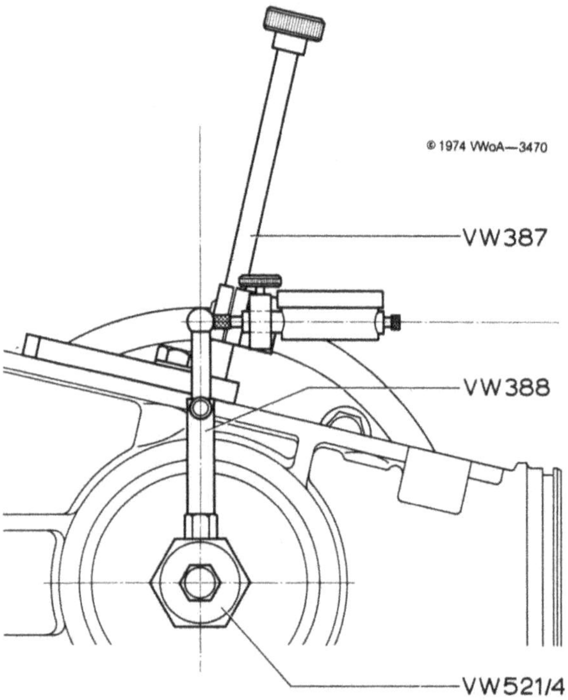

Fig. 12-12. Relative positions of measuring bar and dial indicator pin. It is important that they meet at a 90° angle on the vertical axis of the ring gear (broken lines).

9. Clamp the drive pinion as shown in Fig. 12-13 to keep it from turning.

Fig. 12-13. Spline clamp VW 386 bolted to the final drive housing to prevent the pinion from turning.

10. Hand-turn the ring gear away from the dial indicator until it is stopped by the locked drive pinion. Then zero the dial indicator.

11. From the zeroed position, hand-turn the ring gear toward the dial indicator until it is stopped by the locked drive pinion. Write down the reading. This is the backlash, sometimes abbreviated $\mathbf{Sv_0}$.

12. The $\mathbf{Sv_0}$ should be within a tolerance range of 0.15 to 0.25 mm. If it is not, turn the differential bearing adjusting rings in opposite directions by exactly the same amount until $\mathbf{Sv_0}$ is within this range.

13. Repeat the $\mathbf{Sv_0}$ measurement procedures at three other points 90° apart around the ring gear.

14. Add the four measurements, then divide the sum by four. The quotient is the $\mathbf{Sv_0}$ average.

CAUTION —
The difference between backlash readings must always be smaller than 0.06 mm. If there is a greater variation, something is wrong with the gearset or its installation. Left in this condition, the final drive will be noisy and wear rapidly.

15. If the $\mathbf{Sv_0}$ average is within the 0.15 to 0.25-mm tolerance range, install the lockplates for the adjusting rings. Two lockplates are available. One or the other always fits.

16. Recheck the total turning torque to make sure that it has not been altered by the backlash adjustments.

13. SELECTOR LEVER, SELECTING RODS, AND CABLE

The selector lever used with the automatic transmission shares no parts with the gearshift lever used with the manual transmission. A cable couples the selector rods, which are moved by the selector lever, with a lever on the automatic transmission. This cable is not repairable, and must be replaced as a unit if either the moving part of the cable or its outer housing is worn or damaged.

The automatic transmission was first introduced on 1972 Type 2 VW vehicles. Since that time there has been only one significant modification to the selector lever. On 1974 vehicles, the selector lever has a press button on the side of the knob to unlatch the selector lever. This modification required a different latching segment, selector lever clevis, latching cam, and spring. An exploded view of the new selector lever is given in Fig. 13-1.

Fig. 13-1. Selector lever introduced on 1974 models.

1. M 8 nut
2. Spring washer
3. Bolt
4. Washer
5. Spacer
6. Rubber bellows
7. Bolt (2)
8. M 8 bolt with spring washer (2)
9. Spring (2)
10. Thrust pin (2)
11. Ball (2)
12. Selector cable
13. Pop rivet
14. Rear selecting rod
15. Spring pin
16. Selector lever release
17. Selector lever
18. Cover strip
19. Plastic bushing (2)
20. M 8 bolt (2)
21. Latching segment
22. Rubber bellows
23. M 10 clamp bolt and washer
24. Clamp
25. M 10 nut with washer
26. Spring pin (2)
27. Spring
28. Latching cam
29. Contact
30. Front selecting rod
31. Selector lever clevis
32. Guide plug
33. M 8 nut (2)
34. Washer (2)
35. Circlip
36. Knob
37. Screw (2)
38. Screw (2)
39. Indicator plate
40. Indicator
41. Bulb and holder (2)
42. Console cover
43. Washer (2)
44. Screw (2)
45. M 8 nut
46. Washer
47. Phillips head screw (2)
48. Contact block
49. Selector lever bracket
50. Pivot bolt
51. Clevis pin

13.1 Removing and Installing Selector Lever

The entire drive range selecting mechanism is shown in Fig. 13-2. The selector lever assembly shown in the illustration is the one used on 1972 and 1973 models.

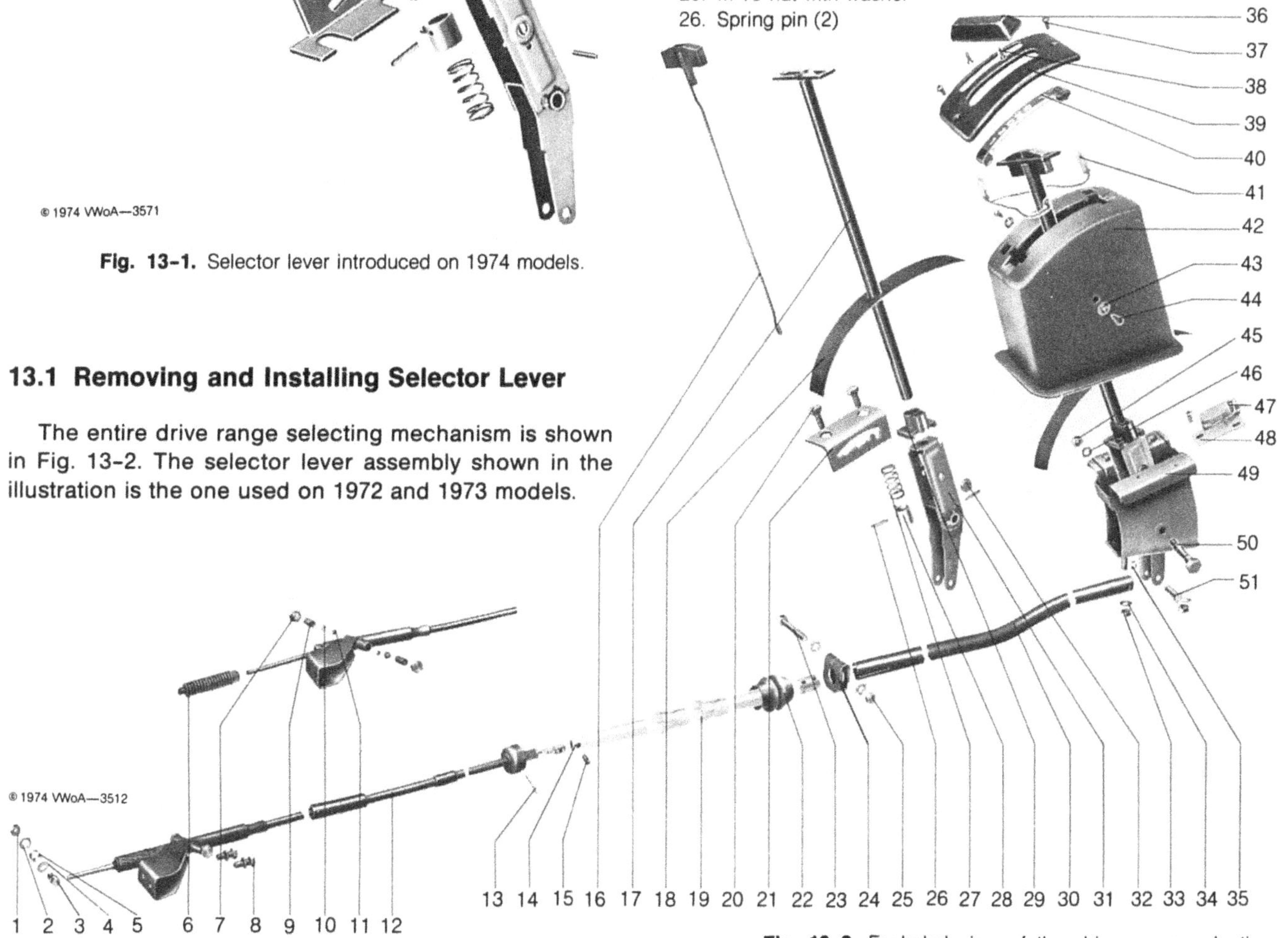

Fig. 13-2. Exploded view of the drive range selecting mechanism.

7

To remove selector lever:

1. Working under the vehicle, remove the cover plate that is beneath the hand controls.
2. Remove the circlip and the clevis pin that hold the selector lever clevis on the front selecting rod.
3. Remove the nuts that hold the selector lever assembly on the floor panel.
4. Working inside the vehicle, lift up the selector lever assembly and disconnect the electrical wires.
5. Remove the selector lever assembly and, if necessary, disassemble it using Fig. 13-1 and Fig. 13-2 as guides.

Installation is the reverse of removal. If you have disassembled the selector lever, adjust the contact block so that the engine can be started in both **N** and **P**. Make sure that the pin on the latching cam is engaged fully in the V-notch in the latching segment when the selector lever is at **R**. If necessary, adjust the latching segment on 1972 and 1973 models by placing a washer between the segment and the selector lever bracket.

To remove selector cable:

1. Working under the vehicle, loosen the clamp nut on the selector cable lever (at transmission). Then disconnect the cable.
2. Remove the selector cable bracket from the transmission case. Cut or drill out the pop rivets in the cable cap (Fig. 13-3).

© 1974 VWoA—3513

Fig. 13-3. Pop rivet (arrow) that holds cable cap on selecting rod guide tube.

3. Loosen the clamp that holds the front and rear selecting rods together. Then pull the rear selecting rod partially out of the selecting rod guide tube.
4. Drive out the spring pin indicated in Fig. 13-4, then remove the selector cable.

© 1974 VWoA—3514

Fig. 13-4. Spring pin (arrow) that holds the cable on the rear selecting rod.

Installation is the reverse of removal. Lightly grease all parts as you install them. Make sure that the pins in the plastic bushings engage the holes in the rear selecting rod before you slide the selecting rod into the guide tube. Install a new pop rivet—3.2 mm in diameter × 7.4 mm long (or 1/8 in. in diameter × 5/16 in. long)—as shown in Fig. 13-5. Then adjust the selector cable.

© 1974 VWoA—3516

Fig. 13-5. Pop rivet being installed to hold the cable cap on the selecting rod guide tube.

13.2 Adjusting Selector Lever Cable

The selector lever cable must never be kinked or bent. Make sure that the rubber bellows is tight-fitting and not cracked or torn.

To adjust:

1. Move the selector lever to **P**. Make sure that the latching cam engages properly.

2. Loosen the clamp that joins the front and rear selecting rods, then pull the rods apart.
3. Hand-press the cable lever on the transmission fully to the rear against spring tension as shown in Fig. 13-6.

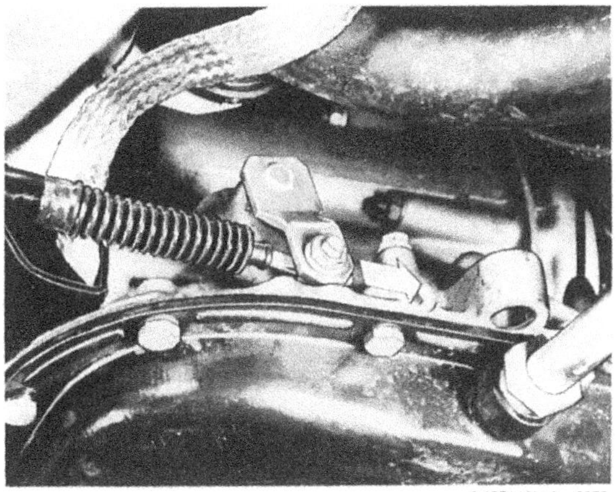

© 1974 VWoA—3277

Fig. 13-6. Cable lever movement toward rear of vehicle (arrow). Moving the lever places the manual valve against the stop in the valve body.

4. While holding the cable lever to the rear with a screwdriver, torque the cable clamp nut to 1.0 mkg (7 ft. lb.). Use the wrench in such a way that the leverage used in tightening the nut aids you in keeping the lever fully to the rear.
5. Push the front selecting rod forward (Fig. 13-7). Then, while holding the rear selecting rod stationary, tighten the clamp.

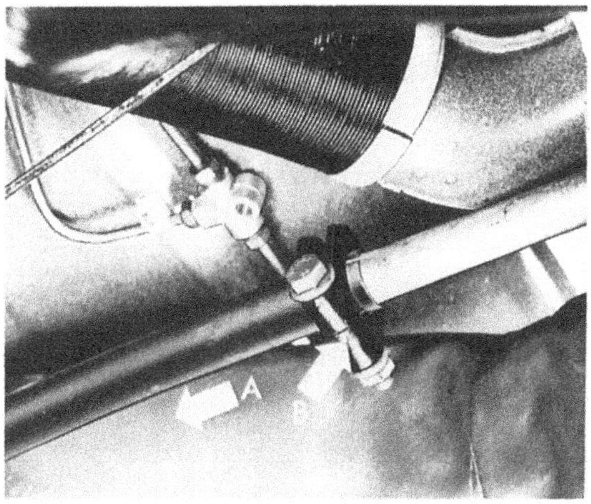

© 1974 VWoA—3278

Fig. 13-7. Front selecting rod being pushed forward (arrow **A**) before clamp (arrow **B**) is tightened.

14. AUTOMATIC TRANSMISSION TECHNICAL DATA

The tables of technical data that appear on the following pages contain all the specifications needed for rebuilding the automatic transmission.

I. Tolerances and Settings

Designation	New part
1. Adjusting planetary gear end play	0.45-1.05 mm (0.018-0.042 in.)
2. Adjusting brake bands—Have transmission horizontal. Before adjusting, center the bands by tightening the screws to 1.0 mkg (86.8 in. lb.) and then loosening them.	
a. 2nd gear	
Tighten screw to 1.0 mkg (86.8 in. lb.), then back off 1¾ to 2 turns and retighten to specification . . . final torque	0.5 mkg (43 in. lb.)
b. 1st gear	
Tighten screw to 1.0 mkg (86.8 in. lb.) then back off 3¼ to 3½ turns and retighten to specification . . . final torque	0.5 mkg (43 in. lb.)
3. Clutches	
a. Forward clutch—Use only lined plates with a concentrically-grooved surface, a pressure plate 6.10-5.85 mm (.240-.230 in.) thick. Note thickness of original circlip. End play **a**	0.8-1.2 mm (0.032-0.048 in.)
b. Direct and reverse clutch (with 2 lined plates)—Use only lined plates with a waffle surface, a pressure plate 6.30-6.15 mm (.248-.242 in.) thick, and a circlip 1.70 mm (.067 in.) thick. Circlip . . . thickness	1.7 mm (0.067 in.)
c. Direct and reverse clutch (with 3 lined plates)—Use only lined plates with a waffle surface, a pressure plate 6.30-6.15 mm (.248-.242 in.) thick. Adjust to correct play of 1.70-2.20 mm (.067-.087 in.) thick. Axial play	1.7-2.2 mm (0.067-0.087 in.)
4. Preload of pinion bearings (turning torque) . . . new	14-20 cmkg (12-18 in. lb.)
bearings used more than 30 miles	2 cmkg (1.7 in. lb.)
5. Total preload (turning torque)	
Pinion and differential . . . new	16-24 cmkg (14-22 in. lb.)
bearings used more than 30 miles	3-5 cmkg (2.6-4.3 in. lb.)

7

II. Constant Velocity Joints (Fig. 14-1)

Designation	Specification
through chassis No. 210 2300 000 from chassis No. 211 2000 001	Part No. 211 501 331 A (groove for metal cap on flange end) 211 501 331 B (annular groove on outside diameter must be toward flange)
Diameter **a**	100.00 − 0.20 mm (3.937 − .007 in.)
Width **b**	32.00 ± 0.30 mm (1.259 ± .012 in.)
Ball diameter through chassis No. 210 2300 000 from chassis No. 211 2000 001	 17.46 mm (.687 in.) 19.05 mm (.750 in.)
Grease per joint	90 g (3.2 oz.) of molybdenum grease
Axial shaft runout (maximum)	0.50 mm (.019 in.)

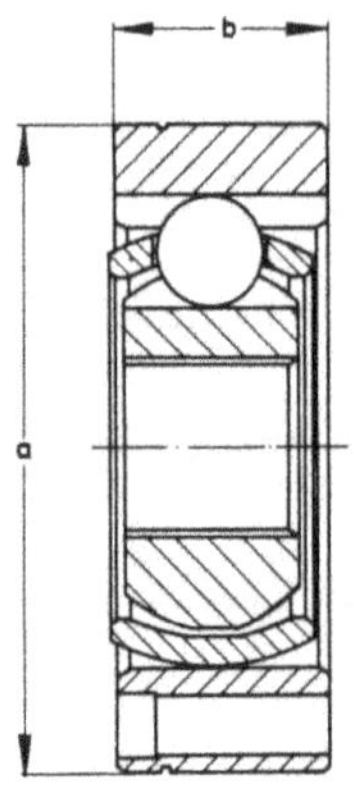

© 1974 VWoA—2581

Fig. 14-1. Constant velocity joint.

III. Double-jointed Axle Driveshafts (Fig. 14-2)

Transmission type .... Manual/Automatic
Code number on end of shaft 2
Shaft dimension **b**:
Manual transmission.................... 476 mm (18¾ in.)
Part No. 211 501 211
Automatic transmission
left side.............................. 457 mm (18 in.)
Part No. 211 501 211 B left
right side........................... 505 mm (19⅞ in.)
Part No. 211 501 211 B right
(The shafts for the automatic transmission have a ridge in the middle.)

IV. Automatic Transmission Test Data

Stall Speed (1900-2000 rpm) Pressure Table			
Selector lever position	**Pressure**	**psi (kg/cm^2)**	**Remarks**
N	Primary throttle pressure Main pressure	45.5 (3.2) 92.0 (6.5)	Increase idle speed to 1000 rpm with vacuum hose *off and plugged*
	Primary throttle pressure Main pressure	5-6 (0.35-0.42) 47-50 (3.3-3.5)	Increase idle speed to 1000 rpm with vacuum hose *on*
R	Main pressure	142-156 (10-11)	
D	Primary throttle pressure Main pressure	31-44 (2.2-3.1) 92 (6.5)	At stall torque speed (full throttle) with vacuum hose *on*
R	Main pressure	256-327 (18-23)	
D	Main pressure	92 (6.5)	At full throttle with a road speed of over 19 mph (30 kph)

Full-throttle gearshift speeds in mph (kph)		
Gears	**"1700" engine**	**"1800" engine**
1-2	18-25 (29-40)	19-25 (30-40)
1-2 kickdown	27-33 (43-53)	27-34 (43-54)
2-3	41-46 (66-74)	43-47 (69-76)
2-3 kickdown	47-51 (76-82)	48-52 (77-83)
3-2	29-25 (47-40)	31-25 (49-40)
3-2 kickdown	48-44 (77-71)	49-45 (78-72)
2-1	16-14 (26-23)	17-14 (27-23)
2-1 kickdown	30-24 (48-39)	31-24 (49-39)

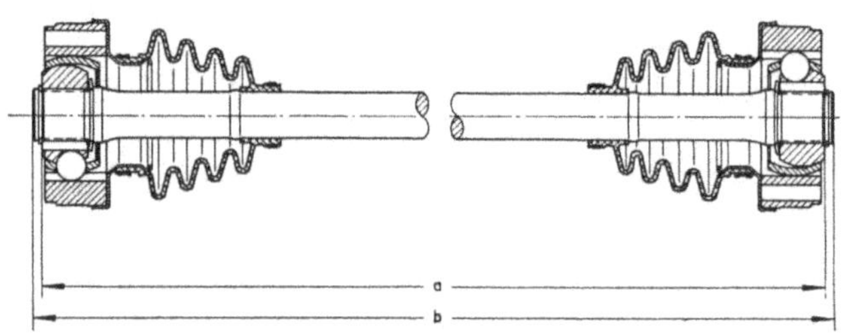

© 1972 VWoA—894

Fig. 14-2. Double-jointed axle driveshaft.

V. Valve Body Springs (Dual-carburetor Engines)

Description	Part No.	Dimensions			
		No. of Coils	Wire Thickness mm (in.)	Free Length, approximate mm (in.)	Coil Inner Dia. ± 0.3 mm (± .012 in.)
Main pressure valve spring (also used in 3rd/2nd accumulator)	003 325 131	16.5	1.50 (.0590)	68.5 ($2\frac{11}{16}$)	11.90 (.469)
Secondary throttle pressure valve spring	003 325 157 A	12.5	0.85 (.0334)	29.1 ($1\frac{5}{32}$)	7.35 (.289)
Kickdown valve spring	003 325 175	10.5	0.63 (.0248)	23.8 ($\frac{15}{16}$)	7.70 (.303)
Modulator valve spring	003 325 185	11.5	0.80 (.0315)	28.5 ($1\frac{1}{8}$)	7.75 (.305)
2nd/3rd shift valve spring	003 325 207	8.5	1.00 (.0393)	25.4 (1)	9.00 (.354)
1st/2nd shift valve spring	003 325 217	9.5	1.00 (.0393)	23.6 ($\frac{29}{32}$)	8.00 (.315)
Throttle pressure limiting valve spring	003 325 227 A	12.5	1.00 (.0393)	32.4 ($1\frac{9}{32}$)	7.70 (.303)
Converter pressure valve spring	003 325 247	9.5	1.25 (.0492)	27.3 ($1\frac{1}{16}$)	8.13 (.320)
Pressure relief valve spring	003 325 267	15.5	0.80 (.0315)	27.7 ($1\frac{3}{32}$)	4.70 (.185)
2nd/3rd valve spring (also used in 3rd/2nd valve in transfer plate)	003 325 269	4.5	0.20 (.0079)	5.8 ($\frac{7}{32}$)	4.30 (.169)
Primary throttle pressure valve spring	003 325 295	10.5	0.63 (.0248)	36.3 ($1\frac{7}{16}$)	9.00 (.354)
3rd/2nd control valve spring	003 325 119 A	15.0	0.56 (.0220)	24.5 ($\frac{31}{32}$)	6.40 (.252)
Part throttle valve spring	003 325 129	6.5	0.40 (.0157)	18.1 ($\frac{23}{32}$)	6.10 (.240)

NOTE —
Because fuel injection engines have different intake manifold vacuum characteristics from dual-carburetor engines, the valve body was slightly modified when fuel injection was introduced on the 1975 models. Owing to these modifications, some of the spring dimensions given in **Table V** may not be applicable to transmissions used with fuel injection engines.

VII. Ratios

Gears	Ratio
Transmission:	
1st gear	2.65
2nd gear	1.59
3rd gear	1.00
Reverse	1.80
Converter:	
Maximum torque multiplication	2.5
Final drive:	
"1700" engine	4.45
"1800" engine	4.36

VI. Automatic Transmission Markings

Code Letters	Final drive, no. of teeth	Engine Displacement	Valve body Code letters	Introduction
NA	11:49	"1700"	none	Sept. 1972
NB	11:48	"1800"	K	Nov. 1973
NC	11:48	"1800"	R	Fuel injection models
ND	11:48	"1800"	S	Fuel injection models

VIII. Tightening Torques

Location	Designation	mkg	ft. lb.
ATF pump to transmission case	bolt	0.4	2.8
Transfer plate on valve body	Phillips head screw	0.3	2.1
Valve body to transmission case	bolt	0.4	2.8
Valve body to transmission case	screw	0.35	2.5
Pan to transmission case	bolt	1.0	7.0
ATF strainer to valve body	slotted screw	0.3	2.1
Manual valve lever/cable lever	nut	0.6	4.3
Cable lever to transmission case	retaining screw	0.5	3.5
Operating lever on transmission case	threaded pin	0.6	4.3
Cable bracket on transmission case	bolt	1.5	11.0
Filler tube on transmission case	bolt	0.5	3.5
Plug for pressure connection/ transmission case	socket or hex. head	1.0	7.0
Vacuum unit/transmission case	—	2.5	18
Lock nut for band adjusting screw	nut	2.0	14
Air deflector on final drive housing	bolt	1.0	7.0
Differential carrier on final drive housing	nut	0.8	6.0
Side cover/final drive housing	nut	0.8	6.0
Starter/final drive housing	nut	2.5	18
Bearing cap/differential carrier	bolt	6.0	43
Ring gear/differential housing	bolt	5.0	36
Transmission case/final drive housing	nut	2.0	14
Converter to drive plate	bolt	2.5	18
Drive shaft/flange	socket head screw	3.5	25
Engine/transmission mounting	nuts and bolts	3.0	22
Transmission/ transmission carrier	bolt	3.5	25
Front transmission mount limiting stop to bonded rubber mounting	bolt	4.5	33
Bonded rubber mounting to transmission case	nut	2.5	18
Selector lever cable clamp nut	nut	1.0	7.0
Adjusting screws for brake bands (have the transmission horizontal. Tighten the screws to 1.0 mkg [86.8 in. lb.]. Loosen them, then retighten to 0.5 mkg [43 in. lb.]. From this position, loosen the front brake band screw by ¾ to 2 turns; the rear brake band screw by 3¼ to 3½ turns.)			

Section 8

BRAKES AND WHEELS

Contents

2 BRAKES AND WHEELS

TABLES

Brakes and Wheels

The 1968 through 1970 Type 2 VWs have four-wheel drum brakes, while 1971 and later models have disc brakes at the front and drum brakes at the rear. On models with front disc brakes, the master cylinder is operated through a vacuum powered brake servo. On all models, each axle has its own hydraulic circuit. On vehicles with front disc brakes, there is a pressure regulator in the rear brake circuit that prevents the rear brakes from locking when they are used to their maximum.

A dual-chamber master cylinder provides operating pressure to both brake circuits. The system is so designed that leaks in one circuit cannot affect the other. An electrical warning system in the master cylinder causes a red warning indicator in the instrument panel to light up if hydraulic pressure is too low in either brake circuit. If you see this light while you are driving, it is imperative that the brake system be given a thorough check, even though braking action may still seem satisfactory. Complete loss of pressure in one of the brake circuits will cause the pedal to fall closer to the floor during braking and will result in abnormally long stopping distances.

Because safe vehicle operation depends very heavily on the brakes, all brake system service and repair work must be carried out with extreme cleanliness, careful attention to specifications, and proper working procedures. All necessary information is given here, although some of the operations that are described may be of practical value only to professional mechanics.

If you lack the skills, special tools, or a clean workshop for servicing the brake system, we suggest you leave such repairs to an Authorized VW Dealer or other qualified shop. We especially urge you to consult your Authorized VW Dealer before attempting repairs on a car still covered by the new-car warranty.

The pages devoted to wheels and tires should have practical interest to all drivers, whether they service their own cars or not. A great many cases of abnormal tire wear or poor vehicle handling are the direct result of improperly fitted tires, tires incorrectly inflated, or driving practices that damage the tires. It is our belief that following the advice offered here will not only save you money but also make your driving safer.

4 Brakes and Wheels

1. General Description

The following major parts make up the brake system:

Master Cylinder Actuated by the foot pedal and connected to the wheel cylinders by brake lines and hoses, the master cylinder generates the hydraulic pressure needed to operate the brakes. The master cylinder has two pistons, one operating the front brakes and the other the rear brakes. On 1971 and later vehicles—those equipped with front disc brakes—the pedal operates the master cylinder through a vacuum powered brake servo.

Fluid Reservoirs The refill reservoir, under the driver's seat, supplies brake fluid via a hose to the twin-chamber reservoir on the master cylinder. In turn, the twin-chamber reservoir supplies the master cylinder.

Wheel Cylinders These are hydraulic cylinders that contain pistons that press the brake shoes against the brake drums. There are two single-piston cylinders in front drum brakes and one cylinder with two opposed pistons in each rear wheel brake.

Brake Shoes Moved by the wheel cylinder pistons, two leading shoes at each front wheel and one leading and one trailing shoe at each rear wheel work against the inside of each brake drum.

Brake Calipers The 1971 and later models have front disc brakes. Each disc brake assembly has a brake caliper which houses the two opposed pistons that press the friction pads against the brake disc.

Hydraulic Lines These are steel tubes and hoses that connect the master cylinder to the front brake calipers or wheel cylinders and to the rear wheel cylinders.

Foot Brakes

The dual-circuit hydraulic foot brakes operate on all four wheels. Pedal pressure on disc brake models is reduced by incorporating a vacuum powered servo.

Parking Brakes

The cable-operated parking brake works on the rear wheels only. The hand lever is held or released by a ratchet and is centrally located under the dashboard.

2. Maintenance

The following routine maintenance operations are covered briefly in **LUBRICATION AND MAINTENANCE**. Additional information can be found in this section under the headings listed after each maintenance check.

1. Checking and changing brake fluid. **9.1**
2. Checking brake linings. **8.3.**
3. Checking brake adjustment. **8.1**
4. Checking brake lines, hoses, and brake lights. **4.4**
5. Checking wheels, tires, and tire pressures. **13.**

3. Brake Fluid Reservoirs

The refill reservoir is mounted above the clutch pedal on 1968 through 1970 models, behind the driver's seat on 1971 and 1972 models, or under the driver's seat on later models. A hose and tubing connect the refill reservoir to the twin-chamber reservoir (Fig. 3-1).

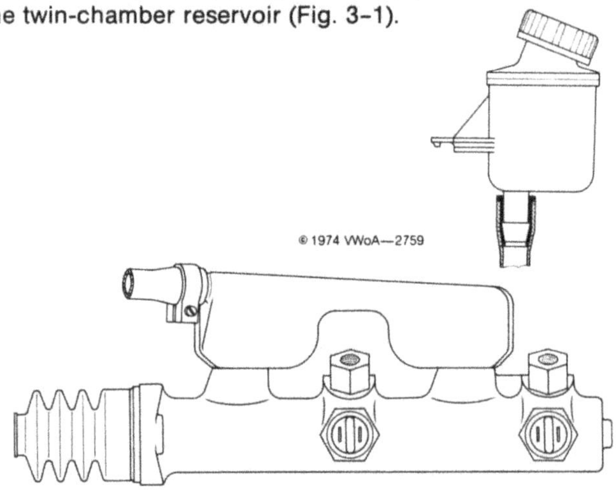

Fig. 3-1. Brake fluid reservoirs. The twin-chamber reservoir is mounted directly on the master cylinder.

3.1 Filling Reservoir

Clean the cap and reservoir top before removing the cap. While the cap is off, check to see that the vent is open. On 1971 and later models, maintain the level between the upper and lower edges of the window in the front of the refill reservoir; on earlier models, fill to within 20 mm (3/4 in.) of the threads for the cap.

> ***CAUTION—***
> *Use only new, unused brake fluid that meets SAE recommendation J 1703 and conforms to Motor Vehicle Safety Standard 116.*

3.2 Removing and Installing Reservoirs

Empty the refill reservoir with a syringe before you remove either reservoir or the line or hose that connects the reservoir.

> ***WARNING—***
> *Do not start a siphon with your mouth or spill fluid on the car. Brake fluid is both poisonous and damaging to paint.*

To remove the refill reservoir, remove the line from the reservoir outlet pipe. Then remove the screw that holds the reservoir in place and remove the reservoir.

During installation, attach the line to the refill reservoir, making sure that all parts fit tightly and are correctly positioned. Then mount the refill reservoir on the vehicle body.

To remove twin-chamber reservoir:

1. Working under the vehicle, remove the cover plate that is beneath the pedal cluster.
2. After removing as much fluid as possible from the refill reservoir, there will still be fluid in the twin-chamber reservoir. So place a container or absorbent cloth under the master cylinder to catch escaping fluid.
3. Press the twin-chamber reservoir out of the rear sealing plug as indicated in Fig. 3-2. Then press the reservoir out of the front sealing plug.

© 1974 VWoA—2769

Fig. 3-2. Twin-chamber reservoir removal. First press the reservoir out of the rear sealing plug as indicated by the arrow.

4. Pull the reservoir slightly to the rear, and then disconnect the refill line at the point indicated in Fig. 3-3.

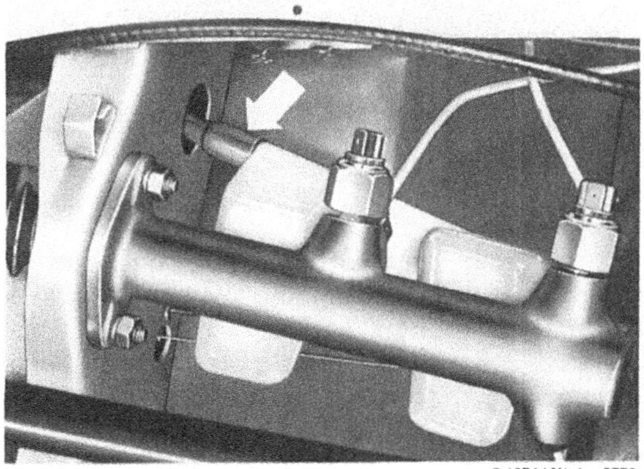

© 1974 VWoA—2770

Fig. 3-3. Point (arrow) where refill line should be disconnected from the twin-chamber reservoir.

Installation is the reverse of removal. On 1969 and later models, a plastic hose replaces the metal tube used as a refill line on 1968 models. Push the rubber seal onto the wider part of the plastic refill line. Then push the line and seal onto the reservoir connection. Following installation, bleed the brakes as described in **9. Bleeding Brakes.**

4. MASTER CYLINDER

The master cylinder has two pistons, one behind the other. One piston supplies hydraulic pressure to the front brakes; the other piston supplies hydraulic pressure to the rear brakes.

4.1 Pushrod Adjustment

On 1968 through 1970 vehicles, the master cylinder pistons are moved by a pushrod connected to the brake pedal. On 1971 and later models, which are equipped with front disc brakes, the pushrod for the master cylinder is part of the vacuum powered brake servo. It is very important that the correct clearance exist between the pushrod and the master cylinder piston (Fig. 4-1).

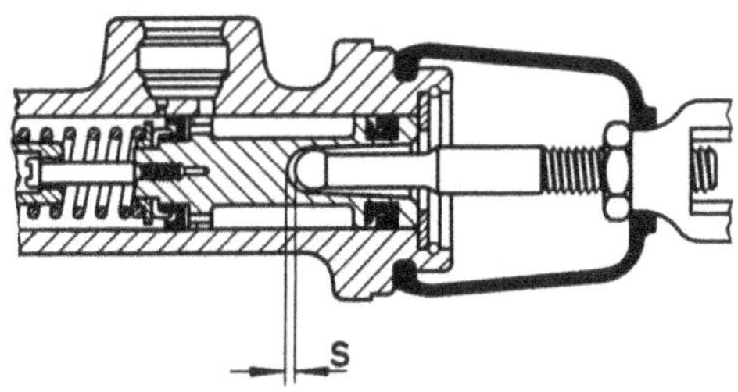

© 1974 VWoA—1453

Fig. 4-1. Required clearance **S** between the end of the pushrod and the master cylinder piston. It should be 1 mm (.040 in.).

Notice that in Fig. 4-1 there is a small compensating port just ahead of the piston, as well as a larger port behind it. The compensating port can admit brake fluid to, or receive fluid from, the working side of the piston only when the piston is in its rest position. A similar pair of ports are positioned adjacent to the other piston in the master cylinder.

The compensating ports are, in many ways, the most important parts of the master cylinder. Their job is to permit surplus brake fluid to return to the reservoirs or to allow the reservoirs to refill the master cylinder. If the pistons block the compensating ports when the brake pedal is fully released, neither of the functions can be fulfilled. It is therefore necessary to maintain the clearance indicated in Fig. 4-1.

Dragging brakes, an abnormally high pedal, and brakes that lock up while driving and fail to release are symptoms of blocked compensating ports. Pushrod clearance should be checked and, if necessary, adjusted. If it is already correct, the master cylinder probably needs to be rebuilt or replaced or, on vehicles with front disc brakes, there is trouble in the vacuum powered brake servo. The clearance must also be checked after servicing the master cylinder, the brake servo, or removing and installing the brake pedal.

On 1968 through 1970 models, the pushrod clearance can be measured directly. Working under the vehicle, remove the cover plate that is beneath the pedal cluster. Operate the pedal lever by hand and see whether the pushrod travels into the master cylinder the prescribed 1-mm (.040 in.) distance before encountering resistance. If not, adjust the clearance by moving the brake pedal stop indicated in Fig. 4-2.

© 1974 VWoA—2774

Fig. 4-2. Brake pedal stop for 1968 through 1970 vehicles (arrow).

If the proper clearance cannot be obtained by moving the brake pedal stop on 1968 through 1970 models, remove the pushrod and check dimension **a** given in Fig. 4-3. It must be 106 mm (4.173 in.). If not, adjust and install the pushrod. Then adjust the pushrod clearance by moving the pedal stop.

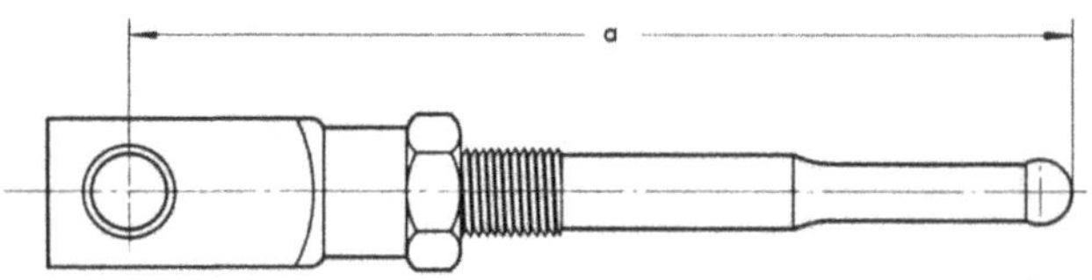

© 1974 VWoA—2773

Fig. 4-3. Correct pushrod length. Dimension **a** must be 106 mm (4.173 in.).

To adjust clearance on vehicle with brake servo:

1. Working under the vehicle, remove the cover plate that is beneath the pedal cluster.
2. Check to see that the pedal lever is being held against the brake pedal stop by the brake pedal return spring.
3. If the pedal lever is against the stop, check the clearance between the connecting rod and the brake servo. The connecting rod should not exert pressure on the servo valve housing when the pedal is all the way up.
4. If necessary, remove the cotter pin from the clevis pin. Remove the clevis pin, and disconnect the connecting rod clevis from the pedal lever.
5. Adjust the connecting rod by screwing it in or out of the threaded part of the servo (Fig. 4-4).
6. If necessary, loosen the locknut and make fine adjustments by turning the connecting rod connection until the clevis pin can easily be inserted as you install the connecting rod on the pedal lever.

© 1974 VWoA—2775

Fig. 4-4. Connecting rod length adjustment. Turn rod as indicated by the curved arrow (left) near the clevis. Make fine adjustments by turning the flat-sided connecting rod connection indicated by the right arrow.

7. Install a new cotter pin in the clevis pin. Tighten the locknut for the connection on the servo.

Because the length of the one-piece pushrod in the brake servo is fixed, its clearance with the master cylinder piston cannot change. Adjusting the connecting rod so that it does not exert pressure on the servo valve ensures that the servo pushrod is fully retracted to maintain the proper clearance. Because close tolerances are maintained during manufacture, the master cylinder or the servo can be replaced without upsetting the clearance between the servo pushrod and the master cylinder piston.

4.2 Removing and Installing Master Cylinder

On 1968 through 1970 vehicles, the master cylinder is bolted to a bracket on the frame. On later models, the master cylinder is mounted on two studs on the brake servo.

To remove the master cylinder, first remove the twin-chamber brake fluid reservoir as described in **3.2 Removing and Installing Reservoirs**. Disconnect the wires from the two brake light/warning light switches, attaching tags to them so they can be returned to their correct terminals during installation. Then disconnect the brake lines and seal them with new brake bleeder valve dust caps. The master cylinder may then be unbolted from the frame or brake servo.

Installation is the reverse of removal. On vehicles with brake servo, install a new O-ring between the servo and the master cylinder. Make sure that replacement master cylinders are either the same type as originally installed or the correct type listed for the vehicle in VW parts list. Torque the master cylinder mounting nuts to 2.5 mkg (18 ft. lb.) on 1968 through 1970 models; torque them to a maximum of 1.3 mkg (9.4 ft. lb.) on 1971 and later models. Torque the brake line unions to 1.5 to 2.0 mkg (11 to 14 ft. lb.). Check the pushrod clearance as described in **4.1 Pushrod Adjustment**. Then fill the reservoirs with fresh brake fluid and bleed the brakes as described in **9. Bleeding Brakes.**

CAUTION —
Use only new, unused brake fluid that meets SAE recommendation J 1703 and conforms to Motor Vehicle Safety Standard 116.

4.3 Repairing Master Cylinder

Faulty master cylinders can be replaced with new or rebuilt units, or the existing master cylinder can be rebuilt. Replacement is fairly easy; rebuilding demands know-how and some special tools. The repair kits available for the master cylinder contain all the seals, cups, pistons, valves, and springs needed for rebuilding. An exploded view of a typical master cylinder is shown in Fig. 4-5.

Fig. 4-5. Exploded view of master cylinder.

1. Cylinder body
2. Sealing plug (2)
3. Plug
4. Seal
5. Spring (2)
6. Cup (2)
7. Warning device piston (2)
8. Brake light switch (2)
9. Brake light/warning light switch (2)
10. Stop screw and seal
11. Front brake piston spring
12. Spring support ring (2)
13. Spring seat (integral with support ring on some late master cylinders) (2)
14. Cup (4)
15. Primary cup washer
16. Front brake piston
17. Stroke limiting screw
18. Stop sleeve
19. Rear brake piston spring
20. Rear brake piston
21. Secondary cup (1968-1970)
22. Stop washer (1968-1970)
23. Lockring (1968-1969)
24. Rubber boot (1968-1969)

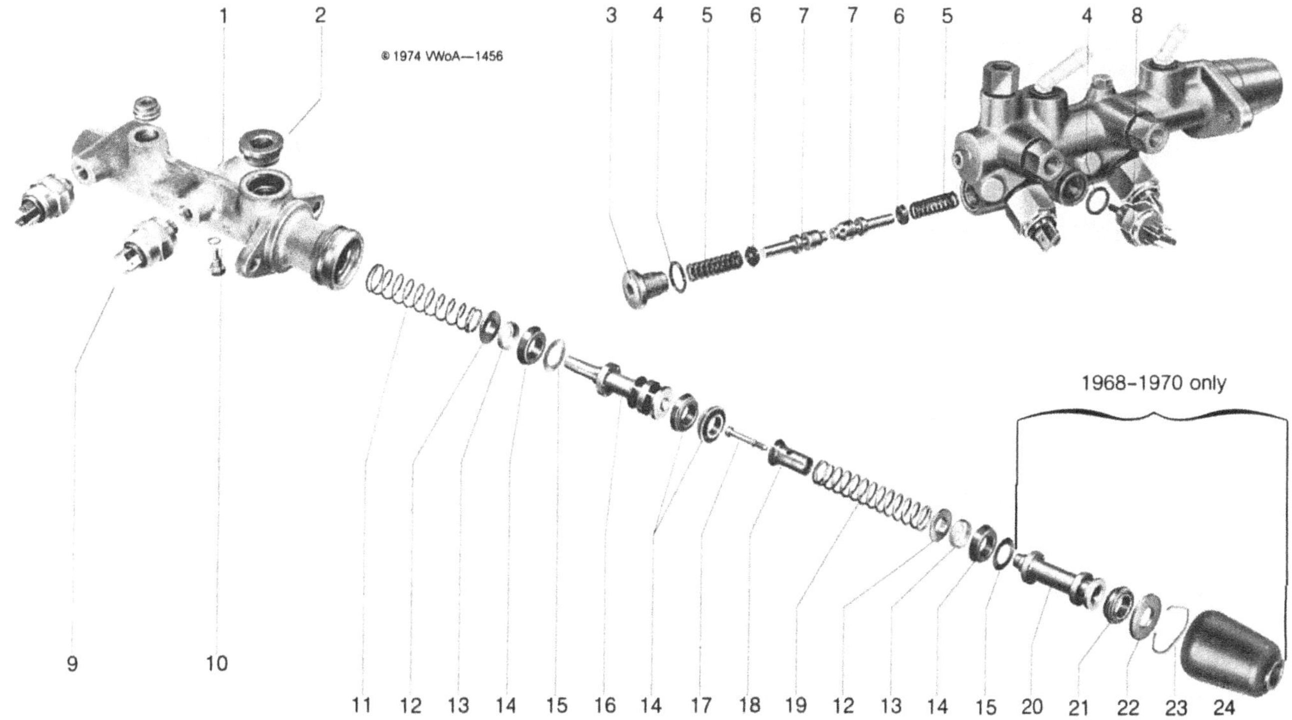

The master cylinders used on 1971 and later models have additional sealing parts because they are mounted against the vacuum chamber of the brake servo. The piston engaged by the servo pushrod is also modified to provide a sealing surface for the added seal cups (Fig. 4-6).

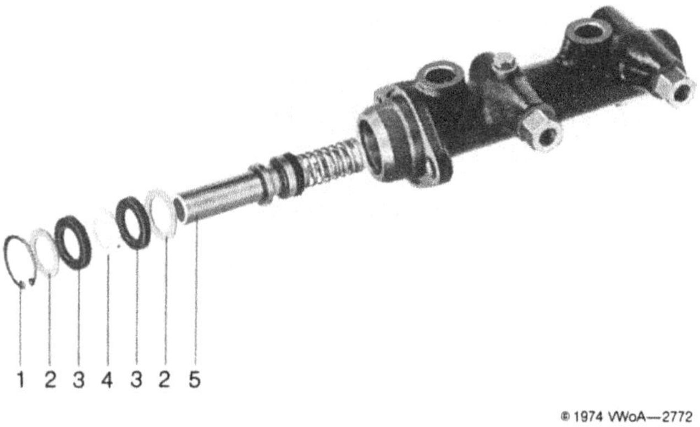

1. Circlip
2. Stop washer
3. Seal cup
4. Plastic washer
5. Pushrod

Fig. 4-6. Sealing parts of master cylinder used with vacuum powered brake servo.

The master cylinder with the separate warning device cylinder was installed during the 1968 and early 1969 model years only. Although the master cylinders in Fig. 4-5 and Fig. 4-6 may not completely resemble the unit on the vehicle, the internal parts of all master cylinders are identical in number and position, with the exception of those additional parts shown in Fig. 4-6.

Rebuilding the master cylinder requires a tool called a brake cylinder hone. Do not install new parts in a master cylinder that has a worn or scored bore. If wear cannot be corrected by honing—without oversizing the bore—a new master cylinder should be obtained.

To disassemble master cylinder:

1. On 1968 through 1970 models, remove the rubber boot.
2. Remove the lockring or circlip over the stop washer. Then remove the stop screw from the center of the master cylinder.
3. Remove the internal parts from the cylinder body.
4. Unscrew the brake light/warning light switches and the residual pressure valves (on vehicles with front drum brakes only).
5. On 1968 and early 1969 master cylinders, remove the plug and then take out the internal parts of the brake warning device cylinder.
6. Remove the sealing plugs.

Inspect the moving parts. If any are worn, replace them all from a repair kit. Replace the entire master cylinder if the cylinder bore is deeply scored. If the bore shows only moderate irregularities from normal wear, remove the irregularities with a brake cylinder hone. The master cylinder with the separate warning device cylinder is not available as a replacement part and must be replaced by a later-type master cylinder, together with combined brake light/warning light switches.

When using a brake cylinder hone, lubricate the stones with brake fluid only, never with mineral oil or kerosene. While the hone is spinning, move it in and out of the bore rapidly to achieve an even polish over the entire cylinder wall.

After honing, check the fit between the pistons and the cylinder bore. If the clearance exceeds 0.10 mm (.004 in.), the entire master cylinder should be replaced. The clearance can be measured by inserting a feeler gauge between the piston and the cylinder wall. However, a brake cylinder bore measuring tool or snap (telescope) gauge and micrometer are preferable tools.

> ***WARNING —***
> *After honing the master cylinder, clean any burrs from the compensating ports. Make certain the ports are not blocked. Otherwise, brake lockup or failure could result.*

To assemble:

1. Clean all parts in brake fluid. Make certain the compensating ports are clear.

> ***WARNING —***
> *Never use kerosene, gasoline, or other petroleum-based solvents for cleaning. These substances are damaging to brake parts.*

2. Using a conical cup sleeve as shown in Fig. 4-7, install the cups on the pistons. A smaller sleeve is required for installing the cups on the warning device pistons of 1968 and early 1969 master cylinders.

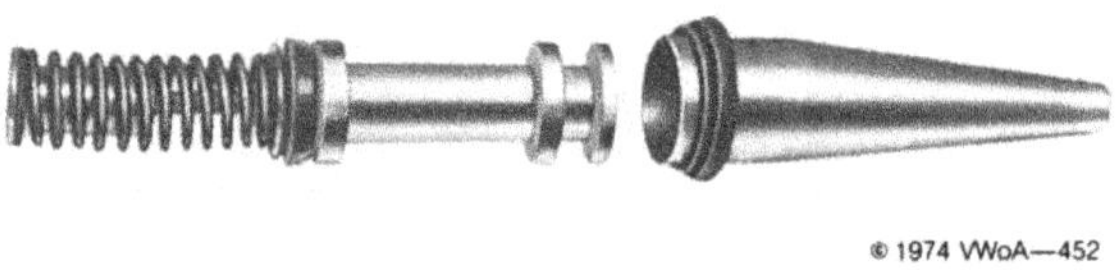

Fig. 4-7. Installing cup on piston. The cup has been slipped over the conical sleeve from the small end. The sleeve is then placed over the piston, and the cup slipped off the sleeve into its groove.

CAUTION —
If the cups are put on the pistons without a conical cup sleeve, extreme care is required to prevent damage to the cups.

3. Hold the master cylinder with its closed end down. Lubricate all internal parts with VW brake cylinder paste (see **9.5 Brake Cylinder Paste**) or with fresh brake fluid. Install the parts in the order shown earlier in Fig. 4-5 and Fig. 4-6.

NOTE —
When you assemble a master cylinder used with the vacuum powered brake servo, lubricate the shaft of the pushrod piston with a light coat of silicone grease (provided in the repair kit). Also fill the annular grooves in the seal cups (Fig. 4-6) with silicone grease. The sealing lips of both seal cups should be toward the master cylinder.

4. Push the pistons into the cylinder against spring tension, then install the stop washer and lockring, or, on 1971 and later models, the stop washer and circlip.
5. Install the stop screw and its seal. Torque to 0.5 to 1.0 mkg (3.5 to 7.0 ft. lb.).

CAUTION —
Be sure the stop screw hole is clear. If blocked by the piston, damage will result as the screw is installed.

6. On 1968 and early 1969 master cylinders with the separate warning device cylinder, install the internal parts using the same procedures used in installing parts in the main cylinder. Then install the plug.
7. Screw the brake light/warning light switches and residual pressure valves (where fitted) into the master cylinder. Torque to 2.0 mkg (14 ft. lb.).
8. Install the sealing plugs. On 1968 through 1970 models, install the rubber boot with the vent hole down.

4.4 Testing and Replacing Brake Light/ Warning Light Switches

The brake light/warning light electrical circuit for 1972 and later vehicles is shown in Fig. 4-8. It contains all the connections found on earlier cars plus several additions and modifications. With this version, the functional check no longer requires that the light lens be pushed in by hand. The warning lamp lights up when the ignition is switched on and goes out in the same manner as the generator and oil pressure warning lamps once the engine has started.

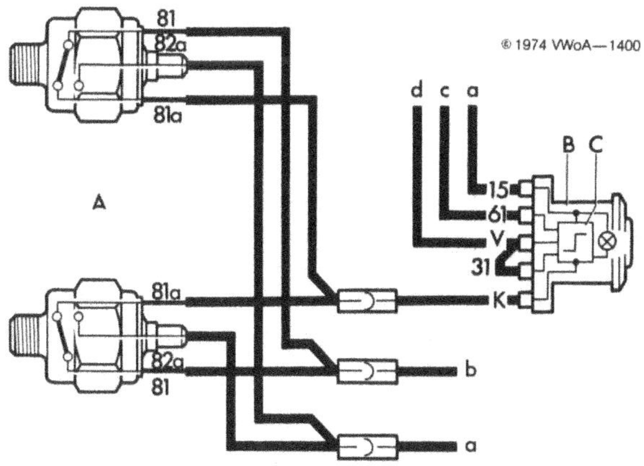

A. Brake light switches
B. Dual circuit brake warning lamp
C. Electronic switch
a. To terminal 15
b. To brake lights
c. From regulator switch terminal 61
d. To ground

Fig. 4-8. Electrical diagram of brake light/warning light switches. Terminal **V** was discontinued in early 1973.

To test brake light switch contacts:

1. Check the brake light bulbs. Replace if necessary.
2. Disconnect the front brake wires from the front brake light/warning light switch (**81** and **81a**, black-red and black wires; see Fig. 4-8).
3. Switch the ignition on and depress the brake pedal. The brake lights should go on. If they do, reconnect the wires to the front switch and remove the rear brake wires (**81** and **81a**) from the other switch. Repeat the test. The brake lights should go on.
4. If the brake lights do not work in one of the tests, replace the defective switch (the one that remained connected during the test).

To replace switch:

1. Disconnect all wires from the defective switch.
2. Unscrew the switch from the master cylinder. Keep the sealing washer.
3. Install the sealing washer and the new switch. Torque to 2.0 mkg (14 ft. lb.). Then connect the wires as indicated in Fig. 4-8.

NOTE —
For additional information about the electrical circuits, consult the wiring diagrams in **ELECTRICAL SYSTEM.**

8

To test brake warning light contact:

1. Check the socket and the light bulb. If necessary, replace them.
2. Switch the ignition on. The warning light should come on.

   NOTE —
   On pre-1972 models, it is necessary to push the light lens to test the light.

3. Open a bleeder valve in the front brake circuit. (See **9. Bleeding Brakes**).
4. Start the engine and depress the brake pedal. The brake warning light should come on.
5. Close the bleeder valve in the front brake circuit and open a bleeder valve in the rear brake circuit. Repeat the test described in Step 4.
6. Check the fluid level in the brake fluid reservoir. If necessary, add fluid.

   CAUTION —
   Use only new, unused brake fluid that meets SAE recommendation J 1703 and conforms to Motor Vehicle Safety Standard 116.

4.5 Removing, Repairing, and Installing Vacuum Powered Brake Servo

The master cylinder on 1971 and later vehicles is operated through a vacuum powered brake servo. The system fails safe, so if the servo ceases to function, the foot brakes can still be applied. However, the pedal pressure required will be somewhat greater than normal.

Before assuming that there is trouble in the master cylinder or the vacuum powered brake servo, check the vacuum hoses carefully. If they are disconnected from either the engine or the servo or if they are cracked and leaking, the servo will not operate.

To remove servo:

1. Thoroughly clean the master cylinder and the brake servo. Then remove the master cylinder as described in **4.2 Removing and Installing Master Cylinder.**
2. Disconnect the vacuum hoses from the brake servo.
3. Remove the cover plate that is beneath the pedal cluster. Then disconnect the air hose from the brake servo's air connection.
4. Loosen the locknut on the connecting rod (Fig. 4-9).
5. Remove the cotter pin from the clevis pin. Remove the clevis pin and disconnect the connecting rod clevis from the pedal lever.

© 1974 VWoA—2775

Fig. 4-9. Connecting rod removal. Loosen locknut at right-hand arrow. Disconnect the clevis from the pedal lever, then unscrew the connecting rod from the servo as indicated by left-hand arrow.

6. Unscrew the connecting rod from the brake servo as shown in Fig. 4-9.
7. Remove the four nuts that hold the brake servo on the bracket, then pull the servo out of the bracket to the rear.

   CAUTION —
   Prior to installation, it is necessary to replace the filter and damping ring, the rubber boot, and the sealing ring inside the air connection as described under the heading **Brake Servo Repairs.** *These replacements are required to ensure that the brake servo will continue to give satisfactory service after it is installed.*

Brake Servo Repairs

Brake Servo repairs are limited to the replacement of the filter and damping ring, the rubber boot, and the sealing ring inside the air connection. These parts should be replaced any time the brake servo is removed or when the brake servo's efficiency has been affected.

The vacuum powered brake servo is shown disassembled in Fig. 4-10. Replacements are available only for those parts mentioned above and no further disassembly should be attempted.

You must remove the vacuum powered brake servo from the vehicle and clean its outside thoroughly before disassembly. Do not use solvents to clean any of the internal parts. Doing so could damage the diaphragm or seals.

To disassemble:

1. Remove the retaining ring from the rubber boot.

Fig. 4-10. The vacuum powered brake servo disassembled. Notice also the O-ring between the vacuum chamber of the servo and the master cylinder. This part should also be replaced whenever the master cylinder is removed and installed.

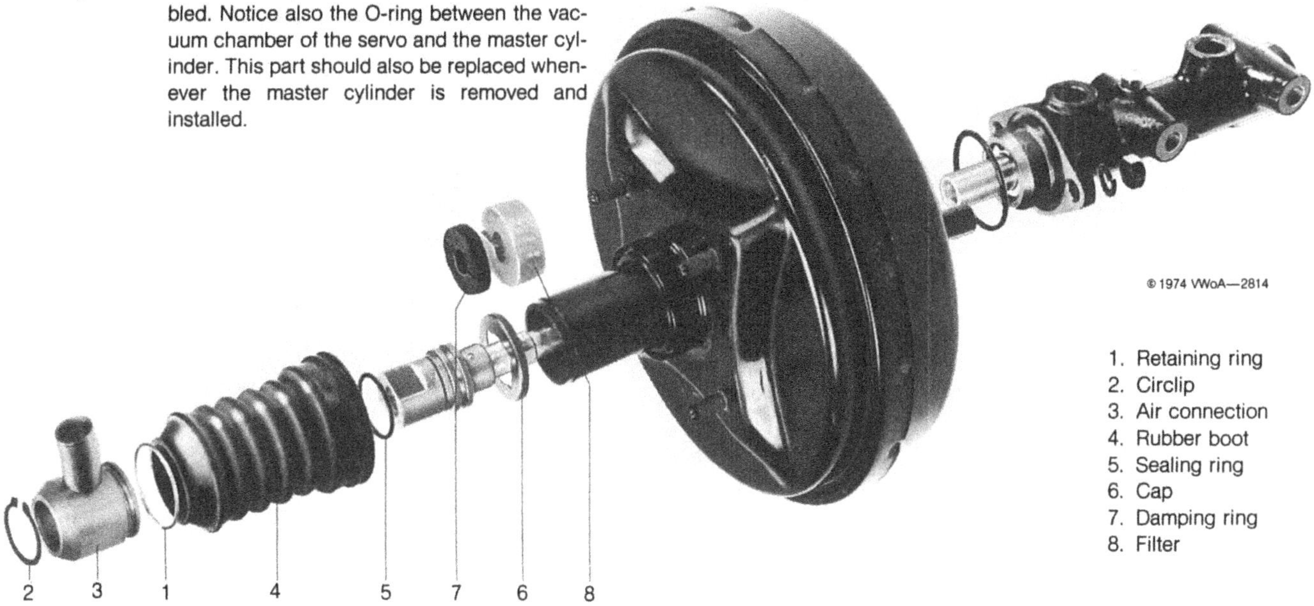

1. Retaining ring
2. Circlip
3. Air connection
4. Rubber boot
5. Sealing ring
6. Cap
7. Damping ring
8. Filter

2. Remove the circlip. Then pull the air connection off the pushrod connection.
3. Pull the rubber boot off its seat on the vacuum chamber.
4. Using a screwdriver, carefully pry the cap off the valve housing on the vacuum chamber. Then pull the damping ring and the filter out of the valve housing.

Installation is the reverse of removal. Make certain that the slits in the new filter and damping ring are positioned 180° apart before you install the cap.

To install brake servo:

1. Install the brake servo on its bracket and torque the nuts to a maximum of 1.3 mkg (9.4 ft. lb.).

> ***WARNING —***
> *Overtightening the mounting nuts can crack the vacuum chamber or pull off the mounting studs. Brake servos that are thus damaged are unsafe and must be replaced.*

2. Using a new O-ring, install the master cylinder as described in **4.2 Removing and Installing Master Cylinder**. Torque the mounting nuts to a maximum of 1.3 mkg (9.4 ft. lb.).
3. Screw the connecting rod into the brake servo. If necessary, make fine adjustments by turning the connection on the servo until the clevis pin can easily be inserted as you install the connecting rod on the pedal lever. See **4.1 Pushrod Adjustment.**
4. Install a new cotter pin in the clevis pin. Tighten the locknut for the connecting rod.
5. Bleed the brakes as described in **9. Bleeding Brakes.**

Testing the Vacuum Check Valve

There is a check valve installed in the vacuum line from the engine to the brake servo. The purpose of this valve is to prevent an engine backfire from producing pressure rather than vacuum in the brake servo vacuum chamber.

To test the vacuum check valve, remove it from the vacuum line. Blow into the valve in the direction indicated by the arrow. The valve should lift from its seat and permit pressure to escape from the top. The valve must seal if you blow into its opposite end. Install the valve with the arrow toward the brake servo.

5. REAR BRAKE PRESSURE REGULATOR

On 1971 and later models—those with front disc brakes and a vacuum powered brake servo—there is a pressure regulator in the rear brake circuit that prevents the rear brakes from locking when they are used to their maximum. This ensures that braking effort is proportioned correctly between the front and rear wheels.

5.1 Testing Brake Pressure Regulator

No repairs are possible to the rear brake pressure regulator. If testing shows the pressure regulator to be faulty, it must be replaced.

8

To test pressure regulator:

1. Install a hydraulic pressure gauge that has a range of 0 to 2300 psi (0 to 160 kg/cm²) in the front brake circuit as shown in Fig. 5-1.

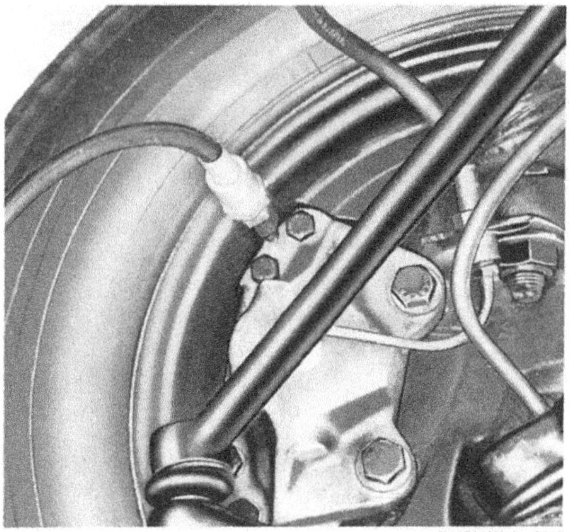

© 1974 VWoA—2810

Fig. 5-1. Pressure gauge hose installed in place of the bleeder valve on the front brake caliper.

2. Install a hydraulic pressure gauge that has the same range in the rear brake circuit as shown in Fig. 5-2.

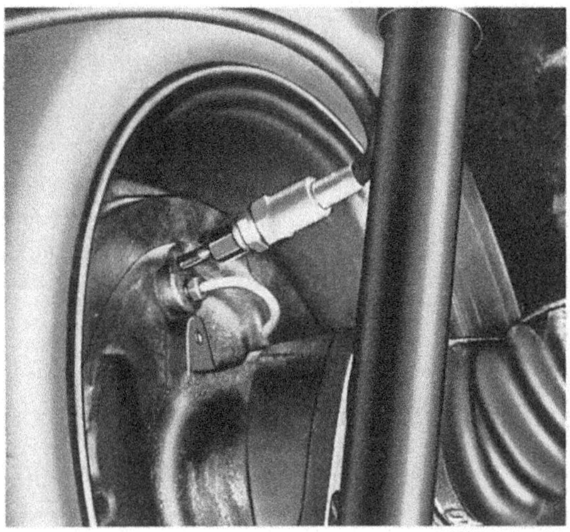

© 1974 VWoA—2811

Fig. 5-2. Pressure gauge hose installed in place of the bleeder valve on the rear wheel cylinder.

NOTE —
If the hoses on the gauges are at least 800 mm (about 3 ft.) long, you can position the gauges as shown in Fig. 5-3. Otherwise, you will need someone under the vehicle to observe the gauges.

© 1974 VWoA—2809

Fig. 5-3. Gauge positioned so that it can be seen from the driver's seat.

3. Bleed both hoses and the pressure gauges via the bleeder valves on the pressure gauges. (See **9. Bleeding Brakes**).
4. Firmly depress the brake pedal several times so that you have a pressure of 1420 psi (100 kg/cm²) in both the front and rear brake circuits.
5. Release the brakes. Then remove both mounting bolts from the pressure regulator.
6. Apply enough pressure to the brake pedal so that both gauges indicate a pressure of 710 psi (50 kg/cm²).
7. While under this pressure, and without producing permanent bending in the brake lines, tilt the front of the pressure regulator 30° downward from its installed position (Fig. 5-4).

© 1974 VWoA—2813

Fig. 5-4. Pressure regulator angled downward from its mounted position. Alpha equals 30°.

8. With the pressure regulator still angled, increase the load on the brake pedal until the front brake circuit pressure reaches 1420 psi (100 kg/cm²). At this time there should be a pressure of 786 to 929 psi (55 to 65 kg/cm²) in the rear brake circuit.

WARNING —
If the pressure in the rear brake circuit is not within the correct tolerance range, the regulator is unsafe for use and must be replaced as described in **5.2 Removing and Installing Brake Pressure Regulator.**

9. If the regulator is in satisfactory condition, return it to its installed position. Install the mounting bolts and torque them to 1.5 mkg (11 ft. lb.).
10. Disconnect the pressure gauges. Then bleed the brake system as described in **9. Bleeding Brakes.**

5.2 Removing and Installing Brake Pressure Regulator

Replace the brake lines along with the rear brake pressure regulator if the unions are locked by corrosion.

To remove:

1. Disconnect the input and output lines from the pressure regulator. Then remove the mounting screws (Fig. 5-5).

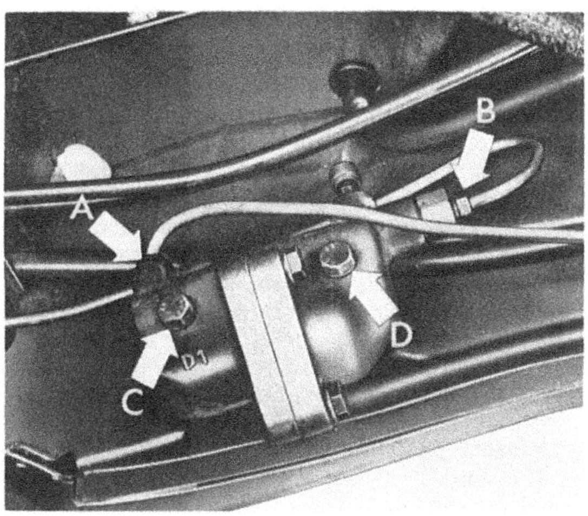

© 1974 VWoA—2808

Fig. 5-5. Brake pressure regulator removal. The input line is at arrow **A**, the output line at arrow **B**. The mounting bolts are indicated by arrows **C** and **D**.

NOTE —
The bleeder valve on the brake pressure regulator and the adaptor between the union and the regulator were discontinued early in 1973. On these late models, the outlet line is installed as shown in Fig. 5-6.

© 1974 VWoA—2817

Fig. 5-6. Late pressure regulator. Output line is connected at Arrow **B** without an adaptor. Arrow **A** indicates front of vehicle.

2. Remove the pressure regulator from the vehicle, being careful not to lose the spacers that are installed between the frame and the regulator.
3. On vehicles manufactured prior to early 1973, remove the adaptor from the output connection.

To install:

1. Install the brake pressure regulator and spacers on the frame sidemember so that the output connection is toward the front of the vehicle. Torque the mounting bolts to 1.5 mkg (11 ft. lb.).

CAUTION —
On replacement pressure regulators, the tapped holes for the brake line unions are plugged by plastic screws. To avoid the entry of dirt, remove these screws only just before installing the unions.

2. On early-type pressure regulators, install the output adaptor and torque it to 2.0 to 2.2 mkg (14 to 16 ft. lb.).
3. Install the input and output lines. Torque the unions to 1.5 to 2.0 mkg (11 to 14 ft. lb.).
4. Bleed the rear brake circuit as described in **9. Bleeding Brakes.**

8

6. BRAKE LINES AND HOSES

The brake lines are steel tubes mounted on the car's frame. They carry brake fluid from the master cylinder to the flexible brake hoses that serve the wheel cylinders.

The brake lines are so routed that they are not exposed to moisture and to the hazard of flying stones. The steel clips that secure the lines to the frame at short intervals prevent vibration and chafing that might weaken the tubing.

6.1 Removing and Installing Brake Lines

The brake lines should be inspected regularly, certainly whenever there is brake trouble or the brakes are being serviced. Look for signs of corrosion, leaks around the unions, leaks in the lines themselves, and dents or cracks that may soon cause trouble.

Replacement lines can be obtained from your Authorized VW Dealer. The unions are factory-installed on the replacement lines, and the lines themselves are preformed to the correct shape for immediate installation.

To remove brake line:

1. Unscrew the unions on the line ends (Fig. 6-1).
2. Remove the spring steel clips that hold the line to the frame.
3. Remove the brake line from the vehicle.

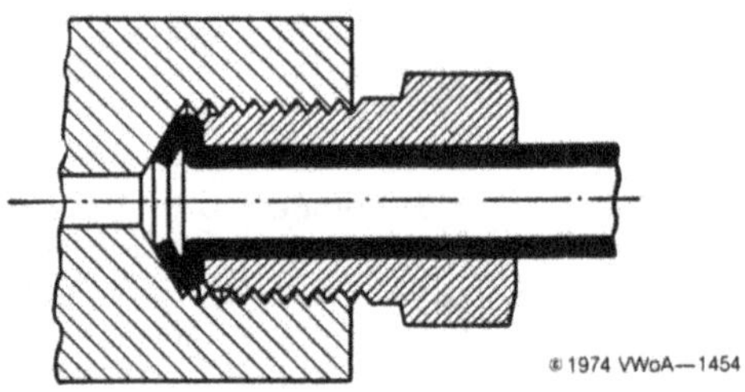

Fig. 6-1. Cross section of brake line union. Notice the double flare that holds the union on the tubing.

To install brake lines:

1. Route the new line so that it follows the routing of the old line.
2. Lubricate the flared ends of the brake lines with brake fluid, then insert the unions and torque them to 1.5 to 2.0 mkg (11 to 14 ft. lb.).

NOTE —
Use a properly fitting wrench to avoid rounding off the union.

3. Carefully install the clips to hold the new line.
4. Bleed the brakes as described in **9. Bleeding Brakes.**

WARNING —
When installing brake lines, be very careful not to dent, flatten, or bend the tubing enough to collapse it. The resulting restriction can upset brake balance and will create stress points in the tubing that may later cause it to crack. Never attempt to straighten a bent or dented brake line.

6.2 Removing and Installing Brake Hoses

Being flexible, brake hoses are much more subject to wear than brake lines. The hoses should, therefore, be inspected very carefully every time routine maintenance is being carried out.

To remove brake hose:

1. Remove the road wheel.
2. Unscrew the union that holds the hose to the brake line.
3. Remove the spring steel hose clip from the bracket on the frame or axle.
4. Pull the hose off the line and plug the line with a new brake bleeder dust cap.
5. Unscrew the hose from the brake caliper or wheel cylinder.

To install brake hose:

1. Obtain a new hose of the correct length.

WARNING —
If the hose is too long, it may rub the wheel or moving suspension parts. If too short, it could break when drawn tight by wheel travel or steering movements. In either case, partial brake failure could result. Hoses must never be painted, and they can be damaged by grease, oil, gasoline, or kerosene. Brake hoses that bulge or appear oil-soaked or cracked must be replaced immediately.

2. Install the hose, following the removal steps in reverse. The hose must hang down and be free of twists.
3. Torque the hose ends to 1.5 mkg (11 ft. lb.) in disc brake calipers and to 1.5 to 2.0 mkg (11 to 14 ft. lb.) in other locations.
4. Check the hose position and the routing in all steering and suspension travel positions.

7. DISC BRAKES

(1971 and later models only)

The principal parts of disc brake assembly are shown in Fig. 7-1. Notice that the brake disc is separate from the front wheel hub.

Fig. 7-1. Principal parts of disc brake assembly.

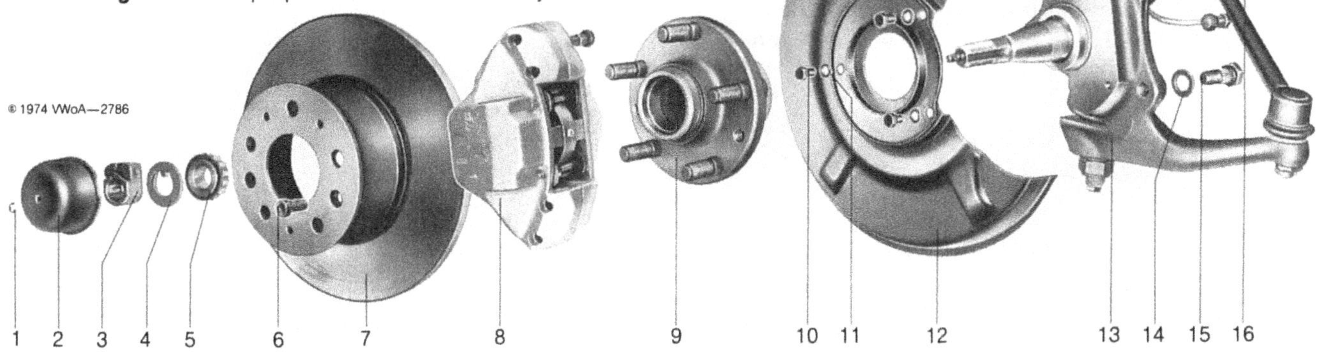

1. E-clip
2. Dust cap
3. Clamp nut
4. Thrust washer
5. Bearing race (2)
6. Socket head screw (2)
7. Brake disc
8. Brake caliper
9. Front wheel hub
10. Bolt (3)
11. Lock washer
12. Splash shield
13. Steering knuckle
14. Spring washer (2)
15. Locating bolt
16. Bolt

The brake caliper is mounted behind the steering knuckle's stub axle. The caliper consists of inner and outer housings, each holding a hydraulic piston for one side of the brake disc. The pistons cause the linings—called pads—to squeeze the disc during braking. Four bolts join the two halves of the housing. Two additional bolts secure the caliper assembly to the steering knuckle.

Three types of brake calipers have been used since the introduction of front disc brakes on the 1971 models. The first two types (before 1973) are identical, except that, beginning in March 1971, a second bleeder valve was added at the bottom of the inner housing (Fig. 7-2).

The 1973 and later models have the larger brake caliper shown in Fig. 7-3. This brake caliper has pads 14 mm ($\frac{35}{64}$ in.) thick rather than 10 mm ($\frac{25}{64}$ in.) thick as in the earlier caliper. The brake disc, steering knuckle, and splash shield were modified accordingly. The pistons of the early calipers are 29 mm ($1\frac{9}{64}$ in.) long. The pistons for the late calipers are 33 mm ($1\frac{19}{64}$ in.) long. The two piston types must not be interchanged.

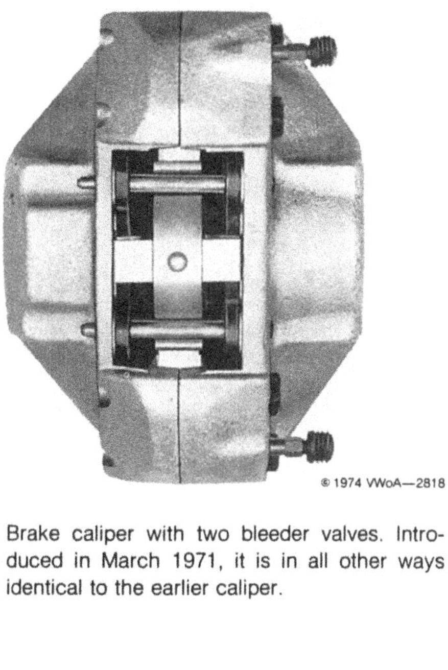

Fig. 7-2. Brake caliper with two bleeder valves. Introduced in March 1971, it is in all other ways identical to the earlier caliper.

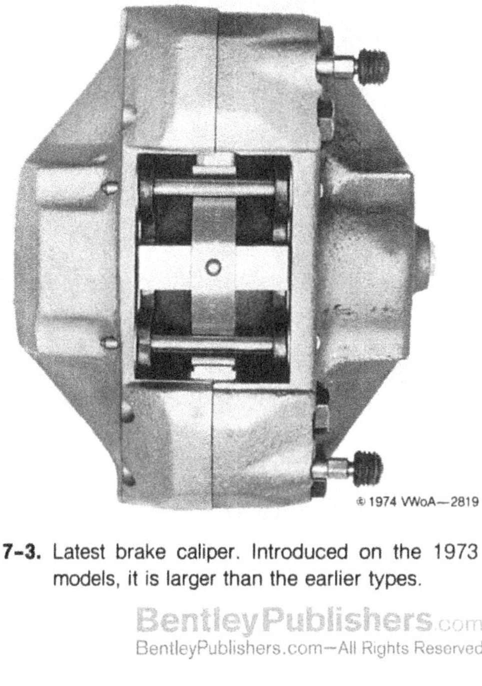

Fig. 7-3. Latest brake caliper. Introduced on the 1973 models, it is larger than the earlier types.

8

7.1 Removing and Installing Brake Pads

The procedure for replacing pads on both the early and late caliper is the same. On both types, the pads should be replaced when the friction material has worn to a remaining thickness of 2.00 mm (.080 in.).

To remove pads:

1. Remove the front wheel. Then drive out the friction pad retaining pins and take out the pad spreader spring (Fig. 7-4).

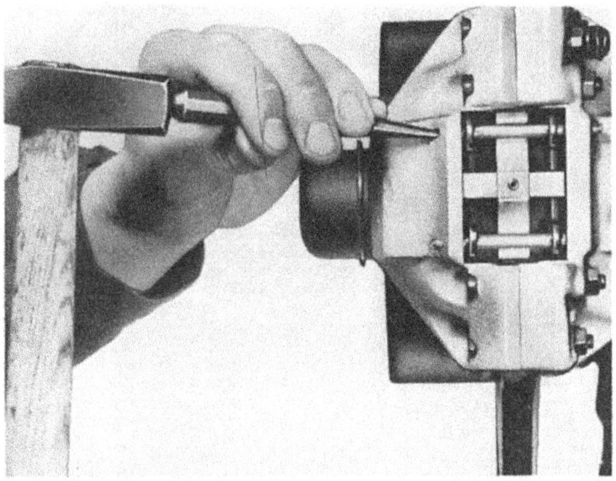

© 1974 VWoA—2249

Fig. 7-4. Driving out pad retaining pin. The spreader spring is the cross-shaped piece behind the pins.

2. Pull the pads out of the caliper. The extractor shown in Fig. 7-5 makes the job easier.

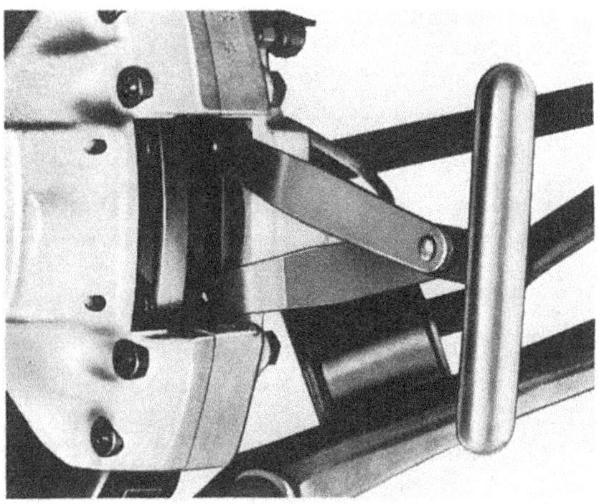

© 1974 VWoA—2250

Fig. 7-5. Pad being pulled out of caliper with extractor.

NOTE —
If the discs are deeply scored, it may be necessary to press the pistons slightly into the caliper to free the pad.

WARNING —
If the friction pads are to be reused, mark each pad and its original position in the caliper. Changing the location of used pads will result in uneven braking.

It is important that certain preparatory steps be carried out prior to actual installation of the pads. First, make sure there are no hard accumulations of dirt and rust on the pad sliding surfaces inside the caliper. If there are, the pads may stick and cause dragging brakes or the car may tend to pull to one side. It is also necessary to note the following points concerning the pads:

1. If the pads are oily, have deep cracks, or are detached from their metal plate, they must be replaced.
2. If the pads are acceptable for reuse, remove any dirt from the radial groove.
3. Always replace all four pads at the same time, even if only one is faulty.

To install pads:

1. Push both pistons into the caliper with a piston retaining device as shown in Fig. 7-6.

© 1974 VWoA—1410

Fig. 7-6. Pistons being pushed into caliper bores.

NOTE —
As you push in the pistons, brake fluid will be forced back into the reservoir. So first remove some fluid to prevent the reservoir from overflowing.

WARNING —
Do not start a siphon with your mouth or spill fluid on the car. Brake fluid is both poisonous and damaging to paint.

2. Remove the piston retaining plates. Scrape clean the pad seating and sliding surfaces in the caliper, then blow out the dirt with compressed air.
3. Check the rubber dust seal shown in Fig. 7-7. It must not be cracked, hard, or swollen. If necessary, remove the caliper from the car as described in **7.2 Removing and Installing Brake Caliper**. Then replace the seals as described in **7.3 Brake Caliper Repair.**

© 1974 VWoA—1411

Fig. 7-7. Rubber dust seal that should be checked before installing the pads.

4. Make sure the piston is positioned to accept the retaining plate as shown in Fig. 7-8.

© 1974 VWoA—2790

Fig. 7-8. Retaining plate that determines piston position. The two pressed-out projections (lower arrow) must engage the cutout in the piston so that the plate will lie flush with the piston face. Upper arrow indicates direction of forward wheel rotation.

5. If necessary, correct the position of the piston using piston rotating pliers.
6. Check the brake disc for wear as described in **7.4 Checking Brake Disc**. If necessary, replace the disc or machine it as described in **7.7 Reconditioning Brake Disc.**
7. Insert the pads in the brake caliper.

> ***WARNING —***
> *Install used pads in the positions you marked for them during removal. If the pads are in the wrong locations, or bind in the caliper, uneven braking will result.*

8. Insert the lower pad retaining pin and install a new spreader spring.

> **NOTE —**
> The split clamping bushing must move freely on the pin. Replace corroded pins.

9. While depressing the top of the spreader spring with your thumb, insert the upper pad retaining pin.

> ***WARNING —***
> *Do not grease the retaining pins. Heat produced by braking can melt the lubricant and cause it to flow onto the pads or disc.*

10. Using a hammer only, drive the retaining pins all the way in so that they will be locked in place by the clamping bushings.

> ***CAUTION —***
> *Do not drive the retaining pin into the caliper with a punch. Doing so can cause the pin shoulder to be sheared off by the clamping bushing.*

11. Install the front wheel. Torque the nuts to 13 mkg (94 ft. lb.).
12. To ensure that the pads are seated against the disc, depress the brake pedal several times while the car is stationary.
13. Check the level of the brake fluid in the reservoir. If necessary, add fresh fluid.

> ***CAUTION —***
> *Use only new, unused brake fluid that meets SAE recommendation J 1703 and conforms to Motor Vehicle Safety Standard 116.*

8

7.2 Removing and Installing Brake Caliper

Never attempt to remove a brake caliper until it has cooled. If the brake caliper is to be completely removed from the vehicle, unscrew the brake hose from the caliper and seal it with a clean bleeder valve dust cap. Support the caliper as it is being unbolted.

If the caliper is only being partially removed, for example to obtain clearance for removal of the brake disc, the brake line need not be disconnected. Simply hang the caliper by a stiff wire hook from one of the steering tie rods. Doing so will eliminate the need for bleeding the brakes, which is necessary if the hose is removed. Never allow the caliper to hang by its hose.

When installing the caliper, make sure that the pressed-out projections on the piston retaining plate are at the bottom. First install the locating (lower) bolt and then the upper bolt. Torque both bolts to 10.0 mkg (72 ft. lb.) on 1971 and 1972 models. Torque the bolts to 16 mkg (116 ft. lb.) on 1973 and later models.

To ensure that the pads are seated against the disc, depress the brake pedal several times while the car is stationary. If the hose has been removed and installed, bleed the brakes as described in **9. Bleeding Brakes.**

NOTE—
If a replacement caliper is being installed, install the pads and related parts as described in **7.1 Removing and Installing Brake Pads.**

7.3 Brake Caliper Repair

The pistons can be taken out of the caliper without separating the two halves of the caliper housing. The two halves of the caliper housing should only be separated if there are signs of leakage. In such cases, the O-rings should be replaced.

Fig. 7-9 is an exploded view that shows the components of a brake caliper. When disassembling the caliper, be very careful not to damage the piston bores, pistons, rubber seals, and the mating surfaces between the caliper halves. The caliper should never be disassembled needlessly.

CAUTION—
If you lack the skills, tools, or a clean workshop for servicing the brake calipers, we suggest you leave such repairs to an Authorized VW Dealer or other qualified shop. We especially urge you to consult your Authorized VW Dealer before attempting repairs on a car still covered by the new-car warranty.

1. Brake disc
2. Caliper outer housing
3. Spreader spring
4. Friction pad (2)
5. Piston retaining plate (2)
6. Rubber dust seal (2)
7. Piston (2)
8. Piston seal (2)
9. Caliper inner housing
10. Bleeder valve (2 after March 1971)
11. O-ring (2)
12. Pad retaining pin
13. Bleeder valve dust cap
14. Housing connecting bolts (4)

Fig. 7-9. Exploded view of the brake caliper along with the brake disc. Note that there are two bleeder valves installed beginning in March 1971. Also, the latest calipers are shaped somewhat differently, despite having similar working parts.

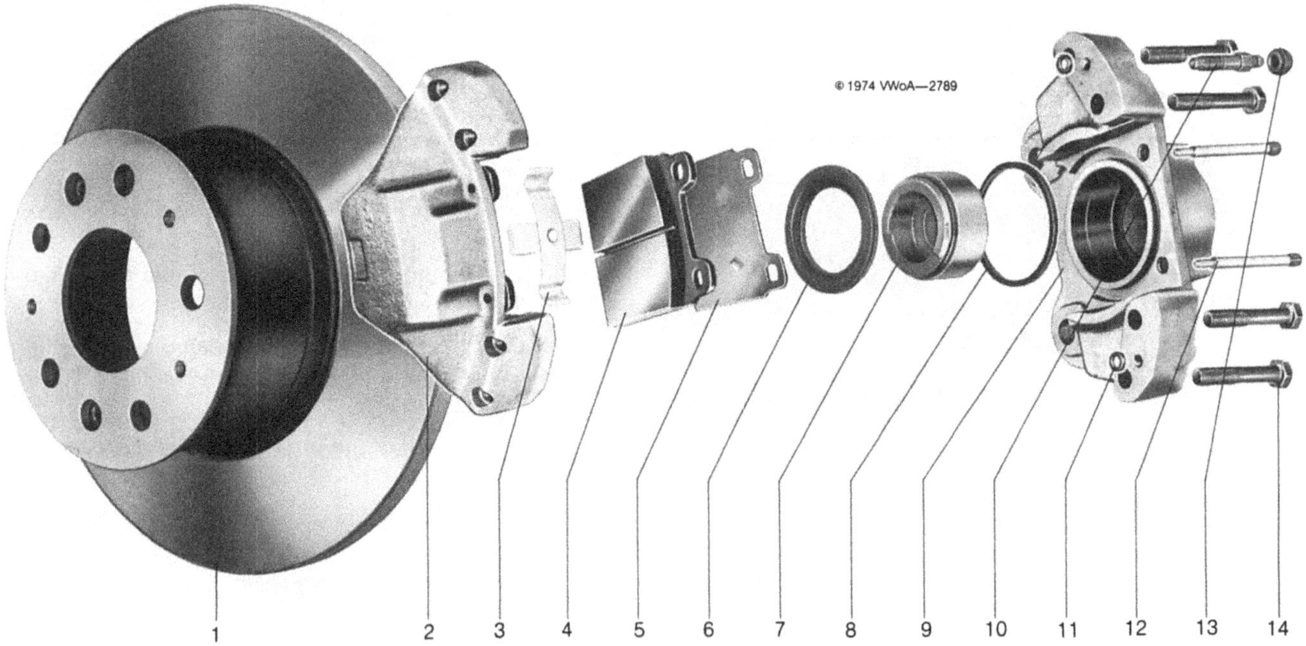

To replace piston seals:

1. Remove the brake caliper from the vehicle and clamp its flange in a vise.

CAUTION —
Install soft jaws in the vise or clamp the flange between two hardwood blocks. Otherwise, damage to the mounting flange could result that would upset the caliper's alignment with the brake disc.

2. Remove the pads and the piston retaining plates. Mark the original positions of the pads if they are to be reused.
3. Pry off the rubber dust seals.

NOTE —
If the seal is simply being lifted to check for fluid leakage, use a plastic rod to avoid puncturing the seal. Use a sharp tool only if the seal is to be replaced. The rubber dust seals and piston seals should be replaced whenever the pistons are removed. All necessary parts are contained in the VW repair kit.

4. Using compressed air as shown in Fig. 7-10, remove one piston from the brake caliper while the second piston is held by the piston retaining device.

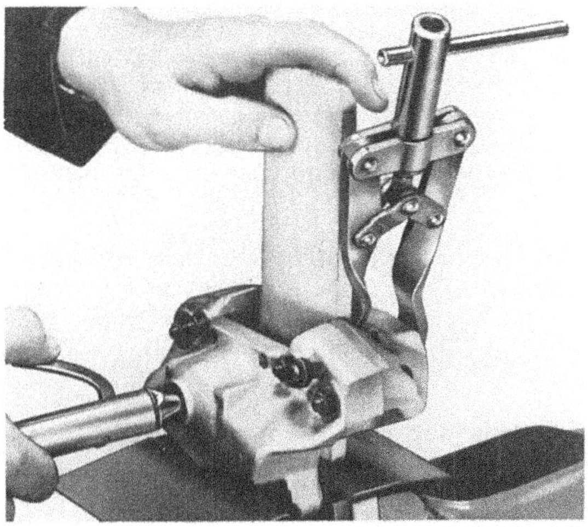

© 1974 VWoA—1465

Fig. 7-10. Removing piston with compressed air. Place a 5-mm to 10-mm (¼-in. to ⅜-in.) hardwood board in the housing to catch the piston. Otherwise, the piston may be damaged.

NOTE —
The cylinders must be serviced one at a time because, with one piston removed, no air pressure can be built up in the brake caliper to expel the other piston.

5. Remove the piston seal from the cylinder with a plastic rod as shown in Fig. 7-11.

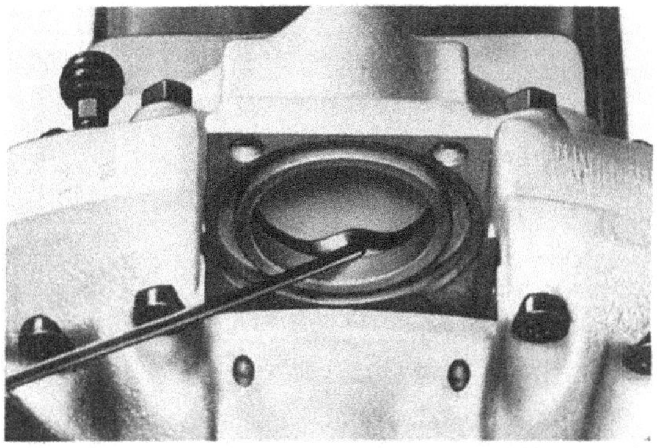

© 1974 VWoA—1466

Fig. 7-11. Piston seal being removed from the groove in the cylinder bore.

To install pistons:

1. Clean all parts with brake fluid only.
2. Check the parts for wear. If the cylinder is damaged, replace the entire brake caliper. Do not hone.
3. Apply a thin coat of VW brake cylinder paste (see **9.5 Brake Cylinder Paste**) to the piston and the new piston seal. Install the seal. Then install the piston with the installing clamp shown in Fig. 7-12. Next, using the installing clamp shown in Fig. 7-12, press the piston in only far enough so that its head is past the seal.

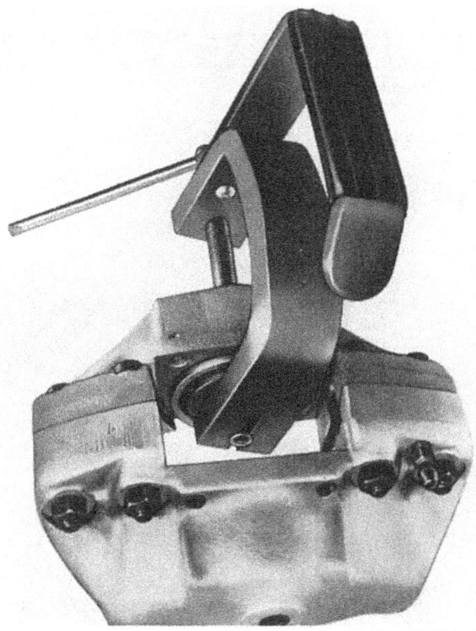

© 1974 VWoA—1467

Fig. 7-12. Piston being pressed past seal with an installing clamp. Do not use the piston retaining device, as it may let the piston tilt, thereby damaging the seal, piston, or caliper housing.

8

4. Lightly rub VW brake caliper paste into the inside of the rubber dust seal. Then install it on the partially pressed-in piston as shown in Fig. 7-13.

© 1974 VWoA—2791

Fig. 7-13. Rubber dust seal partially installed on piston so that its inside lip is seated in groove around piston. The piston has not yet been pressed fully into the cylinder.

5. Using a clamp and tool VW 442, carefully press the seal and piston fully into the caliper housing (Fig. 7-14).

CAUTION —
The rubber dust seal cannot be pushed in far enough by hand. If it is not installed as described here, the seal may not seat properly and will allow dirt and moisture to enter the cylinder.

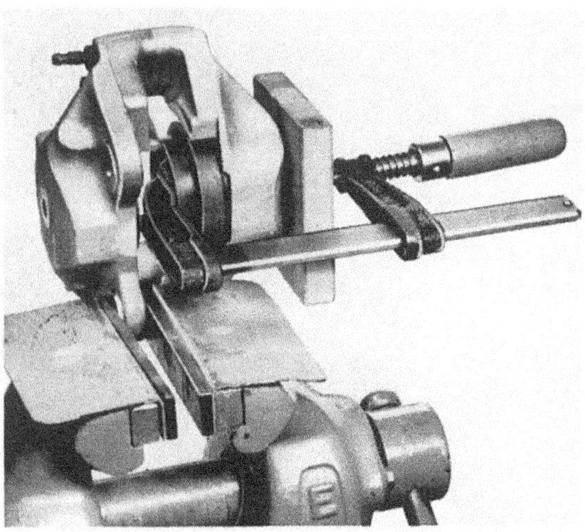

© 1974 VWoA—2792

Fig. 7-14. Rubber dust seal and piston being pressed into caliper housing with tool VW 442 and clamp.

6. Install the piston retaining plate as shown in Fig. 7-15. If necessary, turn the piston in its bore with piston rotating pliers.

CAUTION —
On late calipers with two bleeder valves, it is possible to install accidentally a caliper upside down on the opposite side of the vehicle. The pressed-out projections in the piston retaining plate must be at the bottom when the caliper is installed.

© 1974 VWoA—2793

Fig. 7-15. Retaining plate that determines piston position. The two pressed-out projections (arrow **A**) must engage the cutout in the piston so that the plate will lie flush with the piston face. Arrow **B** indicates direction of forward wheel rotation.

7. Repeat the procedure on the other piston.

The two halves of the brake caliper housing should be separated only if there are leaks in the joint between them. In this case, disassemble the caliper and replace the O-rings surrounding the fluid channels.

To disassemble:

1. Remove the four housing connecting bolts that join the two halves of the housing.
2. Remove the outer housing half.
3. Remove the old O-rings from the chamfers around the fluid channels.

To assemble:

1. Clean the mating surfaces with brake fluid only.
2. Install the two new fluid channel O-rings.

3. Using new connecting bolts, loosely join the two housing halves. Check their alignment.
4. Tighten the two inner bolts and then the two outer bolts to 1 mkg (7 ft. lb.). Again check the alignment and, if necessary, correct it.
5. When the housing halves are properly aligned, torque the two inner bolts and then the two outer bolts to 3.5 mkg (25 ft. lb.).

7.4 Checking Brake Disc

The brake discs should be checked for wear each time repair work is done on the front brakes. Replace the discs if they are worn, scored with sharp ridges, or cracked. Also replace brake discs that have worn down, or have been machined, to a thickness of 11.50 mm (.453 in.) or less.

Check the discs for excessive runout. If a low-speed front end shimmy goes away when you release the brakes, excessive brake disc runout is probably the cause of the shimmy.

To measure runout:

1. Remove the front wheel.
2. Adjust the front wheel bearing axial play to 0.03 to 0.12 mm (.001 to .005 in.) as described in **FRONT AXLE.**
3. Install the measuring appliance on the pad retaining pins in the caliper as shown in Fig. 7-16. Tighten the wing nut to hold it solidly in position.

© 1974 VWoA—1461

Fig. 7-16. Brake disc runout being measured. A dial indicator on a magnetic base can also be used.

4. Install the dial indicator on the appliance with its measuring pin against the disc surface. Then zero the gauge.
5. Slowly hand-turn the brake disc to check the runout. The maximum allowable runout is 0.10 mm (.004 in.).
6. If the brake disc runout exceeds specifications, replace the brake disc.

7.5 Removing and Installing Brake Disc

The brake caliper must be removed from the steering knuckle before the brake disc can be taken off. Hang the caliper from the steering tie rod with a stiff wire hook. Leaving the brake hose attached will save you the job of bleeding the brakes.

To remove the brake disc from the front wheel hub, remove the two socket head screws indicated in Fig. 7-17. Because the brake disc is separate from the front wheel hub, it is unnecessary to remove the hub and wheel bearings from the steering knuckle's stub axle.

© 1974 VWoA—2787

Fig. 7-17. Socket head screws (arrows) that hold brake disc to front wheel hub.

8

CAUTION —
Normally, the disc can be easily pulled off the hub. However, you should not use a hammer to drive the disc off if it is rusted tight. Doing so could ruin the disc. Instead, install three M 8 × 40 bolts as indicated in Fig. 7-18. Evenly tighten the bolts about a quarter turn at a time until the disc is separated from the hub.

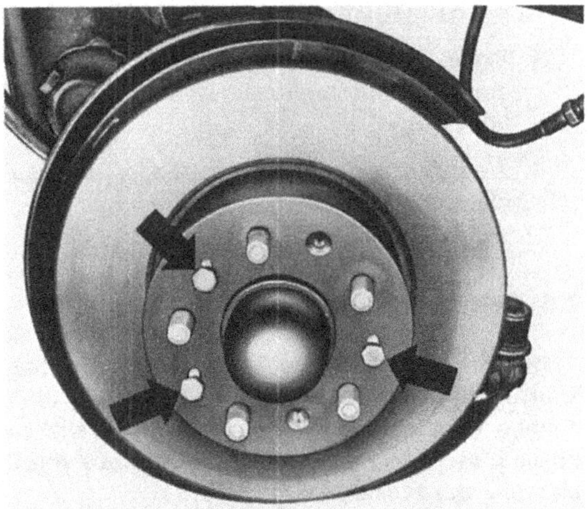

© 1974 VWoA—2788

Fig. 7-18. Bolts (arrows) installed in tapped holes in brake disc. Note that these are not the holes for the socket head screws.

Installation is the reverse of removal. Torque the socket head screws to 2.0 to 2.5 mkg (14 to 18 ft. lb.). Install the brake caliper as described in **7.2 Removing and Installing Brake Caliper**. Torque the caliper mounting bolts to 10.0 mkg (72 ft. lb.) on 1971 and 1972 models. Torque the bolts to 16 mkg (116 ft. lb.) on 1973 and later models. Install the front wheel and torque the nuts or bolts to 13 mkg (94 ft. lb.). Before driving, depress the brake pedal several times to ensure that the pads are seated against the disc.

7.6 Removing and Installing Brake Disc Splash Shield

The brake disc splash shield must be removed whenever the steering knuckle is replaced or if the splash shield itself is damaged. Over the years, minor changes have been made to the shape of the splash shield to accommodate different brake calipers and discs. If the splash shield is to be replaced, be sure to replace it with one that is suitable to the brake caliper on the vehicle.

To remove splash shield:

1. Remove the front wheel.
2. Detach the brake caliper from the steering knuckle and suspend it from the tie rod with a wire hook.
3. Remove the brake disc and the front wheel hub. For details, see the discussion of front wheel bearings given in **FRONT AXLE.**
4. Remove the three bolts that hold the splash shield on the steering knuckle.

To install splash shield:

1. Clean the mounting surface for the splash shield on the steering knuckle. Replace damaged splash shields. Do not attempt to repair them.
2. Install the splash shield. Torque the bolts to 1.0 mkg (7 ft. lb.).
3. Install the front wheel hub and adjust the wheel bearings (see **FRONT AXLE**). Axial play should be 0.03 to 0.12 mm (.001 to .005 in.).
4. Check the brake disc for wear, and install the disc if it is still serviceable. Torque the socket head screws to 2.0 to 2.5 mkg (14 to 18 ft. lb.).
5. Install the brake caliper as described in **7.2 Removing and Installing Brake Caliper**. Torque the mounting bolts to 10.0 mkg (72 ft. lb.) on 1971 and 1972 models; torque the bolts to 16 mkg (116 ft. lb.) on 1973 and later models.
6. Install the front wheel. Torque the nuts to 13 mkg (94 ft. lb.).
7. To seat the pads against the disc, depress the brake pedal several times while the car is stationary.

7.7 Reconditioning Brake Disc

The brake discs should be checked for wear whenever brake repairs are made. Replace discs that have worn below the minimum thickness, are scored with sharp ridges, or are cracked. If thickness is less than 11.50 mm (.453 in.), the disc must be replaced. Brake discs can be reconditioned by an Authorized VW Dealer or a qualified automotive machine shop if the following restrictions are observed:

1. The minimum allowable thickness after rework is 12.00 mm (.472 in.). New brake discs are between 12.80 and 13.00 mm (.504 and .512 in.) thick.

CAUTION —
Never rework brake discs to a lesser thickness. Doing so will allow the pistons to travel farther out in their cylinders. This may severely damage the calipers and pistons.

2. After reworking a brake disc, its thickness should not vary by more than 0.02 mm (.0008 in.) measured at several locations on the disc.
3. The brake disc must be reworked equally on both sides to prevent squeaking, chattering, or brake pedal pulsation.
4. The maximum allowable runout of reworked brake discs is 0.10 mm (.004 in.).

8. BRAKE DRUMS

Fig. 8-1 shows the parts of a rear brake. On 1968 through 1970 models, there are two return springs, and the rear wheel hub and brake drum are integral.

WARNING —
Do not re-use nuts, bolts, washers, circlips, cotter pins, etc. that are worn or deformed in normal use. Always use new parts.

Fig. 8-1. Components of a rear brake (1971 and later).

© 1974 VWoA—2798

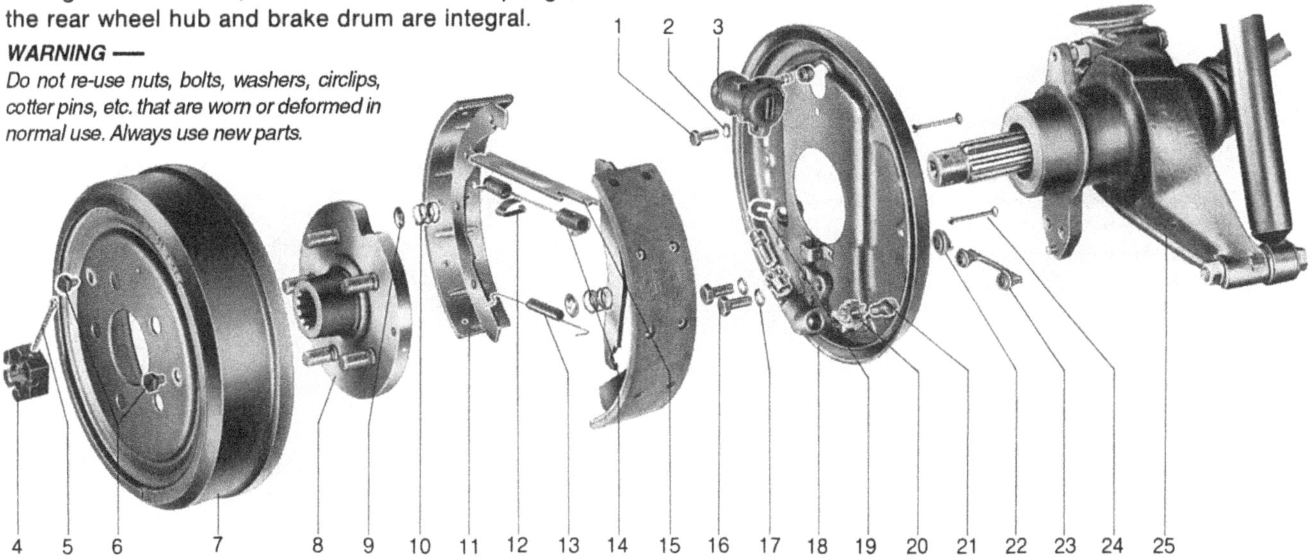

1. Bolt
2. Serrated washer
3. Wheel cylinder
4. Castellated nut
5. Cotter pin
6. Socket head screw (2)
7. Brake drum
8. Rear wheel hub (integral with brake drum on earlier models)
9. Spring cup (2)
10. Shoe retaining spring (2)
11. Brake shoe (one with brake lever)
12. Clip (modified from June 1972)
13. Return spring (2 on early models)
14. Return spring
15. Connecting link
16. Bolt (2)
17. Lock washer (2)
18. Brake backing plate
19. Brake cable clamp
20. Adjuster (2)
21. Adjuster screw (2)
22. Plug
23. Bridge plug
24. Shoe retaining pin
25. Wheel shaft and bearing housing

The front brake assembly used on 1968 through 1970 models is shown in Fig. 8-2. It has two single-piston wheel cylinders and two identical return springs.

© 1974 VWoA—2761

Fig. 8-2. Front drum brake assembly (drum removed).

8.1 Adjusting Drum Brakes

The clearance between the brake linings and the drum gradually increases due to normal wear. This change is indicated by increased pedal travel in applying the brakes. When pedal travel becomes excessive, the brake shoes must be adjusted to position the linings nearer the drums. These adjustments are made at the individual wheels. The disc brakes used on the front of 1971 and later vehicles require no adjustment.

To adjust:

1. Raise the car and fully release the parking brake.
2. Depress the brake pedal as far as it will go several times. This centers the brake shoes in the drums.

NOTE —
If the brakes are far out of adjustment, it may be necessary to recenter the shoes once or twice during the course of adjustments.

3. Remove the rubber plugs from the holes in the backing plate.

8

4. Using a screwdriver or an adjusting lever as shown in Fig. 8-3, turn the adjuster until a slight drag is noted when the wheel is turned by hand. Then back off the adjuster three or four clicks so that the wheel turns freely.

> ***CAUTION —***
> *On 1972 and later models, turn both of the rear wheel adjusting nuts alternately three clicks at a time. If the procedure for earlier models is followed, only one shoe will be adjusted and the wheel cylinder pistons and the brake shoes will not be centered in the drum. When both shoes drag as the wheel is hand-turned, back both adjusters off three or four clicks.*

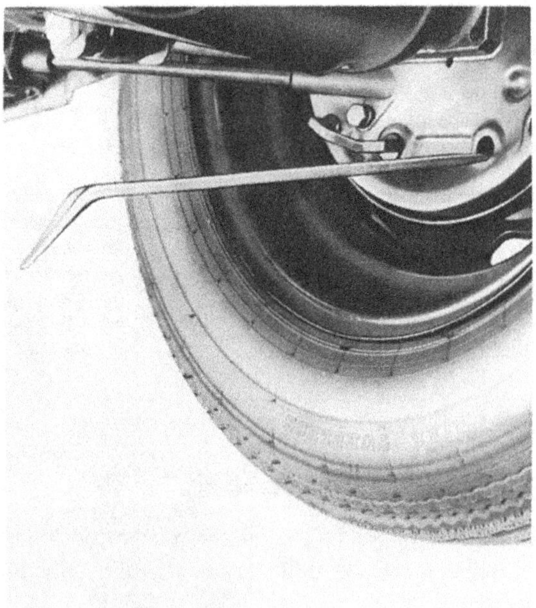

© 1974 VWoA—1652

Fig. 8-3. Lever in position to adjust drum brake.

> **NOTE —**
> There are no holes in the backing plates of vehicles through chassis No. 218 109 823. On these early models you must adjust the brakes through a hole in the face of the brake drum.

5. On 1971 and earlier models, repeat step 4 on the other adjuster on the same wheel's brake assembly.
6. Repeat the entire procedure on the other two or three wheels that have drum brakes.
7. Install the rubber plugs. Replace plugs that are cracked or that no longer fit tight.
8. Road-test the car to check the pedal travel.

8.2 Removing and Installing Brake Drums

Removal and installation of the front brake drums is described in conjunction with front wheel bearings in **FRONT AXLE**. The procedure given here is for the rear brake drums only.

To remove:

1. Fully release the parking brake. Then back off the adjusters slightly.
2. On 1971 and later models only, remove the rear wheel. Then remove the two socket head screws and take the brake drum off the rear wheel hub.

> **NOTE —**
> If the rear wheel hub must be removed from 1971 and later models, as it must when the brake backing plate is removed, leave the brake drum mounted on the hub and follow the additional instructions given here for removing the early brake drum.

3. Remove the cotter pin from the castellated nut on the rear wheel shaft. Remove the nut, then raise the car.

> ***WARNING —***
> *Loosen the wheel shaft nuts while the car is on the ground. The leverage needed for this job is enough to topple a car off the lift.*

4. Using a puller, as shown in Fig. 8-4, remove the brake drum or the brake drum and hub from the splines on the rear wheel shaft.

© 1974 VWoA—2801

Fig. 8-4. Puller, held by wheel lug nuts, being used to remove rear brake drum and wheel hub. Turn the puller screwshaft as indicated by the arrow.

To Install:

1. On 1971 and later models, slide the brake drum onto the rear wheel hub. Then install the socket head screws with a torque of 1.0 to 1.8 mkg (7 to 13 ft. lb.). Install the rear wheel, torquing the nuts to 13 mkg (94 ft. lb.). Then adjust the brake.
2. On 1968 through 1970 models—or when installing the rear brake drum together with the wheel hub on later vehicles—slide the brake drum or brake drum with wheel hub onto the rear wheel shaft.
3. Install the castellated nut finger-tight. Then loosely install the road wheel.
4. With the vehicle on the ground, torque the castellated nut to 35 mkg (253 ft. lb.) and install a new cotter pin. Torque the wheel lugs to 13 mkg (94 ft. lb.).

> ***WARNING —***
> *Do not re-use fasteners such as nuts, bolts, washers, self-locking nuts, circlips, cotter pins that are worn or deformed in normal use. Always replace.*

> ***CAUTION —***
> *The castellated nut must be torqued to specifications to avoid destructive spline wear.*

8.3 Removing and Installing Brake Shoes

Any brake lining that has become oil-soaked or saturated with brake fluid must be replaced. A lining must also be replaced if it is worn at any point to a thickness of 2.50 mm (.100 in.). The lining thickness can be checked by removing the rubber plugs from the brake backing plate.

Although the relative positions of the brake components are the same on all models, individual parts are different. So make certain that the correct parts are obtained for the particular vehicle model and year when making replacements. Bonded brake linings are now commonly used on VW cars instead of the riveted type seen in some of the illustrations.

To remove brake shoes:

1. Remove the road wheel and brake drum.
2. Check the wheel cylinder for sticking pistons by having someone slowly depress the brake pedal while you watch to see whether the brake shoes move out the same distance at a uniform rate.

> **NOTE —**
> Insert two screwdrivers behind the backing plate flange. Press them against the shoes to limit their travel. Also lift the wheel cylinder boot to check for fluid leakage. If the cylinder is sticking, leaking, or if the bleeder valve is rusted tight, rebuild or replace the cylinder.

3. On rear brakes only, remove the spring cups and shoe retaining springs. Press the cup in, rotate it 90°, then release it. If necessary, reach behind the backing plate and hold in the shoe retaining pin. (The parts named are shown earlier in Fig. 8-1.)
4. On rear brakes only, remove the shoe retaining pins.
5. On front drum brakes only, insert a screwdriver as shown in Fig 8-5. Then, being careful not to damage the spring steel adjusting nut detent, pry the brake shoes out of their slots in the adjusting screws.

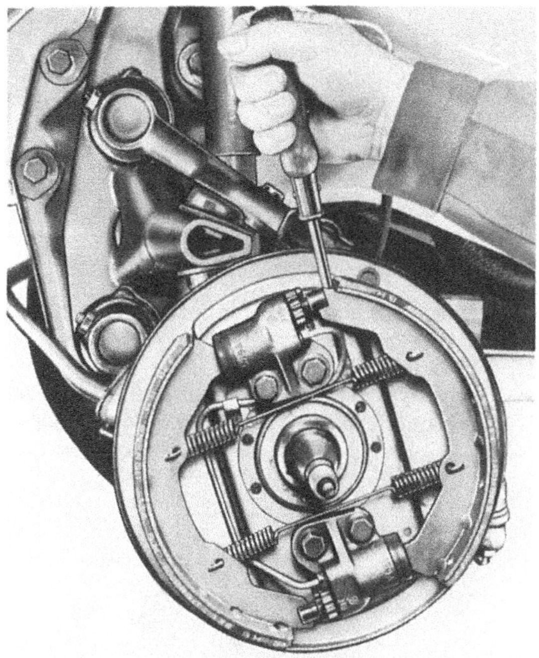

© 1974 VWoA—2782

Fig. 8-5. Front brake shoe being pried out of its slot in the adjusting screw.

6. On front drum brakes only, remove the shoes from the brake backing plate assembly, together with the return springs. Install wire clips over the wheel cylinder pistons to hold them in while the shoes are off.
7. On the rear brakes only, use brake spring pliers to remove the return springs.
8. On rear brakes only, unhook the parking brake cable from the brake lever on the rearmost shoe. Push the lever forward, then pull the cable eye down to unhook it.
9. On rear brakes only, remove the brake shoes. Then install a wire clip over the wheel cylinder pistons to hold them in while the shoes are off.

10. If the rear brake shoes are to be replaced, take off the spring clip (1968 through early 1972 models) or the horseshoe clip (from June 1972). Then remove the anchor pin and parking brake lever from the rearmost brake shoe.

To Install:

1. Inspect the brake drum and compare it to the specifications given in **8.7 Reconditioning Brake Drums**. Make sure that the same type linings are used at both wheels on the axle. Oversize linings are available to fit reconditioned drums on 1968 through 1970 models only.

> ***WARNING —***
> *Using linings of different size or composition on opposite sides of the car can cause dangerously uneven braking.*

2. Disassemble the adjusters and adjuster screws. Lightly coat the threads and the bearing surface on the adjuster with multipurpose grease.
3. Install the adjusters and screws. Fully back them off if new brake shoes are being installed.
4. On rear brakes only, install the brake lever for the parking brake cable as shown in Fig. 8-6.

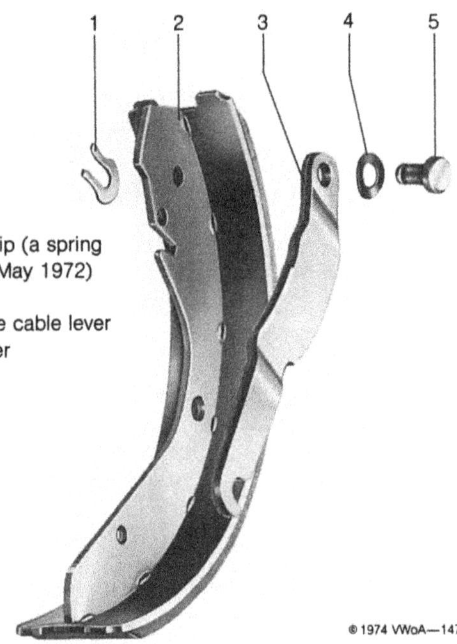

Fig. 8-6. Brake lever installation on rear brake shoe.

5. On rear brakes only, install the connecting link and the upper return spring. On front brakes, install both return springs.
6. On rear brakes only, remove the wire clip from the wheel cylinder pistons. Then install the brake shoes on the pistons and adjuster screws.
7. On rear brakes only, install the lower return spring(s). Then install the shoe retaining pins, the shoe retaining springs, and the spring cups. Hook the parking brake cable eye over the lever.
8. On the front brakes only, position the front brake shoe in the adjusting screw slot and in the upper wheel cylinder piston. Install the rear brake shoe in the lower wheel cylinder piston (Fig. 8-7).

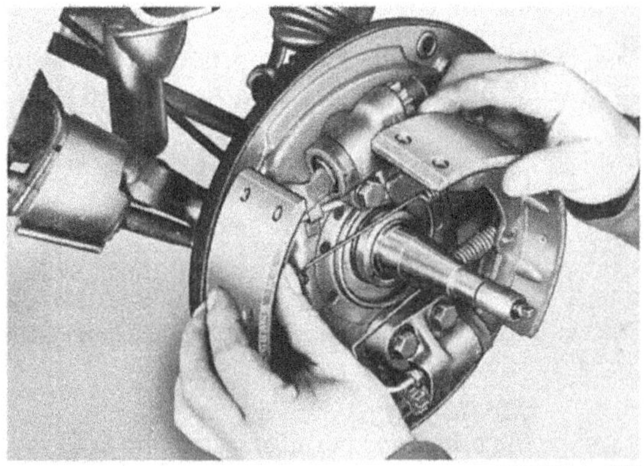

Fig. 8-7. Front brake shoes being installed. Note that the rear shoe has been inserted into the lower wheel cylinder piston.

9. On front brakes only, insert a screwdriver behind the top of the rear brake shoe as you did during removal. Pry the rear shoe rearward against spring tension until you can slip it into the slot in the upper adjuster screw.
10. Install the brake drum over the brake shoes.
11. On front wheel brakes, adjust the front wheel bearings as described in **FRONT AXLE**. Then loosely install the road wheel.
12. On rear wheel brakes through 1970 models—or on later vehicles if the wheel hub has been removed—install the castellated nut, but do not torque it. Loosely install the road wheel.
13. Adjust the brakes.
14. Lower the vehicle to the ground. Torque the wheel lugs to 13 mkg (94 ft. lb.). On rear wheels only, torque the castellated nut to 35 mkg (253 ft. lb.), then install a new cotter pin.

> ***WARNING —***
> *Tighten the wheel lugs and rear wheel shaft nuts with the car on the ground. The leverage needed for these jobs is enough to topple a car off the lift.*

8.4 Removing and Installing Wheel Cylinders

To remove a wheel cylinder, first carry out the procedure for removing the brake shoes as given in **8.3 Removing and Installing Brake Shoes**. Disconnect the brake hose from the rear of the brake assembly and seal it with a clean bleeder valve dust cap. Working behind the backing plate on the rear brakes, remove the wheel cylinder mounting bolt and take the wheel cylinder off to the front. On front brakes, disconnect the brake line that joins the two wheel cylinders. Then remove the two bolts that hold each wheel cylinder to the steering knuckle.

Installation is the reverse of removal. Torque the mounting bolt for the rear wheel cylinders to 2.0 to 3.0 mkg (14 to 22 ft. lb.); the mounting bolts for the front wheel cylinders to 5.5 to 6.0 mkg (40 to 43 ft. lb.). Torque the brake hose and, on front brakes, the connecting line unions to 1.5 to 2.0 mkg (11 to 14 ft. lb.). Carry out the procedure for installing brake shoes as given in **8.3 Removing and Installing Brake Shoes**. Then bleed the brakes as described in **9. BLEEDING BRAKES.**

8.5 Wheel Cylinder Repair

Because replacement wheel cylinders are inexpensive, it is usually more economical to replace them as a unit. However, repair kits are available. Other than a very light honing to remove tarnish or gummy deposits, no machine work should be done.

Many different wheel cylinders have been used on VW vehicles. So when you buy replacement parts, take the old cylinder with you for comparison. Fig. 8-8 shows the components of a rear wheel cylinder; Fig. 8-9 shows a front wheel cylinder.

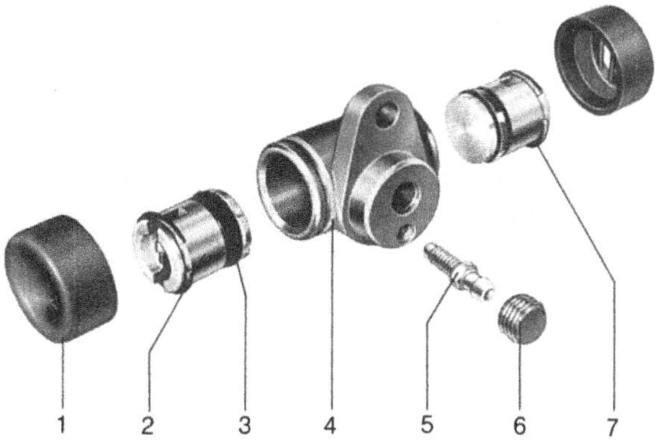

1. Boot (2)
2. Piston (2)
3. Cup (2)
4. Housing
5. Bleeder valve
6. Dust cap
7. Circlip (1968 through 1971 models only) (2)

Fig. 8-8. Exploded view of rear wheel cylinder.

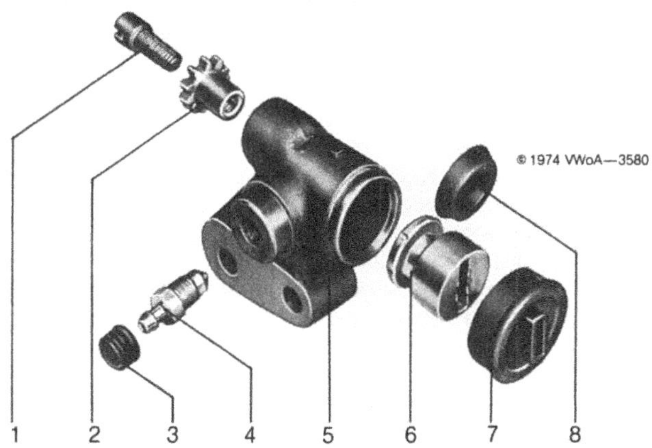

1. Adjuster screw
2. Adjuster
3. Dust cap
4. Bleeder valve
5. Housing
6. Piston (2)
7. Boot
8. Cup

Fig. 8-9. Exploded view of front wheel cylinder.

The internal parts can be hand-pressed out of the housing once the boots are removed. Prior to assembly, clean all parts with brake fluid only.

Check the cylinder for wear. Do not machine or hone metal from the cylinder bore. A new, lubricated piston must be an airtight fit. If it is not, replace the cylinder.

NOTE —
Vacuum should keep a new piston (lubricated with brake fluid) from falling out of the cylinder when the bleeder valve and brake hose holes are sealed and, on rear wheel cylinders, when you cover the opposite end of the cylinder with your thumb.

Lubricate the cups with brake fluid during installation. Coat the pistons with VW brake cylinder paste (see **9.5 Brake Cylinder Paste**) and insert them in the cylinder, then install the remaining parts.

8.6 Removing and Installing Backing Plate

If the backing plate is bent, or if the raised areas that the brake shoes ride against are badly worn, replace the plate. Otherwise, the brake linings will not line up properly with the drum and will wear to a taper.

On front wheel brakes, simply follow the procedure given in **8.4 Removing and Installing Wheel Cylinders**. Once the wheel cylinders have been removed, the backing plate can easily be lifted off.

On rear wheel brakes, remove the brake shoes as described in **8.3 Removing and Installing Brake Shoes**. Remove the wheel cylinder as described in **8.4 Removing and Installing Wheel Cylinders**. Then unbolt the parking brake cable from the backing plate.

> **CAUTION ―**
> *If the bolt that holds the parking brake cable is locked in place by corrosion, treat it with rust solvent or penetrating oil before attempting to force it loose. If the bracket is allowed to turn with the bolt, the flexible cable housing may be damaged.*

The rear backing plate can be taken off the rear wheel bearing housing after the backing plate mounting bolts have been removed.

Installation is the reverse of removal. Prior to installation, however, make sure that the mating surfaces on the backing plate and the steering knuckle or wheel bearing housing are completely clean. Torque the M 8 rear brake backing plate mounting bolts to 2.0 to 3.0 mkg (14 to 22 ft. lb.); torque the M 10 rear brake backing plate mounting bolts to 3.0 to 4.0 mkg (22 to 29 ft. lb.). Then install the wheel cylinder(s), brake shoes, and brake drum as described in **8.4 Removing and Installing Wheel Cylinders** and **8.3 Removing and Installing Brake Shoes.**

8.7 Reconditioning Brake Drums

Check the brake drums whenever new linings are installed. Taper, scoring, or other wear must, if possible, be corrected on a special machine by an Authorized VW Dealer or a qualified automotive machine shop. Both drums on an axle must be machined to the same dimensions. Use oversize linings on 1968 through 1970 models. On later models the linings must be ground to the new radius with a special machine by an Authorized VW Dealer or other qualified shop. Unless oversized linings— or linings radiused to fit the reconditioned drums—are used, uneven or ineffective braking will result.

The accuracy of the machine work will be improved if the road wheel can be mounted on the brake drum during machining and the wheel lugs tightened to the prescribed torque. This means that 1971 and later rear brake drums must be mounted on the brake drum lathe with the rear wheel hubs attached as well as the road wheel.

> **NOTE ―**
> Because of changes in manufacturing methods, rear brake drums installed from June 1973, beginning with chassis No. 213 2215 901, must be machined with the road wheel in place. If this is not possible, drums with excessive radial runout must be replaced.

The specified brake drum dimensions, wear limits, and lining dimensions are given in **Table a.**

Table a. Brake Drum Specifications

Brake drum inside diameters mm (in.)	
1968 through 1970 models	
New diameter	
Front and rear	250.00 + 0.20 (9.842 + .008)
wear limit	251.50 (9.900)
Permissible turned diameter	
Front and rear	251.00 + 0.20 (9.882 + .008)
wear limit	251.50 (9.900)
1971 and later models	
New diameter	
Rear	252.00 + 0.20 (9.920 + .008)
wear limit	253.50 (9.980)
Permissible turned diameter	
Rear	253.00 + 0.20 (9.960 + .008)
wear limit	253.50 (9.980)
Brake drum irregularity tolerances	
Taper	0.10 (.004) max.
Out-of-round	0.10 (.004) max.
Lateral runout	0.25 (.010) max.
Radial runout	0.10 (.004) max.
Brake lining dimensions	
Width, 1968 through 1970 models	
Front and rear	45.00 (1.770)
Width, 1971 and later models	
Rear	55.00 (2.165)
Lining thickness, 1968 through 1970 models	
New standard	5.0–4.8 (.200–.190)
wear limit	2.5 (.100) remaining
New oversize	5.5–5.3 (.220–.210)
wear limit	2.5 (.100) remaining
Lining thickness, 1971 and later models	
New standard	6.0–5.8 (.235–.230)
wear limit	2.5 (.100) remaining
New oversize	not available

9. BLEEDING BRAKES

Bleeding the brakes removes air from the hydraulic system. This task must be performed whenever the brake lines have been disconnected or after a brake cylinder has been replaced or repaired. If the brake pedal feels spongy when you apply the brakes, it is an indication that air has entered the system. If bleeding fails to correct the problem, there are probably leaks to be fixed.

Brake Fluids

Additional fluid must be added to the system when it is bled. The quality of the new brake fluid is important. All VW brake fluids have similar chemical and performance characteristics and may be mixed regardless of differences in color. Using a brake fluid that does not conform with SAE recommendation J 1703 and Motor Vehicle Safety Standard 116 can cause brake failure, premature wear, or erratic operation.

9.1 Changing Brake Fluid

Change the brake fluid in your VW every two years. Brake fluid tends to absorb moisture from the air, and water can initiate corrosion. Water can also cause the fluid to boil when the brakes are used very hard.

It is particularly important that the brake systems of vehicles with disc brakes have a fluid with a high boiling point. Since the brake calipers surround the friction linings, they pick up a great deal of heat from them.

On cars with front drum brakes it is possible to remove all old brake fluid by using the bleeder valve on the lower of the two wheel cylinders. All newer VW disc brake calipers have a second bleeder valve on the lower part of their housing. This makes it possible to remove completely all fluid when it has to be changed.

To change fluid:

1. Attach a suitable hose to the bleeder valves for draining the fluid into containers.
2. Open the rear wheel bleeder valves and pump the pedal until fluid ceases to flow out of them.
3. At the front wheels, open the lower bleeder valves on late cars with front disc brakes; open the single bleeder valve on early cars with front disc brakes; or open the bleeder valve on the lower front wheel cylinder of cars with four-wheel drum brakes. Pump the pedal until fluid ceases to flow out of the bleeder valves.
4. On late models with front disc brakes, open the upper bleeder valves. Leave them open until the fluid stops draining from the lower bleeder valves.
5. Close all bleeder valves.

   NOTE —
   On early models with front disc brakes there is also a bleeder valve on the rear brake pressure regulator. However, this bleeder valve must not be opened either during draining or during bleeding operations.

6. Fill the fluid reservoir with new, unused brake fluid that meets SAE recommendation J 1703 and conforms with Motor Vehicle Safety Standard 116. Then bleed the system by either of the methods described.

   CAUTION —
   It is possible that an air bubble can form in the line that connects the refill reservoir with the twin-chamber reservoir on the master cylinder. This will cause the twin-chamber reservoir to fill very slowly. Make sure that both reservoirs are filled before bleeding the brakes. Otherwise, air will enter the system.

9.2 Bleeding with Pressure Bleeder

Whenever possible, brake bleeding should be done with a pneumatic pressure bleeder similar to that shown in Fig. 9-1. Connect this device to the brake fluid reservoir. The bleeder fills the system with fluid under pressure and will complete the job in a very short time. Simply open the bleeder valve, quickly depress and slowly release the pedal several times, and then close the bleeder valve. A fluid receptacle supplied with the bleeding device must be fitted to the wheel being bled.

© 1974 VWoA—1479

Fig. 9-1. Pressure bleeder for bleeding brake system.

9.3 Bleeding by Pumping

For car owners, the pumping method of bleeding the brakes is usually more practical, even though it requires two persons. Have your helper sit in the car to pump the brake pedal. You will then be free to move from wheel to wheel to perform the actual bleeding.

To bleed:

1. Fully fill the fluid reservoirs with brake fluid that meets SAE recommendation J 1703 and conforms with Motor Vehicle Safety Standard 116.
2. Take the dust cap off the bleeder valve at the right-hand front wheel. Slip a 4-mm ($\frac{5}{32}$-in.) I.D. hose over the bleeder valve and submerge the other end in a clear glass jar partially filled with clean brake fluid. The jar must be clear so that you can see air bubbles coming out of the hose.
3. Open the bleeder valve a half turn. Have your helper slowly depress the brake pedal until it reaches the floor and keep it there while you close the bleeder valve.

4. Have your helper slowly release the pedal until it is completely up. Repeat the preceding step until no more air bubbles emerge from the hose.
5. Repeat the entire bleeding procedure on the other three wheels in the sequence left front, right rear, and left rear.

NOTE —
Refill the reservoir after bleeding each wheel cylinder. If the system contains a great deal of air or if the brake fluid is being changed, it will be necessary to add more fluid once or twice while bleeding each wheel cylinder. Never let the reservoir be emptied completely, or you will have to start bleeding the brakes all over again.

CAUTION —
Do not allow brake fluid to come in contact with painted surfaces. Brake fluid contains a solvent damaging to most finishes.

WARNING —
Do not use soft drink bottles or other food containers to store brake fluid or to bleed the brakes. Brake fluid is poisonous.

9.4 Flushing Brake System

Never use anything but brake fluid to flush the brake system. Alcohol must not be used since it will destroy residual lubrication and will encourage the accumulation of water in the system.

NOTE —
Do not rely on flushing alone to clean a brake system contaminated by dirt or rust. To remove all foreign matter, you must disassemble the system and clean the parts individually.

9.5 Brake Cylinder Paste

Brake cylinders will give longer service when lubricated with VW brake cylinder paste than when lubricated by brake fluid alone. When cleaning or rebuilding a brake cylinder, wash all parts thoroughly in brake fluid. Coat the pistons and the cylinder walls with VW brake cylinder paste, then assemble the cylinder.

VW brake cylinder paste is available from Authorized VW Dealers and should always be the preferred form of lubrication when brake cylinders are repaired. Brake fluid should be considered a satisfactory substitute only if you are unable to reach a VW Dealer.

10. PARKING BRAKE

The components of the parking brake lever are shown in Fig. 10-1.

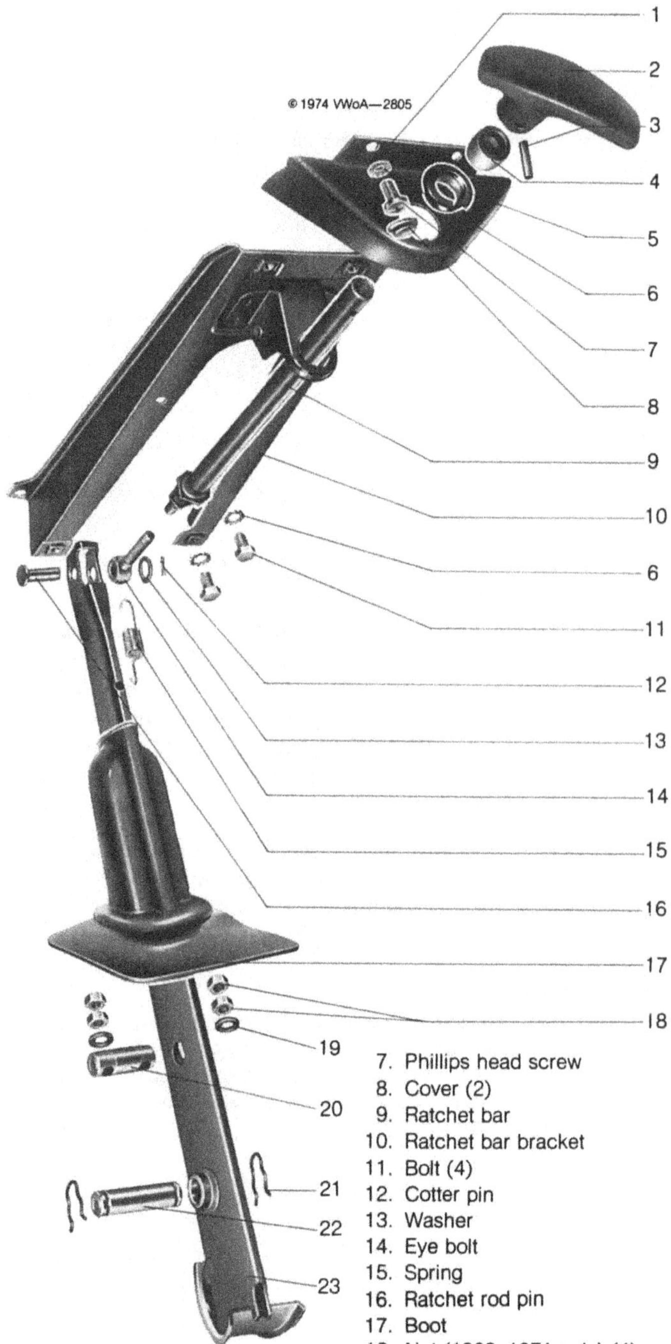

1. Trim plate
2. Handle
3. Rolled pin
4. Ratchet bar stop
5. Guide sleeve
6. Serrated washer (6)
7. Phillips head screw
8. Cover (2)
9. Ratchet bar
10. Ratchet bar bracket
11. Bolt (4)
12. Cotter pin
13. Washer
14. Eye bolt
15. Spring
16. Ratchet rod pin
17. Boot
18. Nut (1968-1971 only) (4)
19. Washer (1968-1971 only) (2)
20. Cable pin (1968-1971 only)
21. Spring clip (2)
22. Pivot pin
23. Parking brake lever

Fig. 10-1. Components of parking brake lever.

The parking brake operates only on the rear wheels. It is completely mechanical and independent of the hydraulic brake system. Once the parking brake handle has been pulled, it is held in position by a ratchet. The lever will remain in the same position until the ratchet is released by twisting the handle and allowing the parking brake lever to move forward.

Two cables extend from the parking brake lever to the left and right rear wheels. Pulling the parking brake handle rearward tightens the cables and moves the rear brake shoes into contact with the rear brake drums. The cable operates on a lever attached to the rear brake shoe of each rear brake. The movement of the lever is transmitted to the front shoe of each rear brake by a flat steel bar called the connecting link.

10.1 Adjusting Parking Brake

The parking brake should be adjusted whenever the rear brake linings have worn enough so that it is possible to pull the brake handle out six clicks without noticeable braking action.

To adjust:

1. Raise the vehicle. Then adjust the rear brakes as described in **8.1 Adjusting Drum Brakes.**
2. On 1968 through 1971 models, slide the boot up the parking brake lever for access to the cable adjusting nuts (Fig. 10-2).

© 1974 VWoA—3629

Fig. 10-2. Parking brake cable adjustments (arrows) on 1968 through 1971 models.

NOTE —
On 1972 and later models, the cable adjustments are on an equalizer bar under the vehicle (Fig. 10-3). On these models the cables are adjusted from under the vehicle.

© 1974 VWoA—3630

Fig. 10-3. Cable adjustments and equalizer bar on 1972 and later models.

3. Insert a screwdriver in the slotted cable ends so that the cables will not be damaged by twisting. Then loosen the locknuts.
4. Pull the parking brake handle out six clicks. Hold the cable ends with a screwdriver and tighten the adjusting nuts until each rear wheel can just be turned by hand. Braking must be equal on both sides of the vehicle.
5. Check whether the equalizing lever on 1972 and later models is horizontal or, on earlier models, that the cable ends extend about the same distance beyond the nut. If not, see **10.3 Removing and Installing Brake Cable.**
6. If the equalizing bar is horizontal, or the cable ends approximately equal, tighten the locknuts and, on 1968 through 1971 models, refit the boot.

8

10.2 Removing and Installing Parking Brake Lever

Before removing the parking brake lever from 1968 through 1971 models, remove the nuts from the parking brake cable ends. On later models, disconnect the equalizer bar connecting rod from the bottom of the parking brake handle. To do this, pry the spring clip off the end of the pin.

To remove lever:

1. Remove the ashtray. Then drive the rolled pin out of the handle and remove the handle and ratchet bar stop.
2. Detach the spring and remove the ratchet rod pin. Then remove the ratchet bar and trim plate.
3. Remove the ratchet bar bracket bolts indicated in Fig. 10-4. Remove the brackets.

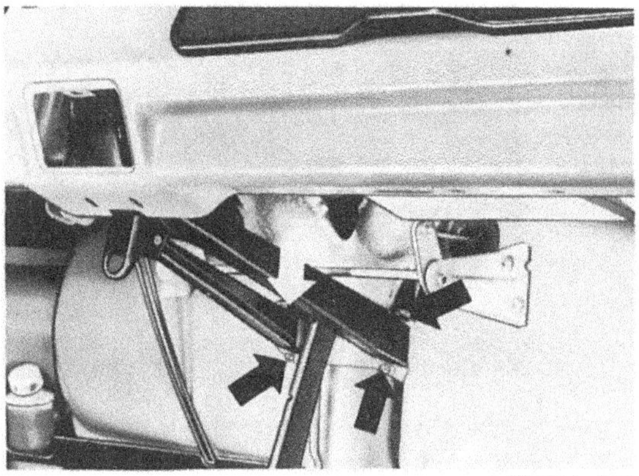

© 1974 VWoA—2806

Fig. 10-4. Bolts (arrows) that hold ratchet bar bracket to body.

4. Slide the boot up and off the parking brake lever. On 1968 through 1971 models, pull the brake cables out downward.
5. Remove the spring clips from the pivot pin. Then remove the pivot pin and take the lever down through the floorboards and out of the vehicle.

Installation is the reverse of removal. When installing the ratchet rod, the eye bolt must be adjusted so that the parking brake lever does not contact the heater pipe when the brake handle is fully released. To do this, screw the eye bolt into the ratchet rod until the parking brake lever must be moved away from the heater pipe in order to install the ratchet rod pin and the spring. Install a new cotter pin through the ratchet rod pin.

10.3 Removing and Installing Brake Cable

If the rear wheel brake shoes are correctly adjusted, but the equalizer bar is not horizontal after adjusting the parking brake (or if one cable end projects considerably farther beyond the nuts on 1968 through 1971 models), one cable may be stretching, and will eventually break.

NOTE —
Replacement parts are available that permit the installation of an equalizer bar on 1968 through 1971 models. However, this conversion should never be used to compensate for a weak and failing cable.

To replace a parking brake cable, take off the locknut and adjusting nut that hold the cable in the equalizer bar or cable pin. Remove the rear brake drum. Then unhook the cable from the lever on the rearmost brake shoe. Unbolt the bracket that holds the cable in the backing plate and take out the cable and flexible housing.

CAUTION —
If the bolt that holds the parking brake cable is locked in place by corrosion, treat it with rust solvent or penetrating oil before attempting to force it loose. If the bracket is allowed to turn with the bolt, the flexible cable housing may be damaged.

Installation is the reverse of removal. Pull the old cable out to the rear. Then grease the new cable and insert it. On 1968 through 1971 models, use a long screwdriver to loop the cable around the curved guide channel on the parking brake lever and into the cable pin. Adjust the parking brakes as you attach the cable. Check the adjustment again after about 300 mi. (500 km).

11. BRAKE TROUBLESHOOTING

Table b lists brake problems, probable causes, and suggested remedies. The numbers in bold type in the Remedy column refer to the numbered headings in this section of the Manual under which the suggested repairs are described.

Table b. Brake Troubleshooting

Problem	Probable Cause	Remedy
1. Pedal goes all the way to floor in braking	a. Linings worn b. No fluid, low fluid	a. Adjust brake shoes (never adjust at pedal). See **8.1**. b. Find and repair leaks. Fill and bleed system. See **9**.
2. Low pedal even after adjustment and bleeding	Master cylinder defective	Replace or rebuild master cylinder. See **4.2**, **4.3**.
3. Pedal spongy or brakes work only after pedal is pumped	a. Insufficient fluid in reservoir b. Air in system c. Spring weak in master cylinder	a. Top up fluid and bleed system. See **9**. b. Check for leaks and bleed system. See **9**. c. Replace or rebuild master cylinder. See **4.2**, **4.3**.

Table b. Brake Troubleshooting (cont.)

Problem	Probable Cause	Remedy
4. Braking action decreases after shoes have been adjusted	a. Brake lines leaking b. Defective master or wheel cylinders	a. Tighten connections or fit new lines and hoses. See **6**. b. Replace or rebuild faulty cylinder. See **4.2**, **4.3**, **7.2**, **7.3**, **8.4**, **8.5**.
5. Brakes overheat	a. Compensating port blocked b. Inadequate pushrod clearance c. Brake shoe return springs weak d. Rubber parts swollen	a. Clean master cylinder. See **4.2**, **4.3**. b. Adjust pushrod clearance at pedal stop. See **4.1**. c. Fit new return springs. See **8.2**, **8.3**. d. Flush system, replace fluid and all rubber parts. See **9**.
6. Brakes inefficient despite high pedal pressure	a. Linings oiled up b. Unsuitable brake linings c. Loose or leaking vacuum hose to power brake servo d. Servo diaphragm leaking e. Poor seal between servo and master cylinder f. Check valve in vacuum line not working properly	a. Clean drums. Replace linings and oil seals. See **8.2**, **8.3**, **TRANSMISSION AND REAR AXLE**. b. Fit new linings. See **7.1**, **8.3**. c. Replace leaking vacuum hoses or tighten hose clamps. See **4.5**. d. Replace the servo. See **4.5**. e. Replace large sealing ring. Check piston rod for damage and, if necessary, replace it. See **4.2**. f. Replace faulty valve. See **4.5**.
7. Brakes bind while car is in motion	a. Compensating port blocked b. Brake fluid unsuitable c. Inadequate pushrod clearance	a. Disassemble master cylinder and clear port. See **4.2**, **4.3**. b. Flush system and refill. See **9**. c. Adjust pushrod clearance at pedal stop. See **4.1**.
8. Brakes chatter and tend to grab	a. Linings worn b. Drums out-of-round	a. Fit new linings. See **7.1**, **8.3**. b. Recondition or replace drums. See **8.2**, **8.7**.
9. Drum brakes squeak	a. Unsuitable or badly fitted linings b. Brake linings dirty c. Backing plates distorted d. Brake shoe return springs weak e. Poor lining contact pattern due to shoe distortion	a. Fit new linings properly. See **8.3**. b. Clean brakes. See **8.2**. c. Check backing plates for distortion and fit new parts if necessary. See **8.2**, **8.6**. d. Fit new return springs. See **8.3**. e. Align shoes with backing plate with 0.20 mm (.008 inch) clearance at lining ends and contact across full width. See **8.3**.
10. Disc brakes squeak	a. Unsuitable pads b. Spreader spring faulty or missing c. Pad guide surfaces dirty or rusted d. Pads dirty or glazed e. Lining loose on pad	a. Fit new pads. See **7.1**. b. Install new spreader spring. See **7.1**. c. Clean pads and calipers. See **7.1**. d. Clean and replace pads. See **7.1**. e. Replace pads. See **7.1**.
11. Brakes give uneven braking	a. Oil or grease on linings b. Brake pressure regulator defective c. Poor contact between lining and drum due to brake shoe distortion d. Brake shoes too tight in the adjusting screw slots or in wheel cylinder pistons e. Different types of linings on same axle f. Incorrect tire pressures or unevenly worn tires g. Drums or discs out-of-round or scored h. Disc brake pads sticking in caliper i. Disc brake pads reinstalled in wrong location j. Brake shoes not in contact with backing plate k. Pistons tight in wheel cylinders l. Dirt in brake lines or hoses	a. Clean drums. Fit new linings and seals or wheel cylinders if necessary. See **8.2**, **8.3**, **8.4**. b. Replace brake pressure regulator. See **5**. c. Shape shoes to leave 0.20 mm (.008 inch) clearance at lining ends. See **8.3**. d. Free up shoes. See **8.3**. e. Fit new shoes or pads. See **7.1**, **8.3**. f. Correct pressures or replace worn tires. See **13.2**, **13.6**. g. Recondition or replace discs or drums. See **7.5**, **7.7**, **8.2**, **8.7**. h. Clean caliper and pads. See **7.1**. i. Install new pads. See **7.1**. j. Reposition shoes or align or replace backing plate. See **8.3**, **8.6**. k. Free up pistons. See **8.5**. l. Clean system and replace defective parts. See **6**, **9**.
12. Brakes pulsate	a. Drums out-of-round (drum brakes only) b. Excessive disc runout or thickness variations (disc brakes only) c. Mating surface between drum and hubs dirty	a. Recondition or replace drums. See **8.2**, **8.7**. b. Recondition or replace discs. See **7.5**, **7.7**. c. Clean surface. Reinstall drum. See **8.2**.
13. Foot pressure on brake pedal must be increased when the pedal reaches a certain position	Groove worn in piston rod that allows air from ventilation drilling to enter servo as groove goes past sealing cup	Replace or rebuild master cylinder. See **4.2**, **4.3**.

12. REMOVING, REPAIRING, AND INSTALLING BRAKE PEDAL

The components of the brake pedal are shown in Fig. 12-1.

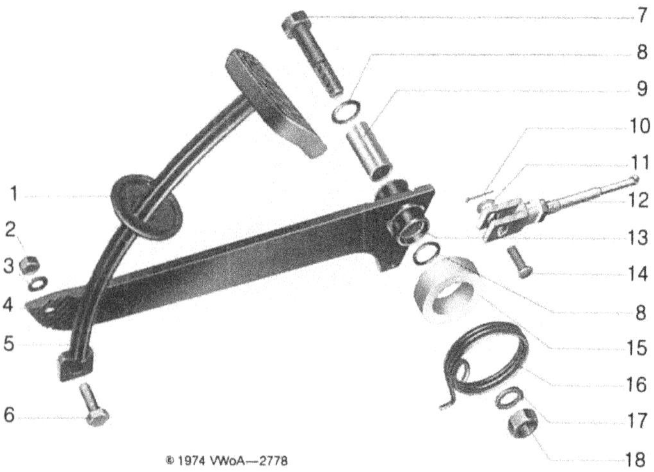

1. Seal
2. Nut
3. Lock washer
4. Brake pedal lever
5. Brake pedal
6. Bolt
7. Mounting bolt
8. Seal (2)
9. Mounting tube
10. Cotter pin
11. Washer
12. Master cylinder pushrod
13. Bushing
14. Clevis pin
15. Plastic ring
16. Return spring
17. Lock washer
18. Nut for mounting bolt

Fig. 12-1. Components of the brake pedal.

WARNING —
Do not re-use fasteners such as nuts, bolts, washers, self-locking nuts, circlips, cotter pins that are worn or deformed in normal use. Always replace.

To remove:

1. Working under the vehicle, remove the cover plate that is beneath the pedal cluster. Then remove the bolt indicated in Fig. 12-2.

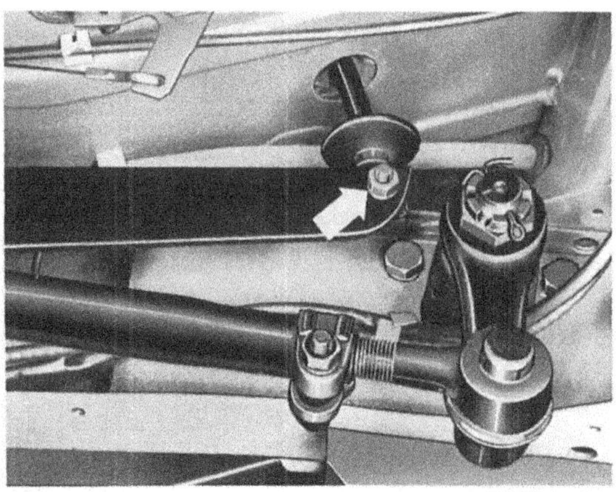

Fig. 12-2. Nut and bolt (arrow) that hold brake pedal on pedal lever.

2. Take the brake pedal out upward into the passenger compartment.
3. To disconnect the master cylinder pushrod or the brake servo connecting rod from the pedal lever, remove the cotter pin and the clevis pin.
4. Using a screwdriver as shown in Fig. 12-3, pry the return spring off the tongue on the pedal lever.

Fig. 12-3. Return spring being pried off pedal lever.

5. Remove the mounting bolt and nut. Then remove the brake pedal lever from its bracket together with the mounting tube, plastic ring, and return spring.

Inspect the bushing inside the pedal lever and, if necessary, replace it. Drive out the old bushing, then press in the new bushing until it is flush with the shorter end of the tube on the brake pedal lever. Inspect the other parts and, if necessary, replace them.

Installation is the reverse of removal. Assemble the mounting tube, seals, plastic ring, and return spring on the pedal lever, then install the pedal lever on the bracket and insert the mounting bolt. Torque the mounting bolt and nut to 3.0 to 4.0 mkg (22 to 29 ft. lb.). Torque the bolt that holds the pedal on the pedal lever to 2.0 mkg (14 ft. lb.). Use a new cotter pin through the clevis pin and adjust the master cylinder pushrod or brake servo connecting rod as described in **4.1 Pushrod Adjustment.**

The clutch pedal and accelerator pedal are mounted separately from the brake pedal. Removal and installation of the accelerator is covered in **FUEL SYSTEM**. Removal, repair, and installation of the clutch pedal is covered in **ENGINE AND CLUTCH.**

13. WHEELS AND TIRES

Tires are subject to many stresses. If they are to perform as intended, they must be inflated to specifications and correctly balanced. Properly maintained, the factory-installed tires will provide long service with comfort and safety. But they must never be kept in service when worn out or damaged by accidents or careless driving.

13.1 Wheels

The 1968 through 1970 models are equipped with 5JK × 14 rims. All later models have 5½J × 14 rims. The 5JK × 14 wheels have an offset of 47.5 mm (1⅞ in.), while the later 5½J × 14 wheels have an offset of 41.0 mm (1⅝ in.). Because the bolt hole circle is not the same, the two types cannot be interchanged.

Offset is the difference between the center of rim width and the mounting face that bolts to the brake drum or brake disc. The use of wheels other than those standard for a given year and model is discouraged. Many of the wider wheels sold by accessory companies do not have the correct offset dimension. This may impose excessive stress on the wheel bearings or alter steering behavior.

The wheels on all models covered by this Manual are suitable for use with tube or tubeless tires of the correct size, including radials. Tires of nonstandard size may be used only if the tire manufacturer specifies them for your particular make and model VW.

13.2 Tire Types and Pressures

Conventional (bias ply) tires are factory-installed on all models through 1970. All 1971 and later models are factory-equipped with radial ply tires.

> ***CAUTION —***
> *Not all bias ply or radial ply tires are the same. When you buy replacement tires, select only bias-ply tires that have an 8 ply rating; select radial ply tires designated "reinforced" or—for Delivery Vans and Station Wagons—with the suffix "C". Other tires of identical size may have inadequate load-carrying capability.*

Conventional Tires

Conventional bias ply tires offer very good riding characteristics and are well suited to highways and rough roads alike. Bias-constructed tires can be made extremely strong through the use of an almost infinite number of plies. The basic simplicity of bias ply design also results in lower construction costs.

Radial Tires

There is less friction between the fabric layers in radial ply tires, so they generate less heat. This makes them especially suitable for long-distance, high-speed driving. They also do not heat up so much when carrying heavy loads. Their rigid tread improves wet-weather traction, but tends to produce a harsher ride and increased road noise on some kinds of surfaces.

Winter Tires

Although inferior to regular tires for dry-road wear and handling, winter (mud and snow) tires can greatly improve operation on snowy or slushy roads. Studded winter tires improve traction on icy surfaces, but can be damaged by fast driving on dry roads and may damage some road surfaces. They should be used only if icy conditions predominate throughout the winter months. Also check your local vehicle laws. The use of studded tires may be restricted in your area.

Tire Pressures

For prolonged high speeds, we recommend that you increase the pressures in conventional bias ply tires by 3 psi over the specifications given here. Winter tires will also work better when inflated an additional 3 psi. However, cold-tire pressure must never exceed the maximum inflated pressure marked on the tire. The following inflation pressures are prescribed:

Conventional (bias ply) Tires

Model	Up to ¾ payload		Fully loaded	
	Front	Rear	Front	Rear
Station Wagon and Campmobile, 1968–1969	28 psi	36 psi	28 psi	36 psi
Trucks, 1968–1971, Station Wagon and Campmobile, 1970–1971	28 psi	36 psi	28 psi	40 psi

Radial (radial ply) Tires

Model	Up to ¾ payload		Fully loaded	
	Front	Rear	Front	Rear
Station Wagon, from August 1971 through March 1972	30 psi	34 psi	30 psi	40 psi
Delivery Van, from August 1971 through April 1972; Campmobile from August 1971	30 psi	37 psi	30 psi	40 psi
Station Wagon and Delivery Van from April 1972	30 psi	40 psi	30 psi	44 psi

8

Because steel and textile cord radial tires have different traction characteristics, it is important that all four tires on the car have the same cord material. Tire effectiveness under various road conditions is shown in the following chart:

Operating conditions	Dry	Wet	Snow	Ice
Snow w/studs radial ply	0	-	X	X
Snow w/studs bias ply	0	-	X	X
Snow radial ply	X	X	X	0
Snow bias ply	X	X	X	0
Standard radial ply	X	X	0	0
Standard bias ply	X	X	0	-

X = Effective 0 = Restricted effectiveness - = Noneffective

13.3 Normal Tire Wear

The original equipment tires on your VW have built-in tread wear indicators. These indicators are molded into the bottom of the tire tread grooves. The indicators eventually appear as the result of normal wear. They are about 13 mm (½ in.) wide in visible bands when the tire tread depth gets down to 1.5 mm (1/16 in.).

When these indicators appear in two or more adjacent grooves of a tire tread, as shown in Fig. 13-1, the tire must be replaced. It is recommended that tires be replaced well before the indicators are as visible as shown. Worn tires cannot grip even a dry road surface properly and are almost completely ineffective on a wet road surface.

WARNING —
Do not assume that a tire is sound merely because the tread wear indicators have not yet appeared. Always check for cuts, cracks, rubber separation, and internal damage. Normal wear is only one factor in determining tire serviceability.

© 1974 VWoA—3539

Fig. 13-1. Indicator showing on worn-out tire.

For best all-round handling, always replace all four tires at the same time. If this is not possible, replace both tires on one axle. Do not combine tires of different ply construction, size, or tread pattern.

WARNING —
Break in new tires by driving at moderate speeds for the first 60 to 100 mi. (100 to 160 km). New tires do not have full traction when first installed.

Normal tire wear is accelerated by higher speeds. Wear at a constant 35 mph (56 kph) is only about a third of that produced at a constant 70 mph (112 kph)—not half, as might seem logical.

Weather also affects normal tire wear. Hot weather is the most damaging and when heat is combined with high speeds and underinflation, tire structure is seriously endangered. Cold weather prolongs tire life; so does wet weather, which reduces the friction between tire and road.

13.4 Removing and Installing Wheels

The VW Owner's Manual supplied with the car lists the proper procedures for this job. The following points, however, deserve mention to mechanics not familiar with the VW Owner's Manual or with VW cars.

First, use the hubcap remover. This is a wire hook that can be slipped into holes in the edges of the hubcaps. It prevents scratches on the painted wheel and makes the job quicker and easier.

Next, use only the VW jack supplied with the vehicle. Never attempt to lift a VW with an ordinary bumper jack. The cross section of VW bumpers is not contoured for the lifting hook on such jacks, and the vehicle cannot be lifted safely by them. If a hydraulic-type floor jack is used, be certain to position it carefully. Place blocks ahead and behind the wheels that remain on the ground to prevent the vehicle from rolling.

CAUTION —
Under no circumstances must the car be lifted by placing a jack under the engine, transmission, or floor pan. Serious damage can result from these practices.

All wheel lugs are removed by turning them counterclockwise. When tightening the wheel lugs, torque them to 13 mkg (94 ft. lb.). Use a torque wrench. Pneumatic tools are seldom capable of attaining the prescribed torque with accuracy.

13.5 Wheel Rotation

Although the tires will develop a normal wear pattern under most conditions, abnormal road surfaces or variations in driving technique may produce unequal wear of the four tires on a car. If, after a period of service, the tires on your vehicle show uneven wear, all four wheels can be rotated as shown in Fig. 13-2.

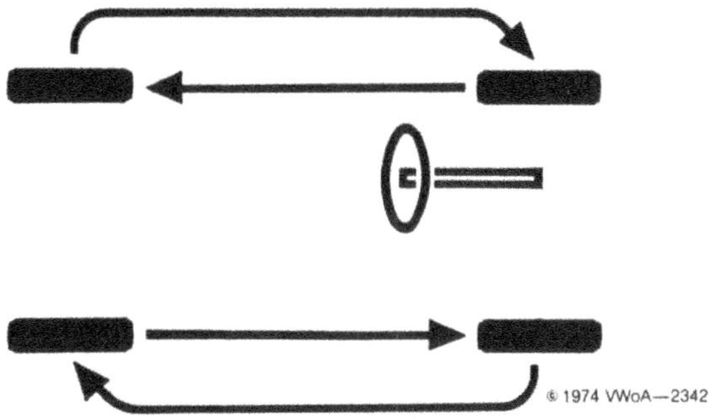

Fig. 13-2. Pattern of recommended wheel rotation. Wheel rotation is not required but may be desirable if tires are wearing unevenly.

13.6 Changing Tires

Dismounting or mounting tires on the rims requires a tire appliance with power enough to force the tire bead over the inner hump on the rim. When carrying out these operations, be sure that the rubber lining on the inner wall of the tire and tire beads is not damaged.

To dismount tire:

1. Take off the valve cap, carefully unscrew the valve core, and let the air out of the tire. Then press the tire bead off the rim as shown in Fig. 13-3.

© 1974 VWoA—488

Fig. 13-3. Pressing tire bead off rim.

2. Pry the tire sidewalls, one after the other, over the rim edges as shown in Fig. 13-4.

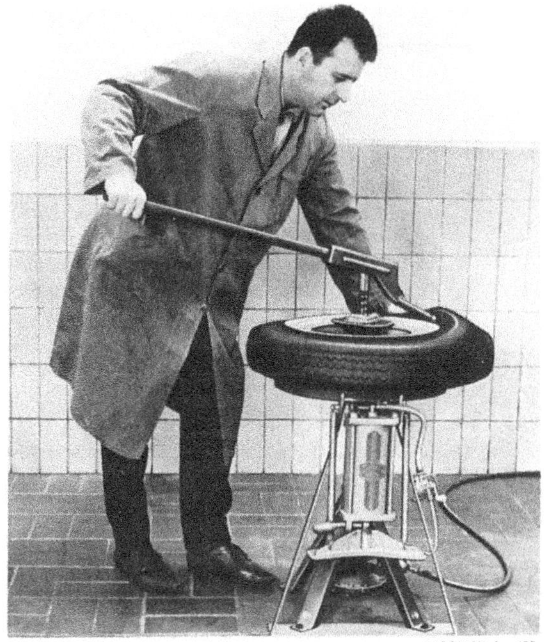

© 1974 VWoA—489

Fig. 13-4. Prying tire sidewalls over rim edges.

3. Check the airtight lining inside the tire for damage and bruises between the lining and casing. Carefully inspect the outside of the tire for embedded stones, cuts, grease, and signs of uneven wear.

4. Check the rubber part of the valve for cracks and damage. If the valve is faulty, remove it from the rim. Lubricate a new valve with soapy water. Then install the new valve as shown in Fig. 13-5.

© 1974 VWoA—490

Fig. 13-5. A valve tool being used to install a new valve in the rim.

8

To mount tire:

1. Check the rim for damage. Radial runout must not exceed 1.25 mm (.050 in.); lateral runout must not exceed 1.50 mm (.060 in.).

> ***WARNING —***
> *Never attempt to straighten bent, dented, or otherwise distorted rims. Doing so could weaken them. Make certain that all wheels are as originally fitted with J-type rims before installing tubeless radial tires. Other rim types are unsuitable for use with radial tires unless tubes are installed.*

2. Using a wire brush, remove any dirt from the rim shoulders and flanges. Smooth any sharp edges before mounting the tire on the wheel.
3. Insert the valve with the valve installing tool.

> **NOTE —**
> Use inner tubes with radial tires if the vehicle will be used on rough roads or off road.

4. Mount the tire on the rim as shown in Fig. 13-6. If there is a red dot on the sidewall, position it toward the valve.

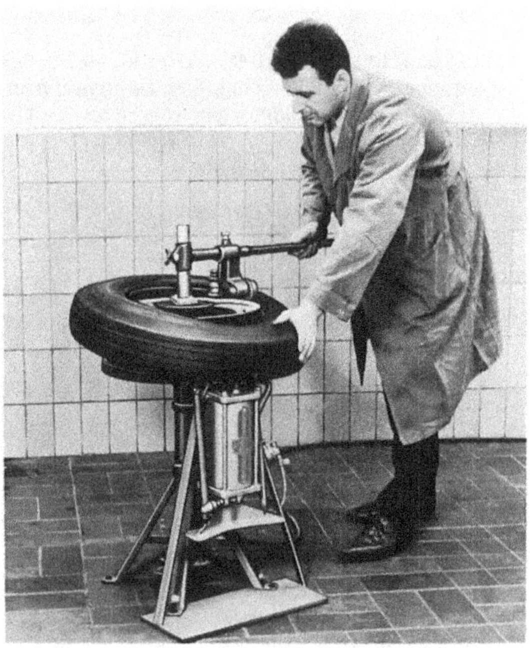

© 1974 VWoA—491

Fig. 13-6. Tire being mounted on rim. Use soft soap or a special rubber lubricant on the rim for safe, easy mounting.

5. Remove the valve core, if not already removed.
6. Inflate the tire to at least 43 psi (3 kg/cm^2).

> **NOTE —**
> Use a bead expander strap, if necessary, to obtain an airtight seal. When inflating the tire you should hear the bead snap over the inner hump on the rim.

7. Install the valve core and inflate the tire to the correct running pressure, as listed earlier in **13.2 Tire Types and Pressures.**
8. Immerse the wheel in water. Check for leaks.
9. Balance the wheel and install it on the car.

13.7 Abnormal Wear

Following are the six most common causes of abnormal tire wear (extra wear):

1. Underinflation or overinflation
2. Hard driving (high speed driving, violent braking, sudden cornering)
3. Rough or abrasive road surfaces, high crown roads, and very uneven road surfaces
4. Poorly aligned wheels (front or rear)
5. Poorly balanced wheels (front or rear)
6. Vehicle overloading or carrying too much weight for tire capacity.

Improper Inflation

Tire life depends greatly on correct inflation. Unfortunately, there are many ways for a tire to lose air. Every tire normally loses some air pressure due to the diffusion of air molecules through the rubber. Although tubeless tires hold air pressure longer than tires with inner tubes, it is recommended that you check the pressure even of tubeless tires once a week.

Tire pressures should be checked before driving, when the tires are still cold. If tire pressures are checked after driving, the pressures will have increased from the heat of road friction and internal flexing. If air is bled from a warm tire to obtain the pressure recommended for a cold tire, the tire will actually be underinflated. A tire that is driven while underinflated will overheat because of increased tire flexing and will rapidly lose its road-holding ability as tire strength begins to diminish.

There is always a reason when a tire loses a significant amount of air in a short period of time. Aside from a hole in the tire, the possible causes are a leaky rim or valve, loose-fitting tire beads, foreign matter in the rim, or an uneven surface between the rim shoulder and the tire.

Fig. 13-7 shows three tire inflation conditions. The shape of the tire is changed by the degree of inflation. This can cause abnormal wear. An underinflated tire wears at its edges; an overinflated tire wears at its center. The profile of an underflated tire is similar to that of an overloaded tire. The same kind of wear will result from overloading, with the added possibility of severe heat damage and possible structural failure.

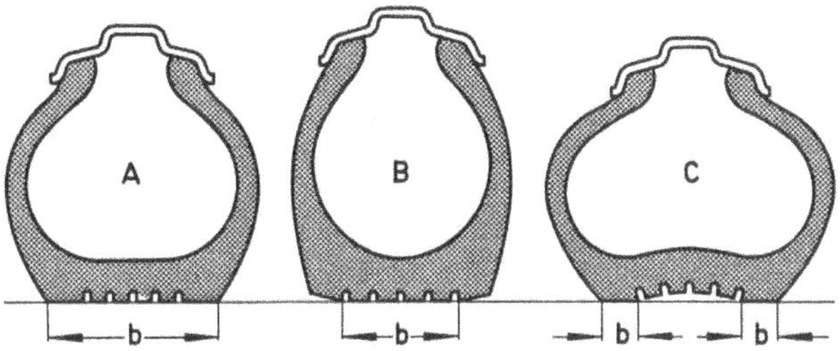

Fig. 13-7. Tire inflation. Condition **A**, normal inflation; condition **B**, overinflation; condition **C**, underinflation. Dimension **b** is the width of the tire tread in contact with the road.

Fig. 13-8 shows typical tire wear due to overinflation. Because a narrower portion of the tread is in contact with the road, wear occurs faster than normal. This tire is worn out despite the good tread remaining at the edges.

© 1974 VWoA—1438

Fig. 13-8. Overinflation wear. Tire worn out by running it with excessive air pressure.

Fig. 13-9 shows the tread wear pattern of an underinflated tire. This worn-out tire still has deep tread in its center, although the edges are completely bald. Such tires should be removed from service immediately. As with tires worn unevenly by overinflation, they provide very limited adhesion in wet weather and almost no traction in deep snow. In addition, side grip for cornering is seriously impaired.

© 1974 VWoA—1435

Fig. 13-9. Underinflation wear. Tire worn out by running it with too low an air pressure.

Hard Driving

Tires may wear abnormally because of excessive speed, heavy braking (see Fig. 13-10), high speed cornering, and similar violent or abrupt maneuvers. Conservative driving preserves tire life.

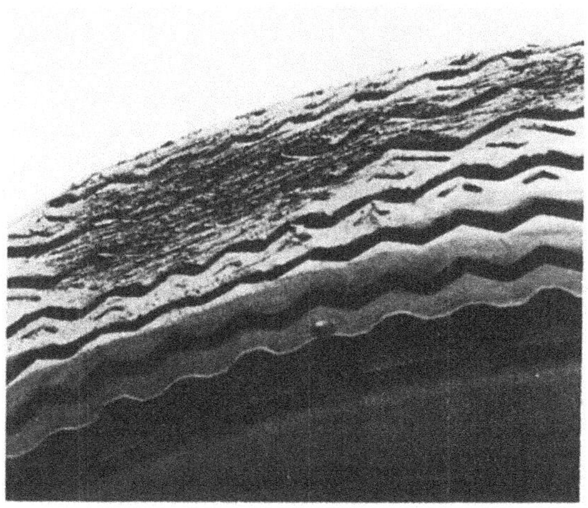

© 1974 VWoA—1439

Fig. 13-10. Abnormal tire wear due to locking the brakes. Uneven braking forces can also cause heavy tire wear.

8

Road Surfaces

Some road surfaces tend to increase tire wear. Rough, anti-skid surfacing materials abrade the tread and accelerate normal wear. Roadways with high crowns necessitate constant steering correction to keep the car headed parallel to the highway. This will eventually produce the kind of abnormal wear usually associated with improper front wheel alignment. Obstructions and road hazards can cause severe tire damage and can even break the casing internally, as shown in Fig. 13-11.

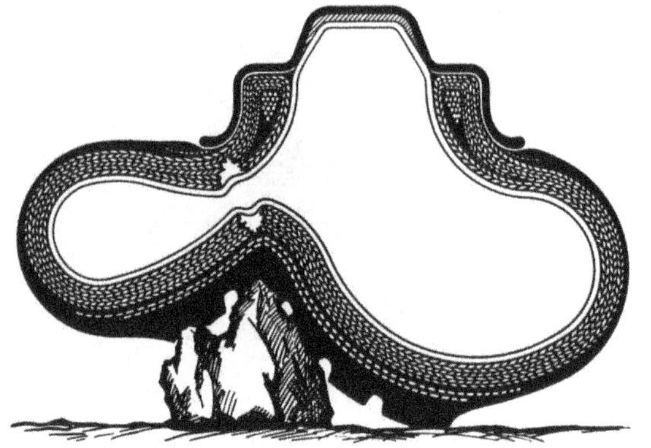

© 1974 VWoA—1441

Fig. 13-11. Internal damage. Striking an obstacle in the road may tear the ply layers. Such damage renders the tire unfit for use. You may have to take the tire off its rim to see such defects.

Faulty Wheel Alignment

Excessive tire wear will result from misalignment of any of the four wheels on the car. If irregular tire wear appears, check the front wheel track, the front wheel angles at full steering lock, the ride height and rear wheel track, the position of the axles relative to each other, the wheelbase on both sides of the car, the front and rear wheel camber, the setting of the rear spring plates, and the condition of the shock absorbers. The abnormal wear caused by misalignment usually takes the form of greater wear at one edge of the tread than on the other. Alignment specifications are given in **TRANSMISSION AND REAR AXLE and FRONT AXLE.**

Wheels Out of Balance

Wheels that are out of balance cause abnormal tire wear and constitute a driving hazard. Unbalanced wheels bounce, tramp, and wobble. The faster you drive, the more dangerous these wheel vibrations are.

Wheel imbalance is always more evident on the front wheels, and usually more dangerous. Fig. 13-12 shows the condition that static balancing of the wheel and tire can detect. The effect of such imbalance is shown in the right-hand part of the illustration.

Fig. 13-13 shows the type of imbalance that only dynamic balancing can detect. Although in perfect static balance, the concentrations of mass are not in line with one another or with the tire centerline. The wheel wobbles sideways when it spins, as shown on the right-hand side of the illustration. This causes deep wear in the form of cupped areas on the tread. Check that wheel runout does not exceed 1.25 mm (.050 in.) either radially or laterally before dynamic balancing the wheel. Replace the wheel if it does.

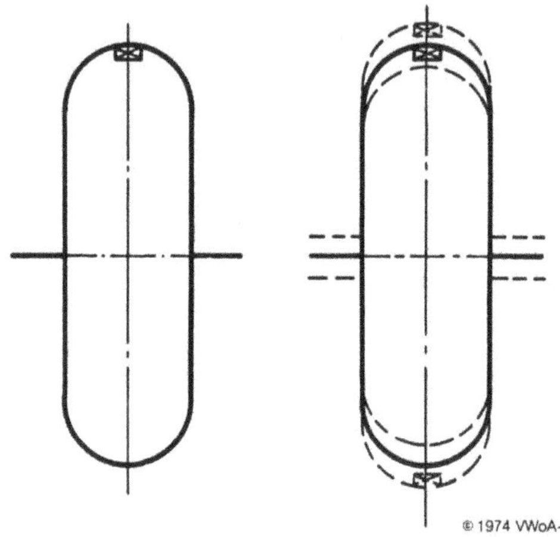

© 1974 VWoA—1445

Fig. 13-12. Imbalance that can be cured by static balancing. Vibration of the spinning wheel is shown at the right.

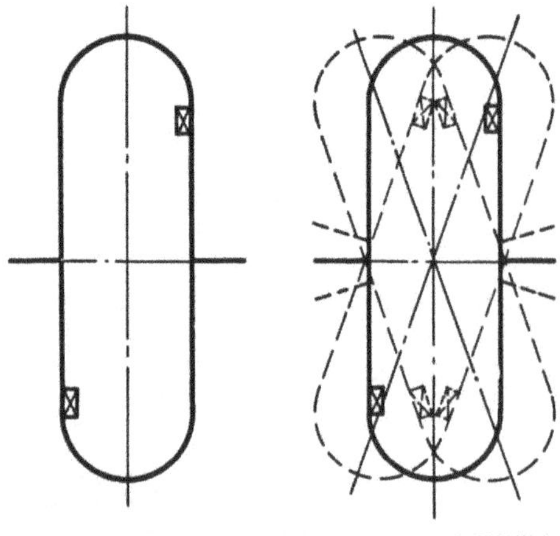

© 1974 VWoA—1446

Fig. 13-13. Imbalance that can be cured only by dynamic balancing. Vibration of the spinning wheel is shown at the right.

Overloading

Overloading the tires causes damage similar to that produced by underinflation. The excessive heat generated by an overloaded tire weakens the tire and causes difficult vehicle handling. The maximum load capacity (given in the Owner's Manual), added to the weight of the vehicle, must never exceed the total load capacity of the four tires (marked on their sidewalls).

14. BRAKES AND WHEELS TECHNICAL DATA

The following tables give all specifications related to the brake system, tires, and wheels. Wheel alignment specifications are given in **TRANSMISSION AND REAR AXLE** and **FRONT AXLE.**

I. Tightening Torques

Location	Part	Thread	mkg	ft. lb.
Master cylinder				
Stop screw in housing	screw	M 6	0.5–1.0	3.5–7
Residual pressure valve on housing	—	M 12 x 1	2.0	14
Brake light switch to housing	—	M 10 x 1	2.0	14
Master cylinder to frame	bolt	M 8	2.5	18
Brake line to master cylinder	union nut	M 10 x 1	1.5–2.0	11–14
Push rod (locknut for clevis)	nut	M 9 x 1	1.5–2.0	11–14
Master cylinder on frame (1968–1970)	bolt	M 8	2.5	18
Master cylinder on servo (1971 on)	nut	M 8	max. 1.3	max. 9.4
Front wheel brakes				
Wheel cylinder and backing plate on steering knuckle (1968–1970)	bolt	M 10	5.5–6.0	40–43
Caliper to steering knuckle, 1971–1972	bolt	M 12 x 1.5	10.0	72
1973 on	bolt	M 14 x 1.5	16.0	116
Bleeder valve in cylinder/caliper	—	M 6/M 7	max. 0.5	max. 3.5
Hose to cylinder/caliper	—	M 10	1.5–2.0	11–14
Screw for clamp nut	socket head screw	M 7	1.5–max. 2.0	11–max. 14
Brake disc to wheel hub	socket head screw	M 8	2.0–2.5	14–18
Splash shield to steering knuckle	bolt	M 7	1.0	7
Brake caliper halves	bolt	—	3.5	25
Rear wheel brakes				
Backing plate on bearing housing/wheel cylinder	bolt	M 8	2.0–3.0	14–22
	bolt	M 10	3.0–4.0	22–29
Cover to bearing housing	bolt	M 10	6.0	43
Brake drum/rear wheel hub to shaft	castellated nut	M 30/M 30	35.0	253
Brake drum to wheel hub	bolt	M 7	1.0–1.8	7–13
Wheels				
Wheel to drum/disc (1968–1970)	bolt	M 14 x 1.5	13.0	94
Wheel to drum/disc (1971 and later)	nut	M 14	13.0	94
Pedals				
Brake pedal on pin	bolt	M 12 x 1.5	3.0–4.0	22–29
Brake pedal to pushrod	bolt	M 8	2.0	14
Clutch pedal on pin	nut	M 10	2.5	18
Support/clutch pedal on frame	bolt	M 8	2.0	14
Brake servo				
Brake servo on front axle	nut	M 8	max. 1.3	max. 9.4
Brake proportioning valve				
Proportioning valve on side member	bolt	M 8	1.5	11
Output adaptor in regulator	—	—	2.0–2.2	14–16

II. Tolerances, Wear Limits, and Settings

Designation	From Chassis No.	New Part mm (in.)	Wear Limit mm (in.)
A. Cylinders			
1. Tandem master cylinder			
Front wheel circuit stroke	218 000 001	24.00 (.944)	—
	211 2000 001	19.00 (.748)	—
Rear wheel circuit stroke	218 000 001	14.00 (.551)	—
	211 2000 001	13.00 (.512)	—
Diameter	218 000 001	22.20 (.874)	—
2. Wheel cylinders			
Front ... diameter	218 000 001	25.40 (1.000)	—
Cylinder in front caliper ... diameter	211 2000 001	54.00 (2.126)	—
Rear ... diameter	218 000 001	22.20 (.874)	—
	212 2000 001	23.81 (.937)	—
B. Brake drums/brake discs			
1. Brake drums			
Front ... inside diameter	218 000 001	250.00 + 0.20 (9.842 + .008)	251.50 (9.900)
Rear ... inside diameter	218 000 001	250.00 + 0.20 (9.842 + .008)	251.50 (9.900)
	211 2000 001	252.00 + 0.20 (9.921 + .008)	253.50 (9.980)

8

continued on next page

II. Tolerances, Wear Limits, and Settings (continued)

Designation	From Chassis No.	New Part mm (in.)	Wear Limit mm (in.)
Front and rear . . . . . wall thickness	—	—	4.00 (.157)
out-of-round	—	0.10 (.004) max.	—
taper	—	0.10 (.004) max.	—
lateral runout	—	0.25 (.010) max.	—
radial runout (at friction surface)	—	0.10 (.004) max.	—
2. Brake discs			
Front only . . . . . thickness	211 2000 001	13.00–.20 (.512–.008)	11.5 (.452)
Thickness after machining	—	12.00 (.472) min.	11.5 (.452)
Machining depth per side	—	0.50 (.020) max.	—
Thickness variation	—	0.02 (.0008) max.	—
Disc runout	—	0.10 (.004) max.	—
C. Brake linings/friction pads			
1. Brake linings			
Front . . . . . width	218 000 001	45.00 (1.770)	—
Rear . . . . . width	218 001 001	45.00 (1.770)	—
	211 2000 001	55.00 (2.165)	—
Front and rear . . . . . thickness	218 000 001	5.0–4.8 (.200–.190)	2.5 (.100) remaining
Oversize . . . . . thickness	218 000 001	5.5–5.3 (.220–.210)	2.5 (.100) remaining
Rear only . . . . . thickness	211 2000 001	6.0–5.8 (.235–.230)	2.5 (.100) remaining
Oversize . . . . . thickness	211 2000 001	not available	—
2. Friction pad (for front disc brakes) . . . . . thickness	211 2000 001	10.00 (.393)	2.0 (.079) remaining
	213 2000 001	14.00 (.551)	2.0 (.079) remaining

III. Tire Data

Original equipment, tires	All models through July 1971	Campmobile from August 1971; Delivery Van and Station Wagon from August 1971 through March 1972	Delivery Van and Station Wagon from April 1972
Tire (tubeless)	7.00 x 14 8 PR	185 SR 14 Reinforced †	185 R x 14 C ‡
Rim x wheel size	5 JK x 14 (5½ J x 14 from Aug. 1970)	5½ J x 14	5½ J x 14
Radial runout	Maximum 1.25 mm (.050 in.)	Maximum 1.25 mm (.050 in.)	Maximum 1.25 mm (.050 in.)
Lateral runout	Maximum 1.5 mm (.060 in.)	Maximum 1.5 mm (.060 in.)	Maximum 1.5 mm (.060 in.)

Inflation pressures in psi (kg/cm²)
***(Under no circumstance must the cold tire pressure exceed the maximum inflated pressure marked on the tire)**

Models with conventional bias ply tires	Up to ¾ payload		Fully loaded	
	front	rear	front	rear
Station Wagon, 1968–1969 Campmobile, 1968–1969	28	36	28	36
Delivery Van, 1968–1971 Station Wagon, 1970–1971 Campmobile, 1970–1971	28	36	28	40
Prolonged high speeds (bias ply tires only) Add to the above specifications*	3 (0.2)	3 (0.2)	3 (0.2)	3 (0.2)
Models with radial ply tires	**Up to ¾ payload**		**Fully loaded**	
	front	rear	front	rear
Delivery Van from August 1971 up to March 1972	30	37	30	40
Station Wagon from August 1971 up to March 1972	30	34	30	40
Campmobile from August 1971	30	37	30	40
Delivery Van and Station Wagon from April 1972	30	40	30	44

† Correct replacement tire must bear word "REINFORCED". The maximum inflation pressure listed on the tire should be 40 psi.

‡ Correct replacement tire must bear the letter "C" immediately after the number 14 in the size designation—and not elsewhere. The maximum load inflation pressure listed on the tire should be 55 psi.

Section 9

LUBRICATION AND MAINTENANCE

Contents

TABLES

Lubrication and Maintenance

The service life of your car depends on the kind of maintenance it receives. The procedures described in this section of the Manual include all periodic checks and maintenance steps necessary for long and reliable operation. You will also find instructions for basic car care. Washing your VW and cleaning its interior need not be done at specific intervals of course, but doing so keeps your VW looking good.

The Owner's Manual and the VW Maintenance Record—or Owner's Manual and Warranty & Maintenance booklet—originally supplied with the car contain the maintenance schedules that apply to your VW. Following these schedules will ensure safe and dependable operation. Because several model years are covered in this Manual, some of the procedures described may not apply to your vehicle. If you are in doubt, always take the Owner's Manual, the Maintenance Record, or the Warranty & Maintenance booklet as your guide.

Some maintenance procedures, such as oil change service, require no special tools and can be carried out by almost all car owners, regardless of their mechanical experience. However, certain other diagnosis and maintenance operations require tools and equipment specifically designed for them. Wheel alignment checks, ignition timing, generator testing, and idle speed adjustments are a few examples. If you lack the skills, tools, or a suitable workshop for performing any of the service steps described, we suggest you leave such work to an Authorized VW Dealer or other qualified shop. We especially urge you to consult your Authorized VW Dealer before attempting any repairs on a car still covered by the new-car warranty.

1. TEST EQUIPMENT

The test equipment shown in this Manual is the kind of equipment that is often found in VW workshops. The methods described for using these pieces of equipment apply only to the equipment that is shown or described. Many car owners and mechanics will have other kinds of test equipment. Their equipment should be connected to the vehicle in the manner that is recommended by the test equipment's manufacturer.

Expensive and complicated equipment is not necessary for routine maintenance. However, it is necessary to use a stroboscopic timing light when you check the engine's ignition system. Static timing is not suitable to the kinds of distributors used on the vehicles covered by this Manual. Additional information on timing lights and their applications can be found in **ENGINE AND CLUTCH.**

Many of the vehicles covered by this Manual are equipped with a system of sensors and test wiring that terminates in a central control socket located in the engine compartment. This socket is designed to receive a plug from the cable of the VW Computer Analysis system. The socket should not be used in conjunction with other kinds of test equipment.

CAUTION —
Never connect any device other than the test plug of the VW Computer Analysis system to the test network central socket in the engine compartment. Incorrect equipment may damage the plug connections, test sensors, or the vehicle components containing them.

2. LUBRICANTS

Because of the recent improvements in the quality of commercially available lubricating oils, completely new oil recommendations for VW cars became effective in 1975. These new oil recommendations, given below, should be applied to all the VWs covered by this Manual.

The superseded oil recommendations, given in the Owner's Manuals supplied with 1968 through 1974 VWs, can still be used in servicing 1968 through 1974 models. You will find, however, that there are many advantages to be gained in adopting the new recommendations. Only the new oil recommendations, given in this Manual, should be applied to 1975 and later VWs.

The lubricants used in your VW have a vital influence on its operation. Use only name brand oils labeled "For Service SD" (or "For Service SE" or both) in the engine. Oils used in 1975 and later cars must be labeled "For Service API/SE". Automatic transmission fluid (ATF) must be labeled Dexron®. The hypoid oil used in the manual transmission must meet specification MIL-L 2105 API/GL4. Use a hypoid oil that meets specification MIL-L 2105 B API/GL5 for the final drive of the automatic transmission. No additives should be used in the engine oil, hypoid oil, or ATF. Experience has shown that name brand lubricants of the correct specification and viscosity meet all operating needs of VW engines and transmissions.

Oil viscosity must be suitable to climatic conditions. Viscosity is a term used to describe how readily a liquid flows. High viscosity oils seem thicker and pour more slowly at room temperature than do low viscosity (thinner) oils. When heated, however, oil loses some of its viscosity. A high viscosity oil heated to 93°C (200°F) may pour as readily as a low viscosity oil at room temperature. If an oil has too low a viscosity, it will not maintain an adequate lubricating film between moving parts. A thin, low viscosity oil may maintain this film at low temperatures but become so much thinner after it has warmed up that it leaves the engine parts unprotected.

It might seem that a high viscosity oil is all that is necessary to properly lubricate an engine. Unfortunately, this is not true. If a high viscosity oil is used during cold weather, it will become so thick and resistant to flow that it cannot properly circulate and reach the parts of the engine requiring lubrication. A thick, high viscosity oil will also become so gummy in cold weather that the starter cannot turn the engine fast enough to start it. The proper viscosity oil will remain fluid enough after the engine has cooled to permit easy starting, yet, after the engine has reached operating temperature, will retain sufficient viscosity to maintain an adequate lubricating film.

Single-grade engine oils, such as SAE 30, were formerly recommended for use in VW engines because of the unreliable quality of the then available multi-grade engine oils. The new high standard for engine oils that conform to the API (American Petroleum Institute) service rating API/SE has made such multi-grade oils suitable for use in the VW engine. Car owners will find that these high-quality multi-grade oils offer many advantages in convenience, performance, and economy—even in older model VWs.

For example, a single-grade oil may have to be discarded after a short period of service owing to the early arrival of winter temperatures. A multi-grade oil, suitable for both summer and winter temperatures, can be left in the engine until the normal oil change mileage has been reached. This feature of multi-grade oils can save the expense of oil changes necessitated by climatic conditions.

Also to be considered is that some oil grades and ratings, as originally recommended for 1968 through 1974 models, may eventually become unavailable. As old stocks of motor oil become depleted, new containers bearing the identifying marks of the new rating system will replace those bearing the identifying marks of older systems. By becoming familiar with the new VW oil recommendations, you can be sure that the oil you buy is the correct kind for your car. Inferior lubricants, no matter how attractively priced, are not a good investment; using the wrong oil will greatly shorten the service life of your VW.

Oil Viscosities

The viscosity grade of oil is designated by an SAE (Society of Automotive Engineers) standard number. An oil designated SAE 40 has a higher viscosity (greater resistance to flow) than an oil designated SAE 30. Multi-grade oils have an extended viscosity range and can be used in place of a number of single-grade oils. For example, an SAE 10W-30 oil is suitable for use within a range of temperatures that would require three different single-grade oils in order to cover it (SAE 10W, SAE 20W/20, and SAE 30).

Table a lists the proper oil viscosity for VW engines under specific climatic conditions. The viscosity SAE number of the oil should be selected for the lowest anticipated temperature at which engine starting will be required, and not for the temperature at the time of the oil change. Because the temperature ranges of the different oil grades overlap, brief variations in outside temperatures are no cause for alarm.

Table a. Engine Oil Viscosity Specifications

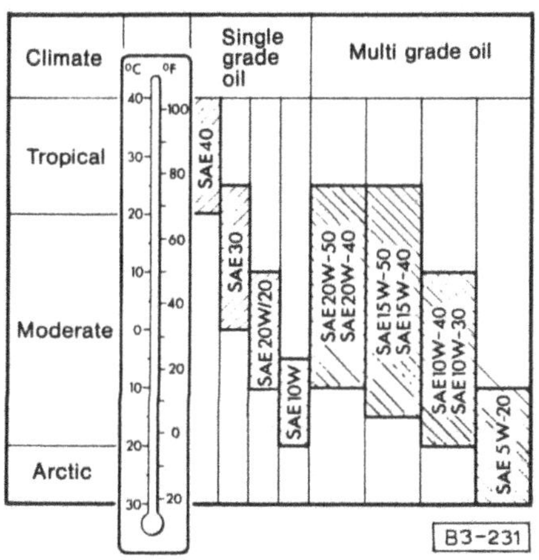

The viscosity of transmission oil is also designated by SAE numbers. Use SAE 80W or SAE 80W/90 for general year-round service in the manual transmission. Always use SAE 90 hypoid oil in the final drive of the automatic transmission. ATF is not graded for viscosity and requires no seasonal change when used in the automatic transmission. However, in arctic areas where temperatures remain consistently below −25°C (−13°F), you can also use ATF in place of hypoid oil in the manual transmission. The ATF should be replaced promptly with hypoid oil if summer temperatures exceed −25°C (−13°F).

Greases

Two types of grease are used for lubrication of chassis and driveline parts. Multipurpose grease (lithium grease) has a wider temperature tolerance range than ordinary grease and should be used for most lubrication purposes. The additives in multipurpose grease give it increased pressure resistance and anticorrosion capabilities. It is suitable for use both in plain bearings and in roller bearings. Molybdenum grease is lithium grease with a friction-reducing molybdenum disulfide additive. It is used in the constant velocity joints of the rear axle. Use dry stick lubricant in the hood locks and on the sliding surfaces of the door striker plates. The battery terminals should be coated with either silicone spray or petroleum jelly.

3. VW MAINTENANCE SCHEDULES FOR 1975 AND LATER CARS

The VW Maintenance schedules given here should be followed carefully even if your car is not being serviced by your Authorized VW Dealer.

3.1 Oil Change Service

Change the engine oil every 5000 mi. (8000 km) on 1975 cars or 7500 mi. (12,000 km) on 1976 and later cars. Change the oil every six months if you drive less than the specified distance during that time. Under severe driving conditions, you may have to change the oil more frequently than scheduled. An oil change includes draining the oil and refilling the crankcase with new oil of the correct grade. Every 15,000 mi. (24,000 km) you should clean the oil strainer and replace the oil filter before you refill the crankcase.

3.2 Brake and Valve Adjustment

The brakes and the engine valves of 1975 cars should be checked and, if necessary, adjusted after 10,000 mi. (16,000 km) of driving. On 1976 and later cars perform
these services at 7500 mi (12,000 km). Repeat these 9
services on 1975 and later cars at each additional 15,000 mi. (24,000 km) thereafter. Brake adjustment includes (1) checking the brake system for damage and leaks, (2) checking the fluid level and, if necessary, adding brake fluid, and (3) adjusting both the foot brakes and the parking brake. The valve clearance adjustment specified at these mileage intervals is recommended for maximum engine life but is not necessary for maintaining the VW Emission Control System Warranty.

3.3 Scheduled Vehicle and Emission Control Maintenance

To keep your Emission Control System Warranty in effect, all maintenance services affecting vehicle emission control have to be performed every 15,000 mi. (24,000 km) or once a year—if you drive less than 15,000 mi. (24,000 km) within twelve months. For additional information, please refer to the Emission Control System brochure supplied with the car. The following emission and vehicle maintenance schedule applies to 1975 and later VWs.

1. Engine: Change oil, clean oil strainer, replace oil filter.
2. Valves: Adjust clearance (unnecessary on 1978 models).
3. Spark plugs: Replace.
4. Ignition distributor: Replace ignition points. Adjust dwell angle and timing. Visually check ignition wires, distributor cap, and distributor rotor. Replace damaged or deteriorating parts.
5. Compression: Check.
6. V-belt: Check tension and condition; replace worn out belts.
7. Fuel filter: Replace.
8. EGR (exhaust gas recirculation) system: Check condition and operation visually.
9. Crankcase ventilation hoses: Check visually.
10. Exhaust system: Check for damage.
11. Fuel tank and EEC (evaporative emission control) system: Visually check fuel filler cap to see that it seals correctly; visually check charcoal filter canister and hose connections.
12. Clutch: Check and, if necessary, adjust freeplay.
13. Transmission: Check oil level and, if necessary, add oil.
14. Automatic transmission: Check and, if necessary, correct ATF (automatic transmission fluid) level. Also check oil in final drive.
15. Brakes: Check and, if necessary, adjust. Check brake pressure regulator visually.
16. Door hinges and sliding door mounting points: Lubricate.
17. Front axle: Lubricate.

Road test the car after you have completed the above maintenance work. During the road test you should check brake operation and observe the efficiency of the heating and ventilation systems. After the road test, check and, if necessary, adjust the engine idle speed and exhaust CO level. Check to see that the cylinder head covers are not leaking oil. Where applicable, recheck the ATF level.

In addition every 30,000 mi. (48,000 km):

1. Air cleaner: Replace filter element.
2. Vacuum passages in throttle valve housing: Clean (where applicable).
3. Automatic transmission: Change ATF.
4. Fuel injection system (1979 and later Calif.): Replace oxygen sensor.

In addition every two years:

1. Brakes: Replace brake fluid.
2. Brake warning light switches: Check operation.

3.4 Scheduled Vehicle and Emission Control Checks

A physical checkup is extremely important for determining the amount of additional maintenance your car may need for continuing peak performance. The checkup should include the items listed below.

Engine

1. Compression: Check.
2. Ignition system and timing: Check.
3. Exhaust system: Check for damage.

Rear axle and transmission

4. Driveshafts: Check boots for leaks.

Front axle and steering

5. Front axle: Check dust seals on ball joints and dust seals on the tie rod ends. Check the tie rods.
6. Ball joints: Check play.
7. Steering: Check play.
8. Front wheels: Check camber and toe.

Brakes, wheels and tires

9. Brake system: Check for damage and leaks.
10. Brake linings and pads: Check thickness.
11. Brake fluid: Check level and, if necessary, add fluid.
12. Tires, including spare wheel: Check for wear and damage; check and, if necessary, correct pressures.

Electrical system

13. Starting current: Check with electronic equipment.

14. Charging system: Check with electronic equipment.
15. Lights: Check all bulbs and switches for correct operation. Check horn and rear window defogger.
16. Headlights: Check adjustment.
17. Windshield wipers: Check operation.
18. Windshield washer: Check operation and, if necessary, add fluid.
19. Battery: Check electrolyte level; check voltage under load.
20. Buzzer and ignition/steering lock: Check.
21. Kickdown switch and solenoid: Check operation (where applicable).
22. Interior lights and instrument lights: Check.
23. Warning lights for alternator and oil pressure: Check.

4. VW MAINTENANCE SCHEDULES FOR 1968 THROUGH 1974 MODELS

The schedules given here list all the maintenance tasks specified for your VW. They should be followed carefully even if your car is not being serviced by your Authorized VW Dealer.

4.1 Oil Change Service

The engine in your VW requires little oil. But for long engine life, this oil should be changed every 3000 mi. (5000 km). An oil change at an Authorized VW Dealer includes the services listed below:

1. Engine: Change the oil, clean the oil strainer, check for leaks. (At this time, most mechanics also check the crankcase breather rubber valve on cars built prior to January 1972.)
2. Battery: Check the electrolyte. If necessary, add distilled water. Clean and grease the terminals.
3. Windshield washer: Check the fluid.

4.2 Scheduled Vehicle and Emission Control Checks

(every 6000 mi. or 10,000 km)

A physical checkup of your VW is extremely important for determining the amount of additional maintenance your car may need for continuing peak performance. The following VW schedule applies to the 1968 through 1974 VWs covered in this Manual.

Engine and clutch (including fuel system):

1. Generator or alternator V belt: Check the tension and condition.
2. Air pump V belt: Check the tension and condition (1973 and later models).
3. Ignition system: Check the dwell and the timing with electronic equipment.
4. Throttle positioner: Check for proper functioning (where applicable).
5. Compression: Check.
6. Exhaust system: Check for damage.
7. Manual transmission: Check the clutch pedal freeplay.
8. Engine: Check the engine oil level.

Transmission and rear axle:

Driveshafts: Check the boots for leaks.

NOTE —
At this time check the transmission for leaks, look for loose constant velocity joint screws, and inspect the shock absorbers.

Front axle and steering:

1. Front axle: Check the dust seals and the ball joint plugs (where applicable) for a proper fit.
2. Ball joints: Check the play.
3. Steering: Check the play. Check the tie rods and the dust seals on the tie rod ends.

NOTE —
At this time the shock absorbers and steering damper can be inspected and the steering gear checked for leaks.

Brakes, wheels, and tires:

1. Brake system: Check for damage and leaks.
2. Brake pedal: Check the pedal freeplay and the pedal travel (brake adjustment).
3. Parking brake: Check the adjustment.
4. Brake fluid: Check the level.
5. Brake linings or pads: Check thicknesses.
6. Tires (including spare): Check for wear and damage. Check and correct the pressure.

Electrical system:

1. Cranking system: Check with electronic equipment.
2. Charging system: Check with electronic equipment.

3. Kickdown switch and solenoid: Check operation (automatic transmission only).
4. Check the operation of the headlights, high beam indicator light, parking lights, side marker lights, license plate light, emergency flasher, stop lights, taillights, back-up lights, turn signals, horn, rear window defogger, and brake warning light.
5. Headlights: Check the adjustment.
6. Windshield wipers: Check the operation.
7. Windshield washer: Check the operation and the fluid level.
8. Battery: Check the electrolyte level. Check the voltage under load.

Test Drive

NOTE —
If maintenance or repair is to follow the checkup, delay the test drive until completion of all repairs.

1. Check the braking, clutch, kickdown, steering, heating, ventilation system (including fresh air fan), and overall vehicle performance.
2. Check the interior lights and the instrument lights.
3. Check the ignition/steering lock and the buzzer alarm.
4. Check the safety belt warning light and the buzzer alarm.
5. Check the warning lights for the generator or alternator and for oil pressure.
6. Check the operation of the automatic transmission (where applicable). Check the ATF level following the test drive.

4.3 Scheduled Vehicle and Emission Control Maintenance

The jobs listed here can be performed at the same time as the preceding checkup or following the checkup.

Every 6000 mi. (10,000 km):

1. Engine: Change the oil. Clean the oil strainer. Replace the oil filter (dual-carburetor engine only).
2. Valves: Check and adjust the clearance.
3. Ignition distributor: Check and adjust the dwell angle and the timing.
4. Engine idle: Check and, if necessary, adjust.
5. Door hinges, door checks, sliding door mounting points: Lubricate.

NOTE —
If necessary, the door and lid locks may also be lubricated at this time.

6. Manual transmission: Check the oil level. Add oil if necessary.
7. Automatic transmission: Check the oil level in the final drive housing. Add oil if necessary. Check the torque of the pan bolts.
8. Front axle: Lubricate (1970 through 1972 vehicles with torsion bar front axles).
9. Test drive as described earlier, then check the cylinder head covers for leaks.

In addition every 12,000 mi. (20,000 km):

1. Ignition distributor: Replace the breaker points, then adjust the dwell angle and the timing.
2. Ignition system: Visually check the distributor cap and rotor.
3. Spark plugs: Replace.

In addition every 18,000 mi. (30,000 km):

1. Front axle: Lubricate.
2. Oil bath air cleaner: Clean and refill the lower part with oil. Check the control flaps for preheated air.
3. Dry-type air cleaner: Replace the paper filter element (at least every two years). Check the control flaps for preheated air.
4. Air cleaner for air pump: Replace element (at least every two years).

In addition every 24,000 mi. (40,000 km):

1. Exhaust recirculation filter (1973 and later models only: Replace (at least every two years).
2. Exhaust recirculation valve (automatic transmission only): Check, then replace if necessary.

In addition every 30,000 mi. (50,000 km):

1. Front wheel bearings (1968 through 1972 vehicles): Clean and repack.
2. Rear wheel bearings (1968 through 1972 vehicles): Clean and repack.
3. Automatic transmission: Change the ATF (includes removing and installing the oil pan).

NOTE —
Change the ATF every 18,000 mi. (30,000 km) if the vehicle is used for trailer towing or other heavy-duty service or if it is regularly driven in constant stop-and-go traffic, at high outside temperature, or under continuous mountain conditions.

In addition every 48,000 mi. (80,000 km):

Activated charcoal filter: Replace (1970 vehicles sold in California and all 1971 and later models).

Every two years:

1. Brakes: Replace the brake fluid.
2. Brake warning light switch: Check functioning.

5. LUBRICATION AND MAINTENANCE OPERATIONS

The maintenance schedules given earlier can be used as a checklist when you are servicing your vehicle. For your additional help, specifications and instructions for performing the more complex checks and maintenance tasks follow.

5.1 Engine and Clutch
(including fuel system and fuel injection)

If you drive only short distances during the winter months, change the oil twice as often as recommended in the maintenance schedules. If you drive only a few hundred miles a month under these conditions, change the oil every six to eight weeks. In arctic areas with temperatures generally below −25°C (−13°F), change the oil four times as often as recommended in the maintenance schedules.

In addition to the normal oil change, you should change engine oil at 600 mi. (1000 km) following engine rebuilding. Doing so gets rid of any metallic particles that have accumulated in the oil during break-in.

Changing Engine Oil, Cleaning Oil Strainer, And Replacing Full-flow Oil Filter

The oil should be drained from the engine while it is hot. Drain the oil by placing a pan of at least one gallon capacity beneath the engine. On single-carburetor engines, loosen all six cap nuts on the oil strainer cover (Fig. 5-1). Pry the oil strainer cover loose. When most of the oil has been drained, remove the oil strainer for cleaning.

1. Gasket
2. Oil strainer
3. Gasket
4. Cover plate
5. Cap nut with washer
6. Oil drain plug with washer

Fig. 5-1. Single-carburetor engine oil strainer removal.

On dual-carburetor and fuel injection engines, remove the drain plug (Fig. 5-2). When most of the oil has drained, remove the oil strainer for cleaning.

Fig. 5-2. Dual-carburetor engine drain plug (**A**). The oil strainer is removed by taking out the bolt at **B**.

On single-carburetor engines, use new gaskets and copper washers when you install the oil strainer. Torque the nuts to 0.7 mkg (5 ft. lb.). On dual-carburetor and fuel injection engines, torque the strainer's sealing bolt to no more than 1.3 mkg (9 ft. lb.).

To change the full-flow oil filter on 1700, 1800 and 2000 engines, use a special wrench such as the one shown in Fig. 5-3. The filter should be changed on carburetor engines at every 6000 mi. (10,000 km) or, on fuel injection engines, at every 15,000 mi. (24,000 km).

© 1974 VWoA—3559

Fig. 5-3. Special wrench in position over full-flow oil filter on dual-carburetor and fuel injection engines.

Fill single-carburetor engines with 5.3 U.S. pints (4.4 Imperial pints, 2.5 liters) of oil labeled "For Service SD" (or "SE" or both). Consult **2. Lubricants** to determine the proper viscosity.

When you have installed a new full-flow filter, fill dual-carburetor and fuel injection engines with 7.4 U.S. pints (6.1 Imperial pints, 3.5 liters) of oil labeled "For Service SD" (or "SE" or both). If you do not change the filter, use 6.3 U.S. pints (5.3 Imperial pints, 3.0 liters) of oil. Consult **2. Lubricants** to determine the proper viscosity.

After filling the crankcase, check the oil with the dipstick. Pull the dipstick out and wipe it dry. Insert it all the way and pull it out again. The oil level should be at the upper mark. After running the engine, look closely at the oil strainer cover and the cylinder head covers to see that no oil is leaking past the gaskets. Replace the gaskets if you find leaks. On 1968 through 1971 vehicles, check the condition of the rubber valve over the lower end of the crankcase breather tube. Replace the valve if it is torn or if its slot does not remain closed.

Checking and Adjusting Generator or Alternator Belt

To check the generator belt tension on single-carburetor engines, press down on the belt halfway between the generator and crankshaft pulleys. The belt should deflect about 10 mm (3/8 in.). If the deflection is significantly greater or smaller than this amount, remove the nut from the center of the generator pulley. Wedge a screwdriver between one of the notches in the rear of the pulley and one of the generator through bolts to keep the pulley from turning. Separate the pulley halves. To tighten the belt, remove shim washers from between the pulley halves; to loosen the belt, add shim washers. Store the leftover shim washers between the pulley outer half and the hat-shaped special washer (Fig. 5-4).

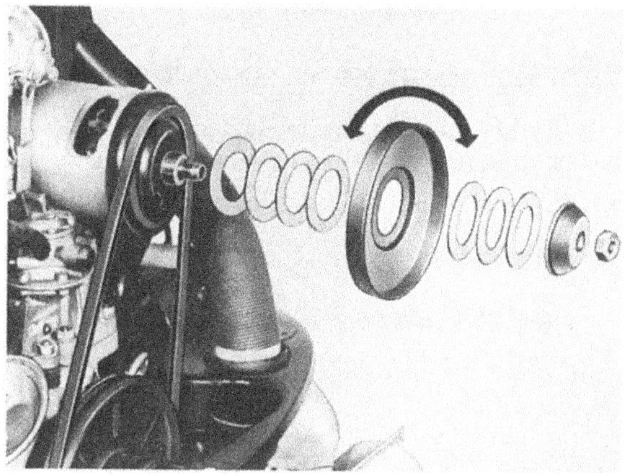

© 1974 VWoA—1995

Fig. 5-4. Single-carburetor engine belt adjustment. Transfer shim washers from one side of the pulley outer half to the other as indicated by the double arrow.

To check the alternator belt tension on dual-carburetor and fuel injection engines, press down on the belt halfway between the alternator and crankshaft pulleys. The belt should deflect about 15 mm (5/8 in.). If the deflection is significantly greater or smaller than this amount, remove the oblong insert from the alternator cover plate. Use a 12-point driver to loosen the socket head bolt shown in Fig. 5-5, then loosen the alternator mounting bolt. Move the alternator as indicated by the arrows until the belt tension is correct. Then tighten the alternator mounting bolt and the socket head bolt.

© 1974 VWoA—3611

Fig. 5-5. Alternator belt adjustment. Arrows indicate the direction in which the alternator can be moved.

Checking Compression

To check the cylinder compression, remove all four spark plugs. Then install a compression testing gauge in one of the spark plug holes. Crank the engine with the starter for a few moments while the accelerator is pressed to the floor or while the throttle valve on the carburetor is held wide open. Note the reading and repeat the procedure with the other cylinders. Compression pressures for the various engines covered by this manual are listed in **ENGINE AND CLUTCH, Tolerances, Wear Limits, and Settings**. If the pressure is below 100 psi (7 kg/cm^2) on 1968 through 1971 models or below 71 psi (5 kg/cm^2) on 1972 and later models, the valves may need to be ground, piston rings replaced, or new cylinders and pistons installed.

Before assuming the problem is in the valves, however, check the valve adjustment to make sure that there is at least a small amount of clearance between the rocker arms and the valve stems. If there is no measurable gap, loosen the adjuster enough to obtain clearance and repeat the compression test. If pressure is still low, the valves need grinding.

Servicing and Replacing Spark Plugs

The spark plugs, which were removed for the compression test, should not be reinstalled until after the valves have been adjusted. This will make it easier to turn the engine by hand. On 1968 through 1974 models, check the gap and the condition of the spark plug electrodes each 6000 mi. (10,000 km). Install new spark plugs each 12,000 mi. (20,000 km). The fuel injection engine introduced in 1975 uses long-life spark plugs. You need only replace such spark plugs each 15,000 mi. (24,000 km).

Adjust the gaps of new spark plugs before you install the plugs in the engine. The correct gap for 1970 through 1974 engines is 0.60 mm (.024 in.). The correct gap for fuel injection engines is 0.70 mm (.028 in.). If the electrodes are rounded, file them to the square profiles of new electrodes before gapping. Additional data on servicing spark plugs can be found in **ENGINE AND CLUTCH**.

The following or similar spark plugs are suitable for use:

Single-carburetor engine
Normal service: Champion L-88-A; Bosch W 145 T1; Beru 145/14
High-speed service: Champion L-85 or L-86; Bosch W 175 T1; Beru 175/14

Dual-carburetor engine
All service: Champion N-88, Bosch W 145 T2; Beru 145/14/3

Fuel injection engine
All service: Champion N-288; Bosch W 145 M2; Beru 145/14/3L.

Checking and Adjusting Valve Clearance

Adjust the valves to the clearances listed in **Table b** with the cylinder in position to fire and the engine cold (oil temperature no more than 50°C (122°F)). See **ENGINE AND CLUTCH** for complete data and instructions.

Table b. Valve Clearance Specifications

Single-carburetor engine or Fuel injection engine	Intake: 0.15 mm (.006 in.) Exhaust: 0.15 mm (.006 in.)
Dual-carburetor engine, through Chassis No. 213 2210 553	Intake: 0.15 mm (.006 in.) Exhaust: 0.15 mm (.006 in.)
Dual-carburetor engine, from Chassis No. 213 2210 554	Intake: 0.15 mm (.006 in.) Exhaust: 0.20 mm (.008 in.)

Checking Distributor

The distributor breaker point dwell (or gap) should be checked when the ignition timing is checked and adjusted. However, there are several service checks that should be made beforehand.

First, remove the distributor cap and wipe it clean inside and out. Then clean the tower of the ignition coil. Check to see that the carbon contact in the center of the cap is not worn away. Carefully inspect the cap and the rotor for carbon tracks, using a magnifying glass if necessary.

Clean away grease, dust, or other foreign matter that may have accumulated inside the distributor. Check to see that the breaker points do not have excessive pitting or buildup. If a feeler gauge is used to check the breaker point gap, first clean the gauge. Otherwise, oil or other foreign matter that may be adhering to it could contaminate the point contacts. Lubricate the distributor as shown in Fig. 5-6.

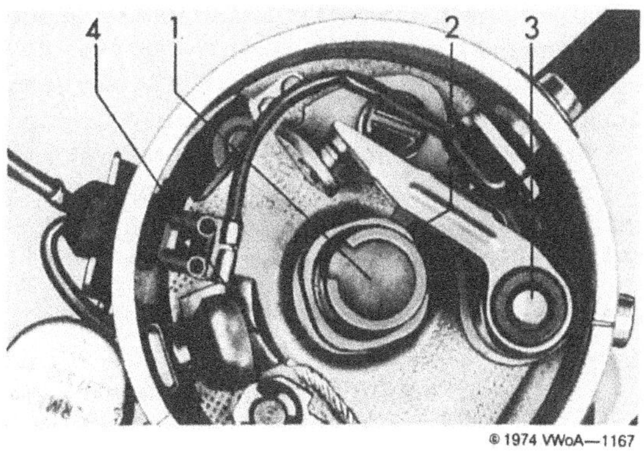

Fig. 5-6. Distributor lubrication. Apply a few drops of engine oil to the felt wick (**1**); one drop to the pivot (**3**). After cleaning away old lubricant, put a small amount of multipurpose grease on the cam and rubbing block (**2**) and on the advance plate guide ball (**4**).

9

Complete information on replacing and adjusting the breaker points, testing the rotor electrically, and checking the high voltage cables can be found in **ENGINE AND CLUTCH**. The section also contains details for disassembling the distributor and checking the spark advance mechanisms.

Checking and Adjusting Engine Idle

For accuracy, adjust the idle speed after the road test. On 1968 through 1974 cars with manual transmissions, set the idle to 800 to 900 rpm. For 1975 and later cars with manual transmissions, set the idle to 850 to 950 rpm. All vehicles with automatic transmissions should idle at 900 to 1000 rpm. To adjust the idle, use only the procedures given in **FUEL SYSTEM** or in **FUEL INJECTION**.

Checking Ignition Timing

The breaker point dwell (or point gap) and the idle speed must be correct before you check or adjust the timing. Set the breaker point dwell to 44° to 50° for new points or 42° to 58° for used points. Alternatively, you can use a feeler gauge to set the point gap to 0.40 to 0.50 mm (.016 to .020 in.).

On 1968 through 1970 cars that have a single vacuum hose attached to the distributor, disconnect and plug the hose. On later models, leave the hose(s) attached. Using a stroboscopic timing light, check the timing with the engine idling at the specified rpm and the oil temperature at 30° to 70°C (86° to 158°F). On single-carburetor engines, the V-notch in the crankshaft pulley (the left V-notch on 1968 through 1970 engines) should be aligned with the central vertical seam in the crankcase.

On dual-carburetor and fuel injection engines, the timing mark on the crankshaft pulley should be aligned with the correct mark on the degree scale. This should be 10° after top dead center on 1973 and 1974 engines used with manual transmissions or 5° after top dead center on all 1975 fuel injection engines, all dual-carburetor engines used with automatic transmissions, and on 1972 dual-carburetor engines used with automatic transmissions. The 1976 fuel injection engine timing mark is at 7.5° BTDC.

Checking Exhaust System

Momentarily cover the tailpipe opening with a pad of rags while the engine is running. Hissing sounds coming from under the car signal leakage. Darker or lighter areas around joints in the system may indicate escaping gases. Check the heater flaps for free movement. Lubricate the pivots with dry stick lubricant. If the flaps are difficult to move, treat them with penetrating oil.

Servicing Air Cleaner

To check the oil bath air cleaner, release the clips that hold the top part of the air cleaner. Do so by raising their wire bails upward. Disconnect any hoses attached to the top part of the air cleaner, carefully noting their position for future installation. Remove the top part of the air cleaner, but do not lay it down with the filter up. Doing so will allow contaminated oil to drain into the upper part of the unit.

Inspect the lower part of the air cleaner. If less than 5 mm (3/16 in.) of oil remains above the sludge layer in the bottom of the lower part of the air cleaner, the air cleaner must be removed, cleaned, and filled with fresh oil.

When you remove an oil bath air cleaner, hold it level to prevent spillage. Removal requires only that the various hoses be disconnected from the air cleaner and the clamp that holds the air cleaner on the carburetor be loosened. Carefully note the location of each hose since accidentally interchanging them will adversely affect the operation of the engine.

On 1968 through 1970 models, the cable for the intake air preheating flap must be disconnected from the air cleaner. First remove the spring clip that holds the cable eye on the warm air control flap arm. Unhook the cable eye from the arm, then loosen the screw in the cable retainer. Pull the cable out of the retainer but leave it attached to the engine. During installation, adjust the cable as described in **FUEL SYSTEM.**

A solvent may be used for final cleaning after the sludge has been wiped out of the air cleaner's lower part. Fill the lower part of the air cleaner to the baffle (Fig. 5-7) or to the mark, using fresh SAE 30 engine oil. SAE 10 W oil should be used in subarctic climates. Use the same viscosity oil all year round.

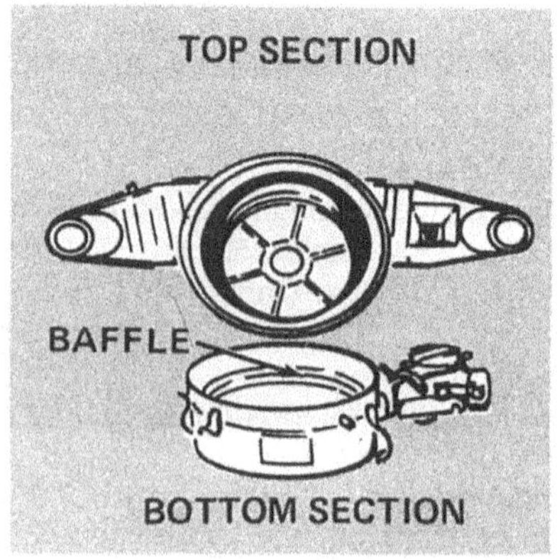

© 1974 VWoA—3723

Fig. 5-7. Air cleaner filling level.

The top part of the air cleaner does not normally require cleaning. However, if the bottom part of the filter has become so dirty that the air inlet holes are partially blocked, they should be cleaned with a scraper. Solvent should never be used to wash the upper part of the air cleaner unless it is allowed to dry completely before reinstallation. Otherwise, dripping solvent will contaminate the oil in the bottom part.

Installation is the reverse of removal. When you service and install any type of air cleaner, check the intake air preheating system as described in **FUEL SYSTEM**. Be sure that the warm air control flaps move freely.

The paper element air cleaner used on 1973 and 1974 cars does not need to be removed from the engine when the filter is cleaned or replaced. To remove the paper element, disconnect the three hoses, pry off the four release clips, and then lift the top section of the air cleaner. Remove the paper element filter and shake the accumulation of dirt from it. Tap the filter—dirty side down—on a hard surface. Install a new paper element each 18,000 mi. (30,000 km). Please consult **FUEL INJECTION** for all information related to the air cleaner used on 1975 and later vehicles.

CAUTION —
Do not clean the paper element with solvents or gasoline or saturate it with oil. These practices will destroy its effectiveness.

With the top section of the air cleaner lifted, as illustrated in Fig. 5-8, install the paper element in the position shown. Replace the original paper element if it is damaged, prematurely clogged with dust, or cannot be cleaned by shaking.

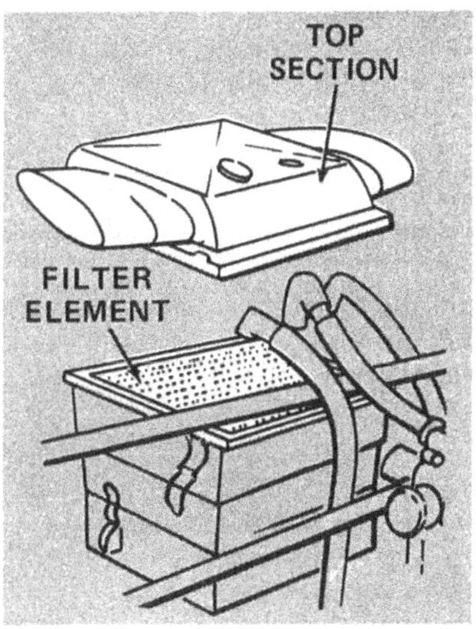

© 1974 VWoA—3724

Fig. 5-8. Paper element filter positioned in bottom section of air cleaner.

Servicing Exhaust Gas Recirculation Filter

(1973 and later models only)

The filter for the exhaust gas recirculation system should be replaced every 24,000 mi. (40,000 km) or two years, whichever comes first. The filter is located under the vehicle, just ahead of the muffler. To remove the filter, remove the two nuts that hold it on the pipes. Install a new filter only. Used filters cannot be cleaned. Filter replacement is not required on 1975 and later vehicles.

Replacing Exhaust Afterburning Air Pump Filter

The air cleaner for the exhaust afterburning air pump has a pleated paper element filter. Remove the hose clamp that holds the air cleaner on the pump. Then remove the air cleaner and disassemble it as shown in Fig. 5-9. Using a new filter element, assemble and install the air cleaner.

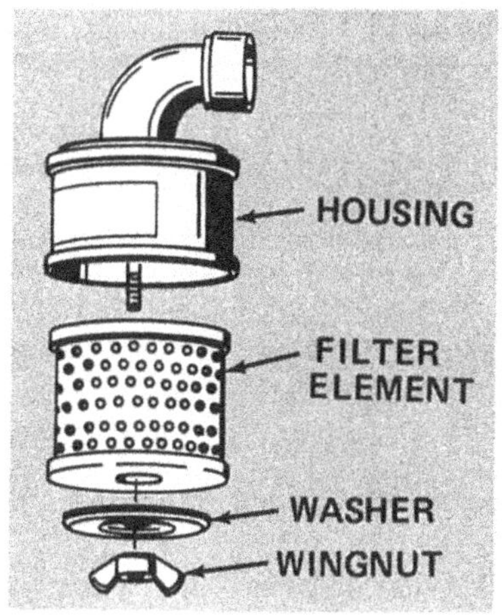

© 1974 VWoA—3725

Fig. 5-9. Components of air pump air cleaner.

Replacing Evaporative Emission Control Canister

Except on 1975 and later models, the activated charcoal filter canister for the evaporative emission control system must be replaced after each 48,000 mi. (80,000 km) or two years, whichever comes first. The filter is located in the upper right-hand side of the engine compartment. To remove the filter for replacement, detach the hoses and unscrew the bolts for the bracket that holds the canister.

Checking Clutch Pedal Freeplay

(manual transmission only)

On 1968 through early 1973 models, it should be possible to depress the clutch pedal 10 to 20 mm (3/8 to 3/4 in.) before encountering working resistance. On vehicles manufactured since February 1973, it should be possible to depress the clutch pedal 15 to 25 mm (5/8 to 1 in.) be-

9

fore encountering working resistance. If the freeplay is outside this range, adjust it as described in **ENGINE AND CLUTCH**. If the longer clutch cable introduced in February 1973 has been installed in an earlier dual-carburetor engined vehicle, use the larger freeplay dimension.

5.2 Manual Transmission and Automatic Transmission

Both the manual transmission and the final drive housing of the automatic transmission must be kept properly filled with hypoid gear oil. Two types of gear oil are commonly available. The first is high pressure transmission oil, commonly designated HP for high pressure. Hypoid transmission oil suitable for VW vehicles is designated EP for extreme pressure. Hypoid oils for use in the manual transmission and in the final drive of the automatic transmission must meet the specifications given in **2. Lubricants**. Oils that meet these specifications have a sulphur-phosphorus additive base that creates a protective coating on gear surfaces.

Checking Hypoid Oil and Cleaning Magnetic Plug

The hypoid oil level in both manual and automatic transmissions should be kept at the lower edge of the filler hole in the side of the transmission case or final drive housing. The level is correct if hypoid oil just barely drips from the hole with the plug removed. The level may be considered satisfactory if you can feel it with your fingertip just below the edge of the filler hole. Hypoid oil is added, if necessary, through the oil filler hole (Fig. 5-10).

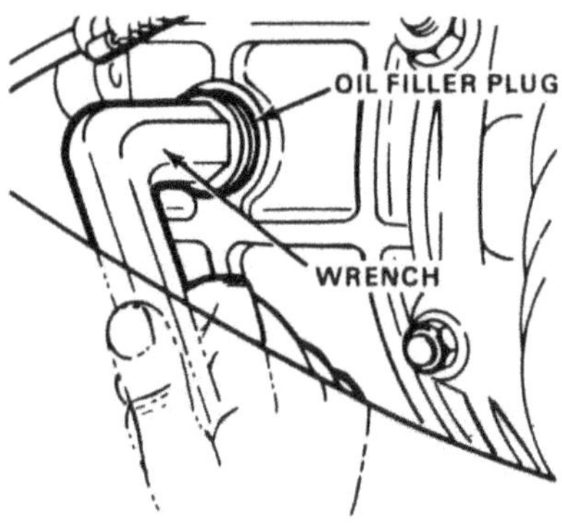

© 1974 VWoA—2758

Fig. 5-10. Oil filler plug being removed from oil filler hole. A 17-mm hexagonal wrench is required.

The drain plug in the bottom of the manual transmission contains a magnet that traps metallic particles as they settle in the oil. The accumulation can be cleaned from the magnet periodically by removing the plug. Have a cork of appropriate size ready to plunge into the transmission drain hole as soon as the magnetic plug is removed. Very little hypoid oil will be lost if the filler plug is left installed so that air cannot enter readily. After cleaning the magnetic plug and reinstalling it, correct the hypoid oil level to make up for any that was lost.

Changing Hypoid Oil

(1968 and 1969 models only)

It is unnecessary to change the hypoid oil on 1970 and later vehicles, unless the oil has become contaminated, or a temperature change makes it necessary to use an oil of a different viscosity. However, the hypoid oil should always be changed 600 mi. (1000 km) after rebuilding the transmission.

The oil will drain faster if it is warm. On manual transmissions, remove both the filler plug and the drain plug. On automatic transmissions, remove the rearmost oil pan. Install the drain plug after drips have slowed to one every 20 seconds. Use a new gasket when you reinstall the oil pan on automatic transmissions. Torque the pan screws to 1.0 mkg (7 ft. lb.).

Refill with hypoid oil of the proper specification and viscosity (see **2. Lubricants**). On manual transmissions, pause occasionally while refilling to give the oil time to flow into the final drive portion of the case. If you attempt to put the oil in too rapidly, it may overflow and give the impression that the case is already full although only 2 or 3 pints have been put in. The manual transmission requires 7.4 U.S. pints (6.1 Imperial pints; 3.5 liters). The final drive of the automatic transmission requires 2.0 U.S. pints (1.8 Imperial pints; 1.0 liter).

Checking Constant Velocity Joint Boots

The rubber boots over the constant velocity joints should not be cracked or torn. Instructions for removing the driveshafts and replacing the boots can be found in **TRANSMISSION AND REAR AXLE**.

Checking Constant Velocity Joint Screws

The socket head screws that hold the constant velocity joints to the flanges should be torqued to 3.5 mkg (25 ft. lb.).

Checking and Correcting ATF Level

(automatic transmission only)

The dipstick for the ATF in the automatic transmission is inserted into the ATF filler tube above the left, forward side of the engine. The dipstick is attached to the cap for the ATF filler tube. The level should be checked with the ATF warm and the engine idling. The transmission selector lever must be in neutral and the parking brake set. Correct ATF level is very important for the proper functioning of the automatic transmission.

Pull out the dipstick and wipe it clean. The ring-shaped handle should be in a vertical position when you reinsert the dipstick to measure the fluid level.

ATF level is correct only if it falls between the two marks on the dipstick. Add ATF, if necessary, but only as much as is needed. Keep in mind that the difference between the lower and upper marks is only 13.5 U. S. oz. (11.25 Imperial oz.; 398 cc). To add ATF, use a clean funnel with a 50-cm (20-in.) hose attached. The ATF added must be labeled Dexron®.

Changing ATF and Cleaning Sump and Strainer

This task must be performed each 30,000 mi. (50,000 km) under normal driving conditions and each 18,000 mi. (30,000 km) under severe conditions, such as trailer towing, stop-and-go driving, or extended mountain driving.

Remove the drain plug from the transmission pan and allow as much fluid as possible to drain. Remove the screws that hold the pan. Pry the pan loose from the bottom of the transmission and lower it carefully. Remove the ATF strainer shown in Fig. 5-11 for cleaning by taking out the central screw.

NOTE ——
The filter-type strainer used on 1976 and early 1977 models cannot be cleaned and should be replaced if it is dirty or clogged. On late 1977 and all later vehicles, a cleanable screen-type strainer was again installed.

1. Manual valve
2. Kickdown solenoid
3. ATF strainer
4. Transfer plate
5. Valve body

Fig. 5-11. Automatic transmission, pan removed. ATF strainer should be removed, cleaned with solvent, and dried carefully with compressed air. Do not wipe it with rags since this could leave lint behind.

CAUTION ——
Do not tow the car or run the engine while there is no ATF in the transmission. This could ruin the bearings in the transmission.

Clean the pan. Remove all traces of old gasket material from both the pan and the transmission case.

CAUTION ——
Never use fluffy rags when cleaning the automatic transmission. Lint from them could clog ATF passages.

Install the strainer. Then install the pan, using a new gasket. Working diagonally, tighten the pan screws to 1.0 mkg (7 ft. lb.) on pans with 13 bolts or to 2.0 mkg (14 ft. lb.) on pans with four bolts. Wait five minutes for the new gasket to compress, then retorque the screws. Repeat this sequence several times until the screws remain at torque.

CAUTION ——
Never exceed the torque specifications in an attempt to cure a leaking gasket. Overtightening will deform the pan and make it impossible to get a good seal. Always install a new gasket to correct leaks.

Refill the transmission with 6.3 U. S. pints (5.3 Imperial pints; 3.0 liters) of ATF. Do not fill above the top mark on the dipstick, as checked with the ATF warm, the engine idling in neutral, and the parking brake set. The ATF used must be labeled Dexron®.

Checking Kickdown Switch Operation
(automatic transmission only)

On 1972 through 1975 models, the kickdown switch is located beneath the accelerator pedal. Operation should be checked as part of road testing. If the kickdown fails to operate, check that it is being operated by the accelerator pedal. If satisfactory, test the switch for electrical continuity. If the switch is sound, test the wiring and solenoid inside the transmission as described in **AUTOMATIC TRANSMISSION.** On 1976 and later vehicles, the kickdown is controlled by the accelerator cable. Adjust the cable as described in **FUEL INJECTION** if the kickdown does not operate correctly.

Lubricating Rear Wheel Bearings
(except 1973 and later models)

After 30,000 mi. (50,000 km) of service, the rear wheel bearings should be removed, cleaned, and packed with fresh multipurpose grease. The rear wheel driveshafts and constant velocity joints must first be removed. At this time the joints should be checked to see that they work smoothly and do not feel gritty. If they are stiff or are suspected of being dirty, they should be disassembled, cleaned, and packed with molybdenum grease as described in **TRANSMISSION AND REAR AXLE.**

Remove the brake drum, wheel hubs (1971 and later models only), and brake backing plates. Drive or press the rear wheel shafts out of the rear wheel bearings.

NOTE —
There is one roller bearing race (outer) and one ball bearing race (inner) at each rear wheel.

Withdraw the ball bearings and the roller bearings from the bearing housing in the rear suspension diagonal arm. This and the preceding operations are described in **TRANSMISSION AND REAR AXLE.**

Carefully clean the bearings in solvent and dry them with compressed air. Clean the bearing housing and other related parts. Pack the bearing races with multipurpose grease. A pressure bearing lubricator is best for this, but the bearings can be packed by hand. Work the grease in between the balls or rollers, then turn the bearing slightly and repeat the process until it is completely filled with grease. Reassemble the parts, replacing the oil seals in the bearing housing so that no grease can reach the brake linings.

Checking Shock Absorbers

If the shock absorbers seem to lack control during the road test, or if the visual inspection shows them to be physically damaged, they should be checked and, if necessary, replaced or repaired.

The best way to check a shock absorber is to remove it from the car and extend and compress it by hand. Damping action should be uniform and extend throughout the entire range of travel. If possible, compare it with a new shock absorber. Faulty shock absorbers should be replaced. However, leakage around the rod does not demand that the shock absorber be replaced unless its action has been impaired. Shock absorbers can be replaced individually as long as the new shock absorber is the same type as the old one.

Checking Transmission and Final Drive For Leaks

A leak from the transmission or final drive should be considered serious if there is wet, fresh oil present at all times. Such leaks should always be corrected by installing new gaskets or seals and never by tightening screws or bolts above their specified torque. The best time to check for leaks is just after a road test.

Checking Pan Screw Torque

(automatic transmission only)

On 1972 through 1974 models, the automatic transmission oil pan screws should be torqued to 1 mkg (7 ft. lb.) each 6000 mi. (10,000 km).

CAUTION —
Never tighten pan screws over 1 mkg (7 ft. lb.) in an attempt to cure a leaking gasket. Overtightening will deform the pan and make it impossible to get a good seal. Always install a new gasket to correct leaks.

5.3 Front Axle and Steering

Most of the maintenance steps involving the front axle can be performed with the car on a lift. It is therefore convenient to do these jobs while you drain the engine oil or service the transmission.

Lubricating Front End

The front axle has five grease fittings (Fig. 5-12). On 1968 through 1970 models, the front axle should be greased every 6000 mi. (10,000 km). On 1971 through 1974 models, lubricate the axle every 18,000 mi. (30,000 km); on 1975 and later models, lubricate it every 15,000 mi. (24,000 km).

Fig. 5-12. Front axle grease fittings. Wheels must be off the ground while axle is being greased.

© 1974 VWoA—3563

The grease fittings should always be wiped clean before pumping grease into them. Keep pumping until fresh grease appears in the joint between the torsion arm or relay lever and the axle beam. Afterward, wipe all surplus grease from the fittings and the suspension parts.

Checking Ball Joint Play

To check the play in the upper and lower ball joints for the front suspension, use the procedures described in **FRONT AXLE**. Play specifications are also given there.

Lubricating Front Wheel Bearings

(except 1973 and later models)

After 30,000 mi. (50,000 km) of service, the front wheel bearings should be removed, cleaned, and packed with fresh multipurpose grease.

CAUTION—
Use only multipurpose (lithium) grease to lubricate the wheel bearings. Other greases will not maintain adequate lubrication and may lead to bearing failure.

Loosen the socket head clamping screw and remove the clamp nut from the stub axle. Then pull the brake drum or brake disc and front wheel hub off the stub axle by hand. Exercise care to avoid dropping the thrust washer and outer wheel bearing race. The outer bearing race can be lifted out of the hub. Before the inner bearing race can be removed, however, it is necessary to pry the oil seal out of the back of the hub. Install a new oil seal during assembly.

Clean the hub cavity with solvent and inspect the condition of the bearing races inside. These steel cups should not be roughened, worn irregularly, or blued by heat. If they are, replace them as described in **FRONT AXLE.** Wash the roller bearing races in solvent and dry them carefully with compressed air.

CAUTION—
Do not use solvents such as gasoline that remove every trace of lubrication. Also, do not spin the races with a blast of compressed air. Without adequate lubrication, the bearings can be damaged by movement.

If a pressure bearing lubricator is not available, get a palmful of multipurpose grease and thrust the edges of the roller bearing races into it, continuing around the bearing until it is completely filled. Coat the races inside the hub with the same lubricant. Do not pack large quantities of grease inside the hub; just coat the interior lightly to prevent corrosion. Clean the stub axle, inspect its bearing seating surfaces to see that they are not roughened or blued by heat, then apply a corrosion-preventing layer of multipurpose grease. Install the bearings and a new oil seal in the hub and install the brake disc on the stub axle.

The wheel bearings must then be adjusted. Install a dial indicator as shown in Fig. 5-13. The brake disc should be turned slightly and axial play measured at several different positions. The readings should not vary greatly and their average should fall within a range of 0.03 to 0.12 mm (.001 to .005 in.). If necessary, bring axial play within this range by turning the clamp nut. See **FRONT AXLE** for additional instructions.

Fig. 5-13. Checking front wheel bearing axial play.

Checking Steering Play

To check the worm and peg steering used on 1968 through 1972 models, the vehicle must be lifted. Turn the steering back and forth through its centered position several times. You should feel resistance as the steering passes through its centerpoint, but it should do so smoothly and without sticking. If there is no resistance or if the steering binds, adjust the steering gearbox as described in **FRONT AXLE.**

To check the worm and roller steering used on 1973 and later models, the vehicle must be lifted. With the front wheels in their straight-ahead position, grip one steering wheel spoke at its extreme outer end. Turn the steering

9

wheel lightly in both directions. The freeplay should not exceed 15 mm (5/8 in.) measured at the wheel rim. If there is excessive freeplay or if the steering binds, adjust the steering gearbox as described in **FRONT AXLE.**

Checking Ball Joint and Tie Rod Seals

The rubber seals on the suspension ball joints and tie rod ends should be checked to make certain that none is torn or cracked. To replace seals, use the procedures given in **FRONT AXLE**. Also check to see that the plastic plugs in the tops of the 1970 through early 1973 suspension ball joints have not been lost. Procedures for lubricating the ball joints are given in **FRONT AXLE**, but this step is required only if broken seals have allowed dirt to enter or if the joints have become stiff.

Checking Wheel Alignment

Check the camber and toe-in of the front wheels. Camber should be 40′ ± 20′. The difference in camber between the wheels should not be more than 30′. Toe-in should be 15′ ± 15′—or 5′ ± 5′ after adding 12 to 18 kg (27 to 39 lb.) of extra weight above the wheel. The maximum toe change should not exceed 25′. If you use a track gauge to check toe with the wheels in their straight-ahead position, the measurement made at the rear of the wheels should be 0.0 to 3.3 mm (0.000 to 0.130 in. or approximately 0 to 1/8 in.) greater than the measurement made at the front of the wheels. If necessary, consult the detailed procedures given in **FRONT AXLE.**

Checking Shock Absorbers

The instructions given for checking shock absorbers in **5.2 Manual Transmission and Automatic Transmission** also apply to the shock absorbers fitted to the front axle.

5.4 Brakes and Wheels

Properly servicing the brake system at the prescribed intervals is vital to safe operation of your vehicle. Skill and approved parts or materials are necessary for satisfactory work.

Checking and Changing Brake Fluid

The fluid level in the brake fluid reservoir on 1968 through 1972 models should be maintained at 15 to 20 mm (about 3/4 in.) below the top (just above the seam near the top of the reservoir). On 1973 and later models, the fluid level must be visible through the window in the front of the reservoir. Brake fluid added to the system must meet SAE recommendation J 1703 and conform to Motor Vehicle Safety Standard 116. The brake fluid should be changed every two years. Detailed procedures for this job are in **BRAKES AND WHEELS**. Test the warning light switch at this time also.

Checking Brake Pedal Travel and Freeplay

To check brake pedal freeplay, use the procedures given in conjunction with master cylinder pushrod adjustment in **BRAKES AND WHEELS.**

If the pedal must be depressed an excessive distance to obtain braking action, the drum brakes require adjustment. For access to adjusters, remove the rubber plugs from the brake backing plates. If necessary, consult the adjustment procedures in **BRAKES AND WHEELS.**

Checking Brake Linings

Brake lining wear should always be checked at the specified mileage. Disc brake pads should not be worn to a remaining thickness of less than 2.00 mm (.080 in.). Drum brake linings should have at least 2.5 mm (3/32 in.) of friction material left above the brake shoe. Except on very early models that have the wheel hub integral with the brake drum, you can check the lining thickness through holes in the backing plates after the rubber plugs have been removed.

Checking Parking Brake Adjustment

You should not be able to pull out the parking brake handle more than five clicks without obtaining noticeable braking action. Make this check only after the rear wheel brakes have been adjusted. If the handle requires more than five clicks to obtain braking, adjust the parking brake as described in **BRAKES AND WHEELS.**

Checking Brake System for Leaks

If the fluid level in the brake fluid reservoir is noticeably low, inspect the brake system for leaks. Check the lines, hoses, unions, and bleeder valves. If there is no sign of leakage, remove the brake drums and lift the rubber boots at the ends of each wheel cylinder. If fluid leaks out, the wheel cylinder is faulty and must be rebuilt or replaced as described in **BRAKES AND WHEELS**. If there is no leakage at the rear brakes on 1971 and later models, make a similar check of the disc brake calipers by lifting the dust seals away from the pistons. If there is leakage past the pistons, rebuild the caliper as described in **BRAKES AND WHEELS.**

Checking Brake Lines and Hoses

The brake lines must not be deeply pitted by corrosion, flattened due to bending, or dented by flying stones. These conditions may soon lead to leaks or may already be interfering with uniform braking action. The brake hoses must not be soft, swollen, or coated with grease and oil. They must not be abraded from rubbing against moving suspension parts or cut due to road hazards. Replace any brake hose that has breaks in its fabric outer cover even though bulges are not yet apparent. Details can be found in **BRAKES AND WHEELS.**

Checking Brake Light Operation

The rear brake lights should come on as soon as the brake pedal encounters resistance when depressed. If not, test the system as described in **BRAKES AND WHEELS.**

Checking Brake Warning Light Switch

Check the brake warning light system every two years (as described in **BRAKES AND WHEELS**) while changing the brake fluid.

Checking Tire Condition and Tire Pressures

Inspect the tires for cuts, separated plies, and abnormal wear. See that the tread wear indicators are not showing. The causes of abnormal tire wear are described in **BRAKES AND WHEELS**. Keep about 42 psi (3.0 kg/cm²) of air in the spare tire. Inflate the tires on the road wheels to the following specifications.

Conventional (bias ply) Tires				
	Up to ¾ payload		**Fully loaded**	
Model	**Front**	**Rear**	**Front**	**Rear**
Station Wagon and Campmobile, 1968-1969	28 psi	36 psi	28 psi	36 psi
Trucks, 1968-1971 Station Wagon and Campmobile, 1970-1971	28 psi	36 psi	28 psi	40 psi

Radial (radial ply) Tires				
	Up to ¾ payload		**Fully loaded**	
Model	**Front**	**Rear**	**Front**	**Rear**
Station Wagon, from August 1971 through March 1972	30 psi	34 psi	30 psi	40 psi
Delivery Van, from August 1971 through April 1972; Campmobile from August 1971	30 psi	37 psi	30 psi	40 psi
Station Wagon and Delivery Van from April 1972	30 psi	40 psi	30 psi	44 psi

NOTE —
Increase the pressures in conventional bias ply tires by 3 psi for prolonged high speeds, but never exceed the maximum tire inflation pressure listed on the tire sidewall.

5.5 Electrical System

Headlight adjustment may be checked following the instructions given in **ELECTRICAL SYSTEM**. However, there are in some states legal specifications for headlight aim that may differ from those recommended by VW. In some areas it is also a legal requirement that headlights be adjusted only by state-authorized shops.

Checking Lights

Checking the lights takes only a few moments. Each light or light system should be switched on and the lights checked visually. Make certain that the generator and the oil pressure warning lights in the instrument panel both light when the ignition is turned on and the engine is not running.

Check the operation of all electrical switches, including those in the door frames. Make certain that the turn signal and emergency flasher switches each produce the appropriate lighting action. Be sure to check the neutral safety switch on vehicles with automatic transmissions. The engine should start only with the selector lever in **N** (neutral) and in **P** (park). Note that the headlight dimmer switch is combined with the turn signal lever. Pulling the lever toward the steering wheel actuates a relay that changes the headlights from high to low beam or from low to high beams.

Checking Windshield Wipers and Washers

On 1968 through 1972 models, the windshield wiper control switch is located on the dashboard. On 1973 and later vehicles, the switch is a lever on the right-hand side of the steering column. Check that the wiper blades are in good condition. A troubleshooting chart for wiper defects can be found in **ELECTRICAL SYSTEM.**

The windshield washers work by compressed air stored in the reservoir. Keep the reservoir filled with the correct quantity of water or a cleaning solution such as VW's Windshield Washer Anti Freeze and Solvent. Mix as directed on the container. The reservoir holds 2.9 U.S. pints (2.4 Imperial pints; 1.4 liters). The air pressure should be maintained at 42 psi (3.0 kg/cm²).

Checking and Filling Battery

The battery is in the engine compartment on the right. If the battery terminals are corroded, correct the condition as described in **ELECTRICAL SYSTEM**.

Check the specific gravity of the electrolyte with a hydrometer. A reading of 1.250 to 1.280 is good; 1.225 to 1.250, fair. Anything below 1.225 is poor. Check the electrolyte level, which can be seen through the translucent battery case. If the level is low, unscrew each filler plug, then bring the electrolyte level up to the indicators with distilled water. Never overfill the battery.

5.6 Body and Frame

There are only a few routine maintenance steps for the body of your car. But they are important steps.

Lubricating Door and Hood Hinges

Pry the plastic plug out of the top of each passenger door hinge. There should be oil in the chamber atop the hinge pin. If not, fill the chamber with SAE 30 engine oil. Press the plug back in and wipe away excess oil. Use the same kind of oil on all hood and lid hinge pins, again being sure to wipe away excess oil. Also apply SAE 30 oil to the notches in the hinges for the sliding door and the rear cargo door (Fig. 5-14).

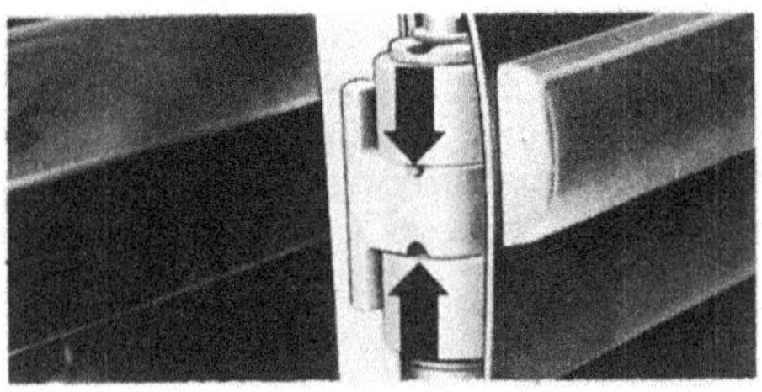

© 1974 VWoA—3565

Fig. 5-14. Lubrication notches (arrows) on sliding door and rear cargo door hinges.

Lubricating Locks

Lubricate the rubbing surfaces of all latches and striker plates with dry stick lubricant. Lubricate door lock cylinders with graphite. Dip the key into graphite powder and turn it in the lock a few times.

6. BASIC CAR CARE

The following brief guide will help you keep your VW looking as good as it runs.

6.1 Care of Car Finish

The longer dirt is left on the paint, the greater the risk of damaging the glossy finish, either by scratching or simply by the chemical effect dirt particles have on the painted surface.

Washing

Never wash the car in direct sunlight. Beads of water not only leave spots when dried rapidly by the sun's heat, but act as tiny magnifying glasses that burn spots into the finish. Use plenty of water, a car-wash soap such as VW's Car Wash and Wax, and a soft sponge.

Begin by spraying water over the dry car to remove all loose dirt. Then apply lukewarm soapy water. Rinse the car after sponging off the soapy water, using plenty of clear water under as little pressure as possible. Wipe the car dry with a chamois or soft terrycloth towel to prevent water-spotting.

Waxing

For a long-lasting, protective, and glossy wax finish, apply a hard wax, such as VW's Classic Car Wax, after the car has been washed and dried. Waxing is not needed after every washing, and a more effortless shine can be obtained by using a car-wash liquid containing wax. You can tell when waxing is required by looking at the finish while it is wet. If the water coats the paint in smooth sheets instead of forming beads that roll off, waxing is in order.

Polishing

Use paint polish, such as VW's Paint Polish, only if the finish assumes a dull look after long service. You can use polish on the car's brightwork to remove tar spots and tarnish, but afterwards apply a coat of wax to protect the clean plating.

Washing Chassis

The best time to wash the underside of the car is just after it has been driven in the rain. Spray the chassis with a powerful jet of water to remove dirt and deicing salt that may have accumulated there.

Special Cleaning

Tar spots can be removed with tar remover. Never use gasoline, kerosene, nail polish remover, or other unsuitable solvents. Insect spots also respond to tar remover. A bit of baking soda dissolved in the wash water will facilitate their removal. This method can also be used to remove spotting from tree sap.

The windshield wiper blades can be removed periodically and scrubbed with a hard bristle brush and alcohol or a strong detergent solution to remove debris. The windows themselves can be cleaned with a sponge and warm water, then dried with a chamois or soft towel. If you use commercial window washing preparations, make certain they are not damaging to automotive finishes.

6.2 Care of Interior

The rubber weatherstrips around windows and doors must be kept pliable if they are to remain effective. Spray these parts with silicone spray, or coat them with talcum powder. Petroleum products are harmful to rubber and should never be used.

Cloth Upholstery and Carpet

Clean cloth interiors with a vacuum cleaner or whisk broom. Dirt spots can usually be removed with lukewarm soapy water. Use spot remover for grease and oil spots. Do not pour the liquid directly on the upholstery, but dampen a clean cloth and rub carefully, starting at the edge of the spot and working inward. Do not use gasoline, naptha, or other flamable substances to clean the upholstery.

Leatherette Upholstery and Trim

Use VW's All Purpose Cleaner or other dry foam cleaner. Grease or paint spots can be removed by wiping with a cloth soaked with VW's All Purpose Cleaner. Use the same cleaner, applied with a soft cloth or brush, on the headliner and side trim panels.

6.3 Tires and Accessories

Never use tar remover, gasoline, or any other petroleum-based substance for cleaning tires. Such liquids damage rubber. Whitewall tires can be cleaned with VW's All Purpose Cleaner. Rubber paints, commonly sold as tire dressing, are largely cosmetic.

Accessories

Most chrome-plated accessories can be polished and waxed along with the rest of the car's bright trim. The radio antenna should be lubricated only if hardened grease and collected dirt are interfering with raising or lowering the antenna. Do not use abrasive polish or cleaners on aluminum trim or accessories. They will destroy the mirror-like shine of anodized surfaces.

The safety belts in your VW should be kept clean. If cleaning is necessary, wash them with a mild soap solution without removing them from the car. Do not bleach or dye safety belts or use any other cleaning agents. They may weaken the webbing.

Carefully check the condition of the webbing while you are cleaning the belts or the interior of the car. Frayed or damaged belts should be replaced.

Section 10

FUEL INJECTION

Contents

Fuel Injection

The electronic fuel injection system covered in this section of the Manual is used only on 1975 and later models. Also covered here are the emission controls used on 1975 and later VWs. The fuel tank, the evaporative emission control, and the fuel gauge sending unit are covered in **FUEL SYSTEM**. These components have remained virtually unchanged from those used on VWs with carburetors. Whether equipped with carburetors or fuel injection, all VWs covered by this Manual are designed to operate on regular (91 octane) gasoline. Lead-free gasoline is required for fuel injection cars that have catalytic converters.

The electronic fuel injection system includes an electric fuel pump, fuel lines, an air cleaner, and a complex system for mixing fuel and air in precisely controlled proportions and for delivering the mixture to the cylinders of the engine. An electronic control unit, commonly called the computer or "brain," monitors various engine operating factors, including the volume of air flowing into the engine, then computes the proper amount of fuel to be delivered for every driving condition. This AFC (air flow controlled) fuel injection system also provides the exacting mixture control necessary for reduced exhaust emissions.

Unlike a carburetor, fuel injection does not depend on the velocity of the incoming air to draw fuel into the engine. Instead, fuel is injected under pressure into the air in the engine's intake ports. The injection pulses are timed to the firing of the engine, ensuring that fully vaporized fuel reaches the combustion chambers. The amount of fuel delivered is primarily determined by the length of the injection pulse. At closed throttle the mixture is made leaner through a reduction in fuel pressure.

Electronic testing requires only a test light and either an ohmmeter or a VOM (a common electrical testing instrument that can be used to measure accurately either ohms resistance or voltage). You should, however, have a thorough basic knowledge of electricity and be familiar with reading wiring diagrams before you attempt such tests. If you lack the skills, instruments, or other equipment necessary for testing the electronic fuel injection system, we suggest you leave such tests or repairs to an Authorized VW Dealer or other qualified and properly equipped shop. We especially urge you to consult your Authorized VW Dealer before attempting repairs on a car still covered by the new-car warranty.

1. GENERAL DESCRIPTION

Fig. 1-1 is a schematic view of the AFC (air flow controlled) electronic fuel injection system that was introduced on the 1975 Type 2 Volkswagens. To help you understand this diagram, a brief explanation is given here of the function of each of the system's components. A detailed description of each component can be found under the heading for the particular component; for example, **4.8 Testing and Replacing Intake Air Sensor**.

For the sake of description, it is convenient to divide the fuel injection system into three subsystems. These subsystems are the fuel circuit, the air system, and the electronic controls.

Fuel Circuit

The electric fuel pump draws fuel from the tank through the filter and into the pressure line to the ring main. The ring main consists of the fuel hoses that supply the four injectors and the cold-start valve. The ring main is connected to a pressure regulator. In order to limit the maximum pressure in the ring main to a predetermined level, the pressure regulator allows surplus fuel from the pump to flow back to the fuel tank. The pressure regulator is not adjustable, its output being controlled by a calibrated spring and by engine vacuum.

Each injector consists of a jet orifice with a solenoid-operated needle valve. All four injectors are triggered simultaneously at every other opening of the ignition distributor's breaker points. The length of the electrical im-

Fig. 1-1. Schematic view of AFC fuel injection.

1. Fuel filter
2. Fuel pump
3. Pressure regulator
4. Cold-start valve
5. Injector
6. Auxiliary air regulator
7. Intake air sensor
8. Throttle valve housing
9. Intake air distributor
10. Temperature sensor I
11. Thermo-time switch
12. Potentiometer with fuel pump switch
13. Throttle valve switch or microswitch (1975–1976)
14. Resistor
15. Temperature sensor II
16. Control unit
17. Ignition distributor

© 1976 VWoA—3851

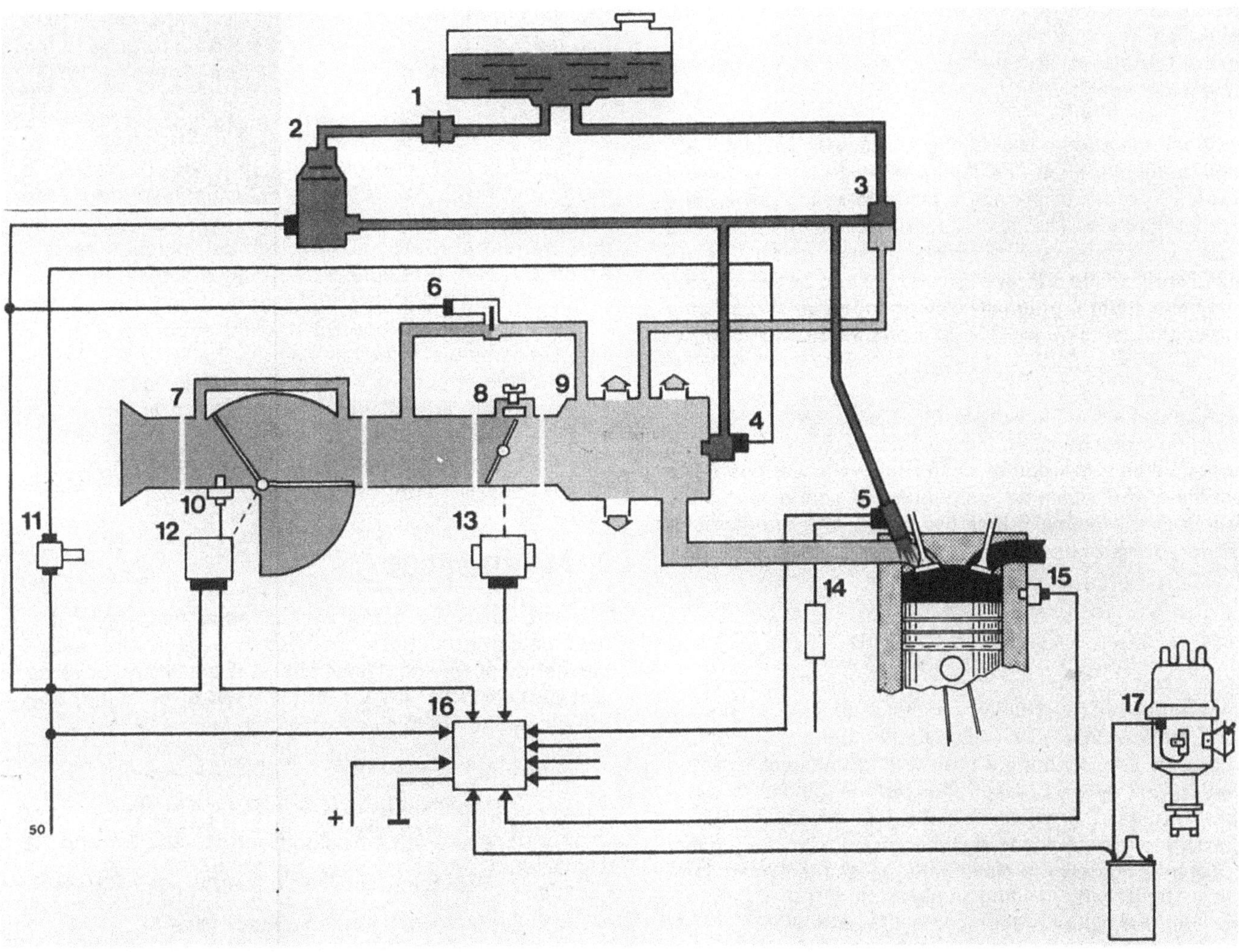

pulse that holds the injector needle valves open is determined by the control unit.

The cold-start valve, sometimes called the fifth injector, injects fuel into the intake air distributor only while the starter is being operated. However, if the thermo-time switch senses that the air surrounding the engine is warm, the cold-start valve does not operate. The thermo-time switch also cuts off current to the cold-start valve if the engine is cranked for a longer than normal period of time in order to get the car started in cold weather. The purpose of the cold-start valve time limit is to prevent flooding.

Air System

The components of the air system are located atop the engine. All air enters the engine through an air cleaner (not shown in Fig. 1-1). From the air cleaner, the air enters the intake air sensor. The intake air sensor consists of a two-vane stator flap that operates a potentiometer and a fuel pump switch. As installed on the car, the intake air sensor is connected by an S-shaped rubber duct to the throttle valve housing which, in turn, is held to the intake air distributor by screws. A bypass screw, located in the throttle valve housing, is used to adjust the engine's idle speed.

When the engine is cold, the auxiliary air regulator admits additional air to the intake air distributor at closed throttle in order to provide a faster idle until the engine has warmed up. The auxiliary air regulator consists of a rotary valve operated by an electrically heated thermostatic spring. The spring begins to heat up as soon as the engine is started, gradually closing the valve and slowing the engine to its normal idle at a predetermined warm-up rate.

As installed on the vehicle, the intake air distributor is a stamped metal housing. It is painted black and located at approximately the center of the top of the engine. The outlets of the intake air distributor are connected to the four intake manifold tubes by pieces of large-diameter fabric-covered hose.

Electronic Controls

The basic fuel metering of the fuel injection system is dependent on the volume of intake air and on engine rpm. Air flow data is transmitted to the control unit by the intake air sensor; engine rpm data is transmitted to the control unit by the breaker points of the ignition distributor. The intake air sensor's stator flap, which moves according to the volume of air being drawn into the engine, constantly adjusts the electrical output of the potentiometer—thereby transmitting an electrical signal to the control unit that varies proportionately to air flow.

In addition to the intake air sensor and the ignition distributor, which are the two main sources of data for the control unit, there are three other electronic controls that signal the control unit—when special operating conditions demand it.

The throttle valve switch (1975 and early 1976 cars) or a microswitch (1976 cars built from February 1976) provides full load enrichment whenever the accelerator pedal is fully depressed, or nearly so. On 1977 and later models there is neither a throttle valve switch nor a microswitch; full load enrichment is handled automatically by the electronic control unit.

The throttle valve switch also controls the operation of the EGR (exhaust gas recirculation) valve; the movement of the throttle valve shaft causes the switch to shut off the EGR valve both at full load and at idle. On 1976 cars that have a microswitch, and on 1977 and later models, the EGR valve is operated by a mechanical linkage attached to the throttle valve shaft.

Temperature sensor I, located inside the intake air sensor, senses intake air temperature and incorporates this data in the output signal of the potentiometer (1975 and early 1976 cars) or transmits the data directly to the electronic control unit (cars built from February 1976).

Emission Controls

The EEC (evaporative emission control) system is covered in **FUEL SYSTEM.** The PCV (positive crankcase ventilation) system and the EGR (exhaust gas recirculation) system are integral with the fuel injection system which, in itself, functions as a highly effective emission control system (the Controlled Combustion System listed in the VW emission control booklet supplied with the car). On some fuel injection VWs, an Emission Afterburning catalytic converter is built into the exhaust system. These cars are identifiable by a fuel filler that will accept only the small pump nozzles used for lead-free gasoline. The 1979 and later California models have an oxygen sensor emission control system.

2. MAINTENANCE

There are only a few maintenance operations that must be carried out at a specified mileage or after a certain period of service. These are listed below and covered in **LUBRICATION AND MAINTENANCE** or under the listed headings in this section of the Manual.

1. Servicing the air cleaner
2. Replacing the fuel filter (see **4.10**)
3. Checking emission controls (see **5.1** and **7.**)
4. Replacing catalytic converter (see **ENGINE**)
5. Replacing oxygen sensor (See **8**).

3. TROUBLESHOOTING FUEL INJECTION

Two things are important prerequisites to successful troubleshooting of the fuel injection system. First, you should be thoroughly familiar with the function of the system as a whole. See **1. General Description**. Before making any electrical tests or fuel circuit tests, make very certain that no hoses in the air system are leaking and that there are no air leaks in the air system components themselves. Further information about the function of each injection system component will be found under the heading for testing and replacing that component.

The second prerequisite to successful troubleshooting is your ability to read the wiring diagram given in Fig. 3-1. If you need help in understanding any of the symbols given in this diagram, please refer to the Wiring Diagrams topic in **ELECTRICAL SYSTEM**. There you will find explanations of all the various symbols. Note especially that the two diodes shown in the double relay in Fig. 3-1 permit current to flow only in the direction indicated by the arrow-like shape of the diode symbol. The part of the wiring diagram that is below the broken line supplements Wiring Diagram B given in **ELECTRICAL SYSTEM**.

Fig. 3-1. Wiring diagram of electronic fuel injection system. The system is also shown in the 1976 and later wiring diagrams given in **ELECTRICAL SYSTEM.** The 1979 California oxygen sensor emission control connection is not shown here.

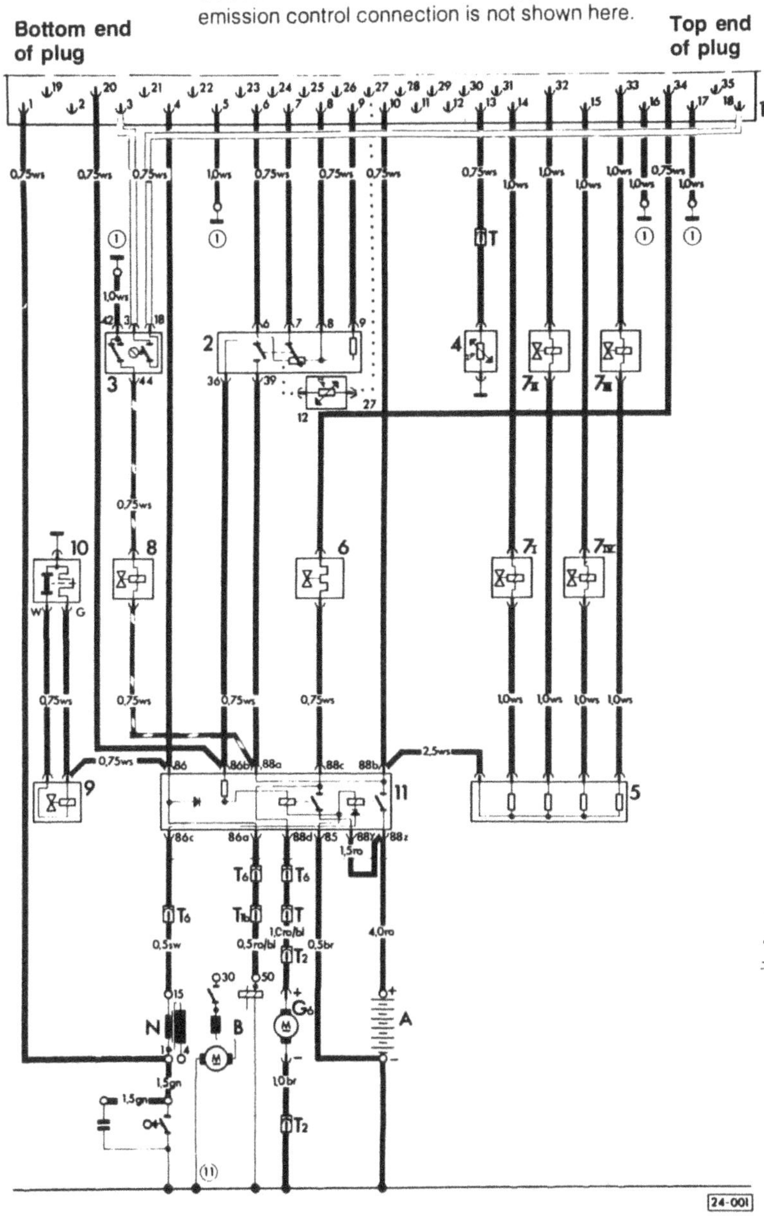

1—Electronic control unit plug
2—Intake air sensor
3—Throttle valve switch
4—Temperature sensor II
5—Series resistance block
6—Auxiliary air regulator
7—Injectors
8—EGR valve (through January 1976)
9—Cold-start valve
10—Thermo-time switch
11—Double relay
12—Temperature sensor I (from February 1976)
N— Ignition coil
G_6— Fuel pump
A— Battery
B— Starter
T— Wire connector
T_{1b}— Wire connector
T_2— Wire connector, double
T_6— Wire connector, 6 point
(1)— Ground connector near alternator
(11)— Ground connector, headlights

Color Codes

ws = white
sw = black
br = brown
ro = red
gn = green
ro/bl = red/blue

··········· Circuit used only from February 1976
— — — Circuit used only through January 1976
═══ Circuit used only on 1975 and 1976 models

Before you begin troubleshooting the fuel injection system, make sure that the engine trouble is not caused by something other than a fuel system problem. As with carburetors, no fuel injection test or adjustment should be made until you are confident that the engine has adequate compression and that the ignition system is not faulty. Because low charging system output can also upset the operation of the fuel injection system, make sure that the voltage in the electrical system is between 12.5 and 14.5 volts.

Erratic fuel metering can be caused by bouncing or arcing ignition distributor breaker points. The fuel injection system control unit, or computer, is programmed to trigger an injection pulse at every other opening of the ignition distributor's breaker points. If the breaker points are bouncing, owing to incorrect dwell or a weak spring, extra injection pulses will be triggered. If the points are arcing, the control unit may fail to sense the opening of the points. Breaker point dwell adjustment, point inspection, and ignition timing are described in **ENGINE**.

Misfiring is not a typical symptom of a faulty fuel injection system. If you encounter misfiring—or if the car starts hard, fails to start, or has inadequate power—check for carbon tracking at the distributor rotor, the distributor cap (Fig. 3-2), and the coil (Fig. 3-3). Replace faulty components. Also check the resistance of the spark plug connectors and the distributor rotor as described in **ENGINE**. Neither connector nor rotor resistance should exceed 10,000 ohms.

© 1976 VWoA—3835 © 1976 VWoA—3836

Fig. 3-2. Typical carbon tracks in distributor cap.

© 1976 VWoA—3837

Fig. 3-3. Typical carbon track on coil tower.

Finally, do not overlook incorrect valve clearances, faulty spark plugs, a dirty air cleaner, or a restricted exhaust system as possible causes of engine trouble. Because the fuel metering is highly dependent on the measurement of intake air volume, any air entering the engine that does not pass through the intake air sensor will cause the engine to operate incorrectly. Typically, it will idle slowly or stall at idle and give symptoms of lean operation at driving speeds. Look for a loose-fitting oil filler cap, leaking or disconnected ventilation hoses, a leaking cylinder head cover gasket, or leaking pushrod tubes on the engine.

Troubleshooting Procedures

The Bosch AFC (air flow controlled) fuel injection system has been designed so that you can make all the necessary electrical tests using only a test lamp and an ohmmeter. In addition to these two electrical instruments, you will need a fuel pressure gauge to test the fuel pump and the pressure regulator.

The double relay is located in the engine compartment on the left-hand side of the fuel tank compartment panel, just in front of the engine's No. 3 cylinder. This relay controls the supply of electrical current to every electrical component of the fuel injection system. Consequently, if current fails to reach the fuel pump or any of the injection system components in the engine compartment, your next step should be to test the double relay.

For example, if the cold-start valve fails to operate during the cranking of a cold engine in cold weather, you would first use a test lamp to determine whether current is reaching the cold-start valve during cranking. If there is no current, you should test the double relay terminals which supply current to the cold-start valve. First, you would check for current to double relay terminal 86a while the starter is being operated. If there is no current, the wire to starter terminal 50 is faulty. If there is current, repeat the test at terminal 86 of the double relay. If there is no current, the double relay is faulty; if there is current, you know that the wire from terminal 86 to the cold-start valve is faulty.

Even though the wiring diagram in Fig. 3-1 shows that many components receive current from the control unit, you should keep in mind that the control unit must first receive this current from the double relay. Therefore, test the double relay thoroughly before you assume that there is trouble in the control unit. Experience has shown that faulty double relays are encountered many times more often than faulty control units.

No attempt should be made to test the electronic control unit itself. Applying test current to its terminals or bridging the terminals could ruin the control unit. The control unit should be judged faulty only if, after careful testing, the wiring and all other components of the fuel injection system are proved to be in good working order.

Before you begin testing the fuel injection system components, select mentally those components that are most likely to be causing the trouble. If the engine is running rich, begin by checking the components of the fuel circuit, then proceed to the intake air sensor and temperature sensor II. If the engine is running lean, concentrate your initial troubleshooting on the components related to the air system—with special attention to the intake air sensor and the auxiliary air regulator.

Once you have (1) determined that there is no engine fault outside the fuel injection system; (2) thoroughly checked the air system for leaking or disconnected hoses; and (3) determined which components are most likely to be causing the trouble, go to the appropriate component topics under **4. Testing and Repairing Fuel Injection System**. There you will find a description of each component's function, the method of testing the component, and instructions for replacing faulty components.

4. TESTING AND REPAIRING FUEL INJECTION SYSTEM

The order in which the components are discussed under the following headings is not intended to indicate which components should be tested first. The sequence in which you test the components should be based upon your own analysis of the operating problem, an understanding of the system as a whole, and on the wiring diagram that is given in **3. Troubleshooting Fuel Injection**.

> ***CAUTION—***
> *If you lack the skills or equipment necessary for testing and repairing the fuel injection system, we suggest you leave such work to an Authorized VW Dealer or other qualified shop. We especially urge you to consult your Authorized VW Dealer before attempting tests or repairs on a car still covered by the new-car warranty.*

4.1 Testing and Replacing Double Relay

The double relay is a small device, approximately 20 mm by 20 mm and about 45 mm in length (¾ in. by ¾ in. and about 2 in. in length). It has eleven numbered terminals. The number of each terminal is printed beside the terminal on the relay. Inside the double relay are two electromagnetic switches (relays), two diodes, and a resistor. These components and their connections were shown earlier in Fig. 3-1.

The double relay is located in the engine compartment on the left-hand side of the fuel tank compartment panel, just in front of the engine's No. 3 cylinder. It is mounted just below the metal cover of the series resistance block. Two white plastic multiple connector plugs connect the double relay to the fuel injection system wiring harness.

The double relay controls the supply of battery voltage to all parts of the fuel injection system—including the control unit. When the engine is stopped, the double relay breaks all the circuits between the fuel injection system and the car's electrical system. The basic sequence of double relay operation is as follows:

1. When the starter is operated, current from terminal 50 on the starter solenoid supplies positive (+) current to terminal 86a of the double relay, causing one of the relays—the fuel pump relay—to close and send battery voltage from terminal 30 on the starter to the electric fuel pump, via double relay terminals 88y and 88d.
2. Because the ignition is turned on when the starter is operating, ignition primary terminal 15 delivers positive (+) current to double relay terminal 86c, thereby closing the power relay that sends positive (+) current from the battery to all electrical parts of the fuel injection system with the exception of the fuel pump, the cold-start valve, and the auxiliary air regulator—which, except for the cold-start valve, receive positive (+) current from the pump relay. The power relay remains closed as long as the ignition is on and will not open again until the ignition is turned off.
3. When the engine has started, the driver moves the key from the start (3) position to the running (2) position. This removes the supply of positive (+) current that was reaching double relay terminal 86a from terminal 50 of the starter—the current that had originally closed the pump relay. However, the pump relay now receives current from the pump switch in the intake air sensor, via terminal 86b and the resistor of the double relay. Whether or not the pump operates depends on whether or not the pump switch in the intake air sensor opens or closes the pump relay.
4. When the driver stops the engine by turning off the ignition, ignition current no longer reaches double relay terminal 86c. The power relay opens, breaking all positive (+) connections between the fuel injection system and the car's electrical system.

The two diodes in the double relay isolate the current that reaches the relay while the engine is running from the current sources that closed the relays during starting. Failure of either diode will prevent its associated relay from closing during starting, causing a malfunction of the fuel injection system. In testing the double relay, make

10

the following two tests before making any other test. In doing so you will check (1) that the double relay is getting current from the car's electrical system and (2) that the diodes are intact and the relays are functioning correctly.

To test current supply:

1. Detach the double relay from the tank compartment panel. Connect one test probe of a test lamp to terminal 85 of the double relay (which is grounded to the negative post of the battery).
2. Apply the second test probe to terminal 88y, as shown in Fig. 4-1. Repeat the test by applying the second test probe to terminal 88z. The test lamp should light during both repetitions of the test. If not, positive (+) battery current is not reaching the terminal to which the second probe is connected.

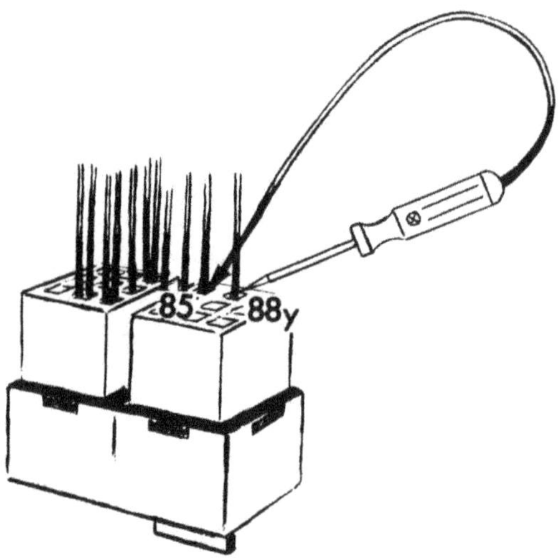

© 1976 VWoA—3852

Fig. 4-1. Test for positive (+) battery current with ignition switch turned off.

3. With the ignition turned to the running (2) position, apply the second test probe to terminal 86c. If the test lamp fails to light, positive (+) battery current is not reaching terminal 86c from terminal 15 of the ignition coil.
4. While running the starter (ignition switch in the 3 position), again apply the second test probe to terminal 86c. The test lamp should light, indicating positive (+) battery current from the ignition coil.
5. Again running the starter, apply the second test probe to terminal 86a. The test lamp should light, indicating that positive (+) battery current is reaching the double relay from terminal 50 on the starter solenoid.

NOTE —

If positive (+) battery current reaches double relay terminal 86a whether the starter is running or not, the wires to terminals 30 and 50 of the starter solenoid have accidentally been reversed. Correct the wire positions. If positive (+) battery current is reaching the double relay correctly, continue with the next test procedure.

To test relay function:

1. Connect one test probe of a test lamp to terminal 85 of the double relay (which is grounded to the negative post of the battery).
2. While running the starter, apply the second test probe to terminal 88d, as shown in Fig. 4-2. If the lamp does not light, the pump relay is not closing and the double relay should be replaced.

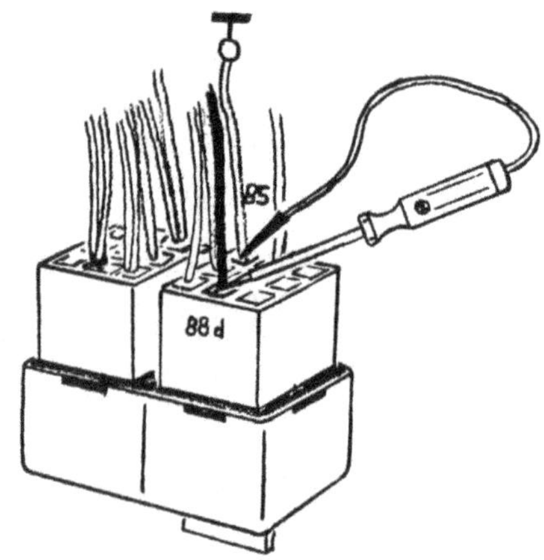

© 1976 VWoA—3853

Fig. 4-2. Test for positive (+) battery voltage to fuel pump during starting.

3. With the ignition turned to the running (2) position, apply the second test probe to double relay terminal 88b. If no positive (+) battery current reaches terminal 88b, the power relay is not closing and the double relay should be replaced.

If the double relay, and the wiring that supplies current to the double relay, pass the two tests that have just been described, you can continue by making tests that will determine whether positive (+) battery current is reaching the other relay terminals at the correct times and under the correct conditions. Further tests should be prompted by the failure of the double relay or the control unit to deliver current to some other component of the fuel injection system. Your tests should be based on a study of the wiring diagram given earlier in **3. Troubleshooting Fuel Injection**.

To replace double relay:

1. Disconnect the battery ground strap.
2. Remove the screw that holds the double relay to the tank compartment panel. Then disconnect the double relay from the two multiple connector plugs.
3. Install the multiple connector plugs on the new double relay, mount the relay on the tank compartment panel, then reconnect the battery ground strap.

4.2 Testing and Replacing Cold-start Valve

The cold-start valve, sometimes called the fifth injector, sprays fuel into the intake air distributor only during the first few seconds of starter operation and only when the engine and the surrounding air are cold. The valve's operation is controlled by the thermo-time switch and by current from terminal 50 of the starter solenoid—current that reaches the valve by way of the double relay. The cold-start valve is located on the rear of the intake air distributor.

If the cold-start valve fails to inject fuel during the cranking of a cold engine, it will be difficult or impossible to start the engine. If the cold-start valve is leaky, the engine may flood during starting—especially if the engine is hot.

To test cold-start valve for leakage:

1. Carefully disconnect the electrical plug from the cold-start valve. Put tape over the plug terminals to prevent accidental electrical sparks.
2. Remove the two screws that hold the cold-start valve to the intake air distributor. Remove the valve from the intake air distributor, but leave the ring main fuel hoses connected to the valve.
3. Disconnect the ignition primary wire from terminal 1 of the ignition coil so that the engine will not start. Tape the terminal of the wire to prevent accidental electrical sparks.
4. Hold the cold-start valve as shown in Fig. 4-3 while someone operates the starter.

> ***WARNING —***
> *Do not smoke or work near heaters or other fire hazards. Have a fire extinguisher handy.*

5. If the cold-start valve leaks fuel while the starter is operated (owing to normal pressure from the fuel pump), replace the cold-start valve.

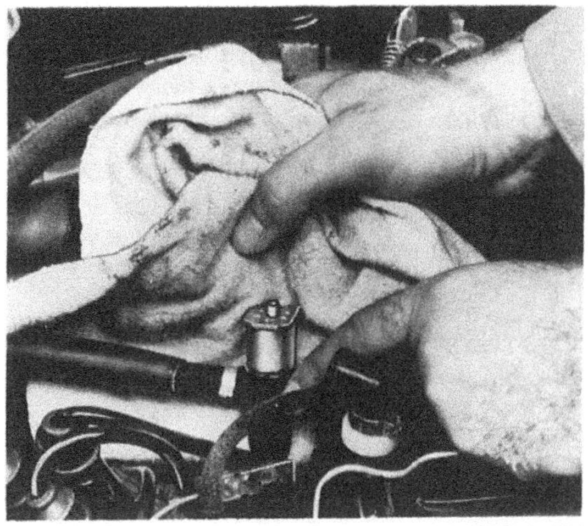

© 1976 VWoA—3823

Fig. 4-3. Cold-start valve being checked for leakage. Use a cloth as shown in order to catch any fuel.

To test cold-start valve operation:

1. Connect a fuel pressure gauge to the pressure testing tap of the ring main. The tap is located in the center of the fuel distribution manifold for the No. 3 and No. 4 cylinder injectors. See **4.4 Testing and Replacing Pressure Regulator**.
2. To prevent the engine from starting, disconnect the ignition primary wire from terminal 1 of the ignition coil. Then operate the starter briefly in order to build up fuel pressure.
3. Pull the electrical plug off the cold-start valve. Using a jumper wire, connect one of the cold-start valve terminals to ground on some clean, unpainted metal part of the engine.
4. Using a second jumper wire, connect the other cold-start valve terminal with terminal 15 of the ignition coil. See Fig. 4-4.

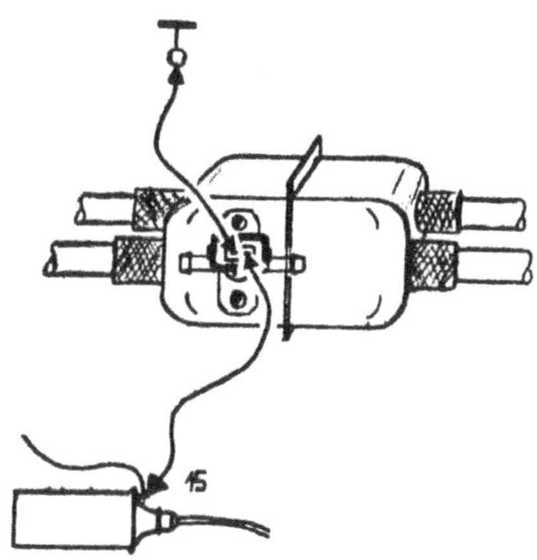

© 1976 VWoA—3861

Fig. 4-4. Hookup for testing cold-start valve operation.

10

5. Turn on the ignition. The fuel pressure should slowly drop, indicating that the cold-start valve is injecting fuel into the intake air distributor. If there is no loss of pressure, the cold-start valve is faulty and should be replaced.

> **NOTE —**
> If the cold-start valve operates, but you suspect that it is not operating when the engine is cold, test the thermo-time switch.

To replace cold-start valve:

1. Disconnect the electrical plug from the cold-start valve. Loosen the hose clamps, then disconnect the ring main fuel hoses from the cold-start valve.

> ***WARNING —***
> *Disconnect the battery ground strap. Do not smoke or work near heaters or other fire hazards. Have a fire extinguisher handy.*

2. Remove the two screws that hold the cold-start valve to the intake air distributor. Then remove the cold-start valve and its gasket.
3. Install the new cold-start valve and a new gasket. Then reconnect the electrical plug, the fuel hoses, and the battery ground strap.

4.3 Testing and Replacing Thermo-time Switch

The thermo-time switch is hidden beneath the intake air distributor, directly under the air outlet for the No. 4 cylinder. The function of the thermo-time switch is to control the operation of the cold-start valve.

The temperature limit and the time limit of the thermo-time switch are marked on the thermo-time switch itself. If the temperature of the air around the engine is below the temperature marked on the thermo-time switch—usually 95°F (35°C)—the thermo-time switch permits operation of the cold-start valve. Above that temperature, the thermo-time switch breaks the circuit to the cold-start valve. In addition to air temperature, an electrical heating element influences the operation of the thermo-time switch. This heating element starts to warm up when the starter begins to operate. After several seconds—usually eight seconds—of starter operation, the heating element will have raised the temperature of the thermo-time switch above its temperature limit, 95°F (35°C) for example, and the thermo-time switch will cut off the activating current to the cold-start valve. Early fuel injection engines had 50°F (10°C)/11-second thermo-time switches. Many of these have been replaced by the higher-temperature units to cure starting problems.

To test thermo-time switch:

1. Make sure that the temperature of the thermo-time switch is below 50°F (10°C) or 95°F (35°C), depending on the temperature marked on the switch. If necessary, cool the thermo-time switch with ice.
2. Carefully disconnect the electrical plug from the cold-start valve. Disconnect the ignition primary wire from terminal 1 of the ignition coil in order to prevent the engine from starting. Position the terminal of the disconnected wire so that it will not accidentally contact any metal part on the engine.
3. Attach the leads of a test lamp to the terminals of the disconnected plug (Fig. 4-5). Operate the starter without interruption for at least 12 seconds. The test lamp should first light brightly and then become noticeably dimmer—or go out—within 8 or 11 seconds, depending on the time marked on the switch. If the lamp lights but does not dim or go out, replace the thermo-time switch.

© 1976 VWoA—3826

Fig. 4-5. Test lamp being used to check thermo-time switch operation.

4. If the test lamp does not light at all, attach one of the test lamp leads to ground on some clean, unpainted metal part of the engine. The other lead should remain connected to one of the plug terminals.
5. Operate the starter. If the test lamp does not light, repeat the test with the test lamp connected to the other terminal of the disconnected plug. If the lamp lights in neither test, the wiring to the double relay is faulty, the double relay itself is faulty or has poor connections, or the wire between terminal 50 on the starter solenoid and double relay terminal 86a is faulty or disconnected.

> **NOTE —**
> If the test lamp lights whether the starter is operating or not, someone has misconnected the wires at the starter solenoid, installing the wire that belongs on terminal 50 on terminal 30 instead.

6. If battery current is reaching the plug, but the test lamp did not light when the starter was operated in step 3, the thermo-time switch is faulty and should be replaced.

To replace thermo-time switch:

1. Remove the intake air distributor as described in **6.3 Removing and Installing Intake Air Distributor**.
2. Disconnect the electrical plug from the thermo-time switch.
3. Using a socket wrench, unscrew the thermo-time switch from the bracket.
4. Install a new thermo-time switch in the bracket and reconnect the plug. Install the intake air distributor as described in **6.3 Removing and Installing Intake Air Distributor**.

4.4 Testing and Replacing Pressure Regulator

When the fuel pump is operating, fuel from the fuel pump enters the pressure regulator and exerts pressure against a spring-loaded diaphragm. Fuel pressure in excess of about 35 psi (2.5 kg/cm²) causes the diaphragm to deflect against spring tension, thereby uncovering a fuel outlet that returns excess fuel to the tank.

In addition to fuel pressure, engine vacuum can act upon the diaphragm, cooperating with fuel pressure in moving the diaphragm against spring tension. When there is high vacuum available to the pressure regulator from the intake air distributor, a fuel pressure in excess of approximately 28 psi (2.0 kg/cm²) will move the diaphragm and permit excess fuel to return to the tank.

The pressure regulator is located in the engine compartment, on the engine's front cover plate, ahead of the No. 1 cylinder. The pressure regulator cannot be adjusted or repaired and must, therefore, be replaced if it is faulty.

To test pressure regulator:

1. If the engine is cold and the air temperature is 50°F (10°C) or less, start the engine and allow it to warm up.
2. With the engine turned off, remove the hex-head plug and its seal from the ring main's pressure testing tap. The tap is located in the center of the fuel distribution manifold for the No. 3 and No. 4 (left) cylinders.

> ***WARNING —***
> *Gasoline may be expelled when you unscrew the plug. Do not smoke or work near heaters or other fire hazards. Have a fire extinguisher handy.*

3. Connect a fuel pressure gauge to the pressure testing tap. Disconnect the vacuum hose from the pressure regulator and plug the hose.

> **NOTE —**
> You do not need a threaded fitting to install the pressure gauge. Slide the gauge hose onto the pressure tap, then lock it in place with a small hose clamp.

4. Start the engine and observe the pressure gauge. The gauge should read approximately 35 psi (2.5 kg/cm²). (33 to 37 psi (2.35 to 2.65 kg/cm²) is considered acceptable by experienced mechanics.) If the pressure is too high or too low, replace the pressure regulator.
5. If the fuel pressure is correct with the vacuum hose disconnected, repeat the test with the vacuum hose connected to the pressure regulator. The gauge should read approximately 28 psi (2.0 kg/cm²). (26 to 30 psi (1.85 to 2.15 kg/cm²) is considered acceptable by experienced mechanics.) If the pressure is too high or too low, replace the pressure regulator.

> **NOTE —**
> Correct pressure is most important at idle, where fuel pressure has the greatest influence on exhaust emissions. However, if the engine seems to lack power, check to see whether the pressure drops considerably below 35 psi (2.5 kg/cm²) at full throttle. If so, look for a kinked or restricted fuel line, a blocked fuel filter, dirt in the fuel tank, or inadequate pump output.

To replace pressure regulator:

1. Disconnect the vacuum hose from the pressure regulator. Disconnect and quickly plug the pressure regulator's two fuel hoses.

> ***WARNING —***
> *Disconnect the battery ground strap. Do not smoke or work near heaters or other fire hazards. Have a fire extinguisher handy.*

2. Working under the vehicle, remove the ring nut that holds the pressure regulator to the engine's front cover plate. The nut is threaded onto the regulator's fuel outlet tube.

3. Remove the pressure regulator. Then install the new pressure regulator on the engine's front cover plate.
4. Reconnect the hoses. The fuel pressure line from the pump should be connected to the side pipe of the pressure regulator.
5. Reconnect the battery ground strap. Because the pressure regulator has an important effect on exhaust emissions, you should test the new regulator to make sure that it maintains the correct fuel pressures.

4.5 Testing and Replacing Auxiliary Air Regulator

The auxiliary air regulator is located atop the engine's crankcase, between the No. 2 cylinder air outlet of the intake air distributor and the box-shaped crankcase breather. It is held on its mounting bracket by two large Phillips head screws.

Fig. 4-6 is a schematic view of the auxiliary air regulator. When open, a rotary valve in the auxiliary air regulator provides additional air—and consequently, additional fuel—during engine warm-up. The auxiliary air regulator's function in the air system can be seen in Fig. 1-1 given in **1. General Description**. On the car, the auxiliary air regulator is located atop the engine, between the intake air distributor and the oil breather.

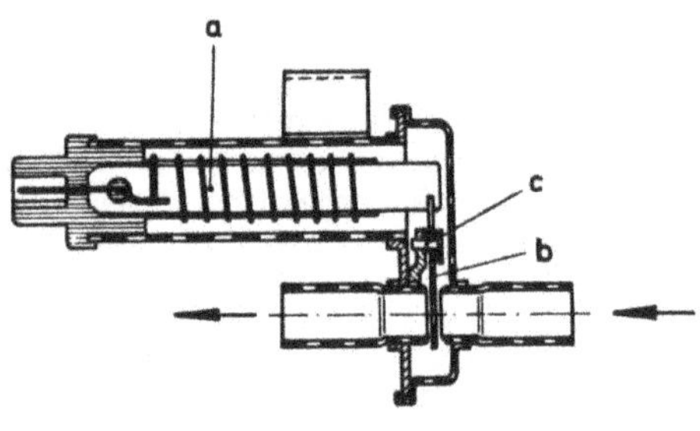

a. Bimetallic strip b. Rotary valve c. Pivot

Fig. 4-6. Schematic view of auxiliary air regulator. Arrows indicate air flow through open valve.

When the engine is started, current from the double relay begins to warm the heating element that is wound around the auxiliary air regulator's bimetallic strip. Heat—whether from the heating element or from hot air surrounding the engine—causes the bimetallic strip to deflect gradually, closing the rotary valve and cutting off the additional air.

To test the auxiliary air regulator, disconnect its fabric-covered intake air hose from the large rubber S-shaped duct that joins the intake air sensor with the throttle valve housing. Then start the engine (the engine must be cold). There should be vacuum at the disconnected hose. If you cover the hose, the engine should slow down. After several minutes of operation, the vacuum should diminish and, eventually, disappear almost completely.

If the auxiliary air regulator does not close after a reasonable period of engine operation (in cold weather, considerable time may be required), stop the engine. Disconnect the electrical plug from the auxiliary air regulator. Using an ohmmeter, measure the resistance between the two terminals on the auxiliary air regulator. The reading should be 30 ohms. If the ohmmeter reads infinity, or if the resistance is considerably less than 30 ohms, replace the auxiliary air regulator.

If the auxiliary air regulator resistance is in the correct range, use a test lamp as shown in Fig. 4-7 to determine whether battery voltage is reaching the heating element while the engine is running. If not, you should make further tests at the double relay. Especially check that someone has not accidentally exchanged the positions of the wires that go to terminals 30 and 50 on the starter solenoid. See the wiring diagram given earlier in **3. Troubleshooting Fuel Injection**.

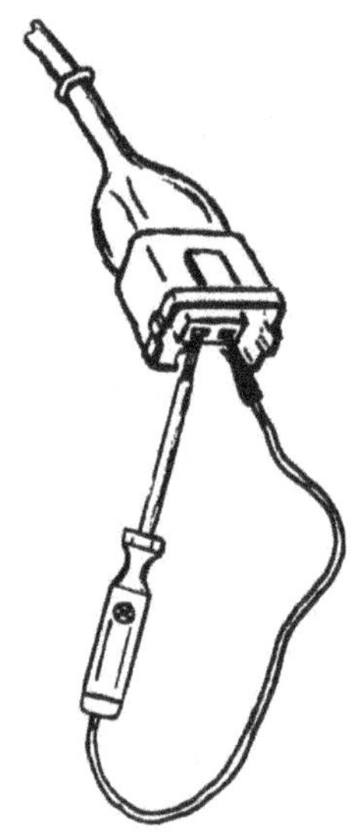

Fig. 4-7. Test lamp being used to check for current at auxiliary air regulator plug while engine is running.

To replace auxiliary air regulator:

1. Disconnect the electrical plug from the auxiliary air regulator.
2. Remove the two Phillips head screws that hold the auxiliary air regulator to its mounting bracket.

3. Lift up the auxiliary air regulator and disconnect both air hoses.

4. Remove the old auxiliary air regulator. Then install the new auxiliary air regulator using a reverse of the removal procedure.

4.6 Testing and Replacing Throttle Valve Switch or Microswitch

On 1976 vehicles manufactured from February 1976, the need for the throttle valve switch was eliminated by changing to a mechanically-controlled EGR valve. An ordinary microswitch was subsequently installed on the throttle valve housing to take over the full load/full throttle enrichment function that had formerly been handled by the throttle valve switch. 1977 and later models have neither a throttle valve switch nor a microswitch.

Testing and Replacing Throttle Valve Switch

(through Chassis No. 226 2077 583)

The throttle valve switch is mounted on the bottom of the throttle valve housing and is both inaccessible and difficult to see with the throttle valve housing installed on the intake air distributor. In addition to its function of signaling the control unit to provide full load enrichment, the throttle valve switch controls the operation of the EGR (exhaust gas recirculation) valve. The throttle valve switch is a black, rectangular device with an electrical plug fastened to it.

Fig. 4-8 shows the arrangement of the throttle valve switch contacts, switch terminals, and switch operating cam. At closed throttle, the control terminal (18) is connected to terminal 2 (42 in the wiring diagram given in **3. Troubleshooting Fuel Injection**). In this position, the EGR valve is closed—preventing exhaust gases from being recirculated. At (or near) full throttle, the control terminal (18) is connected to terminal 3, which signals the control unit to provide full load enrichment and again cuts off the recirculation of exhaust gases. Exhaust gas recirculation occurs only in the mid-range of throttle valve operation.

To test throttle valve switch:

1. With the engine stopped, disconnect the electrical plug from the EGR valve. Then connect a test lamp as shown in Fig. 4-9.

2. Turn on the ignition. With the throttle valve closed, the test lamp should be lit. If not, adjust the switch as described in the replacement procedure. If the lamp still does not light, replace the switch.

3. If the lamp lights at closed throttle, gradually open the throttle valve. The lamp should go out at about 11° of throttle valve shaft rotation and come on again at or near the full-throttle position. If not, replace the throttle valve switch.

4. If the test lamp fails to light at all, ground one of the lamp's test leads against some clean, unpainted metal part of the engine. Then touch the other test lead to one terminal and then the other terminal of the disconnected EGR valve plug. If the test lamp lights in one of these tests, but did not light in step 2, replace the throttle valve switch.

5. If the test lamp does not light when you make the test described in step 4, test the double relay in order to determine whether current is reaching double relay terminal 86a with the ignition on. See the wiring diagram given in **3. Troubleshooting Fuel Injection**.

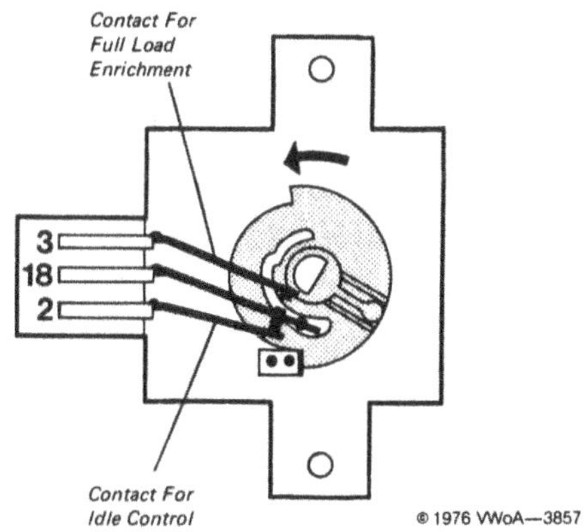

Fig. 4-8. Schematic view of throttle valve switch.

© 1976 VWoA—3827

Fig. 4-9. EGR valve plug being tested after plug has been disconnected from the terminal on the valve (arrow). On Type 2 vehicles, the EGR valve is located on the left side of the engine, right above the No. 4 cylinder.

10

To replace the throttle valve switch, remove the throttle valve housing from the intake air distributor as described in **6.2 Removing and Installing Throttle Valve Housing**. Take out the two screws, then remove the switch from the throttle valve housing.

To avoid damaging the new switch, do not press it against the throttle valve housing until you are sure that the throttle valve shaft is correctly engaged. It may be necessary to rotate the switch slightly in order to engage the cam on the shaft correctly. To adjust the throttle valve switch, hold the throttle valve in its closed position. Turn the switch body in the direction of open throttle until a light resistance is felt. Then tighten the mounting screws.

Testing and Adjusting Microswitch

(1976 cars from Chassis No. 226 2077 584)

At or near full throttle, a cam on the throttle valve shaft operates the microswitch, signaling the electronic control unit to supply a richer mixture for full load operation. The microswitch was eliminated beginning with the 1977 models, as the new control unit handles the enrichment task automatically.

To test and adjust:

1. Connect one wire of a test light to the flat connector of the microswitch that is not covered by insulation (Fig. 4-10). Ground the other test light wire by connecting it to some clean, unpainted metal part of the engine.

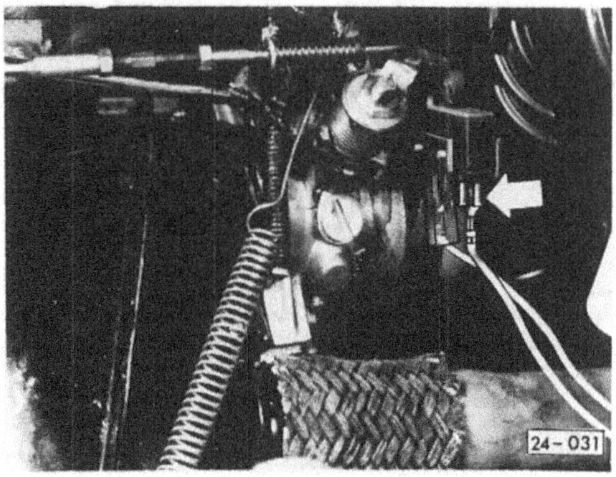

©1977 VWoA-4071

Fig. 4-10. Uninsulated connector on microswitch (arrow).

2. Turn on the ignition.
3. Have someone press the accelerator pedal fully down. The test light should come on shortly before full throttle is reached. If not, adjust the switch as described in the next three steps.
4. Loosen the screws that hold the microswitch to the bracket on the throttle valve housing.
5. With the accelerator pedal fully depressed and the ignition turned on, center the switch roller on the cam, as indicated in Fig. 4-11, then move the switch toward the cam until the test light comes on—but do not move the switch up so far that the roller is under a heavy load.
6. Tighten the screws. Then test the switch as previously described to make sure that it works correctly.

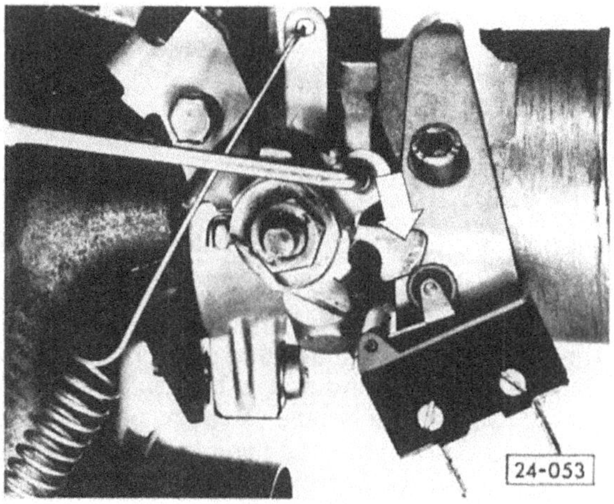

©1977 VWoA-4072

Fig. 4-11. Microswitch roller centered on throttle valve shaft cam (arrow).

4.7 Testing and Replacing Temperature Sensors

Temperature sensor I is built into the intake air sensor. Temperature sensor I is tested as part of the potentiometer test described in **4.8 Testing and Replacing Intake Air Sensor**. Temperature sensor I measures the temperature of the air that enters the engine. This data is incorporated into the potentiometer's data and sent to the control unit.

Temperature sensor II is screwed into the front of the left-hand cylinder head, above the combustion chamber of the No. 3 cylinder. The sensor's rubber air seal is visible in the top of the cylinder cover plate, directly between the intake manifold flange of the No. 3 cylinder and the No. 3 cylinder's spark plug hole. Temperature sensor II provides the control unit with the engine temperature information that is needed for correct starting and warm-up enrichment.

To test temperature sensor II, disconnect its lead wire. Then connect an ohmmeter as shown in Fig. 4-12. At 68°F

(20°C), the resistance should be approximately 2500 ohms. If the resistance is considerably higher or lower, replace temperature sensor II.

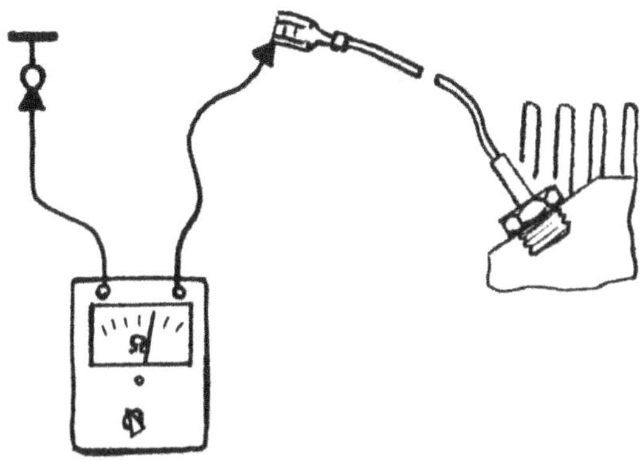

© 1976 VWoA—3858

Fig. 4-12. Ohmmeter being used to test temperature sensor II.

To replace temperature sensor II:

1. Disconnect temperature sensor II's lead wire. Then carefully pry the rubber air seal out of the cylinder cover plate and lift the seal up and off the wire.
2. Using a deep socket wrench (so that the wire can be folded into the top of the socket), remove the temperature sensor.
3. Install the new sensor using a reversal of the removal procedure. Torque the sensor to 1.5 mkg (11 ft. lb.). Make certain that the rubber air seal fits tightly in the cylinder cover plate in order to prevent the loss of cooling air.

4.8 Testing and Replacing Intake Air Sensor

Beginning with vehicles manufactured in February 1976, Temperature sensor I in the intake air sensor is given its own terminal (27). Therefore, the new intake air sensor has a 7-point connector instead of a 6-point connector. The wiring harness has an additional wire that connects terminal 27 of the intake air sensor to terminal 27 of the electronic control unit. Owing to the new temperature sensor I arrangement, the intake air preheating system is eliminated on vehicles built since the end of January 1976. The idle mixture (CO percentage) can be adjusted only after the intake air sensor with seven terminals has warmed to a temperature above 68°F (20°C).

The intake air sensor (Fig.4-13) is a mechanical device that operates a potentiometer and the fuel pump switch. The sensor's mechanical operation depends upon the free movement of the spring-loaded stator flap inside the sensor's main air passage.

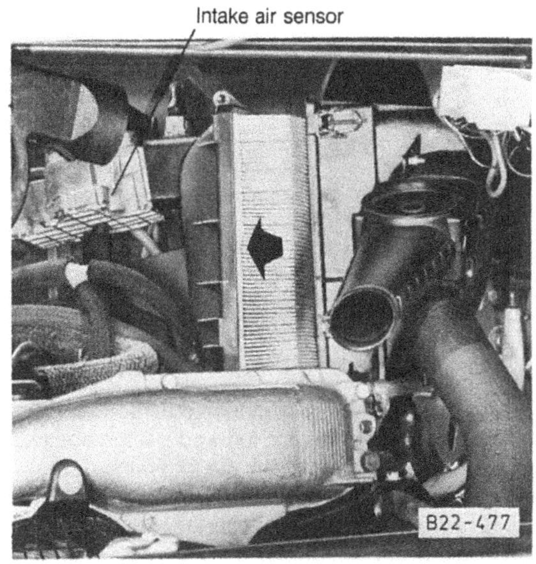

© 1976 VWoA—3847

Fig. 4-13. Location of intake air sensor. Air cleaner cover is partially removed, showing filter element (arrow).

Backfiring can cause the stator flap of the intake air sensor to jam closed—though jamming is normally prevented by the back pressure valve. Dirt or physical impact can cause the stator flap to bind or jam partway open. Therefore, check the stator flap for free movement before you make any electrical tests. To check flap movement, release the four spring clips that hold the air cleaner together. Then separate the two halves of the air cleaner and remove the pleated paper filter element. If necessary, consult **6. Intake Air System**.

Insert a wooden dowel or other suitable probe into the air intake. Using the probe, move the stator flap throughout its entire range. If the flap is merely stuck closed, owing to a backfire, it will usually operate correctly after it is once pressed loose. If the flap binds at any other point, remove the intake air sensor, as described later under this heading, and then carefully clean away any foreign matter that might cause the flap to bind.

If necessary, you can disassemble the intake air sensor by taking out the screws and carefully prying the bottom side of the housing out of the main part of the housing. This bottom part is held in by sealer. If the flap cannot be made to move freely, replace the intake air sensor. In assembling an intake air sensor, be sure to use a suitable sealer at the periphery of the bottom plate and on the screws. Do not wash the sensor with solvent, which could damage the potentiometer.

If the fuel pressure test described in **4.10 Testing and Replacing Fuel Pump and Fuel Filter** shows that the pump is not operating after the engine has started—or if the engine refuses to start—test the fuel pump switch in the intake air sensor. To test the switch, remove the connector plug from the terminals of the intake air sensor.

> ***CAUTION —***
> *To prevent damage to the connector plug and wiring, first remove the rubber boot from the plug and slide the boot down onto the wiring harness. Then, grasping the connector plug (not the wires), pull the plug off the intake air sensor terminals.*

Connect an ohmmeter to intake air sensor terminals 36 and 39. (The numbers are marked on the sensor.) Hand-operate the stator flap. With the flap closed, there should be no electrical continuity between sensor terminals 36 and 39; when the flap is slightly opened, there should be continuity. If there is no continuity, replace the intake air sensor.

To check the potentiometer, connect an ohmmeter to terminals 6 and 9 of the intake air sensor (Fig. 4-14). The resistance should be 200 to 400 ohms on 1975 cars or 100 to 300 ohms on 1976 and later cars. Repeat the test at terminals 7 and 8; the resistance should be 120 to 200 ohms on 1975 cars or 80 to 200 ohms on 1976 and later cars. If the resistance is incorrect for either test, replace the intake air sensor.

© 1976 VWoA—3825

Fig. 4-14. Ohmmeter test probes applied to potentiometer terminals on intake air sensor. (Sensor shown is for a Type 1 VW, but terminal numbers are the same on Type 2.)

Testing Temperature Sensor I

On intake air sensors that have six terminals, temperature sensor I is an "invisible" part that is built into the potentiometer. This temperature sensor is checked when you measure the resistance between intake air sensor terminals 6 and 9. On intake air sensors that have seven terminals, you can test temperature sensor I separately. The location of the sensor is shown in Fig. 4-15.

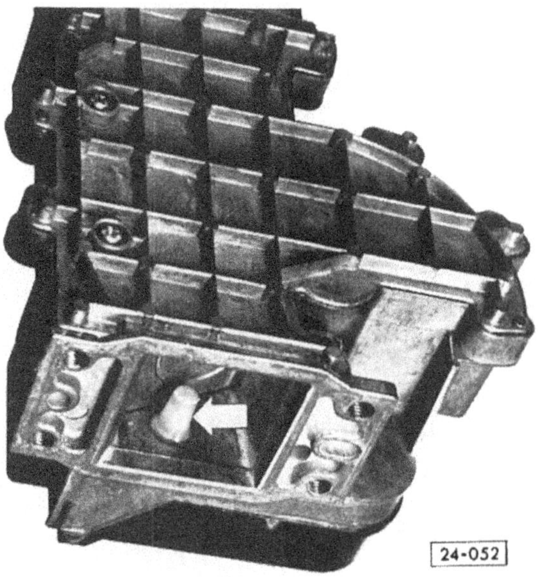

© 1977 VWoA–4060

Fig. 4-15. Location of temperature sensor I (arrow).

The terminal numbers of the seven-terminal intake air sensor are shown in Fig. 4-16. Temperature sensor I is connected to terminal 27. You can test temperature sensor I at the terminals, as described in the procedure that follows. However, if the temperature sensor is found to be faulty, it cannot be replaced without replacing the entire intake air sensor.

© 1977 VWoA–4061

Fig. 4-16. Terminal numbers on intake air sensor introduced in February 1976.

To test:

1. Make sure that the intake air sensor is at 68°F (20°C) or above.
2. Connect an ohmmeter to terminals 6 and 27.
3. If the resistance between terminals 6 and 27 exceeds 2500 ohms, the temperature sensor is faulty and the entire intake air sensor should be replaced.

Replacing Intake Air Sensor

It is important that replacement intake air sensors have the correct part number. The correct intake air sensors for different engines can be determined from **Table a.**

Table a. Intake Air Sensor Part Numbers

Model year	Engine numbers	Intake air sensor Part Number
1975	ED 0 000 001 through ED 0 000 639	022 906 301 C
1975	ED 0 000 640 through ED 0 025 000	022 906 301 A
1976	GD 0 000 001 through GD 0 010 999	022 906 301 A
1976	GD 0 011 000 through GD 0 027 786	022 906 301 B
1977 and later	GD 0 027 787 and later	022 906 301 D
1979 Calif.	GE 0 000 001 and later	039 906 301

To replace intake air sensor:

1. Release the four clips that hold the cover and the body of the air cleaner together. Remove the cover and the pleated paper air filter.
2. Carefully pull back the rubber boot that covers the multiple connector plug on the intake air sensor. Slide the boot down the wiring harness. Then, grasping the connector plug (not the wires), carefully pull the plug off the terminals of the intake air sensor.
3. Loosen the hose clamp that holds the S-shaped rubber duct to the intake air sensor. Then disconnect the rubber duct from the intake air sensor.
4. Remove the body of the air cleaner and the intake air sensor from the car as a unit.
5. Remove the four bolts that hold the body of the air cleaner to the intake air sensor. Then remove the air cleaner body from the intake air sensor. See Fig. 4-17.

Installation is the reverse of removal. If you install a new intake air sensor, or if you have disassembled the old sensor for cleaning or inspection, you should carry out the adjustments described in **5. Adjusting Fuel Injection** after installation is complete.

If you disassemble the intake air sensor, use sealer on the screws and around the periphery of the cover during assembly. The plastic cover can be pried off the potentiometer which, in some cases, may permit you to correct minor mechanical faults. However, the electrical components cannot be replaced or repaired and, if there is an electrical fault in the potentiometer, the entire intake air

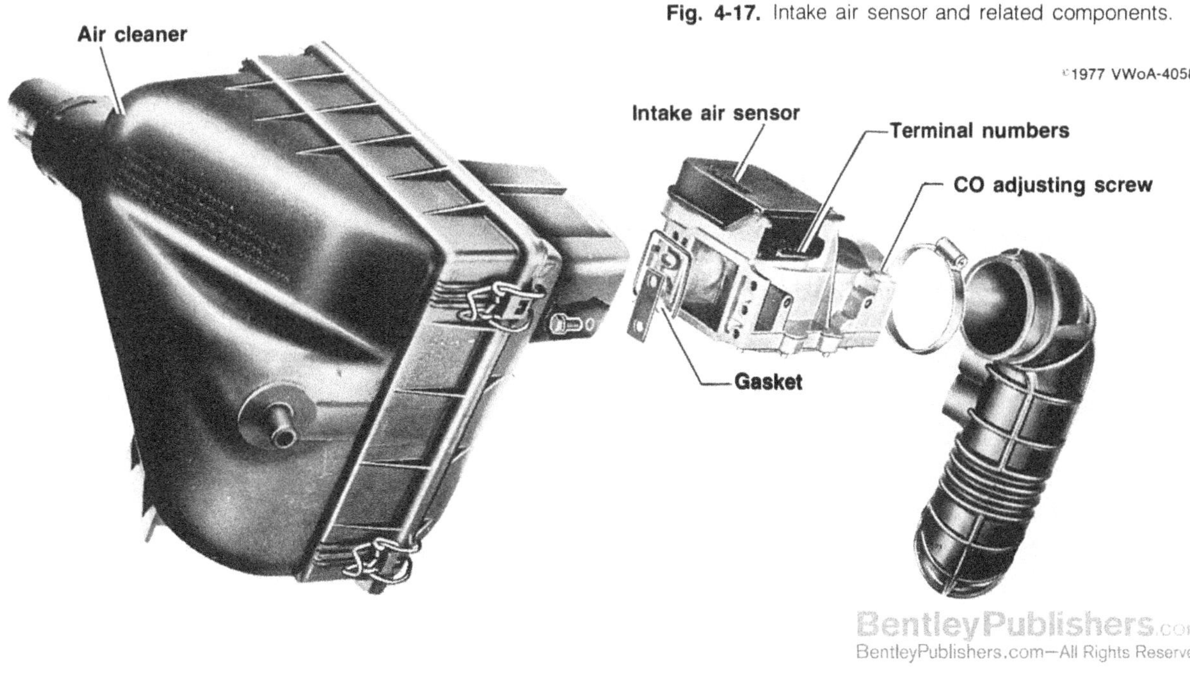

Fig. 4-17. Intake air sensor and related components.

©1977 VWoA-4058

minor mechanical faults. However, the electrical components cannot be replaced or repaired and, if there is an electrical fault in the potentiometer, the entire intake air sensor must be replaced as a unit. Fig. 4-18 is a cutaway view of the intake air sensor.

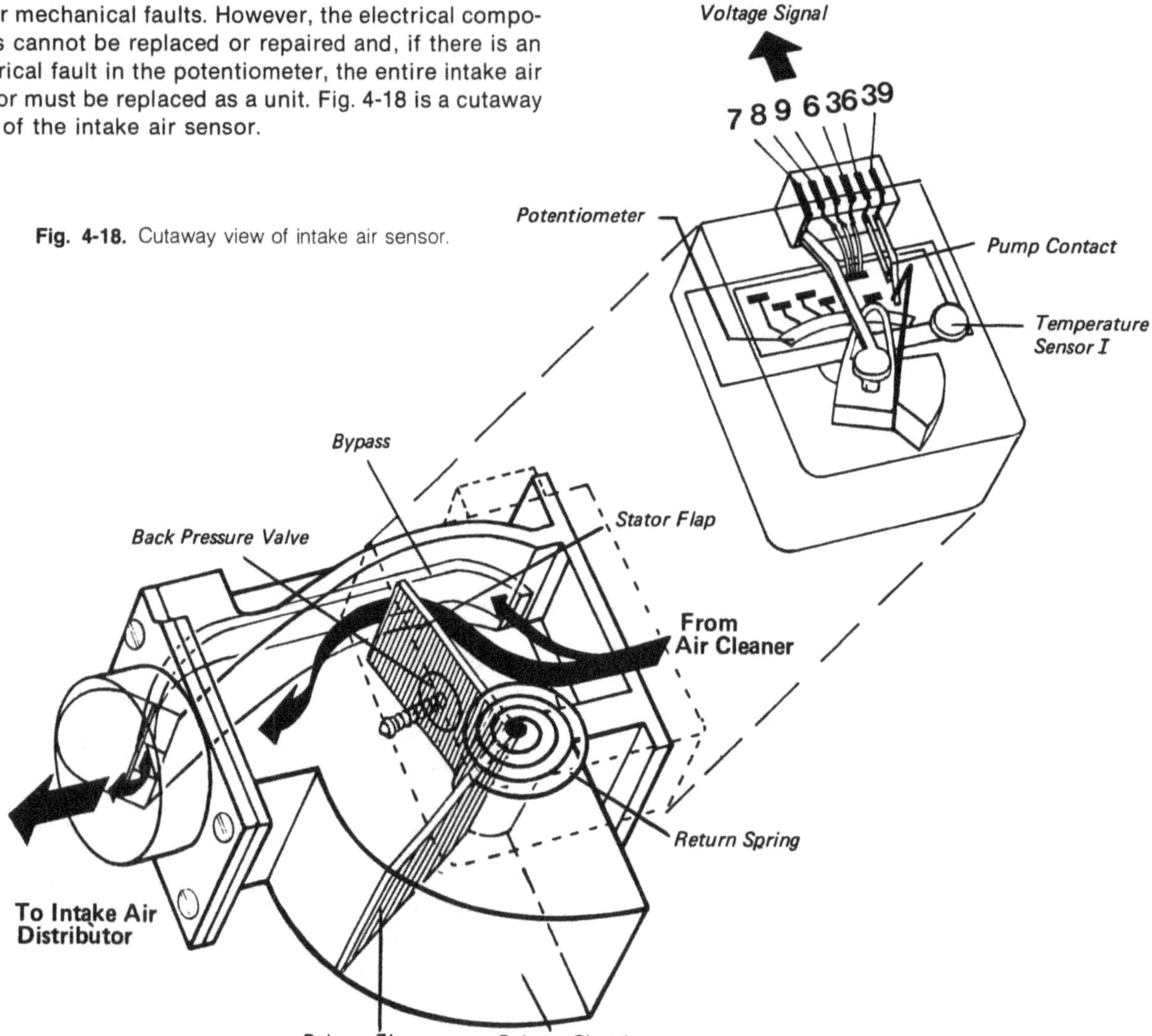

Fig. 4-18. Cutaway view of intake air sensor.

4.9 Testing and Replacing Injectors and Series Resistance Block

A constant supply of positive (+) battery current is delivered to the four injectors. This supply of current comes from terminal 88b of the double relay and is routed through four individual resistors that are contained in the series resistance block. See the wiring diagram given previously in **3. Troubleshooting Fuel Injection**. The purpose of the series resistance is to keep a stable voltage available to the injectors.

The control unit supplies negative (– or ground) current to all four injectors simultaneously at every other opening of the ignition distributor's breaker points. Each time the circuit through the injectors is completed by the application of ground, the solenoids of the injectors open the injector needle valves, allowing fuel from the ring main to spray into the engine's intake ports.

When the engine fails to run or fails to run properly, always troubleshoot the ignition system thoroughly before assuming that the problem is in the fuel injection. If the engine fails to start even though the fuel pump is functioning, it is sometimes best if you determine whether fuel is being injected before you proceed to more detailed troubleshooting.

To determine whether or not fuel is being injected, remove the injectors from the engine, leaving the wires and the fuel hoses attached. Disconnect the ignition high tension cable from coil terminal No. 4.

WARNING ——

Vaporized gasoline will be expelled during the following test. Do not disconnect the cable at its distributor end, which could cause accidental sparks. Do not work near heaters or other fire hazards. Have a fire extinguisher handy.

Run the starter briefly and see whether fuel is discharged from the injectors; the spray pattern should also be correct, as shown in Fig. 4-19. If no fuel is discharged, the fuel injection system is faulty. If one or more injectors fail to discharge fuel, disconnect the electrical plug(s) from the malfunctioning injector(s). Using an ohmmeter, test between the two terminals of each injector that failed to function. If there is no continuity, the solenoid winding of the injector is open-circuited and the injector must be replaced. Also replace injectors that have a faulty spray pattern.

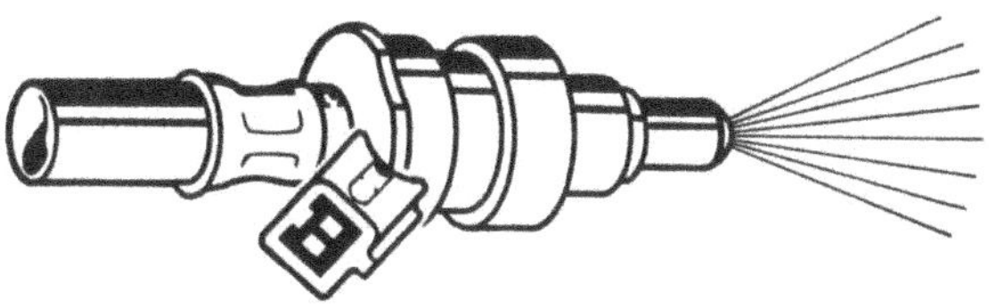

©1977 VWoA-4064

Fig. 4-19. Correct spray pattern. Discharged fuel should form an even cone-shaped spray.

If the injectors are not faulty, use a test lamp as shown in Fig. 4-20 to determine whether positive (+) battery current is reaching the series resistance block from terminal 88b of the double relay when the ignition is on. The series resistance block is located in the left-hand side of the engine compartment, on the fuel tank compartment cover, right above the double relay.

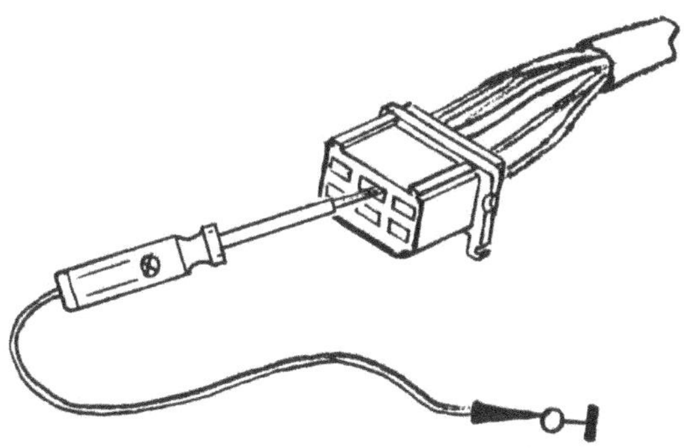

© 1976 VWoA—3862

Fig. 4-20. Series resistance block connector plug being tested for presence of current from double relay terminal 88b.

Once you have determined that positive (+) current is reaching the series resistance block, you can continue testing to determine whether any resistor is faulty. If no current is reaching the series resistance block, test the double relay and the wiring that connects double relay terminal 88b with the series resistance block. See the wiring diagram given in **3. Troubleshooting Fuel Injection**. After pulling the plugs off of the injectors, you can test for voltage at the injector plugs as shown in Fig. 4-21.

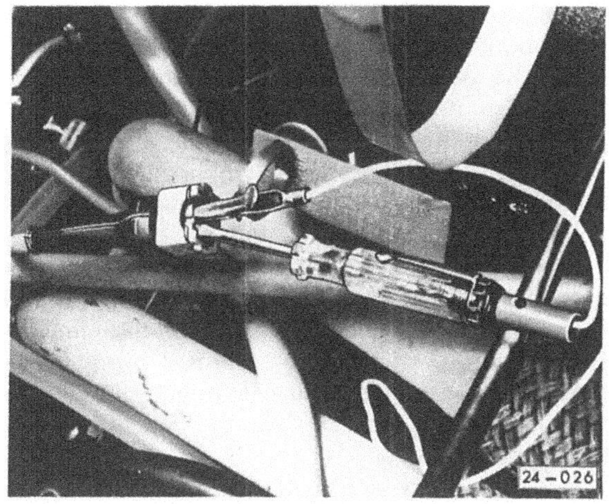

©1977 VWoA-4065

Fig. 4-21. Test lamp being used to determine whether current is reaching injectors. Run starter during test. The light should flicker.

To replace a faulty series resistance block, disconnect its connector plugs from the wiring harness. Remove the two screws that hold the series resistance block cover to the fuel tank compartment cover, then remove the block. Installation is the reverse of removal.

In addition to failing to inject fuel, an injector can become faulty owing to leakage. To check for leakage, wipe clean the ends of the removed injectors (fuel hoses and wires still attached). Then, without operating the starter, turn on the ignition. If the injector jets become wet with fuel but lose no more than two drops of fuel per minute, you should test the pressure regulator in order to determine whether the pressure is excessive. If an injector jet loses more than two or three drops of fuel per minute, the injector itself is faulty and must be replaced.

NOTE —
White exhaust smoke after the engine has warmed up usually indicates leaking injectors or a leaking cold-start valve. High fuel consumption, starting trouble, and oil dilution are more likely to result from continuous operation of the cold-start valve than from leaking injectors. Continuous cold-start valve operation can be caused by reversing the wires at terminals 30 and 50 of the starter solenoid during engine installation or starter replacement.

An additional troubleshooting test should be made in the event of severe engine flooding. This flooding, which may prevent the engine from starting, is probably caused by a faulty injector. The trouble may be intermittent, becoming a major problem in damp weather.

Test each of the fuel injection system's four injectors

10

in order to determine whether the winding of any injector is grounded to the injector case. If one injector winding is grounded, owing to a breakdown of the insulation inside the injector, it will also ground the windings of the other injectors via the connections with the control unit. As a result, all four injectors will open and discharge fuel continuously for as long as the ignition is turned on.

To test an injector for internal ground, disconnect the electrical plug. Apply one test probe of an ohmmeter to the injector case and the other test probe to first one and then the other of the two electrical terminals on the injector. If there is electrical continuity between either terminal and the injector case, the injector is faulty and should be replaced.

> **NOTE —**
> It is not necessary to remove the injectors from the engine in order to test them.

If none of the injectors is faulty, test the wiring by applying one ohmmeter test probe to a clean, unpainted metal part of the car and the other test probe to each of the terminals inside the four injector plugs. With the ignition turned off, there should be no electrical continuity between the injector wires and the car's chassis. Replace the wiring harness if any of the wires is grounded.

To replace injector(s):

1. Because of their attachments to the ring main fuel distributor manifolds, it is easier to remove the injectors in pairs. First, disconnect the electrical plugs from both injectors.
2. Remove the nut that holds each injector retainer to the stud on the intake manifold flange.
3. Being careful not to let the tips of the injectors rub or bump against the cylinder cover plate, withdraw the injectors from the intake manifold. Remove the injector seals and the retainer plates.

> **NOTE —**
> The inner seals sometimes stick inside the injector seat instead of coming out along with the injectors.

4. Loosen the hose clamps, then remove the injector(s) from the ring main fuel hose(s).

> ***WARNING —***
> *Fuel may be expelled. Do not smoke or work near heaters or other fire hazards. Have a fire extinguisher handy.*

Installation is the reverse of removal. Use new injector seals during installation (there should be a small seal around the tip of each injector and a larger seal around the barrel of the injector). Torque the retainer nuts to 60 cmkg (52 in. lb.).

4.10 Testing and Replacing Fuel Pump and Fuel Filter

An electric fuel pump is used with the fuel injection system. The pump and the replaceable in-line fuel filter are located beneath the fuel tank. The operation of the fuel pump is controlled by the pump relay in the double relay. During starting, the pump relay is operated by current from the starter solenoid. With the engine running, the pump relay is operated by current from the intake air sensor.

To test pump:

1. With the engine turned off, remove the hex-head plug from the fuel pressure testing tap. The tap is located on the fuel distribution manifold for the left (No. 3 and No. 4) fuel injectors.

> ***WARNING —***
> *Gasoline may be expelled when you remove the plug. Do not smoke or work near heaters or other fire hazards. Have a fire extinguisher handy.*

2. Slip the hose of a fuel pressure gauge onto the pressure testing tap. Then lock the hose in place with a hose clamp.
3. Operate the starter. The pressure gauge should indicate approximately 35 psi (2.5 kg/cm^2).
4. If the pump supplied pressure during the test but the pressure was far outside the normal range, test the pressure regulator as described in **4.4 Testing and Replacing Pressure Regulator.** If the pump failed to supply pressure, attach a test lamp to the fuel pump connector as shown in Fig. 4-22.

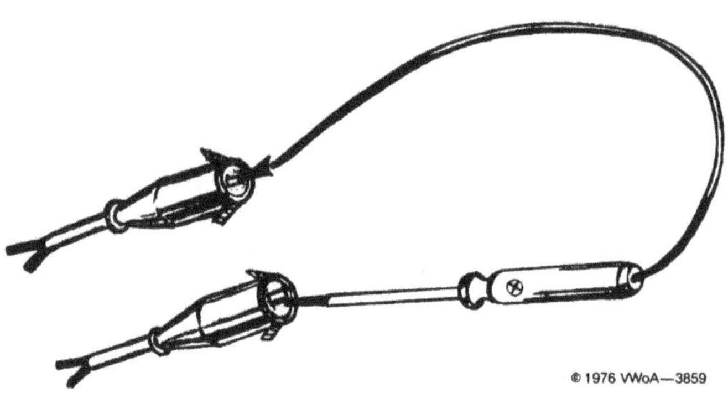

© 1976 VWoA—3859

Fig. 4-22. Test lamp attached to fuel pump connector.

5. Operate the starter. If the test lamp lights, the fuel pump is faulty and should be replaced. If the test lamp does not light up, and there is good ground at the pump via the brown wire, test the double relay as described in **4.1 Testing and Replacing Double Relay**. Also test the wiring using as a guide the wiring diagram given in **3. Troubleshooting Fuel Injection**.

If after replacing the pump there is still no fuel pressure, check for the following possible causes:

1. Fuel filter blocked
2. Rust or other foreign matter in fuel tank
3. Defective pressure regulator
4. Wires for fuel pump or fuel pump connectors interchanged
5. Hoses interchanged at pump or pressure regulator.

To replace fuel pump:

1. Raise the vehicle on a hoist or support it on jack stands.
2. Disconnect the battery ground strap. Then disconnect the connector plugs from the pump.
3. Loosen the hose clamps. Then remove and quickly plug the fuel hoses that are attached to the pump.

WARNING —
Do not smoke or work near heaters or other fire hazards. Have a fire extinguisher handy.

4. Take out the screws, then remove the pump from the car.

Installation is the reverse of removal. The fuel hose from the tank, which contains the filter, should be connected to the rounded end of the pump. The output line that goes to the pressure regulator should be connected to the flat end of the pump which also receives the connector plugs. Use new hose clamps and make sure that the hoses are tight-fitting and in good condition.

To replace fuel filter:

1. Apply pinch clamps to squeeze shut the hoses on each side of the fuel filter.
2. Loosen the hose clamps, then detach the hoses from the old filter.
3. Using new hose clamps, install a new filter; the arrow indicating flow direction should point toward the pump.
4. Remove the pinch clamps and check for leaks.

4.11 Making Test at Control Unit Plug and Replacing Control Unit or Wiring Harness

If, in troubleshooting the fuel injection, you need to test the wiring that connects some component of the fuel injection system with the control unit, you can apply an ohmmeter or a test lamp to the terminals of the disconnected control unit plug. Under no circumstances should you attempt to make tests at the terminals of the control unit itself. Fig. 4-23 shows a typical test being made at the control unit plug.

CAUTION —
In making tests at the control unit plug, you should be careful not to bend or otherwise damage any of the plug connectors. Deformed or misaligned connectors can cause poor electrical contact and could damage the control unit terminals when you install the plug.

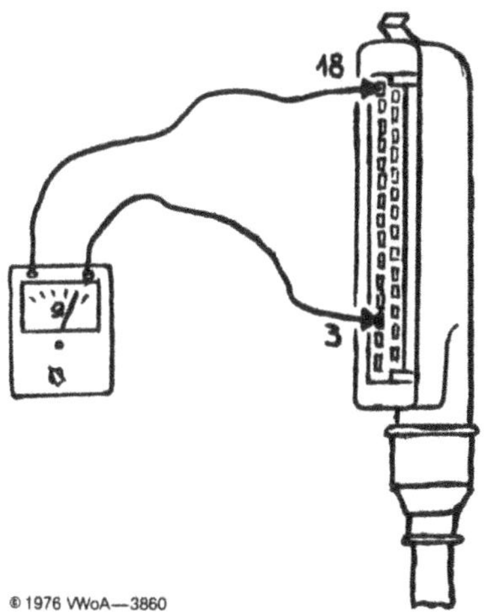

Fig. 4-23. Ohmmeter being used at control unit plug to test wiring to throttle valve switch (early models). Terminal positions and numbers are identified in Fig. 3-1 given in **3. Troubleshooting Fuel Injection**.

In addition to making tests, it may be necessary to disconnect the control unit plug in order to replace the wiring harness. In general, it is best to replace the wiring harness as a unit whenever a problem is discovered in one of the wires for the fuel injection system.

To replace wiring harness:

1. Disconnect the battery ground strap.

NOTE —
As you disconnect the ground strap, you will see the control unit plug, which is near the ground strap, directly in front of the battery.

2. Press down the spring catch that is at the bottom of the control unit, and which retains the lower end of the control unit plug. Then carefully pull out the lower end of the control unit plug in order to free it from the terminals of the control unit. If necessary, use a screwdriver to raise the plug gently. See Fig. 4-24.

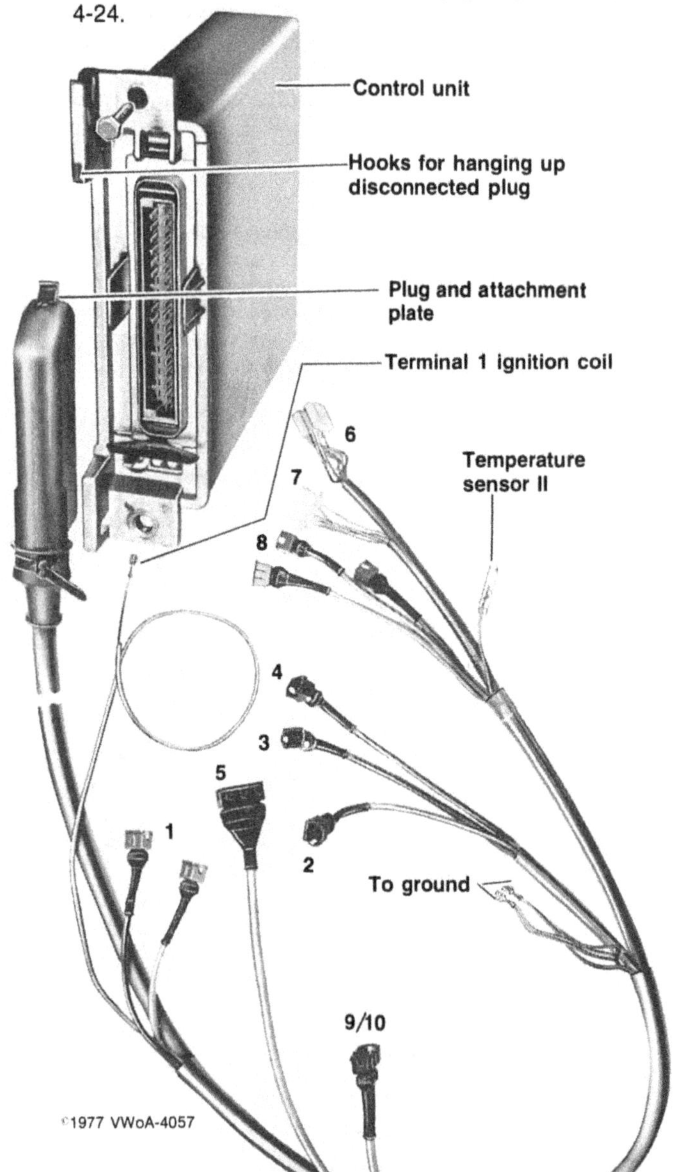

1. Injectors (gray)
2. Cold-start valve (blue)
3. Thermo-time switch (brown)
4. Auxiliary air regulator (black)
5. Intake air sensor (black)
6. Double relay (white)
7. Ballast resistance (white)
8. EGR valve (through Chassis No. 226 2077 583) (green)
9. Throttle valve switch (through Chassis No. 226 2077 583) (black)
10. Microswitch (1976 only from Chassis No. 226 2077 584) (black)

Fig. 4-24. Wiring harness connections.

3. Disconnect the plugs from the various other connecting points that are listed in Fig. 4-24.
4. Install the new wiring harness, using Fig. 4-24 as your guide.

CAUTION —
Be especially careful not to damage the plug connectors when you press the plug onto the terminals of the control unit. Otherwise, the terminals may be bent, causing malfunctions.

The control unit's location is shown in Fig. 4-25. Do not judge the control unit faulty until you have (1) measured the engine's compression and found it within specifications; (2) thoroughly checked the ignition system with special attention to breaker point condition and correct dwell; and (3) tested all fuel injection system components and wiring. If no other fuel injection system fault is found, and the engine still does not run right, replace the control unit.

Fig. 4-25. Location of control unit. To prevent acid spillover in this area of the car, do not fill the battery above the indicated maximum level.

Not all control units are the same, so you should not install a used control unit unless it has the same part number as the original control unit. Because of intake air sensor differences, it is also important that new replacement control units have the correct part number. Beginning with the 1979 cars, the control unit has a built-in device that limits the maximum engine speed to 5400 rpm. **Table b** lists the correct control units.

Table b. Electronic Control Unit Part Numbers

Model year and place where sold	Manual transmission	Automatic transmission
1975 & 1976 Canada and USA (except Calif.)	022 906 021 N	022 906 021 P
1975 & 1976 Calif.	022 906 021 Q	022 906 021 R
1977 Canada and USA	022 906 021 S	022 906 021 T
1978 Canada and USA (except Calif.)	022 906 021 AG	022 906 021 AH
1978 Calif.	022 906 021 AE	022 906 021 AF
1979 Canada and USA (except Calif.)	039 906 021 B	039 906 021 C
1979 Calif.	039 906 021 D	039 906 021 E

To replace control unit:

1. Completely remove the air cleaner as described in **6.1 Removing and Installing Air Cleaner**.
2. Carefully pull out the lower end of the control unit plug in order to free it from the terminals of the control unit. If necessary, use a screwdriver to raise the plug gently and to disengage the hook from the lug on the control unit.
3. Fully remove the plug and position it so that none of its terminals can come into contact with anything that could either damage the system electrically or damage the plug physically.
4. Remove the two bolts that hold the control unit to the body. Then slide the control unit out of its recess toward the rear of the vehicle.

Installation is the reverse of removal. Be especially careful, when installing the plug, to engage all the control unit terminals squarely. Forcing the plug onto the control unit may bend or break a misaligned terminal. If you encounter difficulty, remove the plug and check whether any of the terminals is misaligned or bent before you again attempt to press the plug into position.

5. ADJUSTING FUEL INJECTION

During routine maintenance, you should adjust the idle speed only. Do not attempt to adjust the idle mixture, as described in **5.2 Adjusting Idle Mixture**, without a good quality infrared exhaust gas analyzer such as the Sun® EPA 75 Performance Analyzer.

5.1 Adjusting Idle Speed

Before you attempt to adjust the idle speed, it is imperative that you adjust the valves and the ignition timing to specifications as described in **ENGINE**. Also, for accurate idle speed adjustments, the engine should be thoroughly warmed up by a test drive so that the oil will be hot and the auxiliary air regulator will be fully closed.

To adjust idle speed:

1. When you return from the test drive, stop the engine. Then connect a precision dwell meter/tachometer as specified by the manufacturer or, on cars with breakerless ignitions, as described in **ENGINE AND CLUTCH.** Adjust the instrument to measure rpm for a four-cylinder engine.
2. Start the engine, speed it up, and allow it to return to idle.
3. The idle speed for cars with manual transmissions should be 850 to 950 rpm on 1975 through 1978 models or 800 to 950 rpm on 1979 and later models. The idle speed for cars with automatic transmissions should be 900 to 1000 rpm on 1975 through 1978 models or 850 to 1000 rpm on 1979 and later models. Adjust the idle speed if it is too fast or too slow.
4. To adjust the idle speed, use a screwdriver to turn the bypass screw in the throttle valve housing (Fig. 5-1). Turning the screw clockwise slows the idle; turning the screw counterclockwise increases the idle speed. This adjustment does not alter exhaust emissions so long as the idle rpm is within the prescribed limits.

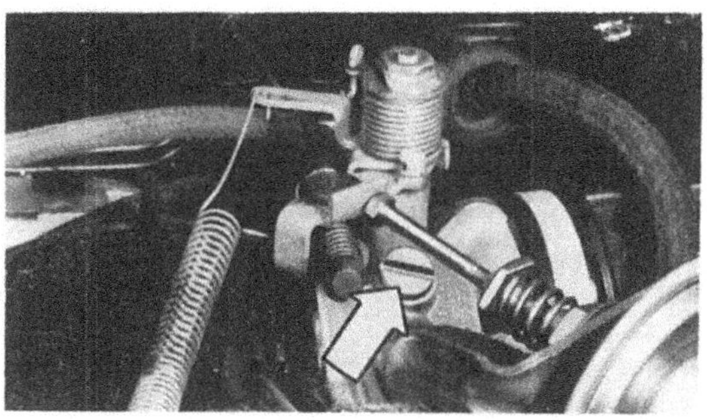

© 1976 VWoA—3828

Fig. 5-1. Bypass screw (arrow) used to adjust idle speed.

5. On cars with automatic transmissions, set the parking brake. Then select a drive range with the engine idling. If the idle speed drops by more than 150 to 200 rpm, adjust the idle speed regulator as described later under a separate unnumbered heading.

10

If turning the bypass screw fully out fails noticeably to increase the idle speed—or if the idle speed was exceptionally slow prior to adjustment—check for air leaks into the crankcase or around the intake manifold connecting hoses. Check especially that the oil filler cap is on tight and that no hoses are disconnected from the intake air distributor. If no air leak is found, check the EGR system as described in **7. Testing and Replacing EGR (Exhaust Gas Recirculation) Valve**.

If the bypass screw must be turned fully in, or nearly so, in order to obtain a slow enough idle—or if the idle speed was exceptionally fast prior to adjustment—check the auxiliary air regulator as described in **4.5 Testing and Replacing Auxiliary Air Regulator**. If the auxiliary air regulator is neither open nor faulty, check the pressure regulator as described in **4.4 Testing and Replacing Pressure Regulator**. Also inspect the pressure regulator's vacuum hose for leaks.

If you find no other faults that could be causing a fast idle on a car with a manual transmission, check the deceleration air enrichment valve (commonly called the decel valve) as described under the following unnumbered heading. If the problem is encountered on a car with an automatic transmission, either check the decel valve (vehicles manufactured from February 1976), as described under the following unnumbered heading, or check the idle speed regulator (vehicles manufactured through January 1976) as described under **Checking and Replacing Idle Speed Regulator.**

Checking and Replacing Decel Valve

The deceleration air enrichment valve (commonly called the decel valve) is located in the center of the front of the engine compartment, approximately between the evaporative emission control charcoal filter canister and the intake air sensor. (Vehicles with automatic transmissions manufactured through January 1976 do not have a decel valve.) The decel valve has a large, flat vacuum chamber that should not be confused with that of the vacuum-controlled EGR (exhaust gas recirculation) valve used through Chassis No. 226 2077 583.

If the decel valve is faulty (fails to close), the idle speed will be too fast. If you suspect this trouble, pinch closed the decel valve's large fabric-covered hose. If the idle speed drops, the decel valve is faulty and must be replaced. Routine functional checks can be made as described in the following procedures.

To check electrically-controlled decel valve (vehicles with automatic transmissions only, through Chassis No. 226 2077 583):

1. Turn on the ignition. Then disconnect the decel valve's wire from the ATF pressure switch.
2. Ground the disconnected wire on a clean, unpainted metal part of the car; listen for an audible click.
3. If the decel valve does not click, replace it.
4. Repeat the test, grounding the wire on its terminal on the ATF pressure switch. If the decel valve does not click, the ATF pressure switch is faulty and should be replaced.

To check vacuum-controlled decel valve (vehicles with manual transmissions and vehicles with automatic transmissions from Chassis No. 226 2077 584):

1. Pull off the large hose (Fig. 5-2) that connects the decel valve to the intake air sensor's S-shaped rubber duct, disconnecting the hose from the duct.
2. Plug the connection in the S-shaped duct with a suitable cork. Then start the engine.
3. Run the engine for a few seconds at about 3000 rpm. Let the throttle valve snap closed suddenly while you feel for suction at the disconnected hose.
4. If there is no suction during engine deceleration, the decel valve is faulty and should be replaced.

©1977 VWoA-4069

Fig. 5-2. Vacuum-controlled decel valve and hoses.

To replace decel valve:

1. Stop the engine. On cars that have the electrically-controlled decel valve, make sure that the ignition is turned off.
2. Disconnect the hoses from the decel valve. Disconnect the wires of electrically-controlled decel valves.
3. Remove the screws that hold the decel valve to the car, then remove the decel valve.
4. Install a new decel valve, using the reverse of the removal procedure.

Checking and Replacing Idle Speed Regulator

(automatic transmission only)

The idle speed regulator is a vacuum diaphragm device located on the throttle valve housing. The regulator can be seen in Fig. 5-1 given earlier. The purpose of the idle speed regulator is to open the throttle valve slightly when a drive range is selected. This maintains the idle rpm within the acceptable speed range for the engine.

When engine vacuum is high at closed throttle, engine vacuum holds in the plunger of the idle speed regulator, allowing the throttle valve to close fully. Under the load of drive range engagement, the engine vacuum drops slightly, reducing the suction on the idle speed regulator's vacuum diaphragm. The exposed spring on the idle speed regulator is then able to extend the plunger, thereby opening the throttle valve slightly.

If the idle speed slows down excessively when a drive range is selected—even though the idle has been adjusted to the correct range—you should check the idle speed regulator. To do this, place the transmission in neutral or park with the engine idling. Then, using feeler gauges, measure the clearance between the end of the plunger and the lever on the throttle valve shaft. If the clearance is not between 0.50 and 1.00 mm (.020 and .040 *in.*), loosen the locknut on the plunger. Screw the plunger in or out until the clearance is correct, then tighten the locknut.

With the clearance correct, pull the vacuum hose off the idle speed regulator. The plunger should move out and slightly open the throttle valve, thereby speeding up the engine's idle. If the plunger fails to move out, or if it fails to retract visibly when you reconnect the vacuum hose, replace the idle speed regulator.

Replacement requires only the removal of the throttle valve housing mounting screw. This screw also holds the idle speed regulator. After you have installed the new idle speed regulator, adjust the clearance as previously described.

5.2 Adjusting Idle Mixture

The idle mixture should be adjusted if you have replaced the intake air sensor or after extensive engine rebuilding. It is unnecessary to adjust the idle mixture during a tune-up or routine maintenance unless exhaust emissions are outside the prescribed range. Professional mechanics should check their state laws to determine whether a shop without state authorization can make adjustments that affect exhaust emissions.

Do not attempt to adjust the idle mixture without a good quality infrared exhaust gas analyzer such as the Sun® EPA 75 Performance Analyzer. The instrument that you use for making idle mixture adjustments should measure parts per million of hydrocarbons and CO percentages.

To adjust idle mixture:

1. Adjust the idle speed as described in **5.1 Adjusting Idle Speed.** On 1979 and later California models, disconnect the oxygen sensor wire at the in-line connector just to the left of the injectors for cylinders No. 3 and No. 4. If necessary, consult **8. Testing and Adjusting Oxygen Sensor Emission Control.**

2. Connect the exhaust gas analyzer to its power source and turn it on. Insert the probe into the car's tailpipe or into the port provided ahead of the catalytic converter on cars that are so equipped (Fig. 5-3).

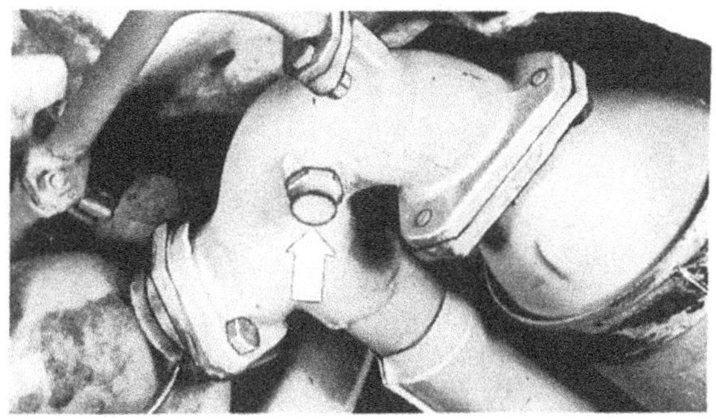

©1977 VWoA-4083

Fig. 5-3. Analyzer probe port in exhaust system of vehicle with catalytic converter. Remove plug, then insert probe. Lubricate plug with graphite paste during installation.

3. On vehicles through Chassis No. 226 2077 583, pull off the hose between the EEC (evaporative emission control) charcoal filter canister and the air cleaner, disconnecting the hose at the air cleaner.

   NOTE ——
   If the vehicle has been parked outside at temperatures below 68°F (20°C), the adjustment must wait until the intake air sensor has reached a temperature of 68°F (20°C). Normally, this temperature—as well as the necessary 122°F (50°C) oil temperature—will have been achieved prior to idle speed adjustment.

4. With the engine running, compare the exhaust analyzer readings with the prescribed ranges for CO emissions and, where prescribed by the state government, for hydrocarbon emissions. The applicable specifications are listed on the engine's emissions decal. (The emissions decal is on the fan housing, just to the rear of the distributor.) If the decal is missing, check to see that the CO volume percentage at idle is within the ranges given in the technical data tables of **ENGINE AND CLUTCH.**

5. If the CO level is outside the prescribed range, carefully pry out the plastic plug that covers the idle mixture adjusting screw on the intake air sensor (Fig. 5-4).

©1977 VWoA-4082

Fig. 5-4. Plastic plug that covers idle mixture adjusting screw. Pry out as indicated by curved arrow.

6. Using a screwdriver, turn the idle mixture adjusting screw one way or the other until emissions, as indicated by the exhaust gas analyzer, are within the correct range. Turning the screw clockwise (screwing it in) makes the mixture richer.
7. If necessary, readjust the idle speed to the prescribed range by turning the bypass screw.
8. On 1979 and later California cars, reconnect the oxygen sensor. If the CO level is not then between 0.4% and 1.2%, check the system as described in **8. Testing and Adjusting Oxygen Sensor Emission Control.**
9. If both the idle speed and the idle mixture are correct, replace the plastic plug that covers the mixture adjusting screw. Then disconnect the exhaust gas analyzer. Where applicable, reconnect the EEC hose and reinstall the plug in the exhaust system.

6. INTAKE AIR SYSTEM

The intake air system includes the air cleaner, the throttle valve housing, the intake air distributor, and the intake manifold tubes. The intake air sensor, which is also a part of the intake air system, is covered separately in **4.8 Testing and Replacing Intake Air Sensor**.

6.1 Removing and Installing Air Cleaner

Use the procedure given here if you intend to remove the air cleaner only. If you must remove the intake air sensor, remove the air cleaner body and the intake air sensor as a unit. See **4.8 Testing and Replacing Intake Air Sensor**.

To remove air cleaner:

1. Remove the air duct hose from the heater air blower.
2. Release the four clips that hold the air cleaner cover to the air cleaner body. See Fig. 6-1.

© 1976 VWoA—3846

A. Heater air blower
B. Clips (2 in front, 2 in rear)
C. Intake air preheating hose connection
D. Air cleaner cover mounting clamp

Fig. 6-1. Components associated with air cleaner removal.

3. If you wish to remove only the pleated paper filter element, push the air cleaner body and the intake air sensor toward the left, as indicated by the arrow in Fig. 6-2. Then withdraw the filter element from between the air cleaner body and the air cleaner cover.

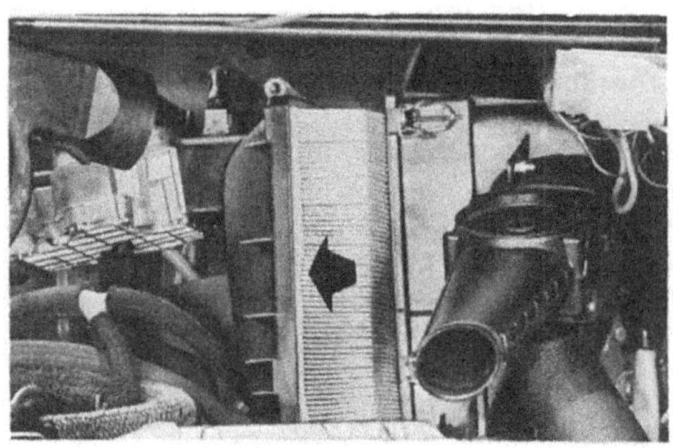

© 1976 VWoA—3847

Fig. 6-2. Preparation for filter removal. Push air cleaner body to left (arrow).

4. To remove the air cleaner cover fully, which is necessary in order to replace the battery or to obtain access to the fuel injection system control unit, disconnect the intake air preheating hose. Then open the air cleaner cover mounting clamp and remove the air cleaner cover.
5. To remove the air cleaner body from the intake air sensor, remove the air cleaner cover and the filter element. Disconnect the evaporative emission control hose and, where applicable, the vacuum hose from the air cleaner body. Then remove the four bolts that hold the air cleaner body to the intake air sensor and remove the air cleaner body.

Installation is the reverse of removal. The air cleaner must be firmly clamped to the car body and engaged with the rubber grommet indicated in Fig. 6-3.

© 1976 VWoA—3848

Fig. 6-3. Mountings for air cleaner cover. Rubber grommet is at **E**; clamp engagement point is at arrow.

Checking Thermostatic Valve

The thermostatic valve, located in the cover part of the air cleaner, can be checked by removing the cover from the intake air duct. The flap of the thermostatic valve should be in position to close off cool air from the intake air duct when the thermostatic bulb is at room temperature. With the engine fully warmed up in warm weather, the flap should close off the supply of warm air from the warm air duct and open the path for cool air from the intake air duct. If the flap is not functioning correctly, replace the thermostatic valve assembly.

Checking EEC Cutoff Valve

An EEC (evaporative emission control) cutoff valve was introduced on the 1976 models. The purpose of the valve is to prevent fuel fumes from entering the air cleaner when the engine is stopped or idling. The cutoff valve is located in the air cleaner body as shown in Fig. 6-4.

© 1976 VWoA—3867

Fig. 6-4. EEC cutoff valve (arrow).

To check the EEC cutoff valve, turn off the engine. Then disconnect the charcoal filter-to-air cleaner hose from the charcoal filter. Blow into the disconnected hose. The valve should be closed. If the cutoff valve is open, allowing air to enter the air cleaner, the valve is faulty and the air cleaner body must be replaced.

If the valve is correctly closed during the first test, start the engine. Accelerate the engine to about 1500 to 2000 rpm, then blow into the disconnected hose. The valve should be open, allowing air to enter the air cleaner. If the cutoff valve is not open, check the vacuum hose that connects the cutoff valve with the throttle valve housing. If the vacuum hose is not leaking or kinked, the cutoff valve is faulty and the air cleaner body must be replaced.

6.2 Removing and Installing Throttle Valve Housing

The throttle valve housing contains the throttle valve, which is cable-operated by the accelerator pedal. The housing also contains the bypass screw used to adjust the idle speed and the inlet port of the EGR (exhaust gas recirculation) system.

To remove throttle valve housing:

1. Remove the intake air sensor and the air cleaner body as a unit. See **4.8 Testing and Replacing Intake Air Sensor**.

2. Loosen the hose clamp and detach the large S-shaped rubber duct from the inlet end of the throttle valve housing. Disconnect the throttle return spring from the lever of the throttle valve shaft.
3. On vehicles through Chassis No. 226 2077 583, remove the two bolts that hold the EGR feed pipe to the throttle valve housing. Move the pipe slightly away from the housing, then slip out the gasket.
4. Disconnect the accelerator cable or linkage from the lever on the throttle valve shaft. Disconnect the vacuum hoses from the throttle valve housing and, on cars with automatic transmissions, from the idle speed regulator.
5. Remove the two screws that hold the throttle valve housing to the intake air distributor. Where applicable, remove the idle speed regulator.
6. On late 1976 cars, disconnect the wires from the microswitch. On 1975 and early 1976 cars, lift the throttle valve housing up, then disconnect the electrical plug from the throttle valve switch.
7. Remove the throttle valve housing from the car. If necessary, take out the two screws, then remove the throttle valve switch or the microswitch from the housing.

Installation is the reverse of removal. Use new gaskets at all points. Torque the screws that hold the throttle valve housing to the intake air distributor to 2.0 mkg (14 ft. lb.). Torque the bolts for the EGR pipe to 1.0 mkg (7 ft. lb.). If it is necessary to clean the EGR ports in the throttle valve housing, be careful not to burr or deform the housing.

6.3 Removing and Installing Intake Air Distributor

The intake air distributor, a black, stamped metal housing on top of the engine, receives intake air from the throttle valve housing. It distributes the air, via the four intake manifold tubes, to the cylinders of the engine. Normally, you will need to remove the intake air distributor only when you are preparing to disassemble the crankcase with the engine removed from the car. Owing to the location of the mounting fasteners beneath the intake air distributor, it is difficult to remove the intake air distributor without first removing other components that are atop the engine.

To remove:

1. If not previously removed, remove the intake air sensor and the air cleaner as described in **4.8 Testing and Replacing Intake Air Sensor**. Remove the large S-shaped rubber duct from the inlet end of the throttle valve housing.
2. Remove the throttle return spring, unhooking it from the lever on the throttle valve shaft and from the bracket on the intake air distributor. Remove the EGR feed pipe. See **7. Testing and Replacing EGR (Exhaust Gas Recirculation) Valve**.
3. Disconnect the accelerator cable or the linkage from the lever on the throttle valve shaft. Disconnect the vacuum hoses and the air hoses from the intake air distributor, the throttle valve housing, and, on cars with automatic transmissions, from the idle speed regulator.
4. If the cylinder heads and intake manifold tubes are installed, slide the four manifold hoses off the intake air distributor and onto the intake manifold tubes.
5. Remove the nuts that hold the intake air distributor to the crankcase. Then lift the intake air distributor off the crankcase and disconnect the electrical plug from the throttle valve switch.
6. Remove the intake air distributor from the car. If necessary, remove the throttle valve housing and the throttle valve switch from the intake air distributor.

Installation is the reverse of removal. Use new gaskets at all points. Torque the nuts that hold the intake air distributor to the crankcase to 2.0 mkg (14 ft. lb.). Torque the nuts and the bolts that hold the EGR pipe to 1.0 mkg (7 ft. lb.).

Removing and Installing Manifold Tubes

You can remove the intake manifold tubes without removing the intake air distributor. This will permit you to remove the cylinder heads and the cylinders without fully disassembling the intake air system.

To remove:

1. If the engine is installed, fully remove the air cleaner as described in **6.1 Removing and Installing Air Cleaner**.
2. Remove the injectors as described in **4.9 Testing and Replacing Injectors and Series Resistance Block**.
3. Remove the nuts that hold each intake manifold tube to the cylinder head. Then pull each intake manifold tube, its gasket, and its connecting hoses off the cylinder head and the intake air distributor. See Fig. 6-5 or Fig. 6-6.

Installation is the reverse of removal. Use new gaskets. Replace loose or leaking connecting hoses. Torque the nuts that hold the manifold tubes to the cylinder heads to 2.0 mkg (14 ft. lb.).

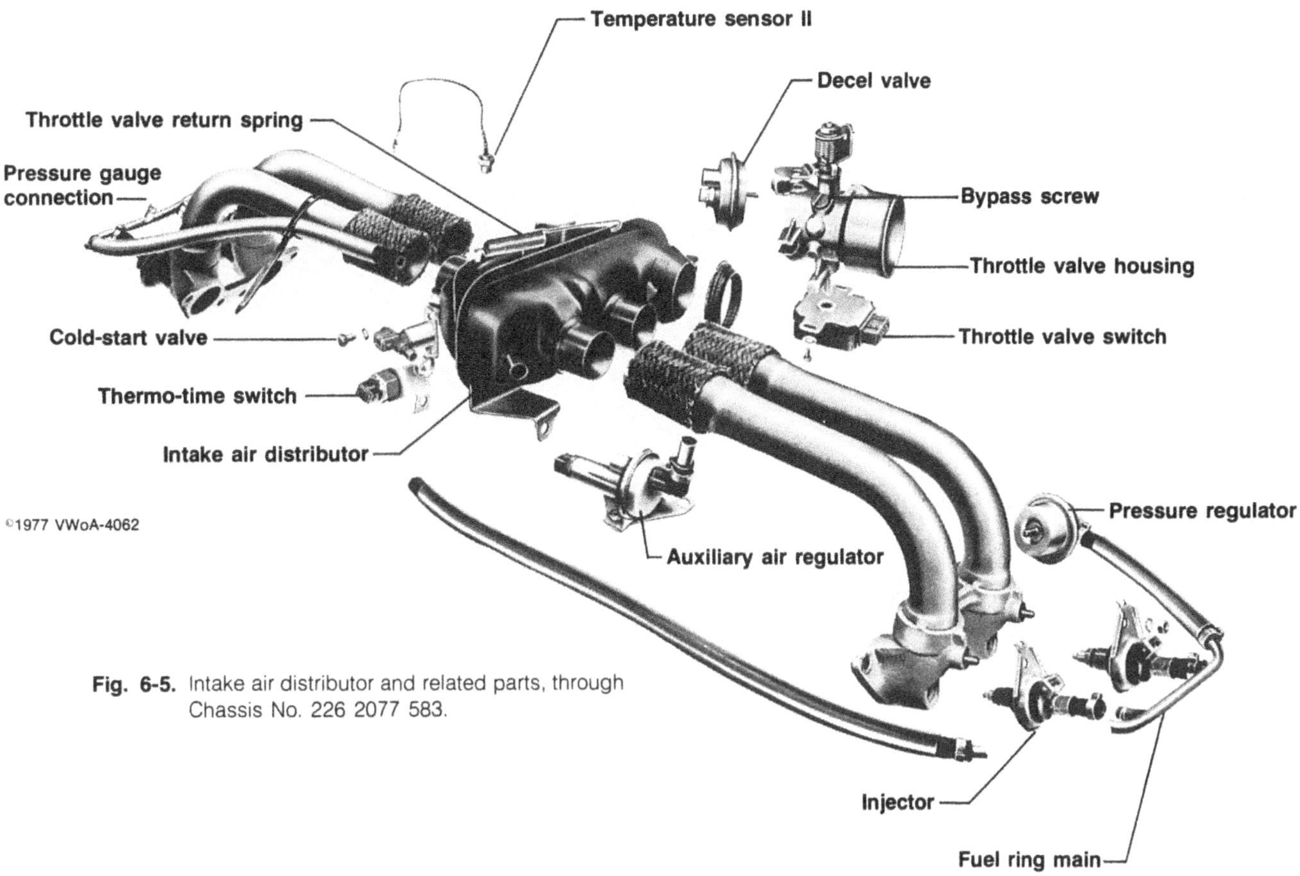

Fig. 6-5. Intake air distributor and related parts, through Chassis No. 226 2077 583.

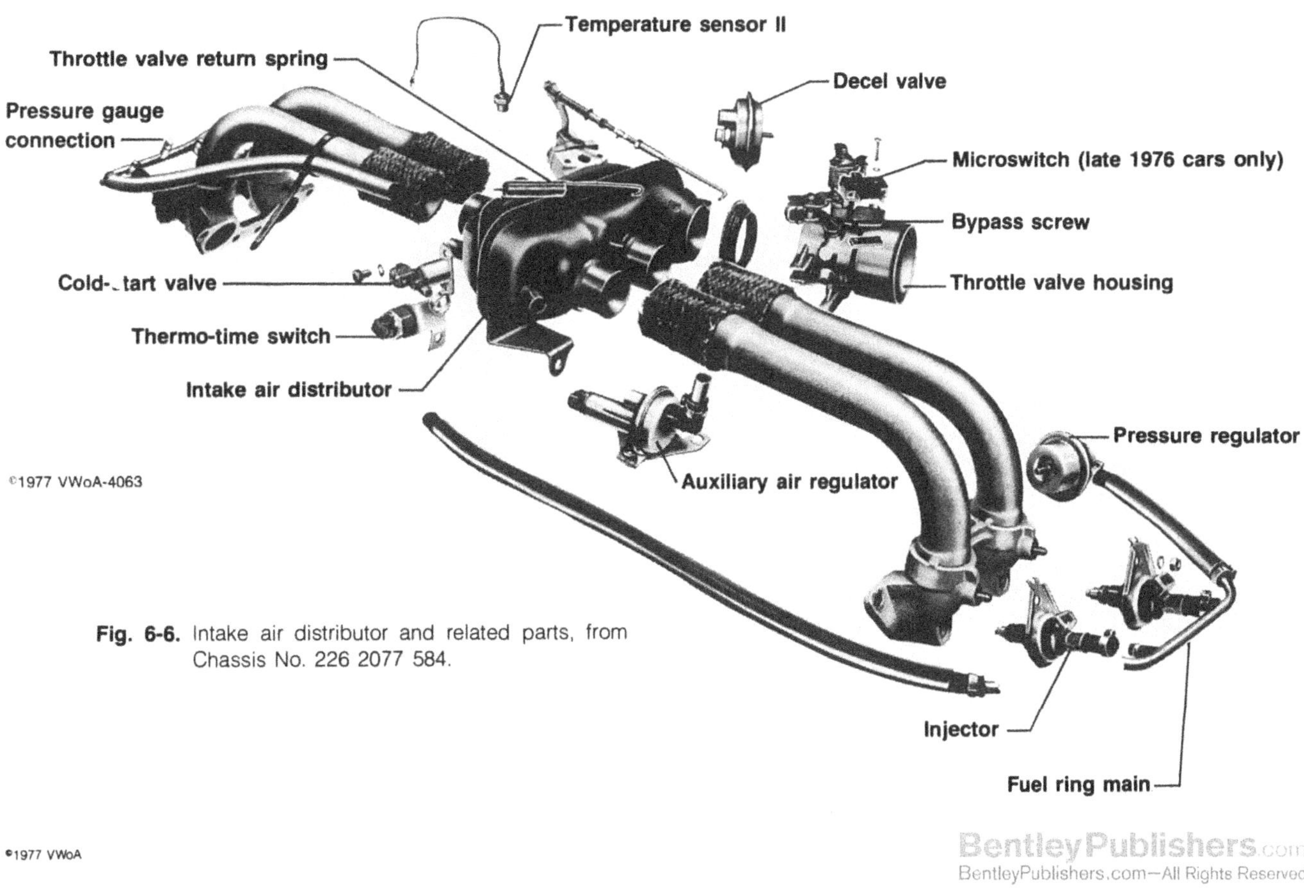

Fig. 6-6. Intake air distributor and related parts, from Chassis No. 226 2077 584.

10

6.4 Automatic Transmission Accelerator Linkage

(1976 and later models only)

Instead of having the accelerator cable connected to the throttle valve arm, as on other models, the 1976 and later cars with automatic transmissions have their accelerator cables connected to an operating lever on the transmission. The operating lever on the transmission is connected to the throttle valve lever on the engine by an adjustable accelerator linkage. Fig. 6-7 is a schematic view of the automatic transmission accelerator linkage.

1. Accelerator pedal
2. Accelerator cable
3. Cable conduit
4. Operating lever (on transmission)
5. Accelerator linkage

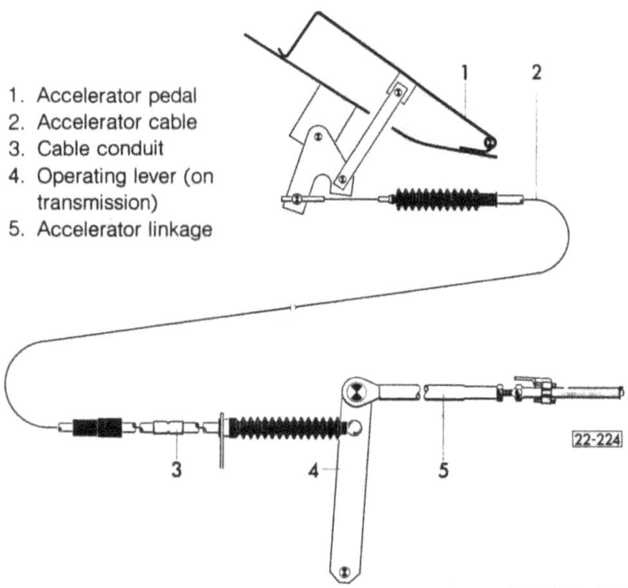

Fig. 6-7. Schematic view of automatic transmission accelerator linkage.

In removing the engine, disconnect the linkage at the operating lever on the transmission. Then pull the linkage out into the engine compartment. After removal and reinstallation of the linkage, the linkage must be adjusted.

To adjust:

1. With the engine turned off, make sure that the accelerator linkage is in the idling position (throttle valve closed).
2. Remove the spring. Then press the linkage in the direction indicated by the arrow in Fig. 6-8 so that the operating lever on the transmission is against its stop.
3. Adjust the linkage end piece by turning the adjusting nut until the flat surface on the pivot pin touches without tension. The parts are identified in Fig. 6-8.
4. Install the spring. Start the engine and check that the idling speed is correct. If necessary, adjust the idle to specifications by turning the bypass screw on the throttle valve housing.
5. Check that the engine returns to idle after the accelerator pedal is released. Then tighten the locknut (Fig. 6-8).

1. Spring
2. Adjusting nut
3. Pivot pin
4. Locknut

Fig. 6-8. Linkage adjustment. During adjustment, pull linkage in direction indicated by arrow.

To adjust accelerator cable:

1. Have someone press the accelerator pedal fully down and hold it in the kickdown position.
2. Measure the play at the operating lever on the transmission (Fig. 6-9).

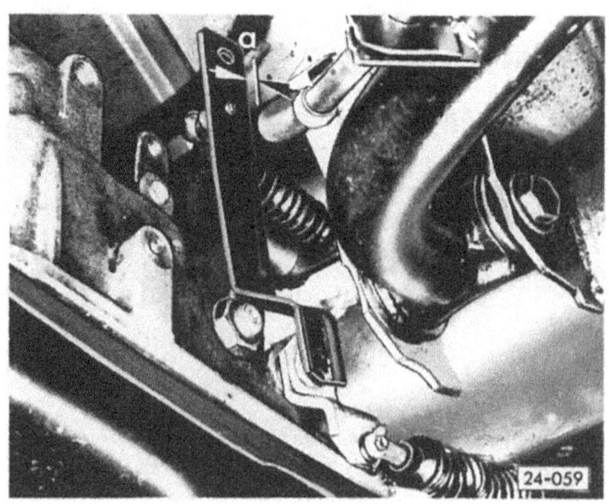

Fig. 6-9. Play at operating lever. Dimension **a** should be 1 to 2 mm (.040 to .079 in.).

3. If the play at the operating lever is not between 1 and 2 mm (.040 and .079 in.), loosen the pivot pin screw at the accelerator and reposition the cable end in order to obtain the correct play. The pedal is the same as on cars with manual transmissions; see Fig. 6-10, given under the next heading.

6.5 Manual Transmission and 1975 Automatic Transmission Accelerator Linakge

Fig. 6-10 is an exploded view of the accelerator pedal and related components. The accelerator cable is connected to the pedal lever and to the throttle valve arm on the engine's throttle valve housing.

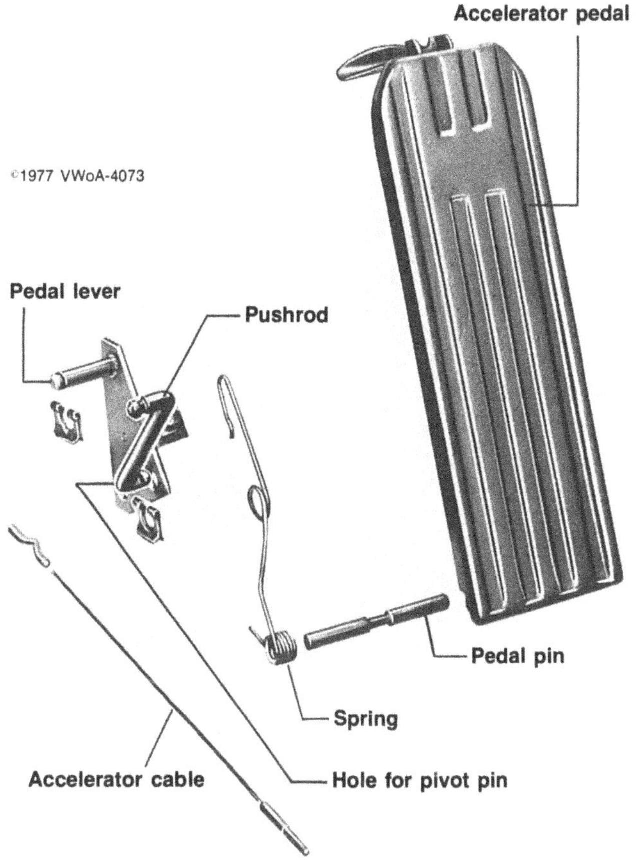

Fig. 6-10. Exploded view of accelerator pedal assembly used on fuel injection vehicles.

To adjust accelerator cable:

1. Have someone press the accelerator pedal fully down. On cars with automatic transmissions, the pedal should be in kickdown position.
2. Measure the clearance between the throttle valve lever and its stop (Fig. 6-11).

Fig. 6-11. Clearance between throttle valve lever and stop on throttle valve housing. Dimension **a** should be 1 to 1.5 mm (.040 to .060 in.) with accelerator pedal fully depressed.

3. If the clearance at the throttle valve lever stop is not between 1 and 1.5 mm (.040 and .060 in.), loosen the pivot pin screw at the accelerator and reposition the cable end in order to obtain the correct clearance.

7. TESTING AND REPLACING EGR (EXHAUST GAS RECIRCULATION) VALVE

Two kinds of EGR valves have been used on the vehicles covered by this Manual. Vehicles manufactured through January 1976 have a vacuum-operated EGR valve that is controlled electrically by the throttle valve switch. Vehicles manufactured from February 1976 have a mechanically-operated EGR valve that is controlled by a linkage connected to the throttle valve lever. With either system, no exhaust gas is recirculated at idle or at full throttle.

Electrically-controlled, Vacuum-operated EGR Valve

The electrically-controlled, vacuum-operated EGR valve used on early cars with fuel injection is located in the engine compartment, approximately on top of the No. 4 cylinder. A long feed pipe links the valve with the throttle valve housing. The EGR valve, when held open by vacuum, permits the recirculation of exhaust gases. At or near full throttle, and at closed throttle, current is applied to a solenoid in the EGR valve by the throttle valve switch. This interrupts the vacuum supply, thereby stopping the recirculation of exhaust gases. If exhaust gases are being recirculated at idle, causing the engine to run poorly or to stall, check the adjustment of the throttle valve switch as

10

described in **4.6 Testing and Replacing Throttle Valve Switch or Microswitch.** Fig. 7-1 is an exploded view of the early-type EGR valve and related components.

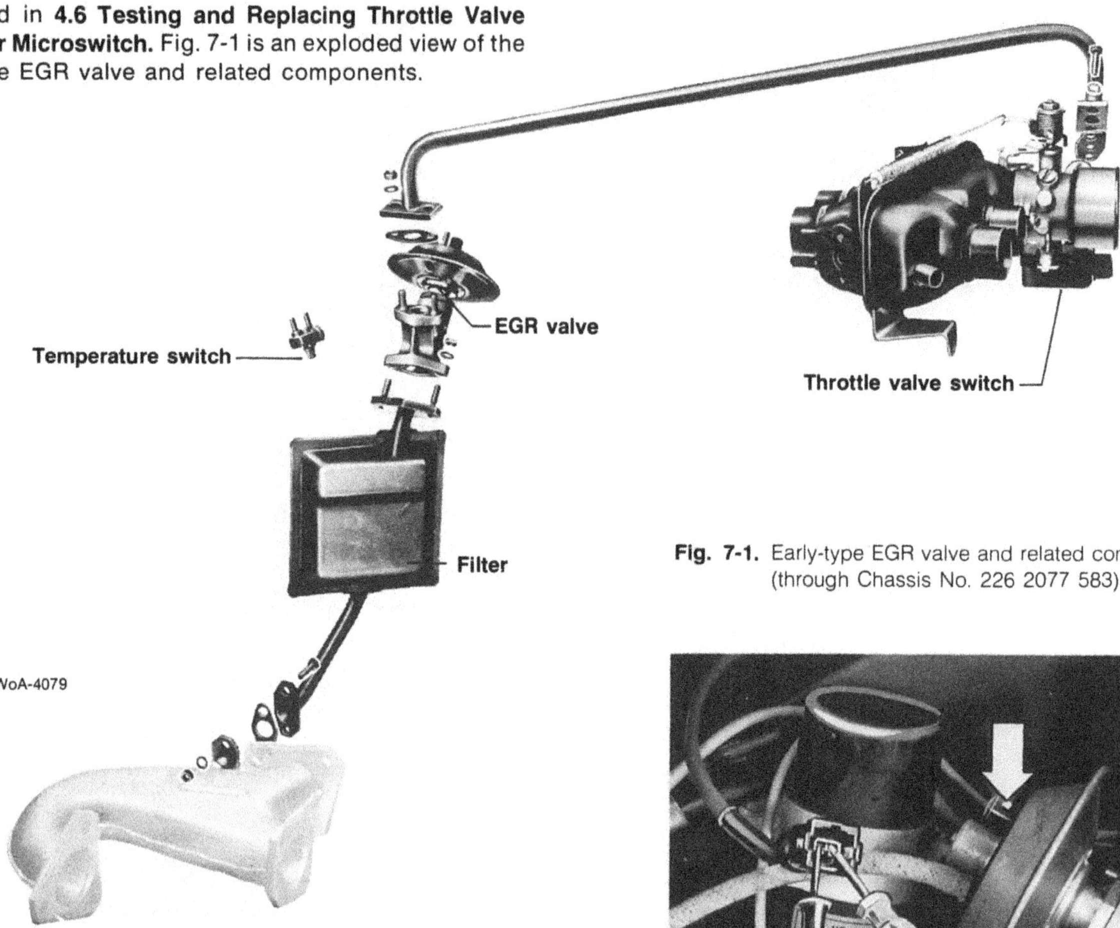

Fig. 7-1. Early-type EGR valve and related components (through Chassis No. 226 2077 583).

Fig. 7-2. Test lamp being used to check whether battery current is reaching EGR valve plug. Arrow indicates installed location of plug on vacuum unit.

To check EGR valve:

1. With the engine idling, pull the plug (Fig. 7-2) off the EGR valve's vacuum unit. The engine should slow down noticeably or stall, indicating that exhaust gases are being recirculated.
2. If there is no change when the plug is pulled off the EGR valve at idle, stop the engine. Then turn the ignition key to the running (2) position without restarting the engine.
3. Connect a test lamp as shown in Fig. 7-2. Then operate the throttle valve by hand, moving the throttle valve shaft from its idle position into the mid-speed range.
4. If the test lamp does not light at idle and at full throttle, test the throttle valve switch as described in **4.6 Testing and Replacing Throttle Valve Switch**. If the throttle valve switch, or the wire to the EGR valve, is not faulty, test the double relay as described in **4.1 Testing and Replacing Double Relay**. Make your tests with the help of the wiring diagram given in **3. Troubleshooting Fuel Injection**.
5. If the test lamp goes off when the throttle is moved off the idle position, and lights at idle and at or near full throttle, the EGR valve is faulty and should be replaced.

To replace early-type EGR valve and feed pipe:

1. To remove the feed pipe only, remove the two nuts that hold the pipe to the EGR valve and the two bolts that hold the pipe to the throttle valve housing. Then remove the pipe and its gaskets. Installation is the reverse of removal. Use new gaskets and torque the nuts to 1.0 mkg (7 ft. lb.).

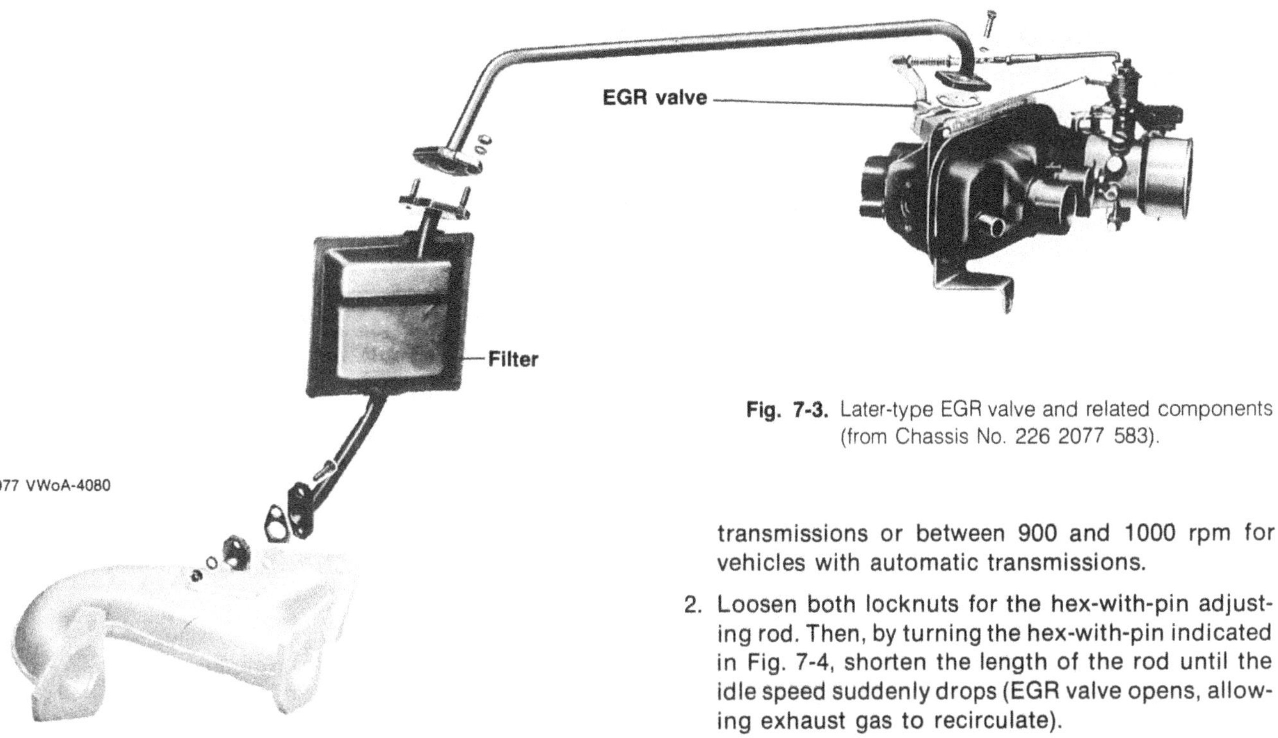

©1977 VWoA-4080

Fig. 7-3. Later-type EGR valve and related components (from Chassis No. 226 2077 583).

2. If you must remove the EGR valve, first remove the feed pipe as described in step 1. Then disconnect the vacuum hose and the electrical plug from the EGR valve's vacuum diaphragm.
3. Remove the two nuts that hold the EGR valve to the cylinder cover plate (and to the EGR inlet pipe, which is beneath the vehicle). Then remove the EGR valve and its gasket.

Installation is the reverse of removal. Use new gaskets at all points. Torque the nuts to 1.0 mkg (7 ft. lb.).

Mechanically-controlled, Mechanically-operated EGR Valve

The mechanically-controlled, mechanically-operated EGR valve used on later cars with fuel injection is located on the intake air distributor (Fig. 7-3). It is connected to the throttle valve lever by an adjustable linkage.

If exhaust gases are being recirculated at idle, causing the engine to run poorly or to stall, check the adjustment of the EGR valve linkage. When correctly adjusted at idle, the mechanical linkage will also cut off exhaust gas recirculation at the correct full-load throttle position without additional adjustment.

To adjust EGR valve operation:

1. Start the engine and allow it to warm up thoroughly. Make sure that the idle speed is correct, between 850 and 950 rpm for vehicles with manual transmissions or between 900 and 1000 rpm for vehicles with automatic transmissions.
2. Loosen both locknuts for the hex-with-pin adjusting rod. Then, by turning the hex-with-pin indicated in Fig. 7-4, shorten the length of the rod until the idle speed suddenly drops (EGR valve opens, allowing exhaust gas to recirculate).

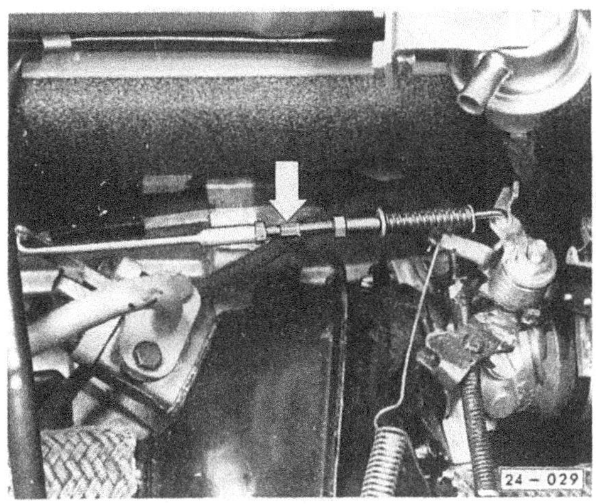

©1977 VWoA-4081

Fig. 7-4. Hex-with-pin (arrow) used to adjust linkage rod length.

3. From this position, turn the hex-with-pin in the opposite direction—by 1 5/6 turns on 1979 and later vehicles with manual transmissions (except California), or by 1 1/6 turns on earlier vehicles with manual transmissions (and on 1979 and later California models with manual transmissions), or by 5/6 turn on all vehicles with automatic transmissions. (The pin on the hex can be used for orientation.).
4. Without moving the adjustment, tighten both locknuts.

10

8. TESTING AND ADJUSTING OXYGEN SENSOR EMISSION CONTROL
(1979 and later California cars only)

The oxygen sensor (Fig. 8-1) generates from 0 to 1 volt depending on sensor temperature and exhaust gas composition. The varying voltage is measured by the injection system's control unit, which alters the injector-open time to give low emissions. The sensor operates neither below 300°C (575°F) nor when turned off at full throttle by a microswitch on the throttle valve housing.

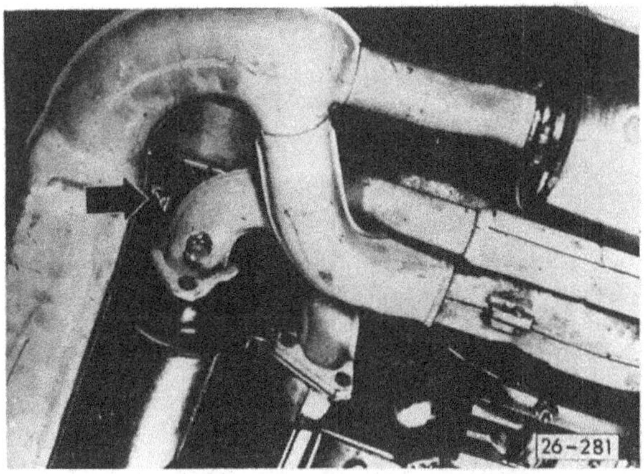

©1979 VWoA—4421

Fig. 8-1. Location of oxygen sensor (arrow) ahead of catalytic converter, and above analyzer probe port.

To troubleshoot:

1. Check and, if necessary, adjust the engine's idle speed and idle mixture. See **5. Adjusting Fuel Injection.** If the CO level is incorrect with the oxygen sensor connected (Fig. 8-2), inspect and if necessary repair the wiring or the microswitch. If no fault is found, replace the oxygen sensor.

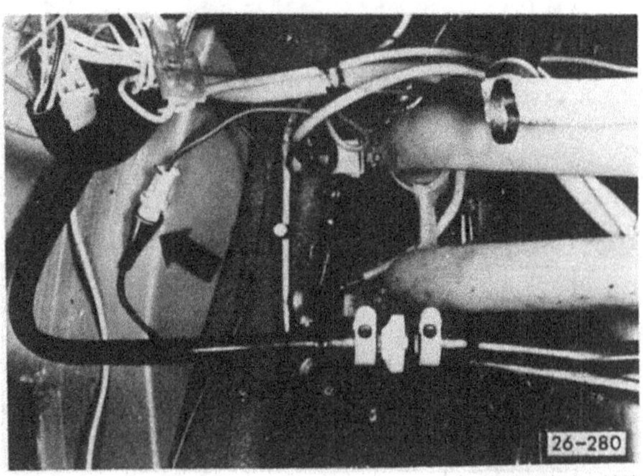

©1979 VWoA—4426

Fig. 8-2. Point at which oxygen sensor can be disconnected (arrow). To remove sensor, disconnect wire and unscrew sensor from exhaust system.

2. If the CO level is correct at idle, recheck it at 2000 rpm. The oxygen sensor system is operating correctly if the CO level is between 0.4% and 0.8%.
3. If the CO level is too low during the above test, there are leaks in the exhaust system ahead of the sensor, or where you have connected the analyzer probe.
4. If the CO level is too high during the test described in step 2, first check the sensor wiring and the microswitch. If no fault is found, replace the sensor. If the CO level remains high after the oxygen sensor has been replaced, the electronic control unit for the fuel injection system is faulty and should be replaced.

To check and adjust microswitch:

1. Stop the engine. Disconnect the wires from the microswitch, then connect an ohmmeter or a battery-powered test light to the switch terminals.
2. Make sure that the ohmmeter indicates infinite ohms (battery-powered test light off) with the throttle valve closed and 0 ohms (battery-powered test light on) with the throttle valve wide open. If the cam on the throttle valve shaft does not correctly operate the switch, check the switch by hand-pressing the roller toward the switch body. If either reading is incorrect, replace the microswitch.
3. Make sure that with the throttle valve wide open, the cam operates the switch when the roller is at the approximate center of the cam (Fig. 8-3). If necessary, loosen the screws and reposition the switch so that it operates correctly. Then tighten the screws and reconnect the wires.

©1979 VWoA—4425

Fig. 8-3. Cam (arrow) on throttle valve shaft in correct contact with roller of microswitch at full throttle. As cam reaches this position, ohmmeter should go from infinite ohms to 0 ohms, or test light should come on.

COLOR WIRING APPENDIX

These color diagrams have been produced from vintage sources and may not reflect the most current editorial content for the components they represent. They have been selected and reproduced here because they contain color, which can be both interesting and helpful.

These color diagrams are intended to supplement rather than replace the corresponding black and white diagrams.

See table below for list of color diagrams and their corresponding black and white diagrams earlier in the manual.

Color Diagram in this appendix	Black & White Diagram from Section 4
A_1	J
Supplement to A	H
A_2	I
Supplement to B	Supplement to G
B	G
C	F
D	E_1
E	D
F	B
G	C

This page intentionally left blank

The current-flow diagrams for 1973 and later cars have symbols for the electrical components rather than pictorial representations as in earlier diagrams. However, pictorial representations do appear separately at the bottom of the diagram to show where the wires attach to the actual component. This is an aid to troubleshooting and component replacement. The symbols used in the diagram itself are explained in Fig. 14-1. The thin black lines in the current-flow diagrams are not actual wires but ground connections via the vehicle chassis.

Along the bottom of each current-flow diagram is a yellow band containing numbers. These numbers (current tracks) will help you find electrical components easily. Appearing after each component listed in the description are numbers in a yellow legend labeled "current track." These numbers indicate which current track in the current-flow diagram contains the component.

Fig. 14-1. Schematic representation of electrical components in current-flow diagrams.

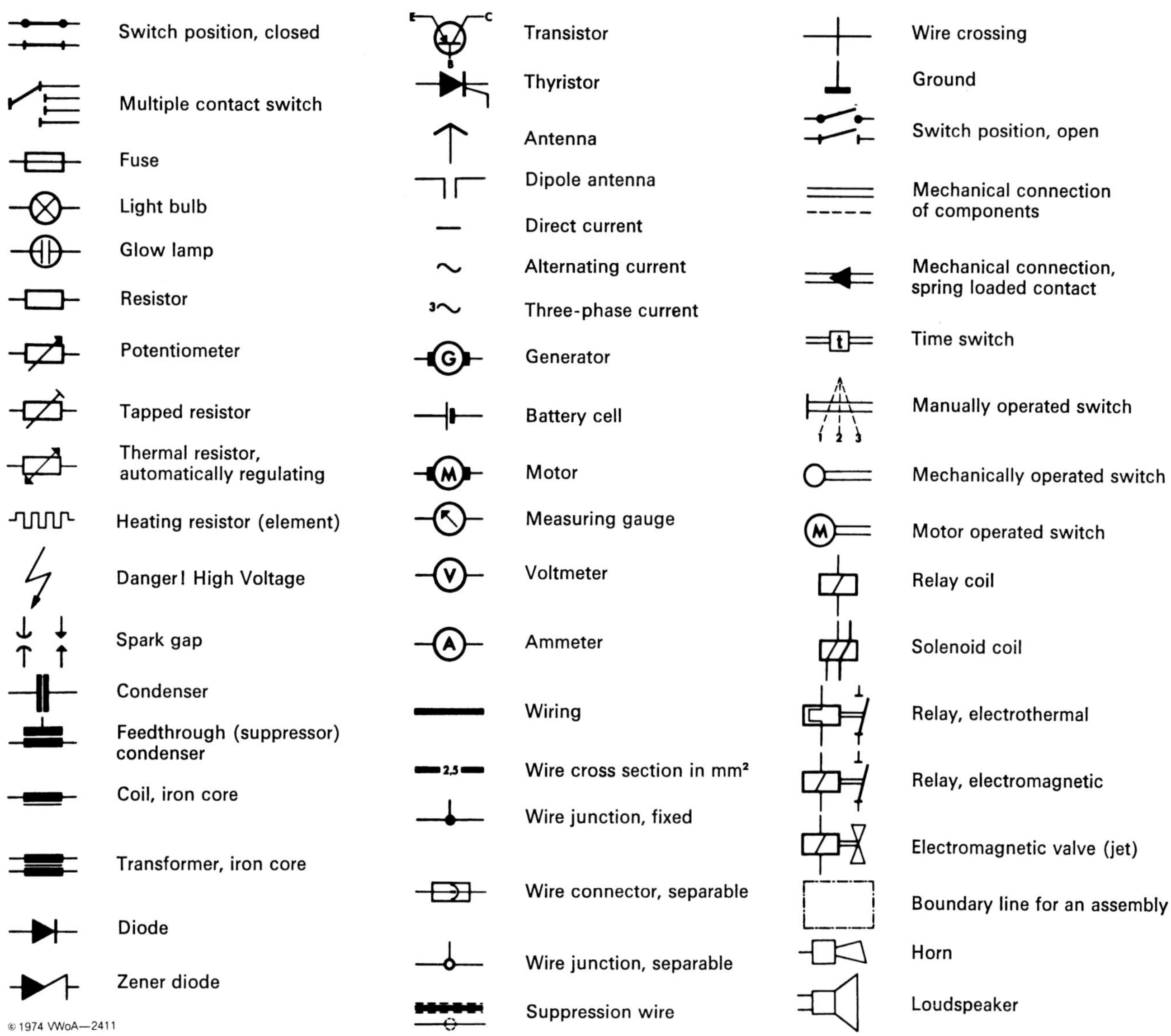

A1.

VW Type 2—from August 1976 (1977 Models)

(Page 1 of 3)

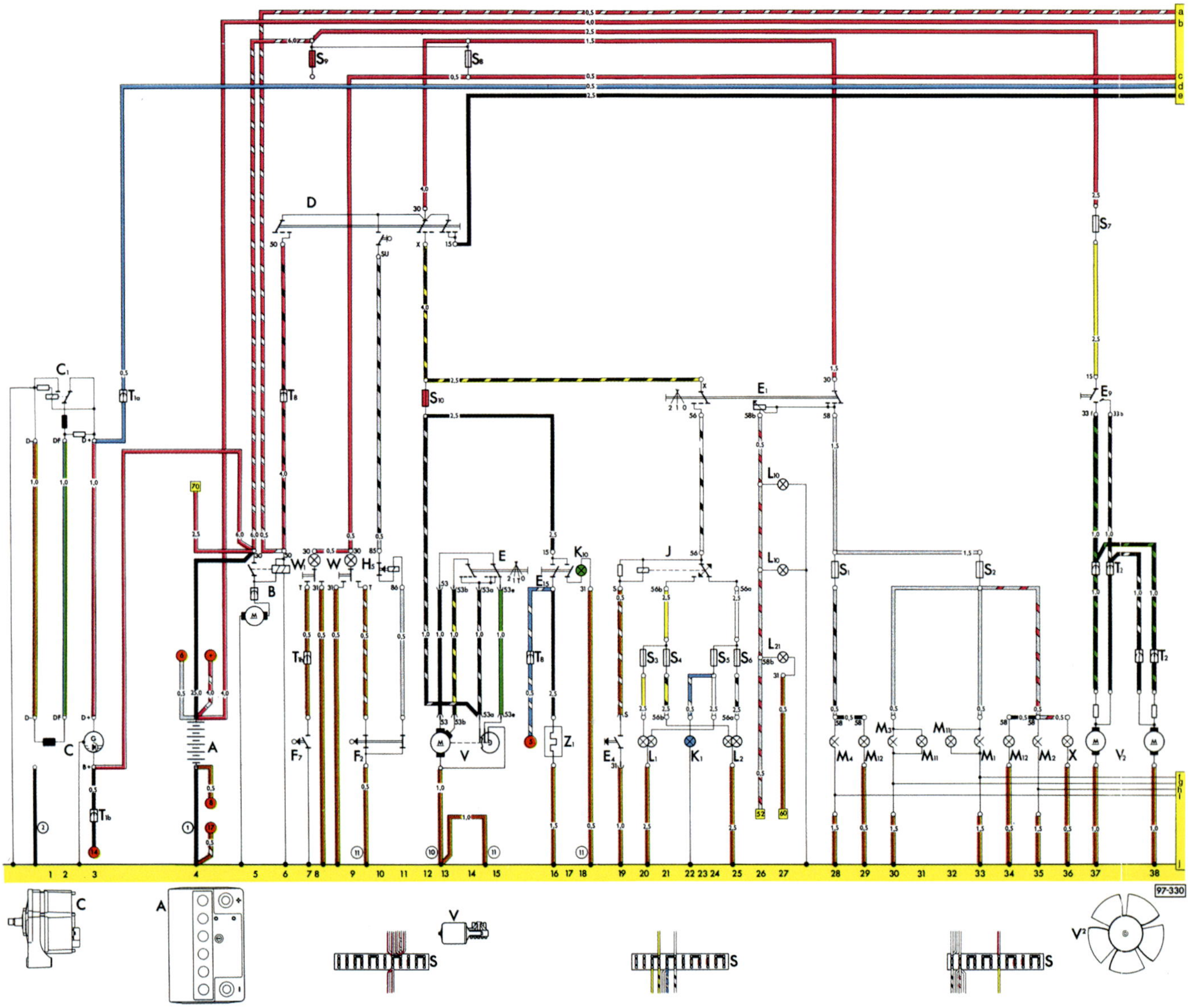

97-330

Description			Current track
A	–	Battery	4
B	–	Starter	5, 6
C	–	Alternator	1, 2, 3
C 1	–	Regulator	1, 2, 3
D	–	Ignition/starter switch	6–13
E	–	Windshield wiper switch	13–15
E 1	–	Light switch	22–28
E 4	–	Headlight dimmer switch	19
E 9	–	Fan motor switch	37
E 15	–	Rear window defogger switch	16–18
F 2	–	Door contact and buzzer alarm switch, left	10
F 7	–	Sliding door switch	7
H 5	–	Ignition key warning buzzer	11
J	–	Headlight dimmer relay	19–25
K 1	–	High beam warning light	22
K 10	–	Rear window defogger warning light	18
L 1	–	Sealed beam unit, left headlight	20
L 2	–	Sealed beam unit, right headlight	25
L 10	–	Instrument panel light	27
L 21	–	Light for heater lever illumination	27
M 1	–	Parking light, left	33
M 2	–	Taillight, right	35
M 3	–	Parking light, right	30
M 4	–	Taillight, left	28
M 11	–	Sidemarker light, front	31, 32
M 12	–	Sidemarker light, rear	29, 34
S 1 to S 10	–	Fuse box	
T 1	–	Wire connector, single a—in engine compartment, right b—in engine compartment, right h—next to interior light front	
T 2	–	Wire connector, double	
T 8	–	Wire connector, eight-point	13–15
V	–	Windshield wiper motor	37, 38
V 2	–	Fan motor	37, 38
W	–	Interior light, front	9
W 1	–	Interior light, rear	8
X	–	License plate light	36
Z 1	–	Rear window defogger heating element	16

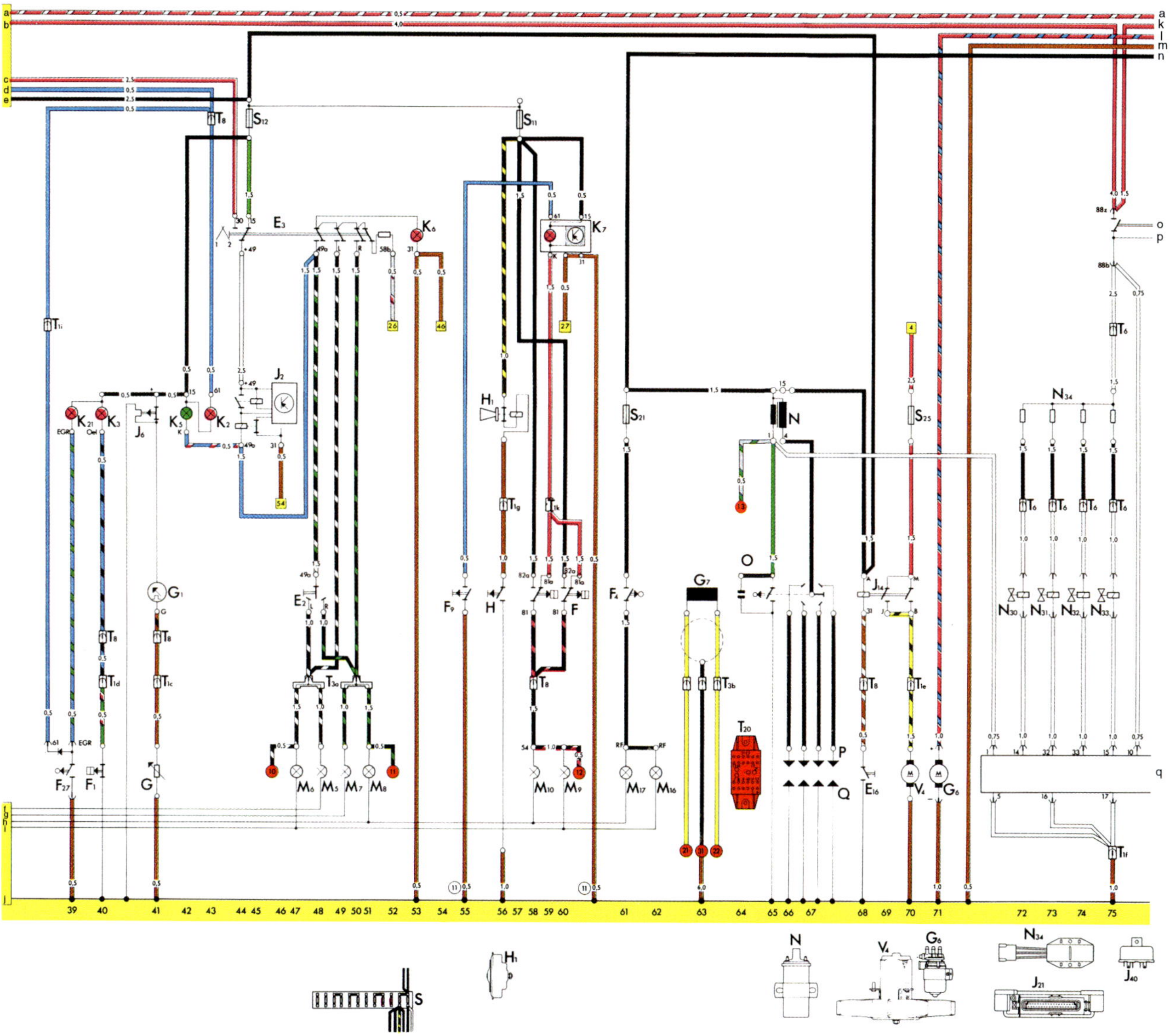

CAUTION —

Disconnect the battery ground strap before starting to work on any part of the electrical system.

① – Ground strap from battery to body

② – Ground strap from transmission to frame

⑩ – Ground connection, dashboard

⑪ – Ground connection, speedometer

A1.

VW Type 2—from August 1976 (1977 Models)

(Page 3 of 3)

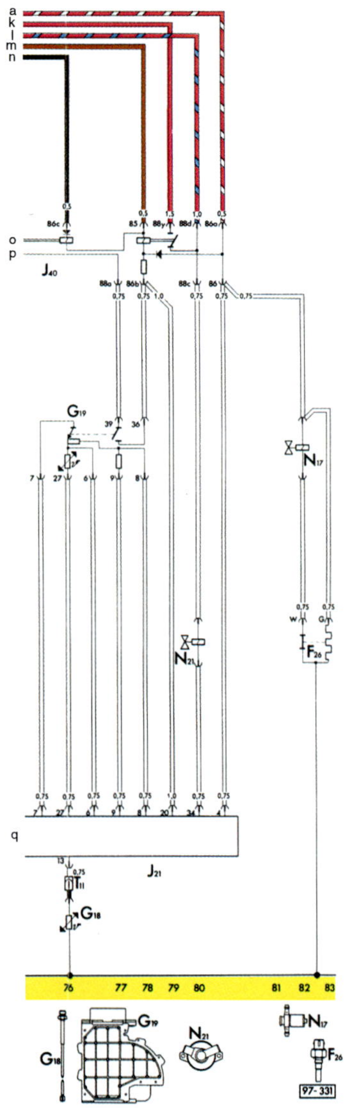

		Description	Current track
E 2	–	Turn signal	48
E 3	–	Emergency flasher switch	44–52
E 16	–	Heater air blower switch	68
F	–	Brake light switch	57–60
F 1	–	Oil pressure switch	40
F 4	–	Back-up light switch	61
F 9	–	Parking brake control light switch	55
F 26	–	Thermo time switch for cold starting valve	82, 83
F 27	–	EGR elapsed mileage switch	39
G	–	Fuel gauge sending unit	41
G 1	–	Fuel gauge	41
G 6	–	Fuel pump	71
G 7	–	TDC marker unit	63
G 18	–	Temperature sensor on cylinder head	76
G 19	–	Intake air sensor	76–78
H	–	Horn button	56
H 1	–	Horn	56
J 2	–	Emergency flasher relay	44–47
J 6	–	Voltage stabilizer	41
J 14	–	Heater air blower relay	68–70
J 21	–	Control unit (electronic)	72–80
J 40	–	Double relay	75–80
K 2	–	Alternator charging warning light	43
K 3	–	Engine oil pressure light	40
K 5	–	Turn signal warning light	42
K 6	–	Emergency flasher warning light	53
K 7	–	Dual circuit brake warning light	59
K 21	–	CAT converter warning light	39
M 5	–	Turn signal, front left	48
M 6	–	Turn signal, rear left	47
M 7	–	Turn signal, front right	49
M 8	–	Turn signal, rear right	51
M 9	–	Brake light, left	60
M 10	–	Brake light, right	58
M 16	–	Back-up light, left	62
M 17	–	Back-up light, right	61
N	–	Ignition coil	65, 66
N 17	–	Cold starting valve	82
N 21	–	Auxiliary air regulator	80
N 30	–	Injector, cyl. 1	72
N 31	–	Injector, cyl. 2	73
N 32	–	Injector, cyl. 3	74
N 33	–	Injector, cyl. 4	75
N 34	–	Resistance for injectors	72–75
O	–	Ignition distributor	64–65
P	–	Spark plug connectors	66–67
Q	–	Spark plugs	66–67
S 11 to S 12	–	Fuse box	
S 21	–	Fuse for back-up light	61
S 25	–	Fuse for heater air blower	70
T 1	–	Wire connector, single	
c	–	fuel tank right side	
d	–	in engine compartment	
e	–	in engine compartment	
f	–	in engine compartment	
g	–	behind dashboard	
i	–	behind dashboard	
k	–	behind dashboard panel	
l	–	in engine compartment, center	
T 3a	–	Wire connector, three-point	
T 3b	–	Wire connector, three-point	
T 6	–	Wire connector, six-point	
T 8	–	Wire connector, eight-point	
T 20	–	Analysis system, analysis socket	
V 4	–	Heater air blower	70
⑪	–	Ground connector, speedometer	

Computer Analysis System ——

The orange-colored spots are connections in the Computer Analysis System that are wired to Socket T 20. The numbers in these spots correspond to terminals in the Computer Analysis socket.

Supplement to Wiring Diagram A_2

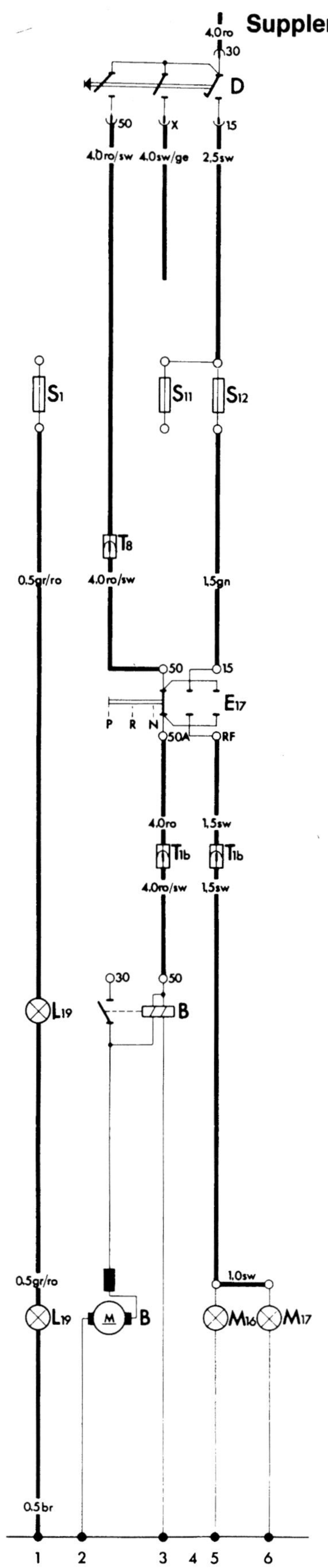

© 1976 VWoA—3869

97-258

	Description	Current track
B	— Starter	2,3
D	— Ingition/starter switch	2,3,5
E^{17}	— Starter cutout and backup lights switch	3,4,5
L^{19}	— Shift console light	1
M^{16}	— Backup light, left	5
M^{17}	— Backup light, right	6
S^1, S^{11}, S^{12}	Fuses in fuse box	
T^{1b}	— Wire connector, single; behind instrument panel	
T^8	— Wire connector, 8 point; behind instrument panel	

Color codes

br = brown
ge = yellow
gn = green
gr = gray
ro = red
sw = black

CAUTION —
Disconnect the battery ground strap before starting to work on any part of the electrical system.

A2.

VW Type 2—from August 1975 (1976 Models)

Description		Current track
A	—Battery	7
B	—Starter	8, 9
C	—Alternator	1, 2, 3
C^1	—Regulator	1, 2, 3
D	—Ignition / starter switch	9, 10, 11, 12, 13, 14, 15, 16, 17
E	—Windshield wiper switch	18, 19
E^1	—Light switch	26, 27, 28, 29, 30, 31
E^2	—Headlight dimmer switch	22
E^9	—Fresh air fan motor switch	40, 41
E^{15}	—Rear window defogger switch	20, 21
E^{16}	—Heater air blower switch	4
F^2	—Door contact and buzzer alarm switch, left	13, 14, 15
F^7	—Sliding door switch	10
H^5	—Ignition key warning buzzer	15
J	—Dimmer relay	22-28
J^{14}	—Heater air blower relay	4-6
K^1	—High beam warning light	25
K^{10}	—Rear window defogger warning light	21
L^1	—Sealed beam unit, left headlight	23
L^2	—Sealed beam unit, right headlight	28
L^{10}	—Instrument panel light	30
L^{21}	—Light for heater lever illumination	30
M^2	—Tail / brake light, right	38
M^4	—Tail / brake light, left	31
M^5	—Parking light, left	32
M^7	—Parking light, right	36
M^{11}	—Sidemarker light, front	34, 35
M^{12}	—Sidemarker light, rear	32, 37
S^1 to S^{10}	Fuse box	
S^{25}	—Fuse for heater air blower	7
T	—Cable adapter behind instrument panel	18
T^1	—Wire connector, single	
	a—in engine compartment, right	4
	b—in engine compartment	6
T^2	—Wire connector, double point behind instrument panel	40, 41
T^{30}	—Wire connector, three point behind instrument panel	19, 17
T^8	—Wire connector, eight point behind instrument panel	47, 20
V	—Windshield wiper motor	17-19
V^2	—Fan motor	40, 41
V^4	—Heater air blower	6
W	—Interior light, front	12
W^1	—Interior light, rear	11
X	—License plate light	39
Z^1	—Rear window defogger heating element	20
(1)	—Ground strap from battery to body	7
(2)	—Ground strap from transmission to frame	1
(10)	—Ground connection, dashboard	17
(11)	—Ground connection, speedometer	13, 19, 21

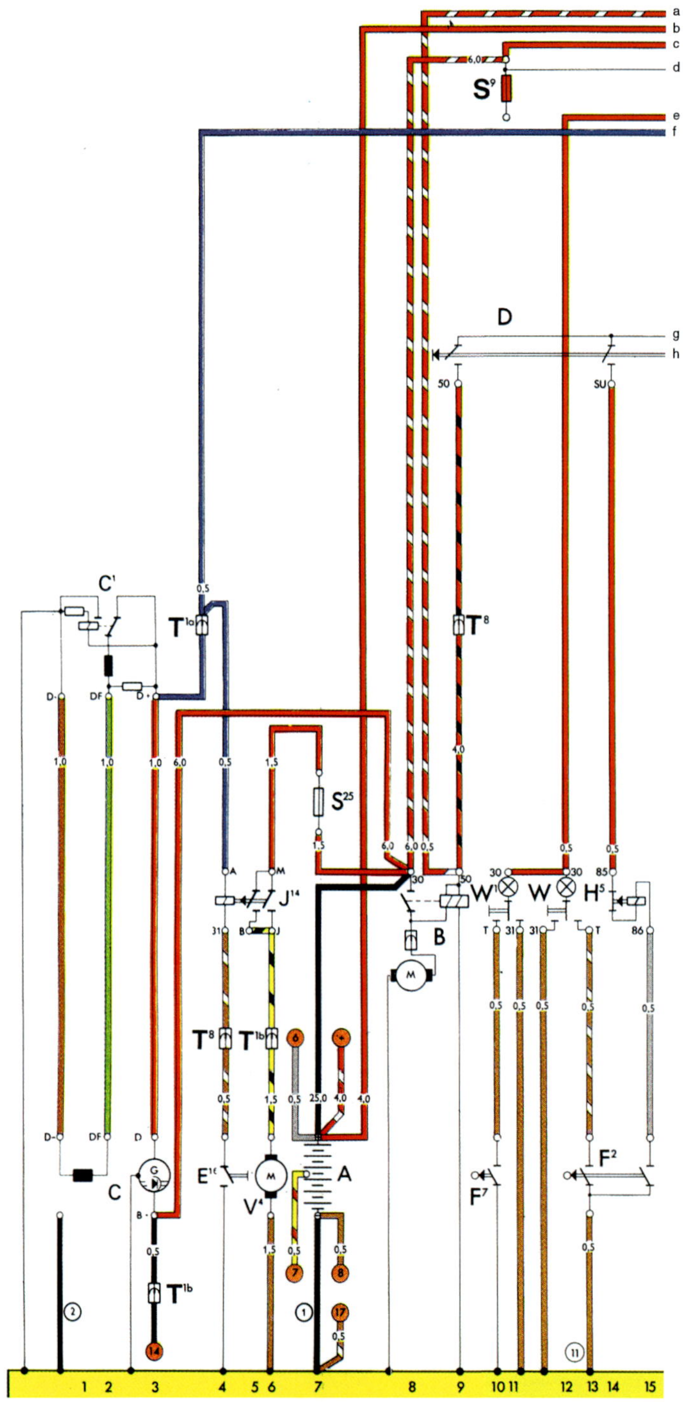

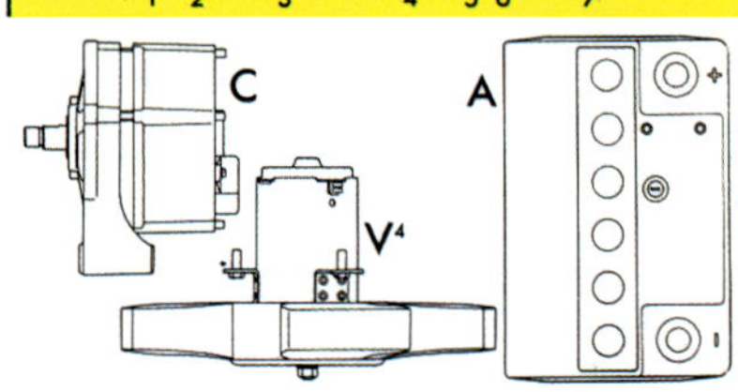

Test Network

The orange colored spots are the connections in the test network that are wired to the test socket (T^{20}). The numbers in the spots correspond to the terminals in the test socket.

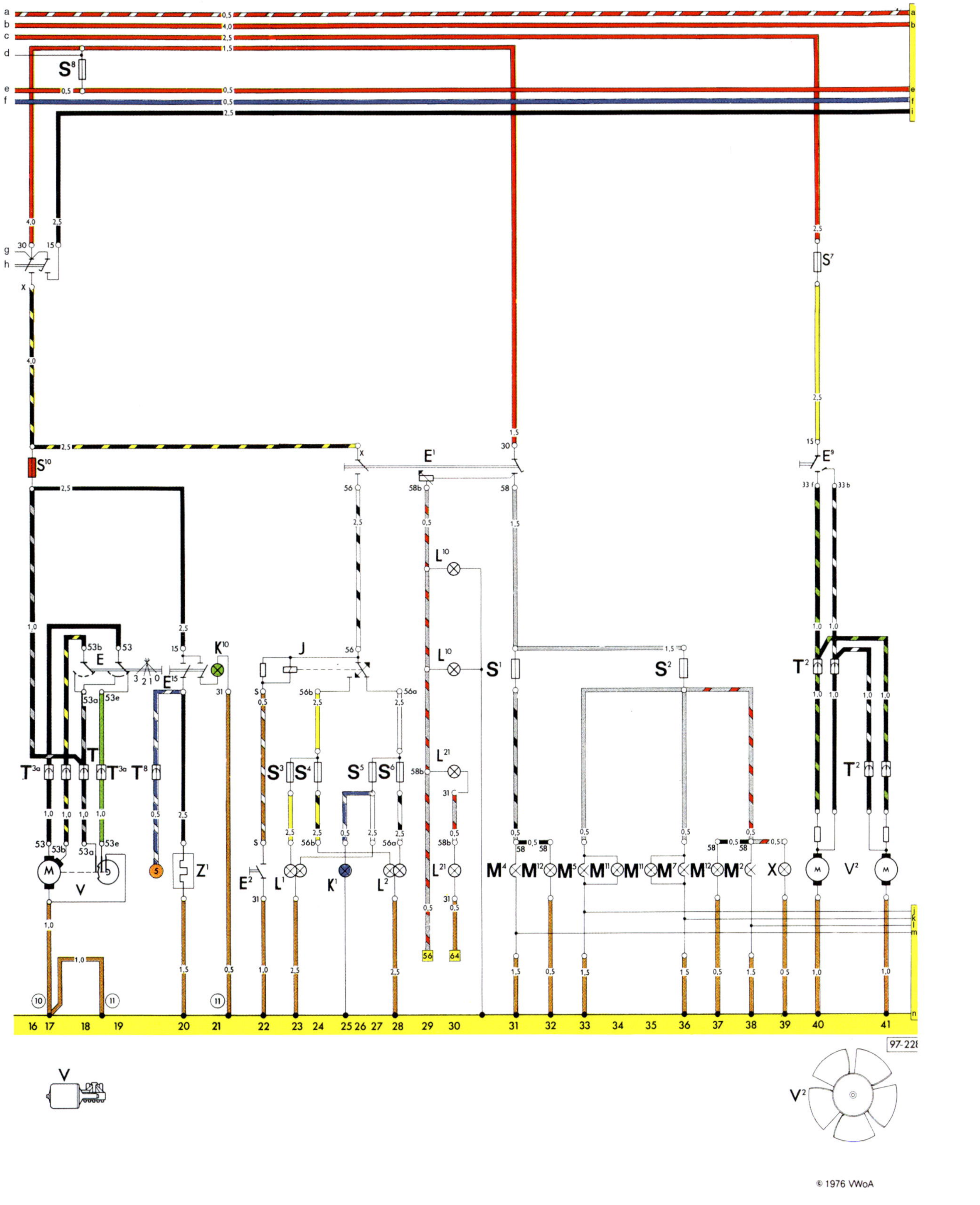
S8
S10
S7
E1
E9
J
E
E2
E15
K10
K1
L1
L2
L10
L21
S1
S2
S3
S4
S5
S6
T
T2
T3a
T8
V
V2
Z1
X
M4
M12
M5
M11
M7
M2
16 17 18 19 20 21 22 23 24 25 26 27 28 29 30 31 32 33 34 35 36 37 38 39 40 41

97-228

A2.

VW Type 2—from August 1975 (1976 Models)

(Page 3 of 4)

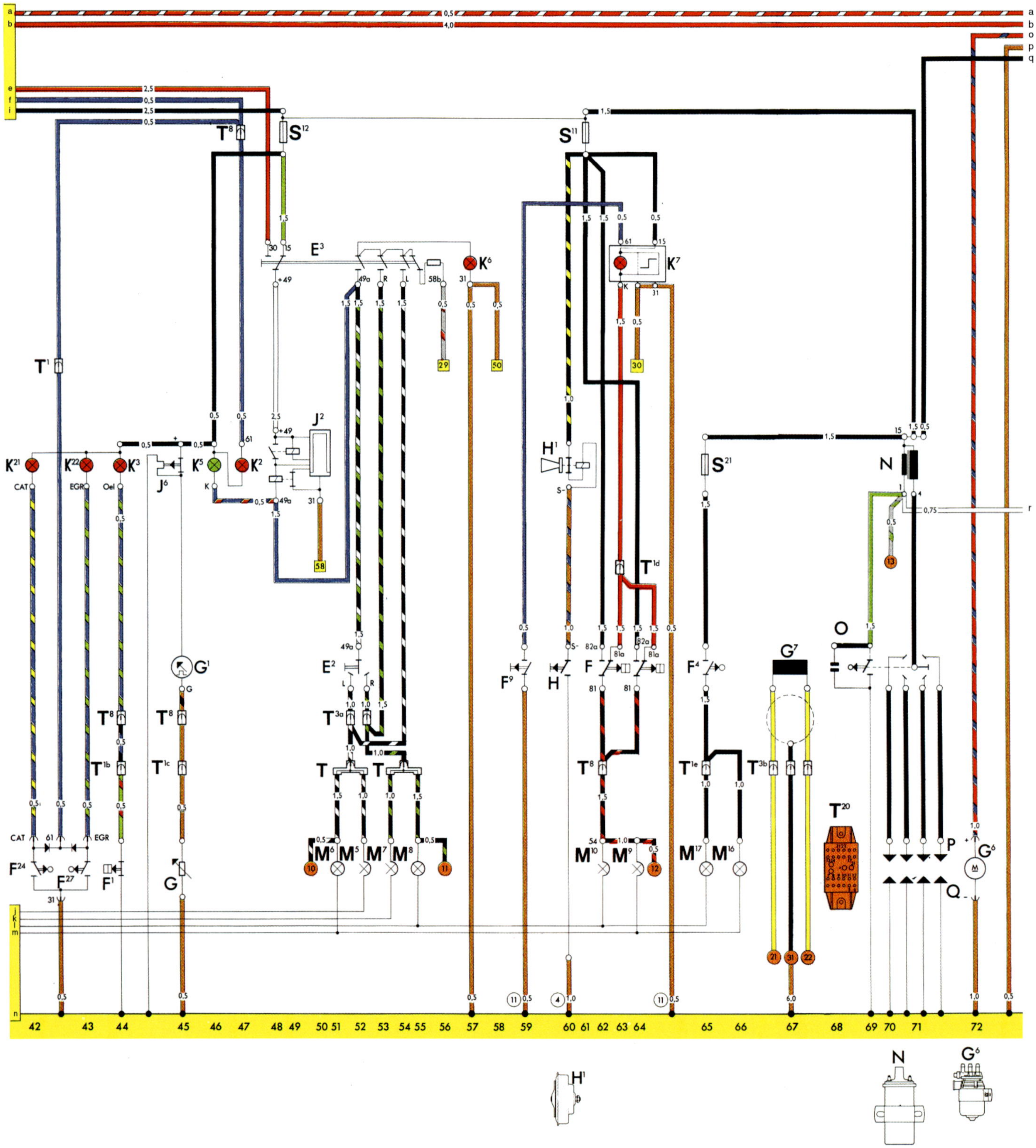

CAUTION —

Disconnect the battery ground strap before starting to work on any part of the electrical system.

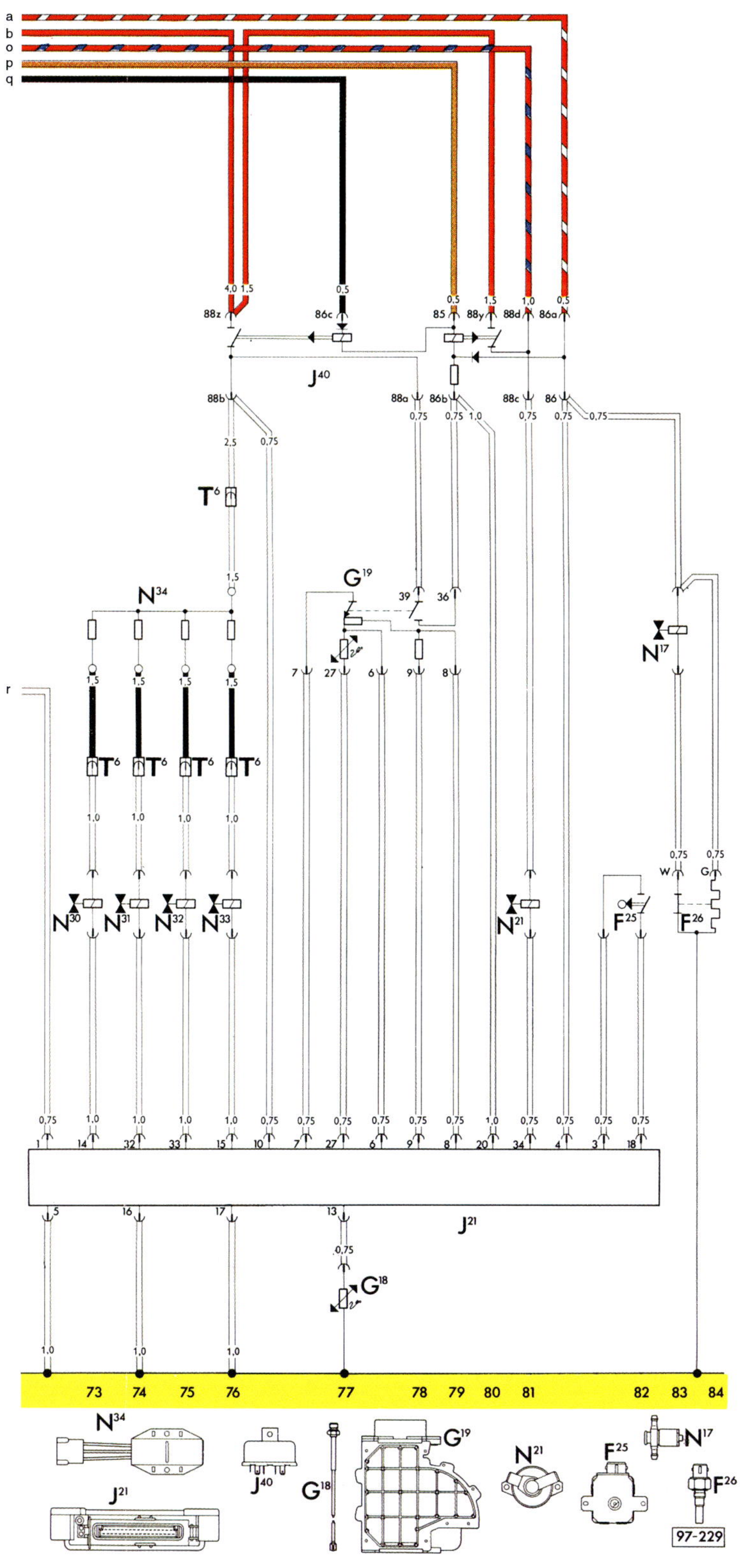

Description		Current track
E^{2}	—Turn signal switch	52
E^{3}	—Emergency flasher switch	48-57
F	—Brake light switch	62-64
F^{1}	—Oil pressure switch	44
F^{4}	—Backup light switch	65
F^{9}	—Parking brake control light switch	54
F^{24}	—Elapsed CAT mileage odometer	42
F^{25}	—Throttle valve switch	82
F^{26}	—Thermo-time switch for cold-start valve	83, 84
F^{27}	—Elapsed EGR mileage odometer	43
G	—Fuel gauge sending unit	45
G^{1}	—Fuel gauge	45
G^{6}	—Fuel pump	72
G^{7}	—TDC marker unit	67
G^{18}	—Temperature sensor on cylinder head	77
G^{19}	—Intake air sensor	77-79
H	—Horn button	60
H^{1}	—Horn	60
J^{2}	—Emergency flasher relay	48-50
J^{6}	—Fuel gauge vibrator	45
J^{21}	—Control unit (electronic)	73-82
J^{40}	—Double relay	76-81
K^{2}	—Alternator charging warning light	47
K^{3}	—Oil pressure warning light	44
K^{5}	—Turn signal warning light	46
K^{6}	—Emergency flasher warning light	57
K^{7}	—Dual circuit brake warning light	63
K^{21}	—CAT converter warning light	42
K^{22}	—EGR warning light	43
M^{5}	—Turn signal, front left	52
M^{6}	—Turn signal, rear left	51
M^{7}	—Turn signal, front right	53
M^{8}	—Turn signal, rear right	55
M^{9}	—Brake light, left	64
M^{10}	—Brake light, right	62
M^{16}	—Backup light, left	66
M^{17}	—Backup light, right	65
N	—Ignition coil	71
N^{17}	—Cold-start valve	83
N^{21}	—Auxiliary air regulator	81
N^{30}	—Injector, cyl. 1	
N^{31}	—Injector, cyl. 2	
N^{32}	—Injector, cyl. 3	
N^{33}	—Injector, cyl. 4	
N^{34}	—Resistance for Injectors	73-76
O	—Distributor	70-71
P	—Spark plug connectors	70-71
Q	—Spark plugs	70-71
S^{11} to S^{12}	} Fuse box	48, 62
S^{21}	—Fuse for backup light	65
T^{1}	—Wire connector, single	42, 43
b	—engine compartment	
c	—beside the fuel tank, right side	45
d	—on the cluster of the instrument panel	64
e	—next to the tail/brake light, left	65
T^{3a}	—Wire connector, three point, behind instrument panel	52, 53
T^{3b}	—Wire connector, three point, in engine compartment	67
T^{8}	—Wire connector, eight point, behind instrument panel	44, 45, 47, 62
T^{20}	—Test network, test socket	68
(4)	—Ground strap, steering joint	60
(11)	—Ground connector, speedometer	

Supplement to Wiring Diagram B

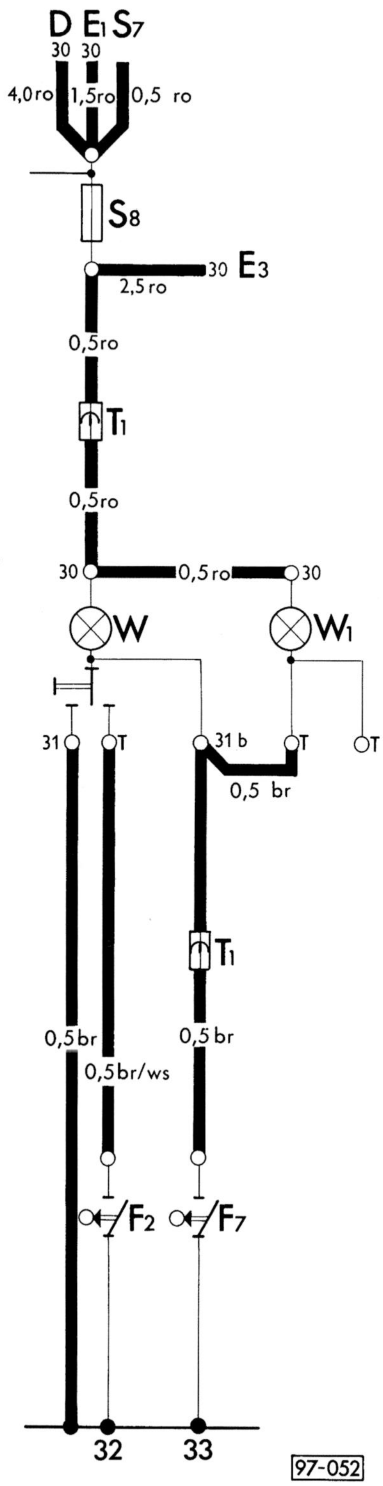

Interior Light Circuit Diagram

The 1974 current flow wiring diagram (Diagram B) remains unchanged for the 1975 model year except for the interior light circuit.

D - to ignition starter switch
E1 - to lighting switch
E3 - to emergency light switch
F2 - Door contact switch (driver's door)
F7 - Door contact switch (sliding door)
T1 - Wire connector—single
S7 - Fuse No. 7 (in fuse box)
S8 - Fuse (in fuse box)
W - Interior light, front
W1 - Interior light, rear

Color Codes

br - brown
ws - white
ro - red

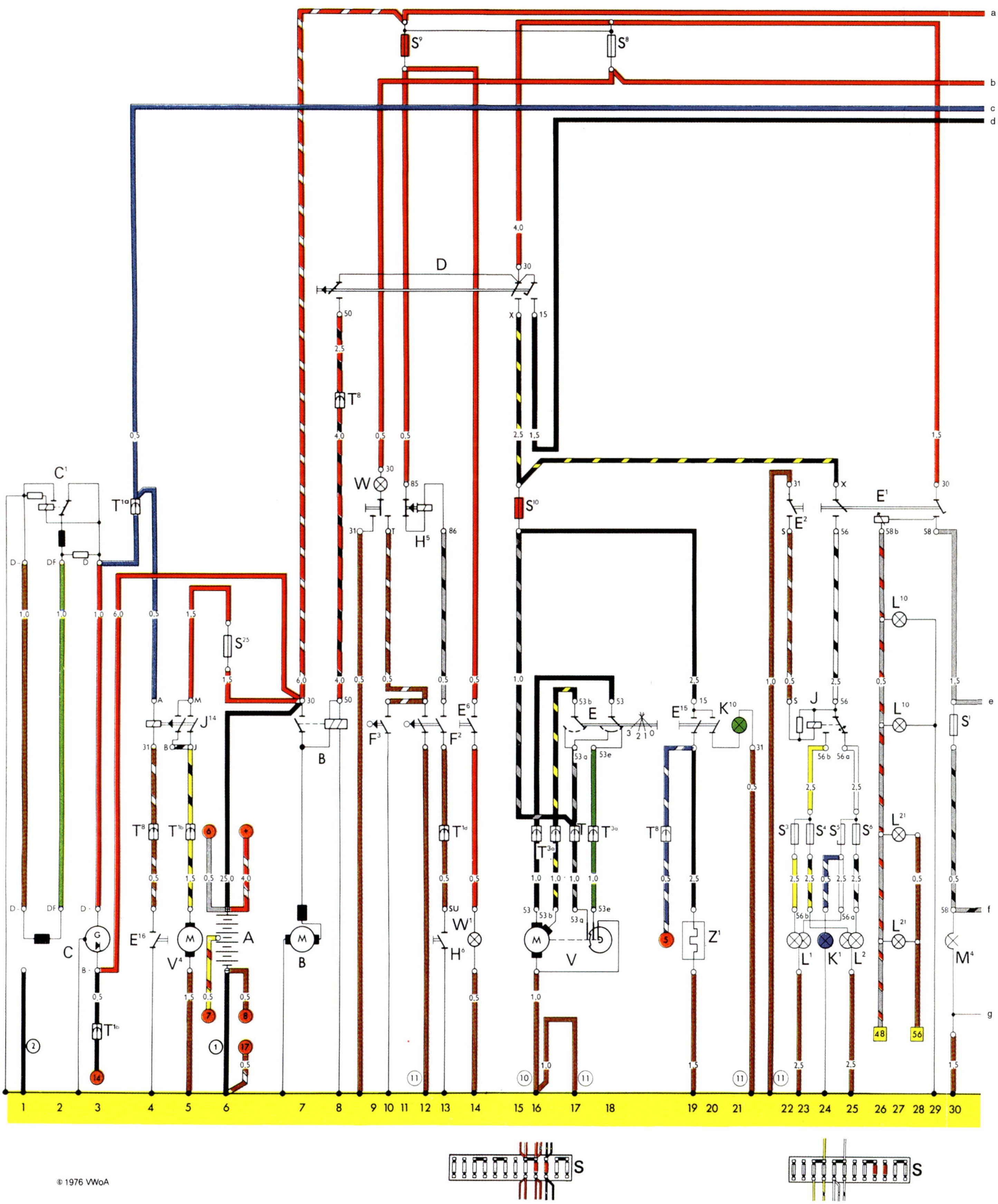

B. **VW Type 2—from August 1972 (1973 through 1975 Models)**
(Page 2 of 3)

	Description	current track
①	– Ground strap from battery to body	6
②	– Ground strap from transmission to frame	1
④	– Ground strap via steering coupling	52
⑩	– Ground connection, dashboard	
⑪	– Ground connection, speedometer	

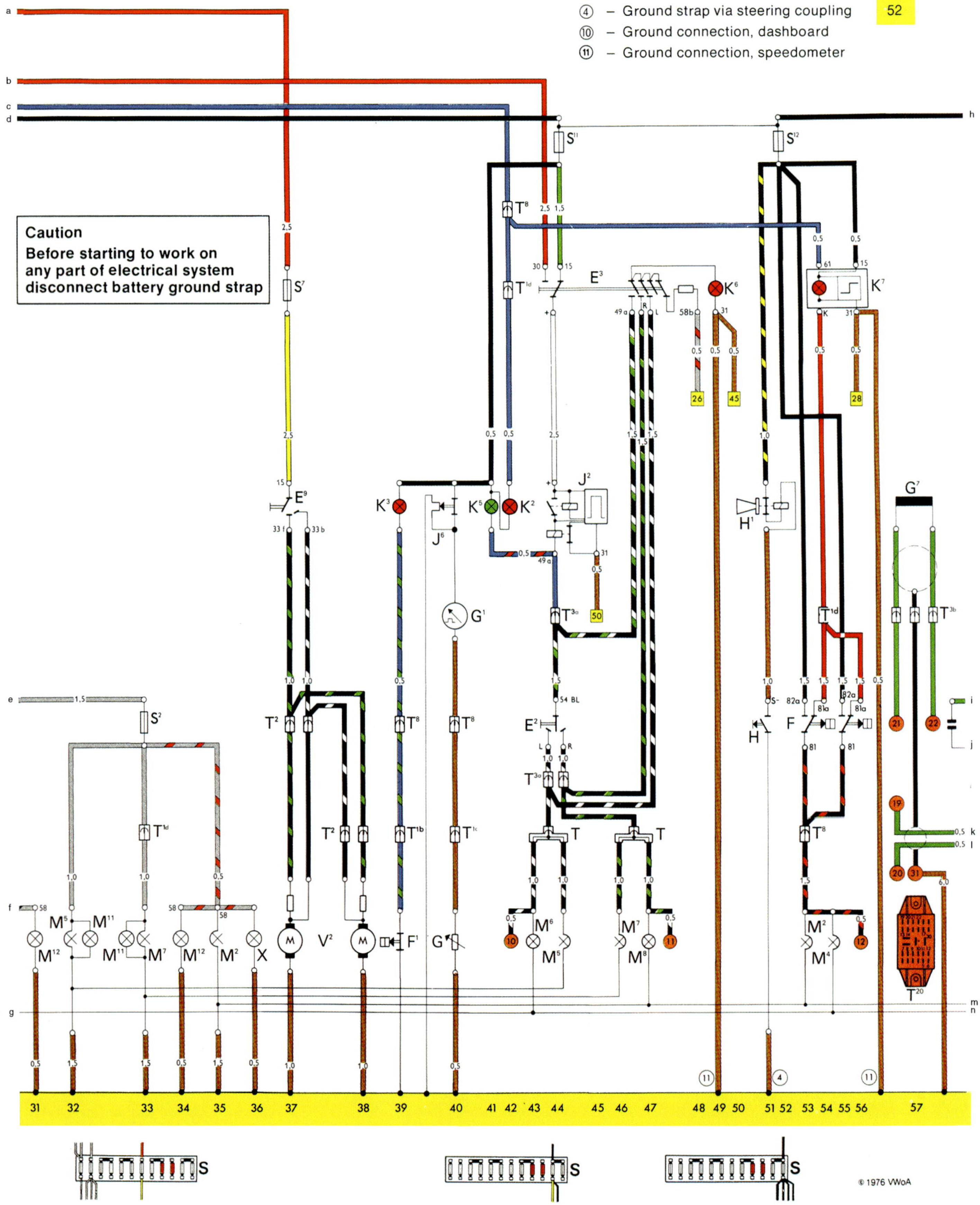

B.

	Description	current track
A	– Battery	6
B	– Starter	7, 8
C	– Alternator	1, 2, 3
C^{1}	– Regulator	1, 2, 3
D	– Ignition/starter switch	8, 15, 16.
E	– Windshield wiper switch	17, 18
E^{1}	– Light switch	25, 26, 29
E^{2}	– Turn signal switch	44
E^{2}	– Headlight dimmer switch	22
E^{3}	– Emergency flasher switch	44, 46, 47, 48, 50
E^{6}	– Interior light switch, rear	14
E^{9}	– Fresh air fan motor switch	37
E^{15}	– Rear window defogger switch	19, 20
E^{16}	– Heater air blower switch	4
F	– Brake light switch	53, 54, 55, 56
F^{1}	– Oil pressure switch	39
F^{2}	– Door contact and buzzer alarm switch, left H 5	12, 13
F^{3}	– Door contact switch, right	10
F^{4}	– Backup light switch	65
G	– Fuel gauge sending unit	40
G^{1}	– Fuel gauge	40
G^{4}	– Ignition timing sensor	59
G^{7}	– TDC marker unit	57
H	– Horn button	51
H^{1}	– Horn	51
H^{5}	– Ignition key warning buzzer	12, 13
H^{6}	– Buzzer alarm contact in ignition / starter switch	13
J	– Dimmer relay	22, 23, 24
J^{2}	– Emergency flasher relay	44, 45
J^{6}	– Fuel gauge vibrator	40
J^{14}	– Heater air blower relay	4, 5
K^{1}	– High beam warning light	24
K^{2}	– Alternator charging warning light	42
K^{3}	– Oil pressure warning light	39
K^{5}	– Turn signal warning light	41
K^{6}	– Emergency flasher warning light	49
K^{7}	– Dual circuit brake warning light	54, 56
K^{10}	– Rear window defogger warning light	21
L^{1}	– Sealed beam unit, left headlight	23
L^{2}	– Sealed beam unit, right headlight	25
L^{10}	– Instrument panel light	27
L^{21}	– Light for heater lever illumination	27
M^{2}	– Tail / brake light, right	35, 53
M^{4}	– Tail / brake light, left	30, 54
M^{5}	– Turn signal and parking light front, left	32, 44
M^{6}	– Turn signal, rear, left	43
M^{7}	– Turn signal and parking light front, right	33, 46
M^{8}	– Turn signal, rear, right	47
M^{11}	– Sidemarker light front	32, 33
M^{12}	– Sidemarker light, rear	31, 34
M^{16}	– Backup light, left	66
M^{17}	– Backup light, right	65
N	– Ignition coil	60
N^{1}	– Automatic choke, left carburetor	62
N^{2}	– Automatic choke, right carburetor	64
N^{3}	– Electromagnetic pilot jet, left carburetor	61
N^{4}	– Electromagnetic pilot jet, right carburetor	63
N^{8}	– Cutoff valve for central idling system	62
O	– Distributor	58, 60
P	– Spark plug connectors	59, 60
Q	– Spark plugs	59, 60
S^{1} to S^{12}	– Fuse box	11, 15, 18, 23, 25, 30, 33, 37, 44, 52
S^{21}	– Fuse for backup lights	65
S^{25}	– Fuse for heater air blower	6
T	– Cable adapter behind instrument panel	
T^{1}	– Wire connector, single a – in engine compartment, right b – in engine compartment c – beside the fuel tank, right side d – on the cluster of the instrument panel e – next to the tail / brake light, left	
T^{2}	– Wire connector, double point behind instrument panel	
T^{3a}	– Wire connector, three point behind instrument panel	
T^{3b}	– Wire connector, three point in engine compartment	
T^{8}	– Wire connector, eight point behind instrument panel	
T^{20}	– Test network, test socket	57
V	– Windshield wiper motor	16, 18
V^{2}	– Fan motor	37, 38
V^{4}	– Heater air blower	5
W	– Interior light, front	10
W^{1}	– Interior light, rear	14
X	– License plate light	36
Z^{1}	– Rear window defogger heating element	19

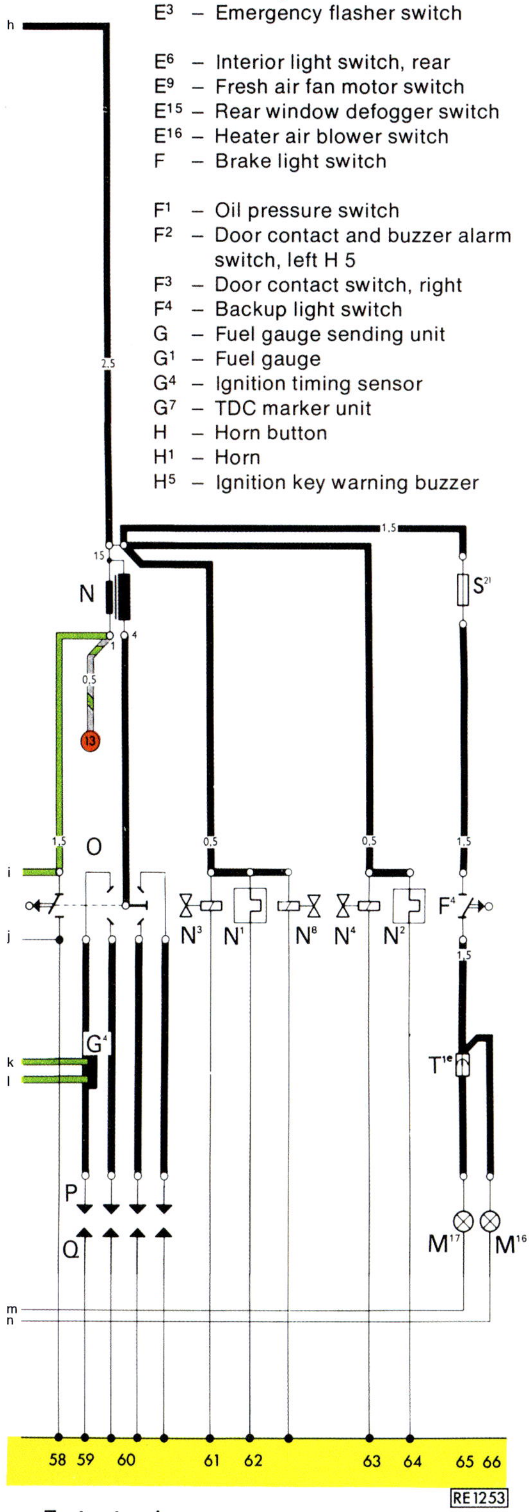

Test network

The orange colored spots are the connections in the test network which are wired to the test socket (T^{20}). The numbers in the spots correspond to the terminals in the test socket.

C. **VW Type 2—from August 1971 (1972 Models)**
(Page 1 of 2)

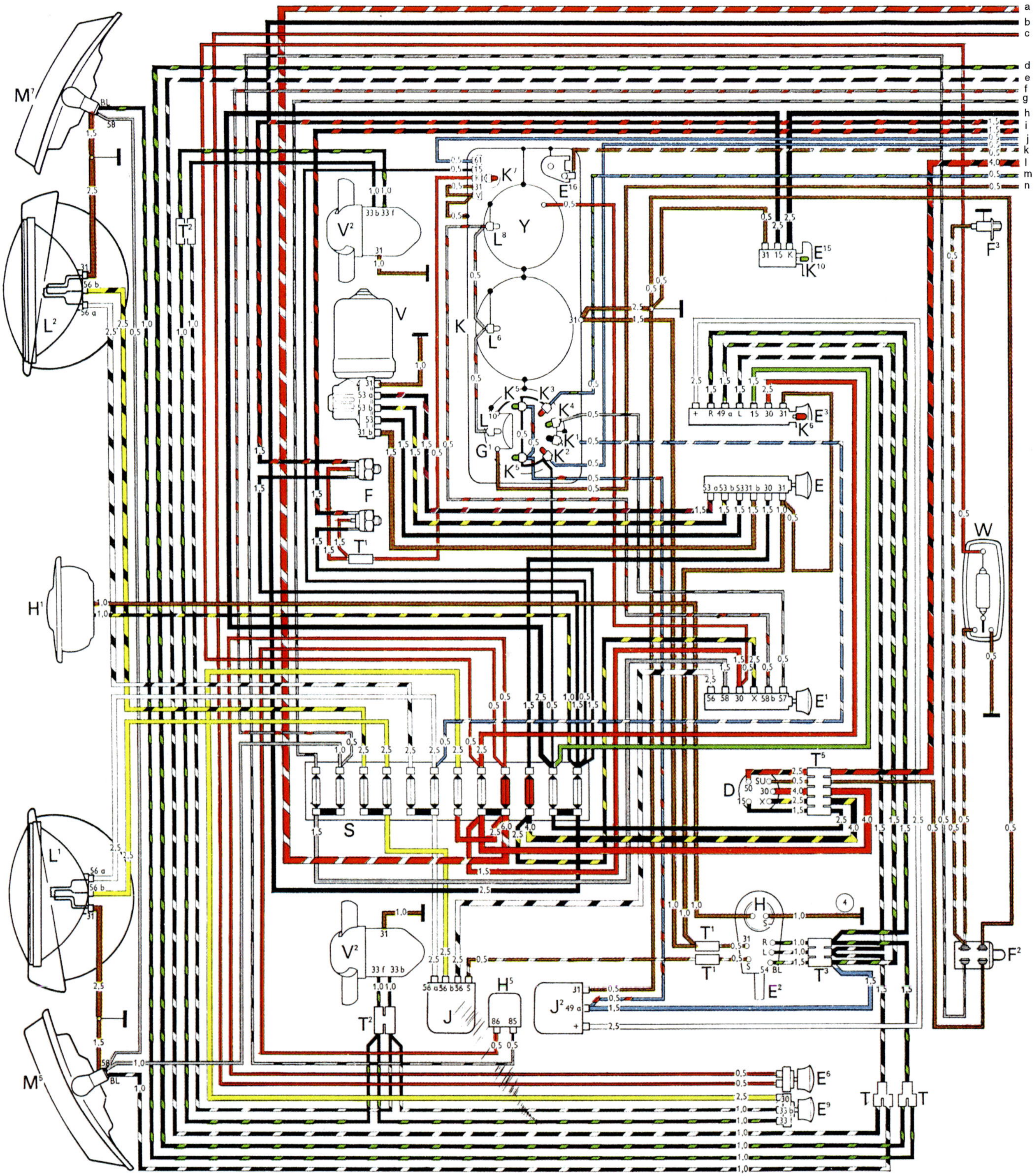

A – Battery
B – Starter
C – Generator
C^1 – Regulator
D – Ignition / starter switch
E – Windshield wiper switch
E^1 – Light switch
E^2 – Turn signal and headlight dimmer switch
E^3 – Emergency flasher switch
E^6 – Interior light switch, rear
E^9 – Switch for fan motors
E^{15} – Rear window defogger switch
E^{16} – Heater air blower switch
F – Brake light and dual / circuit warning light switch
F^1 – Oil pressure switch
F^2 – Door contact switch, left, with contact for buzzer
F^3 – Door contact switch, right
F^4 – Back-up light switch
G – Fuel gauge sending unit
G^1 – Fuel gauge
G^4 – Ignition timing sensor
H – Horn button
H^1 – Horn
H^3 – Ignition key warning buzzer
J – Dimmer relay
J^2 – Emergency flasher relay
J^{14} – Heater air blower relay
K – Instrument panel insert
K^1 – High beam warning light
K^2 – Generator charging warning light
K^3 – Oil pressure warning light
K^4 – Parking light warning light
K^5 – Turn signal warning light
K^6 – Emergency flasher warning light
K^7 – Dual circuit brake system warning light
K^{10} – Rear window defogger warning light
L^1 – Sealed beam unit, left headlight
L^2 – Sealed beam unit, right headlight
L^6 – Speedometer light
L^8 – Clock light
L^{10} – Instrument panel light
M^2 – Tail- and brake light, right
M^4 – Tail- and brake light, left
M^5 – Turn signal and parking light, front, left
M^6 – Turn signal, rear, left
M^7 – Turn signal and parking light, front, right

VW Type 2—from August 1971 (1972 Models)

(Page 2 of 2)

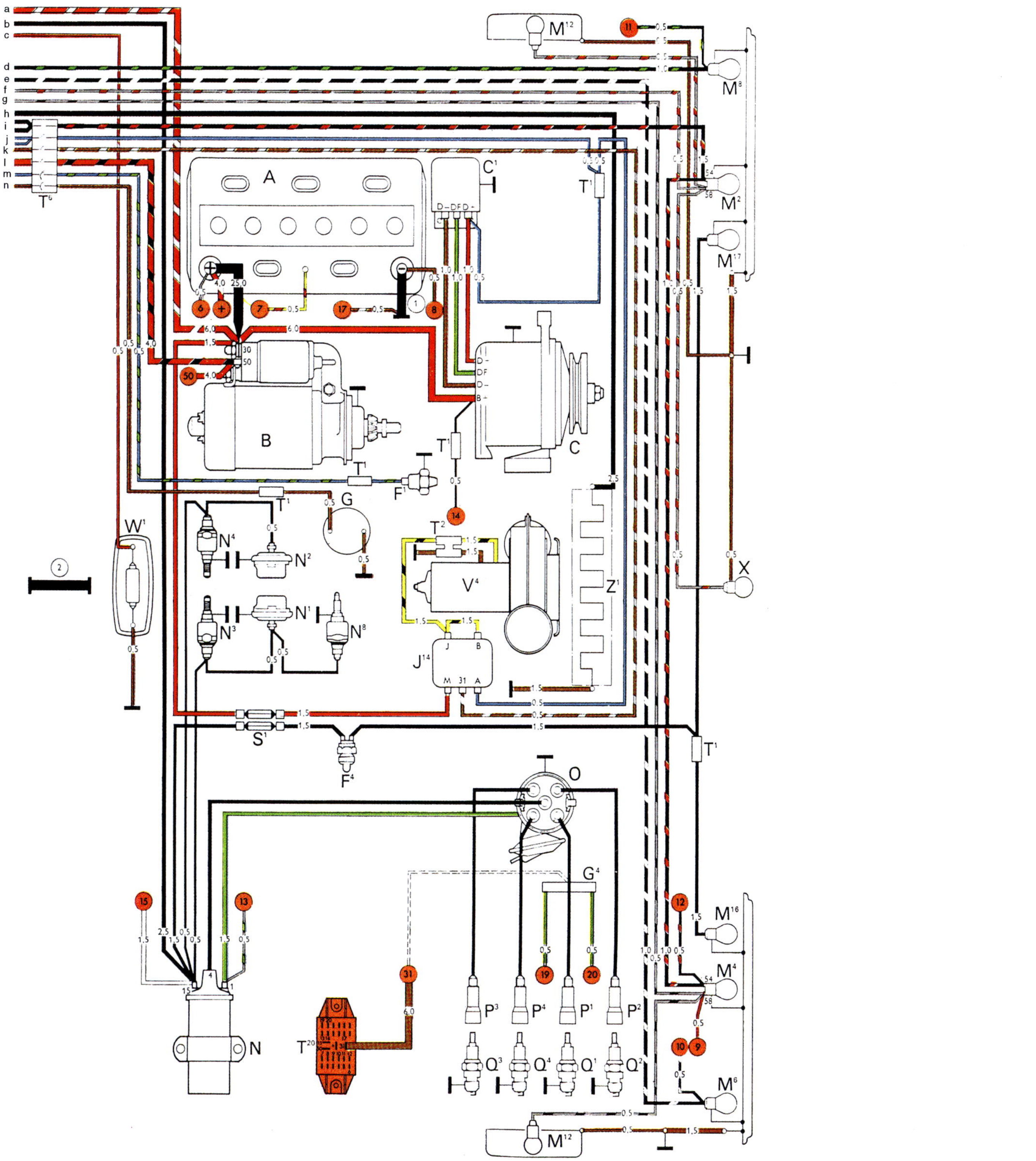

- M^8 – Turn signal, rear, right
- M^{12} – Sidemarker light, rear
- M^{16} – Back-up light, left
- M^{17} – Back-up light, right
- N – Ignition coil
- N^1 – Automatic choke, left carburetor
- N^2 – Automatic choke, right carburetor
- N^3 – Electro-magnetic pilot jet, left carburetor
- N^4 – Electro-magnetic pilot jet, right carburetor
- N^8 – Cut-off valve for central idling system
- O – Ignition distributor
- P^1 – Spark plug connector, No. 1 cylinder
- P^2 – Spark plug connector, No. 2 cylinder
- P^3 – Spark plug connector, No. 3 cylinder
- P^4 – Spark plug connector, No. 4 cylinder
- Q^1 – Spark plug, No. 1 cylinder
- Q^2 – Spark plug, No. 2 cylinder
- Q^3 – Spark plug, No. 3 cylinder
- Q^4 – Spark plug, No. 4 cylinder
- S – Fuse box
- S^1 – Fuse for back-up light
- T – Cable adapter
- T^1 – Cable connector, single
- T^2 – Cable connector, double
- T^3 – Cable connector, triple
- T^6 – Cable connector, eight connections
- T^{20} – Test network, central plug
- V – Windshield wiper motor
- V^2 – Fan motor, front
- V^4 – Heater air blower
- W – Interior light, front
- W^1 – Interior light, rear
- X – License plate light
- Y – Clock
- Z^1 – Rear window defogger heating element

- ① – Ground strap from battery to body
- ② – Ground strap from transmission to the frame
- ④ – Ground cable from horn button to steering coupling

Test network

The orange colored spots are the connections in the test network which are wired to the central plug (T^{20}). The numbers in the spots correspond to the terminals in the central plug.

D.

VW Type 2—from August 1970 (1971 Models)

(Page 1 of 2)

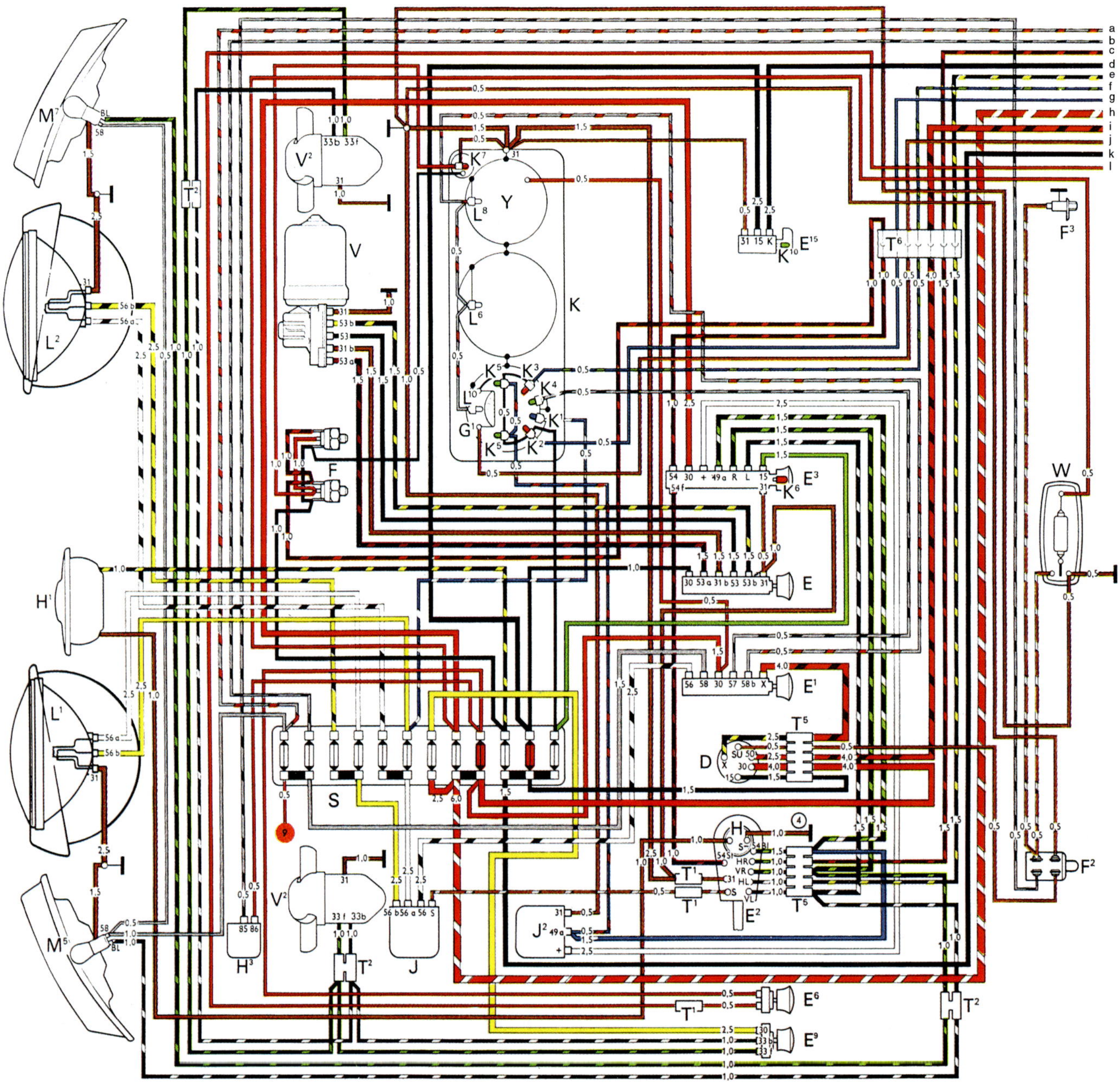

A – Battery
B – Starter
C – Generator
C^1 – Regulator
D – Ignition / starter switch
E – Windshield wiper switch
E^1 – Light switch
E^2 – Turn signal and headlight dimmer switch
E^3 – Emergency flasher switch
E^6 – Interior light switch, rear
E^9 – Switch for fan motors
E^{15} – Rear window defogger switch
F – Brake light switch with warning switch
F^1 – Oil pressure switch
F^2 – Door contact switch, left, with contact for buzzer
F^3 – Door contact switch, right
F^4 – Back-up light switch

G – Fuel gauge sending unit
G^1 – Fuel gauge
H – Horn button
H^1 – Horn
H^3 – Ignition key warning buzzer
J – Dimmer relay
J^2 – Emergency flasher relay
K – Instrument panel insert
K^1 – High beam warning light
K^2 – Generator charging warning light
K^3 – Oil pressure warning light
K^4 – Parking light warning light
K^5 – Turn signal warning light
K^6 – Emergency flasher warning light
K^7 – Dual circuit brake system warning light
K^{10} – Rear window defogger warning light
L^1 – Sealed beam unit, left headlight
L^2 – Sealed beam unit, right headlight

L^6 – Speedometer light
L^8 – Clock light
L^{10} – Instrument panel light
M^5 – Turn signal and parking light, front, left
M^7 – Turn signal and parking light, front, right
M^9 – Tail-, brake-, turn signal light, left
M^{10} – Tail-, brake-, turn signal light, right
M^{12} – Sidemarker light, rear
N – Ignition coil
N^1 – Automatic choke
N^3 – Electro-magnetic pilot jet
O – Ignition distributor
P^1 – Spark plug connector, No. 1 cylinder
P^2 – Spark plug connector, No. 2 cylinder
P^3 – Spark plug connector, No. 3 cylinder
P^4 – Spark plug connector, No. 4 cylinder
Q^1 – Spark plug, No. 1 cylinder
Q^2 – Spark plug, No. 2 cylinder

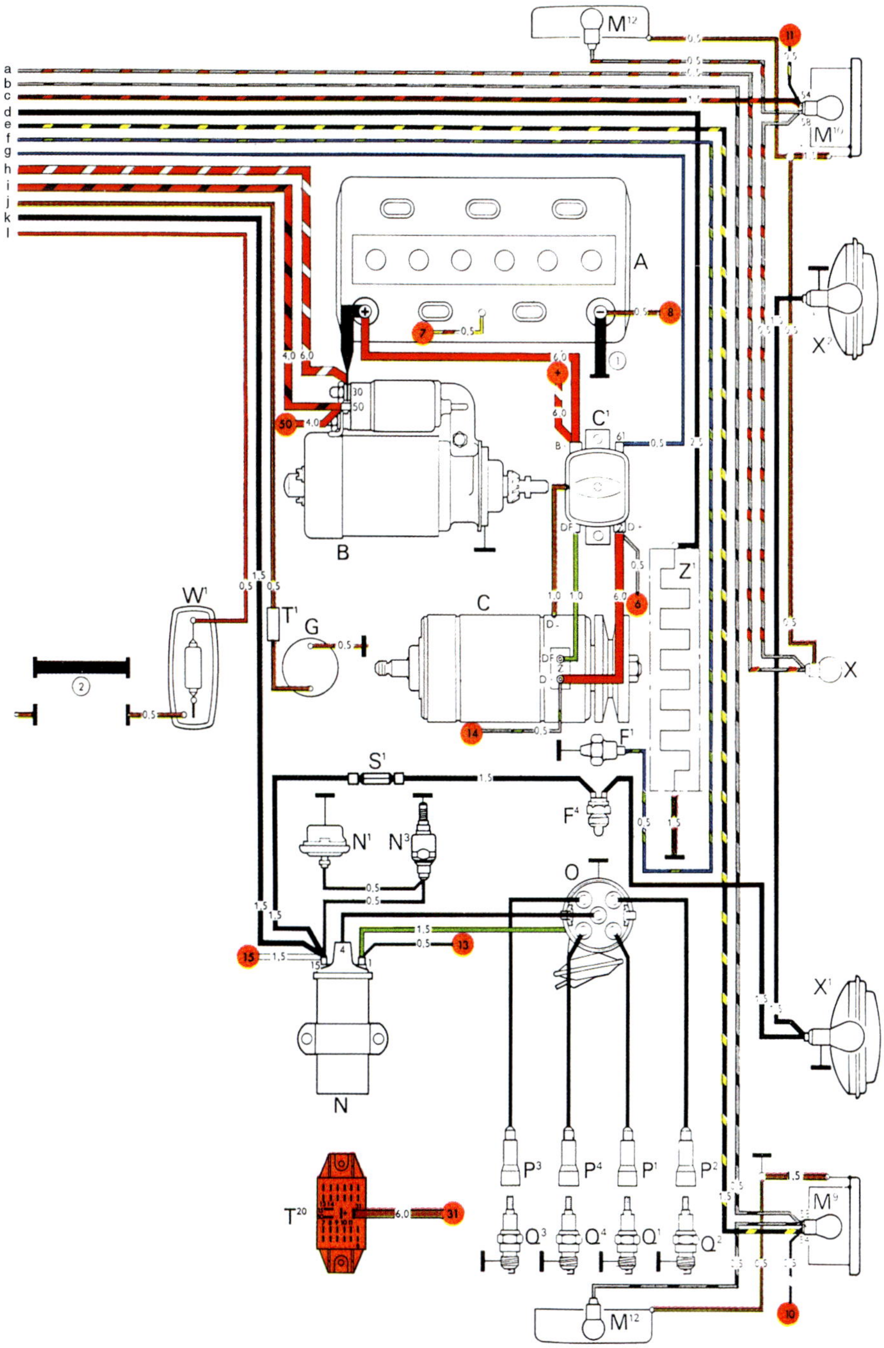

Q^3 – Spark plug, No. 3 cylinder
Q^4 – Spark plug, No. 4 cylinder
S – Fuse box
S^1 – Fuse for back-up light
T^1 – Cable connector, single
T^2 – Cable connector, double
T^5 – Cable connector, five connections
T^6 – Cable connector, eight connections
T^{20} – Test network, central plug
V – Windshield wiper motor
Y^2 – Fan motor, front
W – Interior light, front
W^1 – Interior light, rear
X – License plate light
X^1 – Back-up light, left
X^2 – Back-up light, right
Y – Clock
Z^1 – Rear window defogger heating element

① – Ground strap from battery to frame
② – Ground strap from transmission to the frame
④ – Ground cable from horn button to steering coupling

Test network (from January 1971 only)

The orange colored spots are the connections in the test network which are wired to the central plug. The numbers in the spots correspond to the terminals in the central plug.

Note

Colors of the cables have changed:
cable 13 is grey/green
cable 14 is black

E.

VW Type 2—from August 1969 (1970 Models)

(Page 1 of 2)

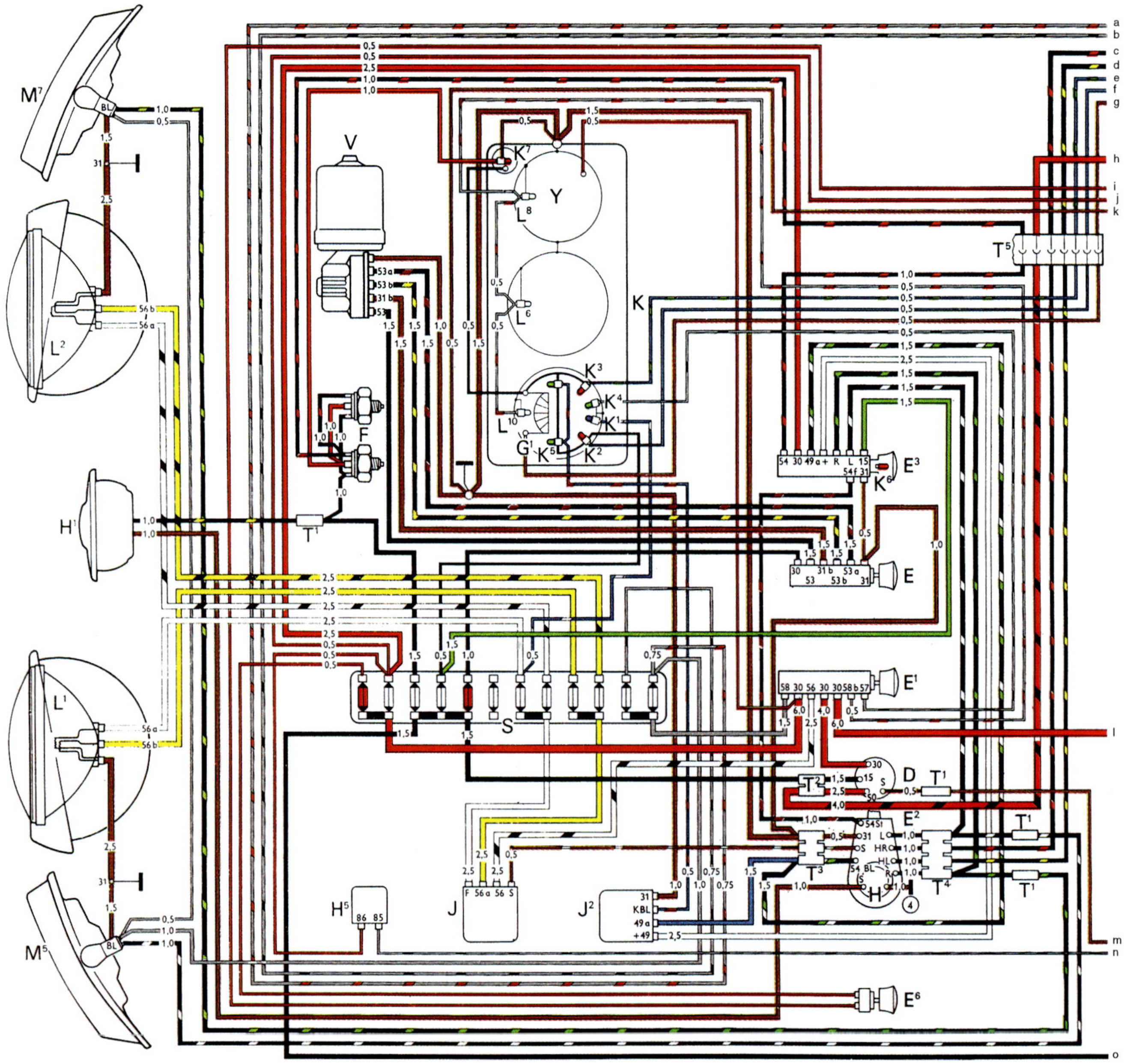

A – Battery
B – Starter
C – Generator
C' – Regulator
D – Ignition / starter switch
E – Windshield wiper switch
E' – Light switch
E² – Turn signal and headlight dimmer switch
E³ – Emergency flasher switch
E⁶ – Switch for interior light, rear
F – Brake light switch
F' – Oil pressure switch
F² – Door contact switch, left with contact for buzzer H⁵
F³ – Door contact, switch right
F⁴ – Switch for back-up lights
G – Fuel gauge sending unit
G' – Fuel gauge
H – Horn button
H' – Horn
H⁵ – Ignition key warning buzzer
J – Dimmer relay
J² – Emergency flasher relay
K – Instrument panel insert
K' – High beam warning light
K² – Generator charging warning light
K³ – Oil pressure warning light
K⁴ – Parking light warning light
K⁵ – Turn signal warning light
K⁶ – Emergency flasher warning light
K⁷ – Dual circuit brake system warning light
L' – Sealed beam light, left headlight
L² – Sealed beam light, right headlight
L⁶ – Speedometer light
L⁸ – Clock light
L¹⁰ – Instrument panel light
M⁵ – Turn signal and parking light, front, left
M⁷ – Turn signal and parking light, front, right
M⁹ – Tail / brake / turn signal light, left
M¹⁰ – Tail / brake / turn signal light, right
M¹² – Side marker lights, rear
N – Ignition coil
N' – Automatic choke
N³ – Electro-magnetic pilot jet
O – Distributor
P' – Spark plug connector, No. 1 cylinder
P² – Spark plug connector, No. 2 cylinder
P³ – Spark plug connector, No. 3 cylinder
P⁴ – Spark plug connector, No. 4 cylinder
Q' – Spark plug, No. 1 cylinder
Q² – Spark plug, No. 2 cylinder

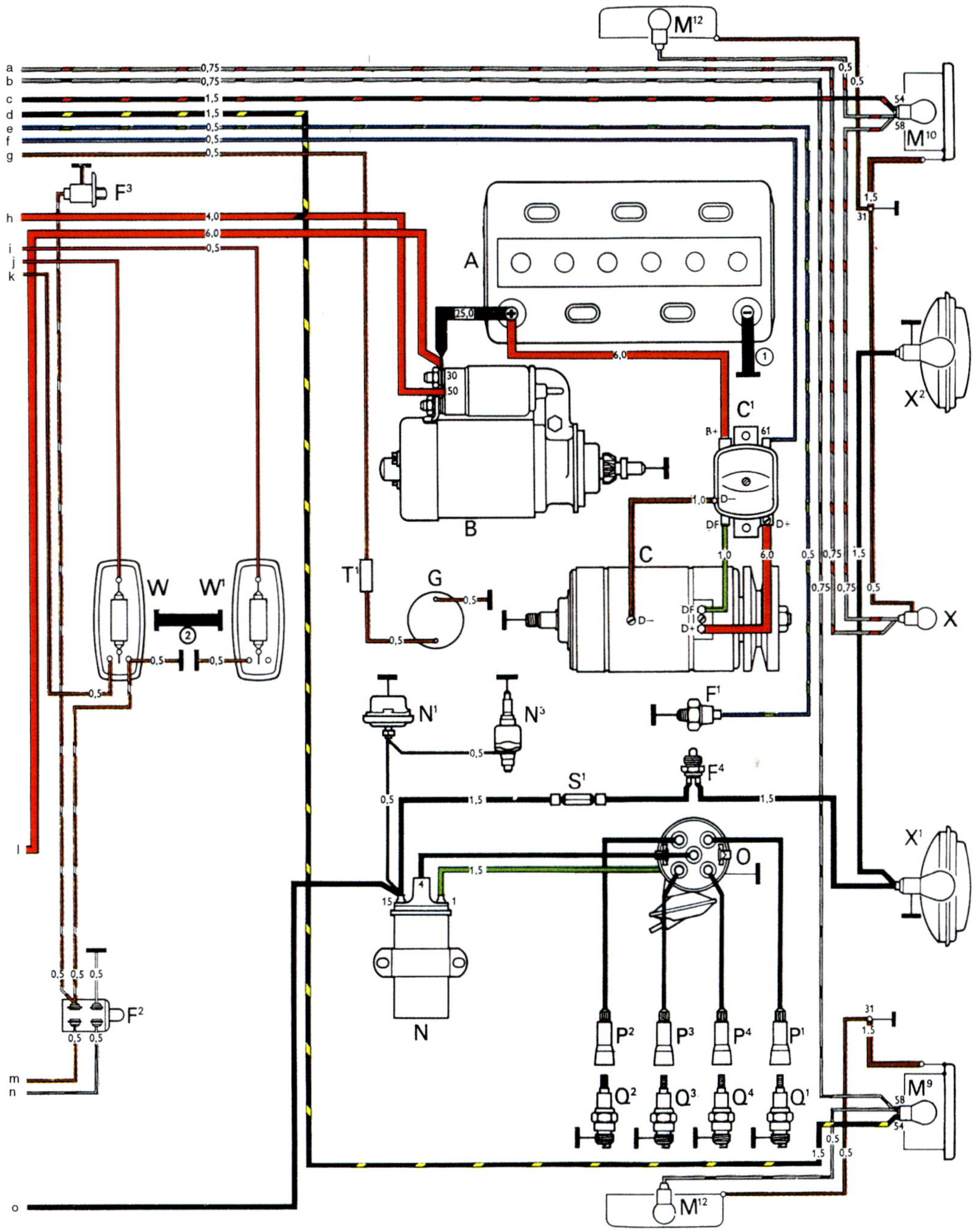

Q³ – Spark plug, No. 3 cylinder
Q⁴ – Spark plug, No. 4 cylinder
S – Fuse box
S¹ – In-line fuse for back-up lights
T¹ – Cable connector, single
T² – Cable connector, double
T³ – Cable connector, triple
T⁴ – Cable connector (four connections)
T⁵ – Push-on connector
V – Windshield wiper motor
W – Interior light, front
W¹ – Interior light, rear
X – License plate light
X¹ – Back-up light, left
X² – Back-up light, right
Y – Clock

(1) – Ground strap from battery to frame
(2) – Ground strap from transmission to the frame
(4) – Ground cable from horn button to steering coupling

NOTE

See also additional wiring diagram G for Type 2 from August 1969 at the end of this section.

F.

VW Type 2—from August 1967 (1969 Models)

(Page 1 of 2)

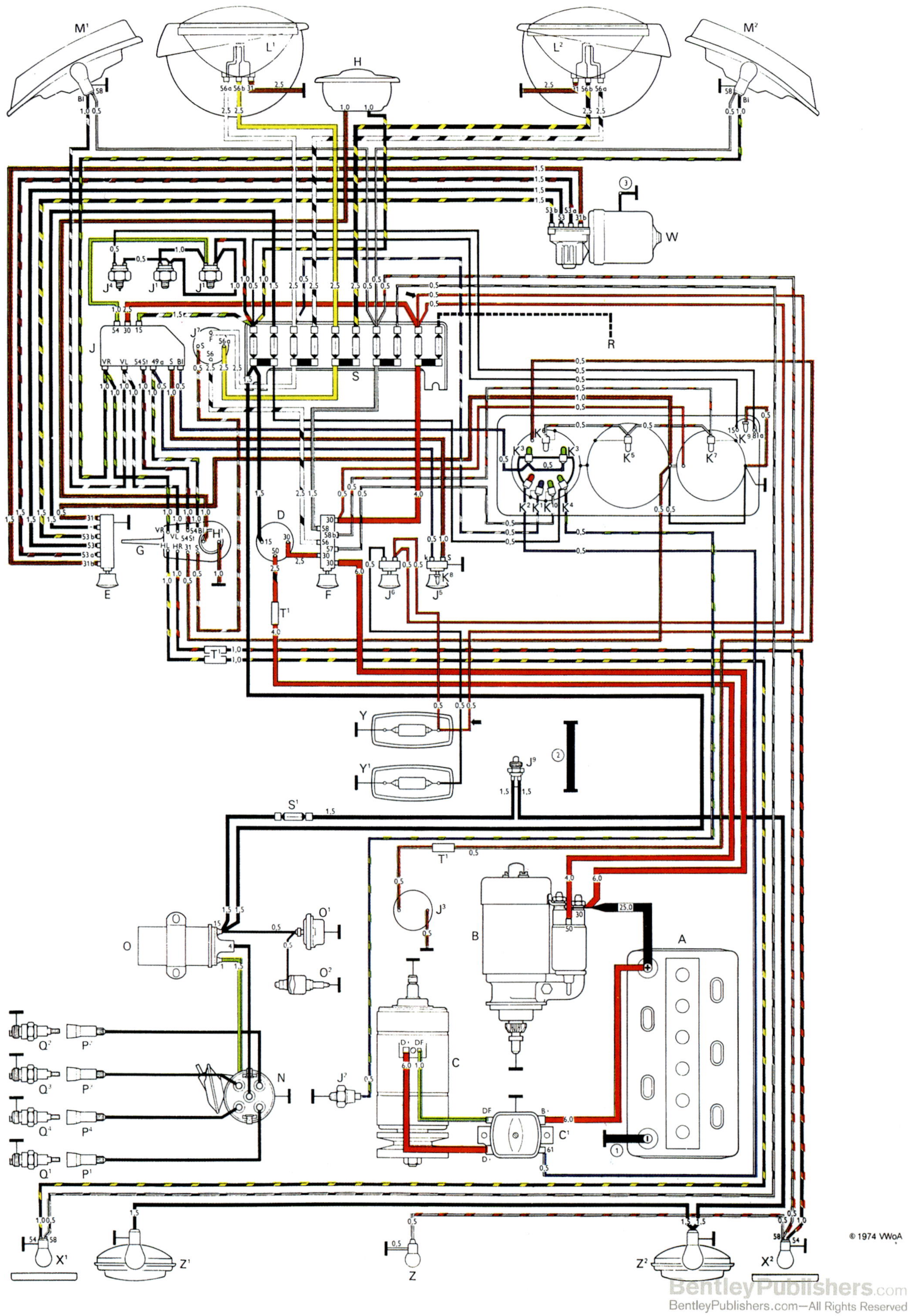

F. VW Type 2—from August 1967 (1969 Models)

(Page 2 of 2)

A – Battery
B – Starter
C – Generator
C¹ – Regulator
D – Ignition/starter switch
E – Windshield wiper switch
F – Light switch
G – Turn signal switch and hand dimmer
H – Horn
H¹ – Horn button
J – Emergency flasher relay
J¹ – Brake light switch (2×)
J² – Oil pressure switch
J³ – Fuel gauge sender unit
J⁴ – Warning switch for brakes
J⁵ – Emergency flasher switch
J⁶ – Interior light switch
J⁷ – Dimmer relay
J⁹ – Back-up light switch
K¹ – High beam warning light
K² – Generator and fan warning light
K³ – Turn signal warning lights
K⁴ – Oil pressure warning lights
K⁵ – Speedometer light
K⁶ – Fuel gauge light
K⁷ – Clock light
K⁸ – Emergency flasher warning light
K⁹ – Brake system warning light
L¹ – Sealed beam unit, left
L² – Sealed beam unit, right
M¹ – Parking light, left
M² – Parking light, right
N – Distributor
O – Ignition coil
O¹ – Automatic choke
O² – Electro-magnetic pilot jet
P¹ – Spark plug connector, No. 1 cylinder
P² – Spark plug connector, No. 2 cylinder
P³ – Spark plug connector, No. 3 cylinder
P⁴ – Spark plug connector, No. 4 cylinder
Q¹ – Spark plug for No. 1 cylinder
Q² – Spark plug for No. 2 cylinder
Q³ – Spark plug for No. 3 cylinder
Q⁴ – Spark plug for No. 4 cylinder
R – Radio
S – Fuse box
S¹ – Back-up light fuse
T¹ – Cable connector
U¹ – Turn signal, front left
U² – Turn signal, front right
W – Windshield wiper motor
X¹ – Brake, turn signal and tail light, left
X² – Brake, turn signal and tail light, right
Y – Interior light, front
Y¹ – Interior light, rear
Z – License plate light
Z¹ – Back-up light, left
Z² – Back-up light, right

Fresh Air Fan Circuit and Rear Window Defogger Circuit for VW Type 2—from August 1969 (1970 Models) G.

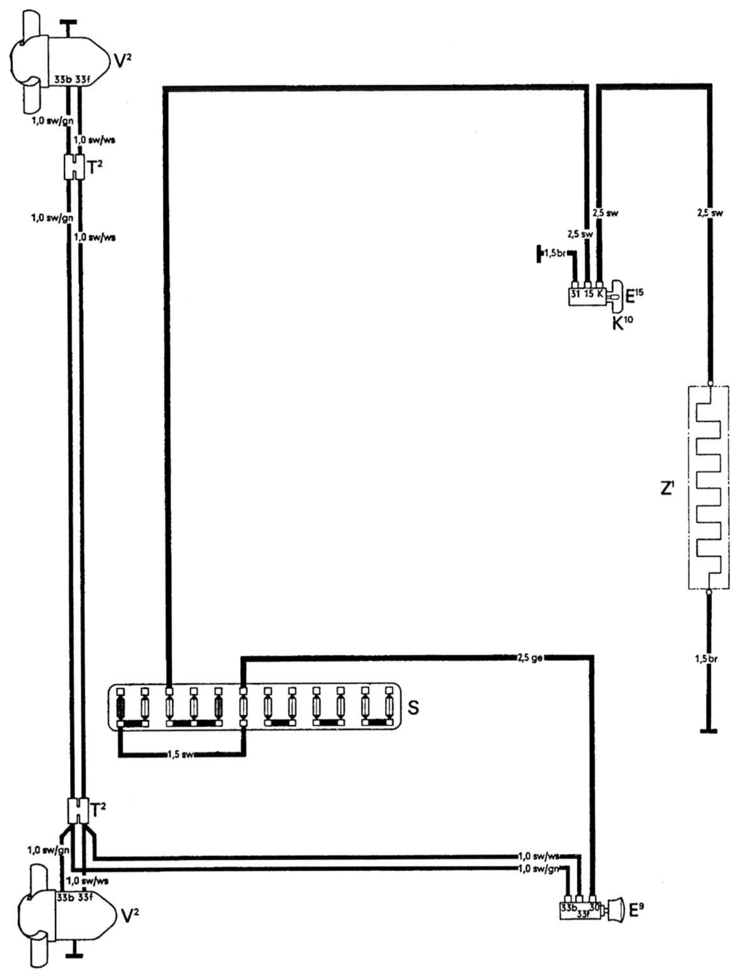

E⁹ – Switch for fan motors
E¹⁵ – Rear window defogger switch
K¹⁰ – Rear window defogger warning light
S – Fuse box
T² – Cable connector, double
Y² – Fan motor, front
Z¹ – Rear window defogger heating element

Color of cables:

sw – black
ge – yellow
br – brown
gn – green
ws – white

① – Battery to body ground strap
② – Transmission to body ground strap
③ – Windshield wiper motor ground cable

Black dotted lines = optional extras
All fuses: 8 amps.

➤ – for Pick up only

Notes